# STUDY GUIDE AND PROBLEMS BOOK

# FOR INTRODUCTION TO

# ORGANIC CHEMISTRY

**BROWN**

**Brent L. Iverson**
**Sheila A. Iverson**
**University of Texas, Austin**

**William H. Brown**
**Beloit College**

SAUNDERS GOLDEN SUNBURST SERIES

**Saunders College Publishing**
Harcourt Brace College Publishers

Fort Worth   Philadelphia   San Diego   New York   Orlando   Austi
San Antonio   Toronto   Montreal   London   Sydney   Tokyo

D0082291

Printed in the United States of America.

Iverson, Iverson & Brown; Study Guide and Problems Book for
Introduction to Organic Chemistry, First Edition.

ISBN 0-03-018304-9

567 202 7654321

To Carina, Alexandra, and Alanna
with love

## PREFACE

This Study Guide and Problems Book is a companion to *Introduction to Organic Chemistry* by William H. Brown. This guide contains a section-by-section *overview* of the major points covered in the text. Reading the overview before and after reading a text chapter should help identify and summarize important elements of the material. The overviews are written in a compact *outline format*. *Key terms* are printed in boldface the first time they appear. Bracketed sentences in italic print provide *hints for studying and pitfalls to avoid*. Especially important ideas are identified with the symbol "**S**". *not ✳ ?*

A *summary of reactions* is presented in tabular form. The starting materials are listed on the vertical axis, and the products of the reaction are listed across the top. Where the two intersect in the table is the section number where a particular reaction can be found in the text. Below the table are generalized descriptions and explanations of the different reactions in the table.

At the back of the book are quizzes of ten multiple-choice problems per chapter with answers. These quizzes were prepared by John A. Miller of Western Washington University. After you study a chapter, you should test your knowledge of the material by trying the quiz.

All of the *problems* from the text have been reprinted in this guide, so there is no need to flip back and forth between the text and the guide. Detailed, stepwise *solutions* to all of the problems are provided. This guide was reviewed for accuracy by Brent and Sheila Iverson and Eric Helms at the University of Texas, Arthur Meyer at Austin Community College and the University of Texas at Austin, and William H. Brown at Beloit College. If you have any comments or questions, please direct them to Professor Brent Iverson, Department of Chemistry and Biochemistry, the University of Texas at Austin, Austin, Texas 78712. E-mail: biverson@utxvms.cc.utexas.edu.

Brent and Sheila Iverson
The University of Texas at Austin
July 1996

# TABLE OF CONTENTS:

# CHAPTER 1: COVALENT BONDS AND SHAPES OF MOLECULES

## 1.0 OVERVIEW
• **Organic Chemistry** is the study of compounds that contain carbon atoms in combination with other types of atoms such as hydrogen, nitrogen and oxygen.
• Chemists have discovered over ten million compounds composed of these elements.
• Organic chemists are concerned primarily with where electrons are located in an atom, molecule, or ion, because then they can understand or predict structure, bonding, and reactivity. ✳

## 1.1 STRUCTURE OF ATOMS
• An atom contains a **nucleus** that is **composed of neutrons** and positively charged **protons**.
   - Most of the mass of an atom is contained in the nucleus. ✳
   - The diameter of an atom's nucleus is in the range of $10^{-14}$ meters to $10^{-15}$ meters.
• The **nucleus is surrounded** by a much larger space that contains **negatively charged electrons**. ✳
   - The diameter of the extranuclear space containing the electrons is approximately $10^{-10}$ meters.
• The electrons are confined to regions of space called **shells**, which are identified by the numbers **1,2,3**, etc. ✳
   - Each **shell contains $2n^2$ electrons** (n is the number of the shell).
   - Electrons in the first shell are held most closely to the nucleus. First shell electrons are thus held the tightest and are referred to as being "low energy."
   - Electrons in higher numbered shells are farther from the nucleus, are held less tightly, and are referred to as being of higher energy.
• Shells are further divided into subregions of space called **atomic orbitals.** ✳
   - An orbital is the region of space where a given electron spends 90%- 95% of its time.
   - An atomic orbital can hold up to two electrons (the **Pauli Exclusion Principle**).
   - Atomic orbitals are classified as s, p, or d.
   - For most of organic chemistry, s and p atomic orbitals are the only types of orbitals that we need to consider.
   - The d atomic orbitals are important for third row elements such as sulfur (S) and phosphorus (P).
• The different types of orbitals are contained within the different shells according to the following rules:
   - There is only one s orbital for a given shell.
      All s orbitals are in the shape of a sphere. A **1**s orbital is the smallest, a **2**s orbital is larger, and a **3**s orbital is larger still.
   - There are three p orbitals for shells of number **2** and higher.
      A p orbital has two lobes arranged in a line with the nucleus at the center. The three p orbitals are orthogonal to each other, so they are designated according to the orthogonal x, y and z axes as $2p_x$, $2p_y$, and $2p_z$.
   - There are five d orbitals for shells with a number of **3** or higher.
   - **All the orbitals of the same type** in a given shell have the **same energy**. For example, all three **2**p orbitals, the $2p_x$, $2p_y$, and $2p_z$, have the same energy. ✳
• Different elements have different numbers of electrons, and a description of the orbitals occupied by the electrons is called an **electron configuration**. ✳
   - The lowest energy electron configuration for a given element is referred to as the **ground-state electron configuration**.
• Ground state electron configurations for atoms are determined according to the following rules:

- Orbitals with the smallest number are lowest in energy and are filled first.  Thus orbitals are filled in the order **1**s, **2**s, **2**p, **3**s, **3**p, **4**s, etc.  Remember there are actually 3 different **2**p orbitals, the **2**$p_x$, **2**$p_y$, and **2**$p_z$ orbitals.
- Each orbital can hold up to two electrons.
- One electron is added to each orbital of the same energy before two electrons are added to any one of them.

• Chemists are primarily interested in the **valence electrons** of an atom.  **Valence electrons** are the electrons in the outermost shell of an atom, the so-called **valence shell**. ✳

- For H, the valence electrons are in the **1**st shell, namely the **1**s orbital.
- For C, N, O, and F the valence electrons are those in the **2**nd shell, namely the **2**s and the three **2**p orbitals.
- For Si, P, S, and Cl the valence electrons are those in the **3**rd shell, namely the **3**s, and the three **3**p orbitals.  For P and S, the **3**d orbitals are important.
- The noble gases have filled valence shells, that is all of the orbitals in their valence shell are filled. ✳

• **Lewis structures** are used to represent atoms and molecules.  In a Lewis structure, the element symbol is surrounded by dots that represent the valence electrons.

## 1.2 THE LEWIS MODEL OF BONDING

• **Chemical bonds** in molecules are made from electrons in the valence shells of atoms. ✳
  - Bonds do not involve electrons from shells below the valence shell.
  - Atoms overwhelmingly prefer to be surrounded by a filled valence shell of electrons (**noble gas configuration**). ✳
  - A filled valence shell for H is 2 electrons, and a filled valence shell for C, N, O, and F is 8 electrons; the "**octet rule**."
  - P and S can have 8 electrons in their valence shell, but the valence shell for S may contain as many as 10 electrons, and the valence shell of P may contain as many as 12 electrons.  This is due to the presence of **3**d orbitals on these third row elements.

• Atoms take part in chemical bonds to fill their valence shells. ✳

• In an **ionic bond** an atom transfers one or more electrons to a different atom creating negatively and positively charged ions that then attract each other.
  - Ionic bonding is only observed when the electron transfer creates filled valence shells for both of the ions.
  - **Negatively charged ions** are called **anions** and **positively charged ions** are called **cations**.

• In a **covalent bond** atoms share pairs of electrons, thus increasing the number of electrons around each atom.  In this way the valence shell of each atom is filled.
  - Sharing of a pair of electrons holds the two atoms together, since the electrons being shared in the bond attract both of the positively-charged nuclei of the bonded atoms.
  - Organic chemistry is concerned primarily with covalent bonds.

• Electron pairs taking part in bonds are not necessarily shared evenly between the atoms.  The more **electronegative element** taking part in a bond attracts the majority of the electron density of the shared electrons. ✳
  - The unequal sharing of electrons (**percent of ionic character**) of a covalent bond can be determined quantitatively using a table of electronegativities of the elements.
  - On the periodic table of the elements, **electronegativity** increases from the bottom left hand corner to the upper right hand corner because atoms with more positively charged nuclei and smaller orbital radii are more electronegative.
  - The most commonly used scale of electronegativity is the Pauling scale of electronegativity in which the most electronegative element, fluorine (F), is given the value of 4.0, and all other elements are assigned values relative to F.

• The different types of bonds can be related to the electronegativity of the elements involved according to the following rules:

- If the elements taking part in a bond have electronegativities that differ by **more than 1.9**, then the bond is **ionic**.
- If the elements taking part in a bond have electronegativities that differ by **less than 1.9**, then the bond is **covalent**.

  If the elements taking part in a bond have electronegativities that differ by **less than 0.4**, than the bond is referred to as **nonpolar covalent**. If the elements taking part in a bond have electronegativities that differ by a value **between 0.4 and 1.9**, than the bond is referred to as **polar covalent.**

• In polar covalent bonds, the more electronegative element taking part in a bond has a partial negative charge, denoted by the symbol δ-, because it attracts a greater proportion of the shared electrons. As a result, the less electronegative element taking part in a bond has a partial positive charge, denoted by the symbol δ+.

• Knowing how to identify the electron-rich and electron-poor regions of molecules is the key to understanding and learning organic chemistry. ✳

- The unequal sharing of electrons in polar covalent bonds forms the basis for reactivity in a molecule, so understanding how electrons are distributed in a given molecule allows chemical reactions to be predicted accurately. *[By the time you have finished studying organic chemistry you should be able to look at the structure of a molecule, and then, based on your understanding of the electron distribution in the molecule, predict chemical reactions. This prediction approach is much more successful than trying to memorize reactions without understanding the reasons they take place.]*

• **Lewis structures** are used to represent molecules. *[Lewis structures are not as hard as they look, but you need a lot of practice to get the hang of them.]*

- In a Lewis structure, a line between two atoms represents a pair of electrons taking part in a bond and a pair of dots represents an unshared or lone pair of electrons.

• **To draw a Lewis structure:**

- **First**, count all of the valence electrons of each atom in the structure. Remember that in the neutral states, H has 1 valence electron, C has 4 valence electrons, N has 5 valence electrons, O has 6 valence electrons, and F has 7 valence electrons. Add a valence electron for each unit of negative charge on an atom or ion, and subtract a valence electron for each unit of positive charge.

- **Second**, draw single bonds (a line) between all of the atoms known to be connected to each other. Usually the connectivity information must be provided for you in the form of a condensed structural formula (Section 1.4 A). For example, for the condensed structural formula $CH_3CHO$ write down:

$$
\begin{array}{ccc}
 & H & O \\
 & | & \| \\
H- & C-C & -H \\
 & | & \\
 & H &
\end{array}
$$

- **Third**, draw any remaining bonds (double or triple bonds) and add any lone pairs of electrons that may be necessary so that each atom in the molecule is surrounded by a **filled valence shell** of electrons. ✳

  For H, a filled valence shell is 2 electrons (1 bond), and for C, N, O, and F; a filled valence shell is 8 electrons distributed as described in the following table:

| Atom | # Of Bonds (Single bonds count as 1 bond, double bonds as 2, and triple bonds as 3) | # Of Lone Pairs Of Electrons |
|:---:|:---:|:---:|
| C | 4 | 0 |
| N | 3 | 1 |
| O | 2 | 2 |
| F | 1 | 3 |

Finishing the example, two lone pairs of electrons are added to the oxygen atom and a double bond is added between the carbon and oxygen atoms to fill all valences and complete the structure:

$$
\begin{array}{c}
\text{H} \quad \ddot{\text{O}} \\
| \quad\quad || \\
\text{H}-\text{C}-\text{C}-\text{H} \\
| \\
\text{H}
\end{array}
$$

In most organic molecules, the halogens Cl, Br, and I are treated the same as F. On the other hand, the neutral atoms B, P, and S have some unusual bonding patterns as described in the following table.

| Atom | # Of Bonds (Single bonds count as 1 bond, double bonds as 2, and triple bonds as 3) | # Of Lone Pairs Of Electrons |
|---|---|---|
| B | 3 | 0 |
| P | 3 | 1 |
| P * | 5 | 0 |
| S | 2 | 2 |
| S | 4 | 1 |
| S * | 6 | 0 |

( * 3d orbitals are involved, so more than 8 valence electrons can be accommodated in the valence shell.)

If there is a negative formal charge (see below) on an atom, add a lone pair of electrons and use one less bond.

If there is a positive formal charge on an atom other than carbon, add a bond and use one less lone pair of electrons.

For carbon with a positive formal charge, you cannot add a bond because carbon already has four. In this case, you should use one less bond and no lone pairs (the carbon is surrounded by only six valence electrons).

• **Some Helpful Hints** for drawing Lewis structures:

- After drawing the single bonds between all connected atoms, draw the lone pairs of electrons. For neutral atoms, the number of lone pairs on a given element does not change from molecule to molecule.

In the neutral, uncharged state, C has zero lone pairs, N has one lone pair, O has two lone pairs and F has three lone pairs. For example, in the molecule $CO_2$, draw the single bonds and lone pairs of electrons around oxygen as follows:

$$\ddot{\text{O}}-\text{C}-\ddot{\text{O}}$$

- After drawing the lone pairs, fill in multiple bonds as necessary. If there are not enough single bonds to other atoms surrounding an atom to give a filled valence shell, then use multiple bonds.

In these cases you should draw the multiple bond(s) to the adjacent atom(s) that also need multiple bond(s) to fill their valence shell. The $CO_2$ example is completed by adding a double bond to each oxygen atom and thereby filling the valence shell of carbon as well.

$$\ddot{\text{O}}=\text{C}=\ddot{\text{O}}$$

- Computation of **formal charges** is a useful bookkeeping method for keeping track of charges on a molecule. ✳
  - For the computation of formal charge, the electrons in bonds are counted as being distributed evenly between the bonded atoms. One electron is counted for each atom taking part in a single bond, two electrons for each atom taking part in a double bond, and three electrons for each atom taking part in a triple bond.
  - Lone pairs of electrons <u>and</u> electrons in shells lower than the valence shell (**1s** electrons for C, N, O, F etc.) are counted as belonging entirely to the atoms to which they are attached.
  - Total formal charge derives from comparing the number of electrons counted as above to the number of protons in the nucleus of the atom (an extra electron results in a formal charge of -1, an extra proton results in a formal charge of +1, etc.).
  - The following tables show the formal charges associated with certain atoms that might be found in organic molecules and reaction intermediates.

## Table Of Atoms With +1 Formal Charge

| Atom | # Of Bonds (Single bonds count as 1 bond, double bonds as 2, and triple bonds as 3) | # Of Lone Pairs Of Electrons |
|------|------|------|
| H | 0 | 0 |
| C | 3 | 0 |
| N | 4 | 0 |
| O | 3 | 1 |
| S | 3 | 1 |
| P | 4 | 0 |

## Table Of Atoms With -1 Formal Charge

| Atom | # Of Bonds (Single bonds count as 1 bond, double bonds as 2, and triple bonds as 3) | # Of Lone Pairs Of Electrons |
|------|------|------|
| C | 3 | 1 |
| N | 2 | 2 |
| O | 1 | 3 |
| S | 1 | 3 |
| F | 0 | 4 |

## 1.3 FUNCTIONAL GROUPS

- Carbon combines with other atoms to form characteristic structural units called **functional groups** such as the hydroxyl group (-OH) of alcohols, the carbonyl group (C=O) of aldehydes and ketones, as well as the carboxyl groups (-CO$_2$H) of carboxylic acids. Thinking about organic molecules in terms of functional groups is important for three reasons: ✳
  - **First,** they are used to divide organic molecules into classes in terms of their physical properties.
  - **Second,** they are sites of chemical reactions, and a particular functional group, in whatever compound it is found, undergoes the same types of chemical reactions.
  - **Third,** they are the basis for naming compounds.
- **Condensed structural formulas** are highly abbreviated versions of Lewis structures that are used to describe molecules.

- The number of hydrogen atoms attached to a given atom are denoted with a subscript. For example -$CH_3$ means there are three hydrogen atoms attached to the carbon atom.
- Lines are used to denote bonds. A single line (-) between two atoms represents a single bond, a double line (=) denotes a double bond, etc. Lone pairs of electrons are not drawn.
- Parentheses are used to denote branching in a molecule. Whole groups of atoms attached to a given atom are placed in parentheses. For example, $(CH_3)_3CH$ indicates that there are three -$CH_3$ groups attached to the remaining carbon atom.

## 1.4 BOND ANGLES AND SHAPES OF MOLECULES

- The **Valence-Shell Electron-Pair Repulsion (VSEPR)** model of molecular structure assumes that *regions* of valence electron density around an atom are distributed to be as far apart as possible in three-dimensional space. ✷
  - When using the VSEPR model, lone pairs of electrons, the two electrons in a single bond, the four electrons in a double bond, and the six electrons in a triple bond are each counted as only a single area of electron density.
  - **Four areas of electron density** around an atom adopt a **tetrahedral shape** with bond angles near 109.5° such as in methane, $CH_4$.
  - **Three areas of electron density** around an atom adopt a **trigonal planar shape** with bond angles near 120° such as in formaldehyde, $H_2C=O$.
  - **Two areas** of electron density around an atom adopt a **linear shape** with bond angles near 180°, such as in acetylene, $HC\equiv CH$.
  - The VSEPR model predicts shape but does not explain why the electrons are located where they are. Thus, it is only a useful model and not a theory.

## 1.5 RESONANCE

- **Resonance theory** is used to depict and understand chemical species for which no single Lewis structure provides an adequate description. ✷ *[This also requires a lot of practice.]*
  - Resonance theory is particularly good at helping to understand cases of partial bonding (for example 1.5 bonds between two atoms, etc.) or when a formal charge is distributed over more than one atom. In these situations, the true structure (referred to as a **resonance hybrid**) is thought of as a composite of two or more **contributing structures**.

    In drawing resonance structures, a straight, double headed arrow (↔) is placed between contributing structures. Curved arrows (⤴) are used to indicate how electrons can be redistributed to make one contributing structure from another. Always draw the curved arrows to indicate where a pair of electrons started (tail of arrow) to where the electrons end up (head of arrow).
  - No atoms are moved between contributing structures, and only *certain* kinds of electrons are moved. ✷

    Lone pairs of electrons <u>or</u> a pair of electrons taking part in a multiple bond are moved, and these electrons can only move a short distance. A lone pair of electrons from an atom can only move to an adjacent bond to make a multiple bond. A pair of electrons in a multiple bond can only move to an adjacent atom to create a new lone pair of electrons or to an adjacent bond to make a multiple bond.
  - Contributing structures cannot have atoms with more than a filled valence shell of electrons.✷

    There can be no more than 2 electrons around H or more than 8 electrons around C, N, O, or F.

    You cannot have more than a filled valence shell because there are no more orbitals in which to place the extra electrons. On the other hand, atoms can have less than the filled valence shell, for example a carbon atom with +1 formal charge only has 6 valence electrons and one empty **2**p orbital.

• Keep in mind that even though we use multiple contributing structures to describe a molecule or ion, the molecule or ion in fact only has <u>one</u> true structure.  It does <u>not</u> alternate between the contributing structures. *

## 1.6 VALENCE BOND MODEL OF COVALENT BONDS

• According to **valence bond theory**, a covalent bond is formed when two atomic orbitals overlap, that is when they occupy the same region in space.  The greater the overlap, the stronger the bond. *
  - Each electron in a covalent bond is delocalized over the two atomic orbitals that overlap to form the bond.  The majority of electron density is thus concentrated between atoms taking part in bonds, consistent with Lewis structures.
• For elements more complicated than hydrogen, it is helpful to combine (**hybridize**) the *valence* atomic orbitals on a given atom before looking for overlap with orbitals from other atoms. *
  - For C, N, and O hybridization means the **2**s atomic orbital is combined with one, two, or all three **2**p atomic orbitals, respectively.
• The results of the orbital combinations are called **hybrid orbitals**, the number of hybrid orbitals are equal to the number of atomic orbitals combined.
  - An **sp$^3$ hybrid orbital** is the combination of the **2**s orbital with three **2**p orbitals.

    Four sp$^3$ orbitals of equivalent energy are created.

    Each sp$^3$ orbital has one large lobe and a smaller one of opposite sign pointing in the opposite direction (with a node at the nucleus).  The large lobes point to a different corners of a tetrahedron.  This explains the tetrahedral structure of molecules like methane, CH$_4$.
  - An **sp$^2$ hybrid orbital** is the result of combining the **2**s orbital with two **2**p orbitals.

    Three sp$^2$ orbitals of equivalent energy are created.

    Each sp$^2$ orbital has one large lobe and a smaller one of opposite sign pointing in the opposite direction (with a node at the nucleus).  The large lobes point to a different corner of a triangle.  This explains the trigonal planar structure of molecules like formaldehyde, CH$_2$=O.

    The left over **2**p orbital lies perpendicular to the plane formed by the three sp$^2$ orbitals.
  - An **sp hybrid orbital** is the combination of one **2**s orbital with one **2**p orbital.

    Two sp orbitals of equivalent energy are created.

    Each sp orbital has two lobes of opposite sign pointing in opposite directions (with a node at the nucleus).

    The lobes with like sign point in exactly opposite directions (180° with respect to each other).  This explains the linear structure of molecules like acetylene, HC≡CH.

    The two left over **2**p orbitals are orthogonal to each other, and orthogonal to the two sp hybrid orbitals as well.
  - Carbon atoms in molecules are either sp$^3$, sp$^2$, or sp hybridized.  The **1**s orbitals are not considered for hybridization with C, N, or O because these **1**s orbitals do not participate in covalent bonding.
• Bonding in complex molecules can be *qualitatively* understood as overlap of hybrid orbitals. *
• Organic chemistry is primarily concerned with two types of covalent bonds, namely sigma bonds and pi bonds.
  - A σ **(sigma) bond** occurs when the majority of the electron density is found on the bond axis.

    A σ bond results from the overlap between an s orbital and any other atomic orbital.

    A σ bond also results from the overlap of other orbitals such as overlap of an sp$^3$, sp$^2$, or sp hybrid orbital with an s, sp$^3$, sp$^2$, or sp hybrid orbital.

    Because rotating a σ bond does not decrease the overlap of the orbitals involved (σ bonds have cylindrical symmetry), a σ bond can rotate freely about the bond axis.
  - A π **(pi) bond** occurs when the majority of the electron density is found above and below the bond axis.

A $\pi$ bond results from the overlap of two **2p** orbitals that are parallel to each other, and orthogonal to the $\sigma$ bond that exists between the two atoms.

Because rotating a $\pi$ bond by 90° destroys the orbital overlap, $\pi$ bonds cannot rotate around the bond axis. *[Understand this before going on.]*

- The hybridization of a given atom ($sp^3$, $sp^2$, or $sp$) determines the geometry and type of bonds made by that atom. The important parameters associated with each hybridization state are listed in the following table.

## Carbon Atom Hybridization State Parameters

| Hybridiztion State | # Of Hybrid Orbitals | # Of 2p Orbitals Left Over | # Of Groups Bonded To Carbon | # Of σ Bonds | # Of π Bonds | Geometry Around Carbon |
|---|---|---|---|---|---|---|
| $sp^3$ | 4 | 0 | 4 | 4 | 0 | Tetrahedral |
| $sp^2$ | 3 | 1 | 3 | 3 | 1 | Trigonal Planar |
| $sp$ | 2 | 2 | 2 | 2 | 2 | Linear |

- To summarize, valence bond theory is based on the approximation that bonding is the result of overlap of atomic orbitals, and that electron density is concentrated between atoms. This approach is intuitively consistent with Lewis structures, and is extremely useful for qualitatively understanding molecular structure and reactivity in organic chemistry.

# CHAPTER 1
## *Solutions to the Problems*

<u>Problem 1.1</u> Write and compare the ground-state electron configurations for
(a) Carbon and silicon

**C (6 electrons) $1s^2 2s^2 2p^2$**
**Si (14 electrons) $1s^2 2s^2 2p^6 3s^2 3p^2$**

**Both carbon and silicon have four electrons in their highest (valence) shells.**

(b) Oxygen and sulfur

**O (8 electrons) $1s^2 2s^2 2p^4$**
**S (16 electrons) $1s^2 2s^2 2p^6 3s^2 3p^4$**

**Both oxygen and sulfur have six electrons in their highest (valence) shells.**

(c) Nitrogen and phosphorus

**N (7 electrons) $1s^2 2s^2 2p^3$**
**P (15 electrons) $1s^2 2s^2 2p^6 3s^2 3p^3$**

**Both nitrogen and phosphorus have five electrons in their highest (valence) shells.**

<u>Problem 1.2</u> Show that the following obey the octet rule.
(a) Sulfur (atomic number 16) forms sulfide ion, $S^{2-}$.

**S (16 electrons): $1s^2 2s^2 2p^6 3s^2 3p^4$**

**$S^{2-}$ (18 electrons): $1s^2 2s^2 2p^6 3s^2 3p^6$**

(b) Magnesium (atomic number 12) forms $Mg^{2+}$.

**Mg (12 electrons): $1s^2 2s^2 2p^6 3s^2$**

**$Mg^{2+}$ (10 electrons): $1s^2 2s^2 2p^6$**

<u>Problem 1.3</u> Judging from their relative positions in the periodic table, which element is more electronegative?
(a) Lithium or potassium

**In general, electronegativity increases from left to right across a row and from bottom to top of a column in the periodic table. Lithium is higher up on the table and thus more electronegative than potassium.**

(b) Nitrogen or phosphorus

**Nitrogen is higher up on the table and thus more electronegative than phosphorus.**

(c) Carbon or silicon

**Carbon is higher up on the table and thus more electronegative than silicon.**

Problem 1.4  Classify these bonds as nonpolar covalent, polar covalent, or ionic.
(a) S-H          (b) P-H          (c) C-F          (d) C-Cl

| Bond | Differences in electronegativity | Type of bond |
|------|------|------|
| S-H | 2.5 - 2.1 = 0.4 | nonpolar covalent |
| P-H | 2.1 - 2.1 = 0 | nonpolar covalent |
| C-F | 4.0 - 2.5 = 1.5 | polar covalent |
| C-Cl | 3.0 - 2.5 = 0.5 | polar covalent |

Problem 1.5  Indicate the direction of polarity in these polar covalent bonds using the symbols δ- and δ+.
(a) C-N

$$\overset{\delta+\quad\delta-}{C-N}$$

**Nitrogen is more electronegative than carbon**

(b) N-O

$$\overset{\delta+\quad\delta-}{N-O}$$

**Oxygen is more electronegative than nitrogen**

(c) C-Cl

$$\overset{\delta+\quad\delta-}{C-Cl}$$

**Chlorine is more electronegative than carbon**

Problem 1.6  Draw Lewis structures, showing all valence electrons, for these covalent molecules.
(a) $C_2H_6$                          (b) $CS_2$                          (c) HCN

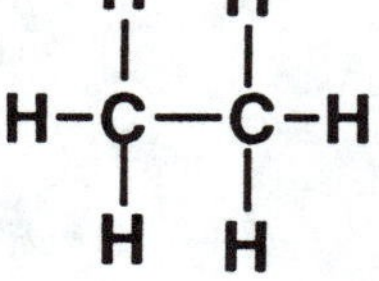

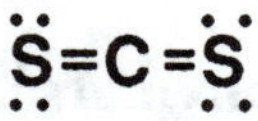

Problem 1.7  Draw Lewis structures for these ions, and show which atom in each bears the formal charge.
(a) $CH_3NH_3^+$                  (b) $CO_3^{2-}$                  (c) $OH^-$
   Methyl ammonium ion            Carbonate ion                  Hydroxide ion

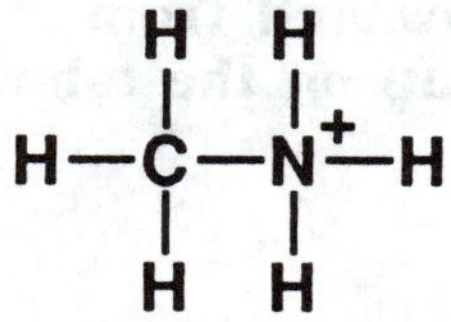

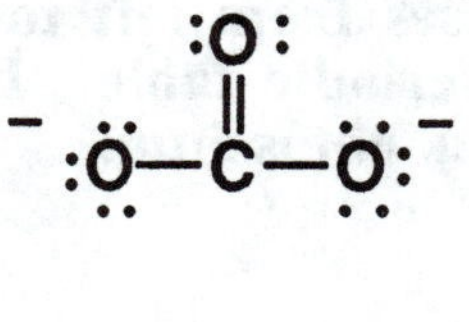

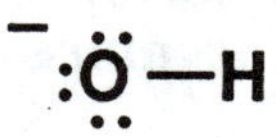

**Problem 1.8** There are four alcohols of molecular formula $C_4H_{10}O$. Draw a condensed structural formula for each one.

CH$_3$-CH$_2$-CH$_2$-CH$_2$-OH

$$\underset{\underset{\overset{|}{\text{OH}}}{}}{CH_3\text{-}CH_2\text{-}CH\text{-}CH_3}$$

$$CH_3\text{-}\overset{\overset{\displaystyle OH}{|}}{\underset{\underset{\displaystyle CH_3}{|}}{C}}\text{-}CH_3$$

$$CH_3\text{-}\overset{\overset{\displaystyle CH_3}{|}}{CH}\text{-}CH_2\text{-}OH$$

**Problem 1.9** Draw condensed structural formulas for the two aldehydes of molecular formula $C_4H_8O$.

$$CH_3\text{-}CH_2\text{-}CH_2\text{-}\overset{\overset{\displaystyle O}{\|}}{CH}$$

$$CH_3\text{-}\overset{\overset{\displaystyle O}{\|}}{\underset{\underset{\displaystyle CH_3}{|}}{CH}}\text{-}CH$$

**Problem 1.10** Draw condensed structural formulas for the two carboxyic acids of molecular formula $C_4H_8O_2$.

$$CH_3\text{-}CH_2\text{-}CH_2\text{-}\overset{\overset{\displaystyle O}{\|}}{C}\text{-OH}$$

$$CH_3\text{-}\overset{\overset{\displaystyle O}{\|}}{\underset{\underset{\displaystyle CH_3}{|}}{CH}}\text{-}C\text{-OH}$$

**Problem 1.11** Predict all bond angles for these molecules.
(a) $CH_3OH$          (b) $CH_2Cl_2$          (c) $H_2CO_3$ (carbonic acid)

**Problem 1.12** Use curved arrows to show the redistribution of electrons in converting contributing structure (a) to (b) and then (b) to (c).

(a)                                        (b)                                        (c)

**Problem 1.13** Describe the bonding in these molecules in terms atomic orbitals involved, and predict all bond angles.

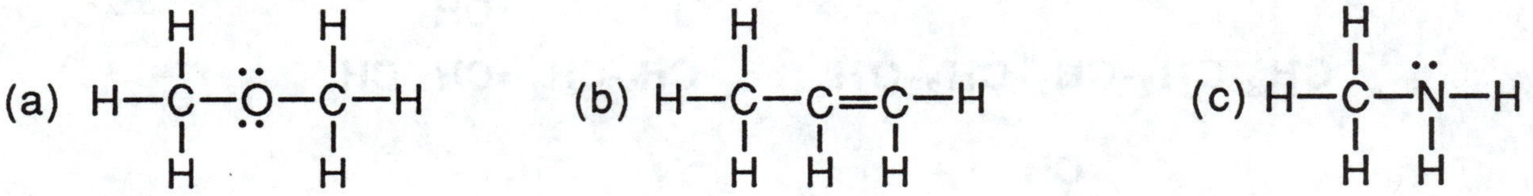

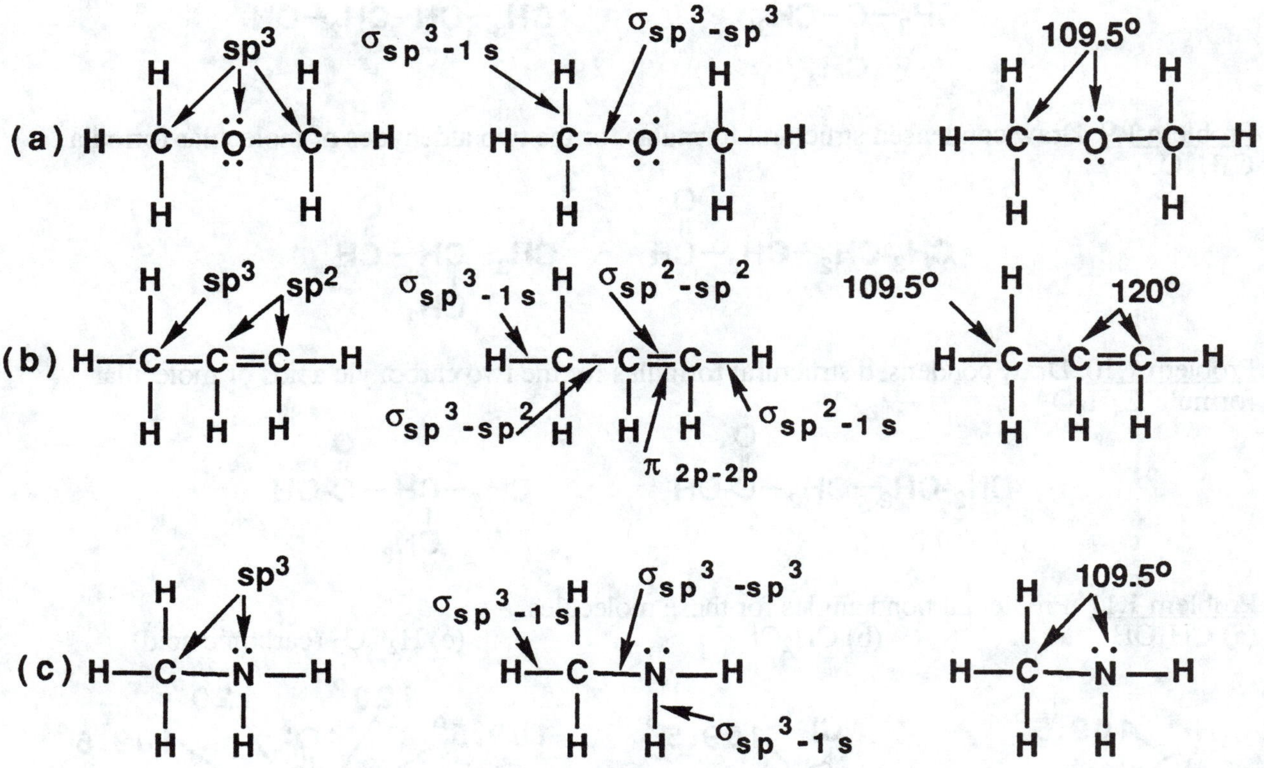

(a) 
```
    H       H
    |       |
H—C—Ö—C—H
    |   ··  |
    H       H
```

(b)
```
    H
    |
H—C—C=C—H
    |   |  |
    H   H  H
```

(c)
```
    H       ··
    |       N—H
H—C—N—H
    |       |
    H       H
```

(a)

(b)

(c)

## Electronic Structures of Atoms

**Problem 1.14** Write ground-state electron configurations for these atoms. After each atom is given its atomic number.

(a) Sodium (11)

$$\text{Na (11 electrons) } 1s^2 2s^2 2p^6 3s^1$$

(b) Magnesium (12)

$$\text{Mg (12 electrons) } 1s^2 2s^2 2p^6 3s^2$$

(c) Oxygen (8)

$$\text{O (8 electrons) } 1s^2 2s^2 2p^4$$

(d) Nitrogen (7)

$$\text{N (7 electrons) } 1s^2 2s^2 2p^3$$

<u>Problem 1.15</u>  Which element has the ground-state electron configuration of
(a) $1s^22s^22p^63s^23p^4$

**Sulfur (16) has this ground-state electron configuration**

(b) $1s^22s^22p^4$

**Oxygen (8) has this ground-state electron configuration**

<u>Problem 1.16</u>  Write the ground-state electron configurations of these elements.
(a) Potassium

**K (19 electrons)  $1s^22s^22p^63s^23p^64s^1$**

(b) Aluminum

**Al (13 electrons)  $1s^22s^22p^63s^23p^1$**

(c) Phosphorus

**P (15 electrons)  $1s^22s^22p^63s^23p^3$**

(d) Argon

**Ar (18 electrons)  $1s^22s^22p^63s^23p^6$**

<u>Problem 1.17</u>  Define valence shell and valence electron.

**The valence shell is the outermost occupied shell of an atom, namely the highest numbered shell.  A valence electron is an electron in the valence shell.**

<u>Problem 1.18</u>  How many electrons are in the valence shell of each atom?
(a) Carbon

**With a ground-state electron configuration of $1s^22s^22p^2$ there are four electrons in the valence shell of carbon.**

(b) Nitrogen

**With a ground-state electron configuration of $1s^22s^22p^3$ there are five electrons in the valence shell of nitrogen.**

(c) Chlorine

**With a ground-state electron configuration of $1s^22s^22p^63s^23p^5$ there are seven electrons in the valence shell of chlorine.**

(d) Aluminum

**With a ground-state electron configuration of $1s^22s^22p^63s^23p^1$ there are three electrons in the valence shell of aluminum.**

### Lewis Structures

**Problem 1.19** Judging from their relative positions in the periodic table, which atom is more electronegative?
(a) Carbon or nitrogen

**In general, electronegativity increases from left to right across a row and from bottom to top of a column in the periodic table. Nitrogen is farther to the right on the table and thus more electronegative than carbon.**

(b) Chlorine or bromine

**Chlorine is higher up on the table and thus more electronegative than bromine.**

(c) Oxygen or sulfur

**Oxygen is higher up on the table and thus more electronegative than sulfur.**

**Problem 1.20** Which of these compounds have covalent bonds and which have ionic bonds?
(a) LiF          (b) $CH_3F$        (c) $MgCl_2$        (d) HCl

**Using the rule that an ionic bond is formed between atoms with an electronegativity difference of 1.9 or greater, the following table can be constructed:**

| Bond | Differences in electronegativity | Type of bond |
|---|---|---|
| Li-F | 4.0 - 1.0 = 3.0 | ionic |
| C-H | 2.5 - 2.1 = 0.4 | nonpolar covalent |
| C-F | 4.0 - 2.5 = 1.5 | polar covalent |
| Mg-Cl | 3.0 - 1.2 = 1.8 | polar covalent |
| H-Cl | 3.0 - 2.1 = 0.9 | polar covalent |

**Based on these values, only LiF has an ionic bond, the other compounds have only covalent bonds.**

**Problem 1.21** Using the symbols $\delta$- and $\delta$+, indicate the direction of polarity, if any, in each covalent bond.
(a) C-Cl

$\delta$+  $\delta$-
**C-Cl**

**Chlorine is more electronegative than carbon**

(b) S-H

$\delta$-   $\delta$+
**S-H**

**Sulfur is more electronegative than hydrogen**

(c) C-S

**Carbon and sulfur have the same electronegativities so there is no direction of polarity in a C-S bond**

(d) P-H

**Phosphorus and hydrogen have the same electronegativities, so there is no direction of polarity in a P-H bond**

Problem 1.22  Write Lewis structures for these molecules.  Be certain to show all valence electrons.  None of these compounds contains a ring of atoms.

(a) $H_2O_2$
   Hydrogen peroxide

(b) $N_2H_4$
   Hydrazine

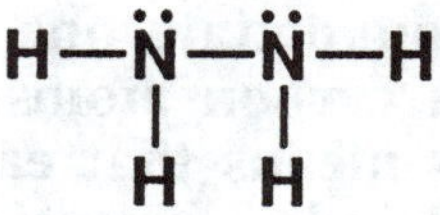

(c) $CH_3OH$
   Methanol

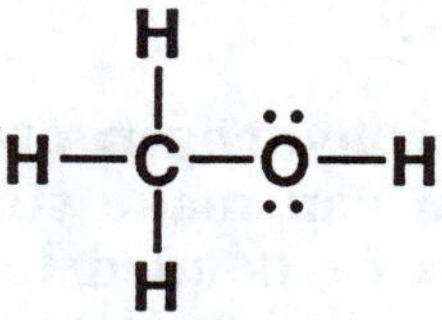

(d) $CH_3SH$
   Methanethiol

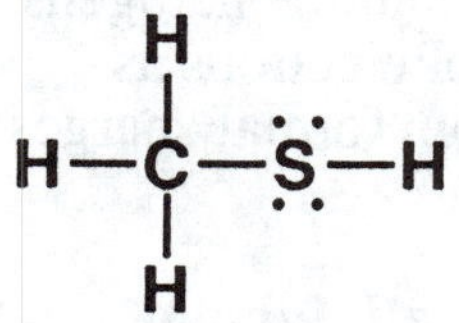

(e) $CH_3NH_2$
   Methylamine

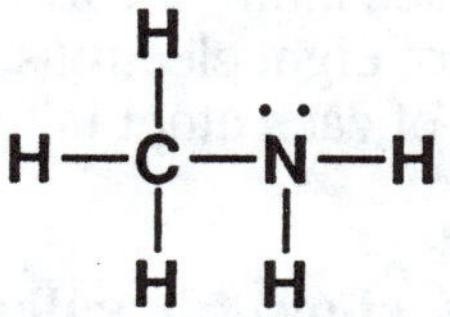

(f) $CH_3Cl$
   Chloromethane

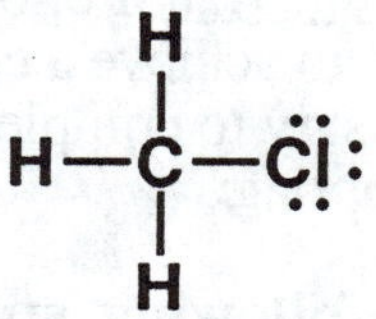

(g) $CH_3OCH_3$
   Dimethyl ether

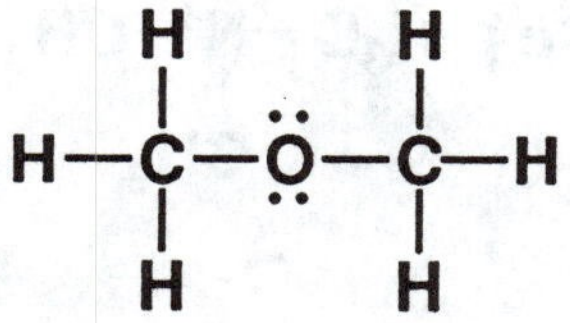

(h) $C_2H_6$
   Ethane

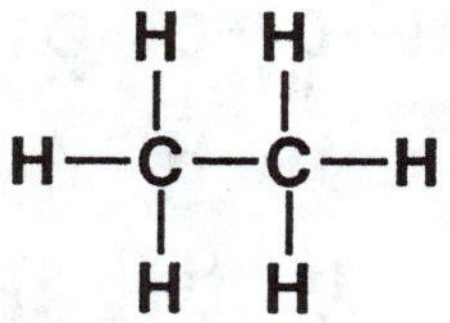

(i) $C_2H_4$
   Ethene

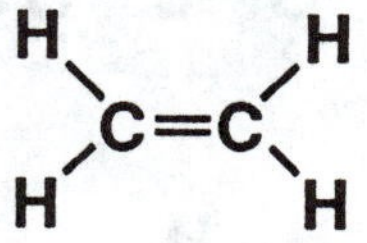

(j) $C_2H_2$
   Ethyne

(k) $CO_2$
   Carbon dioxide

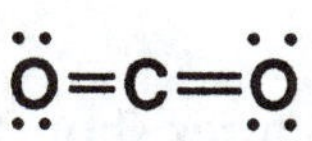

(l) $CH_2O$
   Methanal

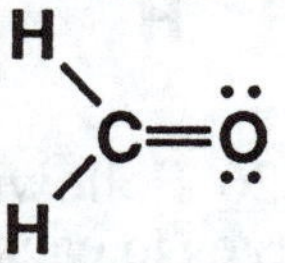

Problem 1.23  Write Lewis structures for these acids.  After each molecular formula is given the arrangement of atoms.

(a) $H_2CO_3$
   Carbonic acid

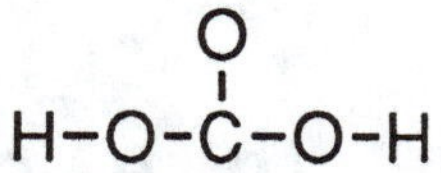

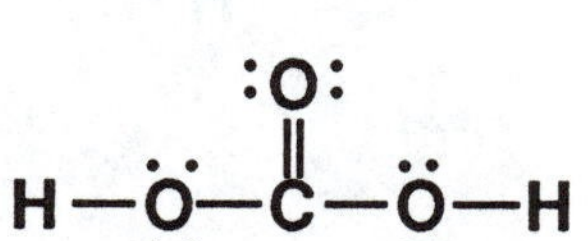

(b) $CH_3CO_2H$
   Ethanoic acid

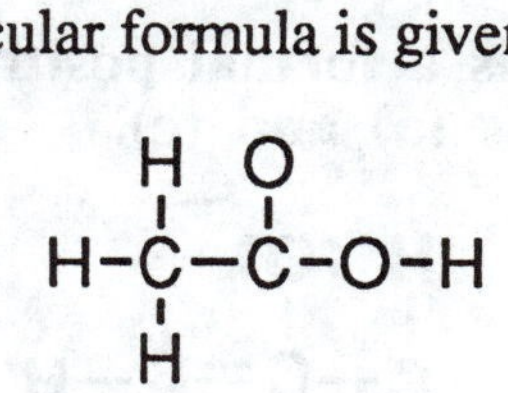

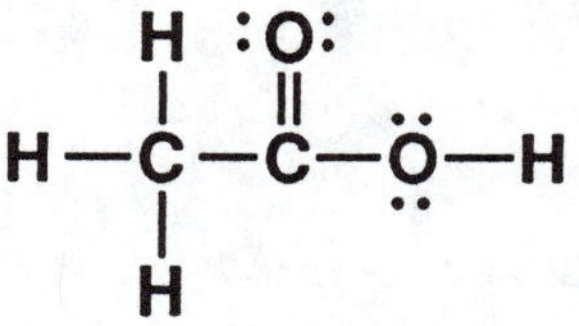

<u>Problem 1.24</u> Why are the following molecular formulas impossible?
(a) $CH_5$

**Carbon atoms can only accommodate four bonds, and each hydrogen atom can only accommodate one bond. Thus, there is no way for a stable bonding arrangement to be created that utilizes one carbon atom and all five hydrogen atoms.**

(b) $C_2H_7$

**Since hydrogen atoms can only accommodate one bond each, no single hydrogen atom can make stable bonds to both carbon atoms. Thus, the two carbon atoms must be bonded to each other. This means that each of the bonded carbon atoms can accommodate only three more bonds. Therefore, only six hydrogen atoms can be bonded to the carbon atoms, not seven hydrogen atoms.**

<u>Problem 1.25</u> Following the rule that each atom of carbon, oxygen, nitrogen, and the halogens reacts to achieve a complete outer shell of eight electrons, add unshared pairs of electrons as necessary to complete the valence shell of each atom in these ions. Then, assign formal charges as appropriate.

**The following structural formulas show all valence electrons and all formal charges.**

(a) H—Ö—C—Ö:⁻ with :O: above C

(b) H—C—C—Ö:⁻ (ethyl group with O)

(c) $H_3C$—N⁺—$CH_3$ with $CH_3$ above and below

(d) H—C—C:⁻ (with H's)

(e) H—N⁺—C—C with :O: and O:⁻

<u>Problem 1.26</u> Following are several Lewis structures showing all valence electrons. Assign formal charges to each structure as appropriate.

**There is a formal positive charge in parts (b). There is a formal negative charge in parts (a) and (c).**

(a) H—C—C=C—H with :Ö:⁻ above middle C

(b) H—C—C⁺—C—H

(c) H—C—C—H with :O: above and charges

**Problem 1.27** Write Lewis structures for these ions. Be certain to show all valence electrons and all formal charges.

(a) OH⁻
   Hydroxide ion

(b) HCO₃⁻
   Bicarbonate ion

(c) CO₃²⁻
   Carbonate ion

(d) Cl⁻
   Chloride ion

**Problem 1.28** Some of these compounds contain only covalent bonds, some contain only ionic bonds, and some contain both ionic and covalent bonds. Draw a Lewis structure for each compound. Show covalent bonds by dashes and ionic bonds by positive and negative charges.

(a) NaCl

(b) NaOH

(c) NaHCO₃

(d) CH₃Cl

(e) CH₃OH

(f) CH₃ONa

(g) MgCl₂

(h) CH₂Cl₂

**Functional Groups**

**Problem 1.29** Draw Lewis structures for the following functional groups. Be certain to show all valence electrons on each.

(a) Carbonyl group

(b) Carboxyl group

(c) Hydroxyl group

**Problem 1.30** Draw the structural formula for a compound of molecular formula C₂H₆O that contains a hydroxyl group.

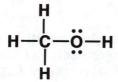

<u>Problem 1.31</u> Draw the structural formula of molecular formula $C_3H_6O$ that is
(a) An aldehyde                                    (b) A ketone

## Polarity of Covalent Bonds

<u>Problem 1.32</u> Which of these statements are true about electronegativity?
(a) Electronegativity increases from left to right in a row of the periodic table.
(b) Electronegativity increases from the top to the bottom in a column of the periodic table.
(c) Hydrogen, the element with the lowest atomic number, has the smallest electronegativity.
(d) The higher the atomic number of an element, the greater its electronegativity.

**Electronegativity increases from left to right across a row and from bottom to top of a column in the periodic table. Thus, statement (a) is true, but (b), (c), and (d) are false.**

<u>Problem 1.33</u> Why does fluorine, the element in the upper right corner of the periodic table, have the largest electronegativity of any element?

**Electronegativity increases with increasing positive charge on the nucleus and with decreasing distance of the valence electrons from the nucleus. Fluorine is that element for which these two parameters lead to maximum electronegativity.**

<u>Problem 1.34</u> Arrange the single covalent bonds within each set in order of increasing polarity (difference in electronegativity).
(a) C-H, O-H, N-H          (b) C-H, C-Cl, C-I          (c) C-S, C-O, C-N
   **C-H < N-H < O-H**          **C-I < C-H < C-Cl**          **C-S < C-N < C-O**
   **(0.4)  (0.9)   (1.4)**          **(0)   (0.4)   (0.5)**          **(0)     (0.5)   (1.0)**

(d) C-Li, C-B, C-Mg
   **C-B < C-Mg < C-Li**
   **(0.5)   (1.3)    (1.5)**

**The difference in electronegativities is shown in parentheses under each bond.**

<u>Problem 1.35</u> Using the values of electronegativity given in Table 1.5, predict which indicated bond in each set is more polar.
(a) $CH_3$-OH or $CH_3$O-H          (b) H-$NH_2$ or $CH_3$-$NH_2$

   **$CH_3$O-H**                              **H-$NH_2$**

(c) $CH_3$-SH or $CH_3$S-H          (d) $CH_3$-F or H-F

   **$CH_3$S-H**                              **H-F**

Problem 1.36 Using the symbols δ+ and δ-, show the direction of polarity in each indicated bond in problem 1.35.

(a)
$\overset{\delta+ \quad \delta-}{H_3C\text{-}OH}$    $\overset{\delta- \quad \delta+}{CH_3O\text{-}H}$

(b)
$\overset{\delta+ \quad \delta-}{H\text{-}NH_2}$    $\overset{\delta+ \quad \delta-}{H_3C\text{-}NH_2}$

(c)
$\overset{\delta+ \quad \delta-}{H_3C\text{-}SH}$    $\overset{\delta- \quad \delta+}{CH_3S\text{-}H}$

(d)
$\overset{\delta+ \quad \delta-}{H_3C\text{-}F}$    $\overset{\delta+ \quad \delta-}{H\text{-}F}$

Problem 1.37 Following are three organometallic compounds. Predict whether each carbon-metal bond is nonpolar covalent, polar covalent, or ionic.

(a) 
$$CH_3CH_2-\underset{\underset{CH_2CH_3}{|}}{\overset{\overset{CH_2CH_3}{|}}{Pb}}-CH_2CH_3$$
Tetraethyllead

(b) $CH_3-Mg-Cl$
Methylmagnesium chloride

(c) $CH_3-Hg-CH_3$
Dimethylmercury

**All of the above carbon-metal bonds are polar covalent.**

Problem 1.38 Identify the most polar bond in each molecule.
(a) $HSCH_2CH_2OH$          (b) $CHCl_2F$          (c) $HOCH_2CH_2NH_2$

**The O-H bond**          **The C-F bond**          **The O-H bond**

## Bond Angles and Shapes of Molecules
Problem 1.39 Use the VESPR model to predict bond angles about each highlighted atom in these molecules.

**Approximate bond angles as predicted by the valence-shell electron-pair repulsion model are as shown.**

Problem 1.40 Use the VSEPR model to predict bond angles about each atom of carbon, nitrogen, and oxygen in these molecules. *Hint*; first add unshared pairs of electrons as necessary to complete the valence shell of each atom. Then, make your predictions of bond angles.

(a) $109.5°$ $CH_3-CH_2-\ddot{O}-H$

(b) $109.5°$ $:O:$ $120°$ $CH_3-CH_2-C-H$

(c) $109.5°$ $120°$ $CH_3-CH=CH_2$

(d) $109.5°$ $180°$ $CH_3-C\equiv C-H$

(e) $109.5°$ $:O:$ $109.5°$ $CH_3-C-\ddot{O}-CH_3$ $120°$

(f) $109.5°$ $109.5°$ $CH_3$ $CH_3-N-CH_3$ $109.5°$

Problem 1.41 Silicon is immediately under carbon in the periodic table. Predict the geometry of silane, $SiH_4$.

**Silicon is in Group 4 of the periodic table, and like carbon, has four valence electrons. In silane, $SiH_4$, silicon is surrounded by four regions of electron density. Therefore, predict all H-Si-H bond angles to be $109.5°$, so the molecule is tetrahedral around Si.**

Problem 1.42 Phosphorus is immediately under nitrogen in the periodic table. Predict the molecular formula for phosphine, the compound formed from phosphorus and hydrogen. Predict the H-P-H bond angle in phosphine.

**Like nitrogen, phosphorus has five valence electrons, so predict that phosphine has the molecular formula of $PH_3$ in analogy to ammonia, $NH_3$. In phosphine, the phosphorus atom is surrounded by four areas of electron density; one lone pair of electrons and single bonds to three hydrogen. Therefore, predict all H-P-H bond angles to be $109.5°$, so the molecule is thus pyramidal.**

## Resonance and Contributing Structures

Problem 1.43 Draw the contributing structure indicated by the curved arrow(s) and assign formal charges as appropriate.

(a) $H-\ddot{O}-C$ (with $:O:$ and $:O:^-$) $\longleftrightarrow$ $H-\ddot{O}-C$ (with $:O:^-$ and $:O:$)

(b) $H_2C=\ddot{O}$ $\longleftrightarrow$ $H_2\overset{+}{C}-\ddot{O}:^-$

(c)

**Problem 1.44** Using the VSEPR model, predict the bond angles about the carbon atom in each pair of contributing structures in problem 1.43. In what way do the bond angles change from one contributing structure to the other?

**As shown on the models below, carbon atoms that are cations are trigonal planar, exhibiting bond angles of 120°, just like the neutral carbon atoms that are sp² hybridized. Therefore, the bond angles do not change from one contributing structure to the other.**

(a)

(b)

(c)

**Problem 1.45** Which of these statements are true about resonance contributing structures?
(a) All contributing structures must have the same number of valence electrons.
(b) All contributing structures must have the same arrangement of atoms. ✓
(c) All atoms in a contributing structure must have complete valence shells.
(d) All bond angles in sets of contributing structures must be the same. ✓

**For sets of contributing structures, electrons (usually π electrons or lone pair electrons) move, but the atomic nuclei maintain the same arrangement in space. Thus, statements (b) and (d) are true. In addition, the total number of electrons, valence and inner shell electrons, in each contributing structure must be the same, so statement (a) is also true. However, the movement of electrons often leaves one or more atoms without a filled valence shell in a given contributing structure, so statement (c) is false.**

Problem 1.46  State the orbital hybridization of each highlighted atom.

(a) [structure with $sp^3$ labeled carbon]

(b) [structure with $sp^2$ labeled carbon]

(c) [structure with $sp$ labeled carbon]

(d) [structure with $sp^3$ labeled nitrogen]

(e) [structure with $sp^2$ and $sp^3$ labeled oxygens]

(f) [structure with $sp^2$ and $sp$ labeled carbons]

Problem 1.47  Describe each highlighted bond in terms of the overlap of atomic orbitals.

(a) [structure] $\sigma$ $sp^3$–$sp^3$

(b) [structure] $\sigma$ $sp^3$–$sp^3$

(c) [structure] $\sigma$ $sp^3$–$sp^3$

Problem 1.48  Describe each highlighted bond in terms of the overlap of atomic orbitals.

(a) [structure] $\sigma$ $sp^2$–$sp^2$  $\pi$ $2p$–$2p$

(b) [structure] $\sigma$ $sp^2$–$sp^2$  $\pi$ $2p$–$2p$

(c) [structure] $\sigma$ $sp^2$–$sp^3$

Problem 1.49  Following is the structural formula of benzene, $C_6H_6$.

[benzene structure]

(a) Predict each H-C-C bond angle in benzene; predict each C-C-C bond angle.

**Each carbon atom in benzene has three centers of electron density around it, so according to the VSEPR model, the carbon atoms are trigonal planar. Predict each H-C-C bond angle to be 120° and each C-C-C bond angle to be 120°.**

(b) State the hybridization of each carbon atom in benzene.

**Each carbon atom is sp$^2$ hybridized because each one makes three σ bonds and one π bond.**

(c) Predict the shape of benzene.

**Since all of the carbon atoms in the ring are sp$^2$ hybridized and thus trigonal planar, predict carbon atoms in benzene to form a flat hexagon in shape, with the hydrogen atoms in the same plane as the carbon atoms.**

# CHAPTER 2: ALKANES AND CYCLOALKANES

## SUMMARY OF REACTIONS

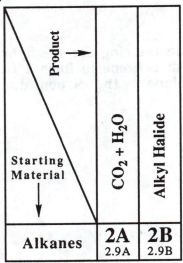

## REACTION 2A: OXIDATION (Section 2.9A)
- Alkanes react with $O_2$ to give $CO_2$, $H_2O$, and heat.

$$CH_4 + 2O_2 \longrightarrow CO_2 + 2H_2O$$

- This reaction is the basis for using alkanes as sources for heat and energy.

## REACTION 2B: HALOGENATION (Section 2.9B)
- Alkanes react with $Cl_2$ and $Br_2$ in the presence of light or heat by a substitution reaction.

$$CH_4 + Cl_2 \longrightarrow CH_3Cl + HCl$$

- In a substitution reaction, one atom or group of atoms is replaced by another atom or group of atoms.
- The reaction can proceed further if more halogen is used.

$$CH_3Cl + Cl_2 \longrightarrow CH_2Cl_2 + HCl$$

-The product **haloalkane** is named by using the same rules used for alkanes (Section 2.3), but the prefixes **fluoro, chloro, bromo,** and **iodo** are used for the halogen that is treated like any other substituent.
• When complex alkanes react with halogens several different products are formed, since the halogen can substitute for different hydrogen atoms. *[This may seem simple enough, but it can get very complicated when complex alkanes are considered]*
  - The substitutions observed in the products are **regioselective**, that is substitution is not statistically random.
      In general, tertiary hydrogen atoms are replaced in preference to secondary hydrogen atoms, and secondary hydrogen atoms are replaced in preference to primary hydrogen

atoms. *[Be sure to review the definitions of tertiary hydrogens, secondary hydrogens, and primary hydrogens in Section 2.3 if necessary.]*
Bromination is more regioselective than chlorination. *[At this point it is not prudent to explain these selectivities, but you will be able to understand them when the mechanism of the reaction is presented in Chapter 5.]*

- The reaction of alkanes with $F_2$ is seldom used because it is highly exothermic and difficult to control.
- Because of their unique physical properties, haloalkanes are widely used as solvents (methylene chloride), refrigerants (Freon-11), non-stick coatings (Teflon), artificial blood replacements (perfluorodecalin), and anesthetics (Halothane).
- Recent concern over possible destructive effects of certain haloalkanes like chlorofluorocarbons (CFCs) in the ozone layer has lead to attempts to limit their use.

## SUMMARY OF IMPORTANT CONCEPTS

### 2.0 OVERVIEW
• A **hydrocarbon** is a molecule that contains only carbon and hydrogen, and an **alkane** is a hydrocarbon that contains only single bonds. ✳

### 2.1 STRUCTURE OF ALKANES
• Alkanes have the general formula $C_nH_{2n+2}$.
• The carbon atoms of alkanes are $sp^3$ hybridized and thus tetrahedral, with bond angles of approximately 109.5°.

### 2.2 CONSTITUTIONAL ISOMERISM IN ALKANES
• **Constitutional isomers** are two or more molecules that have the same molecular formula but the atoms are attached to each other in different ways. (This was discussed previously in Section 1.5).
 - Constitutional isomers have different chemical properties.
 - For methane ($CH_4$), ethane ($C_2H_6$), and propane ($C_3H_8$) there is only one way to attach the carbon atoms to each other, hence there are no constitutional isomers of these alkanes. For alkanes with four or more carbon atoms, the number of constitutional isomers is given by a complex mathematical formula called a generating function.
 - There is no foolproof way to find all constitutional isomers for a given molecular formula, you must use a combination of a systematic method and creativity. *[This is harder than it looks and requires a great deal of practice.]*
 - A reasonable system to answer the constitutional isomer questions is to first write all possible carbon skeletons by starting with the straight chain alkane, then systematically adding appropriate branches.

### 2.3 NOMENCLATURE OF ALKANES
• To name organic compounds, chemists use systematic nomenclature rules established by the **International Union of Pure and Applied Chemistry (IUPAC)**. ✳
• For simple, unbranched alkanes the name consists of a **prefix** and a **suffix**. ✳
 - The prefix indicates the number of carbon atoms. For example, the prefix "**prop**" means three carbon atoms. For a more complete list of prefixes see Table 2.2 in the text. *[It is important to learn these now because the rest of the book assumes you are familiar with these names.]*
 - Following the prefix, the suffix **ane** is used to designate that a compound is an alkane. For example, **propane** is an alkane with three carbon atoms ($CH_3CH_2CH_3$). Later in Section 2.5 you will see that the **ane** suffix is actually composed of the so-called infix "**an**" and the true suffix "**e**."

- For substituted or **branched alkanes**, the nomenclature is based on viewing the molecule as a chain with substituents. ✳
  - To write a **substituent**, the **ane** suffix of the parent hydrocarbon is dropped and is replaced by the suffix **yl**. For example, **propyl** is used to describe a substituent with three carbon atoms ($CH_3CH_2CH_2$-).
- Branched alkanes are named using the following set of rules.
  - The alkane derived from the longest continuous chain of carbon atoms is taken as the **parent chain**. The **root** or **stem name** of the branched alkane is that of the parent chain. *[This can be tricky, especially when the parent chain is drawn in a crooked fashion.]*
  - Each substituent attached to the parent chain is given a name and a number. Certain common names (see below) can be used for naming substituents such as "isopropyl."
  - The **substituent number** shows the carbon of the parent chain to which the substituent is attached. The numbers are designated on the parent chain according to the following rules. *[This numbering scheme is as complicated as it seems, and requires a lot of practice to master thoroughly.]*

    If there is one substituent, number the parent chain from the end that gives the substituent the lower number. For example, a correct name is 2-methylhexane, *not* 5-methylhexane.

    If there are two or more *identical* substituents, number the parent chain from the end that gives the lower number to the substituent encountered first, and the number for each substituent is given in the final name. Indicate the number of times the same substituent occurs by a special set of prefixes. The prefixes **di, tri, tetra, penta**, or **hexa** indicate the presence of 2, 3, 4, 5, or 6 identical substituents, respectively. For example, 2,3-dimethylhexane has methyl groups at positions 2 and 3 on the parent hexane chain.

    If there are two or more *different* substituents, list them in **alphabetical order**, and number the parent chain from the end that gives the lower number to the substituent encountered first. For example, 4-ethyl-3-methyloctane is an acceptable name because ethyl comes before methyl in terms of alphabetization.

    If there are *different* substituents in equivalent positions on the parent chain, give the lower number to the substituent of lower alphabetical order.

    Hyphenated prefixes, for example, *sec-* and *tert-* are not considered when alphabetizing. The prefix **iso** is not a hyphenated prefix, and therefore is included when alphabetizing.
- In spite of the precision of the IUPAC system, an unsystematic set of **common names** is still used for certain compounds. *[These names are deeply rooted in organic chemistry and are still widely used . Remember that it is always correct to use an IUPAC name. However, it is also important to learn how to use the common names, because you will run across them often.]*
  - In the **common nomenclature**, the total number of carbon atoms in an alkane, regardless of their arrangement, determines the name. The following terms are used in common nomenclature to indicate a few selected branching patterns.

    **Normal** or **n-** indicates that all carbons are joined in a continuous chain . For example *n*-butane ($CH_3CH_2CH_2CH_3$)

    **Iso** is used to indicate that one end of an otherwise continuous chain terminates in a $(CH_3)_2CH$- group. For example isobutane ($(CH_3)_2CHCH_3$).

    **Neo** is used to indicate that one end of an otherwise continuous chain terminates in a $(CH_3)_3C$- group. For example, neopentane ($(CH_3)_4C$).

    More complicated patterns of branching cannot be accommodated by common nomenclature, so the IUPAC system must be used.

    Some common names such as *tert*-butyl and isopropyl can be used for substituents, even though the rest of the molecule is named according to IUPAC rules.
- **Atoms are classified** according to their **environment**. ✳
  - Classify a carbon atom in an alkane according to the number of alkyl groups bonded to it. *[This is very important when it comes to understanding relative reactivity.]* A carbon atom bonded to a single alkyl group is a **primary carbon** atom, a carbon atom bonded to two

alkyl groups is a **secondary carbon** atom, a carbon atom bonded to three alkyl groups is a **tertiary carbon** atom, and a carbon atom bonded to four alkyl groups is a **quaternary carbon** atom.
   - **Hydrogen atoms** are also classified as primary, secondary, or tertiary when they are bonded to a primary, secondary, or tertiary carbon atom, respectively.
   - **Equivalent** hydrogen atoms have the same chemical environment. ✳ *[This concept is very important when it comes to spectroscopy, especially  nuclear magnetic resonance (NMR) spectroscopy (Chapter 20)].*
      To determine which hydrogens in a molecule are equivalent, use the following procedure: In your mind, replace each hydrogen with a "**test atom.**" If replacement of two different hydrogens by the "test atom" gives the same compound, then the hydrogens are equivalent. If replacement of two different hydrogens by the "test atom" gives different compounds, then the hydrogens are not equivalent. *[This is very tricky and requires practice.]*

## 2.4  CYCLOALKANES
• Organic chemists use **line-angle drawings** as a simple way to represent complex molecules. In line-angle drawings, each line represents a C-C bond, each double line represents a C=C bond, and each triple line represents a C≡C bond. The vertex of each angle represents a carbon atom. In this way, only the **carbon framework** of the molecule is shown. It is understood that hydrogen atoms complete the tetravalence of the carbon atoms. For example, ⌒⌒ represents pentane, ($CH_3CH_2CH_2CH_2CH_3$).
• A **cycloalkane** is an alkane in which there is a ring of carbon atoms. ✳
   - **IUPAC cycloalkane nomenclature** rules are as follows:
      Use the prefix **cyclo** in front of the name of the alkane with the same number of carbon atoms as the number of carbons in the ring. For example, cyclohexane is a six-member ring, ⬡.
      List substituents on the ring by name and number as you would on an open-chain hydrocarbon.
      If there is only a single substituent on the ring, there is no need to give a number. If there are two or more substituents, give each substituent a number to indicate its location on the ring. Number the atoms of the ring beginning with the substituent of lowest alphabetical order.
• A **bicycloalkane** is an alkane with two rings that share one or more atoms in common.
   - **IUPAC bicycloalkane nomenclature** rules are as follows:
      The parent name of a bicycloalkane is that of the alkane with the same number of carbon atoms as are in the bicyclic ring system.
      Numbering begins at one bridgehead carbon and proceeds along the largest ring to the second bridgehead carbon, and then along the next largest ring back to the original bridgehead carbon, and so forth until all atoms of the bicyclic ring are numbered.
      Ring sizes are shown by counting the number of carbon atom linked to the bridgeheads and placing them in decreasing order in brackets between the prefix **bicyclo** and the parent name. For example, bicyclo[2.2.1]heptane ⬠.
      Name and locate substituents by the rules already described in Section 2.3A.

## 2.5 THE IUPAC SYSTEMS-A GENERAL SYSTEM OF NOMENCLATURE
• The name assigned to any compound consists of at least three parts; **the prefix, the infix** and **the suffix.**
   - The prefix tells the number of carbon atoms in the parent chain. See Table 2.2 in the text for examples.

- The infix (part of the name directly in front of the suffix) tells the nature of the carbon-carbon bonds in the parent chain.
    **an** means the compound has all single bonds, **en** means one or more double bond, and **yn** means there is one or more triple bond.
- The suffix tells the class of the compound to which the substance belongs.
    The class of a compound is determined by the functional groups present.
    Important suffixes include **e** for hydrocarbons, **ol** for alcohols, **al** for aldehydes, **one** for ketones, and **oic acid** for carboxylic acids.

## 2.6 CONFORMATIONS OF ALKANES AND CYCLOALKANES

- The **conformation** of an alkane refers to the *three-dimensional* arrangement of atoms that results from rotation about carbon-carbon bonds. ✳
    - It is convenient to analyze alkane conformations using a **Newman projection.** Although there may be a number of C-C bonds in a molecule to analyze, you can look at only one C-C bond with each Newman projection.
        In a Newman projection, view the molecule along the axis of one C-C bond. ✳
        *[Understanding this statement is the key to using Newman projections.]*
        Thus a Newman projection examines how the different groups are distributed around only the two adjacent carbon atoms involved with the selected C-C bond.
        When drawing a Newman projection, use a large circle to indicate location. Show the three groups bound to the carbon atom that are *nearer* your eye on lines extending from the *center* of a circle at angles of 120°. Show the three groups bound to the carbon atom that is *farther* from your eye on lines extending from the *circumference* of a circle at angles of 120°. *[Make sure you know how to go between a three dimensional molecular model and a Newman projection. Figures 2.5 and 2.6 are especially helpful.]*

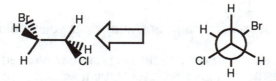

- At room temperature, the C-C bonds can **rotate rapidly.** Thus, an infinite number of conformations are possible around a C-C bond as it rotates. The two extreme conformations are named **eclipsed** and **staggered**.
    - In the **eclipsed conformation**, the groups on the near carbon atom are directly in front of the groups on the back carbon atom.

    - In the **staggered conformation**, the groups on the near carbon atom are as far apart as possible from the groups on the back carbon.

    - **A dihedral angle** (θ), is the angle between a given substituent on the near carbon atom and a given substituent on the far carbon of a Newman projection.
        For eclipsed conformations, the dihedral angles are thus 0°, 120°, 240°, for nearest groups, and for staggered conformations, the dihedral angles are thus 60°, 180°, 300° for nearest groups.

- The staggered conformations have lower **potential energy** than eclipsed conformations, probably due to the **repulsion** of **electron pairs** in the bonds resulting in their preferring to be as far apart as possible. Hydrogen atoms are probably not large enough to "crash" into each other even in the eclipsed conformation. ✳
  - This lower potential energy means that **alkanes spend the majority of their time in a staggered conformation**. ✳
- Taking the bond between the carbons 2 and 3 as reference, there are two types of **staggered conformations for butane**, namely **gauche** and **anti**. ✳
  - The two **gauche** conformations have the two methyl groups adjacent, that is with dihedral angles of 60°.

  - The **anti** conformation has the two methyl groups as far apart as possible, that is with a dihedral angle of 180°. *[If you do not fully understand this, review the preceding sections before going any further.]*

    The methyl groups take up a large amount of space and thus the anti conformation is the most stable (lowest potential energy), because the methyl groups are farthest apart. The gauche conformations are the next most stable, and of course all of the eclipsed conformations are the least stable.
    Other large groups are also more stable in the anti conformation. The larger the groups, the larger the preference for being anti. ✳
- The most stable three-dimensional arrangement of atoms in cycloalkanes minimizes the two types of **steric strain**, namely **angle strain** and **nonbonded interaction strain**. ✳ *[Notice the discussion has turned back to cycloalkanes.]*
  - **Angle strain** arises because the geometry of certain cycloalkanes creates bond angles other than the ideal 109.5°.
  - **Nonbonded interaction strain** arises because the geometry of cycloalkanes forces atoms or groups into close proximity. As expected, this type of strain is proportionately more important for larger atoms or groups.
- To minimize steric strain, the **larger cycloalkanes** exist in a variety of **puckered** nonplanar conformations. ✳

- **An envelope conformation** is the most stable conformation of **cyclopentane**. ⬡.
  There are five possible envelope conformations, each with a different carbon atom that is out of the plane formed by the other four.
  - This puckering relieves nonbonded steric strain by reducing the number of eclipsed atoms in the molecule. In this case the puckering only causes a slight increase in angle strain.
- There are a number of different puckered conformations of **cyclohexane**, by far the most important of which is a remarkably stable **chair conformation** ⬡. ✳ *[Understanding the following details of cyclohexane chair conformations is very important, because you will need to use these ideas in the future when issues like the relative stabilities of reaction intermediates or carbohydrates are discussed.]*

- The chair conformation is dramatically more stable than the planar form, because *all* the groups are perfectly staggered in the chair conformation. Furthermore, the chair conformation has *all* bond angles near the ideal 109.5°. Cyclohexane molecules therefore spend the great majority of their time in the chair conformation.
• In the chair conformation of cyclohexane, the 12 different hydrogen atoms attached to the six carbon atoms of the ring can be classified as one of two types, **axial** or **equatorial**. ✳
    - The **six axial positions** are **perpendicular** to the mean plane of the cyclohexane ring; three axial hydrogens point straight up, and three point straight down.
    - The **six equatorial positions** point roughly outward from the cyclohexane ring. *[Models will help you understand the difference between axial and equatorial positions.]*
        There are two different chair conformations of cyclohexane that are in equilibrium with each other. *[Using models, you should verify that interconverting the two possible chair cyclohexane conformations changes all of the axial hydrogens to equatorial hydrogens, and vice versa.]*
• There are several less stable conformations of cyclohexane such as the **boat** and **twist boat**. These conformations are less stable than the chair conformation because they have either some eclipsed hydrogen atoms or bond angles other than the optimum 109.5°. These less stable conformations are intermediates in the interconversion of the two chair forms of cyclohexane.
• If one or more of the hydrogens of a cyclohexane are replaced by any larger atom or group, the more stable chair conformation is the one that places the larger atom or group in an equatorial position. ✳ *[This is perhaps the most important concept involved with cyclohexane chair conformations, and you will use it over and over again.]*
    - The larger the atom or group, the greater the preference for it to be in an equatorial position. This preference for large atoms or groups to be equatorial derives from a special kind of nonbonded interaction strain called **axial-axial interactions**.
        Atoms or groups that are axial are relatively close to the two other atoms or groups that are also axial, thus groups will crash into these other axial substituents. This contact occurs between axial atoms or groups in the 1 and 3 positions of the cyclohexane ring, hence the name **axial-axial interactions**. *[Confirm this nonbonded interaction strain for yourself by making a model of methylcyclohexane, and make the chair conformation that places the methyl group in an axial position.]*
• Atoms or groups in equatorial positions are pointed away from other groups, thus minimizing nonbonded steric strain. ✳ *[Confirm this by converting your model to the other chair conformation, and notice the methyl group, which is now equatorial, is relatively free from nonbonded interaction strain.]*

## 2.7 *CIS-TRANS* ISOMERISM IN CYCLOALKANES AND BICYCLOALKANES
• Cycloalkanes with substituents on two or more carbons of the ring show a type of isomerism called *cis-trans* isomerism. ✳
    - *Cis-trans* isomerism is a type of **geometric isomerism**, that is isomerism that depends on the placement of substituent groups on the atoms attached to bonds that have inhibited rotation, e.g. part of a ring or on a double bond.
        *Cis-trans* isomerism can be understood by thinking of the cycloalkane as a planar structure. *[This is just a helpful trick, of course the true cycloalkane structures for everything larger than cyclopropane are puckered.]*
        For a cycloalkane with two substituents, the *cis* isomer is the one in which the two substituents are on the same side of the ring plane.
        The *trans* isomer is the one in which the two substituents are on opposite sides of the ring plane. In other words, for a given constitutional isomer such as 1,2 dimethylcyclopentane, the two methyl groups can be either *cis* or *trans* with respect to each other. *[Note that in order for the cis and trans comparison to be valid, the same constitutional isomer must be considered in both cases.]*

No matter how much the cycloalkane ring puckers or interchanges between conformations, the two methyl groups of the *cis* isomer will always be on the same side of the ring plane, and the two methyl groups of the *trans* isomer will always remain on opposite sides of the ring plane. Put another way, no amount of conformational change can convert the *cis* isomer into the *trans* isomer, and vice versa. *[Again, making models will save you a lot of time, and it will also make things more clear .]*

- The analysis of disubstituted cyclohexanes becomes more complicated in the context of chair conformations. For example, *trans*-1,4-dimethylcyclohexane can exist in two chair conformations. *[You should make a model to prove this to yourself.]*

  In one chair conformation, both methyl groups are axial. In the other chair conformation, both methyl groups are equatorial.

  The chair conformation with both methyl groups equatorial is more stable. *[It has fewer axial-axial interactions]*

- For *cis*-1,4-dimethylcyclohexane, each chair conformation has one methyl group axial and one methyl group equatorial, so both of these conformations are equally stable. *[Again, a model will be very helpful here.]*

• **Bicycloalkanes** also exhibit *cis-trans* isomerism. *[You should practice analyzing compounds like cis- and trans-decalin with models using the same ideas just discussed for the cyclopentane and cyclohexane derivatives.]*

## 2.8 PHYSICAL PROPERTIES OF ALKANES AND CYCLOALKANES
• At room temperature, the simple alkanes the size of butane or smaller are **gases**, while pentane through decane are **liquids**.
• At lower temperatures, the alkanes can be frozen into **solids**.
• The fact that these compounds can exist as liquids and solids depends on the existence of **intermolecular forces** that can hold the molecules together.
• In alkanes, the molecules are held together by weak **dispersion forces**, the weakest of all the intermolecular forces.
• Alkanes also have very low **boiling points** because they are held together primarily by nothing but weak dispersion forces.
  - In general, larger alkanes (higher molecular weight) have higher boiling points than smaller alkanes (lower molecular weight), and unbranched alkanes have higher boiling points than branched constitutional isomers. The strength of dispersion forces is proportional to the surface area of contact between molecules. Branched molecules are more compact resulting in smaller surface areas than unbranched constitutional isomers. *[The above ideas explain a large amount of experimental data.]*

## 2.9 REACTIONS OF ALKANES
• Alkanes are relatively unreactive, having small permanent dipole moments and only the relatively strong sigma ($\sigma$) type of bonds.
• On the other hand, alkanes can react with oxygen and halogens under certain conditions. ✳

## 2.10 SOURCES OF ALKANES
• The three natural sources of alkanes are **natural gas, petroleum** and **coal**.
  - **Natural gas** is mostly methane, with some ethane and a small amount of other small alkanes that are gases at room temperature.
  - **Petroleum** is an incredibly complex mixture of compounds, and the commercially important fractions are purified by large scale distillation towers. Presently, petroleum is by far the most important source of organic raw materials for products like fuels, lubricants, nylon, dacron, textile fibers, asphalt, and synthetic rubber.
  - **Coal** has an extremely complex structure, and a great deal of chemistry is required to produce useful alkanes such as fuels from coal.

# CHAPTER 2
*Solutions to the Problems*

<u>Problem 2.1</u> Identify the members of each pair as formulas of identical compounds or as formulas of constitutional isomers.

$$CH_2-CH_3$$

(a) $CH_3-CH-CH-CH_3$    and    $CH_3-CH_2-CH-CH_2-CH-CH_3$
with $CH_3$ and $CH_3$ branches on the right structure, and $CH_2-CH_3$ branch below.

**These molecules are constitutional isomers. Each has six carbons in the longest chain. The first has one-carbon branches on carbons 3 and 4 of the chain; the second has one-carbon branches on carbons 2 and 4 of the chain.**

$$CH_3$$

(b) $CH_3-CH-CH-CH_3$    and    $CH_3-CH-CH-CH_2-CH_3$
with $CH_2-CH_3$ branch below left, and $CH_3$ above and $CH_3$ below right.

**These molecules are identical. Each has five carbons in the longest chain, and one-carbon branches on carbons 2 and 3 of the chain.**

<u>Problem 2.2</u> Draw structural formulas for the three constitutional isomers of molecular formula $C_5H_{12}$.

$CH_3-CH_2-CH_2-CH_2-CH_3$     $CH_3-CH-CH_2-CH_3$ with $CH_3$ above     $CH_3-C-CH_3$ with $CH_3$ above and $CH_3$ below

<u>Problem 2.3</u> Write IUPAC names for these alkanes.

(a) $CH_3-CH-CH_2-CH_2-CH-CH-CH_3$ with $CH_3$ above the second carbon, $CH_3$ above the fifth carbon, and $CH_2-CH_2-CH_3$ below

(b) $CH_3-CH_2-CH_2-C-CH_2-CH_2-CH_3$ with $CH_2-CH_2-CH_3$ above and $CH_3-CH-CH_3$ below

**5-Isopropyl-2-methyloctane**          **4-Isopropyl-4-propylheptane**

Problem 2.4  Write the molecular formula and IUPAC name for each cycloalkane.

(a)

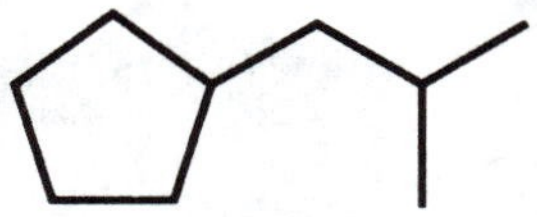

**C₉H₁₈**
**Isobutylcyclopentane**

(b)

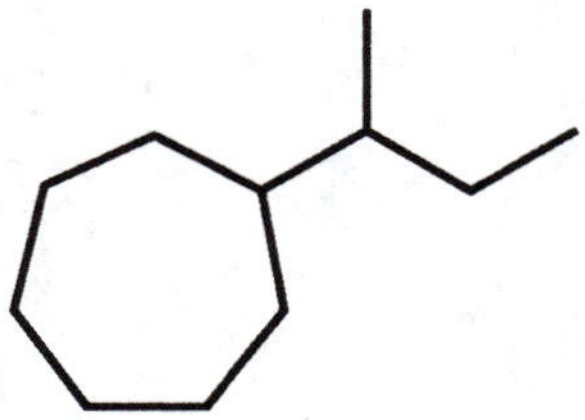

**C₁₁H₂₂**
***sec*-Butylcycloheptane**

(c)

**C₆H₁₂**
**1-Ethyl-1-methylcyclopropane**

Problem 2.5  Write molecular formulas for each bicycloalkane in Figure 2.3, given that hydindrane contains nine carbons, decalin contains ten carbons, and norbornane contains seven carbons.

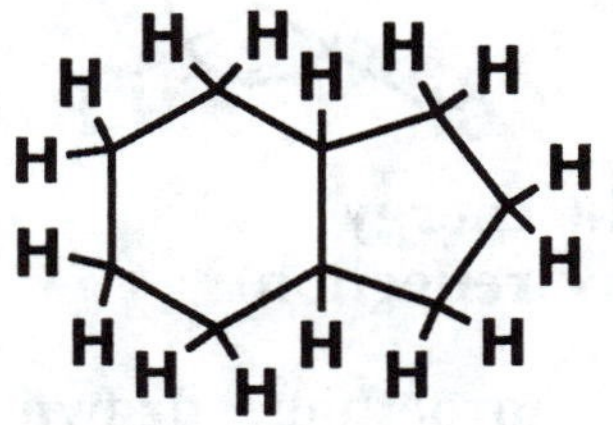

**Hydrindane**
**Molecular Formula C₉H₁₆**

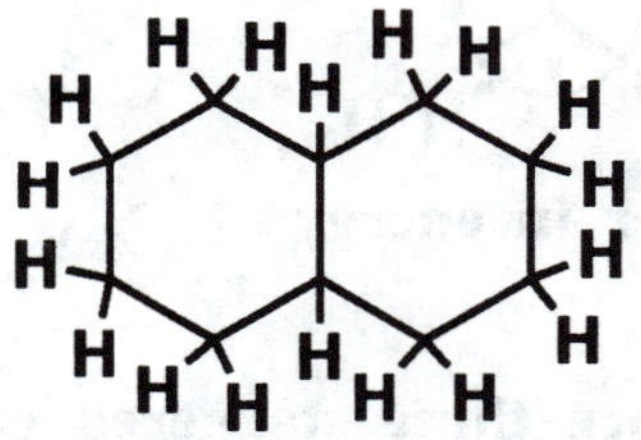

**Decalin**
**Molecular Formula C₁₀H₁₈**

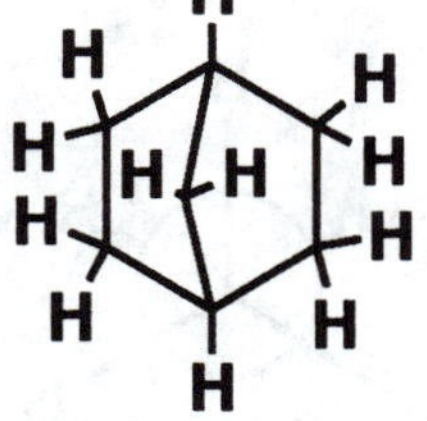

**Norbornane**
**Molecular Formula C₇H₁₂**

Problem 2.6  Combine the proper prefix, infix and suffix and write the IUPAC name for each compound.

(a)

CH₃-C-CH₃

**2-Propanone**

(b) CH₃−CH₂−CH₂−CH₂−C−H

**Pentanal**

(c)

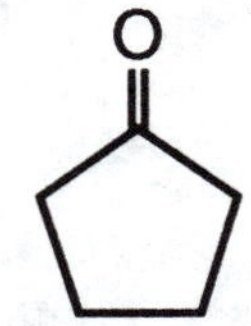

**Cyclopentanone**

(d)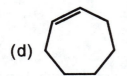

**Cycloheptene**

Problem 2.7
(a) Draw Newman projections for two staggered conformations and two eclipsed conformations of 1,2-dichloroethane.

**Following are three eclipsed conformations of 1,2-dichoroethane drawn as Newman projections. The eclipsed conformation on the left has the highest energy of any of the conformations because the chlorine atoms are closest to each other.**

higher in energy                      lower in energy
                                   (related by reflection)

**Following are three staggered conformations of 1,2-dichloroethane drawn as Newman projections. These staggered conformations are all lower in energy than the eclipsed conformations above. The staggered structure on the left has the lowest overall energy of any of the conformations, because the chlorine atoms are the farthest away from each other.**

lowest in energy                      higher in energy
                                   (related by reflection)

<u>Problem 2.8</u>  Following is a chair conformation of cyclohexane with carbon atoms numbered 1
through 6.

(a)  Draw  hydrogen atoms that are above the plane of the ring on carbons 1 and 2 and below
the plane of the ring on carbon 4.
(b)  Which of these hydrogens are equatorial; which are axial?
(c)  Draw the other chair conformation.  Now, which hydrogens are equatorial; which are axial?

<u>Problem 2.9</u>  The conformational equilibria for methyl, ethyl, and isopropylcyclohexane are all
about 95% in favor of the equatorial conformation, but that for *tert*-butylcyclohexane is virtually
completely on the equatorial side.  Explain by using molecular models and making drawings why
the conformational equilibria for the first three compounds are comparable, but why the
conformational equilibrium for *tert*-butylcyclohexane lies considerably farther toward the equatorial
conformation.

**Rotation is possible about the single bond connecting the axial substituent to the
ring.  Axial methyl, ethyl and isopropyl groups can assume a conformation where
a hydrogen creates the axial-axial interactions.  With a *tert*-butyl substituent,
however, a bulkier -$CH_3$ group creates the axial-axial interaction.  Because of the
increased steric strain (nonbonded interactions) created by the axial *tert*-butyl, the
potential energy of the axial conformation is considerably greater than that for the
equatorial conformation.**

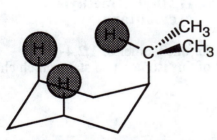

**axial isopropyl group
moderately severe
axial-axial interaction**

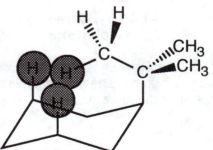

**axial *tert*-butyl  group
very severe axial-axial
interaction**

**Problem 2.10** Following are three cycloalkanes of molecular formula $C_7H_{14}$. State which show *cis-trans* isomerism and for each that does, draw the *cis* and *trans* isomers.

(a) 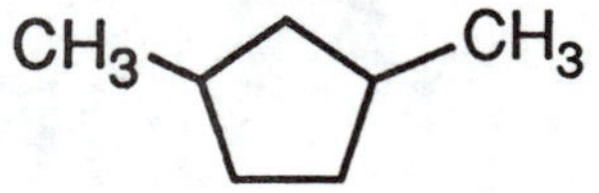  CH$_3$ ⟍⟋ CH$_3$

**1,3-Dimethylcyclopentane shows *cis-trans* isomerism. In the following drawings, the ring is drawn as a planar pentagon with substituents above and below the plane of the pentagon.**

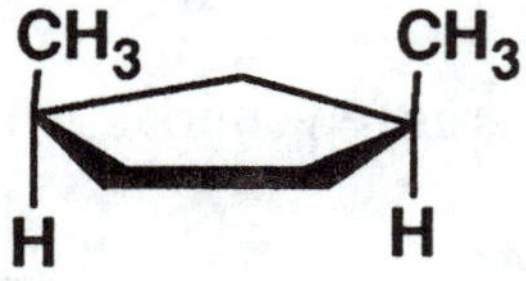

CH$_3$      CH$_3$            CH$_3$      H

H          H                  H          CH$_3$

*cis*-1,3-Dimethyl-            *trans*-1,3-Dimethyl-
cyclopentane                  cyclopentane

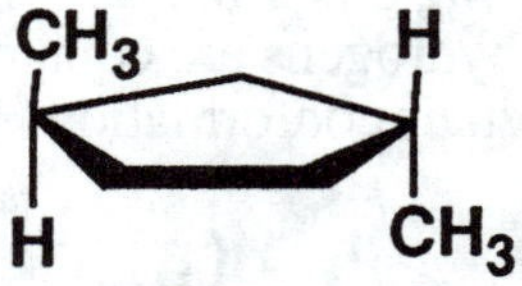

(b) 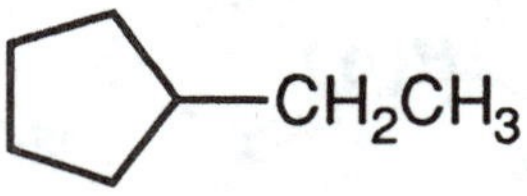  ⬠—CH$_2$CH$_3$

**Ethylcyclopentane does not show *cis-trans* isomerism.**

(c) 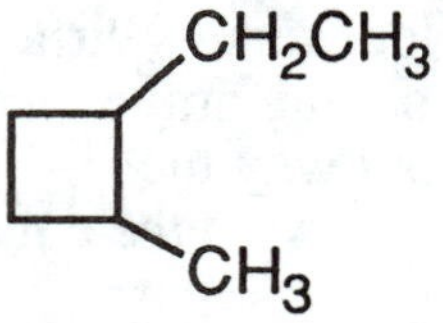  CH$_2$CH$_3$ / ▢ / CH$_3$

**1-Ethyl-2-methylcyclobutane shows *cis-trans* isomerism.**

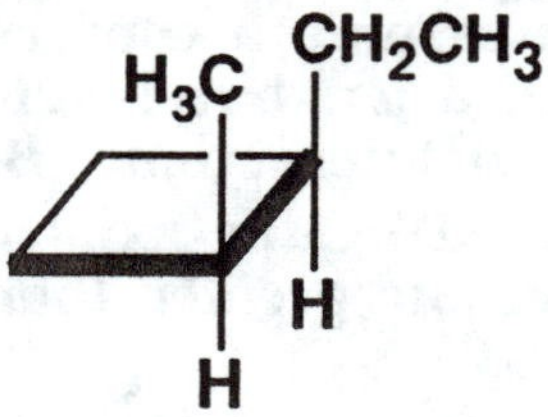

H$_3$C   CH$_2$CH$_3$          H$_3$C   H

H                             CH$_2$CH$_3$

H                             H

*cis*-1-Ethyl-2-methyl-       *trans*-1-Ethyl-2-methyl-
cyclobutane                   cyclobutane

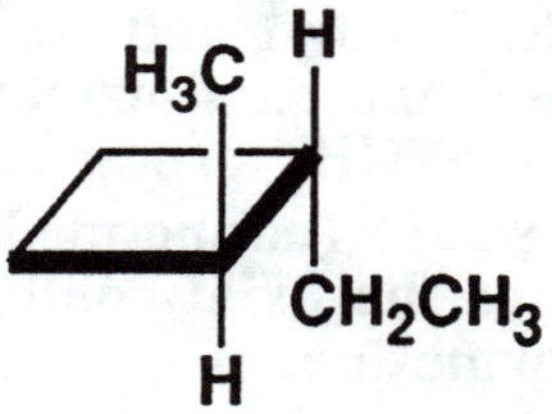

**Problem 2.11** Following is a planar hexagon representation for one isomer of 1,2,4-trimethylcyclohexane. Draw alternative chair conformations of this isomer, and state which chair conformation is the more stable.

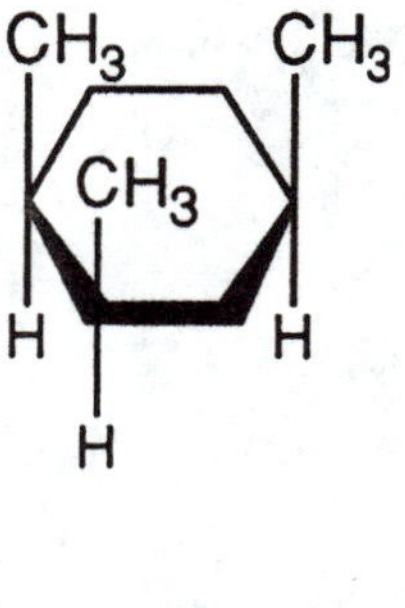

CH$_3$      CH$_3$

CH$_3$

H          H

H

Following are alternative chair conformations for the all *cis* isomers of 1,2,4-trimethylcyclohexane. The alternative chair conformation on the right is the more stable because it has only one axial methyl group.

| | |
|---|---|
| **less stable chair** | **more stable chair** |
| **(two methyl groups axial)** | **(one methyl group axial)** |

<u>Problem 2.12</u> Arrange the alkanes in each set in order of increasing boiling point.
(a)  2-Methylbutane, 2,2-dimethylpropane, and pentane

**2,2-Dimethylpropane,  2-methylbutane,  pentane**
**(bp 9.5°C)          (bp 29°C)      (bp 36°C )**

(b)  3,3-Dimethylheptane, 2,2,4,4-tetramethylpentane, and nonane

**2,2,4,4-Tetramethylpentane,  3,3-dimethylheptane,  nonane**
**(bp 99°C)              (bp 137°C)      (bp 151°C )**

<u>Problem 2.13</u> Name and draw structural formulas for all monochlorination products formed by treatment of butane with $Cl_2$.  Predict the major product.

$$CH_3-CH_2-CH_2-CH_3 \ + \ Cl_2 \ \xrightarrow[\text{heat}]{\text{light or}} \ \text{monochloroalkanes} \ + \ HCl$$

**There are two possible monochlorination products. In butane, there is one set of six equivalent primary hydrogens (on the $-CH_3$ groups) and one set of four equivalent secondary hydrogens (on the $-CH_2-$ groups). Therefore, based on statistical distribution, predict 60% 1-chlorobutane and 40% 2-chlorobutane. However, the known regioselectivity of chlorination (secondary hydrogen > primary hydrogen) predicts that the 2-chlorobutane will be the predominant product in analogy to the chlorination of propane described in the text.**

$CH_3\text{-}CH_2\text{-}CH_2\text{-}CH_2\text{-}Cl$           $CH_3-CH_2-\overset{\overset{\displaystyle Cl}{|}}{C}H-CH_3$

**1-Chlorobutane**              **2-Chlorobutane**
**(Butyl chloride)**           **(sec-Butyl  chloride)**

**Predicted to be the**
**major product**

<u>**Constitutional  Isomerism**</u>
<u>Problem 2.14</u> Which statements are true about constitutional isomers?
(a) They have the same molecular formula.  **True**
(b) They have the same molecular weight.  **True**
(c) They have the same order of attachment of atoms.  **False**

(d) They have the same physical properties.  **False**

<u>Problem 2.15</u>  Which structural formulas represent identical compounds and which represent constitutional isomers?

(a)  $CH_3-CH_2-CH-CH_3$
                          $|$
                          $Cl$

(b)  $CH_3-\overset{\overset{\displaystyle CH_2Cl}{|}}{\underset{\underset{\displaystyle Cl}{|}}{C}}-CH_3$

(c)  $CH_3-\overset{\overset{\displaystyle Cl}{|}}{CH}-\overset{}{CH}-CH_3$
                                              $|$
                                              $Cl$

(d)  $\overset{}{CH_2}-CH_2-\overset{}{CH}-CH_3$
         $|$                    $|$
         $Cl$                   $Cl$

(e)  ⬦—Cl

(f)  $Cl-CH_2-$◁

(g)  $CH_3-\overset{\overset{\displaystyle CH_2-Cl}{|}}{CH}-CH_3$

(h)  $CH_3-CH_2-CH_2-CH_2-Cl$

(i)  $Cl-CH_2-\overset{\overset{\displaystyle CH_3}{|}}{CH}-CH_3$

(j)  $CH_3-\underset{\underset{\displaystyle Cl}{|}}{CH}-CH_2-CH_2-Cl$

(k)  $CH_3-\overset{\overset{\displaystyle CH_2-CH_3}{|}}{CH}-Cl$

(l)  $CH_3-\overset{\overset{\displaystyle CH_3}{|}}{\underset{\underset{\displaystyle Cl}{|}}{C}}-CH_3$

**Following are names and molecular formulas of each**
**(a)  2-Chlorobutane;  $C_4H_9Cl$**
**(b)  1,2-Dichloro-2-methylpropane;   $C_4H_8Cl_2$**
**(c)  2,3-Dichlorobutane;  $C_4H_8Cl_2$**
**(d)  1,3-Dichlorobutane;  $C_4H_8Cl_2$**
**(e)  Chlorocyclobutane;  $C_4H_7Cl$**
**(f)  Chloromethylcyclopropane;  $C_4H_7Cl$**
**(g)  1-Chloro-2-methylpropane;  $C_4H_9Cl$**
**(h)  1-Chlorobutane;  $C_4H_9Cl$**
**(i)  1-Chloro-2-methylpropane;  $C_4H_9Cl$**
**(j)  1,3-Dichlorobutane;  $C_4H_8Cl_2$**
**(k)  2-Chlorobutane;  $C_4H_9Cl$**
**(l)  2-Chloro-2-methylpropane;  $C_4H_9Cl$**

**The following represent identical compounds:  (a),(k)     (d),(j)        (g),(i)**

**The following four compounds represent one set of constitutional isomers: 1-chlorobutane (h), 2-chlorobutane (a)(k), 1-chloro-2-methylpropane (g)(i), and 2-chloro-2-methylpropane (l).**

**Chlorocyclobutane (e) and chloromethylcyclopropane (f) represent a second set of constitutional isomers.**

**1,2-Dichloro-2-methylpropane (b), 2,3-dichlorobutane (c), 1,3-dichlorobutane (d) represent a third set of constitutional isomers.**

<u>Problem 2.16</u> Name and draw structural formulas for all constitutional isomers of molecular formula $C_7H_{16}$.

**There are nine alkanes of molecular formula $C_7H_{16}$.**

$CH_3CH_2CH_2CH_2CH_2CH_2CH_3$

**Heptane**
**(bp 94.8)**

$$CH_3 \overset{\overset{\displaystyle CH_3}{|}}{C}HCH_2CH_2CH_2CH_3$$

**2-Methylhexane**
**(bp 90.0)**

$$CH_3 CH_2 \overset{\overset{\displaystyle CH_3}{|}}{C}HCH_2CH_2CH_3$$

**3-Methylhexane**
**(bp 92.0)**

$$CH_3 \overset{\overset{\displaystyle CH_3}{|}}{\underset{\underset{\displaystyle CH_3}{|}}{C}}CH_2CH_2CH_3$$

**2,2-Dimethylpentane**
**(bp 79.2)**

$$CH_3 \overset{\overset{\displaystyle CH_3}{|}}{C}H\underset{\underset{\displaystyle CH_3}{|}}{C}HCH_2CH_3$$

**2,3-Dimethylpentane**
**(bp 89.8)**

$$CH_3 \overset{\overset{\displaystyle CH_3}{|}}{C}HCH_2\underset{\underset{\displaystyle CH_3}{|}}{C}HCH_3$$

**2,4-Dimethylpentane**
**(bp 80.5)**

$$CH_3CH_2 \overset{\overset{\displaystyle CH_3}{|}}{\underset{\underset{\displaystyle CH_3}{|}}{C}}CH_2CH_3$$

**3,3-Dimethylpentane**
**(bp 86.1)**

$$CH_3 CH_2 \overset{\overset{\displaystyle CH_2CH_3}{|}}{C}HCH_2CH_3$$

**3-Ethylpentane**
**(bp 93.5)**

$$CH_3\underset{\underset{\displaystyle CH_3}{|}}{\overset{\overset{\displaystyle H_3C}{|}}{C}}H\overset{\overset{\displaystyle CH_3}{|}}{C}CH_3$$

**2,2,3-Trimethyl-**
**butane**
**(bp 80.9)**

<u>Problem 2.17</u> Tell whether the compounds in each set are constitutional isomers.

(a) $CH_3-CH_2-OH$ and $CH_3-O-CH_3$

(b) $CH_3-\overset{\overset{\displaystyle O}{||}}{C}-CH_3$ and $CH_3-CH_2-\overset{\overset{\displaystyle O}{||}}{C}-H$

(c) $CH_3-\overset{\overset{\displaystyle O}{||}}{C}-O-CH_3$ and $CH_3-CH_2-\overset{\overset{\displaystyle O}{||}}{C}-OH$

(d) $CH_3-\overset{\overset{\displaystyle OH}{|}}{C}H-CH_2-CH_3$ and $CH_3-\overset{\overset{\displaystyle O}{||}}{C}-CH_2-CH_3$

(e) ⬠ and $CH_3-CH_2-CH_2-CH_2-CH_3$

(f) ⬠ and $CH_2=CH—CH_2-CH_2-CH_3$

**Sets (a), (b), (c), and (f) contain constitutional isomers; sets (d) and (e) do not.**

<u>Problem 2.18</u> Draw structural formulas for all alcohols of molecular formula $C_4H_{10}O$.

$$CH_3-CH_2-CH_2-CH_2—OH \qquad CH_3-CH_2-\overset{\overset{\displaystyle OH}{|}}{C}H-CH_3 \qquad CH_3-\overset{\overset{\displaystyle CH_3}{|}}{\underset{\underset{\displaystyle CH_3}{|}}{C}}—OH$$

$$CH_3-\overset{\overset{\displaystyle CH_3}{|}}{C}H-CH_2-OH$$

<u>Problem 2.19</u> Draw structural formulas for all aldehydes of molecular formula $C_4H_8O$.

$$CH_3-CH_2-CH_2-\overset{\overset{\displaystyle O}{\|}}{C}-H \qquad CH_3-\underset{\underset{\displaystyle CH_3}{|}}{C}H-\overset{\overset{\displaystyle O}{\|}}{C}-H$$

<u>Problem 2.20</u> Draw structural formulas for all ketones of molecular formula $C_4H_8O$.

$$CH_3-CH_2-\overset{\overset{\displaystyle O}{\|}}{C}-CH_3$$

<u>Problem 2.21</u> Draw structural formulas for all ketones of molecular formula $C_5H_{10}O$.

$$CH_3-CH_2-\overset{\overset{\displaystyle O}{\|}}{C}-CH_2-CH_3 \qquad CH_3-CH_2-CH_2-\overset{\overset{\displaystyle O}{\|}}{C}-CH_3 \qquad CH_3-\underset{\underset{\displaystyle CH_3}{|}}{C}H-\overset{\overset{\displaystyle O}{\|}}{C}-CH_3$$

<u>Problem 2.22</u> Draw structural formulas for all carboxylic acids of molecular formula $C_5H_{10}O_2$.

$$CH_3-\underset{\underset{\displaystyle CH_3}{|}}{C}H-CH_2-\overset{\overset{\displaystyle O}{\|}}{C}-OH \qquad CH_3-CH_2-CH_2-CH_2-\overset{\overset{\displaystyle O}{\|}}{C}-OH \qquad CH_3-\overset{\overset{\displaystyle CH_3}{|}}{\underset{\underset{\displaystyle CH_3}{|}}{C}}-\overset{\overset{\displaystyle O}{\|}}{C}-OH$$

$$CH_3-CH_2-\underset{\underset{\displaystyle CH_3}{|}}{C}H-\overset{\overset{\displaystyle O}{\|}}{C}-OH$$

## Nomenclature of Alkanes and Cycloalkanes

Problem 2.23  Write IUPAC names these alkanes and cycloalkanes.

(a) 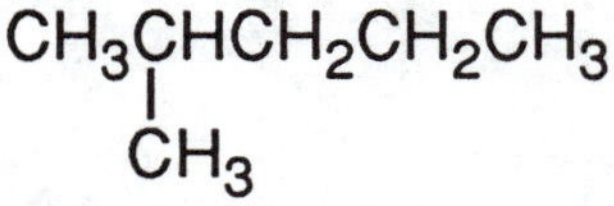  $CH_3CHCH_2CH_2CH_3$
          |
          $CH_3$

**2-Methylpentane (isohexane)**

(b) 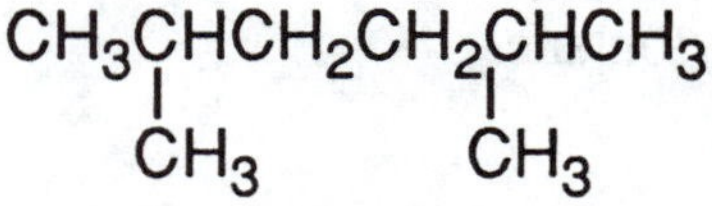  $CH_3CHCH_2CH_2CHCH_3$
          |                    |
          $CH_3$            $CH_3$

**2,5-Dimethylhexane**

(c) 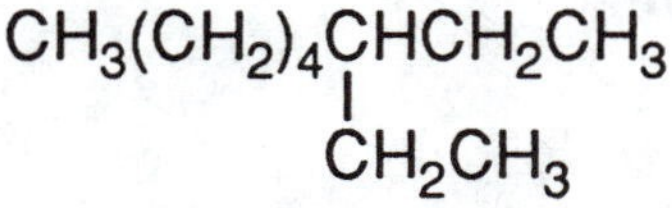  $CH_3(CH_2)_4CHCH_2CH_3$
                    |
                    $CH_2CH_3$

**3-Ethyloctane**

(d)  $(CH_3)_2CHCH_2CH_2C(CH_3)_3$

**2,2,5-Trimethylhexane**

(e) 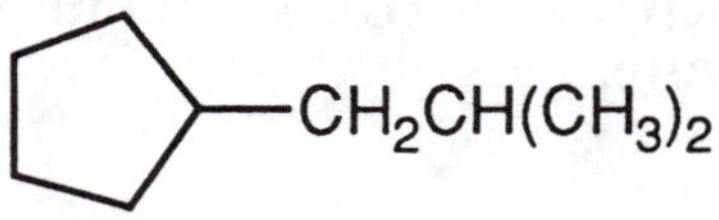  $-CH_2CH(CH_3)_2$

**Isobutylcyclopentane**

(f)

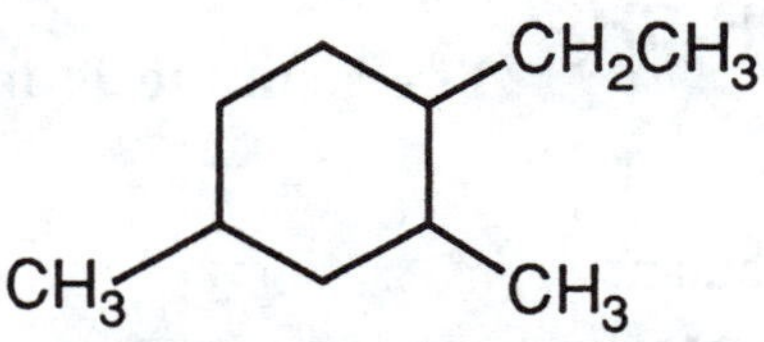

**1-Ethyl-2,4-dimethylcyclohexane**

Problem 2.24  Write structural formulas for these alkanes.

(a)  2,2,4-Trimethylhexane

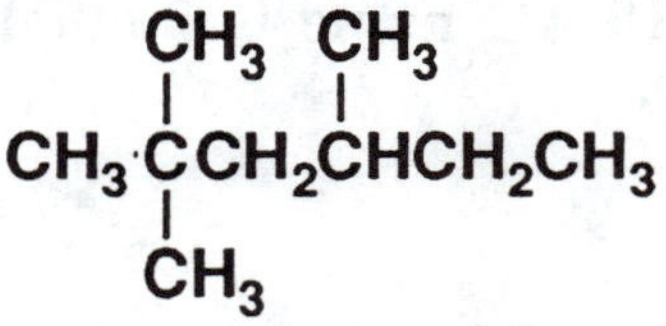

$CH_3$   $CH_3$
   |          |
$CH_3 \cdot CCH_2CHCH_2CH_3$
   |
$CH_3$

(b)  2,2-Dimethylpropane

$CH_3$
   |
$CH_3 CCH_3$
   |
$CH_3$

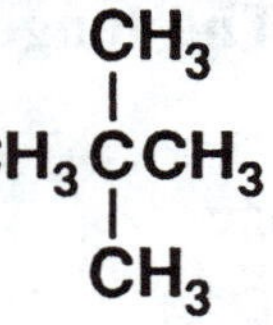

(c)  3-Ethyl-2,4,5-trimethyloctane

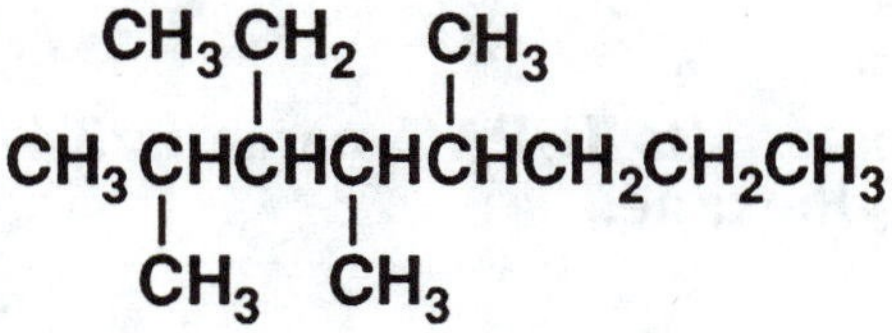

$CH_3 CH_2$   $CH_3$
        |          |
$CH_3 CHCHCHCHCH_2CH_2CH_3$
        |          |
      $CH_3$   $CH_3$

(d)  5-Butyl-2,2-dimethylnonane

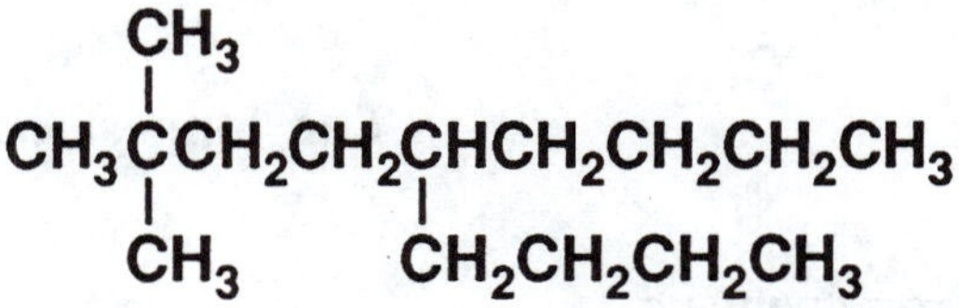

$CH_3$
   |
$CH_3 CCH_2CH_2CHCH_2CH_2CH_2CH_3$
   |                    |
$CH_3$            $CH_2CH_2CH_2CH_3$

(e)  4-Isopropyloctane

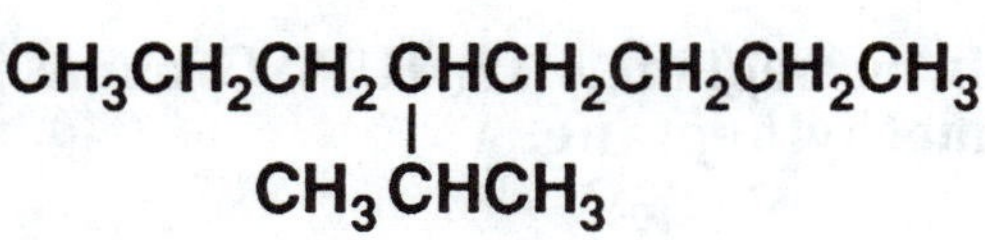

$CH_3CH_2CH_2 CHCH_2CH_2CH_2CH_3$
                    |
            $CH_3 CHCH_3$

(f)  3,3-Dimethylpentane

$CH_3$
   |
$CH_3CH_2 CCH_2CH_3$
   |
$CH_3$

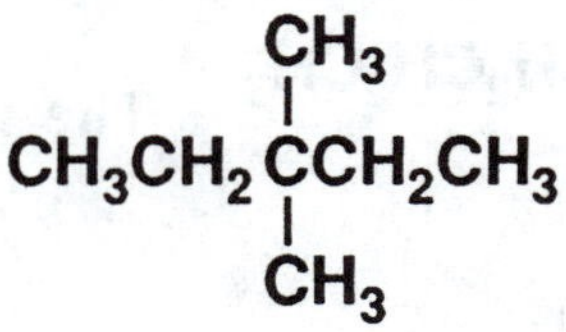

(g)  *trans*-1,3-Dimethylcyclopentane

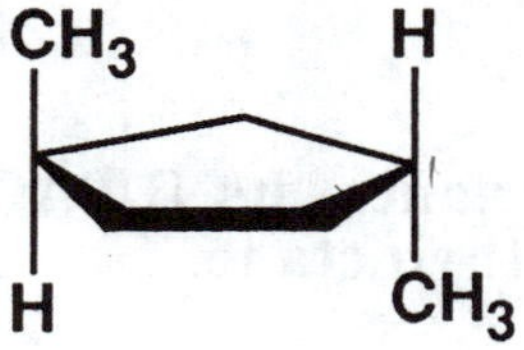

(h)  *cis*-1,2-Diethylcyclobutane

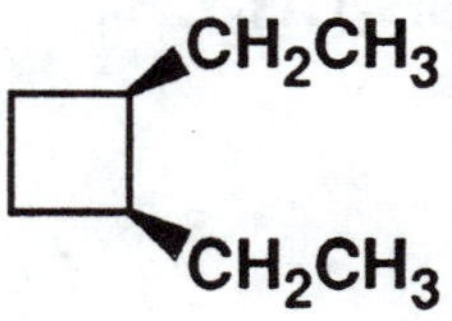

<u>Problem 2.25</u> Explain why each is an incorrect IUPAC name.  Write the correct IUPAC name for the intended compound.

(a) 1,3-Dimethylbutane

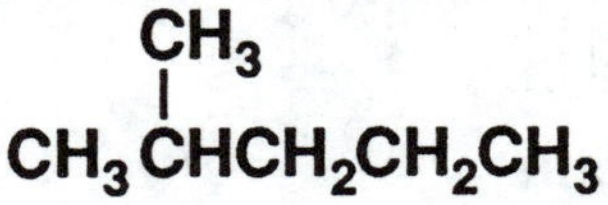

$$CH_3$$
$$|$$
$$CH_3CHCH_2CH_2CH_3$$

**The longest chain is pentane.  Its IUPAC name is 2-methylpentane.**

(b) 4-Methylpentane

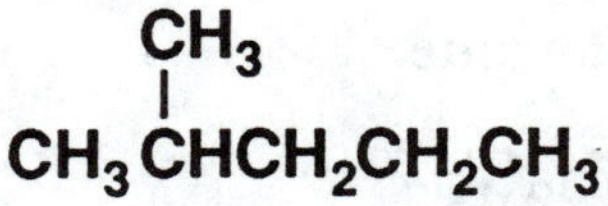

$$CH_3$$
$$|$$
$$CH_3CHCH_2CH_2CH_3$$

**The pentane is numbered incorrectly.  Its IUPAC name is 2-methylpentane.**

(c) 2,2-Diethylbutane

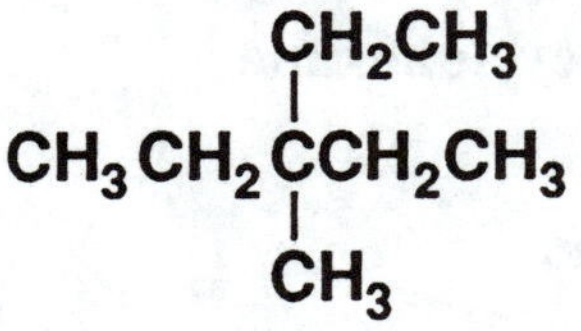

$$CH_2CH_3$$
$$|$$
$$CH_3CH_2CCH_2CH_3$$
$$|$$
$$CH_3$$

**The longest chain is pentane.  Its IUPAC name is 3-ethyl-3-methylpentane.**

(d) 2-Ethyl-3-methylpentane

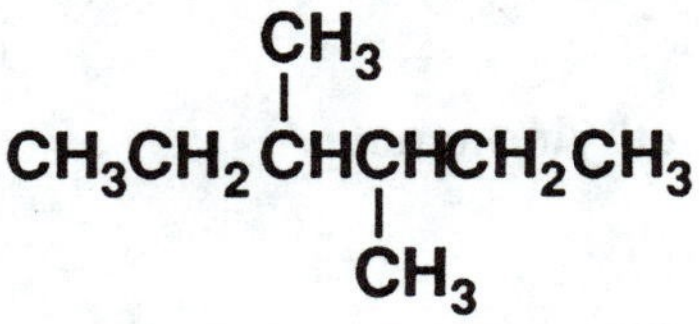

$$CH_3$$
$$|$$
$$CH_3CH_2CHCHCH_2CH_3$$
$$|$$
$$CH_3$$

**The longest chain is hexane.  Its IUPAC name is 3,4-dimethylhexane.**

(e) 2-Propylpentane

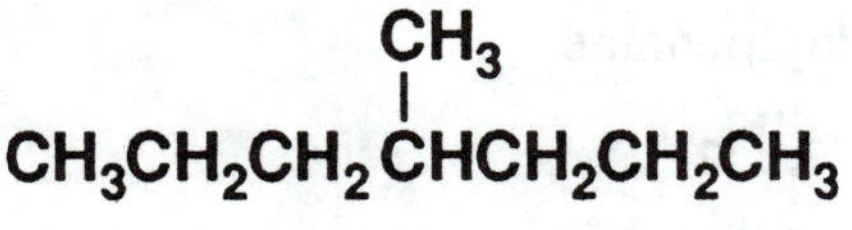

$$CH_3$$
$$|$$
$$CH_3CH_2CH_2CHCH_2CH_2CH_3$$

**The longest chain is heptane.  Its IUPAC name is 4-methylheptane.**

(f) 2,2-Diethylheptane

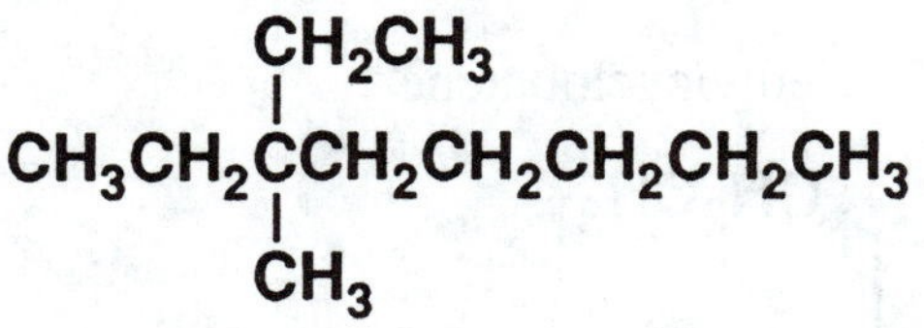

$$CH_2CH_3$$
$$|$$
$$CH_3CH_2CCH_2CH_2CH_2CH_2CH_3$$
$$|$$
$$CH_3$$

**The longest chain is octane.  Its IUPAC name is 3-ethyl-3-methyloctane.**

(g) 2,2-Dimethylcyclopropane

**The ring is numbered incorrectly. Its IUPAC name is 1,1-dimethylcyclopropane.**

(h) 1-Ethyl-5-methylcyclohexane

**The name is numbered incorrectly. Its IUPAC name is 1-ethyl-3-methylcyclohexane.**

## The IUPAC System of Nomenclature

**Problem 2.26** Draw a structural formula for each compound..

(a) 3-Pentanone

$$CH_3-CH_2-\overset{\displaystyle O}{\overset{\|}{C}}-CH_2-CH_3$$

(b) Ethanol

$$CH_3-CH_2-OH$$

(c) Ethanal

$$CH_3-\overset{\displaystyle O}{\overset{\|}{C}}-H$$

(d) Ethanoic acid

$$CH_3-\overset{\displaystyle O}{\overset{\|}{C}}-OH$$

(e) Hexanoic acid

$$CH_3(CH_2)_4-\overset{\displaystyle O}{\overset{\|}{C}}-OH$$

(f) Ethene

$$CH_2{=}CH_2$$

(g) 2-Propanol

$$CH_3-\overset{\displaystyle OH}{\overset{|}{C}H}-CH_3$$

(h) 1-Propanol

$$CH_3-CH_2-CH_2-OH$$

(i) Cyclopentanol

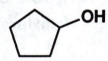

(j) Cyclopentene

(k) Cyclopentanone

**Problem 2.27** Write IUPAC names for these compounds.

(a) $CH_3\text{-}CH_2\text{-}\overset{\displaystyle O}{\overset{\|}{C}}\text{-}CH_3$

**2-Butanone**

(b) $CH_3\text{-}CH_2\text{-}\overset{\displaystyle O}{\overset{\|}{C}}\text{-}H$

**Propanal**

(c) 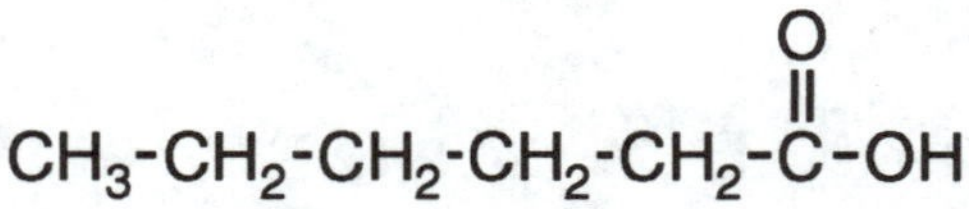 $CH_3-CH_2-CH_2-CH_2-CH_2-\overset{\overset{\displaystyle O}{\parallel}}{C}-OH$

**Hexanoic acid**

(d) 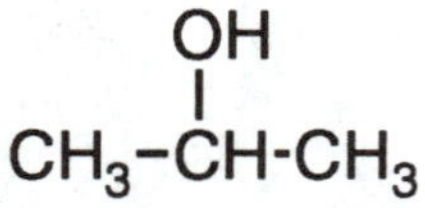 $CH_3-\overset{\overset{\displaystyle OH}{\mid}}{CH}-CH_3$

**2-Propanol**

(e) 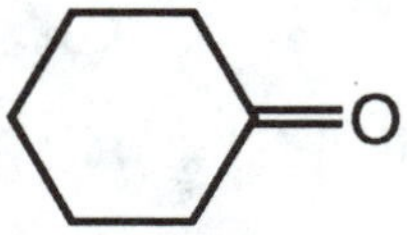 ⬡=O

**Cyclohexanone**

(f) 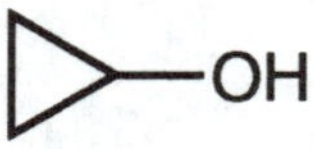 ▷—OH

**Cyclopropanol**

(g) $CH_3-CH=CH_2$

**Propene**

(h)  ⬡

**Cyclohexene**

## Conformations of Alkanes and Cycloalkanes
**Problem 2.28** How many different staggered conformations are there for 2-methylpropane? How many different eclipsed conformations are there?

**Looking down any of the carbon-carbon bonds, there is one staggered and one eclipsed conformation of 2-methylpropane.**

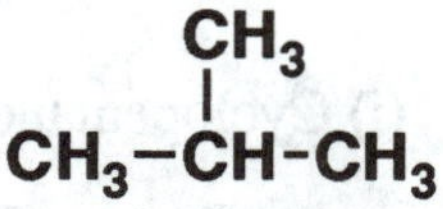

$$CH_3-\overset{\overset{\displaystyle CH_3}{\mid}}{CH}-CH_3$$
**2-Methylpropane**

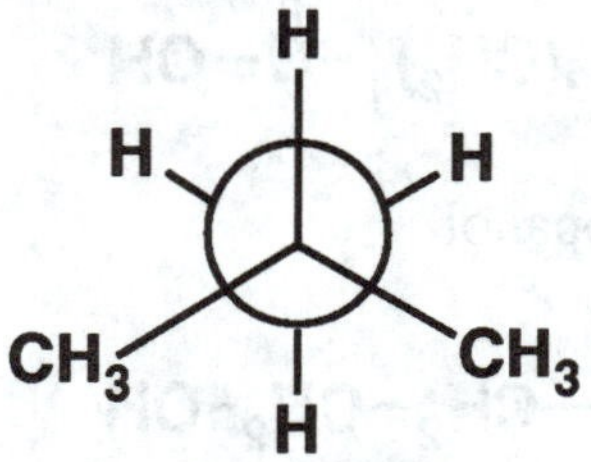

**Staggered** 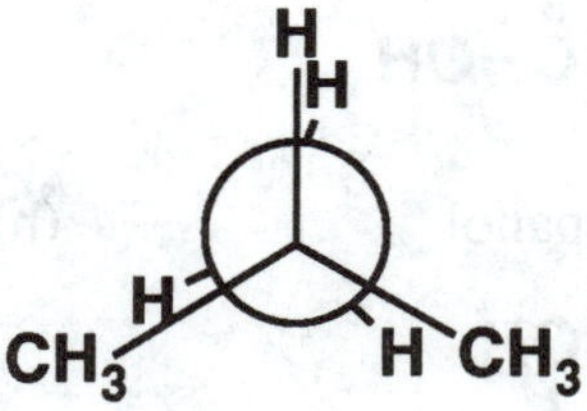          **Eclipsed**

**Problem 2.29** Looking along the bond between carbons 2 and 3 of butane, there are two different staggered conformations and two different eclipsed conformations. Draw Newman projections of each and arrange them in order from most stable conformation to least stable conformation.

**The two staggered conformations are more stable than the eclipsed conformations, with the staggered anti conformation being the most stable. Of the eclipsed conformations, the one with the methyl groups eclipsing each other is the least stable.**

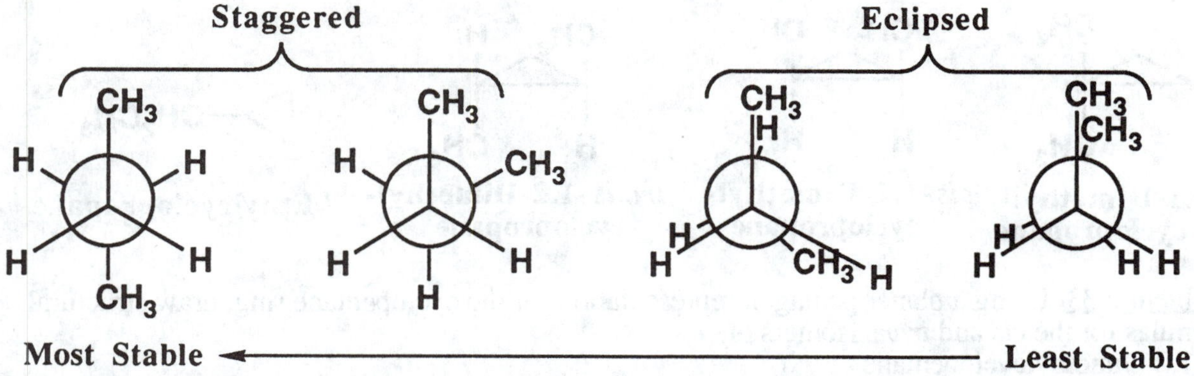
Staggered
Eclipsed
$CH_3$
H
H
H
H
$CH_3$
$CH_3$
H
$CH_3$
H
H
H
$CH_3$
H
H
H
$CH_3$
H
$CH_3$
$CH_3$
H
H
H
H
Most Stable
Least Stable

Problem 2.30 Demonstrate, using molecular models, that in cyclohexane, an equatorial substituent is equidistant from the axial and equatorial groups on an adjacent carbon.

**The best way to see this fact is to draw a Newman projection of one of the carbon-carbon bonds. As can be seen, the axial hydrogen from the carbon atom in the front is in between and thus equidistant to the axial and equatorial hydrogen atoms on the rear carbon atom.**

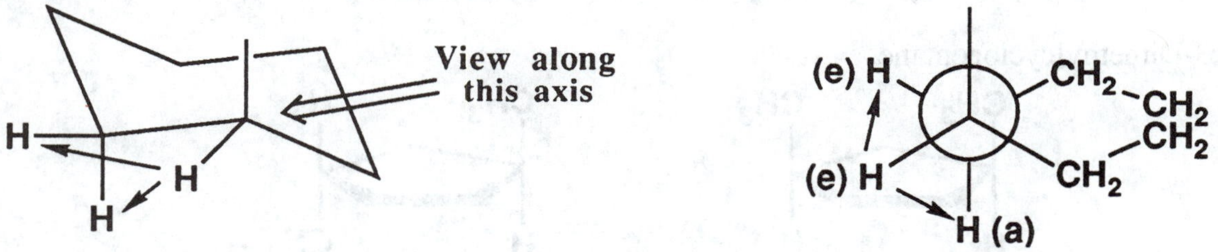
View along this axis
H
H
H
(e) H
(e) H
H (a)
$CH_2$
$CH_2$
$CH_2$
$CH_2$

## *Cis-Trans* Isomerism in Cycloalkanes

Problem 2.31 Name and draw structural formulas for the *cis* and *trans* isomers of 1,2-dimethylcyclopropane.

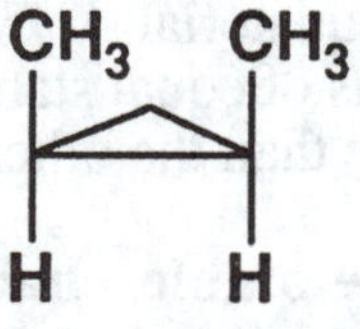
$CH_3$
$CH_3$
H
H

***cis*-1,2-Dimethyl-cyclopropane**

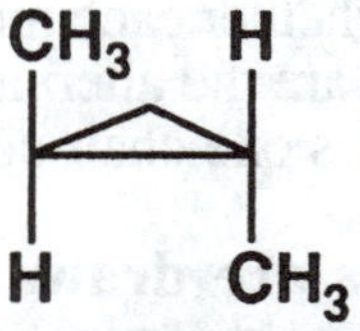
$CH_3$
H
H
$CH_3$

***trans*-1,2-Dimethyl-cyclopropane**

Problem 2.32 Name and draw structural formulas for all cycloalkanes of molecular formula $C_5H_{10}$. Be certain to include *cis-trans* isomers as well as constitutional isomers.

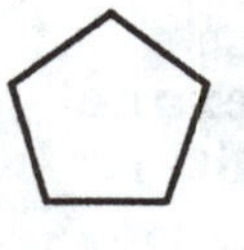

**Cyclopentane**

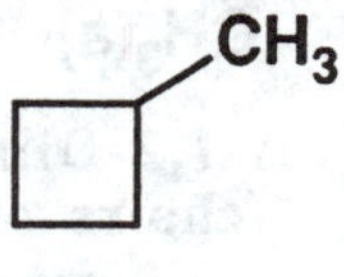
$CH_3$

**Methylcyclo-butane**

**1,1-Dimethyl-
cyclopropane**     ***cis*-1,2-Dimethyl-
cyclopropane**     ***trans*-1,2-Dimethyl-
cyclopropane**     **Ethylcyclopropane**

<u>Problem 2.33</u>  Using a planar pentagon representation for the cyclopentane ring, draw structural formulas for the *cis* and *trans* isomers of:
(a) 1,2-Dimethylcyclopentane

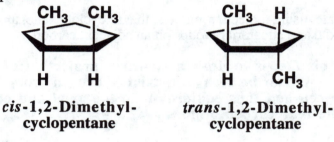

***cis*-1,2-Dimethyl-
cyclopentane**          ***trans*-1,2-Dimethyl-
cyclopentane**

(b) 1,3-Dimethylcyclopentane.

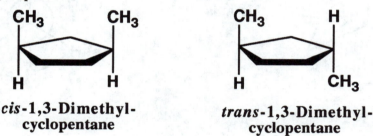

***cis*-1,3-Dimethyl-
cyclopentane**          ***trans*-1,3-Dimethyl-
cyclopentane**

<u>Problem 2.34</u>  Draw the alternative chair conformations for the *cis* and *trans* isomers of 1,2-dimethylcyclohexane;  1,3-dimethylcyclohexane;  and 1,4-dimethylcyclohexane.
(a)  Indicate by a label whether each methyl group is axial or equatorial.
(b)  For which isomer(s) are the alternative chair conformations of equal stability?
(c)  For which isomer(s) is one chair conformation more stable than the other?

**Cis** and **trans** isomers are drawn as pairs.  The more stable chair is labeled in cases where there is a difference.

CH₃(e)     (e)H₃C     CH₃(a)     CH₃(a)

***cis*-1,2-Dimethylcyclohexane
(chairs of equal stability)**

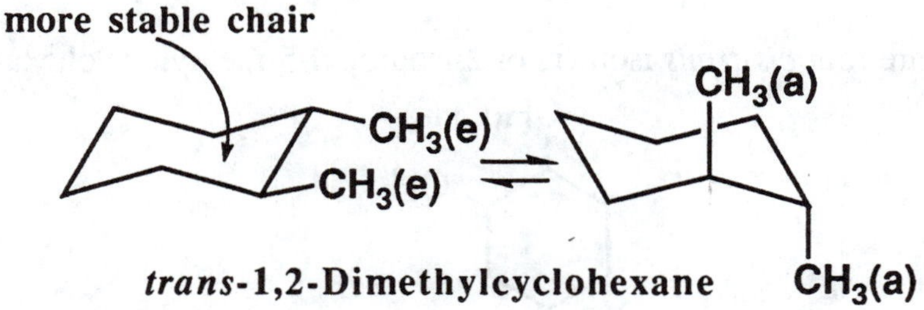

more stable chair

trans-1,2-Dimethylcyclohexane

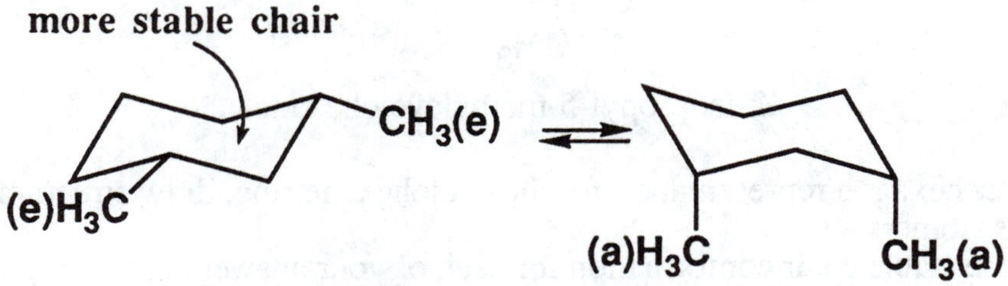

more stable chair

cis-1,3-Dimethylcyclohexane

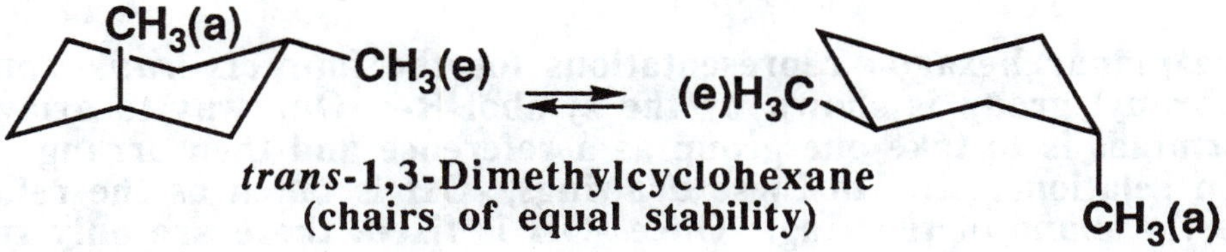

trans-1,3-Dimethylcyclohexane
(chairs of equal stability)

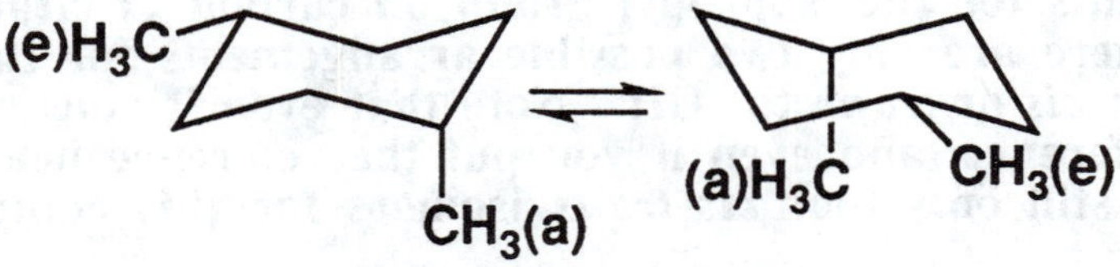

cis-1,4-Dimethylcyclohexane
(chairs of equal stability)

more stable chair

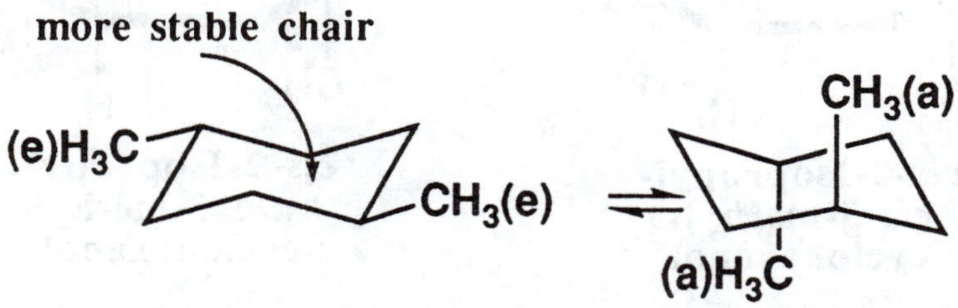

trans-1,4-Dimethylcyclohexane

Problem 2.35 Use your answers from problem 2.34 to complete the table showing correlations between cis, trans and axial, equatorial for the disubstituted derivatives of cyclohexane.

**These relationships are summarized in the following table.**

| Position of Substitution | cis | trans |
|---|---|---|
| 1,4 | a,e or e,a | e,e or a,a |
| 1,3 | e,e or a,a | a,e or e,a |
| 1,2 | a,e or e,a | e,e or a,a |

<u>Problem 2.36</u>  There are four *cis-trans* isomers of 2-isopropyl-5-methylcyclohexanol.

$$CH(CH_3)_2$$

OH

$$CH_3$$

2-Isopropyl-5-methylcyclohexanol

(a)  Using a planar hexagon representation for the cyclohexane ring, draw structural formulas for the four *cis-trans* isomers.
(b)  Draw the more stable chair conformation for each of your answers in part (a).
(c)  Of the four *cis-trans* isomers, which is the most stable?  (If you answered this part correctly, you picked the isomer found in nature and given the name menthol)

**Following are planar hexagon representations for the four *cis-trans* isomers.  In each, the isopropyl group is shown by the symbol R-.  One way to arrive at these structural formulas is to take one group as a reference and then arrange the other two groups in relation to it.  In these drawings, -OH is taken as the reference and placed above the plane of the ring.  Once -OH is fixed, there are only two possible arrangements for the isopropyl group on carbon 2; either *cis* or *trans* to -OH.  Similarly, there are only two possible arrangements for the methyl group on carbon-5; either *cis* or *trans* to -OH.  Note that even if you take another substituent as a reference, and even if you put the reference below the plane of the ring, there are still only four *cis-trans* isomers for this compound.**

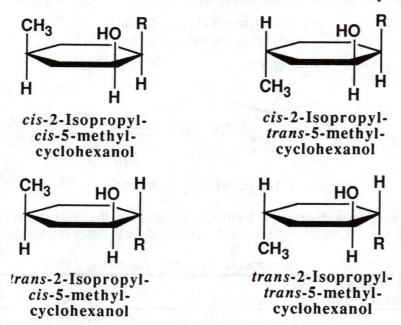

| cis-2-Isopropyl-<br>cis-5-methyl-<br>cyclohexanol | cis-2-Isopropyl-<br>trans-5-methyl-<br>cyclohexanol |
|---|---|
| trans-2-Isopropyl-<br>cis-5-methyl-<br>cyclohexanol | trans-2-Isopropyl-<br>trans-5-methyl-<br>cyclohexanol |

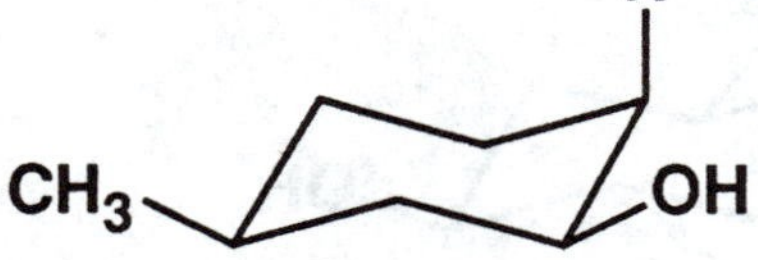

*cis*-2-Isopropyl-*cis*-5-
methylcyclohexanol

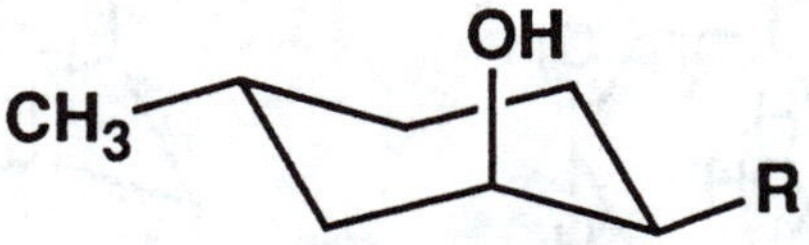

*cis*-2-Isopropyl-*trans*-5-
methylcyclohexanol

most stable chair
(all equatorial)

*trans*-2-Isopropyl-*cis*-5-
methylcyclohexanol

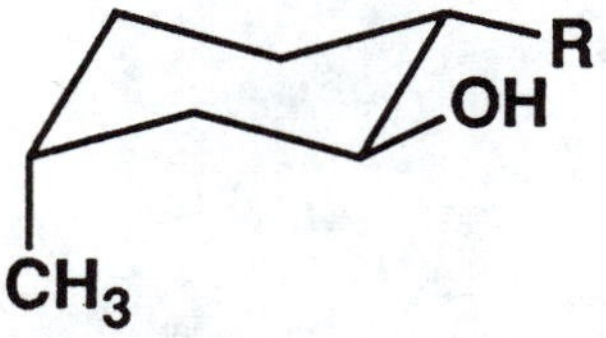

*trans*-2-Isopropyl-*trans*-5-
methylcyclohexanol

**Problem 2.37** Draw alternative chair conformations for each substituted cyclohexane and state which chair is more stable.

(a)

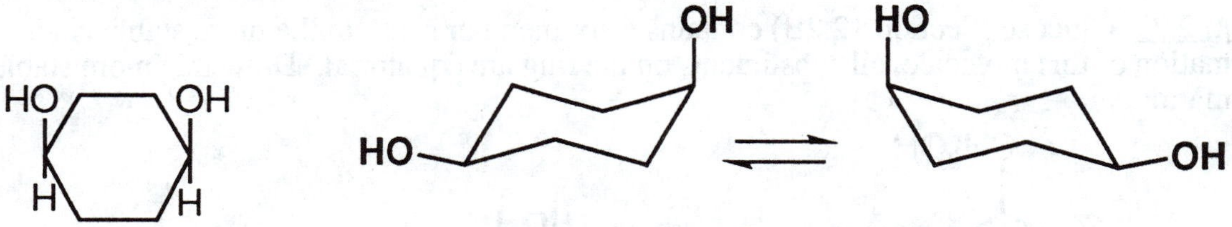

(chairs of equal stability)

(b)

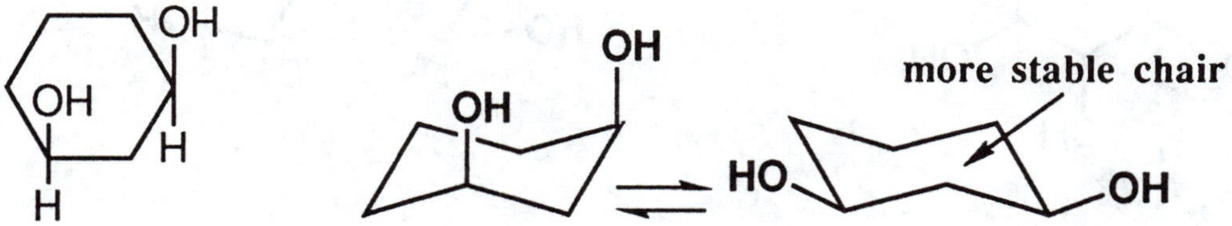

more stable chair

(c)

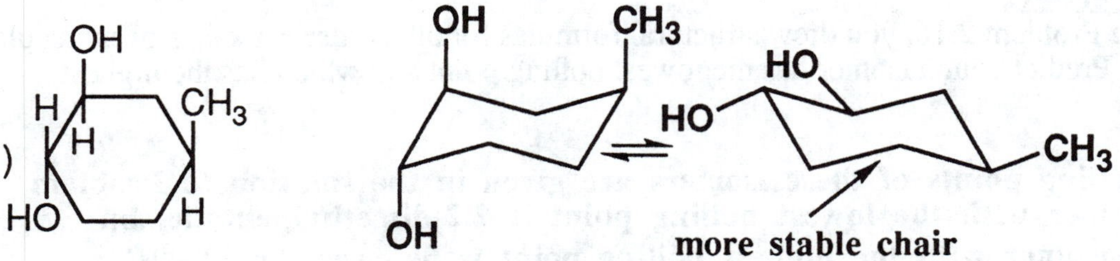

more stable chair

(d)

**Problem 2.38** What kinds of conformations do the six-member rings exhibit in adamantane?

Adamantane

**In adamantane, the cyclohexane rings all have chair conformations.**

**Problem 2.39** Glucose (Section 12.2B) contains a six-member ring. In the more stable chair conformation of this molecule, all substituents on the ring are equatorial. Draw this more stable conformation.

Glucose

## Physical Properties
**Problem 2.40** In Problem 2.16, you drew structural formulas for all isomeric alkanes of molecular formula $C_7H_{16}$. Predict which isomer has the lowest boiling point and which has the highest boiling point.

**Names and boiling points of these isomers are given in the solution to Problem 2.16. The isomer with the lowest boiling point is 2,2-dimethylpentane, bp 79.2°C. The isomer with the highest boiling point is heptane, bp 94.8°C.**

**Problem 2.41** What unbranched alkane has about the same boiling point as water? (see Table 2.4)? Calculate the molecular weight of this alkane and compare it with that of water.

**Heptane, $C_7H_{16}$, has a boiling point of 98.4°C and a molecular weight of 100. Its molecular weight is approximately 5.5 times that of water. Although considerably smaller, the water molecules are held together by the relatively**

strong forces of hydrogen bonding while the much larger heptane molecules are held together only by relatively weak dispersion forces.

## Reactions of Alkanes

Problem 2.42 Write balanced equations for the combustion of each hydrocarbon. Assume that each is converted completely to carbon dioxide and water.

(a) Hexane

$$2\ CH_3(CH_2)_4CH_3\ +\ 19\ O_2\ \longrightarrow\ 12\ CO_2\ +\ 14\ H_2O$$

(b) Cyclohexane

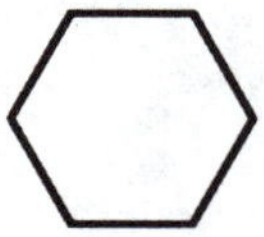

$+\ 9O_2\ \longrightarrow\ 6CO_2\ +\ 6H_2O$

(c) 2-Methylpentane

$$\overset{\displaystyle CH_3}{\underset{\displaystyle |}{2\ CH_3\ CHCH_2CH_2CH_3}}\ +\ 19O_2\ \longrightarrow\ 12CO_2\ +\ 14H_2O$$

Problem 2.43 Methane, propane, and 2,2,4-trimethylpentane are a major sources of energy. On a gram-for-gram basis, which of these hydrocarbons is the best source of heat energy?

| Hydrocarbon | Component of | Heat of Combustion [kcal/mol (kJ/mol)] |
|---|---|---|
| $CH_4$ | natural gas | -212(886) |
| $CH_3CH_2CH_3$ | LPG | -531(2220) |
| $CH_3CCH_2CHCH_3$ (with $CH_3$ groups) | gasoline | -1304(5451) |

On a gram-per-gram basis, methane is the best source of heat energy.

| Hydrocarbon | Molecular Weight | Heat of Combustion (kcal/mol) | Heat of Combustion (kcal/gram) |
|---|---|---|---|
| methane | 16.04 | 212 | 13.3 |
| propane | 44.09 | 531 | 12.0 |
| 2,2,4-trimethylpentane | 114.2 | 1304 | 11.4 |

Problem 2.44 Name and draw structural formulas for all possible monohalogenation products that might be formed in these reactions.

(a) 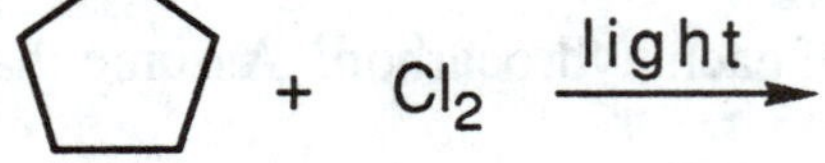 [pentagon structure] + Cl$_2$ →(light)

**Monochlorination of cyclopentane gives only chlorocyclopentane (cyclopentyl chloride).**

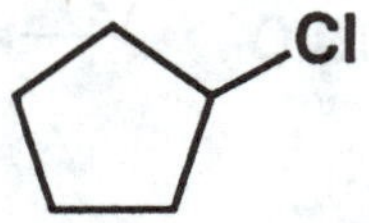

[structure of cyclopentane ring with Cl]

**Chlorocyclopentane**

(b) 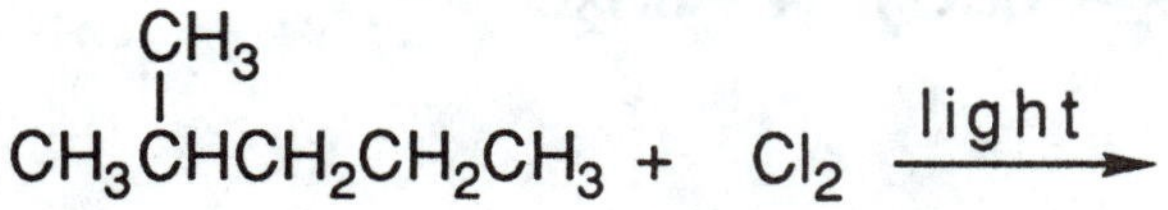 $CH_3CHCH_2CH_2CH_3$ (with CH$_3$ branch) + Cl$_2$ →(light)

**Monochlorination of 2-methylpentane can give a mixture of five constitutional isomers.**

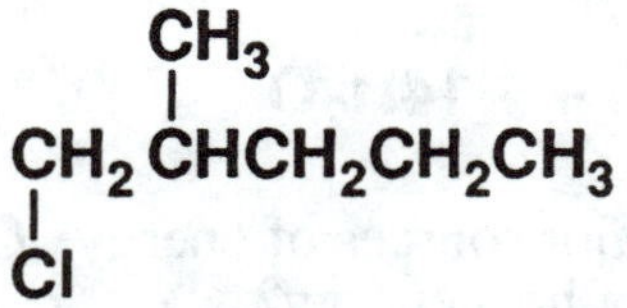

CH$_3$
$CH_2 CHCH_2CH_2CH_3$
Cl
**1-Chloro-2-methyl-pentane**

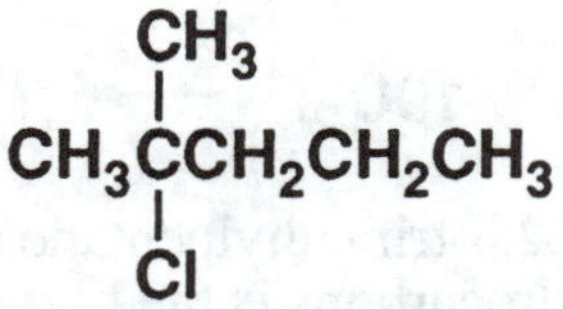

CH$_3$
$CH_3CCH_2CH_2CH_3$
Cl
**2-Chloro-2-methyl-pentane**

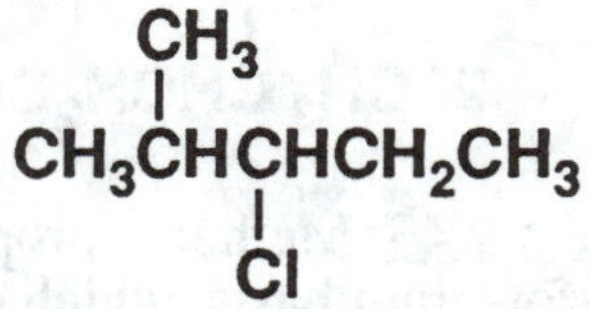

CH$_3$
$CH_3CHCHCH_2CH_3$
Cl
**3-Chloro-2-methyl-pentane**

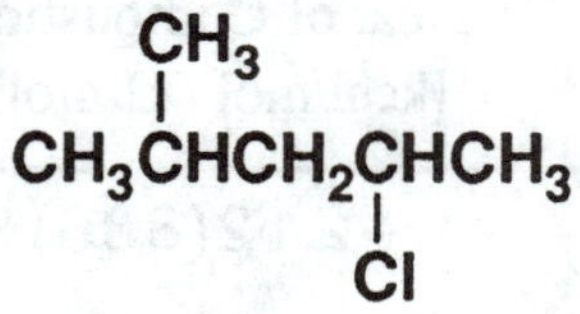

CH$_3$
$CH_3CHCH_2CHCH_3$
Cl
**2-Chloro-4-methyl-pentane**

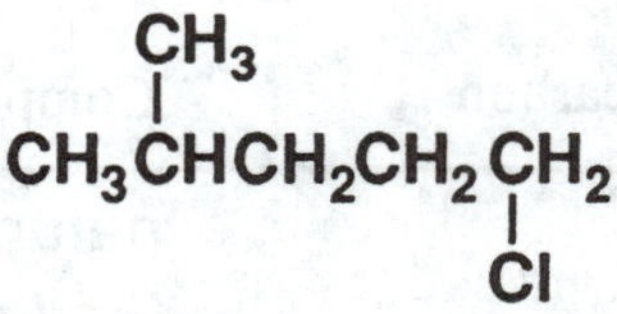

CH$_3$
$CH_3CHCH_2CH_2 CH_2$
Cl
**1-Chloro-4-methyl-pentane**

**2-Chloro-2-methylpentane is predicted to be the major product because it is the only one derived from replacement of a tertiary hydrogen.**

(c) 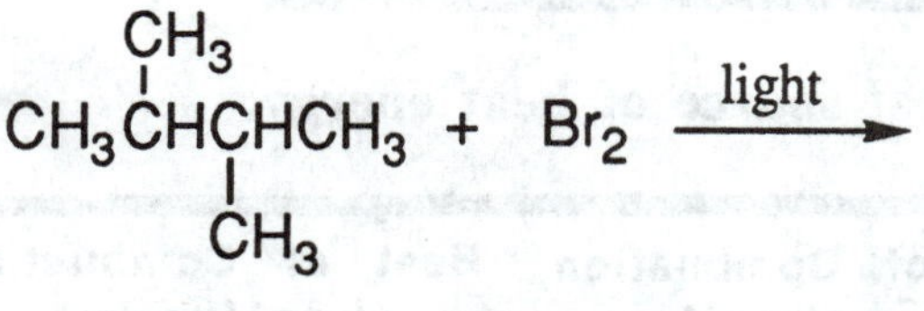 $CH_3CHCHCH_3$ (with CH$_3$ branches) + Br$_2$ →(light)

**Two monobromination products are possible from 2,3-dimethylbutane.**

$$
\begin{matrix}
\text{H}_3\text{C} \quad \text{CH}_3 \\
| \qquad | \\
\text{CH}_2\text{CHCHCH}_3 \\
| \\
\text{Br}
\end{matrix}
\qquad\qquad
\begin{matrix}
\text{H}_3\text{C} \quad \text{CH}_3 \\
| \qquad | \\
\text{CH}_3\text{C}-\text{CHCH}_3 \\
| \\
\text{Br}
\end{matrix}
$$

**1-Bromo-2,3-dimethyl-**
**butane**

**2-Bromo-2,3-dimethyl-**
**butane**

**2-Bromo-2,3-dimethylpentane is predicted to be the major product because it is the one derived from replacement of a tertiary hydrogen.**

(d)   + Br$_2$  $\xrightarrow{\text{light}}$

**Only one monobromination product is possible from cyclopropane.**

$\triangleright$—**Br**

**Bromocyclopropane**

Problem 2.45  Define regioselectivity.

**A regioselective reaction is one in which one direction of bond making and bond breaking occurs preferentially over all other possible directions.  An example would be radical bromination of propane to give primarily 2-bromopropane.**

Problem 2.46  Where you predicted formation of two or more products in Problem 2.44, predict which is the major product.

**The major product is identified in the answer to Problem 2.44**

Problem 2.47  Which compounds could be prepared in high purity (uncontaminated with other isomers) by regioselective halogenation of an alkane?
(a) 2-Chloropentane                          (b) Chlorocyclobutane
(c) 2-Bromo-2-methylbutane               (d) 2-Bromo-3-methylbutane
(e) 2-Bromo-2,4,4-trimethylpentane     (f) Bromocyclohexane

**The compounds that can be prepared in high yields are (b), (c), (e), and (f).  For (b) and (f), there is only one product possible for monocholorination so the yield will be high.  For (c) and (e), the desired products are the result of reaction at the only tertiary carbon atoms in the molecules.  Therefore, the large preference of bromination for substitution at tertiary sites will guarantee that the desired product will predominate.**

**For (a), reaction at the 3 position, also a secondary carbon atom, will lower the purity.  For (d), the preferred reaction with be at the tertiary 3 position, not the secondary 2 position.**

<u>Problem 2.48</u> There are three constitutional isomers of molecular formula $C_5H_{12}$. When treated with chlorine gas at 300°C, isomer A gives a mixture of four monochlorination products. Under the same conditions, isomer B gives a mixture of three monochlorination products, and isomer C gives only one monochlorination product. From this information, assign structural formulas to isomers A, B, and C.

**Structural formulas for the three alkanes of are:**

$$\underset{\substack{\text{A}\\ \text{2-Methylbutane}\\ \text{(Isopentane)}}}{\underset{\displaystyle CH_3-\overset{\displaystyle\overset{\textstyle CH_3}{|}}{CH}-CH_2-CH_3}{}} \qquad \underset{\substack{\text{B}\\ \text{Pentane}}}{CH_3-CH_2-CH_2-CH_2-CH_3} \qquad \underset{\substack{\text{C}\\ \text{2,2-Dimethylpropane}\\ \text{(Neopentane)}}}{CH_3-\overset{\displaystyle\overset{\textstyle CH_3}{|}}{\underset{\displaystyle\underset{\textstyle CH_3}{|}}{C}}-CH_3}$$

**To arrive at the correct assignments of structural formulas, first write formulas for all monochloroalkanes possible from each structural formula. Then compare these numbers with those observed for A, B, and C. Because isomer B gives three monochlorination product, it must be pentane. By the same reasoning, A must be 2-methylbutane, and C must be 2,2-dimethylpropane.**

# CHAPTER 3: ALKENES AND ALKYNES

## 3.0 OVERVIEW
- **Unsaturated hydrocarbons** are hydrocarbons that contain one or more carbon-carbon double or triple bond.
  - **Alkenes** are unsaturated hydrocarbons with one or more carbon-carbon double bond.
  - **Alkynes** are unsaturated hydrocarbons with one or more carbon-carbon triple bond.
  - **Aromatic hydrocarbons** are hydrocarbons that have cyclic structures with special patterns of alternating double bonds.

## 3.1 STRUCTURE
- According to the VSEPR model, the carbon atoms of a double bond should have bond angles of roughly 120°.
- According to the valence bond approach, a carbon-carbon double bond consists of **one sigma bond** formed by the overlap of $sp^2$ hybridized orbitals of adjacent carbon atoms and **one pi bond** formed by the overlap of unhybridized 2p orbitals. * *[This picture of the orbitals involved with the carbon-carbon double bond is crucial to your understanding of the reactions and properties of alkenes.]*
  - The most important implication of this model of carbon-carbon double bonds is that they **can not rotate** because rotation would decrease the extent of 2p-2p overlap. *
  - As correctly predicted by the VSEPR model, each carbon atom in a carbon-carbon double bond is $sp^2$ hybridized, so its geometry is trigonal planar with bond angles near 120°. Notice that this means the carbon atoms and the atoms attached directly to them are in the same plane.
- **Stereoisomers** are molecules that have the same molecular formula, the same connectivity of atoms, but a different arrangement of atoms in space. * *[Stereoisomerism is one of the most important three-dimensional concepts in chemistry, and various types of stereoisomers will be presented throughout the book. The use of molecular models is always helpful in discussions of stereoisomerism.]*
- Because carbon-carbon double bonds cannot rotate, alkenes can display *cis-trans* **isomerization**. For alkenes that have one substituent and one hydrogen atom on each double-bonded carbon:
  - A *cis* alkene is one in which the substituents are on the **same side** of the double bond. For example, *cis*-2-butene. *

$$H_3C \diagdown C{=}C \diagup CH_3$$
$$H \diagup \qquad \diagdown H$$

*cis*-2-Butene

  - A *trans* alkene is one in which the substituents are on **opposite sides** of the double bond. For example, *trans*-2-butene. *

$$H_3C \diagdown C{=}C \diagup H$$
$$H \diagup \qquad \diagdown CH_3$$

*trans*-2-Butene

  - A *trans* **alkene** is **more stable than a corresponding** *cis* **alkene,** because in a *cis* alkenes bulky groups on the same side of a double bond can run into each other, creating strain.
- Alkynes have carbon-carbon triple bonds.
- According to the **valence bond approach**, a **carbon-carbon triple bond** consists of **one sigma bond** formed by the overlap of $sp^2$ hybridized orbitals of adjacent carbon atoms, **one**

**pi bond** formed by the overlap of two parallel, unhybridized $2p_y$ orbitals, and a **second pi bond** formed by overlap of two parallel, unhybridized $2p_z$ orbitals. ✳

## 3.2 NOMENCLATURE

- Form **IUPAC** names of **alkenes** by changing the **an** infix of the parent alkane to the infix **en**. For example, ethene and propene are alkenes with two and three carbon atoms respectively. ✳
  - Form names for more complicated alkenes by choosing the longest carbon chain with the carbon-carbon double bond as the parent alkane.
    **Number the chain** to give the double bond the **smallest number**, indicate the position of the double bond by using the number of the first atom of the double bond.
    **Name branched** or **substituted alkenes** according to the same rules discussed for alkanes.
    For **cycloalkenes**, number the carbon atoms of the double bond 1 and 2 in the direction that gives the substituent encountered first the smallest number.
- Form **IUPAC** names of **alkynes** by changing the **an** infix of the parent alkane to the infix **yn**. For example, ethyne and propyne are alkenes with two and three carbon atoms respectively. ✳
  - Form names for complicated alkynes by finding the longest carbon chain that contains the carbon-carbon triple bond.
  - This chain is numbered so that the carbon atoms of the triple bond are the lower numbers.
  - The location of the triple bond is given as the first carbon of the triple bond.
- Several alkenes and alkynes are known by their common names including ethylene, propylene, isobutylene, butadiene, and acetylene. Substituents are also given common names such as methylene, vinyl, and allyl. *[These must be learned because their use is so widespread.]*
- The *cis-trans* system is used to designate the configuration in an alkene when the parent chain of the molecule can be identified.
  - In *cis* alkenes, the main chain carbon atoms are on the same side of the double bond.
  - In *trans* alkenes, the main chain carbon atoms are on the opposite side of the double bond.
- The **E,Z system** is a more comprehensive system of nomenclature for designating the configuration of an alkene.
- The **E,Z system** uses priority rules to rank the two substituents on each carbon atom of the double bond:
  - Each atom is assigned a priority based on atomic number; the higher the atomic number the higher the priority.
  - If you cannot assign priority differences to the two substituents by comparing the first atoms, then continue down the chains until the first point of difference is reached. *[This can be confusing. Notice the total size of the substituents attached to the double bond is not important, it is the <u>first point of difference in priority</u> that matters. For example, a -$CH_2Cl$ group has a higher priority than $CH_2CH_2CH_2CH_2CH_3$ because the first point of difference, Cl, has a higher priority than any atom on the first carbon of the larger alkyl group.]*
  - In the case of double and triple bonds, count the atoms participating in the double or triple bond as if they are bonded to an equivalent number of similar atoms. For example, a

    -$CH=CH_2$ group is counted as
    $$\begin{matrix} & \overset{C}{|} & \overset{C}{|} & \\ -&C&-C&-H \\ & | & | & \\ & H & H & \end{matrix}$$
    and a -$C{\equiv}CH$ group is counted as
    $$\begin{matrix} & \overset{C}{|} & \overset{C}{|} & \\ -&C&-C&-H \\ & | & | & \\ & C & C & \end{matrix}$$
    *[The only way to get good at this is to practice.]*
  - To assign an alkene as **E** or **Z**, use the following rules:
    If the atoms or groups of atoms of higher priority are on the **same side** of the double bond, it is a **Z alkene**. It is easy to **remember** this as **Z** for **"zame zide."**
    If the higher priority substituents are on **opposite** sides, it is an **E alkene**. *[Caution! Because the letter Z has a zig-zag shape, some students find it tempting to assume that the Z stands for higher priority groups on opposite sides of the alkene making a zig-zag shape*

*around the double bond. This is not the way to assign structures, because E is used for higher priority groups on opposite sides of alkenes. Also, it is important to notice the differences between E and trans as well as Z and cis.]*

higher higher        lower higher
lower lower          higher lower

   **Z**                **E**
("<u>z</u>ame <u>z</u>ide")

For alkenes with more than one double bond, each double bond is named as E or Z as applicable.

- For naming **cycloalkenes**, number the carbon atoms of the double bond 1 and 2 in the direction that gives the substituent encountered first the smallest number.
  - Only cycloalkenes with 8 or more carbon atoms in the ring can have a *trans* geometry, otherwise the angle strain will only allow for *cis* double bonds.
- Alkenes that contain more than one carbon-carbon double bond are called **alkadienes, alkatrienes** and so forth. These molecules may also be referred to by the more general term **polyenes**.
  - For alkenes with **n** double bonds, there are up to $2^n$ possible *cis-trans* isomers. There will be fewer if the molecule contains any symmetry.

## 3.3 PHYSICAL PROPERTIES
- Because alkenes are nonpolar, the only interactions between alkene molecules are dispersion forces, thus their properties are very similar to those of alkanes.

## 3.4 NATURALLY OCCURRING ALKENES-THE TERPENES
- Alkenes are common in nature, and comprise a very important set of biological molecules. **Terpenes** are one group of biological alkene molecules that have some interesting features. **Terpenes** are based on carbon skeletons that can be divided into two or more units that have the same carbon skeleton of **isoprene**.

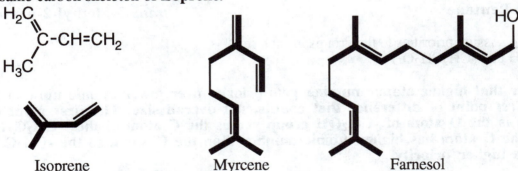

$H_2C$
$\quad$ C-CH=CH$_2$
$H_3C$

   Isoprene                    Myrcene                      Farnesol
(The isoprene units of these terpenes are shown in bold)

- In nature, terpenes are not synthesized from isoprene, but from the pyrophosphate ester of 3-methyl-3-butene-1-ol.

# CHAPTER 3
## *Solutions to the Problems*

<u>Problem 3.1</u>  Give each compound an IUPAC name.

(a) $CH_3CH_2\underset{\underset{CH_3}{|}}{\overset{\overset{CH_3}{|}}{C}}CH=CH_2$

(b) $CH_3(CH_2)_5C{\equiv}CH$

(c) $CH_3\underset{\underset{CH_3}{|}}{\overset{\overset{CH_3}{|}}{C}}C{\equiv}CH$

**3,3-Dimethyl-1-pentene**            **1-Octyne**            **3,3-Dimethyl-1-butyne**

<u>Problem 3.2</u>  Give each compound a common name.

(a) $CH_3\underset{\underset{CH_3}{|}}{CH}C{\equiv}C\underset{\underset{CH_3}{|}}{CH}CH_3$

(b) [cyclohexyl]$-C{\equiv}CH$

(c) $HC{\equiv}CCH_2CH_2CH_2CH_3$

**Diisopropylacetylene**            **Cyclohexylacetylene**            ***n*-Butylacetylene**

<u>Problem 3.3</u>  Which alkenes show *cis-trans* isomerism?  For each alkene that does, draw the *trans* isomer.

(a) 2-Pentene                  (b) 2-Methyl-2-pentene            (c) 3-Methyl-2-pentene

**No *cis-trans* isomers since there two methyl groups on one end of the double bond.**

*trans* **2-Pentene**                                          ***trans*-3-Methyl-2-pentene**

<u>Problem 3.4</u>  Assign priorities to the groups in each set.
(a) $-CH_2OH$ and $-CH_2CH_2OH$

**Remember that higher atomic number gets priority over lower atomic number and it is the *first* point of difference that counts, not overall size. The first point of difference is the O atom of $-CH_2\underline{O}H$ group versus the C atom of the $-CH_2\underline{C}H_2OH$ group.  The O atom has higher atomic number than the C atom, so the $-CH_2OH$ group has higher priority.**

(b) $-CH_2OH$ and $-CH=CH_2$

**Remember that the $-CH=CH_2$ group counts as:**

$$\underset{-CH\cdot CH_2}{\overset{\overset{C}{|}\quad\overset{C}{|}}{}}$$

**The *first* point of difference is the O atom of the first carbon atom of the $-CH_2OH$ group versus a C atom on the first carbon atom of the $-CH=CH_2$ group. Therefore, the $-CH_2OH$ group has higher priority.**

Problem 3.5  Name each alkene and specify its configuration by the E-Z system.

(a)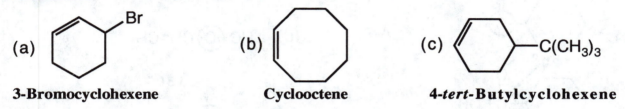

ClCH₂        CH₃
    C=C
CH₃        CH₂CH₃

(E)-1-Chloro-2,3-dimethyl-2-pentene

(b)

Cl        H
    C=C
Br        CH₃

(Z)-1-Bromo-1-chloropropene

(c)

CH₃CH₂CH₂        CH₃
        C=C
    CH₃        CH(CH₃)₂

(E)-2,3,4-Trimethyl-3-heptene

Problem 3.6  Give each cycloalkene an IUPAC name.

(a) [cyclohexene with Br]

3-Bromocyclohexene

(b) [cyclooctene]

Cyclooctene

(c) [cyclohexene with C(CH₃)₃]

4-*tert*-Butylcyclohexene

Problem 3.7  Draw structural formulas for the other two *cis-trans* isomers for 2,4-heptadiene.

H₃C₁  ₂C=C₃  ₄C=C₅  ₆CH₂CH₃⁷  with H's

*cis,trans*-2,4-Heptadiene

*cis,cis*-2,4-Heptadiene

Problem 3.8  The sex pheromone from the silk-worm is (10E, 12Z)-10,12-hexadecadiene-1-ol. Draw a structural formula for this compound.

CH₃CH₂CH₂  ₁₃ ₁₂C=C  ₁₁ ₁₀C=C  CH₂(CH₂)₇CH₂OH  with H's

(10E,12Z)-10,12-hexadecadiene-1-ol

## Structure of Alkenes and Alkynes

**Problem 3.9** Predict all bond angles about each highlighted carbon atom. To make these predictions, use the valence shell electron-pair repulsion model (Section 1.4).

(a)

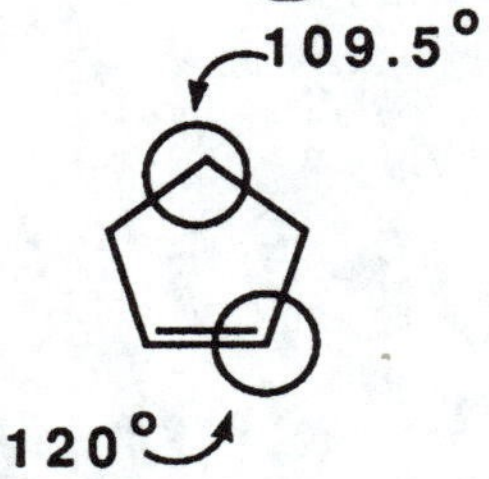

109.5°

120°

(b)

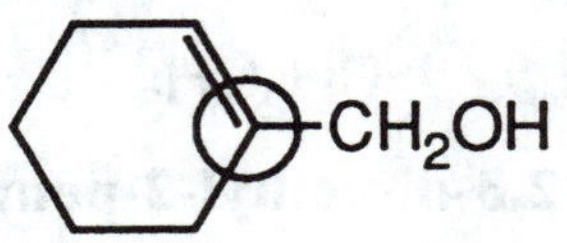

—CH₂OH

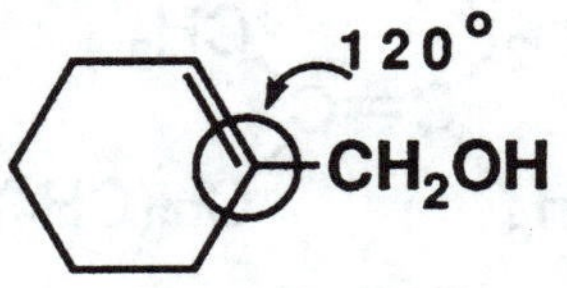

120°

—CH₂OH

(c)

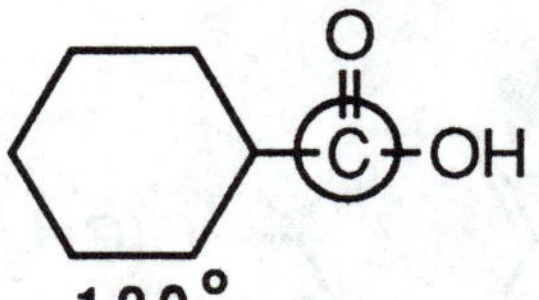

$\overset{O}{\underset{||}{C}}$—OH

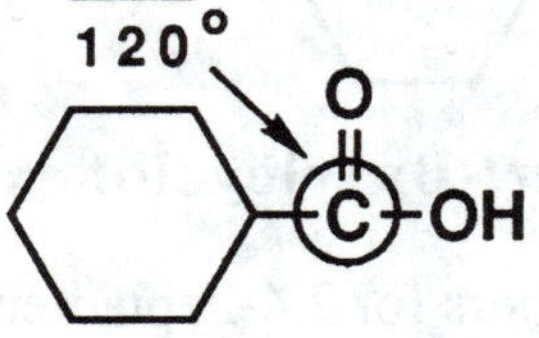

120°

$\overset{O}{\underset{||}{C}}$—OH

(d) 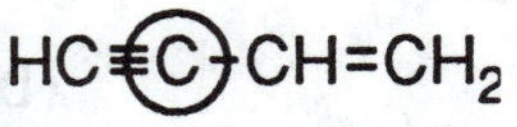    HC≡C—CH=CH₂

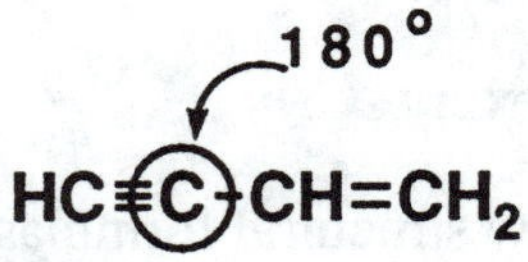

180°

HC≡C—CH=CH₂

**Problem 3.10** For each circled carbon atom in Problem 3.9, identify which atomic orbitals are used to form each sigma bond and which are used to form each pi bond.

**Each bond is labeled sigma or pi and the orbitals overlapping to form each bond are shown.**

(a) 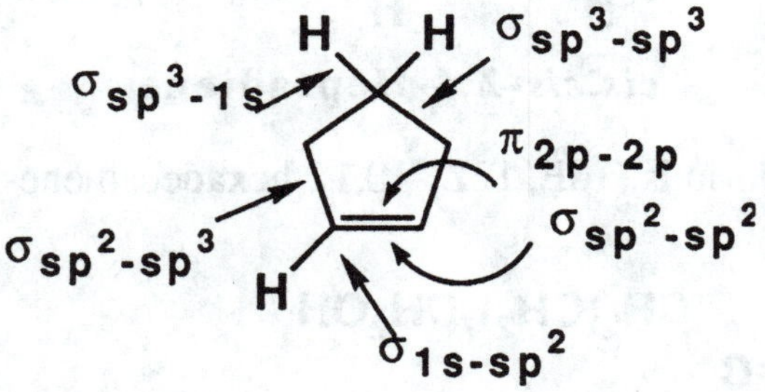   $\sigma_{sp^3-1s}$   $\sigma_{sp^3-sp^3}$   $\pi_{2p-2p}$   $\sigma_{sp^2-sp^3}$   $\sigma_{sp^2-sp^2}$   $\sigma_{1s-sp^2}$

(b) 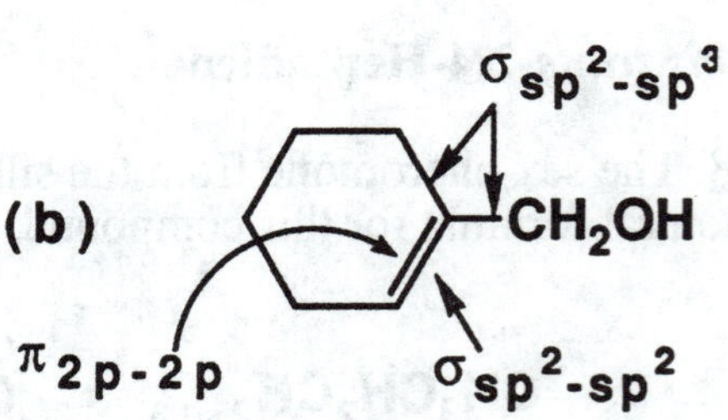   $\sigma_{sp^2-sp^3}$   CH₂OH   $\pi_{2p-2p}$   $\sigma_{sp^2-sp^2}$

(c) 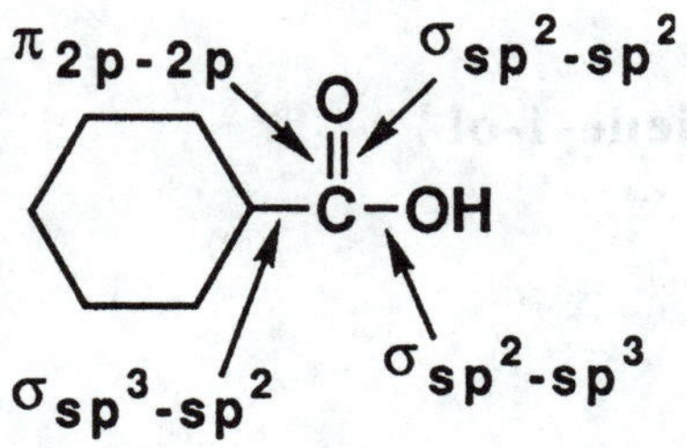   $\pi_{2p-2p}$   $\sigma_{sp^2-sp^2}$   $\overset{O}{\underset{||}{C}}$—OH   $\sigma_{sp^3-sp^2}$   $\sigma_{sp^2-sp^3}$

(d) 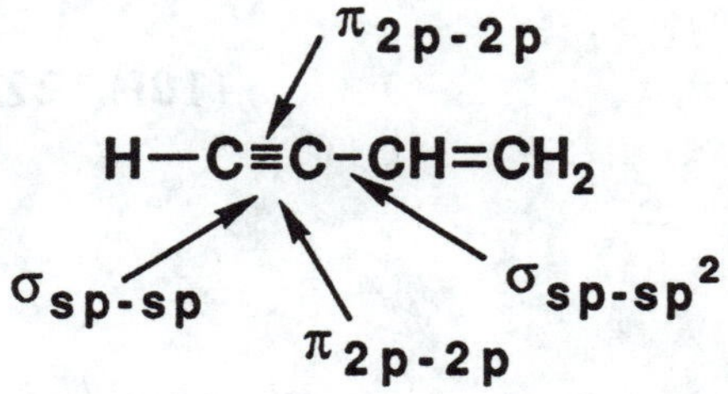   $\pi_{2p-2p}$   H—C≡C—CH=CH₂   $\sigma_{sp-sp}$   $\pi_{2p-2p}$   $\sigma_{sp-sp^2}$

## Nomenclature of Alkenes and Alkynes
Problem 3.11  Write IUPAC names for these compounds.

(a) $CH_3C{\equiv}CCH_3$ with $CH_3$ above and $CH_3$ below the third carbon

**4,4-Dimethyl-2-pentyne**

(b) $HC{\equiv}CCH_2Br$

**3-Bromopropyne**

(c) cyclopentyl$-C{\equiv}CH$

**2-Cyclopentylethyne**

(d) $CH_3(CH_2)_5C{\equiv}CCH_2OH$

**2-Nonyn-1-ol**

(e) $CH_3(CH_2)_6C{\equiv}CH$

**Nonyne**

Problem 3.12  Draw structural formulas for these compounds:
(a) 3-Hexyne

**$CH_3CH_2C{\equiv}CCH_2CH_3$**

(b) 3-Decyne

**$CH_3CH_2C{\equiv}C(CH_2)_5CH_3$**

(c) 3-Chloro-1-butyne

**$HC{\equiv}CCHCH_3$** with $Cl$ above the third carbon

(d) 5-Isopropyl-3-octyne

**$CH_3CH_2C{\equiv}CCHCH_2CH_2CH_3$** with $CH_3-CH$ and $CH_3$ branch

(e) 3-Pentyn-2-ol

**$CH_3C{\equiv}CCHCH_3$** with $OH$ above the fourth carbon

(f) 2-Butyne-1,4-diol

**$HOCH_2C{\equiv}CCH_2OH$**

(g) Diisopropylacetylene

**$CH_3CHC{\equiv}CCHCH_3$** with $CH_3$ above each CH

(g) *tert*-Butylmethylacetylene

**$CH_3C{\equiv}CCCH_3$** with $CH_3$ above and $CH_3$ below the quaternary carbon

Problem 3.13  Which alkenes exist as pairs of *cis-trans* isomers?  For each alkene that does, draw the *trans* isomer.

**For an alkene to exist as a pair of *cis-trans* isomers, both carbon atoms of the double bond must have two different substituents.  Thus, (b), (c), and (e) exist as a pair of *cis-trans* isomers.  The *trans* isomers for these alkenes are drawn under their respective condensed molecular formula.**

(a) $CH_2{=}CHBr$

(b) $CH_3CH{=}CHBr$

$CH_3$ and $H$ on left carbon, $H$ and $Br$ on right carbon (trans C=C)

(c) $BrCH{=}CHBr$

$Br$ and $H$ on left carbon, $H$ and $Br$ on right carbon (trans C=C)

(d) $(CH_3)_2C=CHCH_3$              (e) $(CH_3)_2CHCH=CHCH_3$

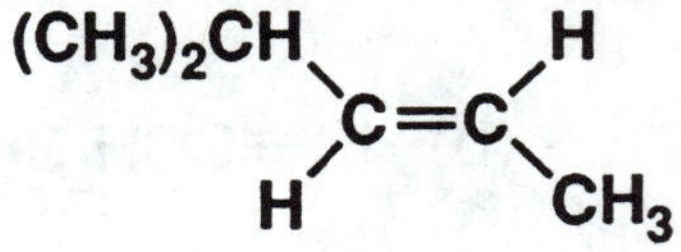

$(CH_3)_2CH$, $H$
$\phantom{(CH_3)_2}C=C$
$H$ , $CH_3$

**Problem 3.14** Name and draw formulas for all alkenes of molecular formula $C_5H_{10}$. As you draw these alkenes, remember the *cis* and *trans* isomers are different compounds and must be counted separately in drawing all alkenes possible for this molecular formula.

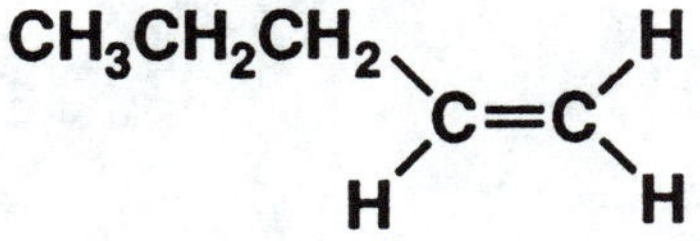

$CH_3CH_2CH_2$, $H$
$\phantom{CH_3CH_2}C=C$
$H$ , $H$
**1-Pentene**

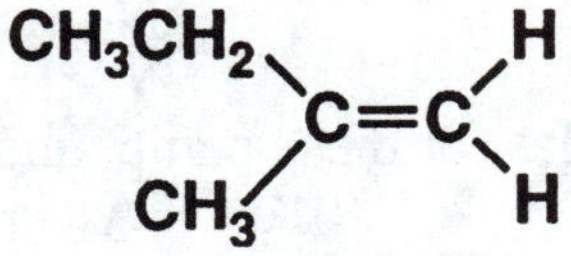

$CH_3CH_2$, $H$
$\phantom{CH_3}C=C$
$CH_3$ , $H$
**2-Methyl-1-butene**

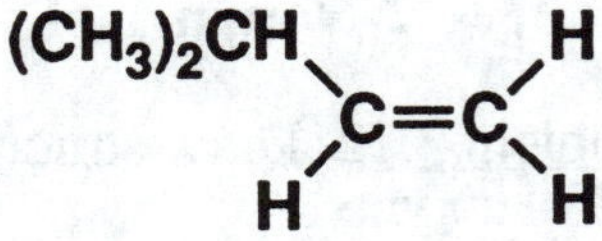

$(CH_3)_2CH$, $H$
$\phantom{(CH_3)_2}C=C$
$H$ , $H$
**3-Methyl-1-butene**

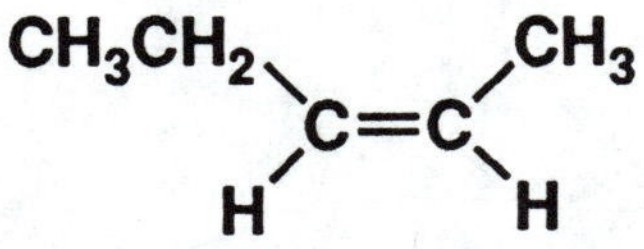

$CH_3CH_2$, $CH_3$
$\phantom{CH_3}C=C$
$H$ , $H$
*cis*-**2-Pentene**

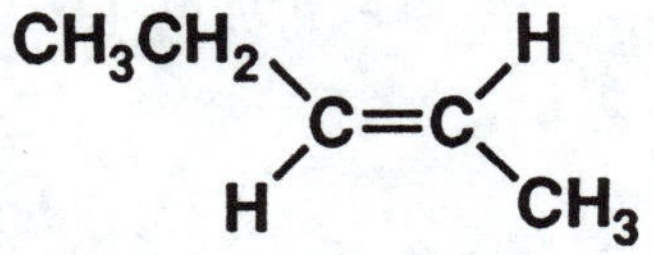

$CH_3CH_2$, $H$
$\phantom{CH_3}C=C$
$H$ , $CH_3$
*trans*-**2-Pentene**

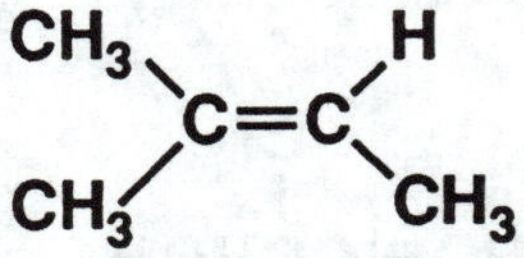

$CH_3$, $H$
$\phantom{CH}C=C$
$CH_3$ , $CH_3$
**2-Methyl-2-butene**

**Problem 3.15** Name and draw structural formulas for all alkenes of molecular formula $C_6H_{12}$ that have these carbon skeletons. Remember *cis* and *trans* isomers!

(a) 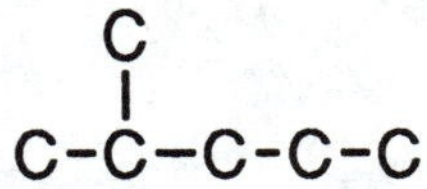 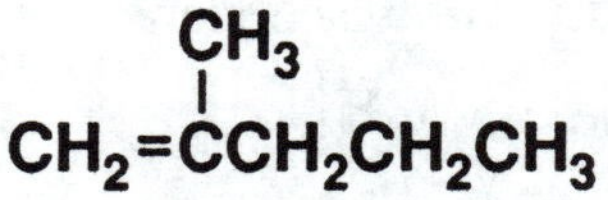
```
      C
      |
C-C-C-C-C
```

$CH_3$
|
$CH_2=CCH_2CH_2CH_3$
**2-Methyl-1-pentene**

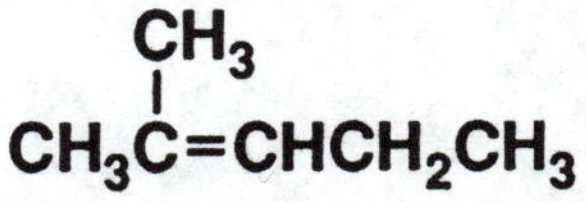

$CH_3$
|
$CH_3C=CHCH_2CH_3$
**2-Methyl-2-pentene**

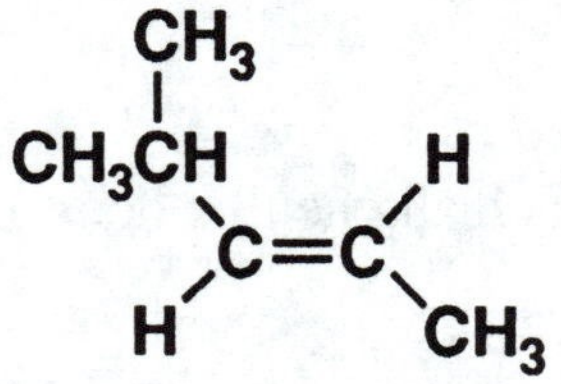

$CH_3$
|
$CH_3CH$, $H$
$\phantom{CH_3}C=C$
$H$ , $CH_3$
*trans*-**4-Methyl-2-pentene**

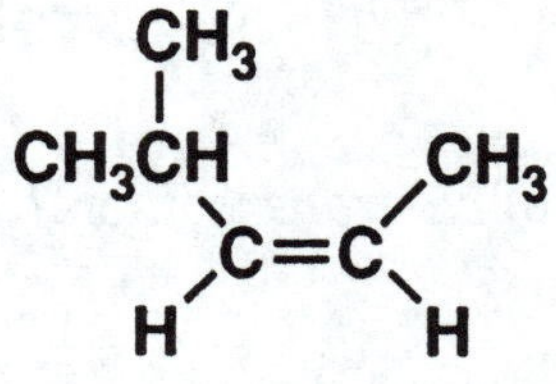

$CH_3$
|
$CH_3CH$, $CH_3$
$\phantom{CH_3}C=C$
$H$ , $H$
*cis*-**4-Methyl-2-pentene**

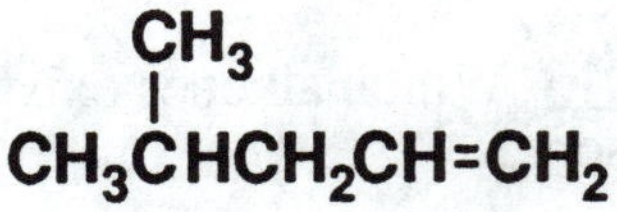

$CH_3$
|
$CH_3CHCH_2CH=CH_2$
**4-Methyl-1-pentene**

(b)

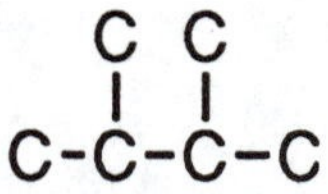

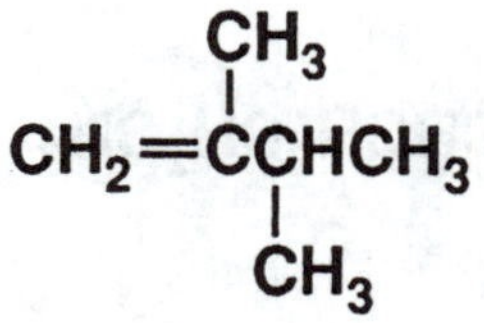

C C
| |
C-C-C-C

CH₃
|
CH₂=CCHCH₃
|
CH₃

**2,3-Dimethyl-1-butene**

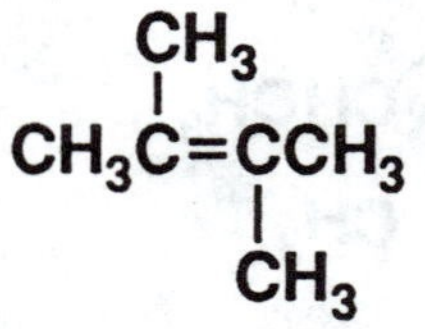

CH₃
|
CH₃C=CCH₃
|
CH₃

**2,3-Dimethyl-2-butene**

(c)

C
|
C-C-C-C
|
C

CH₃
|
CH₃C CH=CH₂
|
CH₃

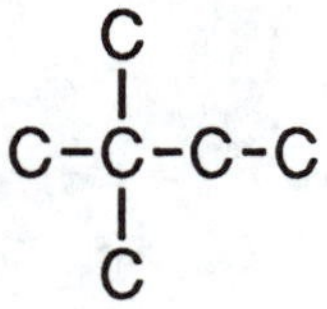

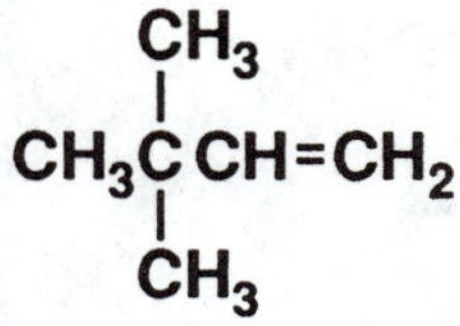

**3,3-Dimethyl-1-butene**

<u>Problem 3.16</u> Arrange the groups in each set in order of increasing priority.
(a) -CH₃  -Br  -CH₂CH₃                    (b) -OCH₃  -CH(CH₃)₂  -CH=CH₂
   **-CH₃ < -CH₂CH₃ < -Br**                     **-CH(CH₃)₂ < -CH=CH₂ < -OCH₃**

(c) -CH₂OH  -CO₂H  -OH
   **-CH₂OH < -CO₂H < -OH**

<u>Problem 3.17</u> Assign an E-Z configuration and a *cis-trans* configuration to these carboxylic acids, each of which is an intermediate in the tricarboxylic acid cycle. Following each is given its common name.

(a)

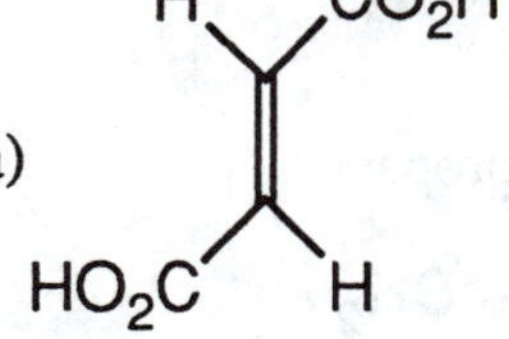

H    CO₂H
 \  /
  C=C
 /  \
HO₂C   H

Fumaric acid

(b)

HO₂C      H
   \     /
    C=C
   /     \
  H    CH₂CO₂H

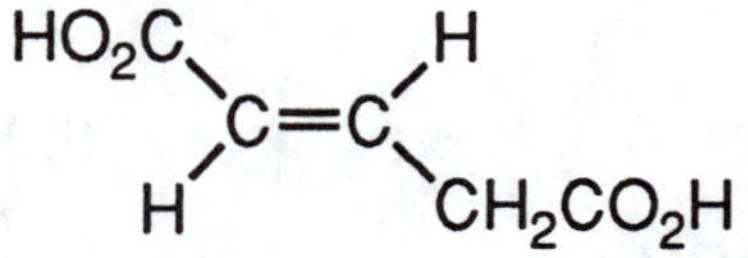

Aconitic acid

**Both of these alkenes have the E or *trans* configuration.**

Problem 3.18  Draw structural formulas for these alkenes.
(a) *trans*-2-Methyl-3-hexene                (b) 2-Methyl-2-hexene

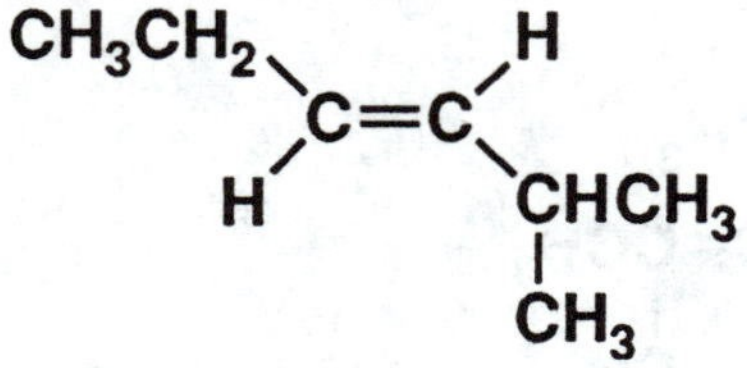

$$CH_3CH_2 \quad \quad H$$
$$C=C$$
$$H \quad \quad CHCH_3$$
$$\quad \quad \quad CH_3$$

$$CH_3$$
$$CH_3C=CHCH_2CH_2CH_3$$

(c) 2-Methyl-1-butene 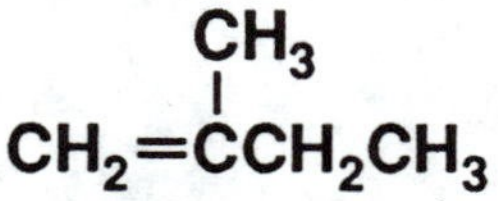                        (d) 3-Ethyl-3-methyl-1-pentene

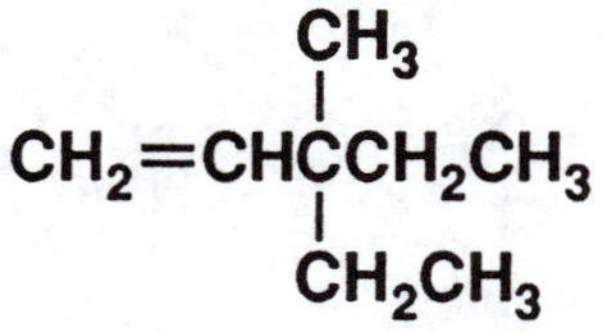

$$CH_3$$
$$CH_2=CCH_2CH_3$$

$$CH_3$$
$$CH_2=CHCCH_2CH_3$$
$$CH_2CH_3$$

(e) 2,3-Dimethyl-2-butene 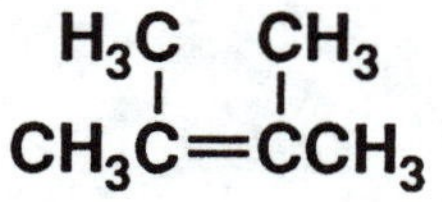                    (f) *cis*-2-Pentene

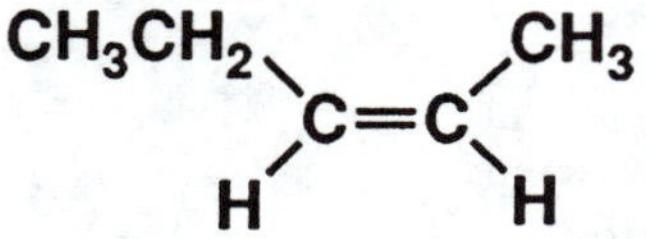

$$H_3C \quad CH_3$$
$$CH_3C=CCH_3$$

$$CH_3CH_2 \quad \quad CH_3$$
$$C=C$$
$$H \quad \quad H$$

(g) (Z)-1-Chloropropene 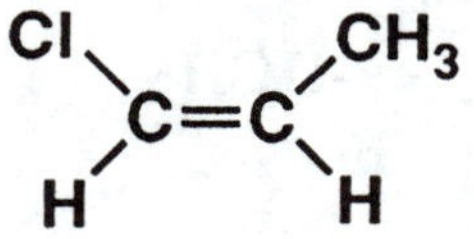                      (h) 3-Methylcyclohexene

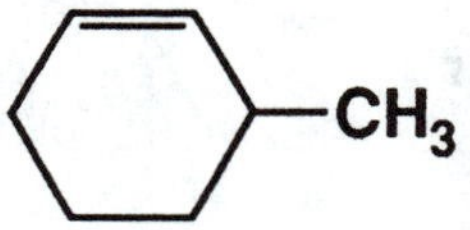

$$Cl \quad \quad CH_3$$
$$C=C$$
$$H \quad \quad H$$

—CH$_3$

(i) 1-Isopropyl-4-methylcyclohexene 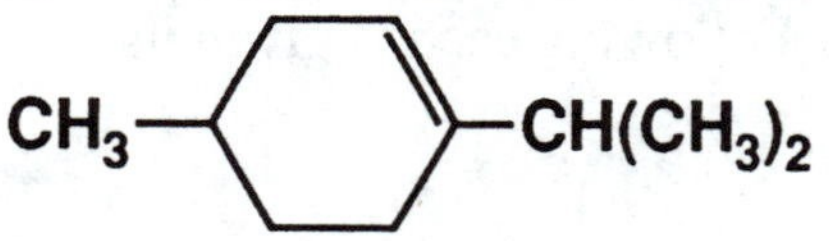          (j) (6E)-2,6-Dimethyl-2,6-octadiene

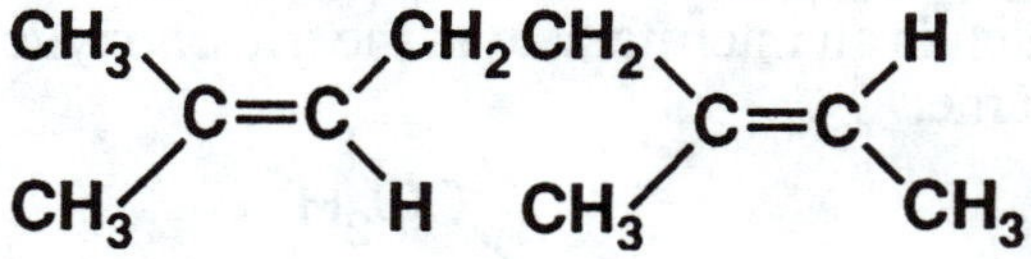

$$CH_3—\text{[ring]}—CH(CH_3)_2$$

$$CH_3 \quad \quad CH_2CH_2 \quad \quad H$$
$$C=C \quad \quad \quad \quad C=C$$
$$CH_3 \quad \quad H \quad \quad CH_3 \quad \quad CH_3$$

(k) Allylcyclopropane 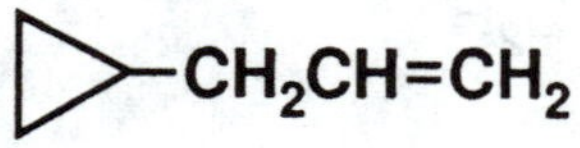                        (l) Vinylcyclopropane

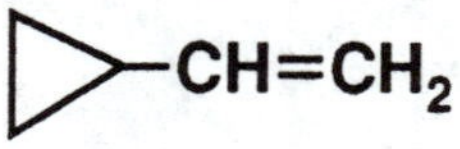

—CH$_2$CH=CH$_2$

—CH=CH$_2$

Problem 3.19  Name these alkenes.

(a)
$$CH_3 \quad \quad CH_2CH(CH_3)_2$$
$$C=C$$
$$CH_3 \quad \quad H$$

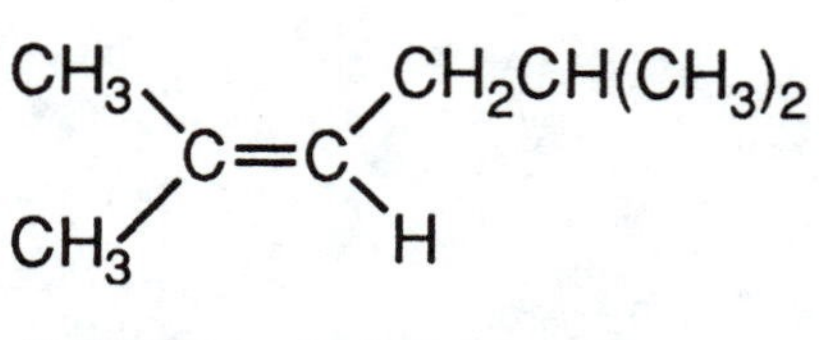

**2,5-Dimethyl-2-hexene**

(b)
$$H \quad \quad H$$
$$C=C$$
$$CH_3CH \quad \quad CH_3$$
$$Cl$$

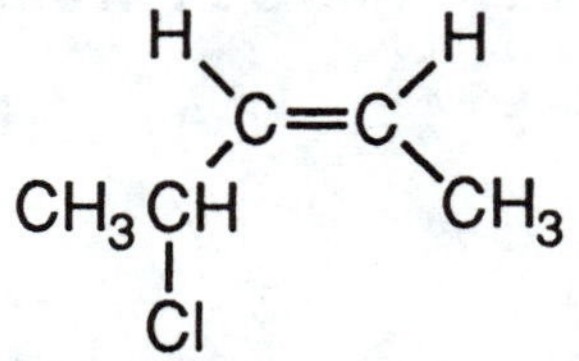

***cis*-4-Chloro-2-pentene**
**(Z)-4-Chloro-2-pentene**

(c)

$$H \diagdown C = C \diagup (CH_2)_4CH_3$$
$$H \diagup \phantom{C=C} \diagdown CH_2CH(CH_3)_2$$

**2-Isobutyl-1-heptene**

(d)

$$Br \diagdown C = C \diagup CH_3$$
$$H_3C \diagup \phantom{C=C} \diagdown Br$$

***trans*-2,3-Dibromo-2-butene**
**(E)-2,3-Dibromo-2-butene**

(e)

$$(CH_3)_2CH \diagdown C = C \diagup CH_3$$
$$H \diagup \phantom{C=C} \diagdown CH_2CH_2CH_3$$

***trans*-2,4-Dimethyl-3-heptene**
**(E)-2,4-Dimethyl-3-heptene**

(f)

$$F \diagdown C = C \diagup F$$
$$F \diagup \phantom{C=C} \diagdown F$$

**Tetrafluoroethene**
**Tetrafluoroethylene**

**Problem 3.20** For each molecule that shows *cis-trans* isomerism, draw the *cis* isomer.

(a)     (b)     (c)     (d)

**Only (b) and (d) show *cis-trans* isomerism. In (a) the two methyl groups are attached to the same carbon atom, and in (c) the two methyl groups are attached to the double bond that must remain *cis* due to strain considerations in a six-member ring.**

**Problem 3.21** Draw structural formulas for all compounds of molecular formula $C_5H_{10}$ that are:
(a) Alkenes that do not show *cis-trans* isomerism.

**Four alkenes of molecular formula $C_5H_{10}$ do not show *cis-trans* isomerism.**

$$CH_2=CHCH_2CH_2CH_3$$

**1-Pentene**

$$\overset{\displaystyle CH_3}{\underset{\displaystyle |}{CH_2=CCH_2CH_3}}$$

**2-Methyl-1-butene**

$$\overset{\displaystyle CH_3}{\underset{\displaystyle |}{CH_2=CHCHCH_3}}$$

**3-Methyl-1-butene**

$$\overset{\displaystyle CH_3}{\underset{\displaystyle |}{CH_3C=CHCH_3}}$$

**2-Methyl-2-butene**

(b) Alkenes that do show *cis-trans* isomerism.

**One alkene of molecular formula $C_5H_{10}$ shows *cis-trans* isomerism.**

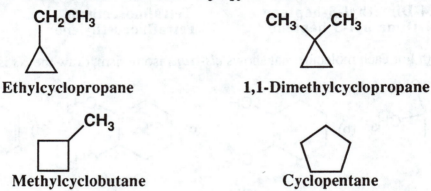

**trans-2-Pentene**                                    **cis-2-Pentene**

(c) Cycloalkanes that do not show *cis-trans* isomerism.

**Four cycloalkanes of molecular formula $C_5H_{10}$ do not show *cis-trans* isomerism.**

**Ethylcyclopropane**                          **1,1-Dimethylcyclopropane**

**Methylcyclobutane**                               **Cyclopentane**

(d) Cycloalkanes that do show *cis-trans* isomerism.

**Only one cycloalkane of molecular formula $C_5H_{10}$ shows *cis-trans* isomerism.**

*cis*-1,2-Dimethyl-                              *trans*-1,2-Dimethyl-
cyclopropane                                       cyclopropane

<u>Problem 3.22</u>  β-Ocimene, a triene found in the fragrance of cotton blossoms and several other essential oils, has the IUPAC name (3Z)-3,7-dimethyl-1,3,6-octatriene.  Draw a structural formula of β-ocimene.

β-Ocimene
(3Z)-3,7-Dimethyl-1,3,6-octatriene

## Terpenes

**Problem 3.23** Show how the carbon skeleton of farnesol can be coiled and cross-linked to give the carbon skeleton of caryophyllene (Figure 3.7).

**The answer to this problem can best be visualized by numbering the atoms along the farnesol chain. Notice how, according to this analysis, the cyclobutane ring of caryophyllene would be formed by cross-linking farnesol from C11 to C1 and from C10 to C2. The structures of caryophyllene and an uncoiled farnesol are drawn above the coiled farnesol structure for comparison.**

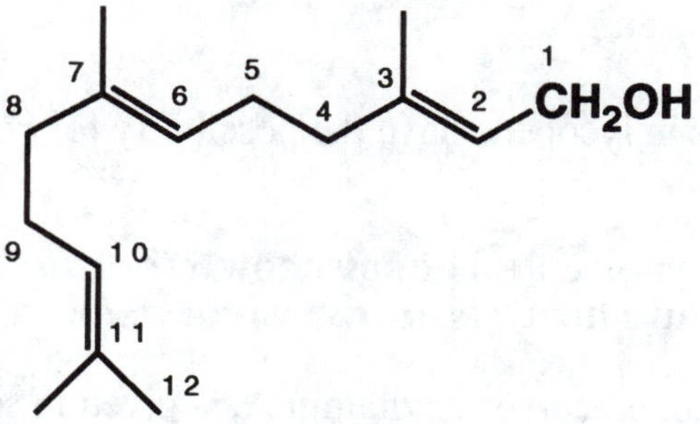

**Uncoiled farnesol** 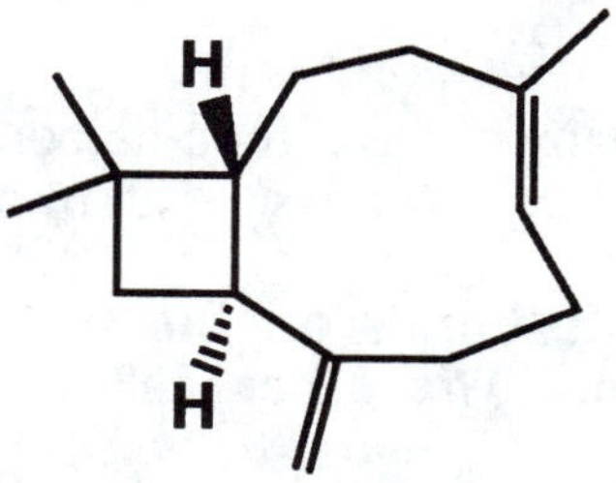                    **Caryophyllene**

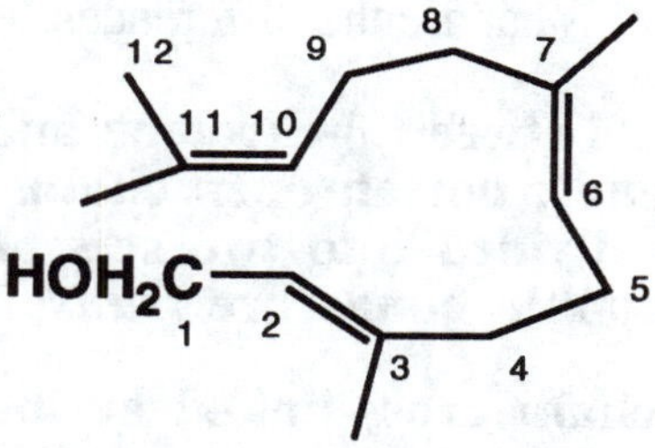

**Coiled farnesol**

**Problem 3.24** Show that the structural formula of Vitamin A (Section 3.2G) can be divided into four isoprene units joined by head-to-tail linkages and cross-linked at one point to form the six-member ring.

isoprene chain cross-linked here

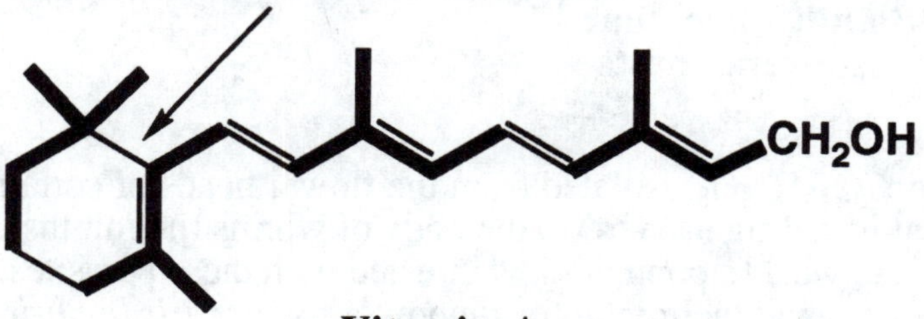

**Vitamin A**

**Problem 3.25** Following is the structural formula of lycopene, a deep red compound that is partially responsible for the red color of ripe fruits, especially tomatoes. Approximately 20 mg of lycopene can be isolated from 1 kg of ripe tomatoes
(a) Show that lycopene is a terpene, that its carbon skeleton can be divided into two sets of four isoprene units with the units in each set joined head-to-tail

**The following structural formula shows the eight isoprene units of lycopene.**

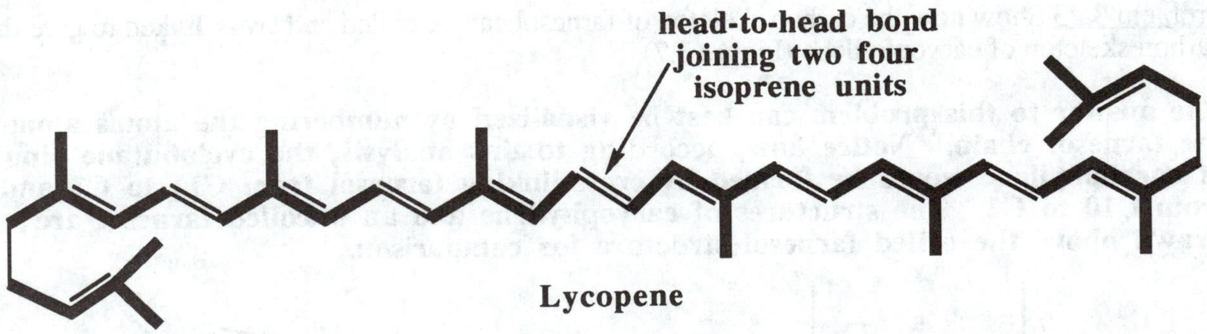

head-to-head bond
joining two four
isoprene units

**Lycopene**

(b) How many of the carbon-carbon double bonds in lycopene have the possibility for *cis-trans* isomerism?  Lycopene is the all *trans* isomer.

**The double bonds on the two ends of the molecule cannot show *cis-trans* isomerism.   The other 11 double bonds can show *cis-trans* isomerism.**

Problem 3.26  The structural formula of β-carotene, precursor to vitamin A, is given in section 16.6A.  As you might suspect, it was first isolated from carrots.  Dilute solutions of β-carotene are yellow, hence its use as a food coloring.  Compare the carbon skeletons of β-carotene and lycopene.  What are the similarities?  What are the differences.

**The main structural difference between β-carotene and lycopene is that β-carotene has six-member rings on the ends, not an open chain.   On the other hand, both β-carotene and lycopene can be divided into two sets of four isoprene units as shown below, and all of the double bonds are *trans* in both molecules.**

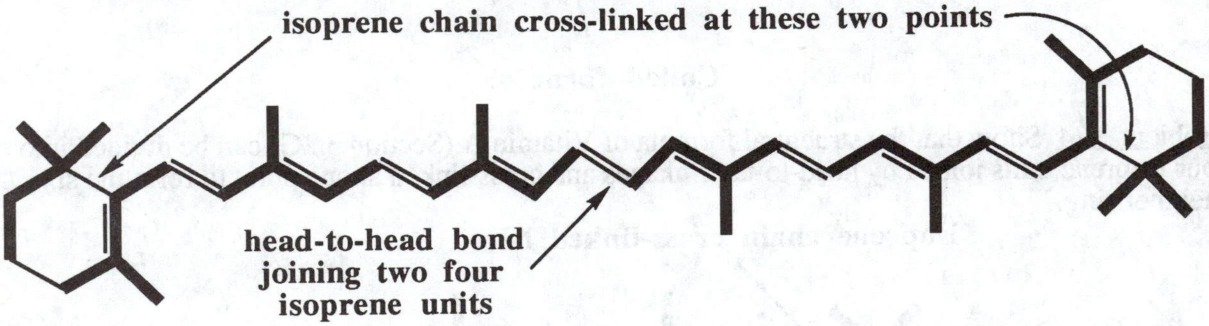

isoprene chain cross-linked at these two points

head-to-head bond
joining two four
isoprene units

Problem 3.27  Santonin, $C_{15}H_{18}O_3$, isolated from the flower heads of certain species of Artemisia, is an anthelmintic, that is, a drug used to rid the body of worms (helminths).  It has been estimated that over one-third of the world's population is infested with these parasites.  Santonin in oral doses of 60 mg is used as an anthelmintic for roundworms (*Ascaris lumbricoides*).

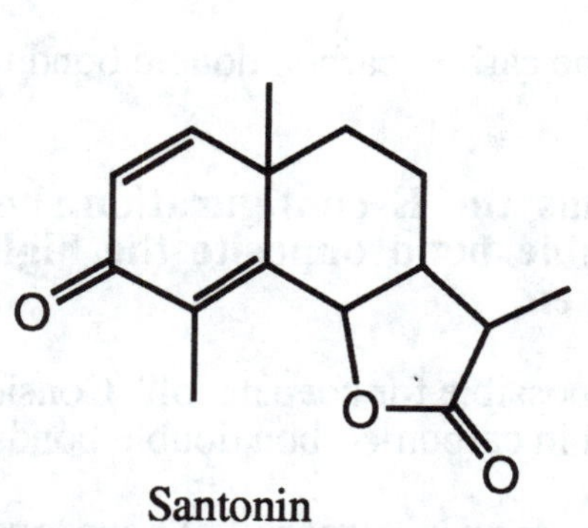

Santonin

Locate the three isoprene units in santonin and show how the carbon skeleton of farnesol might be coiled and then cross-linked to give santonin. Two different coiling patterns of the carbon skeleton of farnesol that could lead to santonin. Try to find them both.

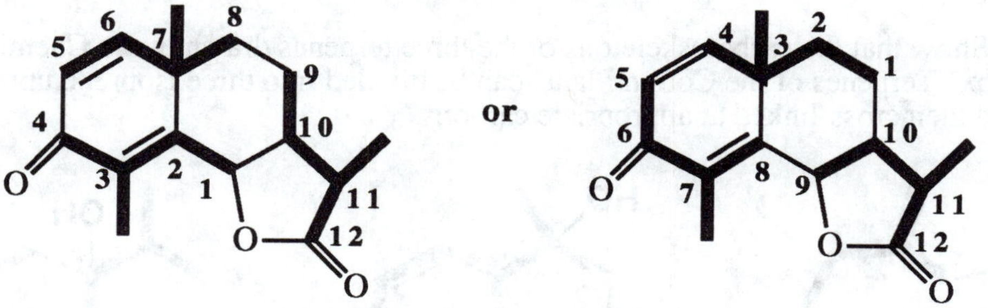

**Problem 3.28** In many parts of South America, extracts of the leaves and twigs of *Montanoa tomentosa* are used as a contraceptive, to stimulate menstruation, to facilitate labor, or to terminate early pregnancy. Phytochemical investigations of this plant have resulted in isolation of a very potent fertility-regulating compound called zoapatanol.

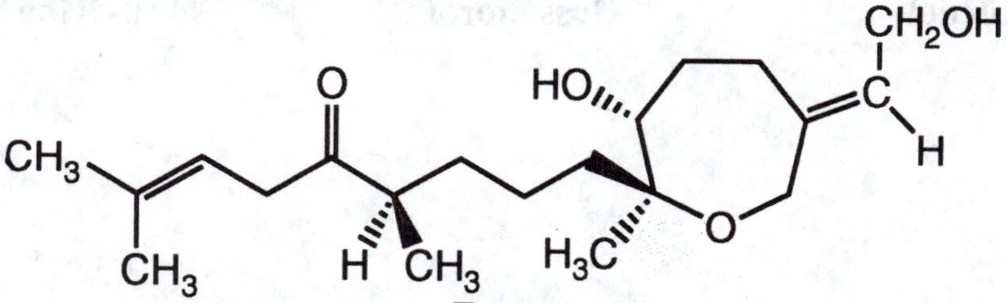

Zoapatanol

(a) Show that the carbon skeleton of zoapatanol can be divided into four isoprene units bonded head-to-tail and then cross-linked in one point along the chain.

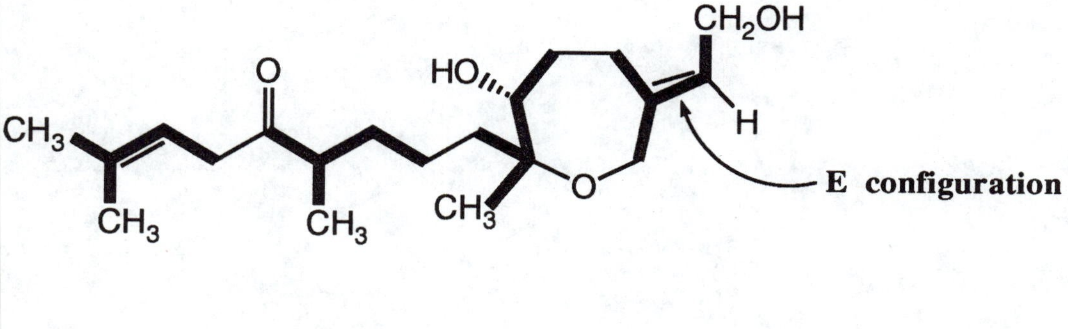

(b) Specify the configuration about the carbon-carbon double bond to the seven-member ring according to the E,Z system.

**The double bond in question has the E configuration, because the hydroxymethyl group is on the side of the double bond opposite the higher priority carbon atom that is linked to the ether oxygen.**

(c) How many *cis-trans* isomers are possible for zoapatanol? Consider the possibilities for *cis-trans* isomerism in both the ring and in carbon-carbon double bonds.

**There are total of 2 x 2 or 4 *cis-trans* isomers. There are two *cis-trans* isomers on the ring, and two *cis-trans* isomers of the double bond attached to the ring. The other double bond has two methyl groups on one carbon atom so it has no *cis-trans* isomers.**

<u>Problem 3.29</u> Show that the carbon skeletons of the three terpenes drawn in the Chemical Connections box "Terpenes of the Cotton Plant" can be divided into three isoprene units bonded head-to-tail and then cross-linked at appropriate carbons.

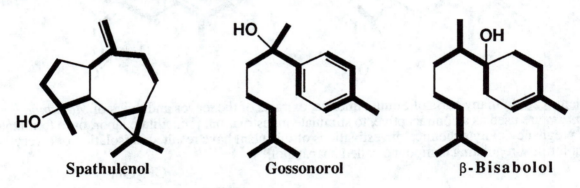

Spathulenol                    Gossonorol                    β-Bisabolol

# CHAPTER 4: REACTIONS OF ALKENES

## SUMMARY OF REACTIONS

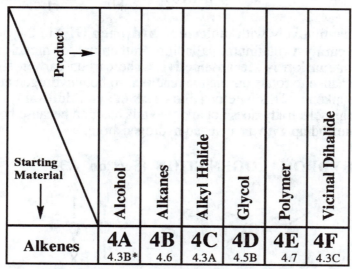

*Section in book that describes reaction.

## REACTION 4A: ACID-CATALYZED HYDRATION (Section 4.3B)

- In the presence of an **acid catalyst** like sulfuric acid, **water adds to alkenes to give alcohols**. ✳
- The reaction mechanism involves formation of a carbocation from protonation of the pi bond, followed by attack of the resulting electrophilic carbocation by water (Section 4.4E) to form a positively-charged **oxonium ion**, that then loses a proton to give the final product. Because a proton is used then produced during the reaction, there is no net change in proton concentration during the reaction. The proton is acting as a catalyst.
- The water can attack from either side of the carbocation, so-acid catalyzed hydration of an alkene is not **stereoselective**. That is, it does not preferentially form **syn** or **anti** products like reactions 4B, 4D and 4F.
- This reaction is **regioselective**, that is one constitutional isomer is produced in preference to other possible constitutional isomers, because **Markovnikov's rule** is followed. Thus, the **hydrogen makes a bond to** the **carbon atom** that **has more hydrogen atoms attached**. This is because the water attacks the more stable carbocation, namely the one that is more highly substituted with alkyl groups. ✳

## REACTION 4B: CATALYTIC HYDROGENATION (Section 4.6)

- **Alkenes** react quantitatively with **molecular hydrogen ($H_2$)** in the presence of a transition **metal catalyst** (platinum, palladium, ruthenium, or nickel) to give **alkanes**. ✳
- **Catalytic hydrogenation is stereoselective**. The carbon-carbon sigma bond usually does not have a chance to rotate during the reaction, so both hydrogen atoms are added to the same face of the alkene. This is referred to as *cis* or **syn addition**.
- The concept of regioselectivity does not apply to this reaction because both carbon atoms of the original alkene end up with an attached hydrogen atom.

## REACTION 4C: HYDROHALOGENATION (Section 4.3A)

- **HF, HCl, HBr, or HI** can add to **alkenes** to give an **alkyl halide**. ✳
- Like reaction 4A, protonation of the pi bond results in formation of a carbocation intermediate, then the halide anion reacts with the carbocation (Section 4.4C).
- The reaction is regioselective in that it **follows Markovnikov's rule**. The **hydrogen makes a bond to** the **carbon atom** that has the **greater number of hydrogens** attached to it.
- Because there is no bridged intermediate, the halide anion can attack from either side of the trigonal planar carbocation. Therefore, the hydrohalogenation of an alkene is **not stereoselective**. That is, it does not preferentially form syn or anti products like reactions 4B, 4D and 4F.

## REACTION 4D: OXIDATION TO VICINAL DIOLS (GLYCOLS) (Section 4.5B)

- **Alkenes** react with **$KMnO_4$ at basic pH (11.8)** to form a **vicinal diol** (glycol). ✳
- In these reactions a cyclic intermediate is formed from the alkene and then hydrolyzed to yield a **vicinal diol** (OH groups on adjacent carbon atoms).
- Note how the cyclic intermediate insures that both oxygen atoms are added to the same face of the alkene (**syn addition**).
- The concept of regioselectivity does not apply to this reaction because both carbon atoms of the original alkene end up with a hydroxyl group attached.

## REACTION 4E: POLYMERS (Section 4.7)

$$CH_2{=}CH_2 \longrightarrow {-}CH_2{-}CH_2{-}CH_2{-}CH_2{-}CH_2{-}CH_2{-}CH_2{-}CH_2{-}$$

Ethylene                                    Polyethylene

- Industrially, an important reaction of alkenes is **polymerization**, the building together of many small molecules into a very large molecular chain. ✶
- Polymerization of ethylene is initiated by trace cations, anions or free radicals. When radicals are used, the ethylene polymerizes by a **radical chain reaction mechanism**. A **radical chain reaction** takes place in the presence of radical initiators such as peroxides.

    In the **initiation step** of a radical chain reaction, an alkoxy radical derived from cleavage of the O-O bond of a peroxide reacts with the alkene to give an alkyl radical.

    In the first **chain propagation step**, the alkyl radical reacts with another alkene to give a new, longer alkyl radical.

    When there is a choice in the reaction to make one of several different radicals, **the more stable radical is formed preferentially. The more stable radical is the one with the unpaired electron on the more highly substituted carbon atom.**

    Tertiary (3°) radicals (with three alkyl groups attached) are more stable than secondary (2°) radicals (with two alkyl groups attached), which are more stable than primary (1°) radicals (with one alkyl group attached), which are more stable than methyl radicals that have no alkyl groups attached. Note that this is the same trend as is observed with relative carbocation stabilities (Section 4.4D).

    In **subsequent propagation steps**, the longer alkyl radical reacts with other alkene molecules to give even longer alkyl radicals. The interesting part is that these even longer alkyl radicals keep reacting with other alkenes and the process is repeated over and over. The number of times a cycle of chain propagation repeats is called **chain length**.

    The chain propagation continues until a **chain termination** step in which two radical species react with one another to make a covalent bond (with no left over unpaired electrons) thereby quenching the radicals.

## REACTION 4F: FORMATION OF VICINAL DIHALIDES (Section 4.3C)

$$\ce{>C=C< ->[X_2] -\underset{X}{\underset{|}{C}}-\overset{X}{\overset{|}{C'}}-}$$

- **Bromination** and **chlorination** involve the addition of $Br_2$ and $Cl_2$ to an alkene, respectively. ✶
- In these reactions, one of the halogen atoms acts as an electrophile and reacts with the pi bond of the alkene. This breaks the halogen-halogen bond, thereby creating a positively-charged intermediate and a halide anion. The positively-charged intermediate has a unique bridged structure and is referred to as a **bridged halonium ion**. The halide anion then completes the reaction by creating a bond to the positively-charged bridged halonium ion from the side of the molecule opposite the halogen bridge.
- The halogen bridge blocks one face of the structure, so the halide anion *must* attack from the side opposite the bridging group. The net result is that the reaction is stereoselective since the two halogens end up on opposite faces of the molecule. This orientation is referred to as **anti addition**.

- The concept of regioselectivity does not apply to this reaction because both carbon atoms of the original alkene end up with an attached halogen atom.

## SUMMARY OF IMPORTANT CONCEPTS

### 4.0 OVERVIEW
• **Reaction mechanisms** describe how chemical bonds are made and broken during the course of a reaction, the order in which the bonds are broken and formed, the rates at which these processes occurs, and the role of solvent or a catalyst if any. Mechanisms provide a theoretical framework within which to organize a great deal of descriptive chemistry.

### 4.1 REACTIONS OF ALKENES
• Alkenes react with a variety of compounds in several characteristic ways. This is in contrast to alkanes, which are relatively unreactive. Two of the most important reactions of alkenes are listed below:
   - **Addition reactions** involve breaking the pi bond of an alkene and replacing it with two sigma bonds.
   - **Polymer addition reactions** involve the formation of polymer chains from monomer alkene molecules.

### 4.2 REACTION MECHANISMS
• **Transition state theory** provides a model for understanding the relationships between reaction rates, molecular structure, and energetics.
   - The **total energy** of any chemical system is always conserved, and is the sum of the **kinetic energy** and **potential energy**. As molecules collide they convert kinetic energy into potential energy in the form of bond vibrational energy.
   - A **reaction coordinate** is a plot of the motion of atoms associated with changing energy as reactants proceed to products during a reaction.
   - The difference in potential energy between reactants and products is called the **heat of reaction**. **Exothermic** reactions have products with lower potential energy than the reactants, and **endothermic** reactions have products with higher potential energy than the reactants.
   - For simple **one step reactions**, reaction occurs if sufficient potential energy becomes concentrated in the proper bonds.
       The **transition state** is the point on the reaction coordinate where the potential energy is a maximum.
       A **transition state** has essentially zero lifetime because it is a maximum on the energy diagram, yet it does have a definite arrangement of atoms and electrons.
       The difference in potential energy between reactants and the activated complex is called the **energy of activation, $E_a$**. A molecule must have more potential energy than the energy of activation to proceed from starting materials to products.
       The greater the energy of activation, the slower the **rate of reaction**, and *vice versa*.
       The rate of reaction also depends on the **collisional frequency between molecules**, the **fraction of collisions with proper orientation** for reaction, and the **fraction of collisions with energy greater than the energy of activation**.
   - In multi-step reactions, each step has its own transition state and energy of activation.
       An **intermediate** is a potential energy minimum between two transition states on a reaction coordinate for a multi-step reaction. Reactive intermediates are never present in appreciable concentrations because the energy of activation for their conversion back to reactants or on to products is so small.
       The slowest step is the one that has the highest energy of activation and is called the **rate-determining step**. *[This is a very important concept in the study of reaction*

*mechanisms. Notice that the overall rate of a multi-step reaction cannot be faster than the rate determining step.]*

## 4.4 ELECTROPHILIC ADDITION TO ALKENES

- The details of **alkene addition reactions** can best be understood by considering the mechanism of the reaction as well as the structure of the alkene.
    - The electrons of the alkene pi bond are located relatively far from the atomic nuclei, so they can react with extremely electron deficient chemical species, referred to as **electrophiles**. *[This is the key idea of the chapter, and addition reactions should be thought of as starting with the alkene pi electrons attacking a highly electrophilic species.]*
    - When the pi electrons react with an electrophile, the pi bond is broken and a new sigma bond is formed with the electrophile. This creates a positively charged intermediate that is itself attacked by an electron rich species (e.g. a chloride or bromide anion or water) to form another new sigma bond, thereby completing the reaction.
        The key to understanding the details of these reactions is to keep track of the electrophile, and the structure of the positively charged intermediate.

    - The bottom line is that the pi bond is replaced by two new sigma bonds, one to an electrophile and one to an electron rich species.
- For addition reactions in which different atoms end up attached to the two carbon atoms of the original alkene (acid-catalyzed hydration, hydrohalogenation), reaction with unsymmetrical alkenes could potentially produce several different products that are referred to as **regioisomers**.
    - For example, during the hydrohalogenation reaction, the halogen atom could potentially end up on either of the two carbon atoms of an alkene. In the case of unsymmetrical alkenes, the halogen atom usually has a preference for only one of the carbon atoms, so one of the regioisomer products predominates.
- In a **regioselective reaction, one regioisomer** is produced or destroyed in **preference to all other possible regioisomers**. For the hydrohalogenation and acid-catalyzed hydration reactions, the regioselectivity of the reaction is predicted using **Markovnikov's rule**.
- **Markovnikov's rule** predicts that for the hydrohalogenation and acid-catalyzed hydration reactions, the hydrogen makes a bond to the carbon atom that has more hydrogen atoms attached.
    - **Markovnikov's rule** can be understood by considering the structure of the **carbocation** intermediate formed during the addition reaction.
        The basic idea is that the more stable carbocation intermediate leads to the predominant (Markovnikov) product. The **more stable carbocation** is the one that has **more alkyl groups attached** to the positively charged carbon atom.
        In other words, a tertiary (3°) carbocation is more stable than a secondary (2°) carbocation, which is more stable than a primary (1°) carbocation, which is more stable than a methyl carbocation.
- In a **stereoselective reaction, one stereoisomer** is formed or destroyed in **preference to all of the other possible stereoisomers**.
    - The bromination and chlorination reactions of alkenes are stereoselective, since only the anti isomer is produced. Similarly, the oxidation of alkenes to form glycols with $KMnO_4$ and the reduction of alkenes with $H_2$ and a transition metal are also stereoselective, giving only syn addition. The acid catalyzed hydration and hydrohalogenation reactions of alkenes are not

stereoselective. Notice how the mechanism of a reaction (the structure of the intermediate, etc.) determines whether a reaction is stereoselective or not.

## 4.5 OXIDATION OF ALKENES

• **Oxidation/reduction reactions** are a very important class of reactions in organic chemistry in which electrons are gained or lost by a reactant during the course of a reaction. Oxidation/reduction reactions can be recognized by writing **balanced half-reactions** that keep track of the organic structures involved.

    - To write a balanced half-reaction, first write a half-reaction showing the organic reactants and products. Complete a material balance using $H^+$ and $H_2O$ for reactions carried out in acid, or $OH^-$ and $H_2O$ for reactions carried out in base. Finally complete the charge balance by adding electrons to one side or the other.

    - An **oxidation** is defined as a reaction in which electrons are lost from a reactant being transformed into products.

    - A **reduction** is defined as a reaction in which electrons are gained by a reactant being transformed into products.

## 4.7 POLYMERIZATION

• Important polymers can be made from a variety of small alkenes including vinyl chloride (makes PVC construction tubing), $CF_2=CF_2$ (makes Teflon), $CH_2=CCl_2$ (makes Saran wrap) and $CH_2=CHC_6H_5$ (makes Styrofoam).

## 4.8 POLYETHYLENE

• Because of the importance of polymers such as polyethylene, vast amounts of ethylene are produced every year. Ethylene is produced by the thermal cracking of hydrocarbons that come from natural gas and petroleum. Ethylene is also converted into other useful compounds such as vinyl chloride and ethylene oxide that are themselves used for a wide variety of industrial applications including polymers.

• Originally, ethylene was polymerized using peroxides and heat in a radical chain process. This produced a soft, tough polymer with highly branched chains called low-density polyethylene (**LDPE**). LDPE can be blow-molded into films for things such as bags or plastic wraps around products.

• Ethylene can also be polymerized by using a catalyst such as $TiCl_3$ combined with $Al(CH_2CH_3)_3$ to produce a higher density polymer called high density polyethylene (HDPE). HPDE has a higher melting point than LDPE, so HDPE can be blow-molded into useful items such as mixing bowls or freezer containers.

# CHAPTER 4
## *Solutions to the Problems*

**Problem 4.1** Name and draw structural formulas for the two possible products of each alkene addition reaction. Use Markovnikov's rule to predict which is the major product.

(a) $CH_3-CH=CH_2$ + HI $\longrightarrow$ $CH_3CHCH_3$ + $CH_3CH_2CH_2I$

                                                     I

                                  **2-Iodopropane**　　**1-Iodopropane**
                                  **Major product**

(b)

**1-Iodo-1-methyl-**
**cyclohexane**
**Major product**

**Iodomethylcyclo-**
**hexane**

**Problem 4.2** Draw structural formulas for the products of these alkene hydration reactions. Predict which is the major product.

(a)
$$CH_3-\underset{\underset{CH_3}{|}}{C}=CH-CH_3 + H_2O \xrightarrow{H_2SO_4} CH_3\underset{\underset{OH}{|}}{\overset{\overset{CH_3}{|}}{C}}CH_2CH_3 + CH_3\underset{\underset{OH}{|}}{\overset{\overset{CH_3}{|}}{C}}HCHCH_3$$

                                                    **major product**

(b)
$$CH_2=\underset{\underset{CH_3}{|}}{C}-CH_2-CH_3 + H_2O \xrightarrow{H_2SO_4} CH_3\underset{\underset{OH}{|}}{\overset{\overset{CH_3}{|}}{C}}CH_2CH_3 + CH_2\underset{\underset{OH}{|}}{\overset{\overset{CH_3}{|}}{C}}HCH_2CH_3$$

                                                    **major product**

**Problem 4.3** Arrange these carbocations in order of increasing stability.

(a)　　(b)　　(c)　

**The order of increasing stability of carbocations is methyl < primary < secondary < tertiary.**

( c )　　( b )　　( a )　

  **primary carbocation**　　**secondary carbocation**　　**tertiary carbocation**

<u>Problem 4.4</u>  Use a balanced half-reaction to show that each transformation involves a reduction.

(a)

**Two hydrogens are required to produce the product alcohol from the ketone.
Therefore, the balanced half-reaction needs two protons and two electrons (for
charge balance) on the left-hand side.  Since the electrons are on the left-hand
side of the equation, the reaction is a two-electron reduction.**

(b)  $CH_3\text{-}CH_2\text{-}\overset{\overset{O}{\|}}{C}OH \longrightarrow CH_3CH_2CH_2OH$

**Two hydrogens are required to produce the product alcohol from the carboxylic
acid.  Therefore, the balanced half-reaction needs two protons and two electrons
(for charge balance) on the left-hand side.  Additionally, the product alcohol has
one less oxygen atom than the carboxylic acid starting material, so there must be
an $H_2O$ molecule added to the right side of the equation to balance the oxygen
atoms.  This $H_2O$ molecule has two more hydrogens that must be balanced by
adding two more protons and electrons to the left-hand side of the equation,
giving a total of four protons and four electrons on the left-hand side.  Since the
electrons are on the left-hand side of the equation, the reaction is a four-electron
reduction.**

$$CH_3\text{-}CH_2\text{-}\overset{\overset{O}{\|}}{C}OH \;+\; 4H^+ \;+\; 4e^- \longrightarrow CH_3CH_2CH_2OH \;+\; H_2O$$

<u>Problem 4.5</u>  Describe the differences between a transition state and a reaction intermediate.

**A transition state is the point on a reaction coordinate in which the potential
energy is a maximum, while an intermediate is a potential energy minimum
between two transition states.  The potential energy of an intermediate is higher
than either reactants or products, so intermediates are highly reactive and can
only rarely be observed directly.**

**Problem 4.6**  Sketch a potential energy diagram for a one-step reaction that is very slow and only slightly exothermic (the energy of the products is less than that of the reactants).  How many transition states are present in this reaction.

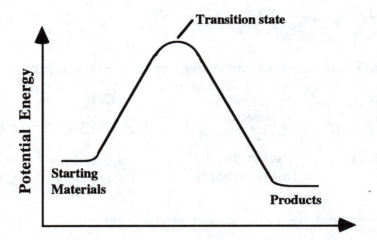

**The reaction described by the above potential energy diagram has one transition state and no intermediates.**

**Problem 4.7**  Sketch a potential energy diagram for a two-step reaction that is endothermic in the first step, exothermic in the second step, and exothermic overall.  How many transition states are present in this two-step reaction?  How many intermediates?

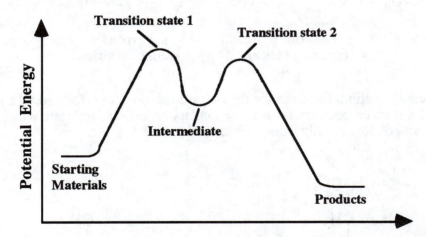

**The reaction described by the above potential energy diagram has two transition states and one intermediate.**

## Electrophilic Additions

Problem 4.8  From each pair, select the more stable carbocation.

(a)   $CH_3CH_2\overset{+}{C}H_2$   or   $CH_3\overset{+}{C}HCH_3$       (b)   $CH_3\overset{CH_3}{\underset{+}{C}H}CHCH_3$ or $CH_3\overset{CH_3}{\underset{+}{C}}CH_2CH_3$

**Cation stability decreases in the order tertiary > secondary > primary > methyl.**

$CH_3CH_2\overset{+}{C}H_2$   or   $CH_3\overset{+}{C}HCH_3$        $CH_3\overset{CH_3}{\underset{+}{C}H}CHCH_3$ o r $CH_3\overset{CH_3}{\underset{+}{C}}CH_2CH_3$

   **primary**            **secondary**                    **secondary**            **tertiary**
                      **(more stable)**                                        **(more stable)**

Problem 4.9  From each pair, select the more stable carbocation.

(a) [ring]$-CH_3$ or [ring]$\overset{+}{-}CH_3$   (b) [ring]$-CH_3$ or [ring]$-CH_2+$

[ring]$-CH_3$ o r [ring]$\overset{+}{-}CH_3$      [ring]$-CH_3$ o r [ring]$-CH_2+$

   **secondary**          **tertiary**                 **secondary**          **primary**
                      **(more stable)**              **(more stable)**

Problem 4.10  Draw structural formulas for the isomeric carbocations formed by addition of H⁺ to each alkene.  Label each carbocation primary, secondary or tertiary, and state which of the isomeric carbocations is formed more readily from each alkene.

(a)   $CH_3-CH_2-\overset{CH_3}{\underset{|}{C}}=CH-CH_3$

$CH_3-CH_2-\overset{CH_3}{\underset{+}{C}}-CH_2-CH_3$   +   $CH_3-CH_2-\overset{CH_3}{\underset{+}{C}}H-CH-CH_3$

   **tertiary**                          **secondary**
**(more stable)**                     **(less stable)**

(b)   $CH_3-CH_2-CH=CH-CH_3$

$CH_3-CH_2-\overset{+}{C}H-CH_2-CH_3$   +   $CH_3-CH_2-CH_2-\overset{+}{C}H-CH_3$

**both secondary carbocations**
**(of equal stability)**

(c)

tertiary
(more stable)

secondary
(less stable)

(d)

primary
(less stable)

tertiary
(more stable)

<u>Problem 4.11</u>  From each pair of alkenes, select the one that reacts more rapidly with HI.

**Reaction rates in these electrophilic additions are a function of the stability of the carbocation intermediate formed in the rate-limiting step.  The more stable the carbocation intermediate (tertiary > secondary > primary) the greater the rate of the reaction.**

(a)   $CH_3-CH=CH-CH_3$  or  $CH_3-\overset{\underset{\displaystyle CH_3}{|}}{C}=CH-CH_3$

$CH_3-CH=CH-CH_3 \longrightarrow CH_3-CH_2-\overset{+}{C}H-CH_3 \longrightarrow CH_3-CH_2-\overset{\underset{\displaystyle I}{|}}{C}H-CH_3$

2-Butene

a secondary
carbocation

2-Iodobutane
(sec-Butyl  iodide)

$CH_3-\overset{\underset{\displaystyle +}{\overset{\displaystyle CH_3}{|}}}{C}-CH_2-CH_3 \longrightarrow CH_3-\overset{\underset{\displaystyle I}{\overset{\displaystyle CH_3}{|}}}{C}-CH_2-CH_3$

a tertiary
carbocation

2-Iodo-2-methylbutane
(major  product)

$CH_3-\overset{\underset{\displaystyle CH_3}{|}}{C}=CH-CH_3$

2-Methyl-2-butene

$CH_3-\overset{\underset{\displaystyle CH_3}{|}}{C}H-\overset{+}{C}H-CH_3 \longrightarrow CH_3-\overset{\underset{\displaystyle CH_3}{|}}{C}H-\overset{\underset{\displaystyle I}{|}}{C}H-CH_3$

a secondary
carbocation

2-Iodo-3-methylbutane
(minor  product)

**The reaction of 2-methyl-2-butene is the only one that can form a tertiary carbocation (upper pathway), so 2-methyl-2-butene is the compound that reacts faster with HI.**

(b) $CH_3-CH=CH_2$ or $CH_2=CH_2$

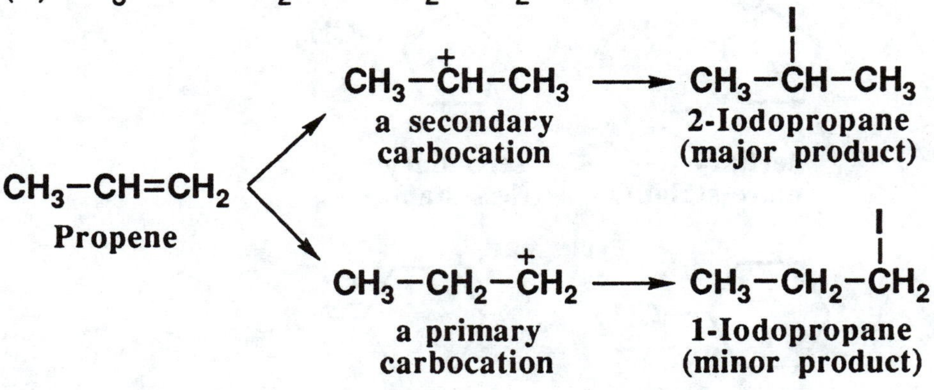

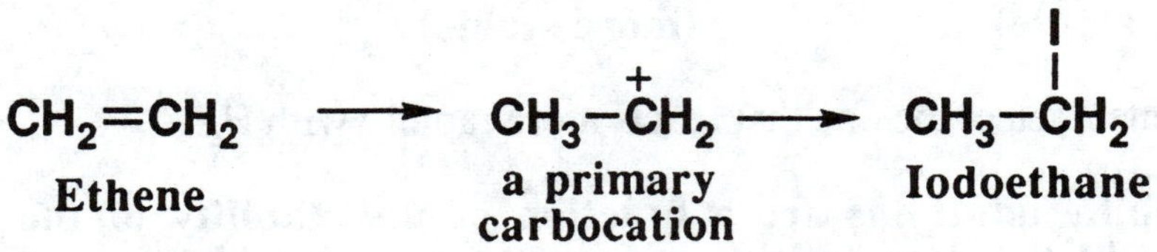

$CH_3-\overset{+}{C}H-CH_3 \longrightarrow CH_3-\overset{I}{\underset{|}{C}H}-CH_3$
a secondary                2-Iodopropane
carbocation                (major product)

$CH_3-CH=CH_2$
Propene

$CH_3-CH_2-\overset{+}{C}H_2 \longrightarrow CH_3-CH_2-\overset{I}{\underset{|}{C}H_2}$
a primary                  1-Iodopropane
carbocation                (minor product)

$CH_2=CH_2 \longrightarrow CH_3-\overset{+}{C}H_2 \longrightarrow CH_3-\overset{I}{\underset{|}{C}H_2}$
Ethene            a primary              Iodoethane
                  carbocation

**The reaction of propene is the only one that can form a secondary carbocation (upper pathway), so propene is the compound that reacts faster with HI.**

<u>Problem 4.12</u> From each pair of cycloalkenes, select the one that reacts more rapidly with HI.

(a) 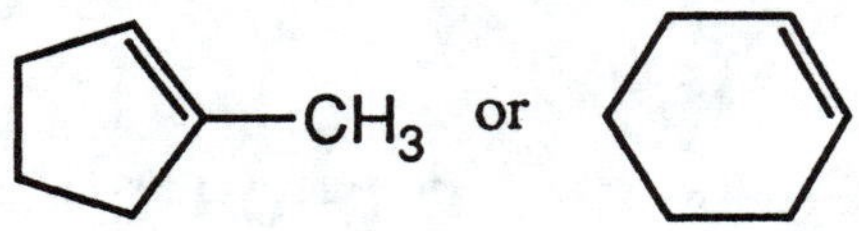 —$CH_3$ or

a tertiary            1-Iodo-1-methylcyclopentane
carbocation           (major product)

1-Methylcyclopentene

a secondary           1-Iodo-2-methylcyclopentane
carbocation           (minor product)

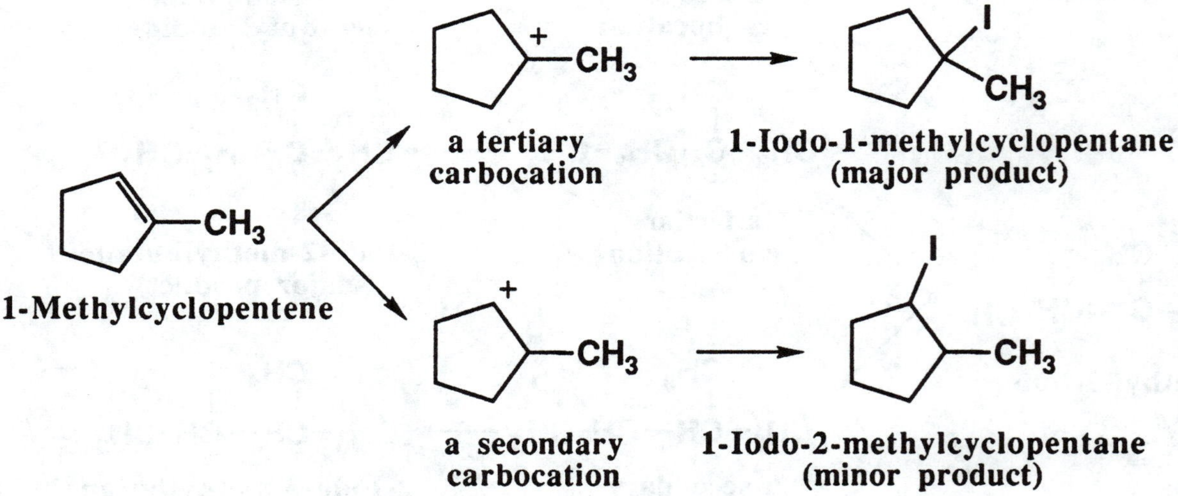

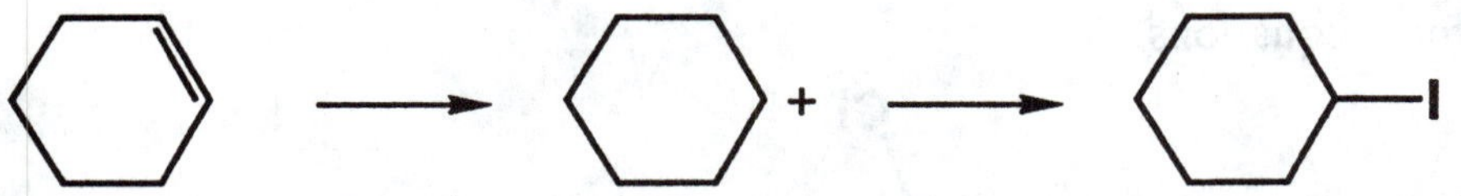

Cyclohexene          a secondary          Iodocyclohexane
                     carbocation

The reaction of 1-methylcyclopentene is the only one that can form a tertiary carbocation (upper pathway), so 1-methylcyclopentene is the compound that reacts faster with HI.

(b) 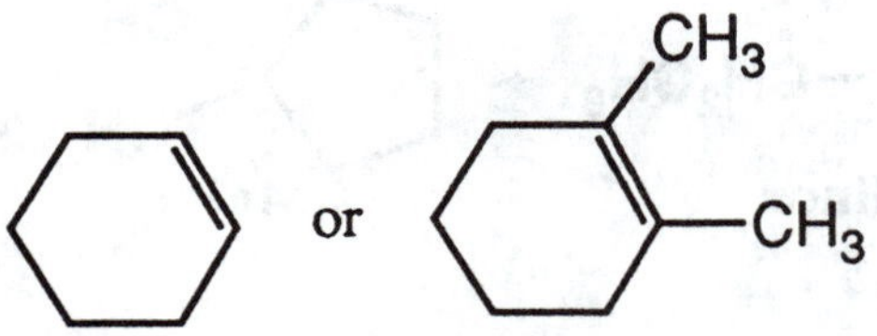 or

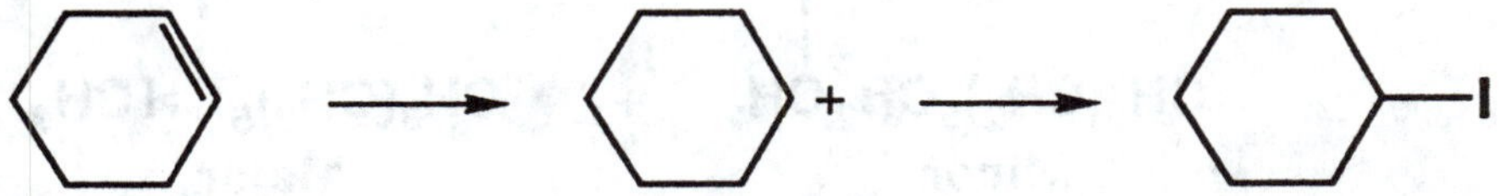

Cyclohexene          a secondary          Iodocyclohexane
                     carbocation

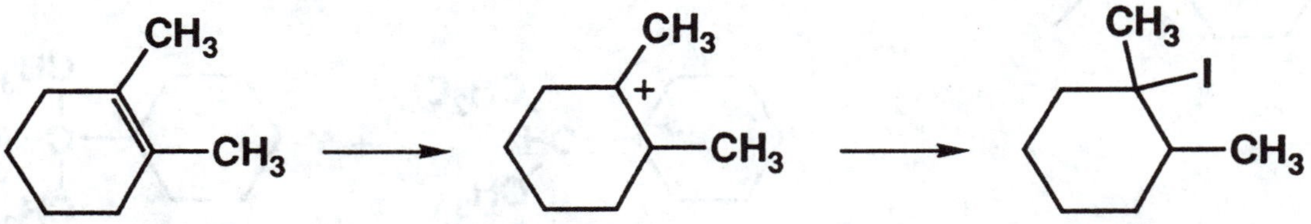

1,2-Dimethylcyclohexene          a tertiary          1-Iodo-1,2-dimethyl-
                                 carbocation          cyclohexane

The reaction of 1,2-dimethylcyclohexene is the only one that can form a tertiary carbocation, so 1,2-dimethylcyclohexene is the compound that reacts faster with HI.

Problem 4.13  Predict the organic product(s) of the reaction of 2-butene with the following reagent(s).

(a) $H_2O$ ($H_2SO_4$)

$CH_3-CH-CH_2-CH_3$
     |
    OH

(b) $Br_2$

$CH_3-CH-CH-CH_3$
     |   |
    Br  Br

(c) $Cl_2$

$CH_3-CH-CH-CH_3$
     |   |
    Cl  Cl

(d) HI

$CH_3-CH-CH_2-CH_3$
     |
     I

Problem 4.14  Complete these equations.

(a) [cyclopentene]—CH$_2$CH$_3$  +  HCl  ⟶  [cyclopentane with Cl and CH$_2$CH$_3$]
**Minor**  +  [cyclopentane with Cl, CH$_2$CH$_3$ on same carbon]
**Major**

(b) [cyclopentene]—CH$_2$CH$_3$  +  H$_2$O  $\xrightarrow{\text{H}_2\text{SO}_4}$  [cyclopentane with OH and CH$_2$CH$_3$]
**Minor**  +  [cyclopentane with OH, CH$_2$CH$_3$ on same carbon]
**Major**

(c) CH$_3$(CH$_2$)$_5$CH=CH$_2$  +  HI  ⟶

CH$_3$(CH$_2$)$_5$CH$_2$CH$_2$ (with I)    +    CH$_3$(CH$_2$)$_5$CHCH$_3$ (with I)
**Minor**                                      **Major**

(d) [cyclohexane with isopropenyl group]  +  HCl  ⟶

[cyclohexane–CH(CH$_3$)CH$_2$Cl]    +    [cyclohexane–C(CH$_3$)$_2$Cl]
**Minor**                                  **Major**

(e)  CH$_3$CH=CHCH$_2$CH$_3$  +  H$_2$O  $\xrightarrow{\text{H}_2\text{SO}_4}$

CH$_3$CHCH$_2$CH$_2$CH$_3$ (with OH)    +    CH$_3$CH$_2$·CHCH$_2$CH$_3$ (with OH)

**Similar amounts of both products will be formed since
they are both derived from a secondary carbocation**

(f)  CH$_2$=CHCH$_2$CH$_2$CH$_3$  +  H$_2$O  $\xrightarrow{\text{H}_2\text{SO}_4}$

CH$_2$CH$_2$CH$_2$CH$_2$CH$_3$ (with OH)    +    CH$_3$CHCH$_2$CH$_2$CH$_3$ (with OH)
**Minor**                                        **Major**

<u>Problem 4.15</u> Reaction of 2-methyl-2-pentene with each reagent is regioselective. Draw a structural formula for the major product of each reaction, and account for your predicted regioselectivity.

(a) HI

(b) $H_2O$ in the presence of $H_2SO_4$

$$CH_3-\underset{\underset{|}{CH_3}}{\overset{\overset{CH_3}{|}}{C}}-CH_2-CH_2-CH_3$$

$$CH_3-\underset{\underset{|}{OH}}{\overset{\overset{CH_3}{|}}{C}}-CH_2-CH_2-CH_3$$

**Both of the above products are explained by Markovnikov's rule.**

<u>Problem 4.16</u> Addition of bromine and chlorine to cycloalkenes is stereoselective. Predict the stereochemistry of the product formed in each reaction and account for your predicted stereoselectivity.

(a) 1-Methylcyclohexene + $Br_2$

**The product has the two bromine atoms _trans_ to each other due to the anti addition geometry observed with these reactions.**

(b) 1,2-Dimethylcyclopentene + $Cl_2$

**The product has the two chlorine atoms _trans_ to each other due to the anti addition geometry observed with these reactions.**

<u>Problem 4.17</u> Draw the structural formula for an alkene of molecular formula $C_5H_{10}$ that reacts with $Br_2$ to give each product.

(a) $CH_3\underset{\underset{Br}{|}}{\overset{\overset{CH_3}{|}}{C}}-\underset{\underset{Br}{|}}{C}HCH_3$

(b) $CH_3\underset{\underset{Br}{|}}{C}H\underset{\underset{Br}{|}}{C}H-CH_2$
    $CH_3$

(c) $\underset{\underset{Br}{|}}{C}H_2\underset{\underset{Br}{|}}{C}CH_2CH_3$
    $CH_3$

$CH_3\overset{\overset{CH_3}{|}}{C}=CHCH_3$

$CH_3\overset{\overset{CH_3}{|}}{C}HCH=CH_2$

$CH_2=\overset{\overset{CH_3}{|}}{C}CH_2CH_3$

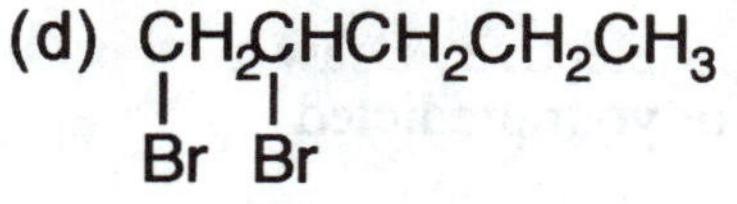

(d)  CH₂CHCH₂CH₂CH₃
      |    |
      Br  Br

**CH₂=CHCH₂CH₂CH₃**

Problem 4.18 Draw the structural formula for a cycloalkene of molecular formula $C_6H_{10}$ that reacts with $Cl_2$ to give each product.

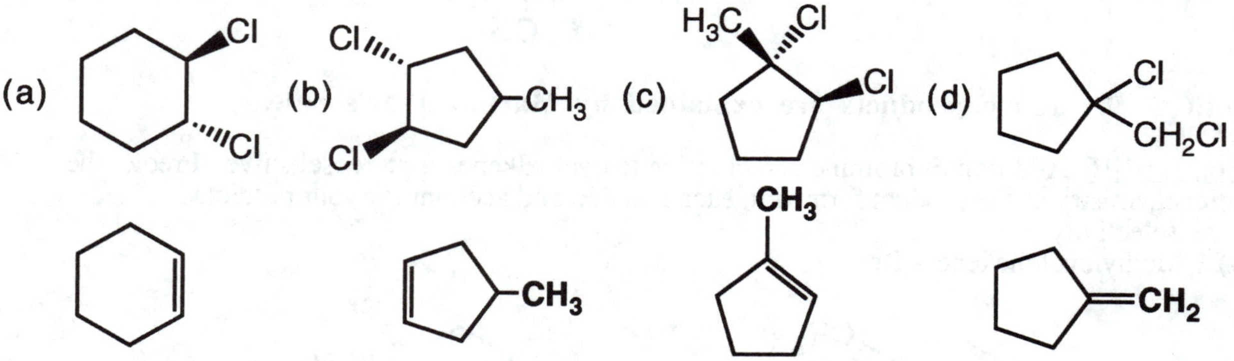

(a)   (b)   (c)   (d)

Problem 4.19 Draw the structural formula for an alkene of molecular formula $C_5H_{10}$ that reacts with HCl to give the indicated chloroalkane as the major product.

(a)  CH₃CCH₂CH₃      (b) CH₃CHCHCH₃      (c) CH₃CHCH₂CH₂CH₃
         |    CH₃              |   CH₃                |
         Cl                    Cl                     Cl

or

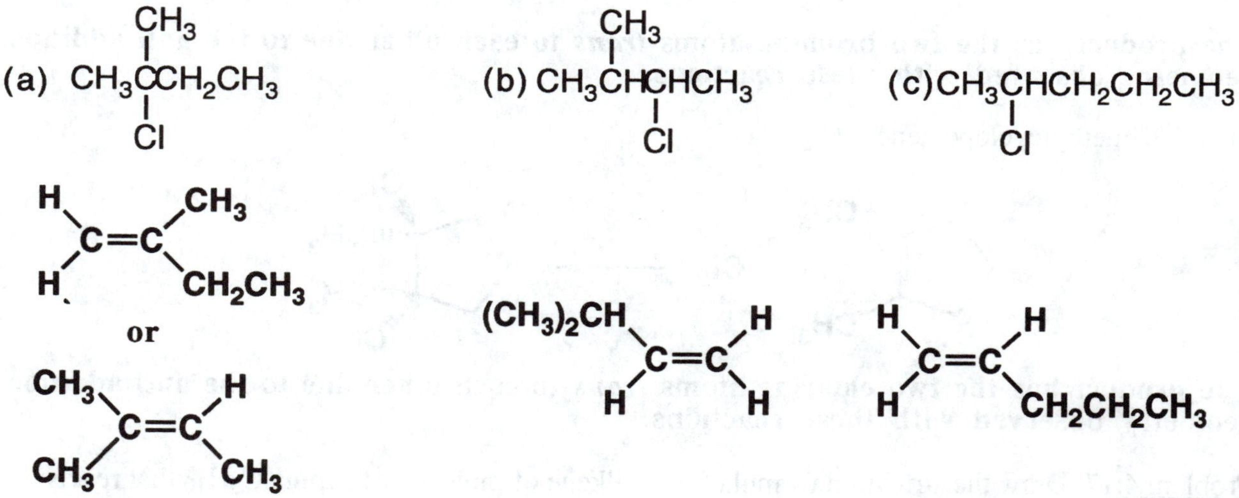

**For part (a) either alkene shown would give 2-chloro-2-methylbutane as the major product as predicted by Markovnikov's rule. For part (b), 2-methyl-2-butene would not work because the predominant product predicted by Markovnikov's rule would be 2-chloro-2-methylbutane (part (a)), not the desired 2-chloro-3-methylbutane. For part (c), *cis* or *trans* 2-butene would not be good choices because 3-chloropentane would be formed in essentially the same amount as the desired product 2-chloropentane.**

**Problem 4.20** Draw the structural formula of an alkene that undergoes acid-catalyzed hydration to give these alcohols as the major product. More than one alkene may give each compound as the major product.

(a) 3-Hexanol

$$CH_3CH_2 \diagdown \underset{C=C}{\overset{H \quad\quad H}{\phantom{C=C}}} \diagup CH_2CH_3$$

or

$$CH_3CH_2 \diagdown \underset{C=C}{\overset{H \quad\quad CH_2CH_3}{\phantom{C=C}}} \diagup H$$

(b) 2-Methyl-2-butanol

$$H \diagdown \underset{C=C}{\overset{\phantom{C}}{\phantom{C=C}}} \underset{CH_2CH_3}{\overset{CH_3}{}}$$
with H below left

or

$$CH_3 \diagdown \underset{C=C}{\overset{}{}} \diagup H$$
$$CH_3 \diagup \phantom{C=C} \diagdown CH_3$$

(c) 1-Methylcyclopentanol

cyclohexene with CH₃

or

cyclohexane with =CH₂

**Note that in part (a) using 2-hexene would give appreciable amounts of 2-hexanol along with the desired 3-hexanol.**

(d) 2-Propanol

$$CH_3 \diagdown \underset{C=C}{\overset{H \quad\quad H}{\phantom{C=C}}} \diagup H$$
with H above left

**Problem 4.21** Draw the structural formula of an alkene that undergoes acid-catalyzed hydration to give these alcohols as the major product. More than one alkene may give each compound as the major product.

(a) Cyclohexanol

cyclohexene

(b) 1,2-Dimethylcyclopentanol

cyclopentene with two CH₃    or    cyclopentane with CH₃ and =CH₂

(c) 1-Methylcyclohexanol

cyclohexene with CH₃    or    cyclohexane with =CH₂

(d) 1-Isopropyl-4-methylcyclohexanol

structure with CH₃ and CH₃ on C, ring with CH₃    or    structure with CH₃ and CH₃ on CH, ring with CH₃

**Problem 4.22** Terpin hydrate is prepared commercially by addition of 2 mol of water to limonene (Figure 3.6) in the presence of sulfuric acid. Terpin hydrate is used medicinally as an expectorant for coughs. It may be given as a mixture of terpin hydrate and codeine.

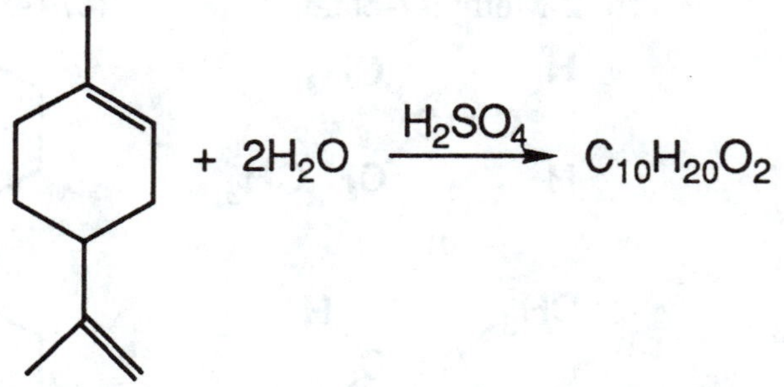

Limonene                                              Terpin hydrate

(a) Propose a structural formula for terpin hydrate and a mechanism for its formation.

**Add water to each double bond by protonation of each double bond to give a 3⁰ carbocation, reaction of each carbocation with water, and loss of the protons to give terpin hydrate.**

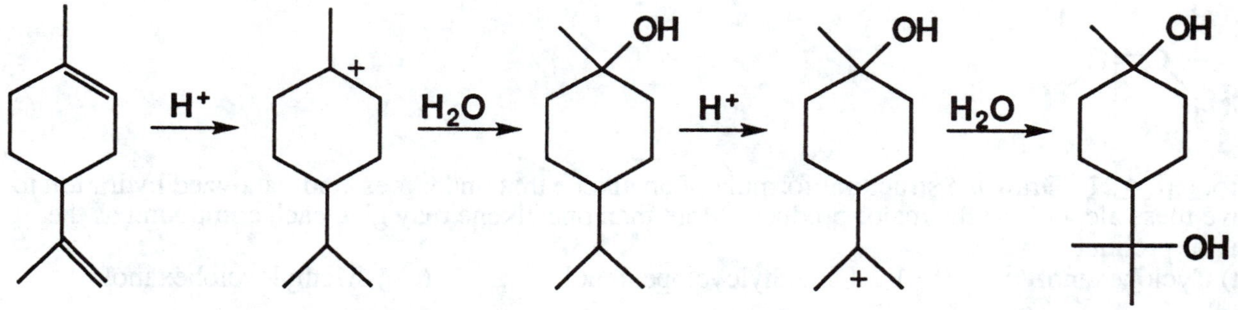

**Limonene**                                                        **Terpin hydrate**

(b) How many *cis-trans* isomers are possible for the structural formula you have proposed?

**There are two *cis-trans* isomers, shown here as chair conformations with the $(CH_3)_2COH-$ side chain equatorial.**

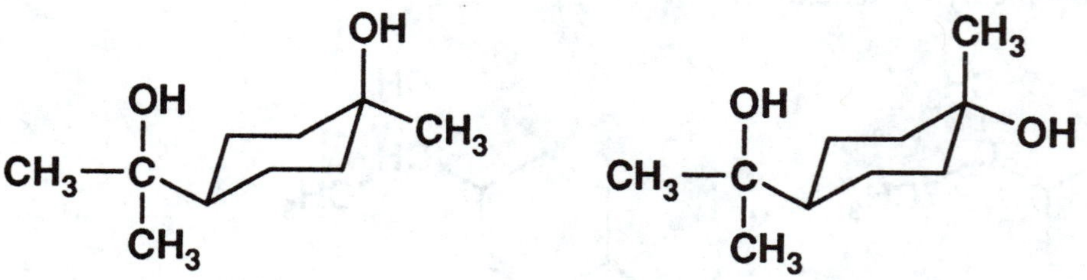

**Problem 4.23** Treatment of 2-methylpropene with methanol in the presence of a sulfuric acid catalyst gives *tert*-butyl methyl ether. Propose a mechanism for formation of this ether.

$$CH_3-\underset{\underset{CH_3}{|}}{C}=CH_2 + CH_3OH \xrightarrow{H_2SO_4} CH_3\underset{\underset{OCH_3}{|}}{\overset{\overset{CH_3}{|}}{C}}CH_3$$

**Reaction of the alkene with a proton gives a tertiary carbocation intermediate. Reaction of this intermediate with the oxygen atom of methanol followed by loss of a proton gives *tert*-butyl methyl ether.**

**Problem 4.24** Treatment of 1-methylcyclohexene with methanol in the presence of a sulfuric acid catalyst gives a compound of molecular formula $C_8H_{16}O$. Propose a structural formula for this compound and a mechanism for its formation.

**Reaction of the alkene with a proton gives a tertiary carbocation intermediate. Reaction of this intermediate with the oxygen atom of methanol followed by loss of a proton gives the product ether.**

## Oxidation-Reduction

<u>Problem 4.25</u>  Use balanced half-reactions to show which transformations involve oxidation, which involve reduction, and which involve neither oxidation nor reduction.

(a) $CH_3\overset{\overset{\displaystyle OH}{|}}{C}HCH_3 \longrightarrow CH_3\overset{\overset{\displaystyle O}{||}}{C}CH_3$

$CH_3\overset{\overset{\displaystyle OH}{|}}{C}HCH_3 \longrightarrow CH_3\overset{\overset{\displaystyle O}{||}}{C}CH_3 + 2H^+ + 2e^-$

(b) $CH_3\overset{\overset{\displaystyle OH}{|}}{C}HCH_3 \longrightarrow CH_3CH=CH_2$

$CH_3\overset{\overset{\displaystyle OH}{|}}{C}HCH_3 \longrightarrow CH_3CH=CH_2 + H_2O$

(c) $CH_3CH=CH_2 \longrightarrow CH_3CH_2CH_3$

$CH_3CH=CH_2 + 2H^+ + 2e^- \longrightarrow CH_3CH_2CH_3$

**Part (a) is an oxidation because the electrons appear on the right-hand side, part (c) is a reduction because the electrons appear on the left-hand side, and part (b) is neither a reduction or an oxidation because no electrons are involved on either side, only $H_2O$.**

<u>Problem 4.26</u>  Write a balanced equation for the combustion of 2-propanol (isopropyl alcohol) in air to give carbon dioxide and water.  The oxidizing agent is $O_2$, which makes up approximately 20% of air.

$2 \; CH_3\overset{\overset{\displaystyle OH}{|}}{C}HCH_3 + 9 \; O_2 \longrightarrow 6 \; CO_2 + 8 \; H_2O$

<u>Problem 4.27</u>  Draw the product formed by treatment of each alkene with aqueous $KMnO_4$ (cold, pH 11.8).
(a) 1-Methylcyclopentene

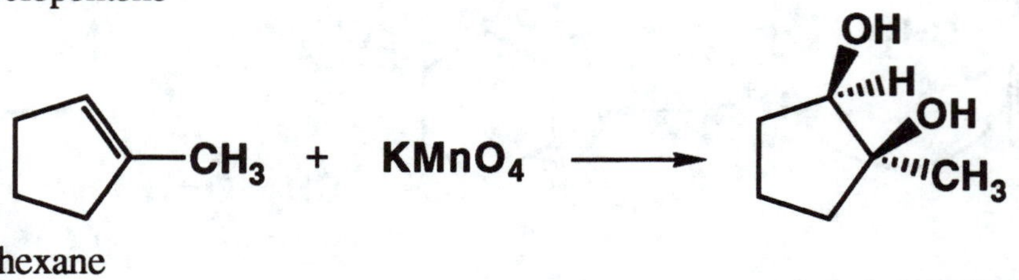

(b) Vinylcyclohexane

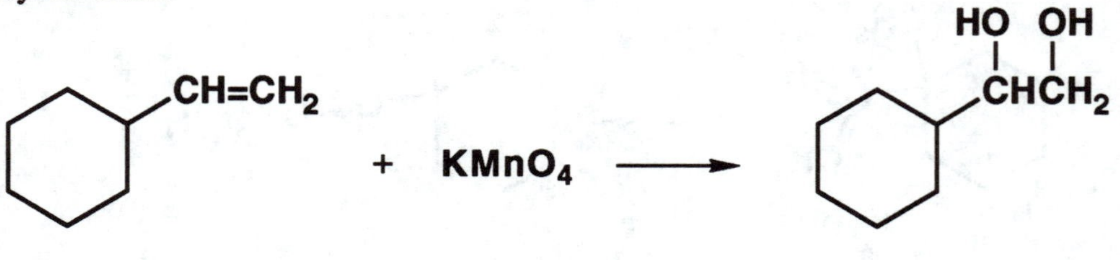

(c) *cis*-2-Pentene

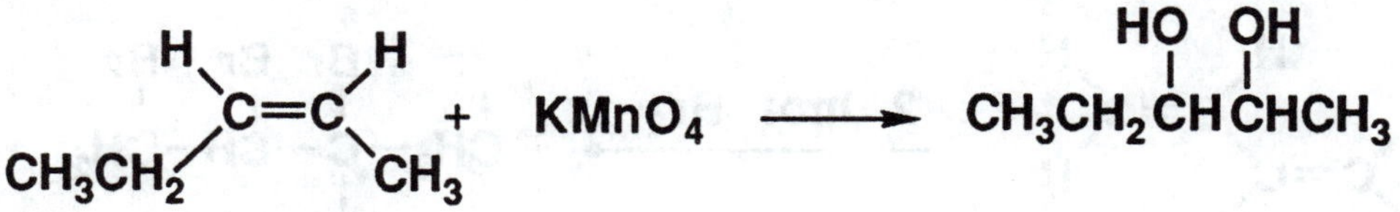

$$\underset{\substack{CH_3CH_2}}{\overset{H}{\diagdown}}C=C\underset{\substack{CH_3}}{\overset{H}{\diagup}} \quad + \quad KMnO_4 \quad \longrightarrow \quad \underset{\substack{\qquad |\ \ \ |}}{CH_3CH_2\overset{HO\ \ OH}{CHCHCH_3}}$$

Problem 4.28  What alkene, when treated with aqueous $KMnO_4$ at pH 11.8, gives each glycol?

(a) 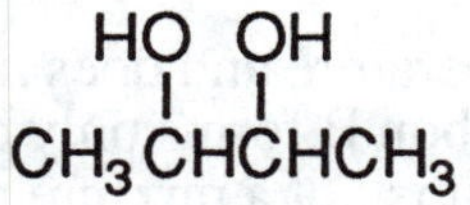 $CH_3\overset{\substack{HO\ \ OH \\ |\ \ \ \ |}}{CHCHCH_3}$

(b)

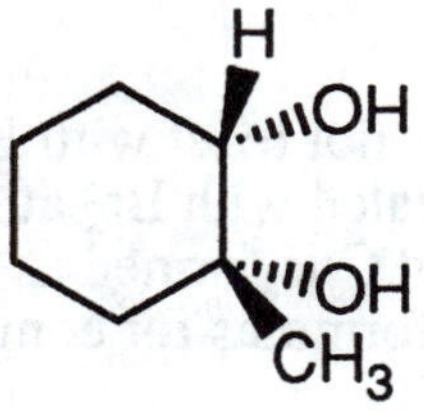

(c) 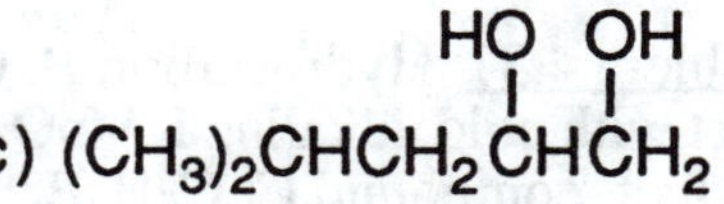 $(CH_3)_2CHCH_2\overset{\substack{HO\ \ OH \\ |\ \ \ \ |}}{CHCH_2}$

$$\underset{\substack{CH_3}}{\overset{H}{\diagdown}}C=C\underset{\substack{CH_3}}{\overset{H}{\diagup}}$$

**or**

$$\underset{\substack{CH_3}}{\overset{H}{\diagdown}}C=C\underset{\substack{H}}{\overset{CH_3}{\diagup}}$$

$$\underset{\substack{(CH_3)_2CHCH_2}}{\overset{H}{\diagdown}}C=C\underset{\substack{H}}{\overset{H}{\diagup}}$$

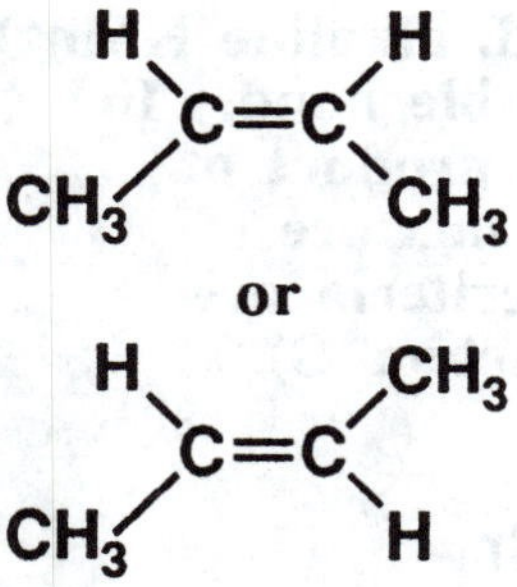

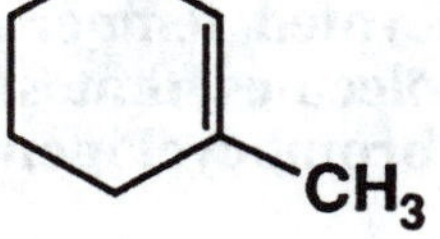

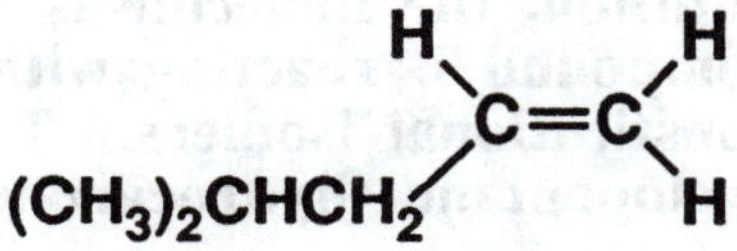

Problem 4.29  Draw the product formed by treatment of each alkene with $H_2/Ni$.

(a) 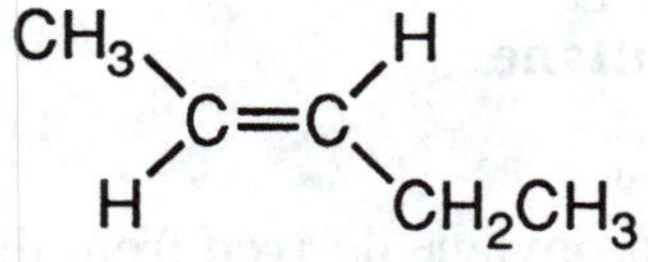 $\underset{\substack{H}}{\overset{CH_3}{\diagdown}}C=C\underset{\substack{CH_2CH_3}}{\overset{H}{\diagup}}$

(b) 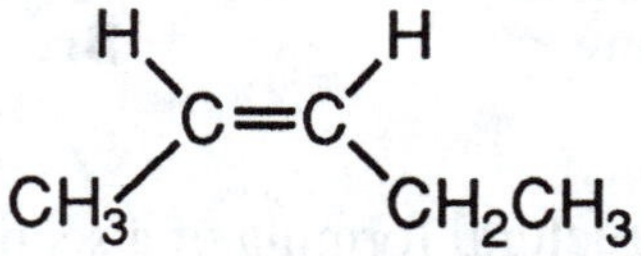 $\underset{\substack{CH_3}}{\overset{H}{\diagdown}}C=C\underset{\substack{CH_2CH_3}}{\overset{H}{\diagup}}$

(c)

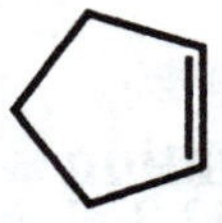

$CH_3CH_2CH_2CH_2CH_3$

$CH_3CH_2CH_2CH_2CH_3$

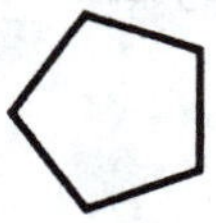

(d)

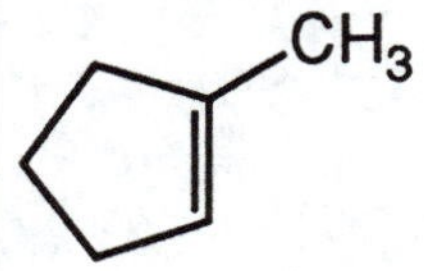

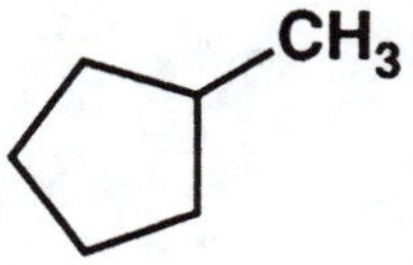

<u>Problem 4.30</u> Hydrocarbon A, $C_5H_8$, reacts with 2 mol of $Br_2$ to give 1,2,3,4-tetrabromo-2-methylbutane. What is the structure of hydrocarbon A?

**Hydrocarbon A ($C_5H_8$)**
**2-Methylbutadiene**

**1,2,3,4-tetrabromo-2-methylbutane**

<u>Problem 4.31</u> Hydrocarbon B, $C_5H_{10}$, does not react with $Br_2$ at room temperature and does not react with cold, alkaline $KMnO_4$. When treated with $Br_2$ at 300°C, hydrocarbon B forms only one product, compound C, $C_5H_9Br$. Compound C is formed as a pure product, it is not a mixture of constitutional isomers. Propose structural formulas for compounds B and C.

**The absence of reactivity with $Br_2$ at room temperature and cold, alkaline $KMnO_4$ indicates that hydrocarbon B does not have a carbon-carbon double bond. In addition, the molecule is symmetrical because compound C, the product of compound B reacting with $Br_2$ at elevated temperature, is not a mixture of constitutional isomers. The only molecules that satisfies these criteria are cyclopentane (hydrocarbon B) and bromocyclopentane (hydrocarbon C).**

**Hydrocarbon B**
**Cyclopentane**

**Hydrocarbon C**
**Bromocyclopentane**

## Polymerization
<u>Problem 4.32</u> Following is the structural formula of a section of polypropylene derived from three units of propylene monomer.

Draw structural formulas for a comparable section of:
(a) Poly (vinyl chloride)

(b) Polytetrafluoroethylene (Teflon)

$$-CF_2-CF_2-CF_2-CF_2-CF_2-CF_2-$$

(c) Poly (methyl methacrylate) (Plexiglass)

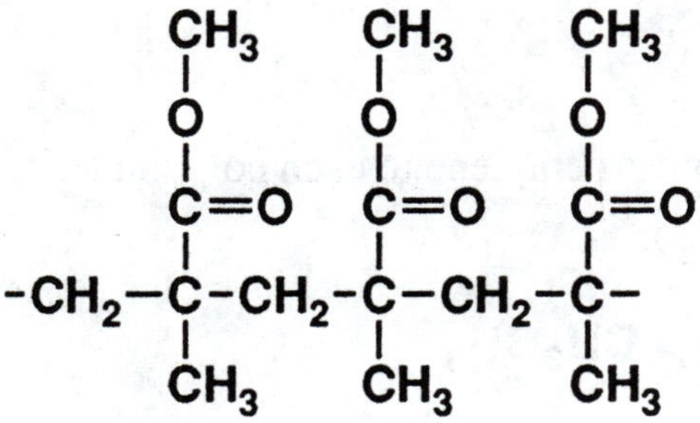

Problem 4.33  Following are structural formulas for sections of three polymers derived from alkene monomers.  From what monomer is each polymer derived?

(a)

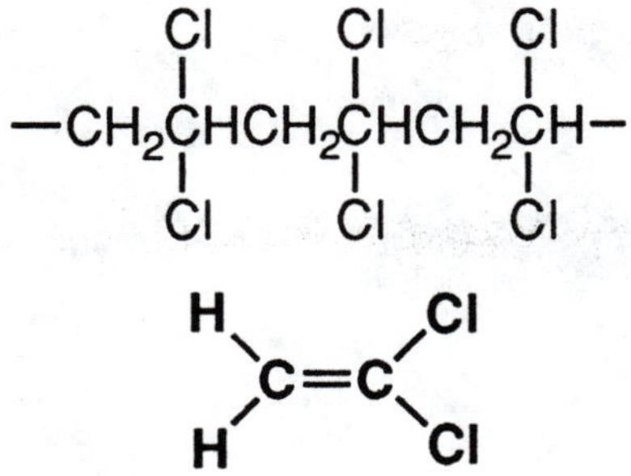

(b)

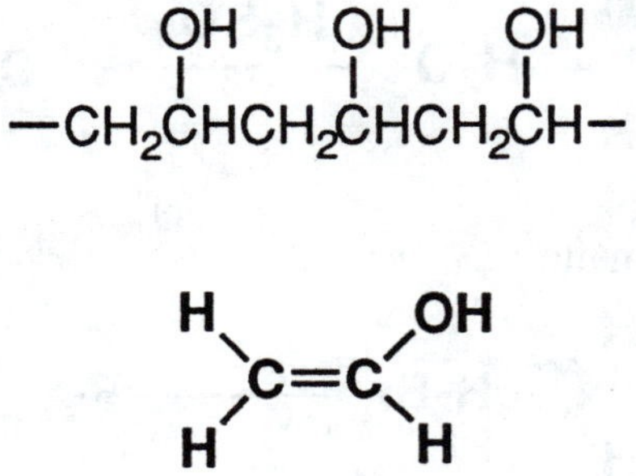

(c)

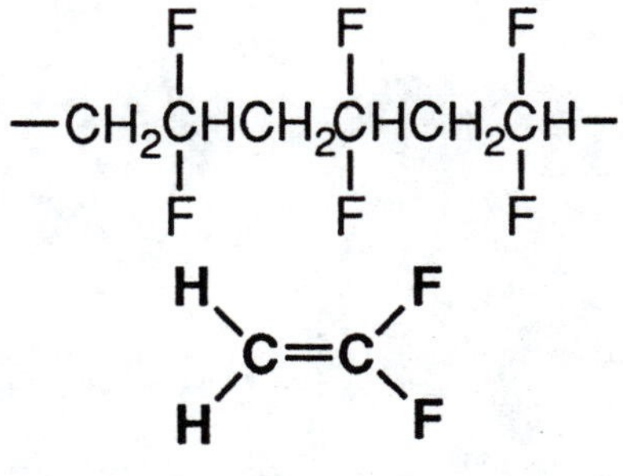

Problem 4.34  Low-density polyethylene (LDPE) has a higher degree of chain branching than high-density polyethylene (HDPE).  Explain the relationship between chain branching and density.

**The branches of LDPE prevent the polymer chains from packing together as tightly as the less-branched HDPE.  Thus, LDPE is less dense than HDPE.**

Problem 4.35  Compare the densities of low-density polyethylene (LDPE) and high-density polyethylene (HDPE) with the densities of the liquid alkanes listed in Table 2.4.  How might you account for the differences between them?

**As stated in the chapter, the density of LDPE is between 0.91 and 0.94 g/cm$^3$, while the density of HDPE is 0.96 g/cm$^3$.  These values are considerably higher than the values of 0.626 to 0.730 g/cm$^3$ for pentane through decane listed in Table 2.4.  A key parameter here is the ratio of hydrogen to carbon atoms in a hydrocarbon.  Hydrogen atoms have such a low atomic weight compared to carbon atoms that a lower hydrogen atom to carbon atom ratio increases the density of a hydrocarbon.  The longer the hydrocarbon, the lower the ratio of**

hydrogen atoms to carbon atoms.  Thus, the polymers, which have the lowest
hydrogen atom to carbon atom ratio, are significantly more dense than the shorter
hydrocarbons.

## Synthesis

Problem 4.36  Show how to convert ethylene to each compound.
(a) Ethane

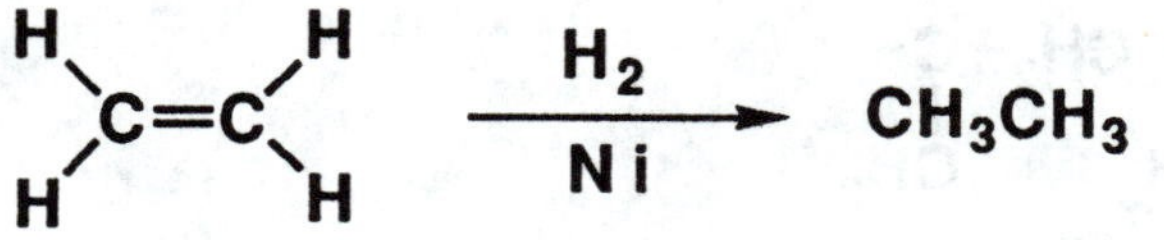

$$H_2C=CH_2 \xrightarrow[\text{Ni}]{H_2} CH_3CH_3$$

(b) Ethanol

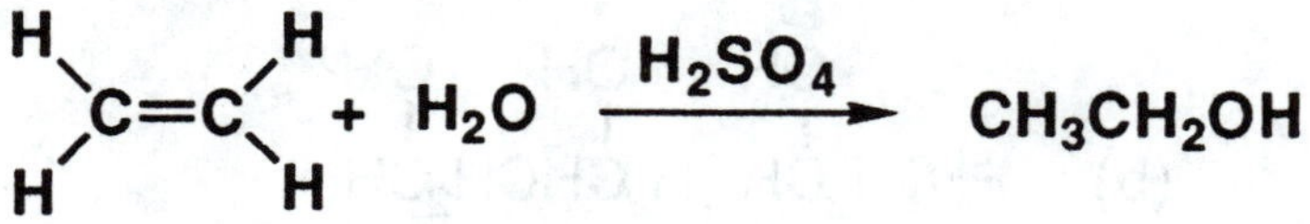

$$H_2C=CH_2 + H_2O \xrightarrow{H_2SO_4} CH_3CH_2OH$$

(c) Bromoethane

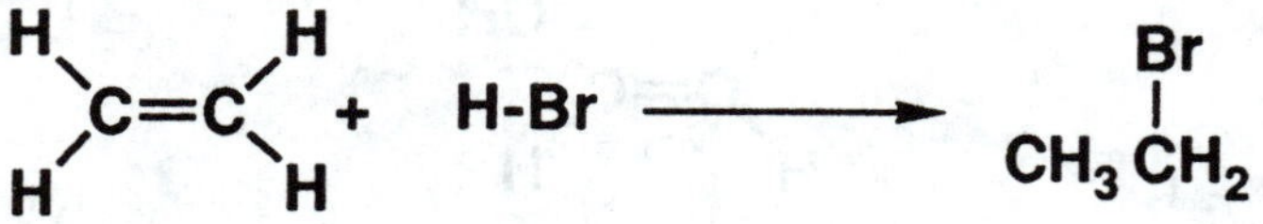

$$H_2C=CH_2 + H\text{-}Br \longrightarrow CH_3CH_2Br$$

(d) 1,2-Dibromoethane

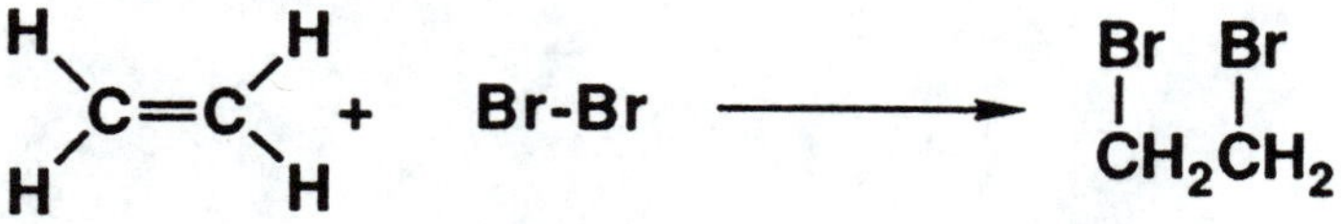

$$H_2C=CH_2 + Br\text{-}Br \longrightarrow BrCH_2CH_2Br$$

(e) 1,2-Ethanediol

$$H_2C=CH_2 + KMnO_4 \xrightarrow[\text{pH 11.8}]{H_2O} HOCH_2CH_2OH$$

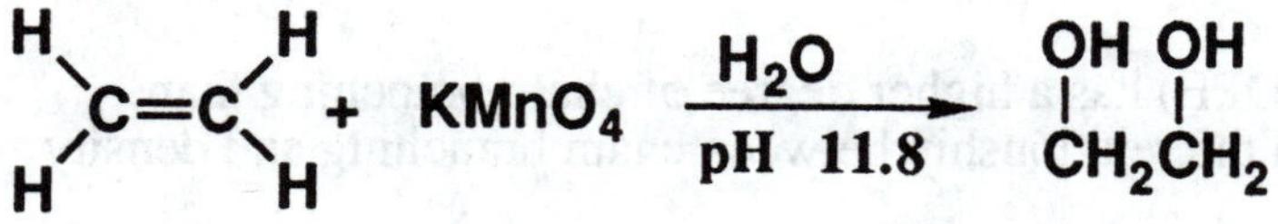

(f) Chloroethane

$$H_2C=CH_2 + H\text{-}Cl \longrightarrow CH_3CH_2Cl$$

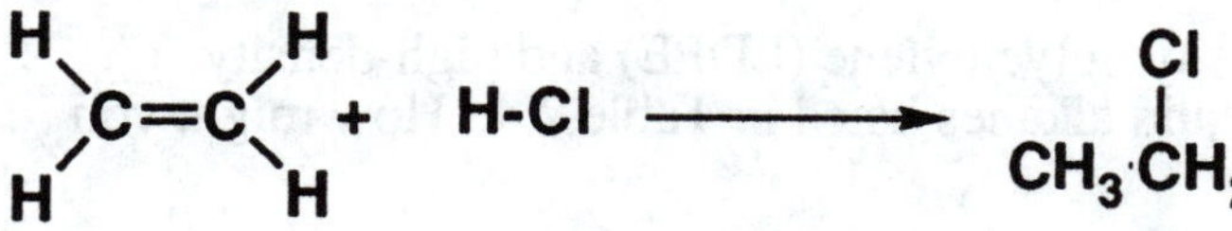

Problem 4.37  Show how to convert cyclopentene into these compounds.

(a)

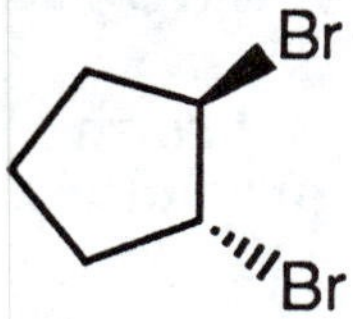

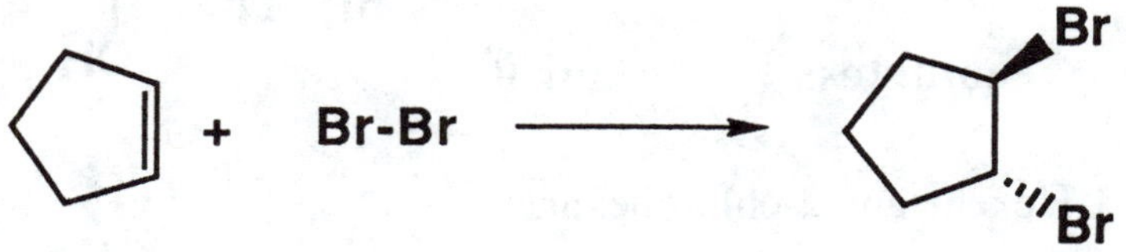

(b)

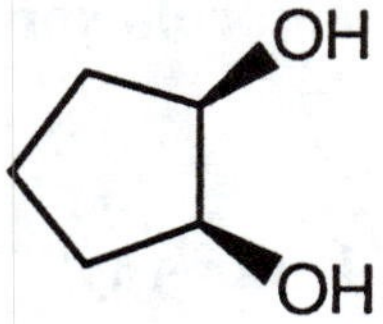

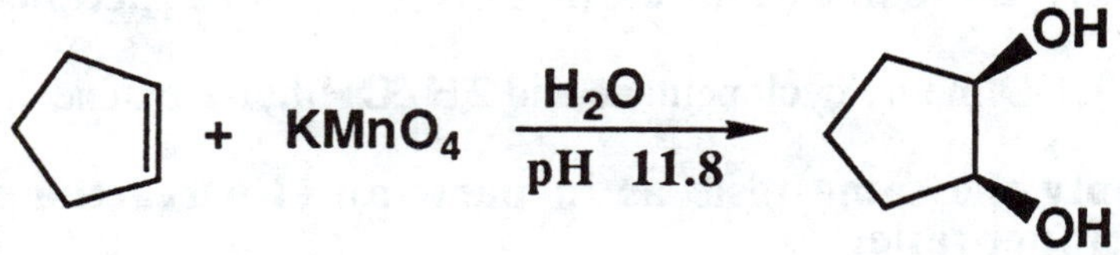

(c)

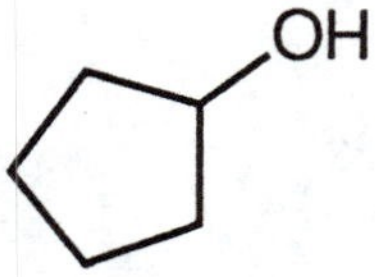

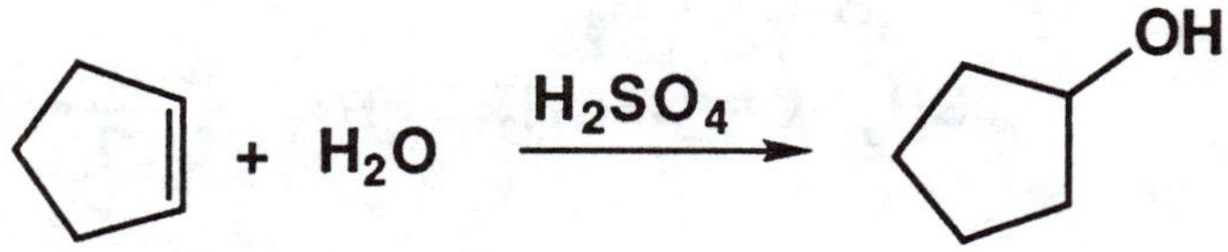

(d)

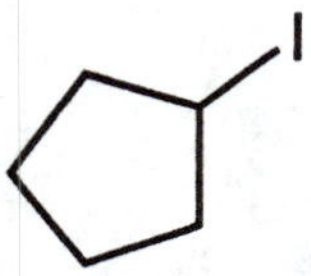

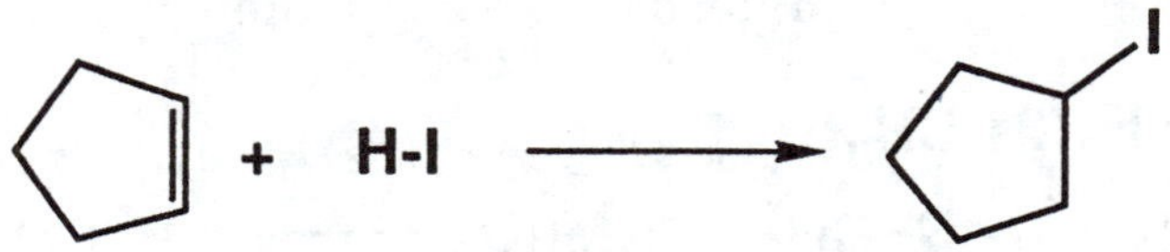

(e)

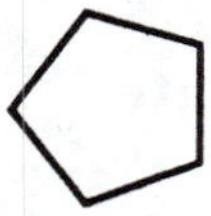

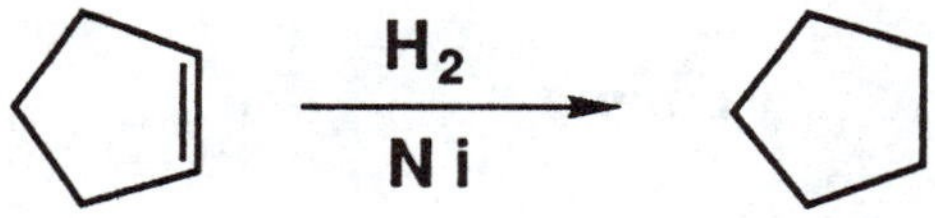

Problem 4.38  Describe how to distinguish between the members of each pair of compounds by a simple chemical test.  For each pair, tell what test to perform and what you expect to observe, and write an equation for each positive test.  For example, to distinguish between cyclohexane and 1-hexene in part (a), you might consider the reaction of each with $Br_2$ and $CCl_4$ or reaction with $KMnO_4$ in basic solution.
(a) Cyclohexane and 1-cyclohexene

**Add a solution of $Br_2$ in $CCl_4$ to each hydrocarbon. No reaction with cyclohexane.  Bromine adds to 1-hexene to form 1,2-dibromohexane. Observations are that a solution of bromine in carbon tetrachloride is deep purple. Addition of this solution to 1-hexene results in loss of the purple color and formation of a colorless solution.**

$$CH_2{=}CHCH_2CH_2CH_2CH_3 \ + \ Br_2 \xrightarrow{\ CCl_4\ } \underset{\underset{\displaystyle Br \quad Br}{|\quad\ |}}{CH_2{-}CHCH_2CH_2CH_2CH_3}$$

**colorless                    purple                          colorless**

**Alternatively, treat each with alkaline $KMnO_4$, pH 11.8.  No reaction with cyclohexane.  1-Hexene reacts to form a glycol and a brown precipitate of manganese dioxide, $MnO_2$.**

$$CH_2=CH(CH_2)_3CH_3 \; + \; KMnO_4 \xrightarrow[\text{pH } 11.8]{} \; CH_2-CH(CH_2)_3CH_3 \; + \; MnO_2$$

colorless        purple        OH  OH   colorless     brown<br>precipitate

(b) 1-Hexene and 2-chlorohexane

**Apply the same tests as in part (a). No reactions are seen with 2-chlorohexane.**

(c) 1,1-Dimethylcyclopentane and 2,3-dimethyl-2-butene

**Apply the same tests as in part (a). No reactions are seen with 1,1-dimethylcyclopentane.**

$$
\begin{array}{ccc}
\underset{\text{colorless}}{\overset{\text{H}_3\text{C}\quad\text{CH}_3}{CH_3-C=C-CH_3}} + \underset{\text{purple}}{Br_2} & \xrightarrow[\text{CCl}_4]{} & \underset{\text{colorless}}{\overset{\text{H}_3\text{C}\quad\text{CH}_3}{CH_3-C-C-CH_3}} \\
& & \quad\quad\quad\text{Br Br}
\end{array}
$$

$$
\begin{array}{ccc}
\underset{\text{colorless}}{\overset{\text{H}_3\text{C}\quad\text{CH}_3}{CH_3-C=C-CH_3}} + \underset{\text{purple}}{KMnO_4} & \xrightarrow[\text{pH } 11.8]{} & \underset{\text{colorless}}{\overset{\text{H}_3\text{C}\quad\text{CH}_3}{CH_3-C-C-CH_3}} + MnO_2 \\
& & \quad\quad\quad\text{OH OH}\qquad\text{brown precipitate}
\end{array}
$$

# CHAPTER 5: CHIRALITY

## 5.0 OVERVIEW
• Molecules are three-dimensional, and this chapter describes very important consequences of that three-dimensionality, namely stereochemistry and chirality.

## 5.1 ISOMERISM
• **Isomers** are *different* molecules that have the same molecular formula. ✳ *[This is the first of several very important definitions that should be learned now to avoid confusion later. They may sound relatively simple, but learning to apply them correctly in cases of complex molecules can be quite challenging.]*
• **Constitutional isomers** are different molecules with the same molecular formula, but with a different order of attachment of atoms. In other words, the atoms are connected to each other differently. For example, pentane ( ⌇⌇⌇ ) and 2-methylbutane ( ⌇⌇ ) are constitutional isomers of molecular formula $C_5H_{12}$.
• **Stereoisomers** are different molecules with the same molecular formula, the same order of attachment of atoms, but a different **orientation** of those atoms or groups in space. ✳ *[This concept is subtle and may be confusing at first, but it is actually a very powerful concept that should be understood before moving on.]*
  - Previously, two types of **stereoisomers** have been presented; E,Z stereoisomers of alkenes and *cis-trans* stereoisomers of cyclic compounds. In general, two different stereoisomers will have the same systematic name, but an additional element is added to the name such as "Z" or "cis" to specify the specific stereoisomer being named. For example, the following two molecules are the same constitutional isomer (2-butene), but they differ in the relative **orientation** of their atoms is space. For the molecule on the left ((E)-2-butene), the two methyl groups are **oriented** away from each other, while for the molecule on the right, (E)-2-butene, the two methyl groups are **oriented** together.

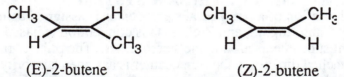

(E)-2-butene                     (Z)-2-butene

  - E,Z isomers of alkenes and *cis, trans* isomers of cyclic molecules are examples of **diastereomers**, that is stereoisomers that are *not* mirror images of each other. This definition is best understood in the context of the definition of enantiomers.
• **Enantiomers** are stereoisomers that are mirror images of each other. ✳
  - A **mirror image** is what you see as the reflection of an object in a mirror.

## 5.2 CHIRALITY
• **Chirality** is a property of three-dimensional objects that is very important in chemistry. An object is **chiral** if it is not superposable on its mirror image. That is, a chiral object and its mirror image cannot be oriented in space so that all of their features (corners, edges, points, etc.) correspond exactly to each other. ✳ *[This is a very difficult concept, but one that is absolutely central to the rest of this chapter and the study of stereochemistry. Chirality should be understood before proceeding.]*
  - Great examples of chiral objects are your hands. They are mirror images of each other (if you hold your left hand up to a mirror you see an image that looks like your right hand in the mirror), yet you cannot orient your two hands in space so that they are superposable. Try

this for yourself if you are having trouble understanding chirality. Because your hands are chiral, they have "handedness", that is the left hand is different in three-dimensional space than your right hand even though they are mirror images of each other and both have four fingers and a thumb. For this reason, a glove that fits your left hand well will not fit your right hand and you must buy one left-handed and one right-handed glove.
- Other examples of chiral objects include your feet, an airplane propeller or ceiling fan, a wood screw or a drill bit. In fact, the vast majority of objects around you are chiral.
• An object is **achiral** if it is superposable on its mirror image. Examples of objects that are achiral include a perfect cube or a perfect sphere. To be achiral, the molecule must have symmetry.
  -Achiral objects can be identified when they possess a **plane of symmetry.**
    A **plane of symmetry** is an imaginary plane passing through an object dividing it such that one half is the mirror reflection of the other half. Only highly symmetric objects such as a perfect sphere have a plane of symmetry. ✳ *[This is an important concept. It may be helpful to carefully examine Figure 5.2 to make sure you understand when an object has a plane of symmetry.]*
• **Molecules can be chiral**. To be chiral, a molecule cannot be superposable upon its mirror image. In addition, a chiral molecule does not have a plane of symmetry. ✳
• Often, molecules are chiral if they contain a tetrahedral carbon atom with four different substituents. This is because a tetrahedral carbon atom that has four different substituents is **not** superposable on its mirror image. ✳ *[This important concept is best understood using models. Try making a model of a tetrahedral carbon atom with four different substituents attached. Next, make a model of its mirror image, and try to superpose the two models. The fact that the two models cannot be superposed confirms that the molecules display chirality. Since the molecules are non-superposable mirror images of each other, they are enantiomers.]*
  - A **stereocenter** is a point in a molecule at which interchange of two atoms or groups of atoms bonded to that point produces a different stereoisomer. A carbon atom with four different substituents is a **tetrahedral stereocenter**. For example, the central carbon atom in lactic acid (2-hydroxypropanoic acid) is a tetrahedral stereocenter. ✳

## 5.3 NAMING ENANTIOMERS : THE R-S SYSTEM
• The absolute configuration of a tetrahedral stereocenter is assigned using the **R-S convention** first introduced by Cahn, Ingold, and Prelog. This convention is based on assigning **priorities** to the four different substituents around the stereocenter. The priority rules are as follows:
  - 1. Each atom attached directly to the stereocenter is assigned a priority based on atomic number. The higher the atomic number, the higher the priority. For example, -Br is assigned a higher priority than -Cl, which is assigned a higher priority than -OH, etc.
  -2. For isotopes, the higher the atomic mass of the isotope, the higher its priority.
  -3. If a priority based on atomic number cannot be assigned to the atoms directly bonded to the stereocenter, then look at the next set of atoms and continue until a priority can be assigned. **It is the first point of difference** that matters here.
  -4. Atoms of double or triple bonds are considered as if they are bonded to an equivalent number of similar atoms by single bonds. *[This is the hardest priority rule, and practice is usually needed to fully understand it.]*
• In order to assign the absolute configuration of a tetrahedral stereocenter according to the R-S convention use the following procedure:
  -1. Locate the tetrahedral stereocenter and identify its four substituents.
  -2. Assign a priority 1,2,3, or 4 to each substituent using the rules listed above.
  -3. Orient the molecule in space such that the group with the lowest priority (4) is directed away from you. In other words, orient the molecule so that when you look at it, the lowest priority group is directly behind the carbon stereocenter. The three remaining groups (1-3) will be arranged like the legs of a tripod directed toward you.

-4. Read the remaining three groups in order from highest (1) to lowest priority (3).

-5. If reading the groups proceeds in a clockwise direction, the absolute configuration is designated as **R**, if reading proceeds in a counterclockwise direction, the absolute configuration is **S**. ✳ *[It is essential that you become very good at assigning absolute stereochemistry, and practice is the best way to become very good.]*

View from this direction

A carbon stereocenter with four different substituents (*a* highest priority, *d* lowest priority)

The view down the C-*d* bond. Groups increase in priority in clockwise direction so the configuration is "R"

## 5.4 FISCHER PROJECTION FORMULAS

• A **Fisher projection formula** is used to indicate the configuration of chiral molecules. To write a Fisher projection formula, orient the molecule so that the **vertical bonds** of a stereocenter are **directed away from you** and the **horizontal bonds** are **directed toward you**. The Fischer projection figure is now drawn as a two dimensional figure, that can be related to the three-dimensional structure by the convention that the two horizontal lines represents bonds directed toward you, and the two vertical lines represent bonds directed away from you.

- You can only manipulate Fischer projection formula in precise ways. For example, you can only rotate a Fischer projection in the plane of the paper by 180°, NOT 90°. If this rule is not followed, then the new Fischer projection formula may inadvertently depict a different stereoisomer than the original Fischer projection formula.

## 5.5 NONCYCLIC MOLECULES WITH TWO OR MORE STEREOCENTERS

• For a molecule with **n** tetrahedral stereocenters, the maximum number of stereoisomers possible is $2^n$. This is because each stereocenter can be either R or S.

• A good way to learn about the stereochemical consequences of having more than one stereocenter in the same molecule is to examine all four possible stereoisomers that arise when there are two stereocenters in the same molecule. For example, consider the four stereoisomers of 2,3,4-trihydroxybutanal drawn below:

|        CHO        |        CHO        |        CHO        |        CHO        |
|:-----------------:|:-----------------:|:-----------------:|:-----------------:|
| H—C—OH (R)        | HO—C—H (S)        | H—C—OH (R)        | HO—C—H (S)        |
| H—C—OH (R)        | HO—C—H (S)        | HO—C—H (S)        | H—C—OH (R)        |
|      CH₂OH        |      CH₂OH        |      CH₂OH        |      CH₂OH        |
|       **a**       |       **b**       |       **c**       |       **d**       |

- From right to left the four isomers are the R,R (**a**); S,S (**b**); R,S (**c**) and S,R (**d**) stereoisomers. When examined in pairs, the following relationships become apparent:

   **a,b** and **c,d** are non-superposable mirror images of each other, thus they are pairs of **enantiomers**. *[This would be a good time to review the definition of enantiomers if necessary.]*

   **a,c** and **a,d** and **b,c** and **b,d** are all pairs of stereoisomers that are not mirror images of each other, thus these are all pairs of **diastereomers**.

- Hint: when asked to identify the stereochemical relationships between pairs of stereoisomers, it is helpful to first assign the absolute configuration (R or S) to each stereoisomer then compare these designations, instead of trying to compare the molecules directly. ✳
- Certain molecules have special symmetry properties that reduce the number of stereoisomers to fewer than the predicted $2^n$. In these cases, some of the stereoisomers contain a plane of symmetry so the molecules are achiral. Molecules or ions that contain two or more tetrahedral stereocenters but are achiral because of a plane of symmetry are called **meso compounds**. ✳
  - **Meso compounds** will always be the R,S/S,R isomer of a molecule. For example, the R,S/S,R isomer of tartaric acid is a meso compound. Of course, not all R,S isomers are meso compounds. To be a meso compound there must be a plane of symmetry in the molecule.

**plane of symmetry** $\Longrightarrow$

**meso-tartaric acid**

## 5.6 CYCLOALKANES WITH TWO OR MORE STEREOCENTERS

- Understanding stereochemical relationships between cycloalkane stereoisomers can be very difficult. The key is being able to identify planes of symmetry in the molecules. If there is a plane of symmetry present then the molecule is achiral, if there is no plane of symmetry then the molecule is chiral. ✳
  - For example, *cis*-3-methylcyclopentanol is chiral because there is no plane of symmetry in the molecule as can be seen by the planar representation. On the other hand, *cis*-1,3-cyclopentanediol is a achiral because there is a plane of symmetry.

Plane of Symmetry

*cis*-3-Methylcyclopentanol
No plane of symmetry; chiral

*cis*-1,3-Cyclopentanediol
Plane of symmetry; achiral

  - With disubstituted cyclohexane derivatives, it is especially important to keep track of the substitution pattern. For example, *trans*-1,4-cyclohexanediol has a plane of symmetry and is thus achiral, while *trans*-1,3-cyclohexanediol does not have a plane of symmetry so it is achiral.

*trans*-1,4-Cyclohexanediol
A plane of symmetry extends
right through both OH groups
perpendicular to ring, so the
molecule is achiral.

*trans*-1,3-Cyclohexanediol
There is no plane of symmetry,
so the molecule is chiral

- Note that sometimes the cyclohexane chair conformations must be considered when looking for planes of symmetry. This is explained further in the text.

## 5.7 PHYSICAL PROPERTIES OF ENANTIOMERS

- In general, **enantiomers have identical physical properties** when those properties are **measured in an achiral environment**. For example, two enantiomers have the same boiling points, melting points, solubilities in achiral solvents, the same $pK_a$ values, etc. On the other hand, two **diastereomers have different physical properties** like melting points or boiling points. ✳
- Although enantiomers have identical physical properties when those properties are measured in an achiral way, they are different compounds. Therefore, they have different physical properties when the measurements are made in a chiral way such as interactions with plane polarized light. Each member of a pair of enantiomers rotates the plane of plane polarized light, so they are called **optically active**.
  - Normal light consists of waves oscillating in all planes perpendicular to its path. Certain materials such as Polaroid sheets only allow waves oscillating in a single plane to pass through. The resulting light is thus called **plane polarized light,** because all of the resulting light is oscillating in the same plane.
  - Samples of enantiomers rotate the plane of plane polarized light. A **polarimeter** is the instrument used in the laboratory to measure the direction and magnitude of rotation of plane polarized light.

    If a sample of an enantiomer rotates plane polarized light in a **clockwise** direction, it is called **dextrorotary**. If a sample rotates plane polarized light in a **counterclockwise** direction, it is called **levorotary**. ✳

    In order to standardize optical rotation data, **specific rotation** has been defined according to the following equation:

    $$[\alpha]_{\lambda}^{T} = \frac{\text{observed rotation (degrees)}}{\text{cell length (dm) x concentration (g/mL)}}$$

    Note that the length of the cell in which the sample is placed is measured in unusual units, namely decimeters. 1 dm = 10 cm. The T stands for the measurement temperature and the $\lambda$ stands for the wavelength of light used to make the measurement (usually the sodium D line at 589 nm).

    By convention, a dextrorotary compound is designated with a plus sign (+) and a levorotary compound is designated with a minus sign (-).

- Two enantiomers rotate plane polarized light by the same number of degrees, but with opposite sign. Meso compounds and all other achiral molecules do not rotate plane polarized light. ✳
- An equimolar mixture of two enantiomers is called a **racemic mixture**. For racemic mixtures, the rotations of plane polarized light exactly cancel and there is no overall rotation. If two enantiomers in a mixture are in unequal amounts, the entire mixture will have a net rotation, and the magnitude and direction of that rotation can be used to determine the exact ratio of the two enantiomers in the mixture according to the following equation:

    $$\% \text{ optical purity} = \frac{[\alpha]_{obs}}{[\alpha]_{pure \ enantiomer}} \times 100$$

  - **Enantiomeric excess**, abbreviated **ee**, is equal to the percent optical purity, and is often referred to when two different enantiomers are made in different amounts in the same reaction. **Enantiomeric excess** is defined according to the following equation:

$$\% \text{ enantiomeric excess (ee)} = \left( \frac{\text{moles of one enantiomer - moles of other enantiomer}}{\text{moles of both enantiomers}} \right) \times 100$$

- There is **no** necessary **relationship between** a **configuration** (R and S) and the **sign of rotation** (+ and -). For some pairs of stereoisomers, the "R" enantiomer has a (+) rotation, while the "S" enantiomer has a (-) rotation, yet for other pairs of stereoisomers, the "S" enantiomer has a (+) rotation and the "R" enantiomer has a (-) rotation. This makes sense since the R and S designations are based on artificial nomenclature rules, while the (+) and (-) rotations are the result of actual measurements. ✴

## 5.9 SEPARATION OF ENANTIOMERS-RESOLUTION
- **Resolution** is the process whereby a racemic mixture is separated into the two enantiomers. A pair of enantiomers are difficult to separate, because they both have identical physical properties such as melting points, boiling points, etc. A common method of resolution involves combining the racemic mixture with another compound that is a single enantiomer. The combination results in the production of two diastereomers that can usually be separated because diastereomers have different physical properties. Following resolution, the enantiomers are recovered. ✴
  - A reaction that lends itself to resolution is salt formation. For example, one enantiomer of a chiral compound such as a positively-charged amine like (+)-cinchonine can be used to form diastereomeric salts with a racemic mixture of negatively-charged species like deprotonated form of chiral carboxylic acids. Following resolution, the pure enantiomers of the carboxylic acid are recovered by acid precipitation.

## 5.10 THE SIGNIFICANCE OF CHIRALITY IN THE BIOLOGICAL WORLD
- Almost all of the molecules of living systems are chiral. Even though all of these chiral molecules could in theory exist as a mixture of stereoisomers, almost invariably only one stereoisomer is found in nature.
  - Enzymes, Mother Nature's molecular machines, are composed of chiral amino acids. As a result, the enzymes are themselves chiral. For that reason, enzymes are able to distinguish substrate enantiomers. A pair of enantiomers can be distinguished in a chiral environment like the active site of an enzyme.

## 5.11 THE STEREOCHEMISTRY OF REACTIONS
- During a reaction, one or more stereocenters in a molecule may be created or destroyed. It is therefore important to keep track of stereocenters during the entire reaction mechanism in order to predict accurately the stereochemistry of the final product. ✴
- In general, optical activity is never produced from optically inactive starting materials, even though the products may be chiral. In other words, if a stereocenter is created from an achiral starting material, then a racemic mixture of the two possible enantiomers is formed (or a meso compound if applicable).
  - Examples of this include the addition of $Br_2$ to *cis*-2-butene to create a racemic mixture of the two enantiomers of 2,3-dibromobutane, and the $KMnO_4$ oxidation of 1-butene to give a racemic mixture of the two enantiomers of 1,2-butanediol.
- Alternatively, optical activity is generated in a reaction only if at least one of the reactants itself is chiral, or if the reaction is carried out in the presence of a catalyst that is itself chiral.

# CHAPTER 5
## *Solutions to the Problems*

<u>Problem 5.1</u> Each of these molecules has one stereocenter? Draw stereorepresentations for each pair of enantiomers.

**The stereocenters are labeled with an asterisk.**

(a)

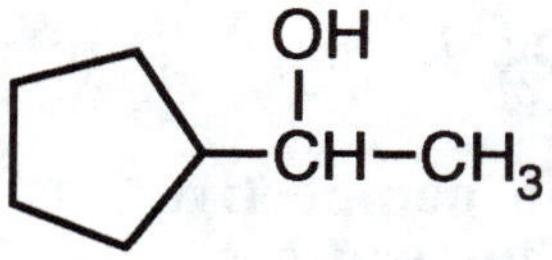

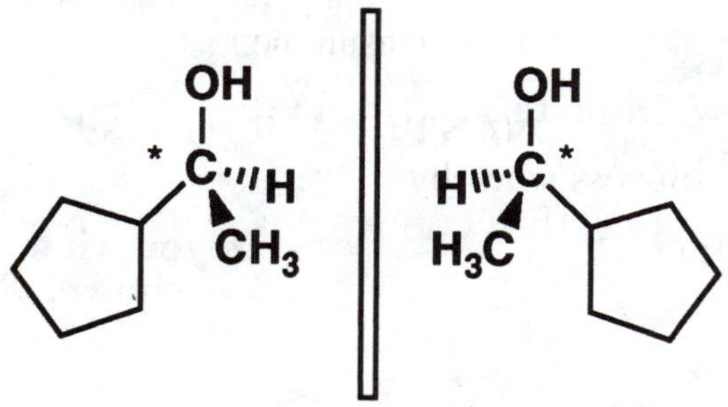

(b)

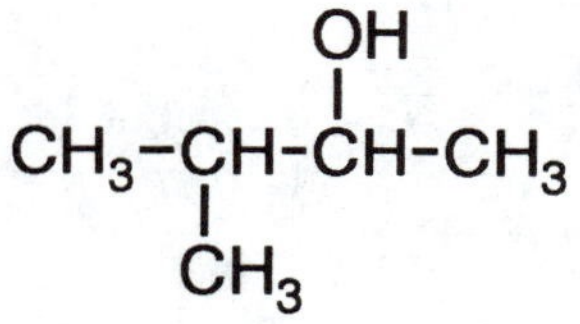

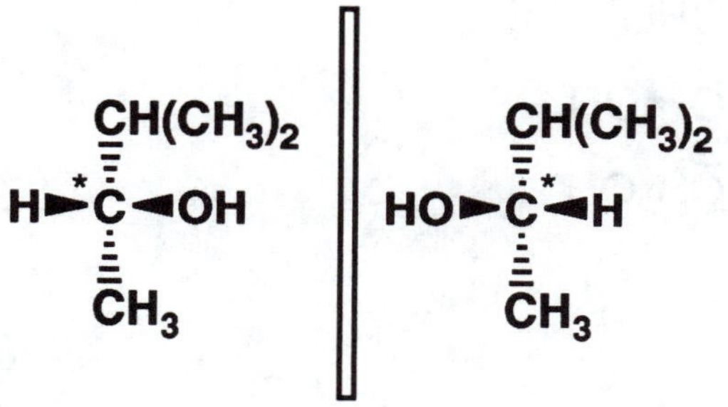

<u>Problem 5.2</u> Assign an R or S configuration to each stereocenter and give each molecule an IUPAC name.

**The drawings underneath each molecule show the order of priority, the perspective from which to view the molecule, and the R,S designation for the configuration.**

(a)

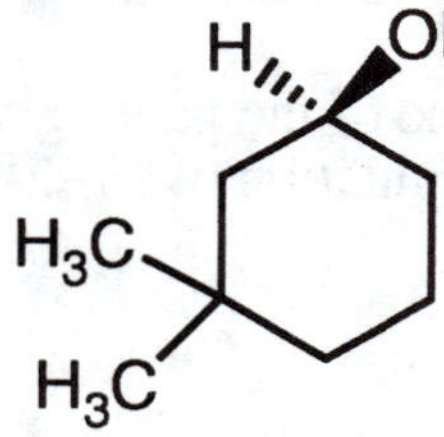

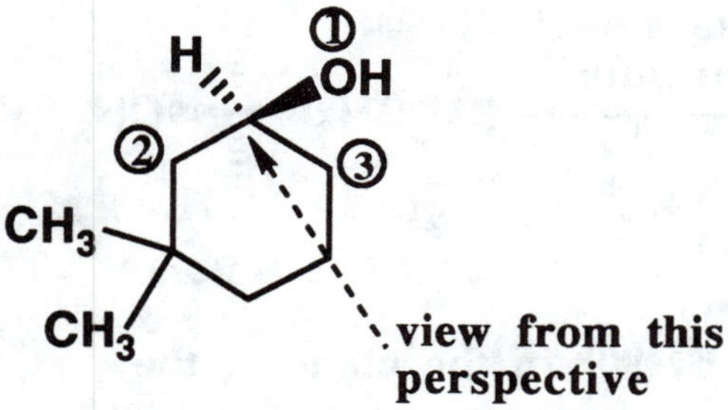

**view from this perspective**

**(1S)-3,3-Dimethylcyclohexanol**

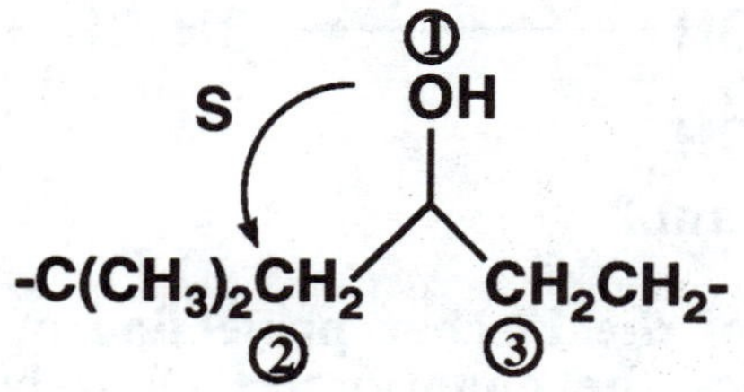

**If you view from the perspective shown, this is what you see**

(b)

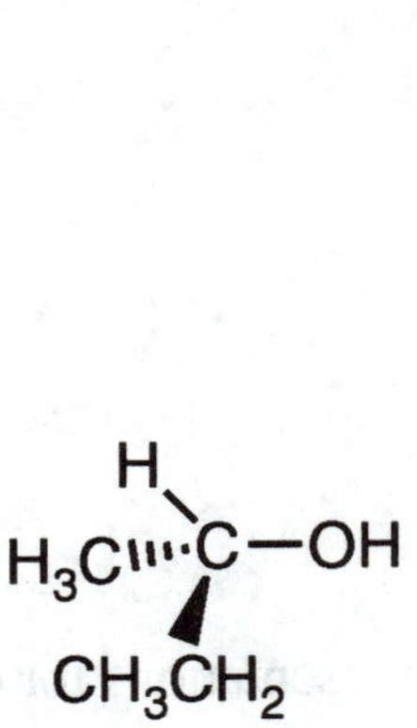

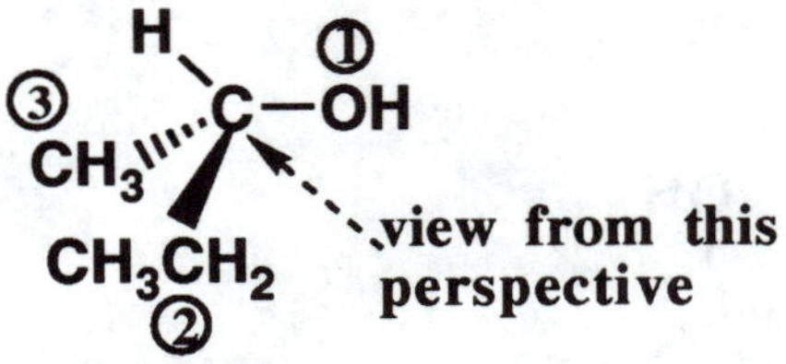

H₃C''''C—OH with H above and CH₃CH₂ below

(S)-2-Butanol

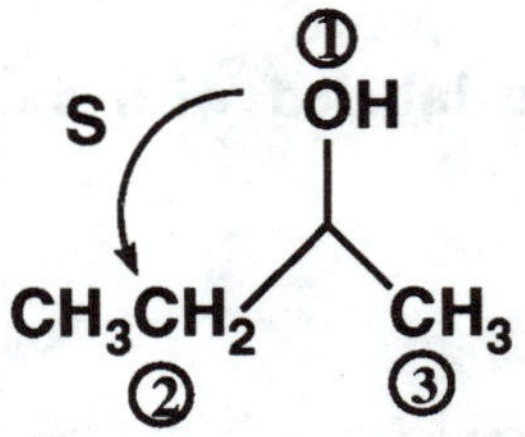

If you view from the perspective shown, this is what you see

(c)

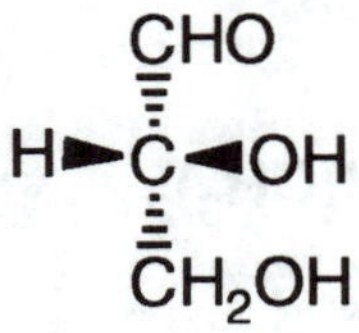

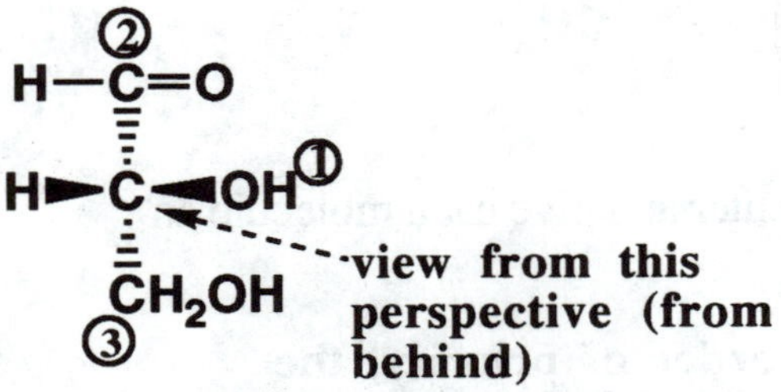

CHO
H►C◄OH
CH₂OH

(2R)-2,3-Dihydroxypropanal

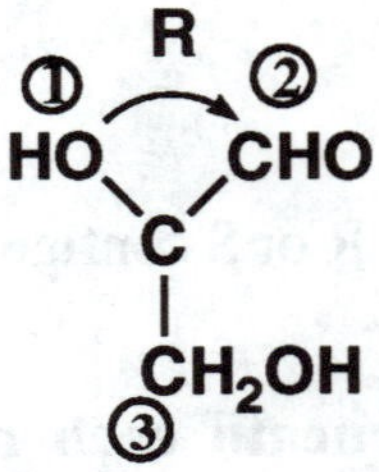

If you view from the perspective shown, this is what you see

Problem 5.3 We said that rotation of a Fischer projection by 90° in the plane of the paper gives a different molecule. What is the different molecule resulting from this manipulation of (S)-2-butanol?

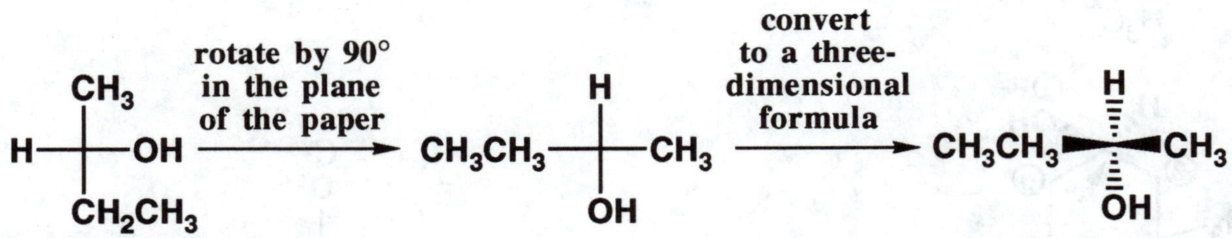

(S)-2-Butanol    rotate by 90° in the plane of the paper    convert to a three-dimensional formula    (R)-2-Butanol

Rotation of the Fischer projection of (S)-2-butanol by 90° in the plane of the paper gives the enantiomer, (R)-2-butanol.

Problem 5.4 Following are Fisher projection formulas for the four stereoisomers of 3-chloro-2-butanol.

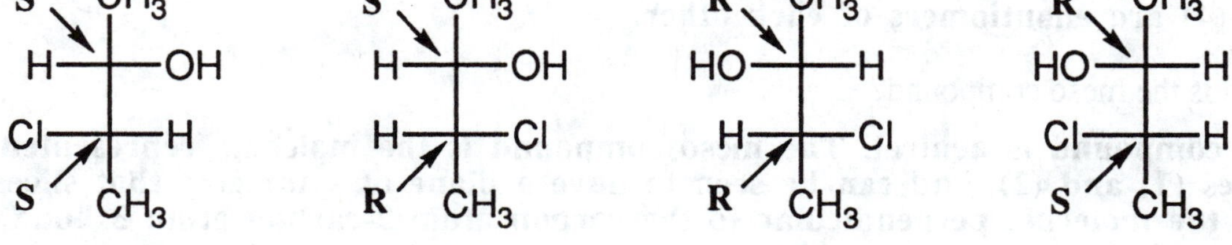

(1)                (2)                (3)                (4)

**The configuration of each tetrahedral stereocenter has been labeled on the above structures. This labeling often helps when trying to establish stereochemical relationships between molecules.**

(a) Which are pairs of enantiomers?

**Enantiomers are stereoisomers that are mirror images of each other. The pairs of enantiomers are structures (1) and (3) (S,S and R,R) as well as structures (2) and (4) (S,R and R,S).**

(b) Which are diastereomers of (1)?

**Diastereomers are stereoisomers that are not mirror images of each other. Therefore, structures (2) and (4) are diastereomers of structure (1).**

(c) Which are diastereomers of (4)?

**Structures (1) and (3) are diastereomers of structure (4).**

Problem 5.5 Following are four Fisher projection formulas for 2,3-dibromobutane.

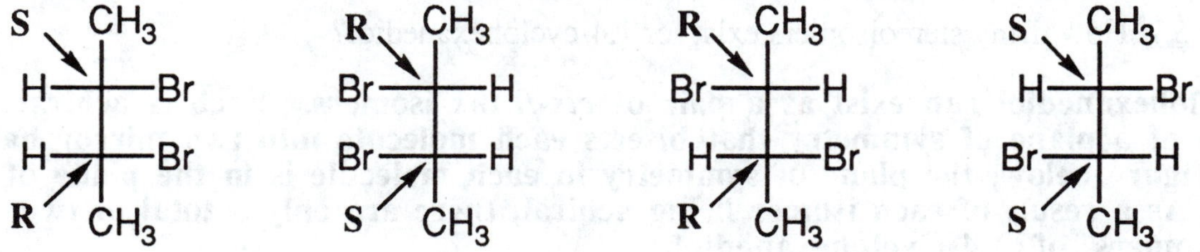

(1)                (2)                (3)                (4)

**The configuration of each tetrahedral stereocenter has been labeled on the above structures. This labeling often helps when trying to establish stereochemical relationships between molecules.**

(a) Which two Fisher projection formulas represent the same compound?

**Structures (1) and (2) represent the same compound, since the two Fisher projections can be interconverted by a 180° rotation in the plane of the paper.**

(b) Which are pairs of enantiomers?

**Enantiomers are stereoisomers that are mirror images of each other. Structures (3) and (4) are enantiomers of each other.**

(c) Which is the meso compound?

**A meso compound is achiral. The meso compound is the molecule represented by structures (1) and (2), and can be seen to have a plane of symmetry that slices through the molecule perpendicular to the carbon atom 2-carbon atom 3 bond.**

(d) Which are diastereomers of (2)?

**Diastereomers are stereoisomers that are not mirror images of each other. Structures (3) and (4) are diastereomers of (2).**

<u>Problem 5.6</u>  How many stereoisomers exist for 1,3-cyclopentanediol?

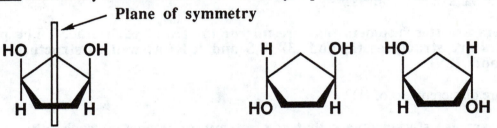

**Plane of symmetry**

*cis*-1,3-Cyclopentanediol
(achiral, a meso compound)

*trans*-1,3-Cyclopentanediol
(a pair of enantiomers)

**1,3-Cyclopentanediol has three stereoisomers. The two *trans* isomers are enantiomers, the *cis* isomer is a meso compound. *Cis*-1,3-cyclopentanediol can be recognized as a meso compound because it is superposable upon its mirror image, and it has a plane of symmetry that bisects it into two mirror halves.**

<u>Problem 5.7</u>  How many stereoisomers exist for 1,4-cyclohexanediol?

**1,4-Cyclohexanediol can exist as a pair of *cis-trans* isomers. Each is achiral because of a plane of symmetry that bisects each molecule into two mirror halves. In the figure below, the plane of symmetry in each molecule is in the plane of the paper. As a result of each isomer being achiral, there are only a total of two stereoisomers of 1,4-Cyclohexanediol.**

*trans*-1,4-Cyclohexanediol          *cis*-1,4-Cyclohexanediol

Problem 5.8  The specific rotation of progesterone, a female sex hormone, is +172°, measured at 20°C.  Calculate the observed rotation prepared by dissolving 300 mg of progesterone in 15 mL of dioxane and placing it in a sample tube 10 cm long.

**The concentration of progesterone, expressed in grams per milliliter is:**
$$300 \text{ mg} / 15 \text{ mL} = 0.020 \text{ g/mL}$$
**Inserting these values into the formula for calculating specific rotation gives:**

$$\text{specific rotation} = \frac{\text{observed rotation (degrees)}}{\text{length (dm) x concentration (g/mL)}}$$

**Rearranging this formula to solve for observed rotation gives:**
**observed rotation (degrees)  =  specific rotation x length (dm) x concentration (g/mL)**
**Plugging in the real values gives the final answer.**

**observed rotation (degrees)  =  +172° x 1.00 dm x 0.020 g/mL  =  +3.4°**

## Chirality
Problem 5.9  Which of these objects are chiral (assume there is no label or other identifying mark)?
(a) Pair of scissors      (b) Tennis ball          (c) Paper clip          (d) Beaker
(e) The swirl created in water as it drains out of a sink or bathtub.

**An object is chiral if it is not superposable upon its mirror image.  Objects that have a plane of symmetry are not chiral.  The tennis ball, paper clip and beaker, (b), (c), and (d), are all achiral because that all have a plane of symmetry and are all superposable on their mirror images.  The pair of scissors and drain swirl, (a) and (e), are chiral because they do not have a plane of symmetry and they are not superposable on their mirror images.**

Problem 5.10  Think about the helical coil of a telephone cord or a spiral binding and suppose that you view the spiral from one end and find that it is a left-handed twist.  If you view the same spiral from the other end, do you find it to be a right-handed twist, or is it a left-handed twist from that end as well?

**A helical coil has the same handedness viewed from either end.**

Problem 5.11  Next time you have the opportunity to view a collection of whelks, augers, or other sea shells that have a helical twist, study the chirality of their twists.  Do you find an equal number of left-handed and right-handed whelks or are they all or mostly all of one chirality?  What about the chirality of whelks compared to augers and other spiral shells?

**This question was just meant to make you think about chirality in nature, but if you do know the answer please share it with your class.**

Problem 5.12  One reason we can be sure that $sp^3$-hybridized carbon atoms are tetrahedral is the number of stereoisomers that can exist for different organic compounds.
(a) How many stereoisomers are possible for $CHCl_3$, $CH_2Cl_2$, and $CHClBrF$ if the four bonds to carbon have a tetrahedral arrangement?

**Both tetrahedral $CHCl_3$ and tetrahedral $CH_2Cl_2$ are achiral, so there is only one stereoisomer of either.**

$$
\begin{array}{c}
Cl \\
| \\
Cl \blacktriangleright C \blacktriangleleft H \\
\vdots \\
Cl
\end{array}
\qquad
\begin{array}{c}
Cl \\
\vdots \\
Cl \blacktriangleright C \blacktriangleleft H \\
| \\
H
\end{array}
$$

**On the other hand, tetrahedral CHBrClF is chiral so there are two stereoisomers possible.**

$$
\begin{array}{c}
Cl \\
\vdots \\
H \blacktriangleright C \blacktriangleleft Br \\
| \\
F
\end{array}
\qquad\Bigg\|\qquad
\begin{array}{c}
Cl \\
\vdots \\
Br \blacktriangleright C \blacktriangleleft H \\
| \\
F
\end{array}
$$

(b) How many stereoisomers are possible for each of these compounds if the four bonds to the carbon have a square planar geometry?

**Even with a square planar geometry (the H and three Cl atoms are in the same plane as the C atom) there is only one stereoisomer possible.**

$$
\begin{array}{c}
H \\
| \\
Cl-C-Cl \\
| \\
Cl
\end{array}
$$

**There are two possible stereoisomers of $CH_2Cl_2$, one with the Cl atoms adjacent to each other, and another with the Cl atoms opposite to each other.**

$$
\begin{array}{c}
H \\
| \\
H-C-Cl \\
| \\
Cl
\end{array}
\qquad\qquad
\begin{array}{c}
H \\
| \\
Cl-C-Cl \\
| \\
H
\end{array}
$$

**There are three possible stereoisomers of a square planar CHBrClF as shown.**

$$
\begin{array}{c}
Br \\
| \\
H-C-Cl \\
| \\
F
\end{array}
\qquad
\begin{array}{c}
Cl \\
| \\
H-C-Br \\
| \\
F
\end{array}
\qquad
\begin{array}{c}
Br \\
| \\
H-C-F \\
| \\
Cl
\end{array}
$$

## Enantiomers

Problem 5.13  Which compounds contain stereocenters?

(a) 2-Chloropentane

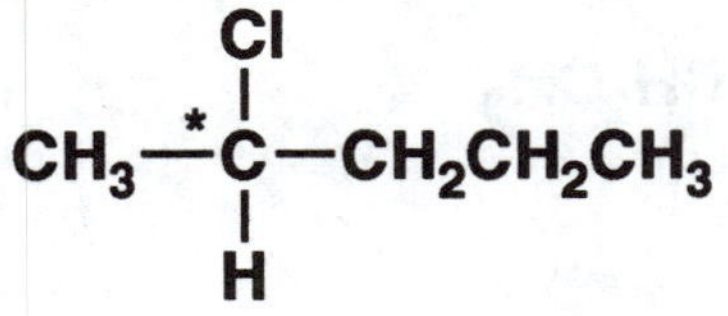

$$CH_3\overset{Cl}{\underset{H}{-\overset{*}{C}}}-CH_2CH_2CH_3$$

(b) 3-Chloropentane

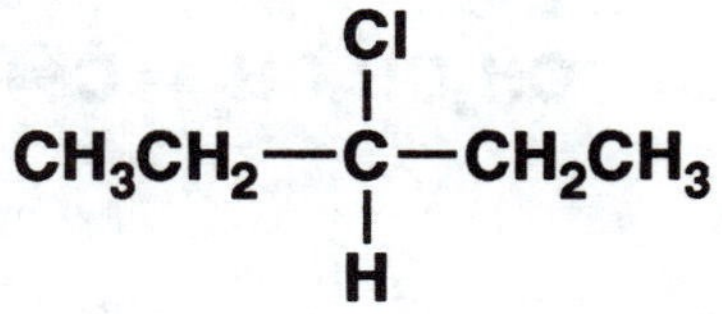

$$CH_3CH_2\overset{Cl}{\underset{H}{-C}}-CH_2CH_3$$

(c) 3-Chloro-1-pentene

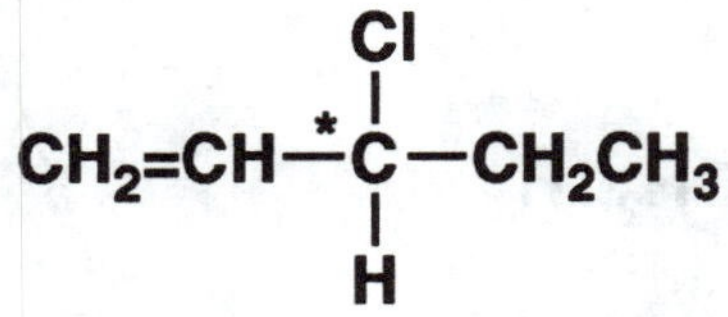

$$CH_2=CH\overset{Cl}{\underset{H}{-\overset{*}{C}}}-CH_2CH_3$$

(d) 1,2-Dichloropropane

$$ClCH_2\overset{Cl}{\underset{H}{-\overset{*}{C}}}-CH_3$$

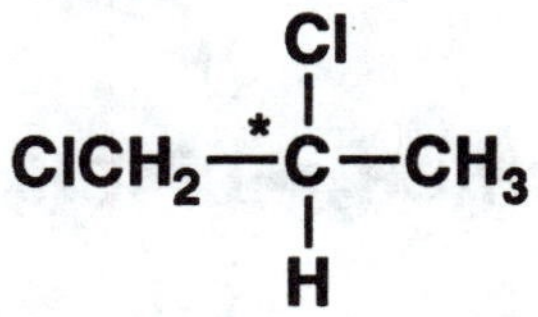

**To be a tetrahedral stereocenter, a carbon atom must have four different groups attached.  Thus, compounds (a), (c), and (d) have stereocenters (marked with an asterisk).**

Problem 5.14  Using only C, H, and O, write structural formulas for
(a) the lowest-molecular-weight alkene that shows chirality

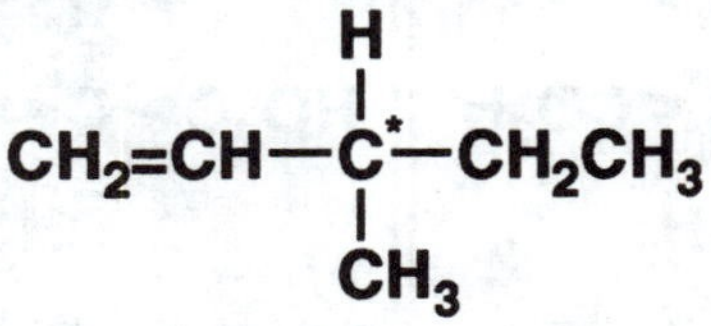

$$CH_2=CH-\overset{H}{\underset{CH_3}{\overset{*}{C}}}-CH_2CH_3$$

**3-Methyl-1-pentene**

(b) the lowest-molecular-weight alcohol that shows chirality

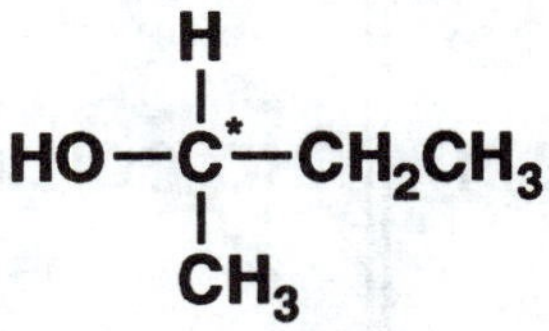

$$HO-\overset{H}{\underset{CH_3}{\overset{*}{C}}}-CH_2CH_3$$

**2-Butanol**

(c) the lowest-molecular-weight aldehyde that shows chirality

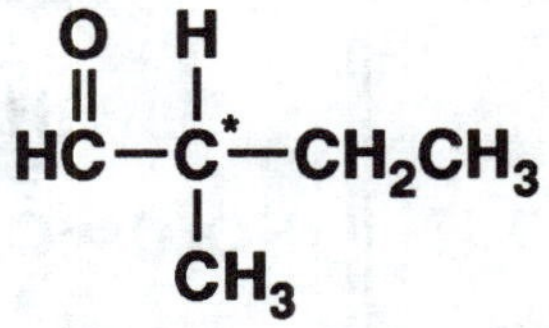

$$HC\overset{O}{\underset{}{\|}}-\overset{H}{\underset{CH_3}{\overset{*}{C}}}-CH_2CH_3$$

**2-Methylbutanal**

<u>Problem 5.15</u>  Which alcohols of molecular formula $C_5H_{12}O$ exist as pairs of enantiomers?

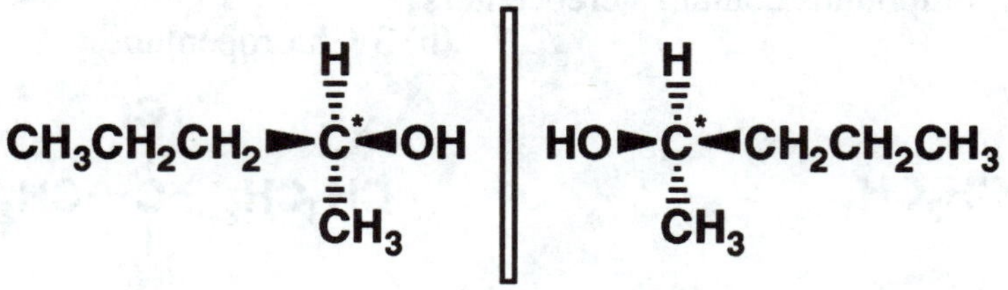

$$CH_3CH_2CH_2-\overset{\overset{H}{|}}{\underset{\underset{CH_3}{|}}{C^*}}-OH \quad \Vert \quad HO-\overset{\overset{H}{|}}{\underset{\underset{CH_3}{|}}{C^*}}-CH_2CH_2CH_3$$

**2-Pentanol**

$$CH_3CH_2-\overset{\overset{H}{|}}{\underset{\underset{CH_3}{|}}{C^*}}-CH_2OH \quad \Vert \quad HOH_2C-\overset{\overset{H}{|}}{\underset{\underset{CH_3}{|}}{C^*}}-CH_2CH_3$$

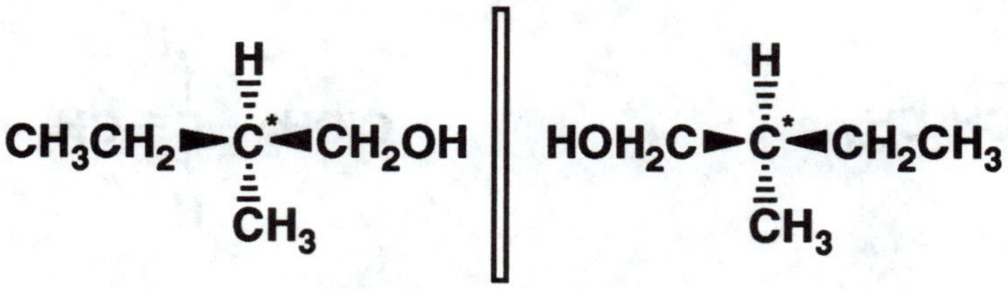

**2-Methyl-1-butanol**

<u>Problem 5.16</u>  Which carboxylic acids of molecular formula $C_6H_{12}O_2$ exist as pairs of enantiomers?

$$CH_3CH_2CH_2-\overset{\overset{H}{|}}{\underset{\underset{CH_3}{|}}{C^*}}-CO_2H \quad \Vert \quad HO_2C-\overset{\overset{H}{|}}{\underset{\underset{CH_3}{|}}{C^*}}-CH_2CH_2CH_3$$

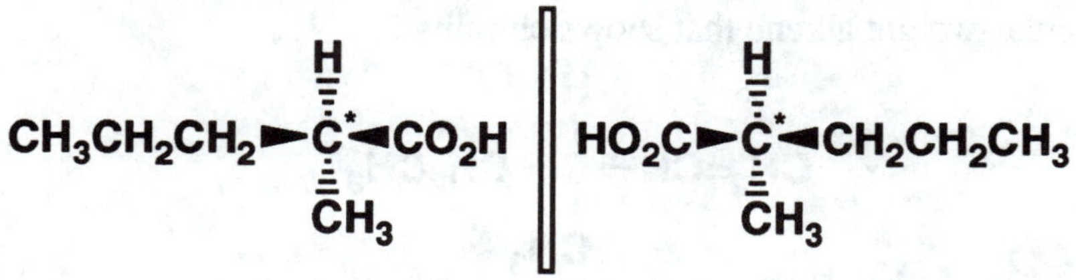

**2-Methylpentanoic  acid**

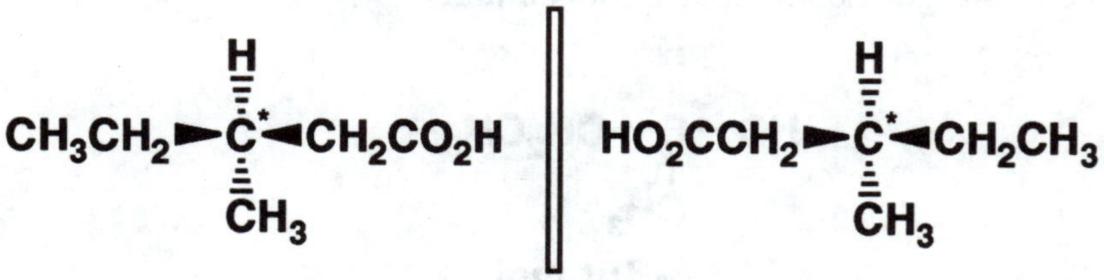

$$CH_3CH_2-\overset{\overset{H}{|}}{\underset{\underset{CH_3}{|}}{C^*}}-CH_2CO_2H \quad \Vert \quad HO_2CCH_2-\overset{\overset{H}{|}}{\underset{\underset{CH_3}{|}}{C^*}}-CH_2CH_3$$

**3-Methylpentanoic  acid**

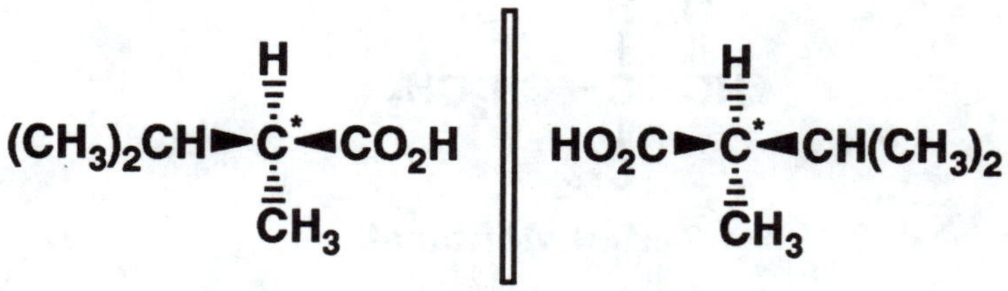

$$(CH_3)_2CH-\overset{\overset{H}{|}}{\underset{\underset{CH_3}{|}}{C^*}}-CO_2H \quad \Vert \quad HO_2C-\overset{\overset{H}{|}}{\underset{\underset{CH_3}{|}}{C^*}}-CH(CH_3)_2$$

**2,3-Dimethylbutanoic  acid**

Problem 5.17 Write the structural formula of an alcohol of molecular formula $C_6H_{14}O$ that contains two stereocenters.

$$CH_3CH_2 - \overset{\overset{\displaystyle H}{|}}{\underset{\underset{\displaystyle H_3C}{|}}{C^*}} - \overset{\overset{\displaystyle OH}{|}}{\underset{\underset{\displaystyle H}{|}}{C^*}} - CH_3$$

**3-Methyl-2-pentanol**

Problem 5.18 Draw the structural formula for at least one bromoalkene of molecular formula $C_5H_9Br$ that shows:
(a) Neither E-Z-isomerism nor enantiomerism.

**Draw any structural formula in which there is no tetrahedral stereocenter and in which at least one carbon of the double bond has two identical atoms or groups of atoms on it. For example:**

$$\underset{\displaystyle CH_3}{\overset{\displaystyle CH_3}{>}}C=C\underset{\displaystyle CH_3}{\overset{\displaystyle Br}{<}}$$

**2-Bromo-3-methyl-2-butene**

(b) E-Z-isomerism but not enantiomerism.

**Each carbon of the double bond must have two different atoms or groups of atoms on it, and there can be no tetrahedral stereocenter. For example:**

$$\underset{\displaystyle CH_3}{\overset{\displaystyle H}{>}}C=C\underset{\displaystyle CH_2Br}{\overset{\displaystyle CH_3}{<}}$$

**(Z)-1-Bromo-2-methyl-2-butene**

(c) Enantiomerism but not E-Z-isomerism.

**One carbon of the double bond must have two identical atoms or groups of atoms on it, and there must be a tetrahedral stereocenter. For example:**

$$\underset{\displaystyle H}{\overset{\displaystyle H}{>}}C=C\underset{\displaystyle H}{\overset{\displaystyle CH_2-}{<}}\overset{\overset{\displaystyle Br}{|}}{\underset{\underset{\displaystyle CH_3}{}}{C}}\cdots H$$

**(S)-4-Bromo-1-pentene**

(d) Both enantiomerism and E-Z-isomerism.

**There is only one constitutional isomer that shows E-Z isomerism and has a tetrahedral stereocenter. Drawn is the (4R), (Z) isomer:**

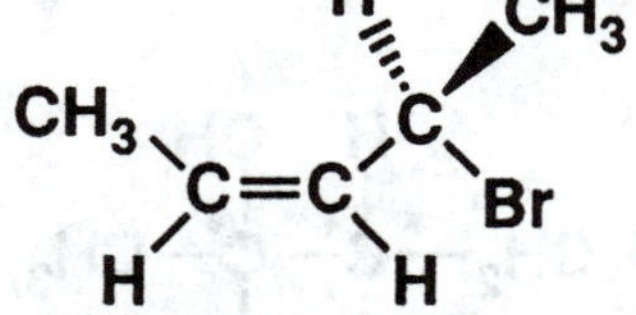

**(R)-4-Bromo-(Z)-2-pentene**

<u>Problem 5.19</u> Draw mirror images for these molecules.

**The mirror images are shown in bold.**

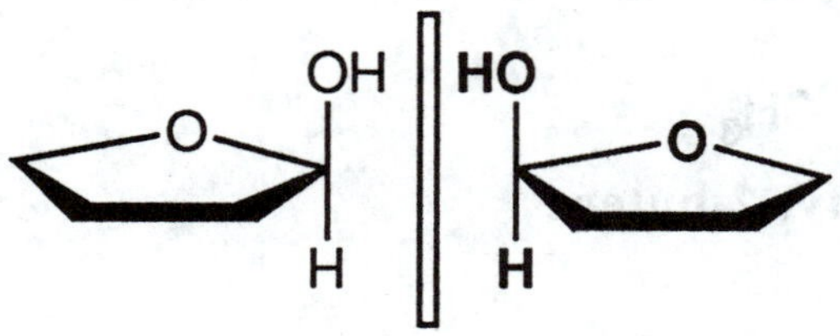

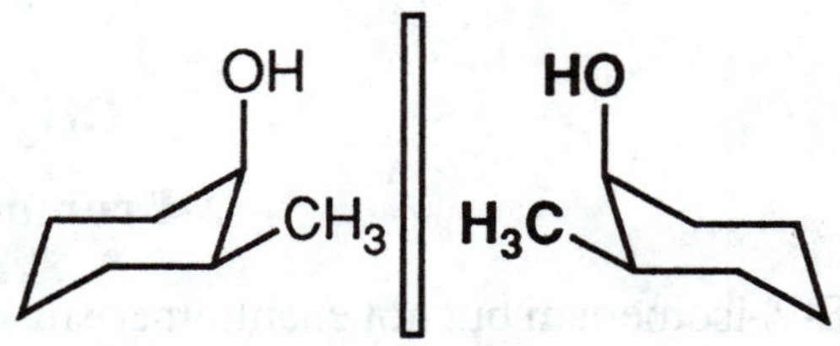

<u>Problem 5.20</u> Following are four stereorepresentations for lactic acid. Take (a) as a reference structure. Which stereorepresentations are identical with (a) and which are mirror images of (a)?

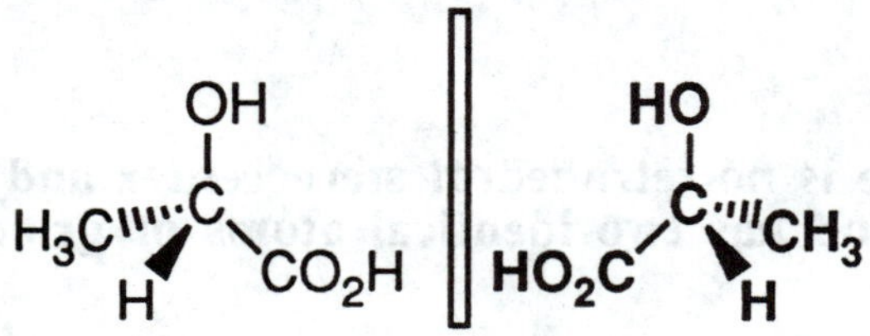

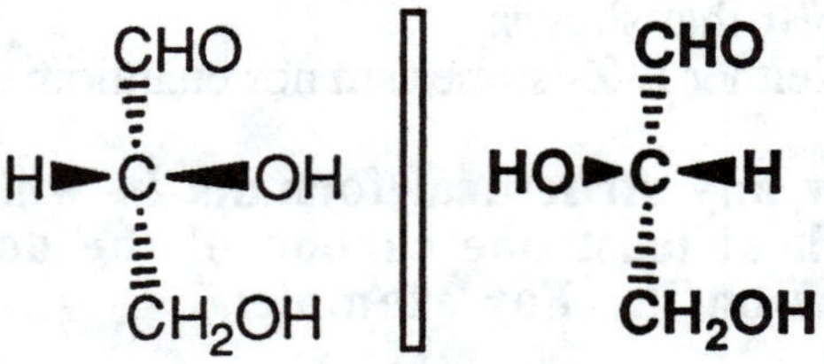

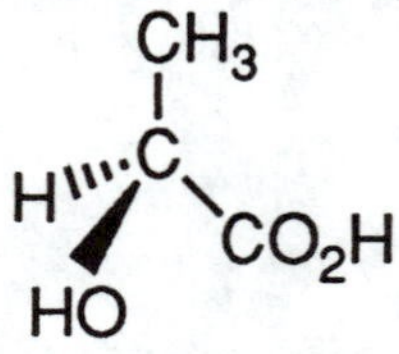

**All stereorepresentations have the (S)-configuration so they are identical.**

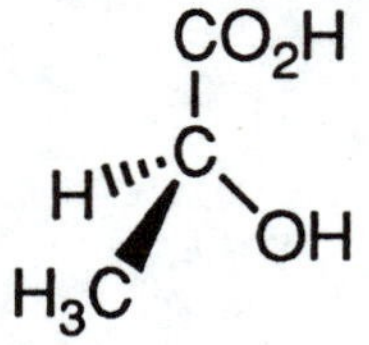

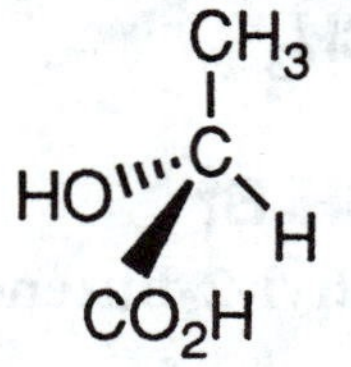

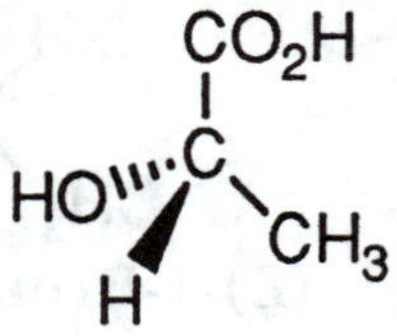

<u>Problem 5.21</u> Label each stereocenter in these molecules with an asterisk. Note that not all of the molecules contain stereocenters.

(a) 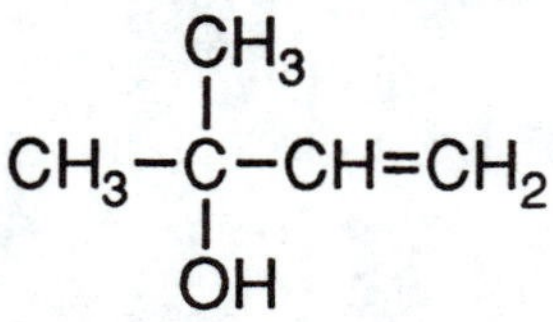   $CH_3-\overset{\overset{\displaystyle CH_3}{|}}{\underset{\underset{\displaystyle OH}{|}}{C}}-CH=CH_2$

**No stereocenters**

(b) 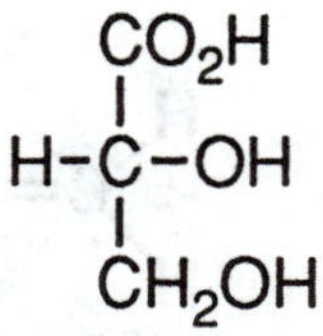   $H-\overset{\overset{\displaystyle CO_2H}{|}}{\underset{\underset{\displaystyle CH_2OH}{|}}{C}}-OH$

$H-\overset{\overset{\displaystyle CO_2H}{|}}{\underset{\underset{\displaystyle CH_2OH}{|}}{\overset{*}{C}}}-OH$

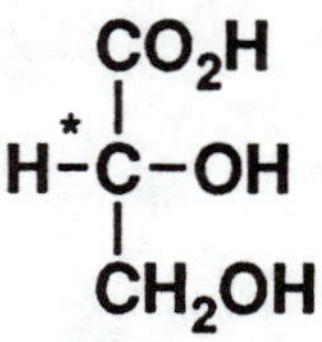

**2 stereoisomers
(a pair of enantiomers)**

(c) 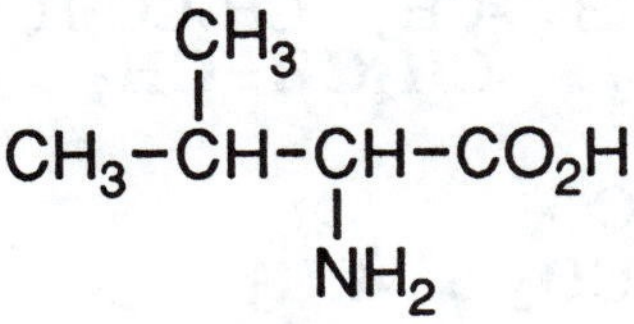  CH$_3$–CH–CH–CO$_2$H
        |      |
       CH$_3$   NH$_2$

(d)
           O
           ||
    CH$_3$–C–CH$_2$–CH$_3$

CH$_3$–CH–*CH–CO$_2$H
    |     |
   CH$_3$  NH$_2$

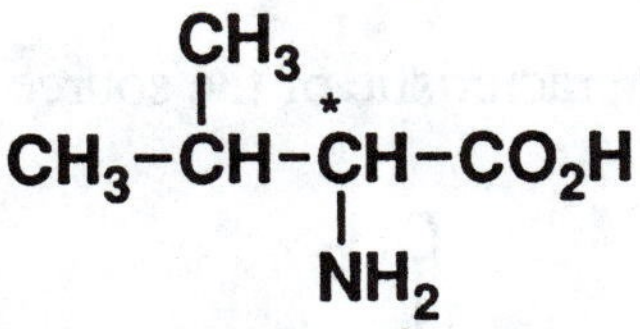

**2 stereoisomers**
**(a pair of enantiomers)**

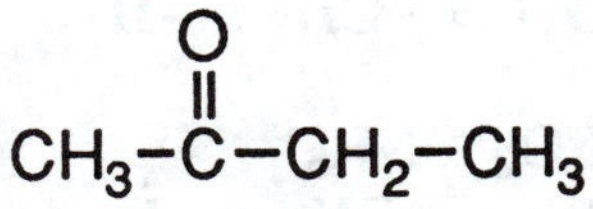

**No stereocenters**

(e) 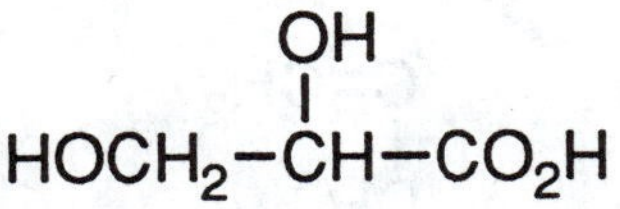 HOCH$_2$–CH–CO$_2$H
         |
        OH

(f)
        CH$_2$OH
        |
    H–C–OH
        |
        CH$_2$OH

HOCH$_2$–*CH–CO$_2$H
        |
        OH

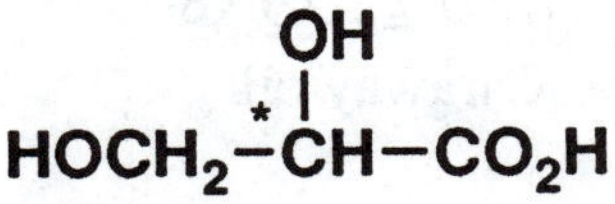

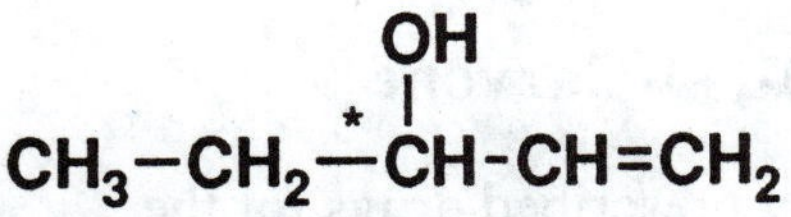

**2 stereoisomers**
**(a pair of enantiomers)**

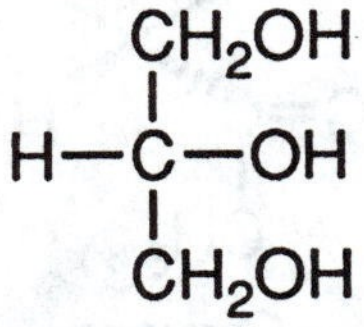

**No stereocenters**

(g) 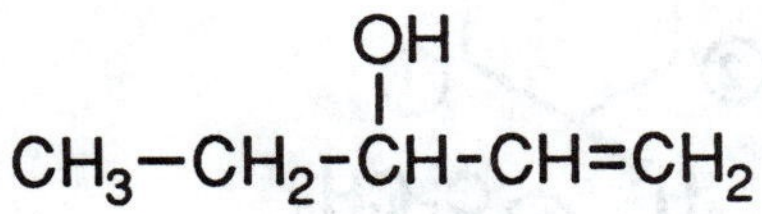  CH$_3$–CH$_2$–CH–CH=CH$_2$
            |
          OH

(h)
        CH$_2$-CO$_2$H
        |
    H–C–CO$_2$H
        |
        CH$_2$-CO$_2$H

CH$_3$–CH$_2$–*CH–CH=CH$_2$
         |
        OH
**2 stereoisomers**
**(a pair of enantiomers)**

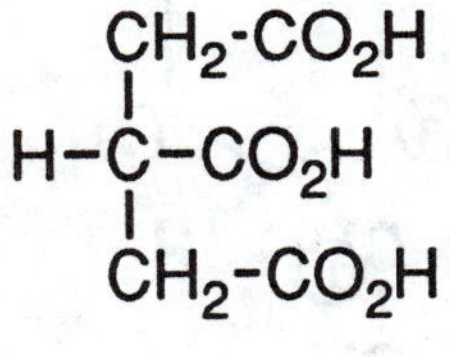

**No Stereocenters**

**Problem 5.22**  How many stereoisomers exist for each molecule in problem 5.21?

**The total number of stereoisomers is written below each structure in the answer to problem 5.21 above.**

## Designation of Configuration: The R-S Convention

**Problem 5.23**  Assign the priorities to the groups in each set.

**The groups are ranked from highest to lowest under each problem. Remember that priority is assigned at the <u>first</u> point of difference.**

*Should be Switched!*

(a) -H  -CH$_3$  -OH  -CH$_2$OH
-OH > -CH$_2$OH > -CH$_3$ > -H

(b) -CH$_2$CH=CH$_2$  -CH=CH$_2$  -CH$_3$  -CH$_2$CO$_2$H
-CH$_2$CO$_2$H > -CH=CH$_2$ > -CH$_2$CH=CH$_2$ > -CH$_3$

(c) -CH$_3$ -H -CO$_2^-$ -NH$_3^+$
-NH$_3^+$ > -CO$_2^-$ > -CH$_3$ > -H

(d) -CH$_3$ -CH$_2$SH -NH$_3^+$ -CO$_2^-$
-NH$_3^+$ > -CH$_2$SH > -CO$_2^-$ > -CH$_3$

<u>Problem 5.24</u>  Each enantiomer of carvone has a distinctive odor characteristic of the source from which it is isolated.  Assign R-S configurations to each enantiomer.

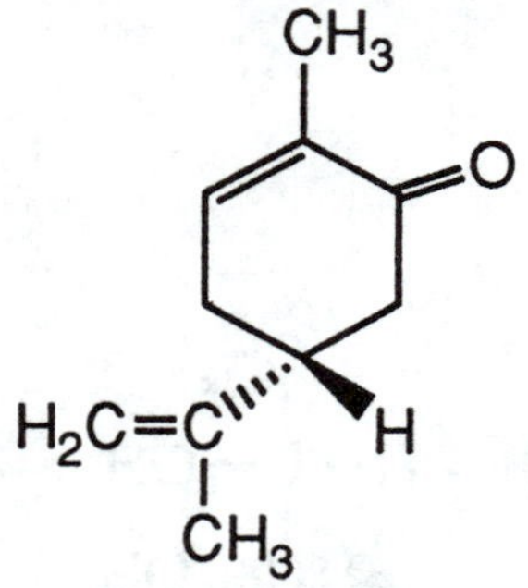

(-)-Carvone
$[\alpha]_D^{20}$ = -62.5°
Spearmint oil

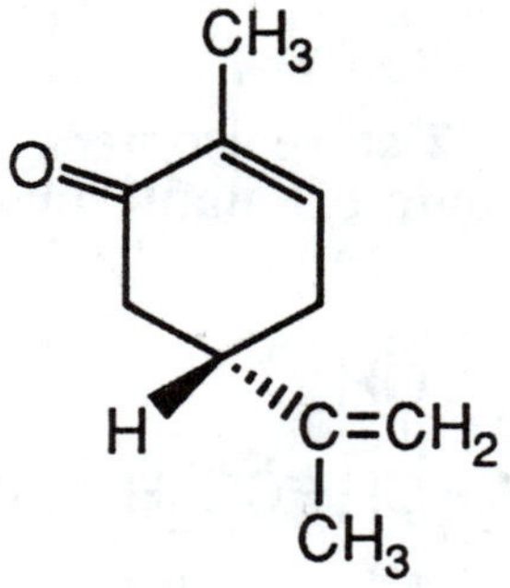

(+)-Carvone
$[\alpha]_D^{20}$ = +62.5°
Caraway oil

**Following are R-S designations for each enantiomer.**

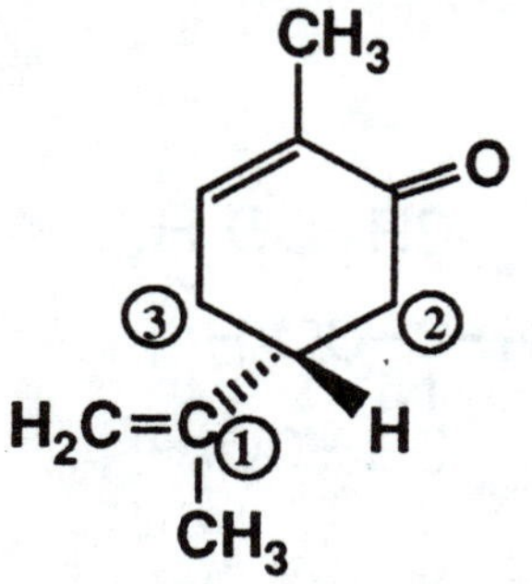

(R)-(-)-Carvone

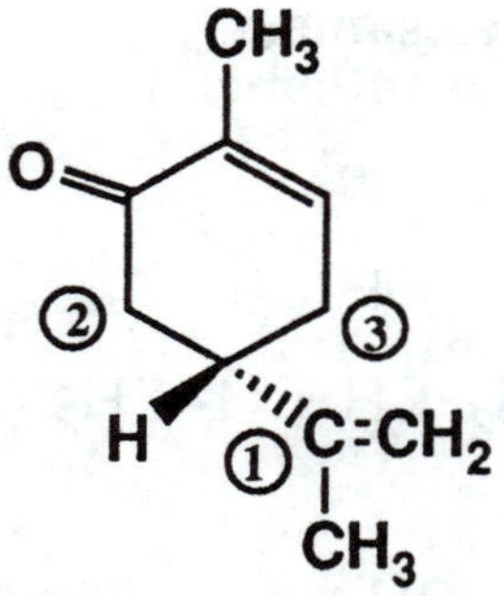

(S)-(+)-Carvone

<u>Problem 5.25</u>  The compound L-DOPA is one of the most widely prescribed drugs for the treatment of Parkinson's disease.  Its enantiomer, D-DOPA, has no therapeutic effect.  DOPA is an abbreviation of a chemical name of this compound, dihydroxyphenylalanine.  Is the active drug the R enantiomer or the S enantiomer?

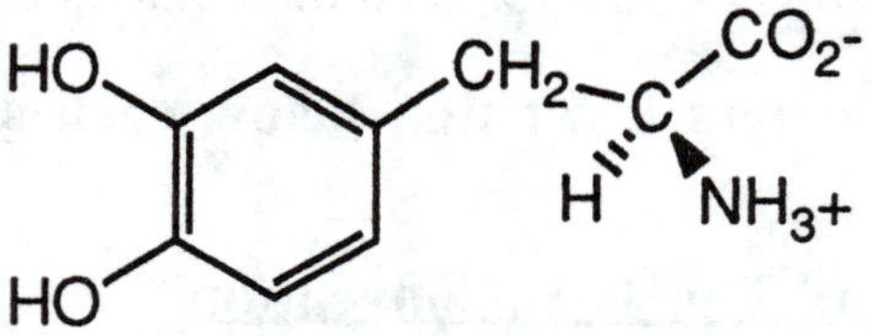

L-DOPA

**L-DOPA is the S enantiomer, as indicated by the rankings shown on the following structure .**

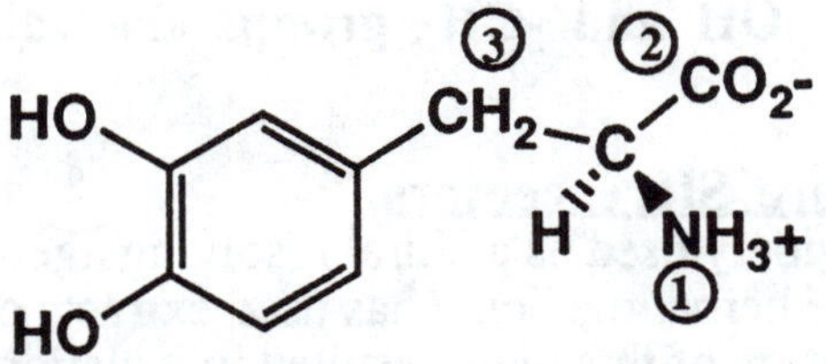

**Problem 5.26** Following is a staggered conformation for one of the enantiomers of 2-butanol.

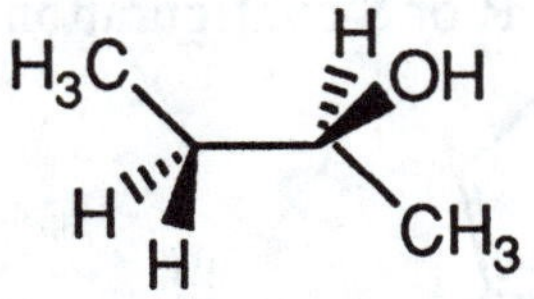

(a) Is this (R)-2-butanol or (S)-2-butanol?

**The structure drawn is (S)-2-butanol.**

(b) Draw a Newman projection for this enantiomer, viewed along the bond between carbons 2 and 3.

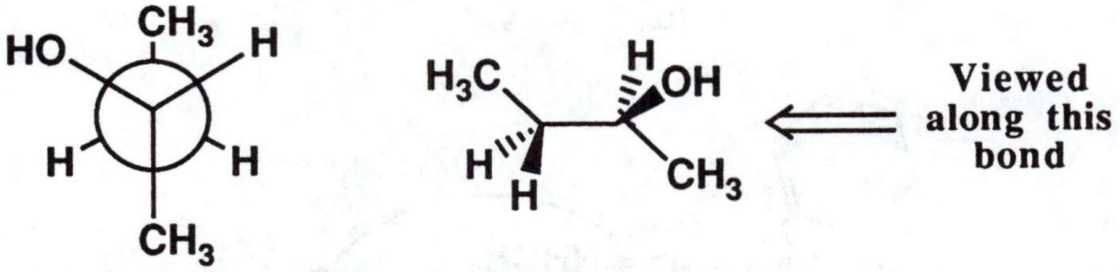

(c) Draw a Newman projection for one more staggered conformations of this molecule. Which of your conformations is the more stable? Assume that -OH and -CH₃ are comparable in size.

**There are actually two more staggered conformations of (S)-2-butanol.**

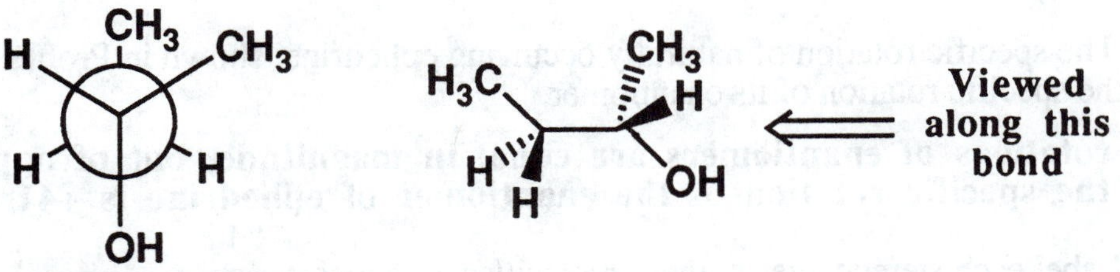

More stable

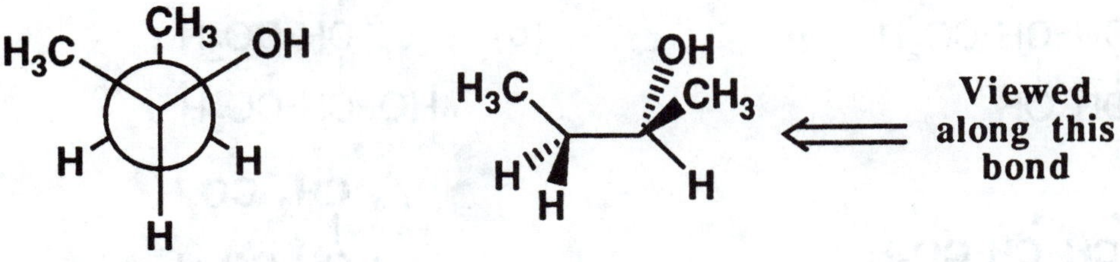

Less stable

**Assuming that -OH and -CH₃ are the same size, then the structure drawn in part (b) and the upper structure shown in part (c) are of equal stability, and these are more stable than the lower structure shown in part (c). This lower structure is**

**less stable because both the -OH and -CH₃ groups are adjacent to the -CH₃ group.**

## Molecules With Two Or More Stereocenters

<u>Problem 5.27</u> An organic base widely used as a chiral resolving agent of organic acids is ephedrine. For centuries, Chinese herbal medicine has used extracts of *Ephedra sinica* to treat asthma. Phytochemical investigation of this plant resulted in isolation of ephedrine, a very potent dilator of the air passages of the lungs. The naturally occurring stereoisomer is levorotatory and has the following configuration. Assign R or S configuration to each stereocenter.

Ephedrine

$$[\alpha]_D^{21} = -41°$$

(1R,2S)-(-)-Ephedrine

<u>Problem 5.28</u> The specific rotation of naturally occurring ephedrine, shown in Problem 5.27 is -41°. What is the specific rotation of its enantiomer.

**The specific rotations of enantiomers are equal in magnitude, but of opposite sign. Thus, the specific rotation of the enantiomer of ephedrine is +41°.**

<u>Problem 5.29</u> Label each stereocenter in these molecules with an asterisk.

(a)

$$CH_3-CH-CH\cdot CO_2H$$
$$\quad\quad | \quad\ |$$
$$\quad\ OH\ OH$$

(b)

$$CH_2-CO_2H$$
$$|$$
$$CH-CO_2H$$
$$|$$
$$HO-CH-CO_2H$$

<br>

$$CH_3-\overset{*}{CH}-\overset{*}{CH}\cdot CO_2H$$
$$\quad\quad | \quad\ |$$
$$\quad\ OH\ OH$$

**4 stereoisomers
(two pairs of enantiomers)**

$$CH_2-CO_2H$$
$$|$$
$$*\ \overset{}{CH}\cdot CO_2H$$
$$|$$
$$HO\overset{*}{-}CH-CO_2H$$

**4 stereoisomers
(two pairs of enantiomers)**

(c)

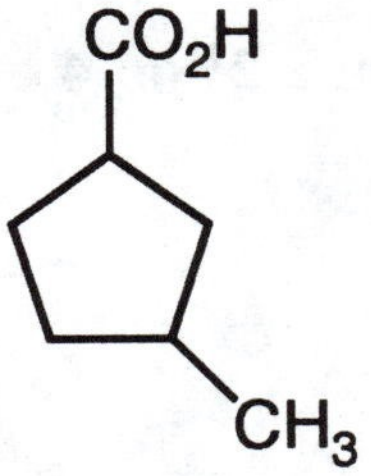

CO₂H

CH₃

(d)

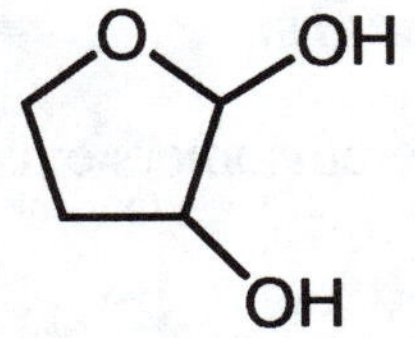

O

OH

OH

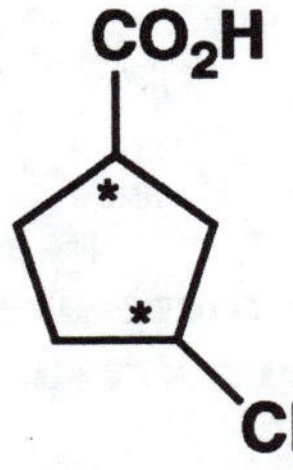

CO₂H

*

*

CH₃

**4 stereoisomers**
**(two pairs of enantiomers)**

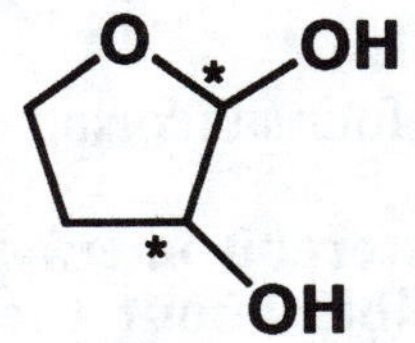

O

*

OH

*

OH

**4 stereoisomers**
**(two pairs of enantiomers)**

(e)

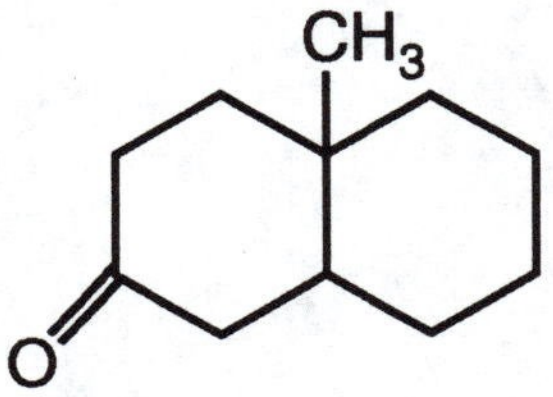

CH₃

O

(f)

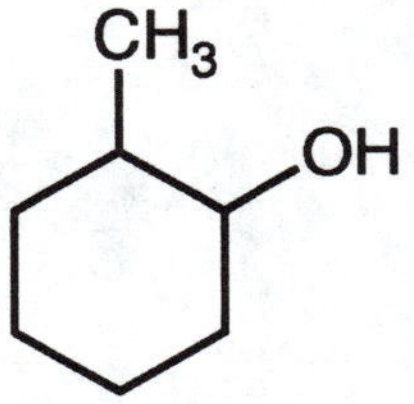

CH₃

OH

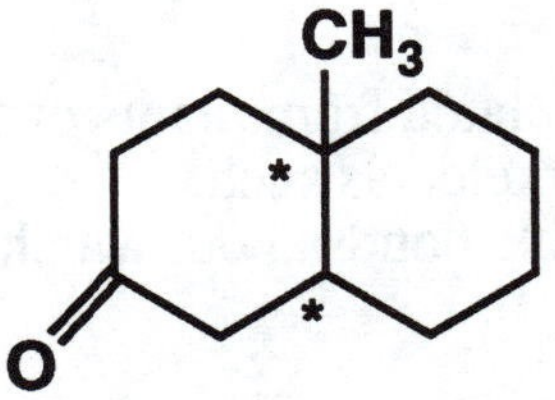

CH₃

*

*

O

**4 stereoisomers**
**(2 pairs of enantiomers)**

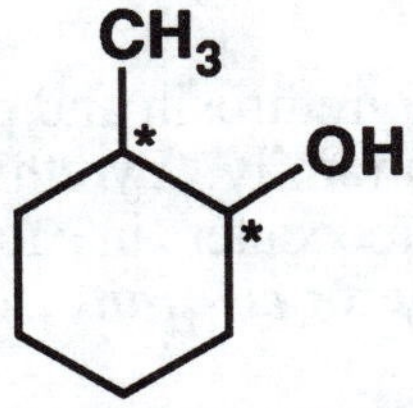

CH₃

*

OH

*

**4 stereoisomers**
**(2 pairs of enantiomers)**

<u>Problem 5.30</u> How many stereoisomers are possible for each compound in Problem 5.29?

**The total number of stereoisomers are listed below the structures that are drawn in the answer to Problem 5.29.**

<u>Problem 5.31</u> How many stereoisomers are possible for each compound, which is the aggregating pheromone for the Norway spruce beetle?

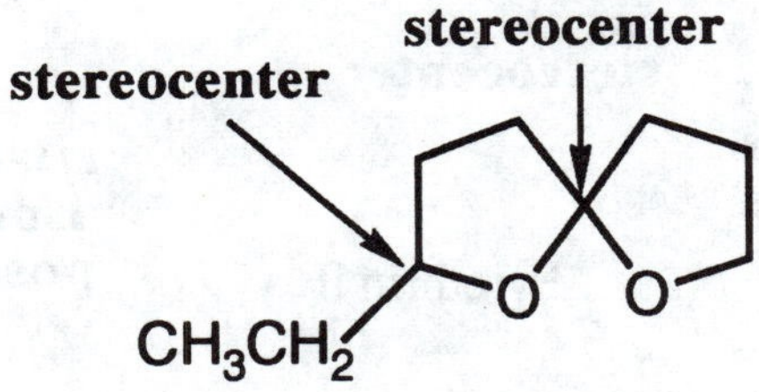

**stereocenter**

**stereocenter**

CH₃CH₂

O    O

**There are two stereocenters in the molecule.  As a result, there are $2^2$ or 4 possible  stereoisomers.**

<u>Problem 5.32</u>  Four stereoisomers exist for 3-penten-2-ol.

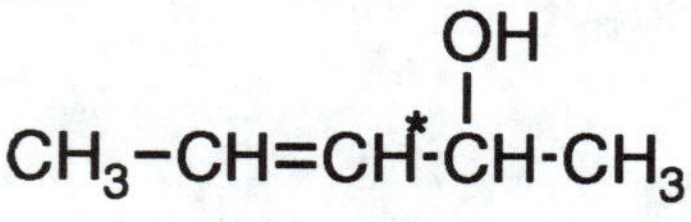

$$CH_3-CH=CH\overset{*}{-}CH-CH_3$$

with OH on the CH

3-Penten-2-ol

(a) Explain how these four stereoisomers arise.

**There are E or Z stereoisomers possible about the double bond and there are two configurations possible about the one stereocenter (carbon 2) so there are a total of 2 x 2 = 4 stereoisomers.**

(b) Draw the stereoisomer having the E configuration about the carbon-carbon double bond and the R configuration at the stereocenter.

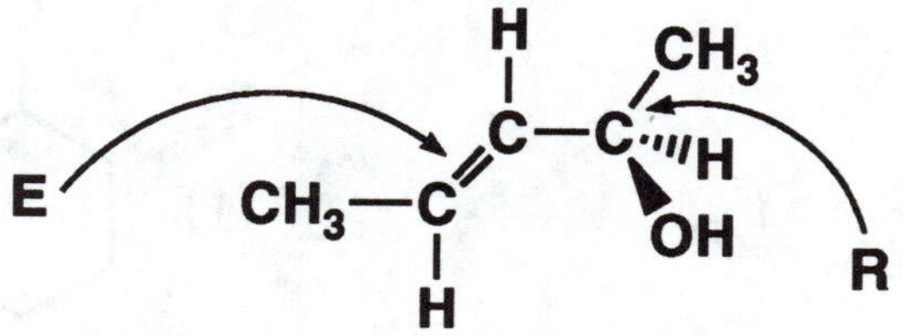

<u>Problem 5.33</u>  Pyrethrosin and pyrethrin II are two natural products isolated from plants of the chrysanthemum family.  Pyrethrin II is a natural insecticide and is marketed as such.
(a) Label all stereocenters in each molecule and all carbon-carbon double bonds about which there is the possibility for *cis-trans* isomerism.

**The stereocenters and carbon-carbon double bonds with possible *cis-trans* isomers are indicated on the structures.**

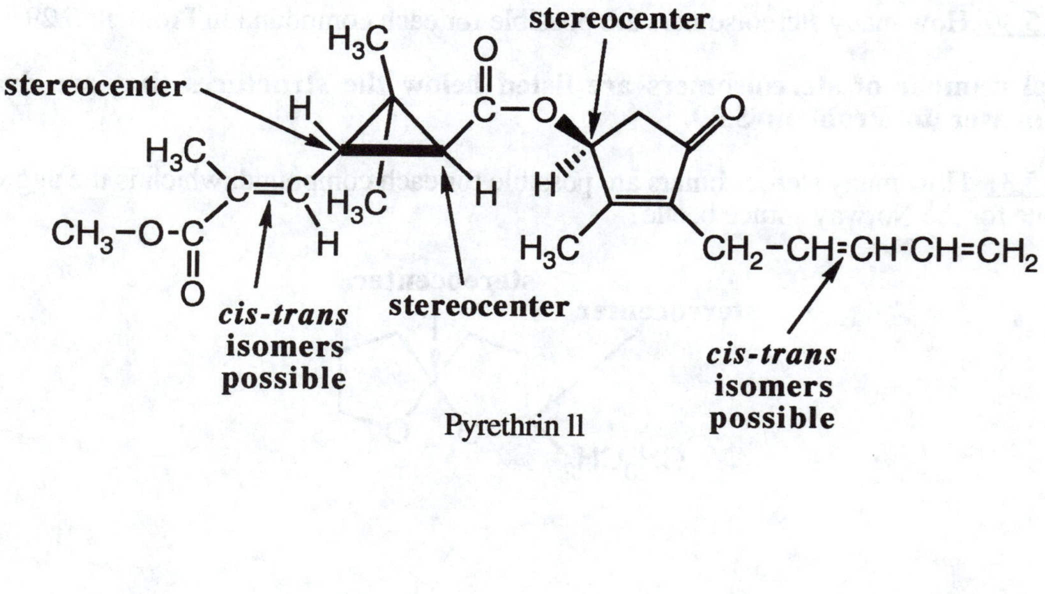

Pyrethrin ll

Pyrethrosin

(b) State the number of stereoisomers possible for each molecule.

**For pyrethrin II there are 2 alkenes with possible *cis-trans* isomers and three stereocenters. Thus, there $2^2 \times 2^3 = 32$ possible stereoisomers of this molecule.**

**For pyrethrosin, there are five stereocenters and one alkene with possible *cis-trans* isomers. Recall that a ten-member ring is large enough to allow the *trans* alkene configuration. There are a total of $2^5 \times 2 = 64$ possible stereoisomers of this molecule.**

(c) Show that the cyclic part of pyrethrosin is a terpene composed of three isoprene units.

**The isoprene units are highlighted on the structure.**

**Problem 5.34** Label the eight stereocenters in cholesterol. How many stereoisomers are possible for a molecule of this structural formula?

**Each stereocenter is marked with an asterisk.**

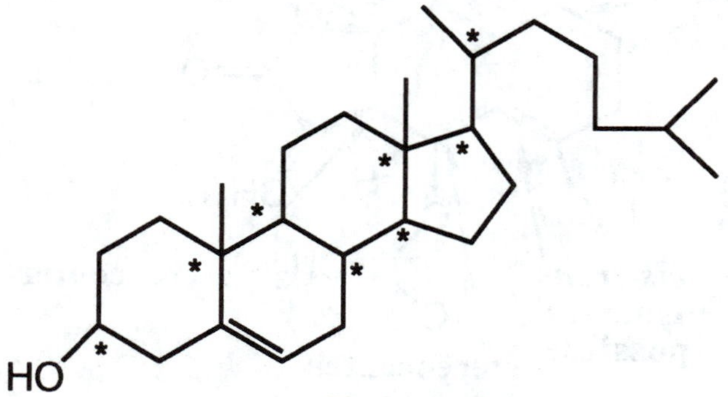

Cholesterol

**There are a total of $2^8 = 256$ stereoisomers possible for a molecule of this structural formula.**

**Problem 5.35** Label the four stereocenters in amoxicillin, which belongs to the family of semisynthetic penicillins.

**Each stereocenter is marked with an asterisk.**

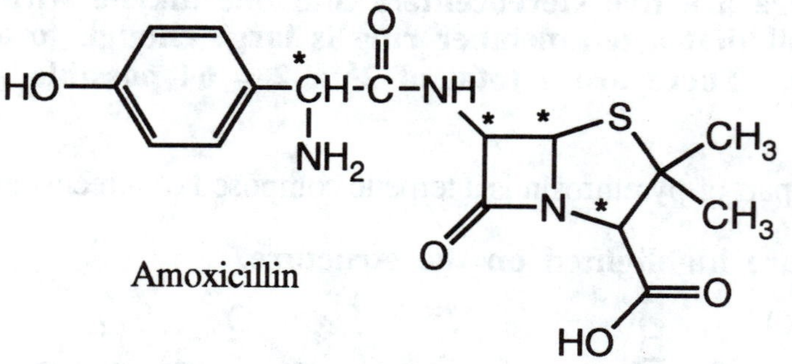

Amoxicillin

**Problem 5.36** Label all stereocenters in the antihistamine terfenadine (Seldane). Terfenadine received FDA approval in May 1985 and by year's end had become the top-selling antihistamine in the United States. It provides relief from allergic disorders, but , unlike so many of the earlier antihistamines, does not cause drowsiness.

**Each stereocenter is marked with an asterisk.**

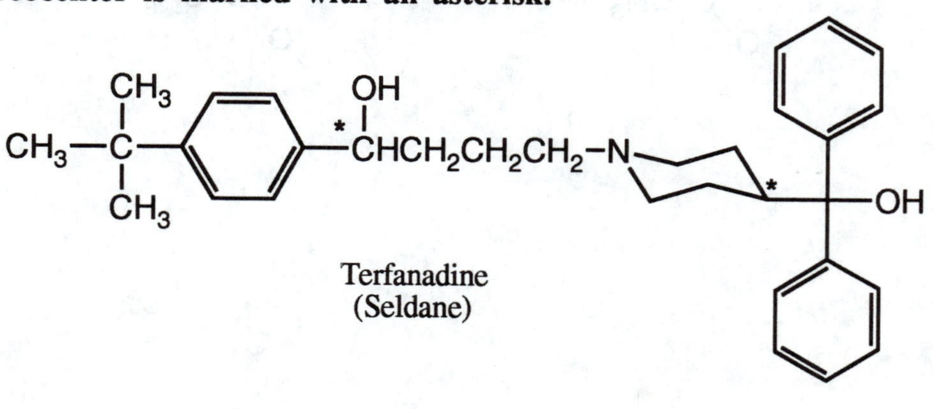

Terfanadine
(Seldane)

Problem 5.37  How many stereoisomers are possible for 2,3-butanediol?

**There are three stereoisomers possible, two enantiomers and a meso compound.**

**(S,S)-2,3-Butanediol          (R,R)-2,3-Butanediol          (R,S)-2,3-Butanediol**
**a pair of enantiomers                                    a meso compound**

Problem 5.38  How many stereoisomers are possible for 3-chloro-2-butanol?

**There are four stereoisomers possible, two pairs of enantiomers.**

**(2S,3S)                (2R,3R)                (2S,3R)                (2R,3S)**
**a pair of enantiomers                        a pair of enantiomers**

Problem 5.39  Which of these are meso compounds?

**The meso compounds are (a), (c), (d), and (f).**

<u>Problem 5.40</u>  Convert the stereorepresentations in parts (f), (g), and (h) of Problem 5.39 to Fischer projections.

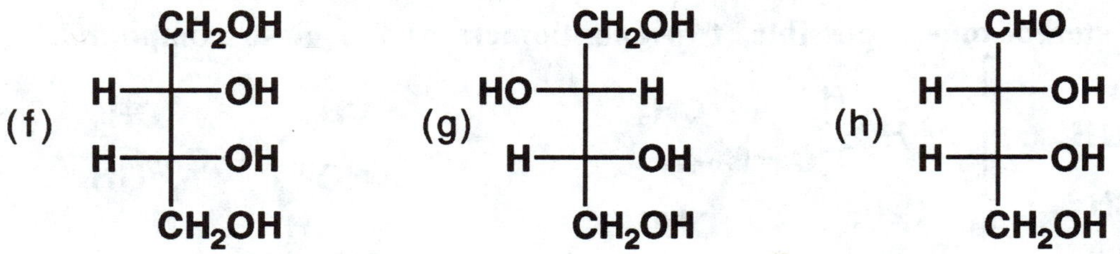

(f)          (g)          (h)

**If your Fischer projections look different than those drawn above, remember that rotation of a Fisher projection 180° in the plane of the paper produces an equivalent structure.**

<u>Problem 5.41</u>  Convert the stereorepresentations in parts (a) and (b) of Problem 5.39 to Fischer projections.  In doing so, orient the carbon chain vertically with the groups attached to it pointing to the left and right.

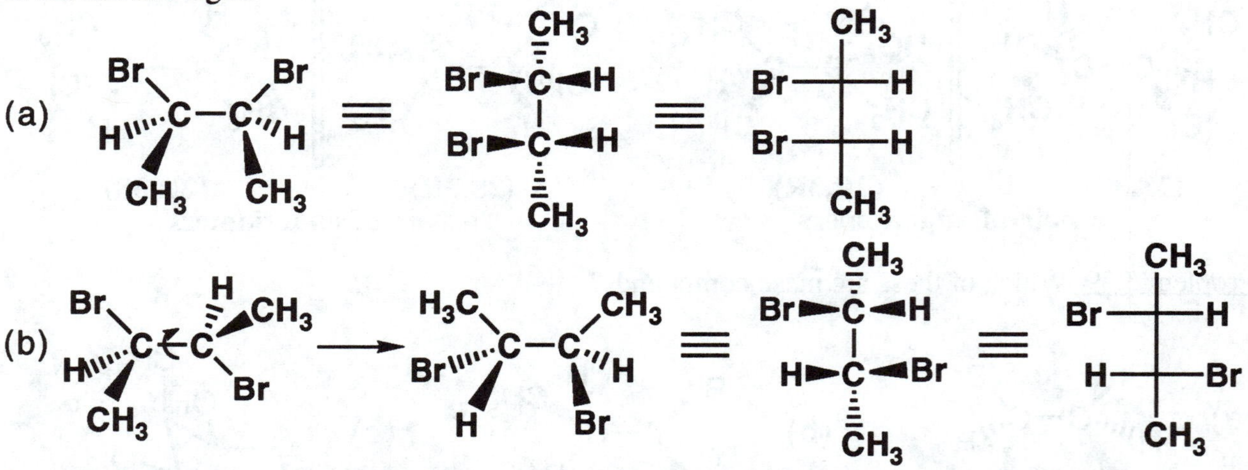

(a)

(b)

**It may help to make molecular models of the molecules.  If your Fischer projections look different than those drawn above, remember that rotation of a Fisher projection 180° in the plane of the paper produces an equivalent structure.**

<u>Problem 5.42</u>  Draw a Newman projection, viewed along the bond between carbons 2 and 3, for both the most stable conformation and the least stable conformation of meso-tartaric acid.

**The carboxylic acid groups are the largest groups.  In the most stable conformer, the carboxylic acid groups are anti to each other.  In the least stable conformer, they are eclipsed.**

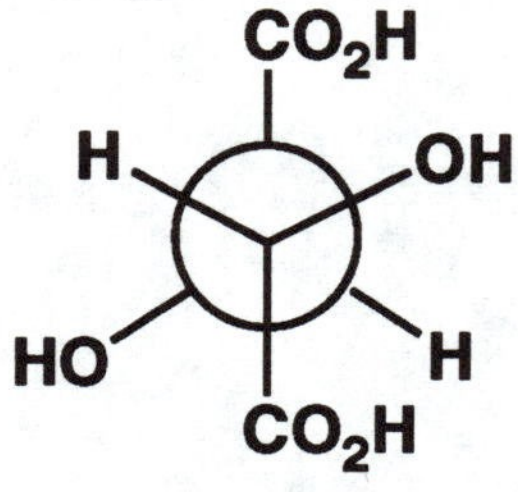

Most stable
conformer

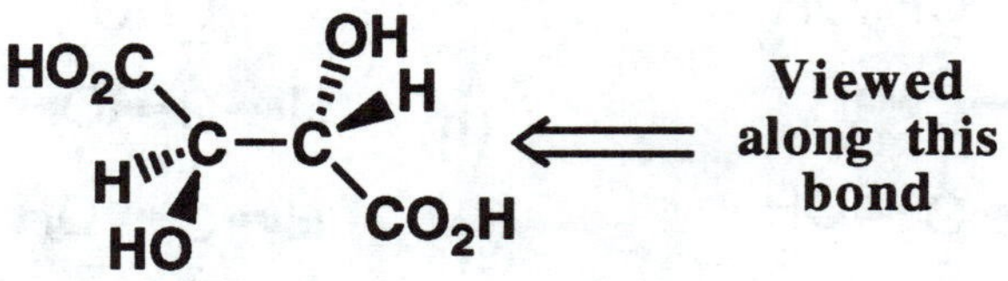

meso-Tartaric acid

Viewed
along this
bond

Least stable
conformer

meso-Tartaric acid

Viewed
along this
bond

**Problem 5.43** Draw Fischer projections for your answers to Problem 5.42.

$$CO_2H$$

HO —— H

HO —— H

$$CO_2H$$

**meso-Tartaric acid**

**Problem 5.44** How many stereoisomers are possible for 1,3-cyclopentanediol? Which are pairs of enantiomers? Which are meso compounds?

**There are three stereoisomers of 1,3-cyclopentanediol, one pair of *trans* enantiomers and a *cis* meso compound.**

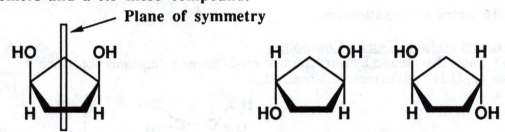

Plane of symmetry

***cis*-1,3-Cyclopentanediol**
**(achiral, a meso compound)**

***trans*-1,3-Cyclopentanediol**
**(a pair of enantiomers)**

**Problem 5.45** How many stereoisomers are possible for 2-chlorocyclopentanol? Which are pairs of enantiomers? Which are meso compounds?

**There are four stereoisomers of 3-methylpentanol, two pairs of enantiomers.**

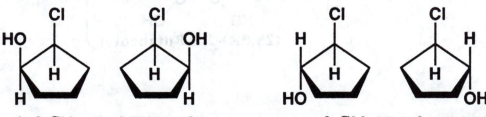

***cis*-2-Chlorocyclopentanol**
**(a pair of enantiomers)**

***trans*-2-Chlorocyclopentanol**
**(a pair of enantiomers)**

Problem 5.46 In answer to Problem 2.39 you were asked to draw the more stable chair conformation of glucose, a molecule in which all groups on the six-member ring are equatorial. Here is a drawing of that conformation.

(a) What is the configuration (R or S) at carbon 1 and carbon 5?

**The configuration at carbon 1 is R, and the configuration at carbon 5 is also R.**

(b) Identify all stereocenters in this molecule.

(c) How many stereoisomers are possible?

**There are $2^5 = 32$ stereoisomers.**

(d) How many pairs of enantiomers are possible?

**There are 16 pairs of enantiomers.**

## Reactions that Produce Chiral Compounds
Problem 5.47 Show that the syn hydroxylation of cis-2-butene by aqueous potassium permanganate at pH 11.8 gives meso-2,3-butanediol.

Problem 5.48  Show that acid-catalyzed hydration of 1-butene gives a racemic mixture of (R)-2-butanol and (S)-2-butanol and that the product is, therefore, optically inactive.

**The reaction mechanism involves an initial protonation to form a carbocation intermediate.**

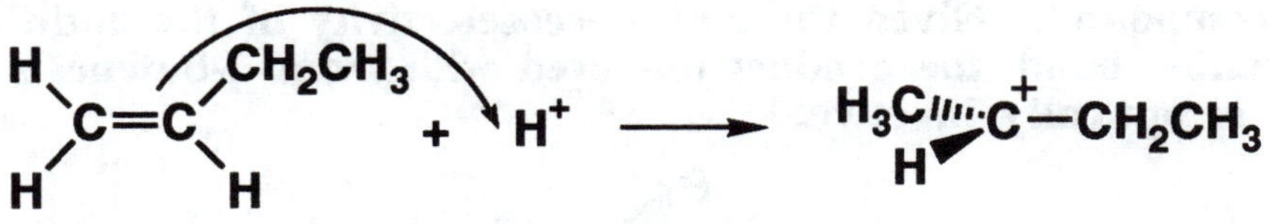

**Water could approach the carbocation from either side with equal probability, so a racemic mixture will be formed.  Thus, the product will be optically inactive.**

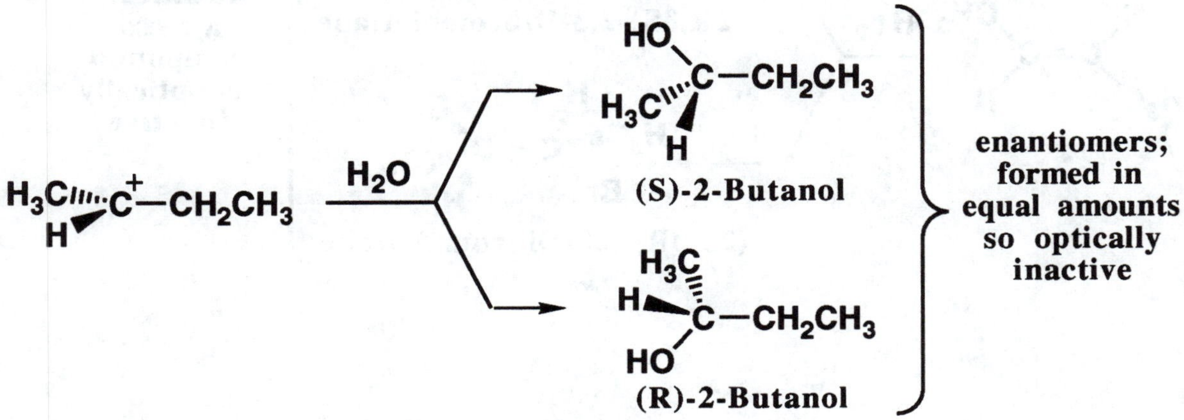

(S)-2-Butanol

(R)-2-Butanol

enantiomers;
formed in
equal amounts
so optically
inactive

Problem 5.49  Addition of bromine to *cis*-2-butene gives 2,3-dibromobutane.  What is the stereochemistry of the product; Is it a meso compound or is it a pair of enantiomers?  Is it optically active?

**Three stereoisomers are possible for 2,3-dibromobutane; one pair of enantiomers and one meso compound.  Given the anti stereoselectivity of the addition of bromine to a double bond, the product observed with *cis*-2-butene is the pair of enantiomers.  These will be formed in identical amounts so the product is optically inactive.**

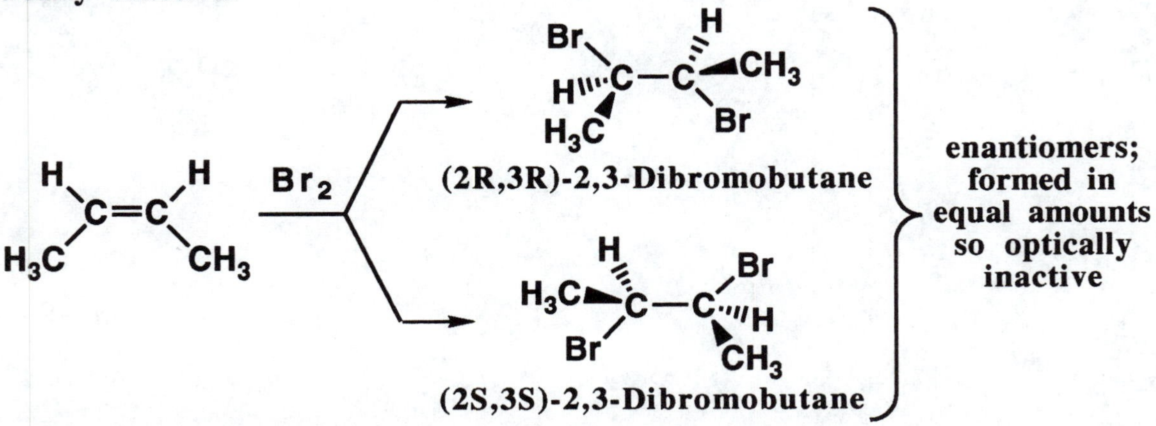

(2R,3R)-2,3-Dibromobutane

(2S,3S)-2,3-Dibromobutane

enantiomers;
formed in
equal amounts
so optically
inactive

<u>Problem 5.50</u> Addition of bromine to *trans*-2-butene also gives 2,3-dibromobutane. What is the stereochemistry of the product; Is it a meso compound or is it a pair of enantiomers? Is it optically active?

**Three stereoisomers are possible for 2,3-dibromobutane; one pair of enantiomers and one meso compound. Given the anti stereoselectivity of the addition of bromine to a double bond, the product observed with *trans*-2-butene is the meso compound and is optically inactive.**

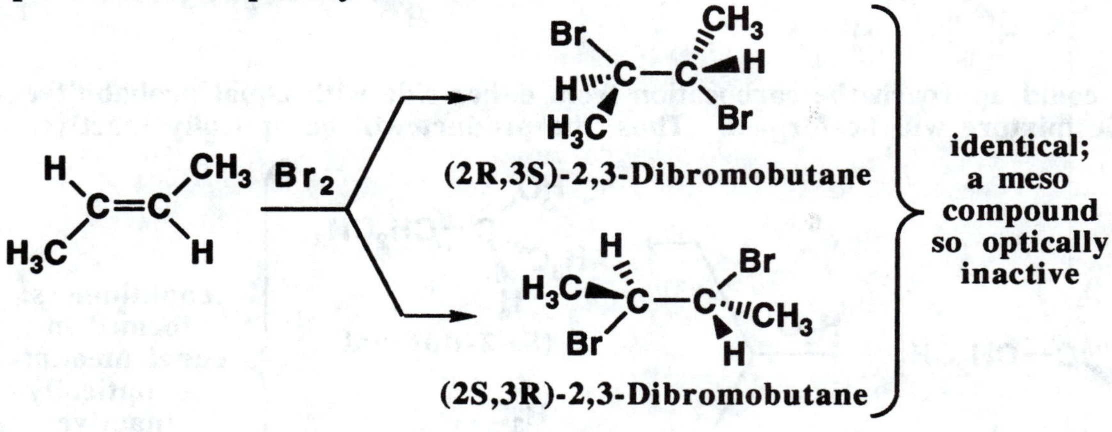

**(2R,3S)-2,3-Dibromobutane**

**(2S,3R)-2,3-Dibromobutane**

identical;
a meso
compound
so optically
inactive

# CHAPTER 6: ACIDS AND BASES

## 6.0 OVERVIEW
- A large number of organic reactions involve acid-base reactions.  These include acid-base reactions of important functional groups such as alcohols and carbonyl compounds.  In addition, many reactions involve catalysis by protons or Lewis acids, such as $AlCl_3$.

## 6.1 BRØNSTED-LOWRY ACIDS AND BASES
- According to **Johannes Brønsted** and **Thomas Lowry**, an **acid** is a **proton donor**, and a **base** is a **proton acceptor**. *
  - The reaction of an acid with a base involves a **proton transfer** from the acid to the base.  When an acid transfers a proton to a base, the acid is converted to its **conjugate base**.  When a base accepts a proton, the base is converted to its **conjugate acid**.

$$H\text{-}A \quad + \quad B^- \quad \Longleftrightarrow \quad A^- \quad + \quad H\text{-}B$$

| Acid | Base | Conjugate Base | Conjugate Acid |

## 6.2 QUANTITATIVE MEASURE OF ACID AND BASE STRENGTH
- A **strong acid** is one that is completely ionized in aqueous solution, in other words there is complete transfer of the proton from the acid to water to form $H_3O^+$.  A **weak acid** is one that is incompletely ionized in aqueous solution.  Most organic acids, such as carboxylic acids, are weak acids. *
- A **strong base** is one that is completely ionized in aqueous solution, in other words there is complete proton transfer from $H_2O$ to the base to form $HO^-$.  A **weak base** is one that is incompletely ionized in aqueous solution.  Most organic bases are weak bases. *
- The strength of a weak acid is described by the acid ionization constant $K_a$ that is defined by the following equation. *

$$K_a = \frac{[H^+][A^-]}{[HA]}$$

Where $A^-$ is the conjugate base of the acid HA.
  - The ionization constants for weak acids have negative exponents, so it is convenient to refer to the $pK_a$ where $pK_a = -\log_{10}K_a$.  In other words, a weak acid with a $pK_a$ of 5.0 has a $K_a$ of $1.0 \times 10^{-5}$.
  - Because $pK_a$ is defined as the *negative* log of the $K_a$, the larger the $pK_a$, the weaker the acid.  Similarly, the smaller the $pK_a$, the stronger the acid.  Note, the strongest acids actually have negative $pK_a$ values.

## 6.3 THE POSITION OF EQUILIBRIUM IN ACID-BASE REACTIONS
- The conjugate base of a strong acid is a weak base, and the conjugate base of a weak acid is a strong base. * *[This relationship will make the following rule easier to use]*
- For any proton transfer reaction, the **position of equilibrium favors** the side of the reaction equation that has the **weaker acid and weaker base**. * *[With this extremely helpful rule, you  can predict the outcome of virtually any proton transfer reaction as long as you know the relevant pKₐ's.]*

$$\text{H-A} \quad + \quad \text{B}^- \quad \longrightarrow \quad \text{A}^- \quad + \quad \text{H-B}$$

| Stronger Acid | Stronger Base | Weaker Base | Weaker Acid |

$$\text{Equilibrium Favors}$$
$$\text{This Side}$$

## 6.4 LEWIS ACIDS AND BASES

• According to the definition first proposed by G.N. Lewis, a **Lewis Acid** is a species that forms a new covalent bond by accepting a pair of electrons. A **Lewis base** is a species that forms a new covalent bond by donating a pair of electrons. ✻ *[This concept is important, because it describes much more than just proton transfer reactions.]*

  - The reaction of a Lewis acid with a Lewis base can be described by the following equation:

$$\text{A} + \text{:B} \longrightarrow \text{A-B}$$

  - Here the "A" represents a Lewis acid and ":B" represents a Lewis base.

• The Lewis definitions of acids and bases are more general than the Brønsted-Lowry definitions. All Brønsted-Lowry acids (proton donors) are also Lewis acids and all Brønsted-Lowry bases (proton acceptors) are also Lewis bases. The Lewis definitions cover reactions other than proton transfer reactions such as water reacting with a carbocation.

# CHAPTER 6
## *Solutions to the Problems*

Problem 6.1 Write these reactions as proton-ransfer reactions. Label which reactant is the acid and which the base; which product is the conjugate base and which the conjugate acid. Use curved arrows to show the flow of electrons in each reaction.

(a)  $CH_3-S-H$  +  $OH^-$  ⟶  $CH_3-S^-$  +  $H_2O$

$$CH_3-\ddot{S}-H \quad + \ddot{:}\ddot{O}-H \quad ⟶ \quad CH_3-\ddot{S}\ddot{:}^- \quad + \quad H-\ddot{O}-H$$

**acid**          **base**                    **conjugate**          **conjugate**
                                              **base**                **acid**

(b)  $CH_3-O-H$  +  $NH_2^-$  ⟶  $CH_3-O^-$  +  $NH_3$

$$CH_3-\ddot{O}-H \quad + \ddot{:}\overset{-}{N}-H \quad ⟶ \quad CH_3-\ddot{O}\ddot{:}^- \quad + \quad H-\overset{H}{\underset{H}{N}}-H$$

**acid**          **base**                    **conjugate**          **conjugate**
                                              **base**                **acid**

Problem 6.2 For each value of $K_a$, calculate the corresponding value of $pK_a$. Which compound is the stronger acid?
(a) Acetic acid, $K_a = 1.74 \times 10^{-5}$          (b) Water, $K_a = 2.00 \times 10^{-16}$

**The $pK_a$ is equal to $-log_{10}K_a$. The $pK_a$ of acetic acid is 4.76 and the $pK_a$ of water is 15.7. Acetic acid, with the smaller $pK_a$ value, is the stronger acid.**

Problem 6.3 Predict the position of equilibrium for these acid-base reactions.

(a)  $CH_3NH_2$  +  $CH_3CO_2H$  ⇌  $CH_3NH_3^+$  +  $CH_3CO_2^-$
     Methylamine    Acetic acid              Methylammonium    Acetate
                                                  ion            ion

**Acetic acid is the stronger acid; equilibrium lies to the right**

$$CH_3NH_2 + CH_3CO_2H \rightleftharpoons CH_3NH_3^+ + CH_3CO_2^-$$

              $pK_a$ 4.76              $pK_a$ 9.64

      **(stronger        (stronger                 (weaker        (weaker
        base)             acid)                      acid)          base)**

(b) $CH_3CH_2O^- + NH_3 \rightleftharpoons CH_3CH_2OH + NH_2^-$

Ethoxide     Ammonia     Ethanol     Amide
ion                                   ion

**Ethanol is the stronger acid; equilibrium lies to the left.**

$$CH_3CH_2O^- + NH_3 \rightleftharpoons CH_3CH_2OH + NH_2^-$$

$$pK_a\ 33 \qquad\qquad pK_a\ 15.9$$

(weaker     (weaker      (stronger    (stronger
base)       acid)        acid)       base)

<u>Problem 6.4</u> Write an equation for the reaction between each Lewis acid/base pair, showing electron flow by means of curved arrows. <u>Hint:</u> Aluminum is in Group IIIA of the periodic table, just under boron. Aluminum in $AlCl_3$ has only six electrons in its valence shell and, like boron in $BF_3$, has an incomplete octet.

(a) $Cl^- + AlCl_3 \longrightarrow$

Lewis      Lewis
base       acid

(b) $CH_3Cl + AlCl_3 \longrightarrow$

Lewis      Lewis
base       acid

## Brønsted-Lowry Acids and Bases

<u>Problem 6.5</u> Complete an equation for these proton-transfer reactions using curved arrows to show the flow of electron pairs in each. In addition, write Lewis structures for all starting materials and products. If you are uncertain about which substance in each equation is the proton donor, refer to Table 6.1 for the relative strengths of proton acids.

(a) $CH_3CO_2H + OH^- \longrightarrow$

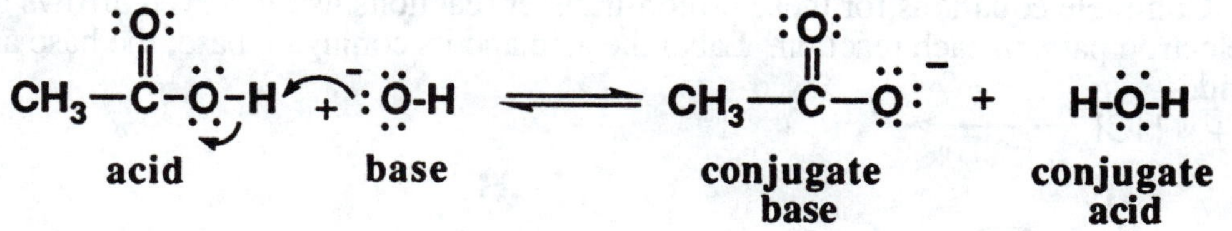

$$CH_3-C-O-H + :\overset{..}{O}-H \rightleftharpoons CH_3-C-\overset{..}{O}:^- + H-\overset{..}{O}-H$$

acid　　　　　　base　　　　　　　conjugate　　　conjugate
　　　　　　　　　　　　　　　　　　base　　　　　acid

(b) $CH_3CO_2H + H_2O \longrightarrow$

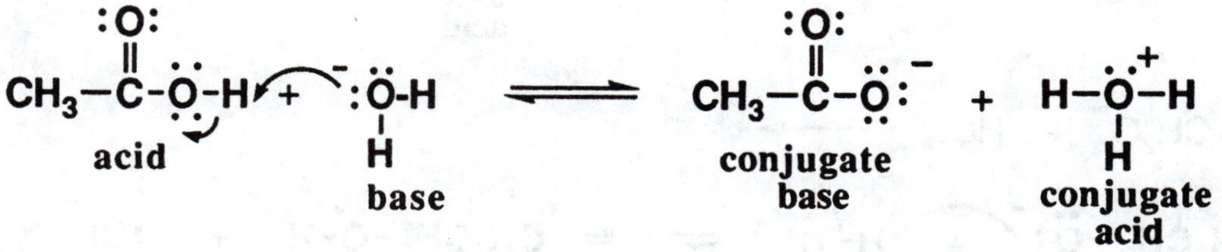

$$CH_3-C-O-H + ^-:\overset{..}{O}-H \rightleftharpoons CH_3-C-\overset{..}{O}:^- + H-\overset{+}{\underset{H}{O}}-H$$
　　　　　　　　　　　　　H

acid　　　　　　　　　conjugate　　　conjugate
　　　　base　　　　　base　　　　　acid

(c) $CH_3CO_2H + HCO_3^- \longrightarrow$

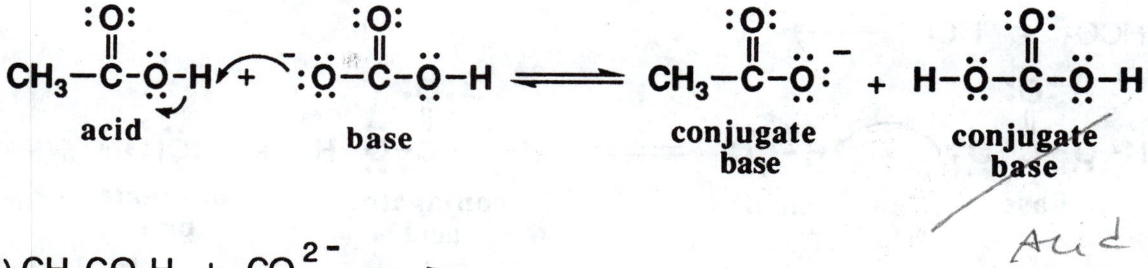

$$CH_3-C-\overset{..}{O}-H + ^-:\overset{..}{O}-C-\overset{..}{O}-H \rightleftharpoons CH_3-C-\overset{..}{O}:^- + H-\overset{..}{O}-C-\overset{..}{O}-H$$

acid　　　　　　base　　　　　　conjugate　　　　conjugate
　　　　　　　　　　　　　　　　base　　　　　　base

Acid

(d) $CH_3CO_2H + CO_3^{2-} \longrightarrow$

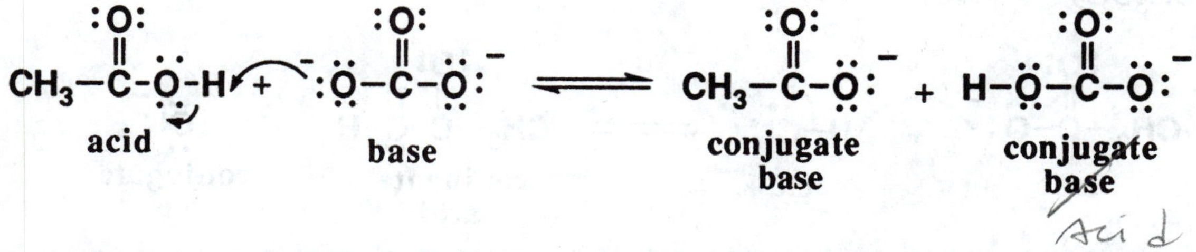

$$CH_3-C-O-H + ^-:\overset{..}{O}-C-\overset{..}{O}:^- \rightleftharpoons CH_3-C-\overset{..}{O}:^- + H-\overset{..}{O}-C-\overset{..}{O}:^-$$

acid　　　　　　base　　　　　　conjugate　　　　conjugate
　　　　　　　　　　　　　　　　base　　　　　　base

Acid

Problem 6.6  For each of the proton-transfer reactions in Problem 6.5, label which starting material is the acid and which product is its conjugate base; which starting material is the base and which product is its conjugate acid.

See the answers to Problem 6.5 for the labels.

Problem 6.7 Complete equations for these proton-transfer reactions using curved arrows to show the flow of electron pairs in each reaction. Label the acid and its conjugate base; the base and its conjugate acid.

(a) $NH_3$ + HCl ⟶

$$H-\overset{\overset{\displaystyle H}{|}}{\underset{\underset{\displaystyle H}{|}}{N}}: \quad + \quad H-\ddot{\underset{\cdot\cdot}{C}}l: \quad \rightleftharpoons \quad H-\overset{\overset{\displaystyle H}{|}}{\underset{\underset{\displaystyle H}{|}}{\overset{+}{N}}}-H \quad + \quad :\ddot{\underset{\cdot\cdot}{C}}l:^{-}$$

base          acid                    conjugate                conjugate
                                        acid                      base

(b) $CH_3CH_2O^{-}$ + HCl ⟶

$$CH_3CH_2-\ddot{\underset{\cdot\cdot}{O}}:^{-} \quad + \quad H-\ddot{\underset{\cdot\cdot}{C}}l: \quad \rightleftharpoons \quad CH_3CH_2-\ddot{\underset{\cdot\cdot}{O}}-H \quad + \quad :\ddot{\underset{\cdot\cdot}{C}}l:^{-}$$

base                    acid                          conjugate              conjugate
                                                        acid                    base

(c) $HCO_3^{-}$ + HCl ⟶

$$H-\ddot{\underset{\cdot\cdot}{O}}-\overset{\overset{\displaystyle :O:}{\|}}{C}-\ddot{\underset{\cdot\cdot}{O}}:^{-} \quad + \quad H-\ddot{\underset{\cdot\cdot}{C}}l: \quad \rightleftharpoons \quad H-\ddot{\underset{\cdot\cdot}{O}}-\overset{\overset{\displaystyle :O:}{\|}}{C}-\ddot{\underset{\cdot\cdot}{O}}-H \quad + \quad :\ddot{\underset{\cdot\cdot}{C}}l:^{-}$$

base                    acid                          conjugate              conjugate
                                                        acid                    base

(d) $CH_3CO_2^{-}$ + HCl ⟶

$$CH_3-\overset{\overset{\displaystyle :O:}{\|}}{C}-\ddot{\underset{\cdot\cdot}{O}}:^{-} \quad + \quad H-\ddot{\underset{\cdot\cdot}{C}}l: \quad \rightleftharpoons \quad CH_3-\overset{\overset{\displaystyle :O:}{\|}}{C}-\ddot{\underset{\cdot\cdot}{O}}-H \quad + \quad :\ddot{\underset{\cdot\cdot}{C}}l:^{-}$$

base                    acid                          conjugate              conjugate
                                                        acid                    base

Problem 6.8 Complete equations for these proton-transfer reactions using curved arrows to show the flow of electron pairs in each reaction. Label the acid and its conjugate base; the base and its conjugate acid.

(a) $NH_4^{+}$ + $OH^{-}$ ⟶

$$H-\ddot{\underset{\cdot\cdot}{O}}:^{-} \quad + \quad H-\overset{\overset{\displaystyle H}{|}}{\underset{\underset{\displaystyle H}{|}}{\overset{+}{N}}}-H \quad \rightleftharpoons \quad H-\ddot{\underset{\cdot\cdot}{O}}-H \quad + \quad :\overset{\overset{\displaystyle H}{|}}{\underset{\underset{\displaystyle H}{|}}{N}}-H$$

base                    acid                          conjugate              conjugate
                                                        acid                    base

(b) $CH_3CO_2^-$  +  $CH_3NH_3^+$  ⟶

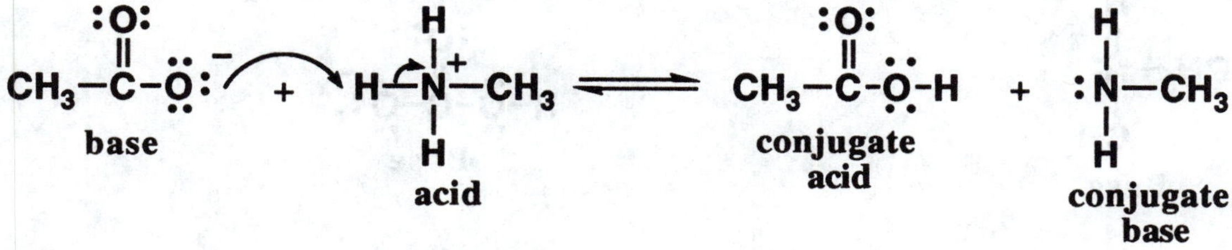

CH₃–C–O: ⁻  +  H–N⁺–CH₃  ⇌  CH₃–C–O–H  +  :N–CH₃

    base               acid             conjugate acid      conjugate base

(c) $CH_3CH_2O^-$ + $NH_4^+$ ⟶

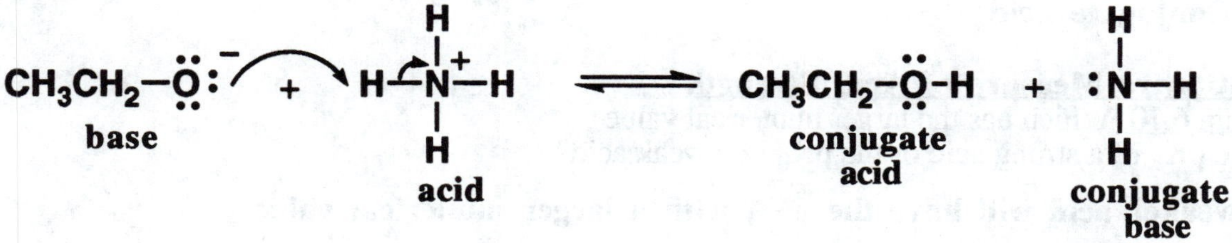

$CH_3CH_2–\ddot{O}:^-$  +  H–N⁺–H  ⇌  $CH_3CH_2–\ddot{O}–H$  +  :N–H

    base                acid             conjugate acid      conjugate base

(d) $CH_3NH_3^+$  +  $OH^-$ ⟶

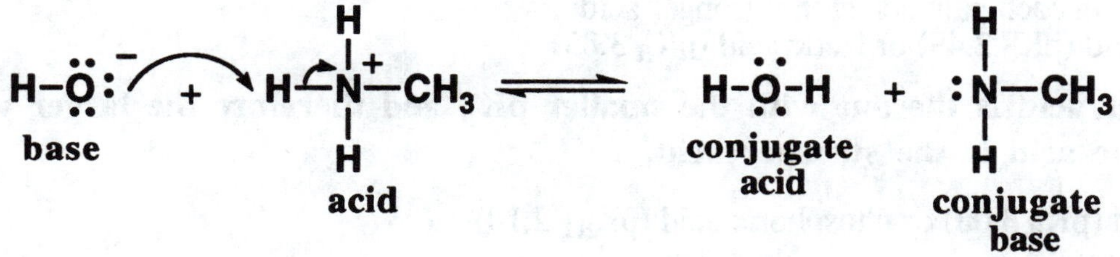

$H–\ddot{O}:^-$  +  H–N⁺–CH₃  ⇌  $H–\ddot{O}–H$  +  :N–CH₃

    base                acid             conjugate acid      conjugate base

**Problem 6.9** Each of these molecules and ions can function as a base. Complete the Lewis structure, showing all valence electrons of each base and write the structural formula of the conjugate acid formed by reaction of each with HCl.

(a)  $CH_3CH_2OH$

(b)   HCH (with O double bonded)

$CH_3–CH_2–\ddot{O}–H$

     **Base**

$CH_3–CH_2–\overset{+}{\underset{H}{\ddot{O}}}–H$

 **Conjugate acid**

H–C–H  (with :O: double bonded)

     **Base**

H–C–H  (with $:\overset{+}{O}$–H)

 **Conjugate acid**

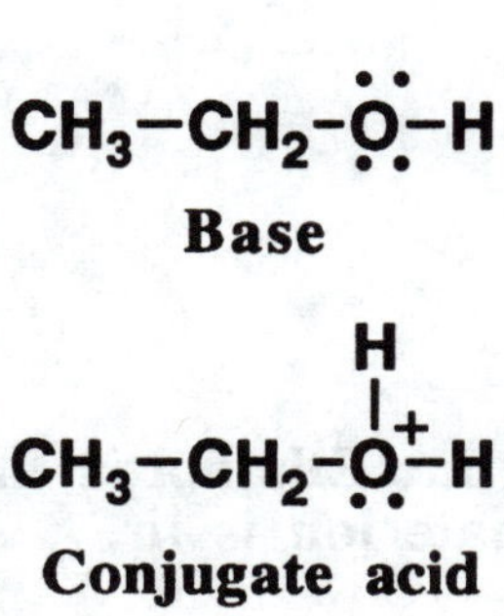

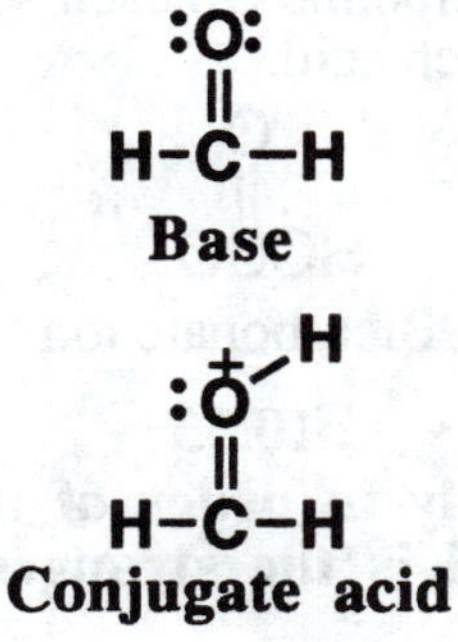

(c)   $(CH_3)_2NH$

H
|
$CH_3-N:$
|
$CH_3$
**Base**

H
|
$CH_3-\overset{+}{N}-H$
|
$CH_3$
**Conjugate acid**

(d)   $HCO_3^-$

:O:
‖
$H-\overset{..}{\underset{..}{O}}-C-\overset{..}{\underset{..}{O}}:^-$
**Base**

:O:
‖
$H-\overset{..}{\underset{..}{O}}-C-\overset{..}{\underset{..}{O}}-H$
**Conjugate acid**

## Quantitative Measure of Acid Strength

**Problem 6.10** Which has the larger numerical value
(a) The $pK_a$ of a strong acid or the $pK_a$ of a weak acid?

**The weaker acid will have the $pK_a$ with a larger numerical value.**

(b) The $K_a$ of a strong acid or the $K_a$ of a weak acid?

**The stronger acid will have the $K_a$ with a larger numerical value.**

**Problem 6.11** In each pair, select the stronger acid:
(a) Pyruvic acid ($pK_a$ 2.49) or lactic acid ($pK_a$ 3.85)

**The stronger acid is the one with the smaller $pK_a$ and therefore the larger value of $K_a$. Pyruvic acid is the stronger acid.**

(b) Citric acid ($pK_a$ 3.08) or phosphoric acid ($pK_{a1}$ 2.10)

**Phosphoric acid is the stronger acid.**

(c) Nicotinic acid (niacin, $K_a$ $1.4 \times 10^{-5}$) or acetylsalicylic acid (aspirin, $K_a$ $3.3 \times 10^{-4}$)

**Acetylsalicylic acid is the stronger acid.**

(d) Phenol ($K_a$ $1.12 \times 10^{-10}$) or acetic acid ($K_a$ $1.74 \times 10^{-5}$)

**Acetic acid is the stronger acid.**

**Problem 6.12** Arrange the compounds in each set in order of increasing acid strength. Consult Table 6.1 for $pK_a$ values of each acid.

(a)   $CH_3CH_2OH$        $HO\overset{O}{\overset{‖}{C}}O^-$        $C_6H_5\overset{O}{\overset{‖}{C}}OH$
      Ethanol            Bicarbonate ion       Benzoic acid

$pK_a$:    15.9                10.33                4.19

**The compounds are already in order of increasing acid strength. Ethanol is the weakest acid, benzoic acid is the strongest acid, and bicarbonate ion is in between.**

(b)    $\overset{\displaystyle O}{\overset{\|}{HOCOH}}$            $\overset{\displaystyle O}{\overset{\|}{CH_3COH}}$            HCl

   Carbonic acid            Acetic acid            Hydrogen chloride

**pKa:   6.36                    4.76                          -7**
**Again, the compounds are already in order of increasing acid strength.  Carbonic acid is the weakest acid, hydrogen chloride is the strongest acid, and acetic acid is in between.**

<u>Problem 6.13</u> Arrange the compounds in each set in order of increasing base strength.  Consult Table 6.1 for $pK_a$ values for the conjugate acid of each base.

**The weaker the conjugate acid (higher pKa), the stronger the base.**

(a)    $NH_3$        $\overset{\displaystyle O}{\overset{\|}{HOCO^-}}$        $CH_3CH_2O^-$

        9.24          6.34              15.9              $pK_a$ of conjugate acid
**Base strength increases in the order:**

$$\overset{\displaystyle O}{\overset{\|}{HOCO^-}} \quad < \quad NH_3 \quad < \quad CH_3CH_2O^-$$

(b)    $OH^-$        $\overset{\displaystyle O}{\overset{\|}{HOCO^-}}$        $\overset{\displaystyle O}{\overset{\|}{CH_3CO^-}}$

        15.7          6.34              4.76              $pK_a$ of conjugate acid
**Base strength increases in the order:**

$$\overset{\displaystyle O}{\overset{\|}{CH_3CO^-}} \quad < \quad \overset{\displaystyle O}{\overset{\|}{HOCO^-}} \quad < \quad OH^-$$

(c)    $H_2O$        $NH_3$        $\overset{\displaystyle O}{\overset{\|}{CH_3CO^-}}$

        -1.74          9.24              4.76              $pK_a$ of conjugate acid
**Base strength increases in the order:**

$$H_2O \quad < \quad \overset{\displaystyle O}{\overset{\|}{CH_3CO^-}} \quad < \quad NH_3$$

(d)   $NH_2^-$        $CH_3\overset{\displaystyle O}{\overset{\|}{C}}O^-$        $OH^-$

      33              4.76            15.7                    $pK_a$ of conjugate acid

**Base strength increases in the order:**

$$CH_3\overset{\displaystyle O}{\overset{\|}{C}}O^- \; < \; OH^- \; < \; NH_2^-$$

## The Position of Equilibrium in Acid-Base Reactions

<u>Problem 6.14</u> For an acid-base reaction, one way to indicate the predominant species at equilibrium is to say that the reaction arrow points to the acid with the higher value of $pK_a$. For example:

$$NH_4^+ \; + \; H_2O \; \longleftarrow \; NH_3 \; + \; H_3O^+$$
$$pK_a \; 9.24 \hspace{4cm} pK_a \; -1.74$$

$$NH_4^+ \; + \; OH^- \; \longrightarrow \; NH_3 \; + \; H_2O$$
$$pK_a \; 9.24 \hspace{4cm} pK_a \; 15.7$$

Explain why this rule works.

**In acid-base reactions, the position of equilibrium favors reaction of the stronger acid and stronger base to give the weaker acid and weaker base. The acid with the higher $pK_a$ is the weaker acid, so the arrow will point toward it.**

<u>Problem 6.15</u> Acetic acid, $CH_3CO_2H$, is a weak organic acid, $pK_a$ 4.76. Which equilibria involving acetic acid lie considerably toward the left? Which lie considerably toward the right? Values of $pK_a$ for other acids in this problem can be found in Table 6.1

(a)   $CH_3CO_2H \; + \; HCO_3^- \; \rightleftharpoons \; CH_3CO_2^- \; + \; H_2CO_3$
$$\text{p}K_a \; \textbf{6.36}$$

(b)   $CH_3CO_2H \; + \; NH_3 \; \rightleftharpoons \; CH_3CO_2^- \; + \; NH_4^+$
$$\text{p}K_a \; \textbf{9.24}$$

(c)   $CH_3CO_2H \; + \; H_2O \; \rightleftharpoons \; CH_3CO_2^- \; + \; H_3O^+$
$$\text{p}K_a \; \textbf{-1.74}$$

**Based on the $pK_a$ values shown, reactions (a) and (b) have equilibria that lie considerably to the right, while reaction (c) has an equilibrium that lies considerably to the left.**

**Problem 6.16** Alcohols are very weak organic acids, $pK_a$ 16-18. The $pK_a$ of ethanol, $CH_3CH_2OH$, is 15.9. Which equilibria involving ethanol lie considerably toward the left? Which lie considerably toward the right? Values of $pK_a$ for other acids in this problem can be found in Table 6.1

(a)    $CH_3CH_2OH$   +   $HCO_3^-$   $\rightleftharpoons$   $CH_3CH_2O^-$   +   $H_2CO_3$

$$pK_a \ \ 6.36$$

(b)    $CH_3CH_2OH$   +   $OH^-$   $\rightleftharpoons$   $CH_3CH_2O^-$   +   $H_2O$

$$pK_a \ \ 15.7$$

(c)    $CH_3CH_2OH$   +   $NH_2^-$   $\rightleftharpoons$   $CH_3CH_2O^-$   +   $NH_3$

$$pK_a \ \ 33$$

(d)    $CH_3CH_2OH$   +   $NH_3$   $\rightleftharpoons$   $CH_3CH_2O^-$   +   $NH_4^+$

$$pK_a \ \ 9.24$$

**Based on the $pK_a$ values shown, reaction (c) has an equilibrium that lies considerably to the right, while reactions (a) and (d) have equilibria that lie considerably to the left. Reaction (b) has an equilibria that lies somewhat to the left.**

**Problem 6.17** Benzoic acid, $C_6H_5CO_2H$, is insoluble in water, but its sodium salt, $C_6H_5CO_2^-Na^+$, is quite soluble in water. Will benzoic acid dissolve in
(a) Aqueous sodium hydroxide?               (b) Aqueous sodium bicarbonate?
(c) Aqueous sodium carbonate?

**The $pK_a$ of benzoic acid is 4.19. The $pK_a$ values for the conjugate acids of sodium hydroxide, sodium bicarbonate, and sodium carbonate are 15.7, 6.36, and 10.33, respectively. Thus, equilibrium will favor reaction of benzoic acid with all three of these bases to give the soluble $C_6H_5CO_2^-Na^+$. Therefore, benzoic acid will dissolve in aqueous solutions of all three bases.**

**Problem 6.18** Phenol, $C_6H_5OH$, is only slightly soluble in water, but its sodium salt, $C_6H_5O^-Na^+$, is quite soluble in water. Will phenol dissolve in
(a) Aqueous sodium hydroxide?               (b) Aqueous sodium bicarbonate?
(c) Aqueous sodium carbonate?

**The $pK_a$ of phenol is 9.95. The $pK_a$ values for the conjugate acids of sodium hydroxide, sodium bicarbonate, and sodium carbonate are 15.7, 6.36, and 10.33, respectively. Thus, equilibrium will favor reaction of phenol with only sodium hydroxide and sodium carbonate to give the soluble $C_6H_5O^-Na^+$. Phenol will dissolve in aqueous solutions of these two bases. Sodium bicarbonate is not a strong enough base to deprotonate phenol, so phenol will not dissolve in an aqueous solution of sodium bicarbonate.**

Problem 6.19 Unless under pressure, carbonic acid in aqueous solution breaks down into carbon dioxide and water, and carbon dioxide is evolved as bubbles of gas. Write an equation for the conversion of carbonic acid to carbon dioxide and water.

$$\underset{\text{HOCOH}}{\overset{\overset{\displaystyle O}{\parallel}}{\phantom{.}}} \longrightarrow \quad H_2O \quad + \quad CO_2 \uparrow$$

Problem 6.20 Will carbon dioxide be evolved when sodium bicarbonate is added to an aqueous solution of these compounds?
(a) Acetic acid                    (b) Ethanol                    (c) Ammonium chloride

**In order for carbon dioxide to be evolved, the sodium bicarbonate must be protonated to give carbonic acid (Problem 6.19). The $pK_a$ of carbonic acid is 6.36. The $pK_a$'s for acetic acid, ethanol and ammonium chloride are 4.76, 15.9, and 9.24, respectively. Thus, acetic acid is the only acid strong enough to protonate sodium bicarbonate and thus evolve carbon dioxide.**

## Lewis Acids and Bases
Problem 6.21 Complete equations for these reactions between Lewis acid-Lewis base pairs. Label which starting material is the Lewis acid and which is the Lewis base, and use a curved arrow to show the flow of the electron pair in each reaction. In solving these problems, it is essential that you show all valence electrons for the atoms participating directly in each reaction.

(d)    CH$_3$-$\overset{+}{C}$H-CH$_3$   +   :$\ddot{B}$$\ddot{r}$:$^-$   $\longrightarrow$   CH$_3$-CH-CH$_3$ with :$\ddot{B}\ddot{r}$: above
       Lewis          Lewis
       Acid           Base

(e)    CH$_3$-CH$_2$-$\ddot{O}$H  +  H$^+$   $\longrightarrow$   CH$_3$-CH$_2$-$\overset{+}{\underset{\cdot\cdot}{O}}$H with H above
       Lewis          Lewis
       Base           Acid

(f)    CH$_3$-$\overset{+}{C}$H-CH$_3$   +   CH$_3$-$\ddot{O}$-H   $\longrightarrow$   CH$_3$-CH-CH$_3$ with CH$_3$-$\overset{+}{O}$-H above
       Lewis              Lewis
       Acid               Base

# CHAPTER 7: ALCOHOLS, ETHERS, AND THIOLS

## SUMMARY OF REACTIONS

| Starting Material → Product | Aldehydes | Alkenes | Alkyl Halides | | Carboxylic Acids | Disulfides | Epoxides | Glycols | Ketones | Metal Alkoxides |
|---|---|---|---|---|---|---|---|---|---|---|
| Alcohols | | **7A** 7.4E* | **7B** 7.4D | **7C** 7.4D | | | | | | **7D** 7.4C |
| Alcohols (Primary) | **7E** 7.4F | | | | **7F** 7.4F | | | | | |
| Alcohols (Secondary) | | | | | | | | | **7G** 7.6B | |
| Alkenes | | | | | | | **7H** 7.4F | | | |
| Epoxides | | | | | | | | **7I** 7.7 | | |
| Thiols | | | | | | **7J** 7.8B | | | | |

*Section of book that describes reaction.

## REACTION 7A: ACID-CATALYZED DEHYDRATION (Section 7.4E)

- **Alcohols** can be heated with $H_3PO_4$ or $H_2SO_4$ to generate an **alkene**. The net result of this process is the removal of $H_2O$ from the alcohol, thus the process is called **dehydration**.
  ✳
- When more than one alkene can be formed from dehydration, the more stable alkene is formed in larger amounts. This generalization is known as **Zaitsev's rule**. In general, the more stable alkene is the one that is more highly substituted, that is, the one with more alkyl groups on the $sp^2$ carbon atoms.
- The mechanism of dehydration involves protonation of the oxygen atom of the -OH group, followed by loss of water to form a carbocation. Since there is no good nucleophile (see Reaction 7B, Section 7.4D for the definition of a nucleophile) to react with the carbocation, a different reaction takes place, namely loss of $H^+$ to give the alkene.

- This mechanism is referred to as an **E1 mechanism**. **E** because a small molecule ($H_2O$) is eliminated, and **1** because there is only one reactant in the slowest step. The slowest step in a reaction is referred to as the **rate-determining step**.
- Because loss of water to form a carbocation occurs in the rate-determining step, the reactivity of different alcohols in this reaction parallels the ease with which carbocations are formed. That is, tertiary alcohols react the fastest in the dehydration reactions, because tertiary carbocations are the most stable.
- Note how this dehydration reaction is just the exact reverse of acid-catalyzed hydration of an alkene (Reaction 4A, Section 4.3B).

- The reaction conditions determine the position of this equilibrium. Large amounts of water favor formation of the alcohol, removing all traces of water favors formation of the alkene.

## REACTION 7B: REACTION WITH H-X (Section 7.4D)

- Tertiary **alcohols** can be converted to **alkyl halides** by treatment with **H-X**, where X = Cl, Br, or I. The mechanism involves initial protonation of the oxygen atom of the -OH group followed by departure of the resulting $H_2O$ to create a carbocation that reacts with X⁻ to give the final product. ✳
    The step involving loss of water is the slowest, the so-called **rate-determining step**. This reaction is considered to follow an **$S_N1$ mechanism**, because it involves **S**ubstitution of X for -OH via a **N**ucleophilic attack by X⁻ onto the carbocation and there is only **1** reactant in the rate-determining step (the protonated oxonium species that simply loses water without reacting with any other molecules).
- In these reactions, the X⁻ is acting as a **nucleophile**. A nucleophile is a molecule or ion that provides a pair of electrons to be shared by another atom in formation of a new covalent bond. For the $S_N1$ mechanism, the carbocation can be considered an electrophile. *[The nucleophile concept is one of the central concepts in all of organic chemistry. A vast number of different reactions can be viewed as involving an electron rich nucleophile reacting with an electron deficient group, that is an electrophile (Section 4.4).]*
- Primary alcohols react much more slowly than tertiary alcohols with H-X reagents. The Lucas reagent ($ZnX_2$) is used to speed up the reaction of secondary and primary alcohols. The mechanism of reaction with the Lucas reagent involves initial complexation of $Zn^{2+}$ with the oxygen atom of the -OH group, followed by displacement of the Zn-O-H by X⁻ to give the alkyl halide product.
    This reaction is classified as an $S_N2$ reaction, because the reaction involves **S**ubstitution of the X for -OH via a **N**ucleophilic attack by X⁻ onto the backside of the C-O bond and the rate determining step involves **2** reagents, the X⁻ and the Zn complex of the alcohol.
- In general, tertiary alcohols react faster by the $S_N1$ mechanism because carbocation formation is involved in the rate determining step, and tertiary carbocations are the most stable. Primary alcohols form the least stable carbocations, so they are the least likely to undergo an

$S_N1$ reaction.  Secondary alcohols fall somewhere in between in both carbocation stability and $S_N1$ reactivity.

- On the other hand, primary alcohols are most likely to react by the $S_N2$ mechanism, because primary alcohols have less steric hindrance to interfere with nucleophilic attack.  Tertiary alcohols have more alkyl groups that can act to screen out nucleophiles trying to react.  Secondary alcohols fall somewhere in between.

## REACTION 7C: REACTION WITH SOCl₂ (Section 7.4D)

$$\underset{}{-\overset{|}{\underset{|}{C}}-\overset{OH}{\underset{|}{C'}}-} \quad \xrightarrow{\text{SOCl}_2} \quad -\overset{|}{\underset{|}{C}}-\overset{X}{\underset{|}{C'}}-$$

- The **-OH group** of **alcohols** can also be **replaced** with a **halide** using **SOCl₂**.  The products of the reaction are the alkyl halide, $SO_2$, and HCl, respectively.

## REACTION 7D: REACTION WITH ACTIVE METALS (Section 7.4C)

$$2 \; -\overset{|}{\underset{|}{C}}-OH \quad \xrightarrow{\text{2 M}} \quad 2 \; -\overset{|}{\underset{|}{C}}-O^- M^+ \; + H_2$$

- **Alcohols** react with **active metals** such as Li, Na, and K to produce **hydrogen gas ($H_2$)** and a **metal alkoxide**. ✱
- The product alkoxides are slightly more basic than hydroxide, HO⁻.  This means that alcohols are slightly weaker acids than water.

## REACTION 7E: OXIDATION OF PRIMARY ALCOHOLS TO ALDEHYDES (Section 7.4F)

$$-\overset{H}{\underset{H}{\overset{|}{\underset{|}{C}}}}-OH \quad \xrightarrow{\text{PCC}} \quad -\overset{O}{\underset{H}{C}}$$

- A special reagent called **pyridinium chlorochromate** reacts with **primary alcohols** and **stops at the aldehyde**, without reacting further to the carboxylic acid. ✱

## REACTION 7F: OXIDATION OF PRIMARY ALCOHOLS TO CARBOXYLIC ACIDS (Section 7.4F)

$$-\overset{H}{\underset{H}{\overset{|}{\underset{|}{C}}}}-OH \quad \xrightarrow{\text{[O]}} \quad -\overset{O}{\underset{OH}{C}}$$

- **Primary alcohols are oxidized** all the way **to carboxylic acids** by aqueous solutions of various forms of chromium(VI) such as $CrO_3$, $K_2Cr_2O_7$, and especially $H_2CrO_4$, a compound called chromic acid.  Chromic acid is prepared from $CrO_3$ or $K_2Cr_2O_7$ and

$H_2SO_4$. In reactions of a primary alcohol with chromic acid, an aldehyde is initially formed, but is oxidized further to a carboxylic acid before it can be isolated.

## REACTION 7G: OXIDATION OF SECONDARY ALCOHOLS TO KETONES
### (Section 7.4F)

- **Secondary alcohols** can be **oxidized to ketones**. Oxidizing agents such as PCC and chromic acid can be used. ✳

## REACTION 7H: OXIDATION OF ALKENES WITH PEROXYACIDS
### (PERACIDS) (Section 7.6B)

- The most common laboratory **synthesis of epoxides** is **from alkenes, using peroxyacids** as oxidizing agents. A commonly used reagent is peroxybenzoic acid. ✳

## REACTION 7.I: ACID-CATALYZED RING OPENING OF EPOXIDES (Section 7.7)

- In the presence of an **acid catalyst**, the oxygen of an **epoxide** is protonated to form a bridged oxonium ion intermediate, that is then susceptible to nucleophilic attack at one of the carbon atoms to generate the **glycol product**. ✳
- The oxygen atom of the oxonium ion intermediate blocks one face of the attached carbon atoms, so the water attacks from the opposite side. For epoxides derived from cyclic alkenes, this means the addition results in formation of a *trans* glycol product. Contrast this with oxidation of an alkene with potassium permanganate to give a *cis* glycol product (Reaction 4D, Section 4.5B).

## REACTION 7J: OXIDATION OF THIOLS TO DISULFIDES (Section 7.8B)

- Even relatively **mild oxidizing agents** such as $I_2$ or $O_2$ **react with thiols** to produce **disulfides**. In fact, thiols are so susceptible to oxidation that they must be protected from contact with air (the $O_2$) to avoid spontaneous disulfide formation. ✳

- Disulfide bonds are especially important in proteins, where they help stabilize three-dimensional structure.

## SUMMARY OF IMPORTANT CONCEPTS

### 7.0 OVERVIEW

- Alcohols, ethers, and thiols are important functional groups that are involved in a number of characteristic reactions, and are very common in nature. Well-known examples of these types of molecules include ethanol (an alcohol), diethyl ether (the first inhalation anesthetic), and ethane thiol (the foul-smelling material added to natural gas so that leaks may be detected).

### 7.1 STRUCTURE

- An **alcohol** is a molecule that contains an -OH (**hydroxyl**) group attached to an $sp^3$ hybridized carbon atom. The oxygen atom of an alcohol is also $sp^3$ hybridized. The oxygen atom forms one sigma bond each to the hydrogen and the carbon atom. There are two lone pairs of electrons on the oxygen atom, each residing in an $sp^3$ orbital. ✳
- An **ether** is a molecule that contains an oxygen atom bonded to two $sp^3$ hybridized carbon atoms. The oxygen atom of an alcohol is also $sp^3$ hybridized. The oxygen atom forms one sigma bond with each of the two carbon atoms. There are two lone pairs of electrons on the oxygen atom, each residing in an $sp^3$ orbital. ✳
- A **thiol** is analogous to an alcohol, except that thiols have an -SH (**sulfhydryl**) group attached to an $sp^3$ hybridized carbon atom. ✳

### 7.2 NOMENCLATURE

- All alcohols, ethers, and thiols can be named according to IUPAC rules, but numerous common names are still used for the simpler ones so these must be learned as well.
- In the **IUPAC system, alcohols are named** by selecting the parent chain as the longest continuous chain that contains the -OH group. The location of the -OH group takes precedence over alkyl groups and halogens when numbering the parent chain. The suffix **e** is changed to **ol** and a number is added to designate the position of the -OH group.
  - When there is a choice, the chain is numbered to give the -OH group the lowest number. This includes cyclic and bicyclic alcohols. Examples of IUPAC names for alcohols include 2-pentanol (not 4-pentanol!) and cyclohexanol.
- Alcohols are characterized according to how many alkyl groups are attached to the carbon atom bearing the -OH group. A **primary alcohol (1°)** is one in which the -OH group is attached to a primary carbon atom, for example 1-propanol. A **secondary alcohol (2°)** is one in which the -OH group is attached to a secondary carbon atom, for example 2-propanol . A **tertiary alcohol (3°)** is one in which the -OH group is attached to a tertiary carbon atom, for example 2-methyl-2-propanol. ✳ *[Characterizing an alcohol as primary, secondary, or tertiary will be the key to understanding some of the reactions of alcohols.]*

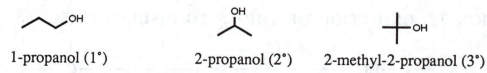

1-propanol (1°)          2-propanol (2°)          2-methyl-2-propanol (3°)

- A molecule that has more than one -OH group is called a **diol** if it has two, or a **triol** if it has three, etc. These molecules are named by adding the suffix **diol** or **triol**, etc., after the suffix **e** and then providing a number for each carbon atom having an -OH group attached. Again the chain is numbered to give the lowest possible numbers to -OH groups.
  - Examples of IUPAC names for diols or triols include 1,2-pentanediol or 2,3,5-octanetriol.

- Diols that contain -OH groups on adjacent carbon atoms are still referred to with the common name of **glycols**. For example ethylene glycol is really 1,2-ethanediol, the major component of antifreeze.
- Compounds that contain an -OH group and a C=C bond are named as an alcohol, with the parent chain numbered so that the -OH group is assigned the lowest number. The infix **en** is used in place of **an** and the suffix **ol** is used in place of the suffix **e**. Numbers are assigned as normal.

   For example, 3-penten-2-ol is the correct IUPAC name for $CH_3$-CH=CH-CHOH-$CH_3$.

- In the **IUPAC system**, **ethers are named** by choosing the longest alkyl chain as the parent chain. The remaining -OR group is named as an alkoxy substituent. For example, ethoxyethane and 3-methoxyheptane are acceptable IUPAC names.

| Ethoxyethane | 3-Methoxyheptane |

- Low-molecular-weight ethers have common names that are often used. The common names are constructed by listing the two alkyl groups in alphabetical order followed by the word ether. For example, diethyl ether or methyl propyl ether are acceptable common names.
- Some important ethers are cyclic, and they are given special names such as tetrahydrofuran or 1,4-dioxane.
- Several small ethers are useful solvents, such as diethyl ether, 2-methoxyethanol, 2-ethoxyethanol, and diethylene glycol dimethyl ether
- According to the **IUPAC system**, **thiols are named** the same as an alcohol, except the suffix **thiol** is placed *after* the suffix **e**. For example 2-butanethiol is a correct IUPAC name. In the common nomenclature rules, the -SH group is referred to as a mercaptan. For example, *n*-butyl mercaptan is the common name for 1-butanethiol.
- The alcohol group has higher priority than the thiol group when it comes to naming complex molecules, so molecules that contain both an -OH group and an -SH group are named as alcohols. In this case, the -SH group is referred to as **mercapto**. For example, 4-mercapto-2-butanol is the correct IUPAC name for $CH_2SH$-$CH_2$-CHOH-$CH_3$. *[This concept of functional group priority for nomenclature will become more important as we learn about more functional groups.]*

## 7.3 PHYSICAL PROPERTIES

- The **electrons are distributed unevenly** in the O-H bond of an **alcohol**. Because oxygen is more electronegative than either carbon or hydrogen, there is increased electron density and thus a partial negative charge on the oxygen atom of an alcohol. Similarly, since hydrogen is so much less electronegative than oxygen, the hydrogen atom has relatively little electron density and thus a partial positive charge. ✳ *[This picture of the alcohol group having partial positive charge on hydrogen and partial negative charge on oxygen is very important and explains most of the reactions of alcohols. Better yet, thinking of the alcohol group in this way will allow you to <u>predict</u> successfully these reactions.]*

$$-\overset{|}{\underset{|}{C}}-\ddot{\underset{\displaystyle H\,\delta+}{\overset{..}{O}}}\!\!:^{\,\delta-}$$

- As the result of the polarization of the O-H bond, there is a permanent dipole moment in the molecule. Two molecules with permanent dipole moments are attracted to each other since the negative end of one dipole is attracted to the positive end of the other dipole, and *vice versa*.

This attraction is called **dipole-dipole interaction**, an example of a weak intermolecular force that holds different molecules together. ✶

• A **hydrogen bond** is a special type of dipole-dipole interaction that occurs when the positive end of one of the dipoles is a hydrogen that is bonded to a very electronegative element (O, N, F). For example, hydrogen bonds that occur between different alcohol molecules involve the hydrogens of one -OH group interacting with the lone pair of electrons on the oxygen atom of another -OH group.

- The strengy of a hydrogen bond is about 5 kcal/mol in water. This is small compared to the strength of covalent bonds that are about 100 kcal/mol, but when several hydrogen bonds are working together, they are strong enough to hold large linear chains in precise, three dimensional structures such as those found in proteins and nucleic acids.
- The alcohol group can take part in hydrogen bonding, so molecules with -OH groups can stick to each other. As a result, alcohols have boiling points that are significantly higher than the corresponding alkanes or alkenes of similar molecular weight.
- Because water ($H_2O$) molecules are hydrogen bonded to each other in solution, only molecules that can themselves take part in dipole-dipole interactions like hydrogen bonds can break up the water molecules and thus dissolve. For that reason, the -OH group on small alcohols allows them to dissolve in water.
    The longer the non-polar alkyl chain on an alcohol, the lower the solubility in water and the higher the solubility in non-polar solvents like hexane and benzene.

• **Ethers** are **polar molecules** since the oxygen atom possesses a partial negative charge and each attached carbon atom possesses a partial positive charge. However, there is only limited dipole-dipole interaction between molecules because there is too much steric hindrance around the carbon atoms with the partial positive charges. As a result, the boiling points of ethers are not that much higher than those of similar hydrocarbons. On the other hand, the oxygen atom of ethers can accept hydrogen bonds, so ethers have higher solubilities in water than the corresponding hydrocarbons. ✶

• The most noticeable property of thiols is their stench. For example, thiols are responsible for the foul scent of skunks.

• The **physical properties of thiols** are completely **different** from those of alcohols, because the **S-H bond** is **not** as **polar** as the O-H bond. Sulfur and hydrogen atoms are similar in electronegativity, so these atoms do not have significant partial charges in the -SH group. As a result, thiols show little association by hydrogen bonding so they have lower boiling points and are less soluble in water compared with analogous alcohols. ✶ *[This correlation between hydrogen bonding ability, water solubility and boiling/melting points is a good illustration of how understanding molecular structure can lead to accurate predictions of physical properties.]*

## 7.4 REACTIONS OF ALCOHOLS

• **Alcohols** are **weakly acidic** in aqueous solution, slightly weaker than water. ✶
• The oxygen atom of an alcohol reacts as a **weak base** and can be protonated by extremely strong acids to generate **oxonium ions**. ✶

- **Primary alcohols** can be **oxidized to aldehydes** (Reaction 7E, Section 7.4F) or **carboxylic acids** (Reaction 7.F, Section 7.4F) and **secondary alcohols** can be **oxidized to ketones** (Reaction 7G, Section 7.6B). In order to be oxidized, an alcohol must have a hydrogen attached to the carbon atom containing the -OH group, thus tertiary alcohols are extremely resistant to oxidization (because they do not have any such hydrogens). ✳

## 7.5 REACTIONS OF ETHERS
- Ethers are very resistant to chemical reactions. Because they are so unreactive and have good solvation properties, ethers such as diethyl ether are good solvents in which to carry out organic reactions.

## 7.6 EPOXIDES
- An **epoxide is a cyclic ether** contained in a **three-member ring**. Epoxides are usually synthesized from an alkene reacting with a peracid such as peroxybenzoic acid (Reaction 7H, Section 7.4F).
- Common names for epoxides can be assigned by using the name of the alkene from which the epoxide could have been derived, followed by the word **oxide**. For example, 1-butene oxide is an acceptable common name.

## 7.7 ACID-CATALYZED RING OPENING OF EPOXIDES
- **Epoxides** are **more reactive than normal ethers** because of the strain present in the three-member ring. In particular, the bond angle in an epoxide is 60° instead of the normal 109.5° for an sp$^3$ atom. Epoxides can undergo ring-opening with a variety of reagents. For an example, please see the acid-catalyzed ring opening of epoxides (Reaction 7I). ✳

## 7.8 REACTIONS OF THIOLS
- **Thiols** are **slightly more acidic than** the corresponding **alcohols**. In other words, they are easier to deprotonate in base.
- **Thiols** are **easily oxidized** to a number of higher oxidation states. As a result, thiols can be considered as mild reducing agents. ✳ *[Recall that when a molecule is oxidized, it gives up electrons. Thus an easily oxidized substance can be considered a reducing agent since it likes to give away electrons.]*
  - Thiols (R-SH) can be oxidized to disulfides (R-S-S-R), sulfinic acids (R-SO$_2$H), and sulfonic acids (R-SO$_3$H).

# CHAPTER 7
*Solutions to the Problems*

**Problem 7.1** Give IUPAC names for these alcohols.

(a)

$$CH_2OH$$
$$H_3C-C\text{···}CH_2CH_3$$
$$H$$

(b)

HO   $CH_3$

(c)   $(CH_3)_3CCH_2OH$

**(S)-2-Methyl-1-butanol**          **1-Methylcyclopentanol**          **2,2-Dimethyl-1-propanol**

**Problem 7.2** Classify each alcohol as primary, secondary, or tertiary.

(a) $CH_3$
$CH_3\overset{\displaystyle CH_3}{\underset{\displaystyle CH_3}{C}}CH_2OH$

(b) ▷—OH

(c) $CH_2{=}CH{-}CH_2{-}OH$

(d)
$CH_3$
$OH$

**Primary**                    **Secondary**                    **Primary**                    **Tertiary**

**Problem 7.3** Write IUPAC names for these unsaturated alcohols:

(a)   $CH_3{-}CH{=}CH{-}CH_2OH$

**2-Buten-1-ol**

(b)   
$$OH$$
$$CH_3\overset{|}{C}HCH{=}CH_2$$

**3-Buten-2-ol**

**Problem 7.4** Write IUPAC and common names for these ethers.

(a)
$CH_3$
$CH_3\overset{|}{C}HCH_2OCH_2CH_3$

**1-Ethoxy-2-methylpropane**
**Ethyl isobutyl ether**

(b)

—$OCH_3$

**Methoxycyclopentane**
**1-Cyclopentenyl methyl ether**

**Problem 7.5** Write names for these thiols and thioethers.

(a)
$CH_3$
$CH_3\overset{|}{C}HCH_2CH_2SH$

**3-Methyl-1-butanethiol**

(b)   $CH_3SCH_2CH_3$

**Ethyl methyl sulfide**

(c)
$SH$
$CH_3CH{=}CH\overset{|}{C}HCH_3$

**3-Pentene-2-thiol**

**Problem 7.6** Arrange these compounds in order of increasing boiling points.

$CH_3OCH_2CH_2OCH_3$          $HOCH_2CH_2OH$          $CH_3OCH_2CH_2OH$

**In order of increasing boiling point they are:**

$$CH_3OCH_2CH_2OCH_3 \qquad CH_3OCH_2CH_2OH \qquad HOCH_2CH_2OH$$

**1,2-Dimethoxyethane**          **2-Methoxyethanol**          **1,2-Ethanediol**
                                 (Methyl cellosolve)          (Ethylene glycol)
**bp 84°C**                      **bp 125°C**                  **bp 198°C**

**Hydrogen bonding, or lack of it, is the key. Although 1,2-dimethoxyethane is a polar molecule, there is little intermolecular association between molecules in the pure liquid because centers of positive and negative charge on adjacent molecules cannot approach each other close enough to develop appreciable dipole-dipole interactions. However, the fact that its boiling point is higher than that of hexane (bp 69°C), a nonpolar hydrocarbon of comparable molecular weight, indicates that there is some dipole-dipole interaction. Both 2-ethoxyethanol and 1,2-ethanediol can associate by hydrogen bonding. Because 1,2-ethanediol has more sites for hydrogen bonding, it has a higher boiling point than 2-methoxyethanol.**

<u>Problem 7.7</u> Predict the position of equilibrium for this acid-base reaction. (<u>Hint</u>: review Section 6.3).

**Acetic acid is the stronger acid; equilibrium lies to the right.**

| (stronger base) | (stronger acid) | (weaker acid) | (weaker base) |

<u>Problem 7.8</u> Draw structural formulas for the alkenes formed by acid-catalyzed dehydration of these alcohols. For each, predict which is the major product.

Problem 7.9 Draw the product of treatment of each alcohol in Example 7.9 with chromic acid.

(a)  $CH_3(CH_2)_4\overset{\overset{\displaystyle O}{\|}}{C}OH$      (b)  $CH_3(CH_2)_3\overset{\overset{\displaystyle O}{\|}}{C}CH_3$      (c)

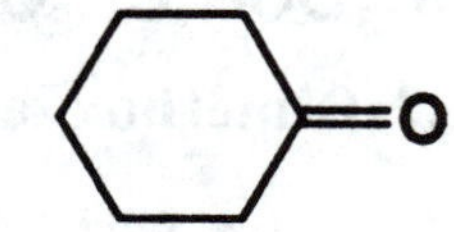

Problem 7.10 Draw the structural formula of the epoxide formed by treating 1,2-dimethylcyclopentene with a peroxycarboxylic acid.

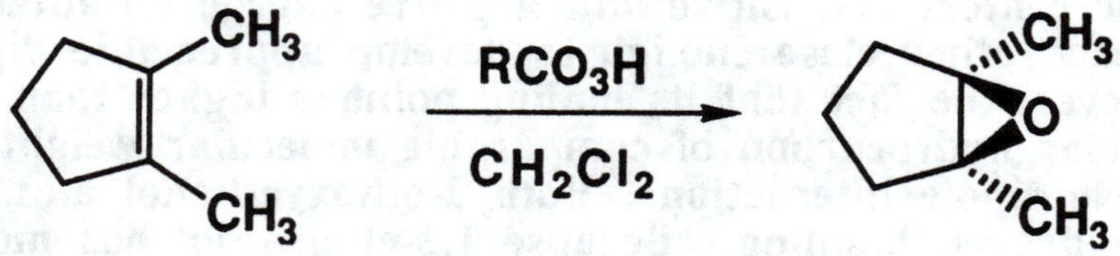

Problem 7.11 Show how to convert cyclohexene to *cis*-1,2-cyclohexanediol.

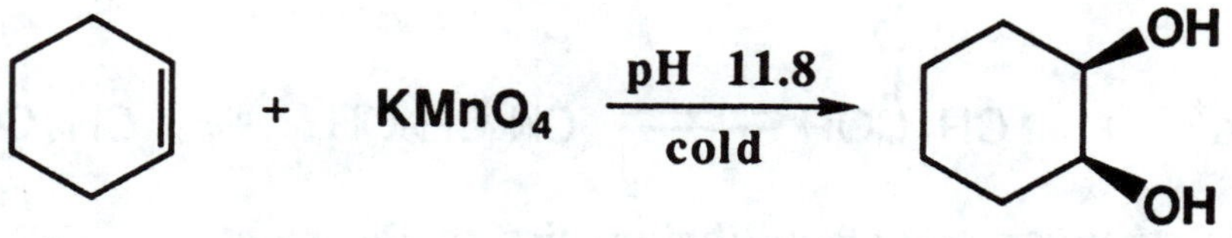

**Cyclohexene**                          ***cis*-1,2-Cyclohexanediol**

### Structure and Nomenclature
Problem 7.12 Which are secondary alcohols?

(a) 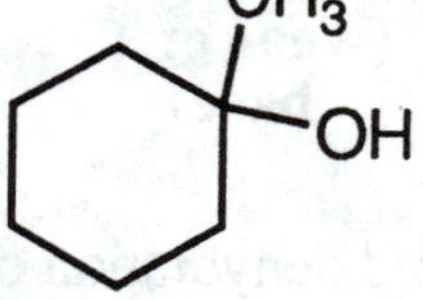      (b)  $(CH_3)_3COH$   (c) 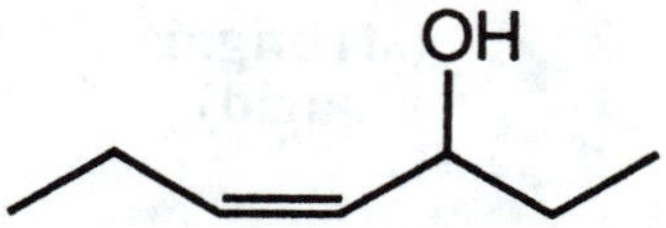      (d)

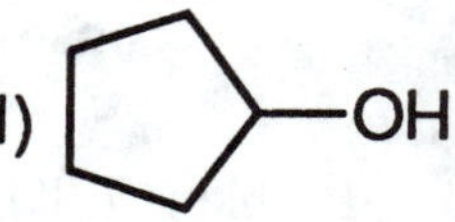

**The secondary alcohols are (c) and (d).  Molecules (a) and (b) are tertiary alcohols.**

Problem 7.13 Name these alcohols and thiol.

(a)  $CH_3CH_2CH_2CH_2CH_2OH$ (b)   $HOCH_2CH_2CH_2OH$   (c)  $CH_2=CHCH_2CH_2OH$
    **1-Pentanol**                      **1,3-Propanediol**                  **3-Buten-1-ol**

(d)  $HOCH_2CH_2\overset{\overset{\displaystyle CH_3}{|}}{C}HCH_3$      (e) 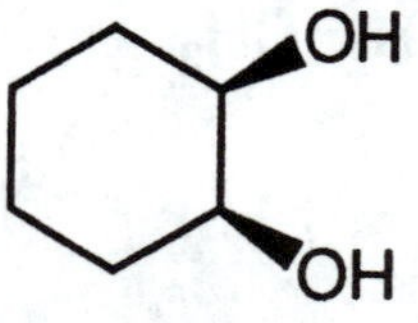                       (f)  $CH_3CH_2CH_2CH_2SH$

  **3-Methyl-1-butanol**          ***cis*-1,2-Cyclohexanediol**                  **1-Butanethiol**
   **(Isopentyl alcohol)**

**Problem 7.14** Write structural formulas for these alcohols:

(a) Isopropyl alcohol

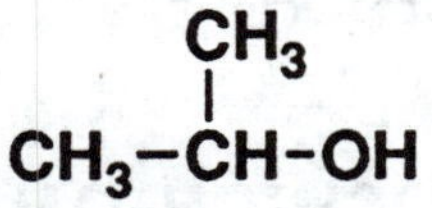

$$CH_3-\underset{\underset{\displaystyle CH_3}{|}}{CH}-OH$$

(b) Propylene glycol

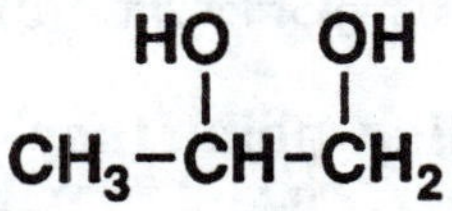

$$CH_3-\underset{\underset{\displaystyle HO}{|}}{CH}-\underset{\underset{\displaystyle OH}{|}}{CH_2}$$

(c) (R)-5-Methyl-2-hexanol

$$H_3C-\underset{\underset{\displaystyle CH_2CH_2CHCH_3}{|}}{\overset{\overset{\displaystyle OH}{|}}{C}}\cdots H \qquad CH_3$$

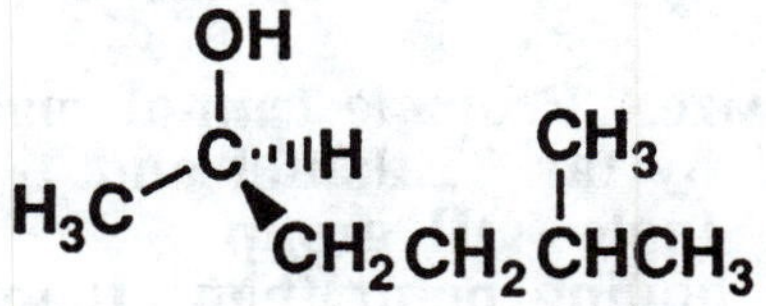

(d) 2-Methyl-2-propyl-1,3-propanediol

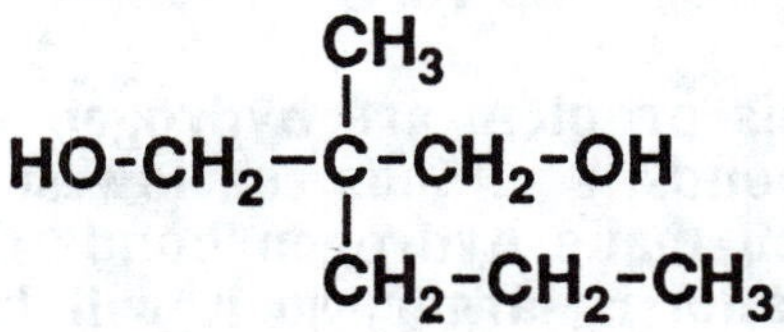

$$HO-CH_2-\underset{\underset{\displaystyle CH_2-CH_2-CH_3}{|}}{\overset{\overset{\displaystyle CH_3}{|}}{C}}-CH_2-OH$$

(e) 2,2-Dimethyl-1-propanol

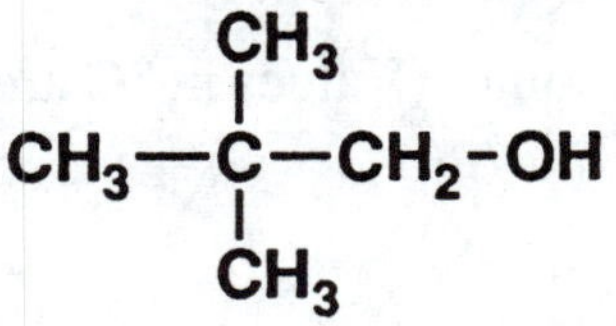

$$CH_3-\underset{\underset{\displaystyle CH_3}{|}}{\overset{\overset{\displaystyle CH_3}{|}}{C}}-CH_2-OH$$

(f) 2-Mercaptoethanol

$$HS-CH_2-CH_2-OH$$

(g) 1,4-Butanediol

$$HOCH_2CH_2CH_2CH_2OH$$

(h) (Z)-5-Methyl-2-hexen-1-ol

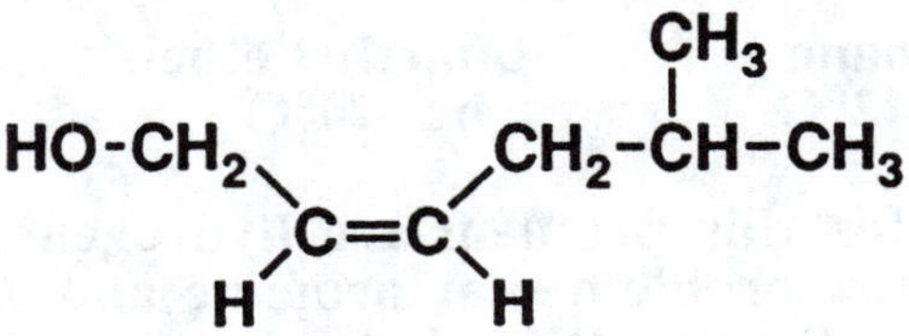

$$\underset{\displaystyle H}{HO-CH_2}\;\;\underset{\displaystyle H}{\overset{\displaystyle CH_2-CH-CH_3}{\underset{\displaystyle}{C=C}}}\;\;CH_3$$

(i) *cis*-3-Pentene-1-ol

$$\underset{\displaystyle H}{HO-CH_2-CH_2}\;\;\underset{\displaystyle H}{\overset{\displaystyle CH_3}{C=C}}$$

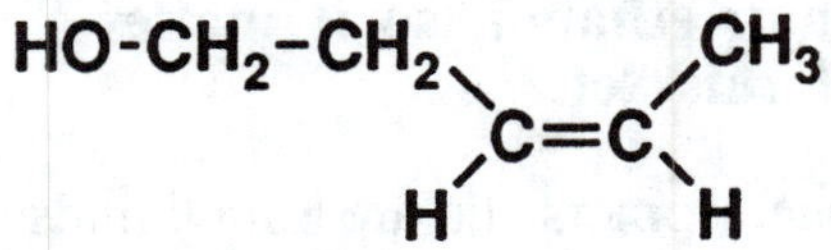

(j) *trans*-1,4-Cyclohexanediol

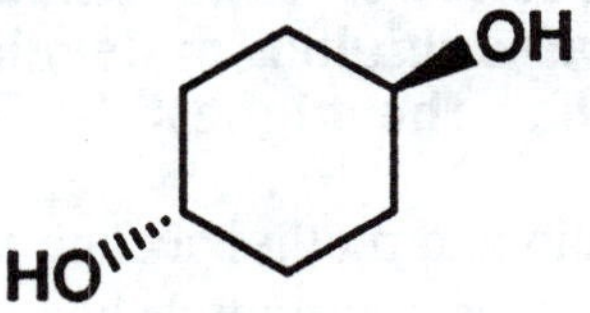

**Problem 7.15** Write names for these ethers.

(a)

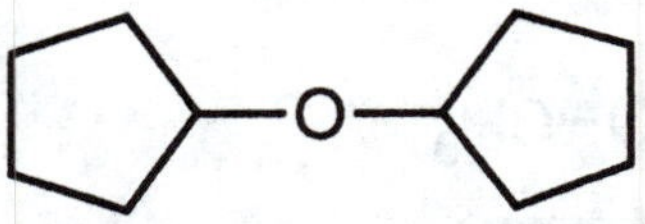

**Cyclopentoxycyclopentane
(Dicyclopentyl ether)**

(b) $[CH_3(CH_2)_4]_2O$

**Pentoxypentane
(Dipentyl ether)**

(c)   $CH_3CH_2OCH_2CH_2OH$

**2-Ethoxyethanol**

## Physical Properties

Problem 7.16 Arrange these compounds in order of increasing boiling point. (Values in °C are -42, 78, 117, and 198)

(a) $CH_3CH_2CH_2CH_2OH$    (b) $CH_3CH_2OH$    (c) $HOCH_2CH_2OH$    (d) $CH_3CH_2CH_3$

**In order of increasing boiling point they are:**

| $CH_3CH_2CH_3$ | $CH_3CH_2OH$ | $CH_3CH_2CH_2CH_2OH$ | $HOCH_2CH_2OH$ |
|---|---|---|---|
| **Propane** | **Ethanol** | **1-Butanol** | **Ethylene glycol** |
| **bp -42°C** | **bp 78°C** | **bp 117°C** | **bp 198°C** |

**The keys for this problem are hydrogen bonding and size. Propane cannot make any hydrogen bonds, so it has the lowest boiling point by far. Ethanol and 1-butanol can each make hydrogen bonds through their single -OH group. However, 1-butanol is larger, so it will have a higher boiling point than ethanol. Ethylene glycol has two -OH groups per molecule with which to make hydrogen bonds, so it will have the highest boiling point of this set.**

Problem 7.17 Arrange these compounds in order of increasing boiling point. (Values in °C are -42, -24, 78, and 118)

(a) $CH_3CH_2OH$    (b) $CH_3OCH_3$    (c) $CH_3CH_2CH_3$    (d) $CH_3CO_2H$

**In order of increasing boiling point they are:**

| $CH_3CH_2CH_3$ | $CH_3OCH_3$ | $CH_3CH_2OH$ | $CH_3CO_2H$ |
|---|---|---|---|
| **Propane** | **Dimethyl ether** | **Ethanol** | **Acetic acid** |
| **bp -42°C** | **bp -24°C** | **bp 78°C** | **bp 118°C** |

**The keys for this problem are hydrogen bonding, polarity, and size. We know from the last problem that propane and ethanol have boiling points of -42°C and 78°C, respectively. Dimethyl ether is polar, but cannot make hydrogen bonds. Therefore, it makes sense that dimethyl ether has a boiling point (-24°C) that is higher than propane, but lower than ethanol. Acetic acid can make strong hydrogen bonds and has a higher molecular weight than ethanol, so it makes sense that acetic acid has the highest boiling point of this set.**

Problem 7.18 Propanoic acid and methyl acetate are constitutional isomers and are both liquids at room temperature. One of these compounds has a boiling point of 141°C, the other has a boiling point of 57°C. Which compound has which boiling point? Explain.

$$CH_3-CH_2-\overset{\overset{\textstyle O}{\|}}{C}-O-H \qquad\qquad CH_3-\overset{\overset{\textstyle O}{\|}}{C}-O-CH_3$$

Propanoic acid                           Methyl acetate

**Propanoic acid has the higher boiling point. Because the carboxyl group of propanoic acid can function as both a hydrogen bond donor (through the O-H group) and a hydrogen bond acceptor (through the C=O and C-O groups) there is a high degree of intermolecular association between molecules of propanoic acid in the liquid state. Methyl acetate, in the pure liquid state, cannot associate by hydrogen bonding.**

CH₃–CH₂–C–O–H
propanoic acid
bp 141°C

CH₃–C–O–CH₃
methyl acetate
bp 57°C

**Problem 7.19** Compounds that contain an N-H group associate by hydrogen bonding.
(a) Do you expect this association to be stronger or weaker than that of compounds containing an O-H group?

**Weaker. The O-H bond is more polar, because the difference in electronegativity between N and H is less than the difference between O and H. Thus, the degree of intermolecular interaction between compounds containing an N-H group is less than that between compounds containing an -OH group.**

| Bond | Difference in electronegativity | % Partial ionic character |
|------|--------------------------------|---------------------------|
| N-H  | 3.0 - 2.1                      | 30%                       |
| O-H  | 3.5 - 2.1                      | 40%                       |

(b) Based on your answer to part (a), which would you predict to have the higher boiling point, 1-butanol or 1-butanamine?

$CH_3CH_2CH_2CH_2OH$                    $CH_3CH_2CH_2CH_2NH_2$

1-Butanol
bp 117°C

1-Butanamine
bp -78°C

**The stronger the hydrogen bonds, the higher the boiling point since hydrogen bonds in the liquid state must be broken upon boiling. Therefore, 1-butanol, with the stronger hydrogen bonds, will have the higher boiling point.**

**Problem 7.20** Following are the structural formulas for 1-butanol and 1-butanethiol. One of these compounds has a boiling point of 98.5° C; the other has a boiling point of 117°C. Which compound has which boiling point? Explain your reasoning.

$CH_3CH_2CH_2CH_2OH$                    $CH_3CH_2CH_2CH_2SH$

1-Butanol
**bp 117°C**

1-Butanethiol
**bp 98.5°C**

**The S-H bond of thiols are much less polar than the O-H bond of alcohols. Therefore, thiols do not make hydrogen bonds that are as strong as the hydrogen bonds made by alcohols. Hydrogen bonds hold molecules of a liquid together, so athiol such as 1-butanethiol has a lower boiling point than a corresponding alcohol such as 1-butanol.**

<u>Problem 7.21</u>  Which of these compounds can participate in hydrogen bonding with water? For each that can, indicate which site(s) can function as a hydrogen bond acceptor and which can function as a hydrogen bond donor.

**The molecules in bold can function as hydrogen bond donors and/or acceptors. Each site is labeled "donor" or "acceptor" as appropriate.**

(a)  $CH_3CH_2CH_2CH_2CH_2OH$

acceptor

$CH_3CH_2CH_2CH_2CH_2-O-H$

donor

(b)  $(CH_3CH_2)_2NH$

acceptor          donor

$CH_3CH_2-N-H$
$\qquad\qquad |$
$\qquad\quad CH_2CH_3$

(c)  $CH_3CH=CHCH_3$

**None**

(d)  $CH_3\overset{O}{\overset{||}{C}}CH_3$

acceptor

$CH_3-\overset{O}{\overset{||}{C}}-CH_3$

(e)  $CH_3\overset{O}{\overset{||}{S}}CH_3$

acceptor

$CH_3\overset{O}{\overset{||}{S}}CH_3$

(f)  $CH_3CH_2\overset{O}{\overset{||}{C}}OH$

acceptor                    acceptor

$CH_3CH_2\overset{O}{\overset{||}{C}}-O-H$

donor

<u>Problem 7.22</u>  From each pair of compounds, select the one that is more soluble in water; explain the basis for your reasoning.

(a)    $CH_2Cl_2$    or    $CH_3OH$

**Methanol, CH$_3$OH, is soluble in all proportions in water. Dichloromethane, CH$_2$Cl$_2$, is insoluble. The highly polar -OH group of methanol is capable of participating both as a hydrogen bond donor and hydrogen bond acceptor with water and, therefore, interacts strongly with water by intermolecular association. No such interaction is possible with dichloromethane.**

(b)

$$\underset{\text{CH}_3\overset{\displaystyle\text{O}}{\overset{\displaystyle\|}{\text{C}}}\text{CH}_3}{} \quad \text{or} \quad \underset{\text{CH}_3\overset{\displaystyle\text{CH}_2}{\overset{\displaystyle\|}{\text{C}}}\text{CH}_3}{}$$

**Propanone (acetone), $CH_3COCH_3$, is soluble in water in all proportions. 2-Methylpropene (isobutylene) is insoluble in water. Acetone has a large dipole moment and can function as a hydrogen bond acceptor from water.**

(c)      $CH_3CH_2Cl$      or      $NaCl$

**$NaCl$ is the more soluble. Chloroethane is insoluble in water. Following is a review of some of the general water solubility rules developed in General Chemistry. For these rules, <u>soluble</u> is defined as dissolving greater than 0.10 mol/L. <u>Slightly soluble</u> is dissolving between 0.01 mol/L and 0.10 mol/L.**
**1. Sodium, potassium, and ammonium salts of halogens or nitrates are soluble.**
**2. Silver, lead, and mercury(l) salts of halogens are insoluble.**
**Thus, applying Rule 1, $NaCl$ is soluble in water. Chloroethane (ethyl chloride) is a nonpolar organic compound and insoluble in water.**

(d)   $CH_3CH_2CH_2SH$   or   $CH_3CH_2CH_2OH$

**Sulfur is less electronegative than oxygen, so an S-H bond is less polarized than an O-H bond. Hydrogen bonding is therefore weaker with thiols than alcohols, so the alcohol will be more able to interact with water molecules through hydrogen bonding. The alcohol will be more soluble in water.**

(e)

$$\underset{\text{CH}_3\text{CH}_2\overset{\displaystyle\text{OH}}{\overset{\displaystyle|}{\text{C}}}\text{HCH}_2\text{CH}_3}{} \quad \text{or} \quad \underset{\text{CH}_3\text{CH}_2\overset{\displaystyle\text{O}}{\overset{\displaystyle\|}{\text{C}}}\text{CH}_2\text{CH}_3}{}$$

**The alcohol group has both a hydrogen bond donor and acceptor (the oxygen and hydrogen atoms of the -OH group, respectively), while the ketone has only a hydrogen bond acceptor (the oxygen atom). Thus, the alcohol will be more able to interact with water through hydrogen bonding, so the alcohol will be more soluble in water.**

<u>Problem 7.23</u> Arrange the compounds in each set in order of decreasing solubility in water; explain the basis for your answers.
(a) Ethanol; butane; diethyl ether

| $CH_3CH_2OH$ | $CH_3CH_2OCH_2CH_3$ | $CH_3CH_2CH_2CH_3$ |
|---|---|---|
| **soluble in all proportions** | **8 g/100 mL water** | **insoluble in water** |

**In general, the more strongly a molecule can take part in hydrogen bonding with water, the greater the molecule will be able to interact with the water molecules and dissolve. Only the ethanol can act both as a donor and acceptor of hydrogen bonds with water. Diethyl ether can act as an acceptor of hydrogen bonds. Butane can act as neither a donor nor an acceptor of hydrogen bonds.**

(b) 1-Hexanol; 1,2-hexanediol; hexane

$$CH_2CHCH_2CH_2CH_2CH_3 \quad CH_3CHCH_2CH_2CH_2CH_2OH \quad CH_3CH_2CH_2CH_2CH_2CH_3$$
$$|\ \ |$$
$$OH\ OH$$

**The 1,2-hexanediol molecules can take part in more hydrogen bonds with water than the 1-hexanol, since the diol has two -OH groups. The hexane has no polar bonds and thus cannot take part in any dipole-dipole interactions with water molecules.**

Problem 7.24 Each compound in this problem is a common organic solvent. From each pair of compounds, select the solvent with the greater solubility in water.

**Solubility in water increases with increasing hydrogen bonding ability and decreases with increasing surface area of hydrophobic groups such as alkyl groups.**

(a) $CH_2Cl_2$ or $CH_3CH_2OH$              (b) $CH_3CH_2OCH_2CH_3$ or $CH_3CH_2OH$

   **$CH_3CH_2OH$**                                                 **$CH_3CH_2OH$**

(c) $CH_3\overset{O}{\overset{||}{C}}CH_3$ or $CH_3CH_2OCH_2CH_3$    (d) $CH_3CH_2OCH_2CH_3$ or $CH_3(CH_2)_3CH_3$

   **$CH_3\overset{O}{\overset{||}{C}}CH_3$**                                    **$CH_3CH_2OCH_2CH_3$**

## Synthesis of Alcohols

We have encountered three reactions for the synthesis of alcohols, including glycols.
(1) Acid-catalyzed hydration of alkenes (Section 4.3B).
(2) Oxidation of alkenes to glycols by alkaline $KMnO_4$ (Section 4.5B).
(3) Acid-catalyzed ring opening of epoxides to give glycols (Section 7.7).

Problem 7.25 Give the structural formula of an alkene or alkenes from which each alcohol or glycol can be prepared.
(a) 2-Butanol

**$CH_3CH_2CH=CH_2$**

*or*

(b) 1-Methylcyclohexanol

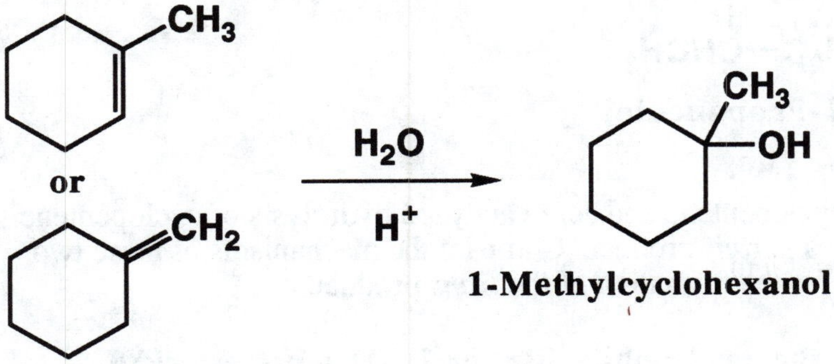

$$\underset{\text{or}}{\text{(structure)}} \xrightarrow[\text{H}^+]{\text{H}_2\text{O}} \text{(structure)}$$

**1-Methylcyclohexanol**

(c) 3-Hexanol

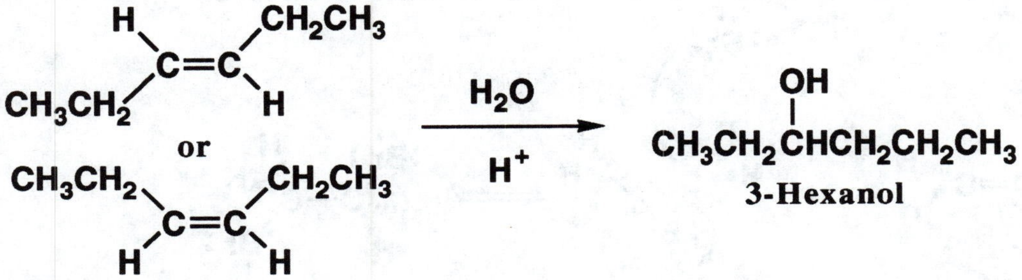

$$\underset{\text{or}}{\text{(structure)}} \xrightarrow[\text{H}^+]{\text{H}_2\text{O}} \underset{\text{3-Hexanol}}{\overset{\text{OH}}{\text{CH}_3\text{CH}_2\text{CHCH}_2\text{CH}_2\text{CH}_3}}$$

(d) 2-Methyl-2-pentanol

$$\underset{\text{or}}{\text{(structure)}} \xrightarrow[\text{H}^+]{\text{H}_2\text{O}} \text{(structure)}$$

$$\text{CH}_2{=}\text{CCH}_2\text{CH}_2\text{CH}_3 \text{ with } \text{CH}_3$$

**2-Methyl-2-pentanol**

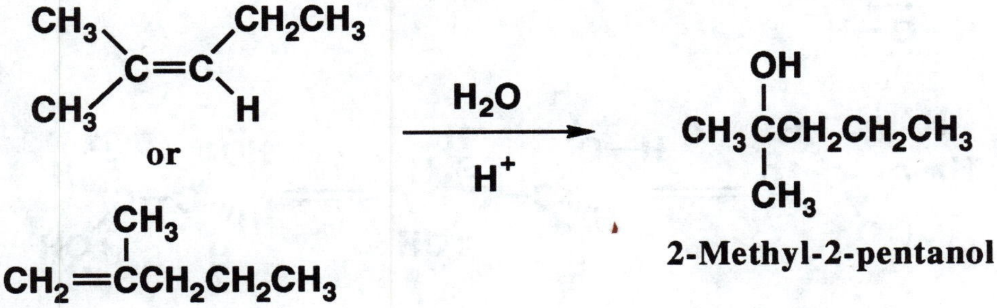

(e) Cyclopentanol

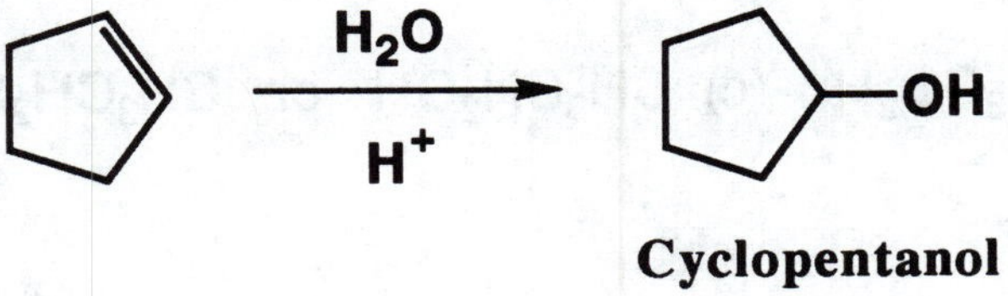

$$\text{(structure)} \xrightarrow[\text{H}^+]{\text{H}_2\text{O}} \text{(structure)}\text{—OH}$$

**Cyclopentanol**

(f) 1,2-Propanediol

$$CH_2{=}CHCH_3 \xrightarrow[\substack{cold \\ pH\ 11.8}]{KMnO_4} \underset{\text{1,2-Propanediol}}{\overset{\overset{\displaystyle HO\quad OH}{|\qquad|}}{H_2C{-}CHCH_3}}$$

<u>Problem 7.26</u>  Addition of bromine to cyclopentene and acid-catalyzed hydrolysis of cyclopentene oxide are both stereoselective; each gives a *trans* product.  Compare the mechanisms of these two reactions and show how each accounts for the formation of the *trans* product.

**Notice how in both mechanisms, the nucleophile (Br⁻ or H₂O) must attack a three-member ring species.  One face of this three member ring is blocked by either the Br or OH atoms, so the nucleophile must attack from the opposite side as shown.  The result in both cases is that a *trans* product is formed.**

<u>**Acidity of Alcohols and Thiols**</u>
<u>Problem 7.27</u>  Select the stronger acid from each pair.  For each stronger acid, write a structural formula for its conjugate base.

(a) $H_2O$  or  $H_2CO_3$  (b) $CH_3OH$  or  $CH_3CO_2H$  (c) $CH_3CH_2OH$  or  $CH_3CH_2SH$

Under each acid is given its $pK_a$. The stronger acid has the smaller value of $pK_a$.

|     | weaker acid | stronger acid | conjugate base of stronger acid |
|-----|-------------|---------------|----------------------------------|
| (a) | $H_2O$ $pK_a$ 15.7 | $H_2CO_3$ $pK_a$ 6.36 | $HCO_3^-$ |
| (b) | $CH_3OH$ $pK_a$ 15.5 | $CH_3CO_2H$ $pK_a$ 4.76 | $CH_3CO_2^-$ |
| (c) | $CH_3CH_2OH$ $pK_a$ 15.9 | $CH_3CH_2SH$ $pK_a$ 8.5 | $CH_3CH_2S^-$ |

Problem 7.28 Arrange these compounds in order of increasing acidity (weakest to strongest).

$$CH_3CH_2CH_2OH \qquad CH_3CH_2\overset{\overset{O}{\|}}{C}OH \qquad CH_3CH_2CH_2SH$$

The compounds listed in order of increasing acidity are given below. The $pK_a$ of alcohols are ~16, the $pK_a$ of thiols are ~8.5 and the $pK_a$ of carboxylic acids are ~4.5.

$$CH_3CH_2CH_2OH \qquad CH_3CH_2CH_2SH \qquad CH_3CH_2\overset{\overset{O}{\|}}{C}OH$$

Problem 7.29 From each pair, select the stronger base. For each stronger base, write the structural formula of its conjugate acid.

(a) $OH^-$ or $CH_3O^-$ (each in $H_2O$)

$$HO^- \longrightarrow HOH$$

stronger       weaker
base           acid

(b) $CH_3CH_2S^-$ or $CH_3CH_2O^-$

$$CH_3CH_2O^- \longrightarrow CH_3CH_2OH$$

stronger       weaker
base           acid

(c) $CH_3CH_2O^-$ or $NH_2^-$

$$NH_2^- \longrightarrow NH_3$$

stronger     weaker
base       acid

**Problem 7.30** For each equilibrium, label the stronger acid, stronger base, weaker acid, and weaker base. Also predict the position of each equilibria. For pKa values, see Table 6.1.

(a)   $CH_3CH_2O^- + HCl \rightleftharpoons CH_3CH_2OH + Cl^-$

$$CH_3CH_2O^- + HCl \longrightarrow CH_3CH_2OH + Cl^-$$

| | pKa -7 | pKa 15.9 | |
|:---:|:---:|:---:|:---:|
| stronger base | stronger acid | weaker acid | weaker base |

(b)   $CH_3\overset{O}{\overset{\|}{C}}-OH + CH_3CH_2O^- \rightleftharpoons CH_3\overset{O}{\overset{\|}{C}}-O^- + CH_3CH_2OH$

$$CH_3\overset{O}{\overset{\|}{C}}-OH + CH_3CH_2O^- \longrightarrow CH_3\overset{O}{\overset{\|}{C}}-O^- + CH_3CH_2OH$$

| pKa 4.76 | | weaker | pKa 15.9 |
|:---:|:---:|:---:|:---:|
| stronger acid | stronger base | weaker base | weaker acid |

**Problem 7.31** Predict the position of each equilibrium.

(a)  $CH_3CH_2OH + Na^+OH^- \rightleftharpoons CH_3CH_2O^- Na^+ + H_2O$

$$CH_3CH_2OH + Na^+OH^- \longleftarrow CH_3CH_2O^- Na^+ + H_2O$$

| pKa 15.9 | | | pKa 15.7 |
|:---:|:---:|:---:|:---:|
| weaker acid | weaker base | stronger base | stronger acid |

(b) $CH_3CH_2SH + Na^+OH^- \rightleftharpoons CH_3CH_2S^- Na^+ + H_2O$

$$CH_3CH_2SH + Na^+OH^- \longrightarrow CH_3CH_2S^- Na^+ + H_2O$$

| pKa 8.5 | | | pKa 15.7 |
|:---:|:---:|:---:|:---:|
| stronger acid | stronger base | weaker base | weaker acid |

## Reactions of Alcohols

**Problem 7.32** Show how to distinguish between cyclohexanol and cyclohexene by a simple chemical test. *Hint:* Treat each with $Br_2$ in $CCl_4$ and see what happens.

**Cyclohexene discharges the color of bromine in carbon tetrachloride and also of cold, alkaline potassium permanganate to produce a brown precipitate of manganese dioxide. Cyclohexanol does not react with either of these reagents under these conditions.**

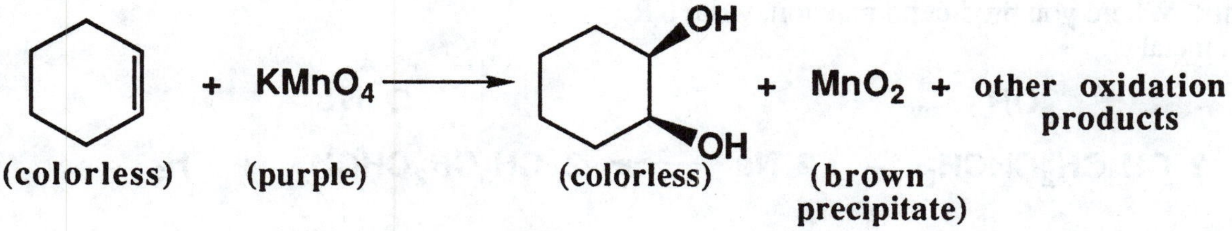
+ KMnO4
OH
OH
+ MnO2 + other oxidation products
(colorless)
(purple)
(colorless)
(brown precipitate)

**Alternatively, only cyclohexanol reacts with active metals such as sodium and potassium to liberate hydrogen gas which then bubbles from the test solution.**

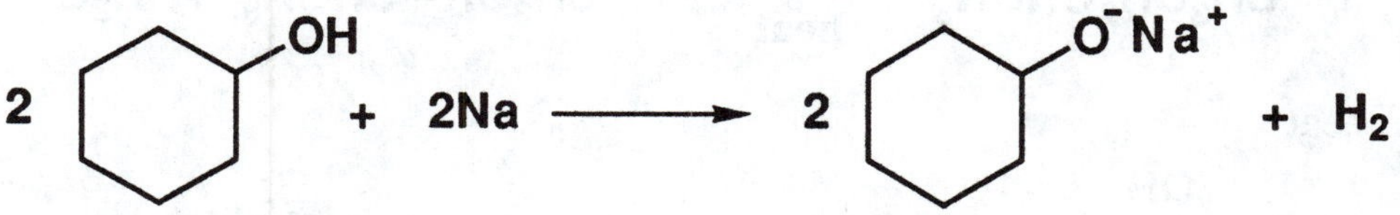
OH
2
+ 2Na
2
O⁻Na⁺
+ H2

Problem 7.33 Write equations for the reaction of 1-butanol, a primary alcohol, with these reagents. Where you predict no reaction, write NR.

(a) Na metal

$$2\ CH_3CH_2CH_2CH_2OH + 2\ Na \longrightarrow 2CH_3CH_2CH_2CH_2O^-\ Na^+ + H_2$$

(b) HCl, $ZnCl_2$, heat

$$CH_3CH_2CH_2CH_2OH + HCl \xrightarrow[\text{heat}]{ZnCl_2} CH_3CH_2CH_2CH_2Cl + H_2O$$

(c) $K_2Cr_2O_7$, $H_2SO_4$, heat

$$CH_3CH_2CH_2CH_2OH + K_2Cr_2O_7 \xrightarrow[\text{heat}]{H_2SO_4} CH_3CH_2CH_2\overset{O}{\overset{\|}{C}}\text{-}OH + Cr^{3+}$$

(d) $SOCl_2$

$$CH_3CH_2CH_2CH_2OH + SOCl_2 \xrightarrow{\text{pyridine}} CH_3CH_2CH_2CH_2\text{-}Cl + SO_2 + HCl$$

(e) Pyridinium chlorochromate (PCC)

$$CH_3CH_2CH_2CH_2OH + PCC \longrightarrow CH_3CH_2CH_2\overset{O}{\overset{\|}{C}}\text{-}H + Cr^{3+}$$

<u>Problem 7.34</u> Write equations for the reaction of 2-butanol, a secondary alcohol, with these reagents. Where you predict no reaction, write NR.
(a) Na metal

$$2 \ CH_3CH_2\overset{\overset{\displaystyle OH}{|}}{C}HCH_3 \ + \ 2 \ Na \longrightarrow 2 \ CH_3CH_2\overset{\overset{\displaystyle O^-Na^+}{|}}{C}HCH_3 \ + \ H_2$$

(b) $H_2SO_4$, heat

$$CH_3CH_2\overset{\overset{\displaystyle OH}{|}}{C}HCH_3 \ \xrightarrow[\text{heat}]{H_2SO_4} \ CH_3CH{=}CHCH_3 \ + \ H_2O$$

(c) HBr, heat

$$CH_3CH_2\overset{\overset{\displaystyle OH}{|}}{C}HCH_3 \ + \ HBr \ \xrightarrow{\text{heat}} \ CH_3CH_2\overset{\overset{\displaystyle Br}{|}}{C}HCH_3 \ + \ H_2O$$

(d) $K_2Cr_2O_7$, $H_2SO_4$, heat

$$CH_3CH_2\overset{\overset{\displaystyle OH}{|}}{C}HCH_3 \ + \ K_2Cr_2O_7 \ \xrightarrow[\text{heat}]{H_2SO_4} \ CH_3CH_2\overset{\overset{\displaystyle O}{\|}}{C}CH_3 \ + \ Cr^{3+}$$

(e) $SOCl_2$

$$CH_3CH_2\overset{\overset{\displaystyle OH}{|}}{C}HCH_3 \ + \ SOCl_2 \ \xrightarrow{\text{pyridine}} \ CH_3CH_2\overset{\overset{\displaystyle Cl}{|}}{C}HCH_3 \ + \ SO_2 \ + \ HCl$$

(f) Pyridinium chlorochromate (PCC)

$$CH_3CH_2\overset{\overset{\displaystyle OH}{|}}{C}HCH_3 \ + \ PCC \ \longrightarrow \ CH_3CH_2\overset{\overset{\displaystyle O}{\|}}{C}CH_3 \ + \ Cr^{3+}$$

<u>Problem 7.35</u> What is the most likely mechanism of this reaction? Draw a structural formula for the reactive intermediate formed during the reaction.

$$CH_3CH_2\overset{\overset{\displaystyle CH_3}{|}}{\underset{\underset{\displaystyle CH_3}{|}}{C}}OH \ + \ HCl \ \longrightarrow \ CH_3CH_2\overset{\overset{\displaystyle CH_3}{|}}{\underset{\underset{\displaystyle CH_3}{|}}{C}}Cl \ + \ H_2O$$

**The tertiary alcohol reacts via an $S_N1$ mechanism, involving a relatively stable tertiary carbocation as shown.**

**Problem 7.36** Complete these equations.

(a)   $CH_3CH_2CH_2OH + H_2CrO_4 \longrightarrow CH_3CH_2\overset{\overset{\textstyle O}{\|}}{C}OH + Cr^{3+}$

(b)   $\underset{\underset{\textstyle CH_3}{|}}{CH_3CHCH_2CH_2OH} + SOCl_2 \longrightarrow \underset{\underset{\textstyle CH_3}{|}}{CH_3CHCH_2CH_2Cl} + SO_2 + HCl$

(c)   + HCl $\longrightarrow$ + $H_2O$

(d)   $HOCH_2CH_2CH_2CH_2OH + HCl \xrightarrow{ZnCl_2} ClCH_2CH_2CH_2CH_2Cl + 2H_2O$

(e)   + $H_2CrO_4 \longrightarrow$ + $Cr^{3+}$

(f)   $\xrightarrow[\text{pH 11.8}]{KMnO_4, H_2O}$

Problem 7.37  In the commercial synthesis of *tert*-butyl methyl ether, an antiknock, octane-improving gasoline additive, 2-methylpropane and methanol are passed over an acid catalyst to give the ether.  Propose a mechanism for this reaction.

2-Methylpropene                Methanol
(Isobutylene)

2-Methoxy-2-methylpropane
(*tert*-Butyl methyl ether)

**The mechanism for this reaction is analogous to acid-catalyzed hydration of an alkene (Section 4.3B).  The first step involves protonation of the alkene to produce a carbocation intermediate.  Methanol reacts as a nucleophile with the carbocation intermediate followed by loss of a proton to give the final product.**

## Reactions of Thiols

Problem 7.38  Predict the position of equilibrium for each acid-catalyzed reaction; that is does each lie considerably to the left, considerably to the right, or are concentrations evenly balanced?

(a)  $CH_3CH_2OH$   +   $CH_3CH_2S^- \, Na^+$   $\rightleftharpoons$   $CH_3CH_2O^- \, Na^+$   +   $CH_3CH_2SH$

$CH_3CH_2OH$  +  $CH_3CH_2S^- \, Na^+$   $\longleftarrow$   $CH_3CH_2O^- \, Na^+$   +   $CH_3CH_2SH$

| pKa 15.9 | | | pKa 8.5 |
|---|---|---|---|
| weaker | weaker | stronger | stronger |
| acid | base | base | acid |

(b)  $CH_3CH_2S^- \, Na^+$   +   $CH_3\overset{\displaystyle O}{\overset{\|}{C}}OH$   $\rightleftharpoons$   $CH_3CH_2SH$   +   $CH_3\overset{\displaystyle O}{\overset{\|}{C}}O^- \, Na^+$

$CH_3CH_2S^- \, Na^+$   +   $CH_3\overset{\displaystyle O}{\overset{\|}{C}}OH$   $\longrightarrow$   $CH_3CH_2SH$   +   $CH_3\overset{\displaystyle O}{\overset{\|}{C}}O^- \, Na^+$

| | pKa 4.76 | pKa 8.5 | |
|---|---|---|---|
| stronger | stronger | weaker | weaker |
| base | acid | acid | base |

## Synthesis

Problem 7.39  Show how to convert 1-propanol to 2-propanol in two steps.

$$CH_3CH_2CH_2OH \xrightarrow[\text{heat}]{H_3PO_4} CH_3CH=CH_2 \xrightarrow[H_2SO_4]{H_2O} CH_3\overset{\overset{\displaystyle OH}{|}}{C}HCH_3$$

**1-Propanol**                              **Propene**                              **2-Propanol**

**Note how the hydration of propene will give 2-propanol as the major (Markovnikov) product.**

Problem 7.40  Show how to convert cyclohexene to cyclohexanone in two steps.

**Cyclohexene**                  **Cyclohexanol**                     **Cyclohexanone**

Problem 7.41  Show how to convert cyclohexanol to *cis*-1,2-cyclohexanediol in two steps.

**Cyclohexanol**                  **Cyclohexene**             ***cis*-1,2-cyclohexanediol**

Problem 7.42 Show how to convert propene to propanone (acetone) in two steps.

$$CH_3CH=CH_2 \xrightarrow[H_2SO_4]{H_2O} CH_3\overset{\overset{\displaystyle OH}{|}}{C}HCH_3 \xrightarrow{H_2CrO_4} CH_3\overset{\overset{\displaystyle O}{||}}{C}CH_3$$

**Propene**                          **2-Propanol**                      **Propanone (Acetone)**

Problem 7.43 Show how to convert cyclohexanol to these compounds:
(a) Chlorocyclohexane

(b) Cyclohexene

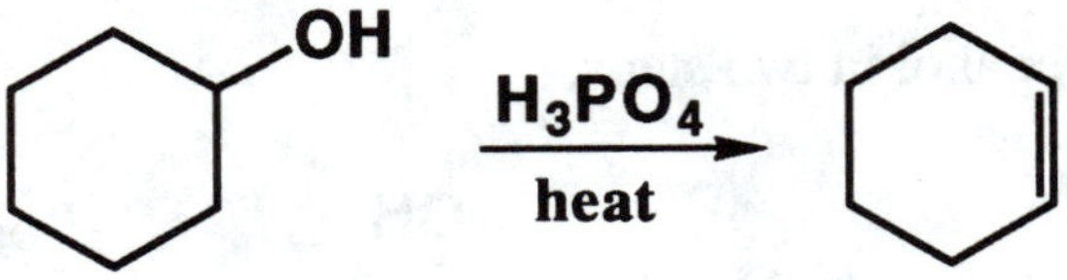

(c) Cyclohexane

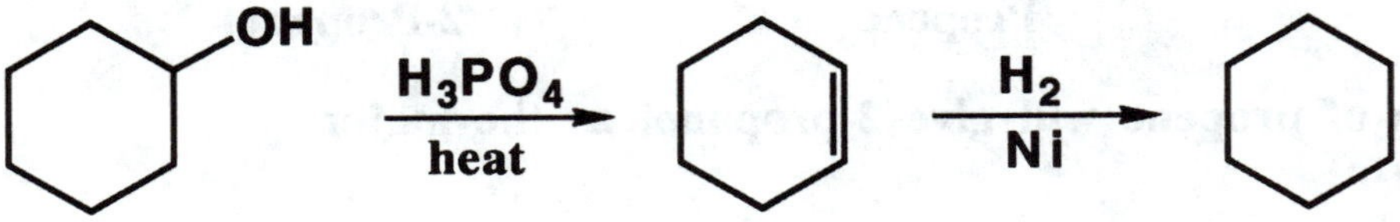

(d) Cyclohexanone

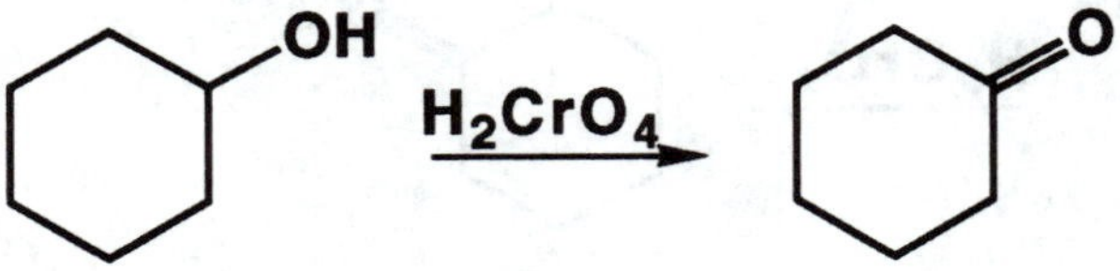

Problem 7.44 Show reagents and experimental conditions to synthesize these compounds from 1-propanol. Any derivative of 1-propanol prepared in an earlier part of this problem may then be used for a later synthesis.

(a) Propanal

$$CH_3CH_2CH_2OH \ + \ PCC \ \longrightarrow \ CH_3CH_2\overset{\displaystyle O}{\overset{\|}{C}}H$$

(b) Propanoic acid

$$CH_3CH_2CH_2OH \ \xrightarrow{H_2CrO_4} \ CH_3CH_2\overset{\displaystyle O}{\overset{\|}{C}}OH$$

(c) Propene

$$CH_3CH_2CH_2OH \ \xrightarrow[\text{heat}]{H_3PO_4} \ CH_3CH{=}CH_2$$

(d) 2-Propanol

$$\underset{\textbf{From (c)}}{CH_3CH{=}CH_2} \ \xrightarrow[H_2SO_4]{H_2O} \ CH_3\overset{\displaystyle OH}{\overset{|}{C}}HCH_3$$

(e) 2-Bromopropane

$$\underset{\textbf{From (d)}}{CH_3\overset{\displaystyle OH}{\overset{|}{C}}HCH_3} \ \text{or} \ \underset{\textbf{From (c)}}{H_2C{=}CHCH_3} \ + \ HBr \ \xrightarrow{\text{heat}} \ CH_3\overset{\displaystyle Br}{\overset{|}{C}}HCH_3$$

(f) 1-Chloropropane

$$CH_3CH_2CH_2OH + HCl \xrightarrow[\text{heat}]{ZnCl_2} CH_3CH_2CH_2Cl$$

(g) Propanone

$$\underset{\text{From (d)}}{CH_3\overset{\overset{\displaystyle OH}{|}}{C}HCH_3} \xrightarrow{H_2CrO_4} CH_3\overset{\overset{\displaystyle O}{||}}{C}CH_3$$

(h) 1,2-Propanediol

$$\underset{\underset{\text{pH 11.8}}{\text{From (c)}}}{CH_3CH{=}CH_2} \xrightarrow[\text{cold}]{KMnO_4} \overset{\overset{\displaystyle HO \quad OH}{|\quad\;\;|}}{CH_3CH{-}CH_2}$$

**Problem 7.45** Show how to bring about the following conversions. For any conversion involving more than one step, show each intermediate compound formed.

(a) $CH_3\overset{\overset{\displaystyle CH_3}{|}}{C}HCH_2OH \xrightarrow[\text{heat}]{H_3PO_4} CH_3\overset{\overset{\displaystyle CH_3}{|}}{C}{=}CH_2$

(b) $CH_3\overset{\overset{\displaystyle CH_3}{|}}{C}HCH_2OH \xrightarrow[\text{heat}]{H_3PO_4} CH_3\overset{\overset{\displaystyle CH_3}{|}}{C}{=}CH_2 \xrightarrow[H_2SO_4]{H_2O} CH_3\overset{\overset{\displaystyle CH_3}{|}}{\underset{\underset{\displaystyle OH}{|}}{C}}CH_3$

(c) $CH_3\overset{\overset{\displaystyle CH_3}{|}}{C}HCH_2OH \xrightarrow[\text{heat}]{H_3PO_4} CH_3\overset{\overset{\displaystyle CH_3}{|}}{C}{=}CH_2 \xrightarrow[\underset{\text{pH 11.8}}{\text{cold}}]{KMnO_4} CH_3\overset{\overset{\displaystyle CH_3}{|}}{\underset{\underset{\displaystyle HO}{|}}{C}}{-}\overset{}{\underset{\underset{\displaystyle OH}{|}}{C}H_2}$

(d) $CH_3\overset{\overset{\displaystyle CH_3}{|}}{C}HCH_2OH \xrightarrow{H_2CrO_4} CH_3\overset{\overset{\displaystyle CH_3}{|}}{C}HCO_2H$

**Problem 7.46** Show how to bring about the following conversions. For any involving more than one step, show each intermediate compound formed.

(a)

(b)

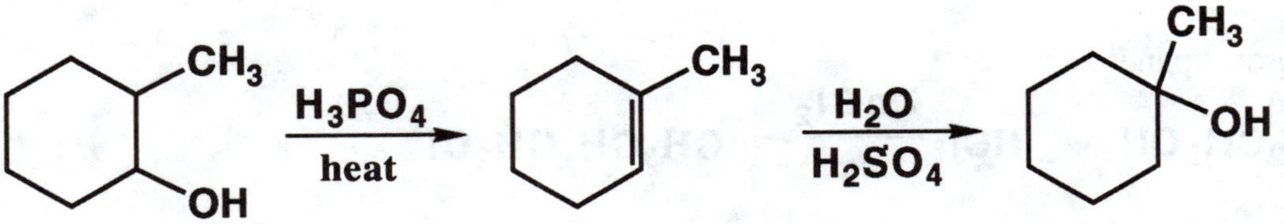

$$\xrightarrow[\text{heat}]{H_3PO_4}$$

$$\xrightarrow[H_2SO_4]{H_2O}$$

(c)

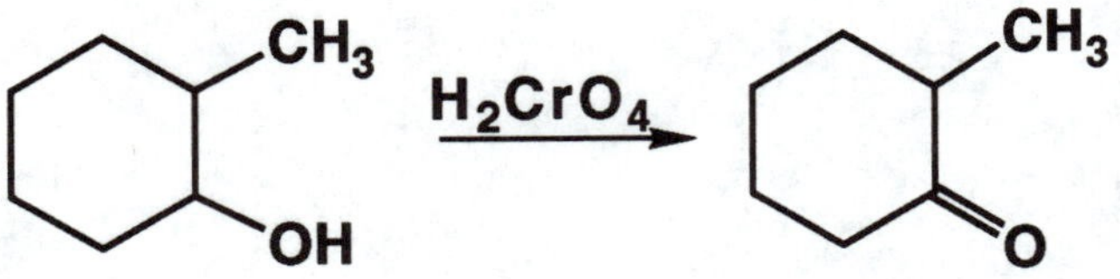

$$\xrightarrow{H_2CrO_4}$$

(d)

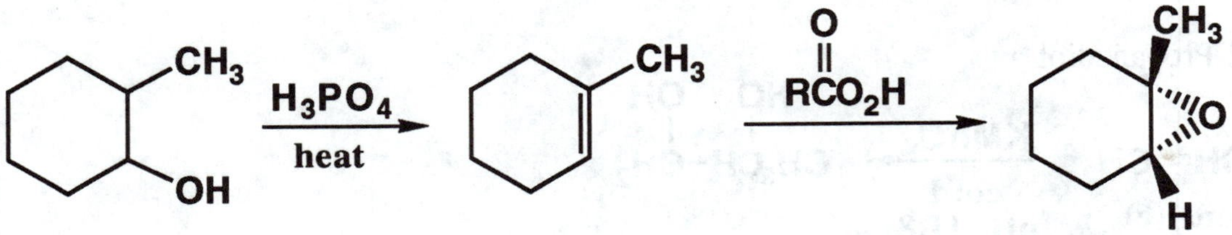

$$\xrightarrow[\text{heat}]{H_3PO_4}$$

$$\xrightarrow{\overset{\overset{\displaystyle O}{\|}}{RCO_2H}}$$

(e)

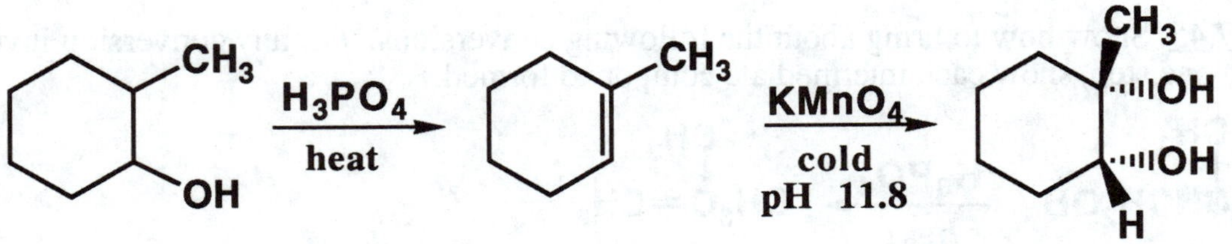

$$\xrightarrow[\text{heat}]{H_3PO_4}$$

$$\xrightarrow[\substack{\text{cold} \\ \text{pH } 11.8}]{KMnO_4}$$

(f)

$$\xrightarrow[\text{heat}]{H_3PO_4}$$

$$\xrightarrow{\overset{\overset{\displaystyle O}{\|}}{RCO_2H}}$$

$$\xrightarrow[H_2O]{H^+}$$

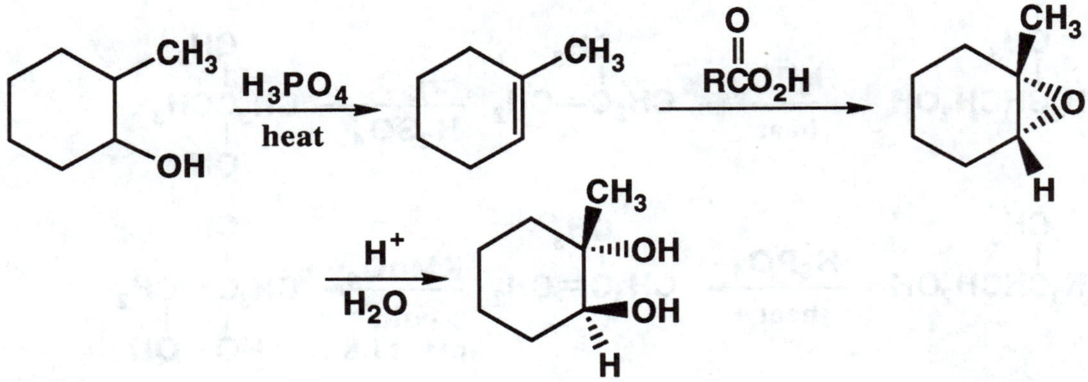

Notice how in parts (e) and (f) it is the stereochemistry of the products that determines which reagents must be used. In the case of (e), the *cis*-diol product is desired, so cold $KMnO_4$ at pH 11.8 can be used. In the case of (f), the *trans*-diol product is desired, and this can be produced by reaction of the epoxide with $H^+$ and $H_2O$.

**Problem 7.47**  Show how to convert the compound on the left to compounds (a), (b), and (c).

**Problem 7.48**  At this point, the following spectroscopy problems may be assigned.  [1]H-NMR and
[13]C-NMR: End-of-Chapter Problems 20.15-20.18
IR Problems: End-of-Chapter Problems 21.5-21.6

# CHAPTER 8: ALKYL HALIDES

## SUMMARY OF REACTIONS

| Starting Material ↓ / Product → | Alcohols | Alkenes | Alkyl Halides | Ammonium Ions | Ethers | Thioethers | Thiols |
|---|---|---|---|---|---|---|---|
| Alkyl Halides | 8A 8.2* | 8B 8.6 | 8C 8.2 | 8D 8.2 | 8E 8.9 | 8F 8.2 | 8G 8.2 |

*Section of book that describes reaction.

## REACTION 8A: FORMATION OF ALCOHOLS: REACTION WITH HYDROXIDE ION AND WATER (Section 8.2)

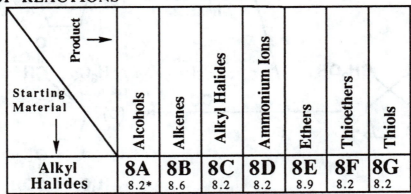

- **Alkyl halides** can be converted into **alcohols** by treatment with either **hydroxide or water**.✱
- For HO ⁻, substitution occurs via an $S_N2$ mechanism with primary alkyl halides, but with secondary and especially tertiary alkyl halides elimination (Reaction 8B) can become important because HO ⁻ is also a relatively strong base.
- The $H_2O$ can react predominantly via an $S_N2$ mechanism with primary alkyl halides, but with secondary and especially tertiary alkyl halides the $S_N1$ mechanism can become important.

## REACTION 8B: FORMATION OF ALKENES: β-ELIMINATION (Section 8.6)

$$H-\overset{|}{\underset{|}{C}}-\overset{|}{\underset{|}{C'}}-X \xrightarrow{\text{base}} \diagdown C = C' \diagup$$

- **Alkyl halides** undergo **β-elimination** in the presence of **base** to produce **alkenes**. ✱
- Primary alkyl halides will only undergo appreciable elimination (E2) with a very strong base, for example RO ⁻.
- Secondary alkyl halides may undergo some E2 elimination with strong bases, or E1 elimination in solvolysis reactions.
- Tertiary alkyl halides readily undergo E2 elimination with base, or E1 in solvolysis reactions.

## REACTION 8C: HALOGEN EXCHANGE (Section 8.2)

$$—C—X \xrightarrow{\;X'^-\;} —C—X' \;+\; X^-$$

- The **halogen** of an alkyl halide can be **exchanged** by using the **halide ion** as a nucleophile in a substitution reaction. ✻
- Like most of the non-basic nucleophiles, the reaction will take place via an $S_N2$ mechanism for methyl, primary, and secondary alkyl halides, but the $S_N1$ mechanism is important for tertiary alkyl halides.

## REACTION 8D: ALKYLATION OF AMINES (Section 8.2)

$$—C—X \xrightarrow{\quad —N: \quad} —C—N^+— \;\; X^-$$

- **Amines** also react with **alkyl halides** to produce **alkylated amines**. ✻

## REACTION 8E: FORMATION OF ETHERS: REACTION WITH ALKOXIDE ANIONS, THE WILLIAMSON ETHER SYNTHESIS (Section 8.9)

$$—C—X \xrightarrow{\quad —C'—O^- \quad} —C—O—C'—$$

- **Alkyl halides** react with **alkoxide anions** to produce **ethers**. ✻
- This reaction is most important for methyl and primary alkyl halides. The alkoxide anions are such strong bases that β-elimination is a competing reaction with secondary and tertiary alkyl halides.

## REACTION 8F: FORMATION OF THIOETHERS: REACTION WITH THIOLATE ANIONS (Section 8.2)

$$—C—X \xrightarrow{\quad —C'—S^- \quad} —C—S—C'—$$

- **Alkyl halides** react with the **thiolate anions** to produce **thioethers**. ✻

## REACTION 8G: FORMATION OF THIOLS: REACTION WITH HS⁻ (Section 8.2)

$$\begin{array}{c}|\\-C-X\\|\end{array} \xrightarrow{\text{HS}^-} \begin{array}{c}|\\-C-SH\\|\end{array}$$

- **Alkyl halides** react with **HS⁻** to produce **thiols**. ✻

## SUMMARY OF IMPORTANT CONCEPTS

### 8.0 OVERVIEW

- **Haloalkanes**, also known as **alkyl halides** in the common nomenclature, are compounds that contain a **halogen atom** attached to an **sp³ carbon atom**. The terms haloalkanes and alkyl halides are used interchangeably for the rest of the chapter. They can be prepared from a variety of different compounds, and they can be converted into a variety of different compounds. Thus, alkyl halides are an important class of molecules for organic synthesis. ✻
- The general symbol for haloalkanes is **R-X**, where **R** is **any alkyl group** (the carbon attached to the halogen must be sp³ hybridized) and **X** can be any of the **halogens**, namely -F, -Cl, -Br, or -I.

### 8.1 NOMENCLATURE

- IUPAC names are derived for haloalkanes by naming the parent hydrocarbon according to normal rules, and treating the halogen atom as a substituent to be listed in alphabetical order like the other substituents. For example, 2-bromobutane or 3-fluoro-4-methylnonane are acceptable IUPAC names.
- When there is a question about numbering the parent chain, number the parent chain so that the first alkyl or halogen substituent is given the lowest possible number.
- Common names of haloalkanes consist of the name of the alkyl group followed by the name of the halide as a separate word. For example, *n*-propyl iodide is the common name for the compound called 1-iodopropane in the IUPAC nomenclature.
  - Haloalkanes in which all of the hydrogens of a hydrocarbon are replaced by halogen atoms are called perhaloalkanes. For example, perchloropropane is the common name for the compound of molecular formula $C_3Cl_8$.

### 8.2 NUCLEOPHILIC ALIPHATIC SUBSTITUTION

- Alkyl halides react with electron rich reagents that are both **Lewis bases** and **nucleophiles**. A **nucleophile** is any species capable of **donating a pair of electrons** to form a new covalent bond. Because nucleophiles can also function as bases, this must always be considered when nucleophiles react with alkyl halides.
- **Nucleophilic substitution reactions** are reactions in which one nucleophile is substituted for another. They are very important reactions for alkyl halides, because a wide variety of different functional groups can be prepared in this way. In these reactions, the **halogen atom** is **replaced by** the **nucleophile**. ✻
  - For example, Table 8.1 in the text lists a number of negatively-charged and neutral nucleophiles that can react with alkyl halides such as methyl bromide to produce numerous types of molecules.

### 8.3 MECHANISMS OF NUCLEOPHILIC ALIPHATIC SUBSTITUTION

- There are two different limiting mechanisms for nucleophilic aliphatic substitution reactions. These are called **S$_N$2 and S$_N$1** mechanisms, and they will now be discussed individually. ✻

• The term $S_N2$ stands for Substitution reaction, Nucleophilic, 2nd order (also called bimolecular). According to the $S_N2$ mechanism, bond-breaking and bond-making occur at the same time. Thus, both the nucleophile and alkyl halide are involved in the rate determining step, hence this is a **bimolecular reaction**. The reaction takes place in a single step, so there is only a single transition state, not any intermediates. In this case, the departing halogen atom is called the **leaving group**. ✳

Transition state in which
Nu-C bond is formed as
C-Br bond is broken.

- A **bimolecular reaction** is one in which two reactants take part in the transition state of the slow or rate-determining step of a reaction. Thus the rates of bimolecular reactions such as $S_N2$ reactions are proportional to the concentration of both the alkyl halide and nucleophile.
- If the halide leaving group is attached to a stereocenter, then the **configuration** of the stereocenter is **inverted** during an $S_N2$ reaction. This is because the nucleophile enters from the **opposite side** of the molecule **as the departing leaving group**, causing the stereocenter to invert analogous to how an umbrella is inverted in the wind.
- $S_N2$ reactions are particularly **sensitive** to **steric factors**, since they are greatly retarded by steric hindrance (crowding) at the site of reaction.
- Since there is no carbocation produced in $S_N2$ reactions, there is no skeletal rearrangement observed.

• The term $S_N1$ stands for Substitution reaction, Nucleophilic, 1st order (also called unimolecular). According to the $S_N1$ mechanism, there are two steps. The carbon-halide bond breaks in the rate-determining first step, creating a carbocation intermediate that then makes a new bond to the nucleophile in the second step. Only the alkyl halide is involved with the rate-determining step, thus the reaction is **unimolecular**. Since the reaction involves two steps, there are two transition states and one intermediate. ✳

- A **unimolecular reaction** is one in which only one reactant takes part in the transition state of the rate-determining-step. Thus the rates of unimolecular reactions such as $S_N1$ reactions are proportional only to the concentration of the alkyl halide.
- Since nucleophiles are not involved in the rate-determining step, stronger nucleophiles do not react faster in $S_N1$ reactions.

- Because the $S_N1$ mechanism involves creation and separation of unlike charges in order to form the carbocation intermediate, polar solvents that can stabilize these charges by solvation greatly accelerate $S_N1$ reactions. For example, $S_N1$ reactions are much faster in water than in ethanol.
- If the leaving group is attached to a stereocenter, then the **configuration** of the stereocenter is **racemized** during an **$S_N1$ reaction**. This is because the carbocation intermediate is planar and therefore achiral, so the nucleophile can approach from either side leading to both possible enantiomers as products. In theory, an $S_N1$ reaction will result in complete racemization, but in fact only partial racemization is observed; with the inversion product predominating. This is accounted for by proposing that while bond-breaking between carbon and the leaving group is complete, the leaving group remains associated for some period of time with the carbocation as an ion pair. To the extent that the leaving group remains associated as an ion pair, it hinders approach of the nucleophile from that face, favoring attack from the opposite face resulting in an excess of inversion.
- **$S_N1$** reactions are greatly accelerated by electronic factors that stabilize carbocations.
- Since there is a carbocation intermediate in $S_N1$ reactions, **skeletal rearrangements** are observed if they produce another carbocation of equal or greater stability.

## 8.4 FACTORS THAT INFLUENCE THE RATE OF $S_N1$ AND $S_N2$ REACTIONS

- Since the nucleophile is involved in the rate-determining-step of the **$S_N2$** reaction, stronger nucleophiles react at a faster rate. Stronger nucleophiles are said to have increased **nucleophilicity**.
- The **structure** of the **alkyl halide** greatly **influences** which **mechanism** will be followed. The order of reactivity for the $S_N2$ mechanism increases in the order: $3° < 2° < 1°$ allylic = $1° <$ methyl, since steric hindrance is highest for $3°$ alkyl halides and least for methyl halides. On the other hand, the order of reactivity for the $S_N1$ mechanism increases in the order: methyl < $1° < 2° < 1°$ allylic < $3°$, since $3°$ carbocations are most stable and methyl carbocations are least stable.✳
  - The net result is that when nucleophilic substitution reactions occur, **methyl** and **1° alkyl** halides react **exclusively** by the **$S_N2$** mechanism. **2° alkyl halides** can react by **either** the **$S_N2$** or **$S_N1$** mechanism. **3° alkyl halides** react **exclusively** by the **$S_N1$** mechanism. Alkyl groups at the β positions of alkyl halides can also inhibit $S_N2$ reactions of even primary alkyl halides due to steric hindrance. ✳
  - Other reaction mechanisms such as β-elimination reactions can take place when a nucleophile/base reacts with an alkyl halide, and the structure of the alkyl halide greatly influences which of these reactions occurs.
- The leaving group develops a partial negative charge as it is departing by either the $S_N1$ or $S_N2$ mechanism. Thus, the lower the basicity, the better a halide is able to function as a leaving group. $I^-$ is the best leaving group, and leaving group ability increases in the order: $F^- < Cl^- < Br^- < I^-$.
- The solvent can have a strong influence on nucleophilic aliphatic substitution reactions. ✳
  - Two different types of solvents are used in substitution reactions, **protic solvents** and **aprotic solvents**. **Protic solvents** are solvents that contain a functional group such as -OH that can act a hydrogen bond donor. **Aprotic solvents** do not have any functional groups that can act as hydrogen bond donors.
  - Solvents are further classified as **polar** and **nonpolar**. **Polar** solvents interact strongly with ions and polar molecules, while **nonpolar** solvents do not interact with ions and polar molecules. **Dielectric constant** is a common measure of solvent polarity and is defined as the amount of electrostatic insulation provided by molecules placed between two charges.
  - Water, formic acid, methanol, and ethanol are considered to be **polar protic solvents**. Important **polar aprotic** solvents include dimethyl sulfoxide, acetonitrile, dimethylformamide, and acetone while **nonpolar aprotic** solvents include

dichloromethane, diethyl ether, and benzene.  It is helpful to categorize solvents this way, because solvents in a given category influence reactions between nucleophiles and alkyl halides in the same ways.

- **Polar protic solvents greatly inhibit $S_N2$ reactions** with negatively charged nucleophiles, because the nucleophile is so highly solvated and thus unreactive.  As a result, $S_N2$ reactions are dramatically faster in polar aprotic solvents such as acetonitrile ($CH_3CN$) compared with polar protic solvents like water. ✳
- On the other hand, **polar sovents favor $S_N1$ reactions**, because the greater the polarity of the solvent, the easier it is to form carbocations. ✳

## 8.6  ß-ELIMINATION

• Most nucleophiles are also bases and alkyl halides are predisposed to ß-elimination, so this must always be considered as a competing reaction.  **ß-Elimination** involves **loss of the halide ion** and a **proton from a ß-carbon atom** (the carbon adjacent to the one with the halide).  This process is also referred to as **dehydrohalogenation**.  The stronger the base, the higher the percentage of ß-elimination product formed in a reaction.  Note that when there is more than one ß carbon atom with a hydrogen atom attached, multiple alkene products are possible. ✳

## 8.7  MECHANISMS  FOR  ß-ELIMINATION

• There are two mechanisms for the ß-elimination reaction of alkyl halides, called **E1** and **E2**.  These are analogous in some ways to the $S_N1$ and $S_N2$ mechanisms discussed above. ✳

- **E1 reactions** involve departure of a leaving group such as a halide ion to create a carbocation (analogous to the first step of the $S_N1$ reaction), followed by departure of a hydrogen atom on a ß-carbon to yield the final product.  Like the $S_N1$ reaction, the carbocation is a true intermediate, and loss of halide ion is the slow step.

The rate-determining step in an E1 reaction is loss of the halide to generate the carbocation.  Thus, the **E1 reaction is unimolecular** (first order) since the rate only depends on the concentration of alkyl halide.

**E1 reactions** give predominantly the **Zaitsev elimination product**, namely the **more highly substituted alkene**.  This is because the transition state of the product determining step has partial double bond character.  The transition state with lower energy is the one with the more stable partial double bond.  A reaction will follow predominantly the reaction path with the rate limiting transition state of lower energy.

- **E2 reactions** are concerted in that the base removes the ß-hydrogen at the same time the C-X bond is broken.

anti and coplanar
geometry of H and X

The only step in the E2 reaction involves both the base and the alkyl halide. Thus, the **E2 reaction is bimolecular** (second order), since the rate depends on both the concentration of base and the concentration alkyl halide.

**E2 reactions** also give predominantly the **Zaitsev elimination product**, since there is significant partial double bond character in the transition state.

## 8.8 SUBSTITUTION VERSUS ELIMINATION

• In the absence of any base, tertiary alkyl halides in polar solvents undergo unimolecular reactions to give a combination of substitution ($S_N1$) and elimination (E1). Although the exact ratios are hard to predict, the amount of substitution can be increased by increasing the concentration of non-basic nucleophile.

• In general, for bimolecular reactions, increased steric hindrance increases the ratio of elimination to substitution products. This is because steric hindrance interferes with the approach of the nucleophile to the backside of the C-X bond, thus impeding the substitution reaction.

- **Tertiary halides** react with all basic reagents to give **elimination products**. There is too much steric hindrance for substitution to compete effectively with elimination.

- **Secondary alkyl halides** have an intermediate amount of steric hindrance and are **borderline**. Substitution or elimination may predominate depending on the particular nucleophile/base, solvent, and temperature of the reaction. Strongly basic nucleophiles such as alkoxides favor E2 reactions, but weakly basic strong nucleophiles favor substitution.

- **Primary alkyl halides** and **methyl halides** have very little steric hindrance, so they react with all nucleophiles, even strongly basic nucleophiles like hydroxide ions and alkoxides ions, to give predominantly **substitution products**.

# CHAPTER 8
## *Solutions to the Problems*

Problem 8.1  Give the IUPAC name for each compound.

(a)   $(CH_3)_2C=CHCH_2Cl$

(b)

(c)   $CH_3-\overset{\overset{\displaystyle OH}{|}}{CH}-CH_2-Cl$

**1-Chloro-3-methyl-2-butene**          **2-Chlorocyclohexanone**          **1-Chloro-2-propanol**

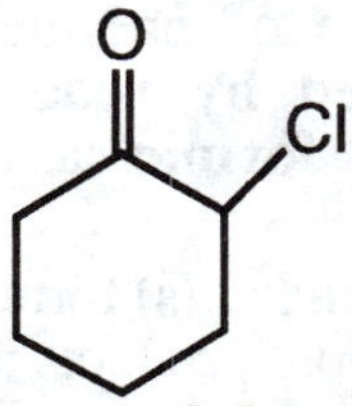

Problem 8.2  Complete these nucleophilic substitution reactions.

(a)   [cyclopentane]—Br  +  $CH_3CH_2S^-$ Na$^+$  $\longrightarrow$  [cyclopentane]—$SCH_2CH_3$  +  Na$^+$B r$^-$

(b)  [cyclopentane]—Br  +  $CH_3CH_2O^-$ Na$^+$  $\longrightarrow$  [cyclopentane]—$OCH_2CH_3$  +  Na$^+$B r$^-$

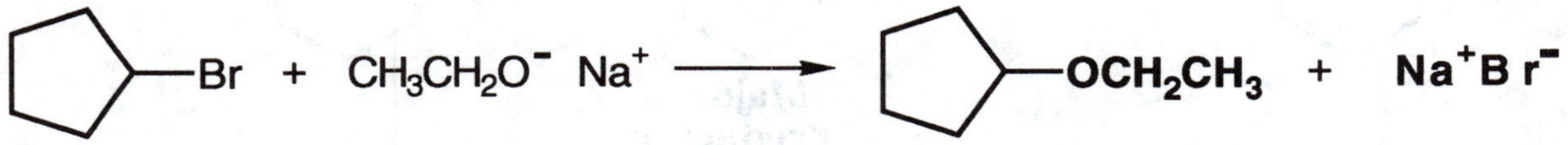

Problem 8.3  Write the expected substitution product(s) for each reaction and predict the mechanism by which each product is formed.

(a) 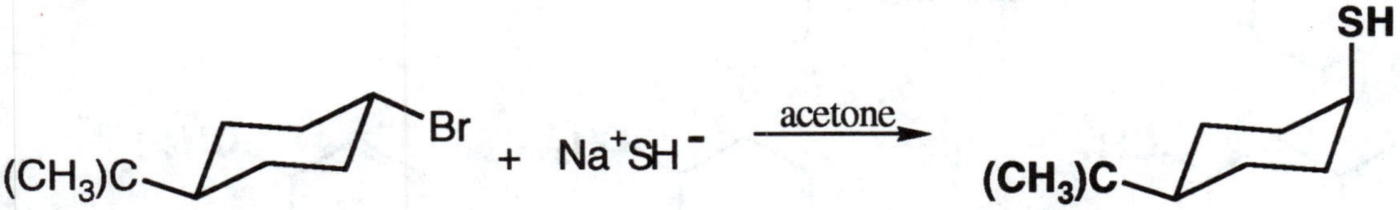  $(CH_3)C$—[cyclohexane]—Br  +  Na$^+$SH$^-$  $\xrightarrow{\text{acetone}}$  $(CH_3)C$—[cyclohexane]—SH

**The SH$^-$ is a very good nucleophile and since the reaction involves a secondary alkyl halide with a good leaving group, the reaction mechanism is S$_N$2.**

(b) 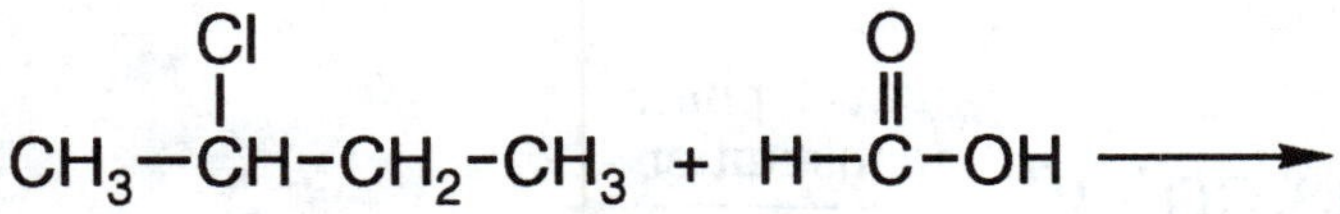  $CH_3-\overset{\overset{\displaystyle Cl}{|}}{CH}-CH_2-CH_3$  +  $H-\overset{\overset{\displaystyle O}{||}}{C}-OH$  $\longrightarrow$

       (R)-Enantiomer

**The alkyl halide is secondary and chloride is a good leaving group.  Formic acid is an excellent ionizing solvent and a poor nucleophile.  Therefore, substitution takes place by an S$_N$1 mechanism and leads to racemization.**

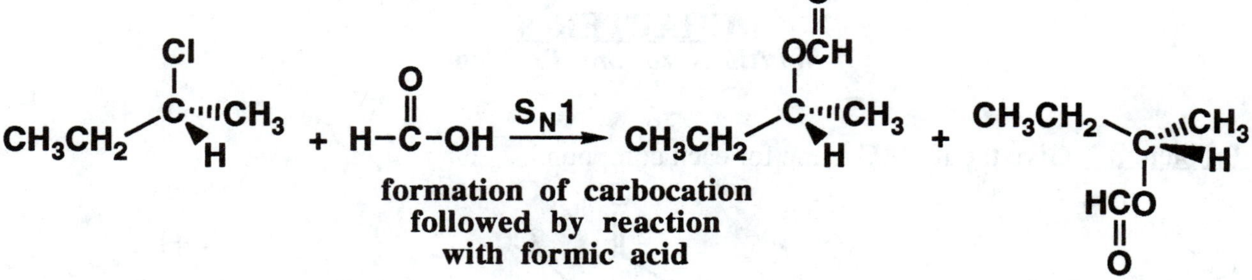

formation of carbocation
followed by reaction
with formic acid

**Problem 8.4** Predict the β-elimination product(s) formed when each chloroalkane is treated with sodium ethoxide in ethanol. If two or more products might be formed, predict which is the major product.

**When there is a choice, the more highly substituted alkene will be the major product as predicted by Zaitsev's rule.**

(a)

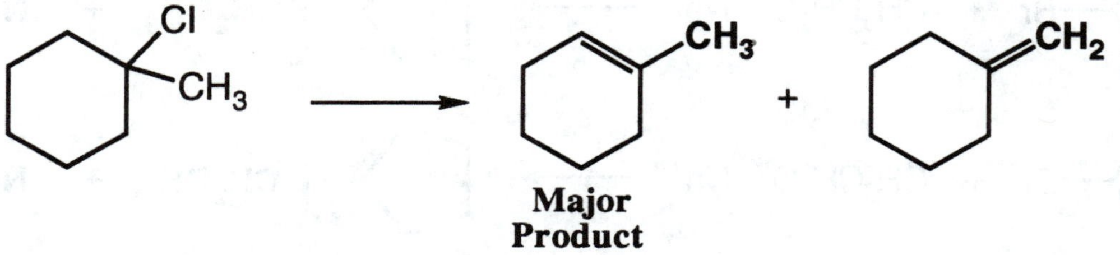

Major
Product

(b)

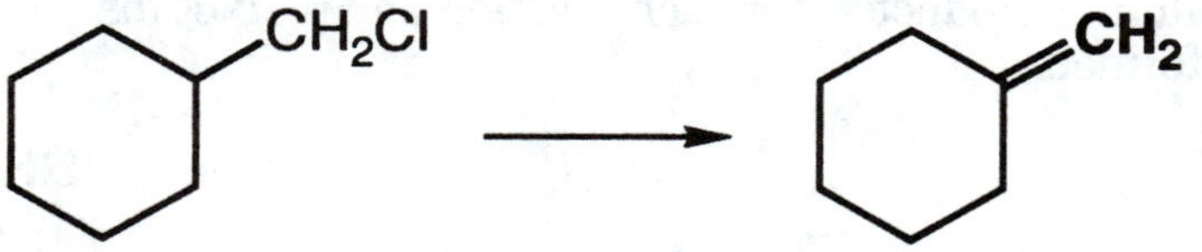

(c)

Similar amounts of each product

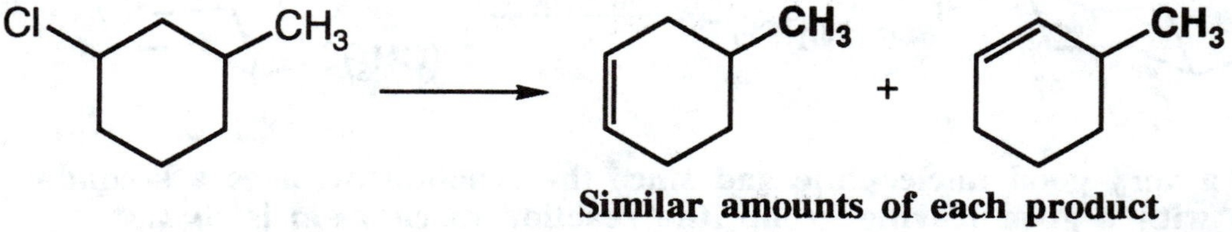

**Problem 8.5** Complete these reactions.

(a)

$$CH_3CHCH_2CH_2Br + (CH_3)_3CO^- K^+ \xrightarrow[\substack{(CH_3)_3COH}]{\substack{nucleophilic \\ substitution}}$$

$$CH_3CHCH_2CH_2OC(CH_3)_3 + K^+Br^-$$

(b)   $CH_3\overset{\overset{\displaystyle CH_3}{|}}{C}HCH_2CH_2Br$  +   $(CH_3)_3C\overset{-}{O}$ $K^+$   $\xrightarrow[\text{(CH}_3\text{)}_3\text{COH}]{\text{β-elimination}}$

$CH_3\overset{\overset{\displaystyle CH_3}{|}}{C}HCH=CH_2$  +  $(CH_3)_3COH$  +  $K^+Br^-$

Problem 8.6  Predict whether each reaction proceeds predominantly by substitution or elimination, or whether the two compete.  Write structural formulas for the major organic product(s).

**According to Table 8.7, E2 will be the main reaction observed when a secondary alkyl halide reacts with a strong base such as sodium methoxide.**

(a) $CH_3CH_2\overset{\overset{\displaystyle |}{|}}{C}H\text{-}CH_2\text{-}CH_3$ + $CH_3\overset{-}{O}$ $Na^+$   $\xrightarrow[\text{methanol}]{}$   $CH_3CH_2CH{=}CHCH_3$ + $CH_3OH$ + $Na^+I^-$

**According to Table 8.7, $S_N2$ will be the main reaction observed when a secondary alkyl halide reacts with a strong nucleophile / weak base such as sodium iodide.**

(b)   

+ $Na^+I^-$   $\xrightarrow[\text{acetone}]{}$   + $Na^+Cl^-$

Problem 8.7  Show how you might use the Williamson ether synthesis to prepare these ethers.
(a)   $(CH_3CH_2CH_2CH_2)_2O$

**Treatment of 1-bromobutane with sodium butoxide gives dibutyl ether.**

$CH_3CH_2CH_2CH_2\overset{-}{O}$ $Na^+$ + $CH_3CH_2CH_2CH_2Br$  $\longrightarrow$  $(CH_3CH_2CH_2CH_2)_2O$
+ $Na^+Br^-$

(b)   $-CH_2\text{-}O\text{-}\overset{\overset{\displaystyle CH_3}{|}}{\underset{\underset{\displaystyle CH_3}{|}}{C}}\text{-}CH_3$

**There is only one combination of alkyl halide and metal alkoxide that gives cyclopentyl *tert*-butyl ether in good yield.**

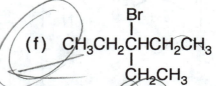

If this synthesis were attempted with the alkoxide derived from cyclopentanol and a *tert*-butyl halide, the reaction would produce an alkene by an E2 pathway.

## Nomenclature
**Problem 8.8** Give IUPAC names for these compounds.

(a) $CH_2{=}CF_2$

**1,1-Difluoroethene**

(b) ⬡—Br

**3-Bromocyclopentene**

(c) $(CH_3)_2CHCH_2CH_2CHCH_3$ with Cl

**2-Chloro-5-methylhexane**

(d) $Cl(CH_2)_6Cl$

**1,6-Dichlorohexane**

(e) $CF_2Cl_2$

**Dichloro-difluoromethane**

(f) $CH_3CH_2CHCH_2CH_3$ with Br and $CH_2CH_3$

**3-Bromo-3-methylpentane**

**Problem 8.9** Give IUPAC names for the following compounds. Be certain to include a designation of configuration in your answer.

(a)

**(R)-2-Bromobutane**

(b)

**trans-1-Bromo-4-methyl-cyclohexane**

(c) $H_3C-C$ with H, $(CH_2)_2CH_3$, Br

**(S)-2-Bromopentane**

(d)

**(E)-1-Chloro-2-butene**

Problem 8.10 Draw structural formulas for these compounds (given are IUPAC names).
(a) 3-Bromopropene                          (b) (R)-2-Chlorobutane

$CH_2\!\!=\!\!CHCH_2Br$

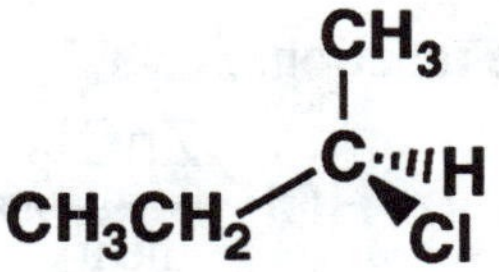

(c) *meso*-2,3-Dibromobutane 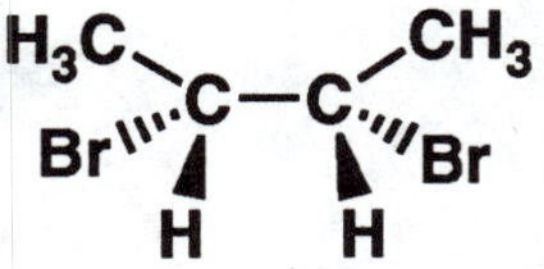                (d) *trans*-1-Bromo-3-isopropylcyclohexane

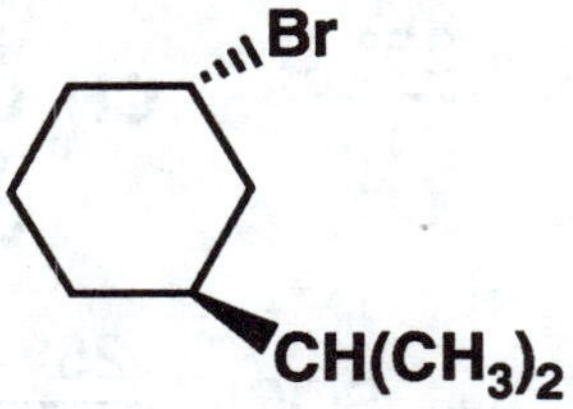

(e) 1,2-Dichloroethane 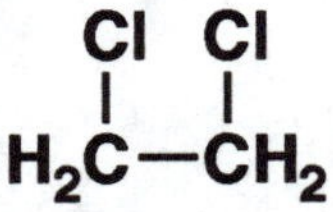                      (f) Bromocyclobutane

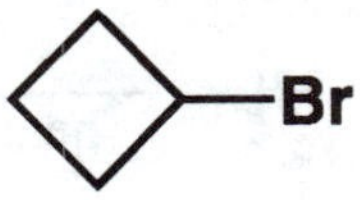

Problem 8.11 Draw structural formulas for these compounds (given are common names).
(a) Isopropyl chloride                      (b) *sec*-Butyl bromide

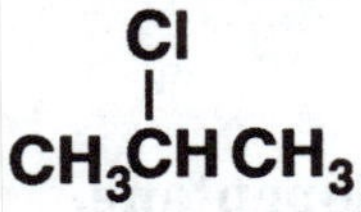

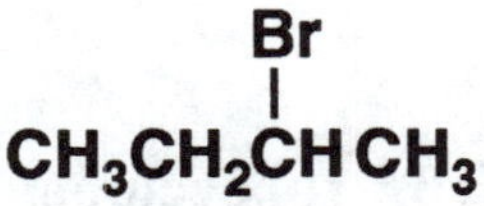

(c) Allyl iodide                            (d) Methylene chloride

$CH_2\!\!=\!\!CHCH_2I$                        $CH_2Cl_2$

(e) Chloroform                              (f) *tert*-Butyl chloride

$CHCl_3$                                    $(CH_3)_3CCl$

(g) Isobutyl chloride

$(CH_3)_2CHCH_2Cl$

Problem 8.12 Which compounds are secondary (2°) alkyl halides?
(a) Isobutyl chloride          (b) 2-Iodooctane              (c) *trans*-1-Chloro-4-
                                                                 methylcyclohexane

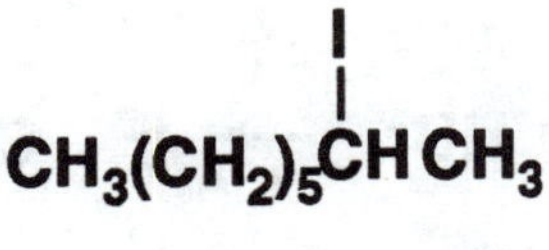

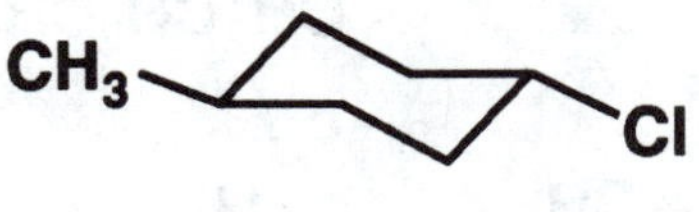

$(CH_3)_2CHCH_2Cl$             $CH_3(CH_2)_5CHCH_3$

**As can be seen in the above structures, (b) and (c) are secondary alkyl halides, while (a) is a primary alkyl halide.**

## Synthesis of Alkyl Halides

Alkyl halides may be synthesized from alkanes (Section 2.9B), from alkenes (Section 4.3A) and from alcohols (Section 7.4D).

Problem 8.13  Complete these reactions.

(a) $CH_3CH_2CH_2CH_2OH$  +  HCl  $\xrightarrow[\text{heat}]{ZnCl_2}$  $CH_3CH_2CH_2CH_2Cl$  +  $H_2O$

(b)  $CH_3\overset{\displaystyle CH_3}{\underset{\displaystyle CH_3}{C}OH}$  +  HCl  $\xrightarrow{25°\ C}$  $CH_3\overset{\displaystyle CH_3}{\underset{\displaystyle CH_3}{C}Cl}$  +  $H_2O$

(c)  $CH_3(CH_2)_5CH_2OH$  +  $SOCl_2$  $\xrightarrow{25°\ C}$  $CH_3(CH_2)_5CH_2Cl$  +  HCl  +  $SO_2$

(d) $CH_3CH_2CH_2CH{=}CH_2$  +  HI  $\xrightarrow{\text{heat}}$  $CH_3CH_2CH_2\overset{\displaystyle |}{C}HCH_3$

Problem 8.14  What alkene or alkenes and reaction conditions give each alkyl halide in good yield?

(a)

**Electrophilic addition of HBr to cyclopentene will give bromocyclopentane.**

(b)  $CH_3\overset{\displaystyle CH_3}{\underset{\displaystyle Br}{C}}CH_2CH_2CH_3$

**Electrophilic addition of HBr to either 2-methyl-1-pentene or 2-methyl-2-pentene will give 2-bromo-2-methylpentane in accord with Markovnikov's rule.**

(c)

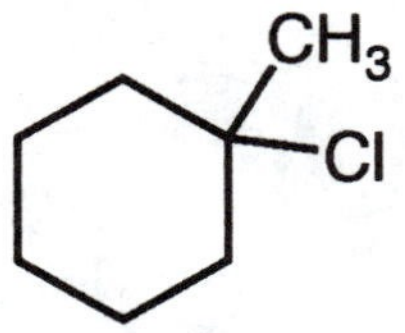

**Electrophilic addition of HCl to either of the two alkenes shown below will give 1-chloro-1-methylcyclohexane in accord with Markovnikov's rule.**

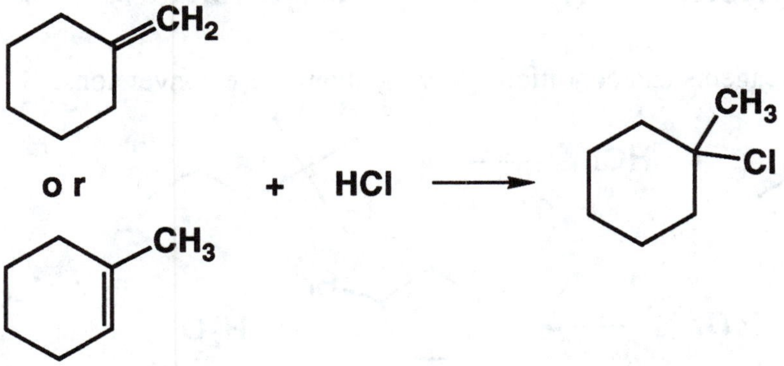

**or** + **HCl** ⟶

<u>Problem 8.15</u> What alcohol and reaction conditions give each alkyl halide in good yield?

(a)

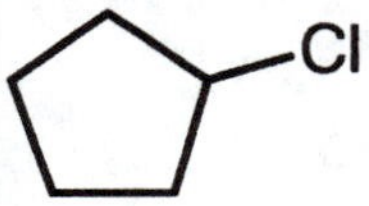

**This transformation can be accomplished by reaction of cyclopentanol with concentrated HCl or with thionyl chloride, SOCl$_2$.**

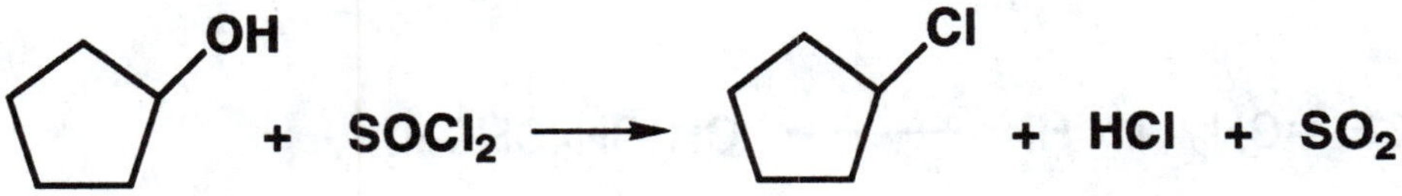

OH + HCl ⟶ Cl + H$_2$O

OH + SOCl$_2$ ⟶ Cl + HCl + SO$_2$

(b) 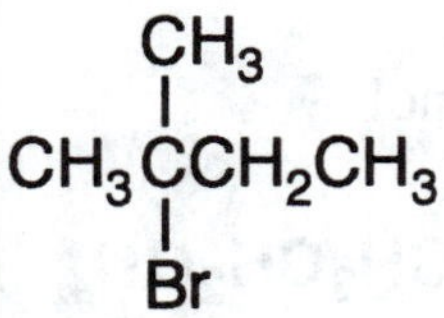 CH$_3$CCH$_2$CH$_3$ with CH$_3$ on top and Br on bottom

**This transformation can be accomplished through the reaction of 2-methyl-2-butanol with HBr to give the intended product.**

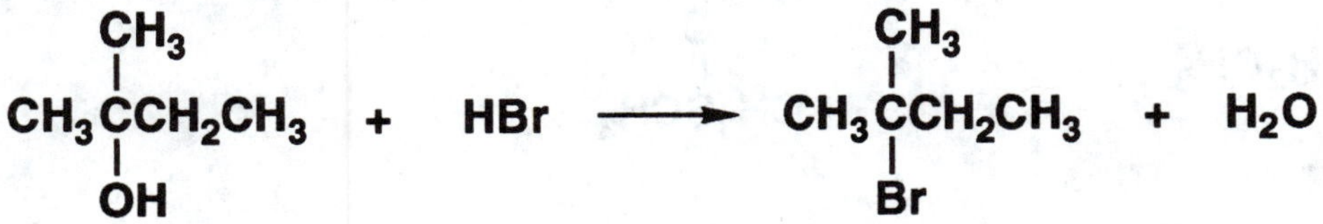

CH$_3$CCH$_2$CH$_3$ (with CH$_3$ on top, OH on bottom) + HBr ⟶ CH$_3$CCH$_2$CH$_3$ (with CH$_3$ on top, Br on bottom) + H$_2$O

$$CH_3CHCH_2CH_2I$$
(c) with $CH_3$ on the CH

**Reaction of 3-methyl-1-butanol with HI will give the desired product.**

$$CH_3CHCH_2CH_2OH \ + \ HI \ \longrightarrow \ CH_3CHCH_2CH_2I \ + \ H_2O$$

(with $CH_3$ groups on each middle carbon)

**Problem 8.16** Show reagents and conditions to bring about these conversions.

(a) 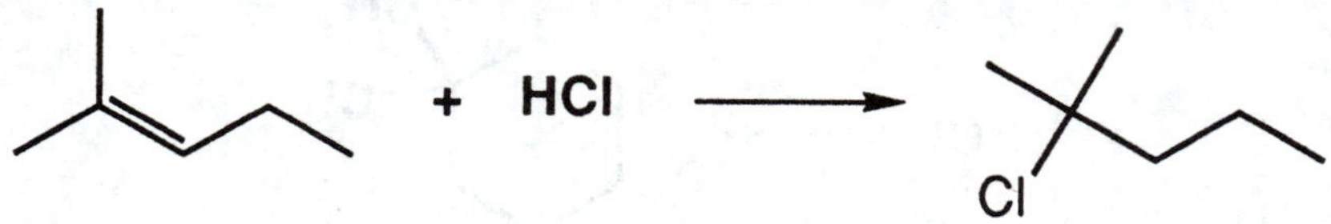 + HCl ⟶

(b)  + HBr ⟶ + H_2O

(c) $CH_3CH{=}CHCH_3 \ + \ HCl \ \longrightarrow \ CH_3CHCH_2CH_3$ (with Cl)

(d) 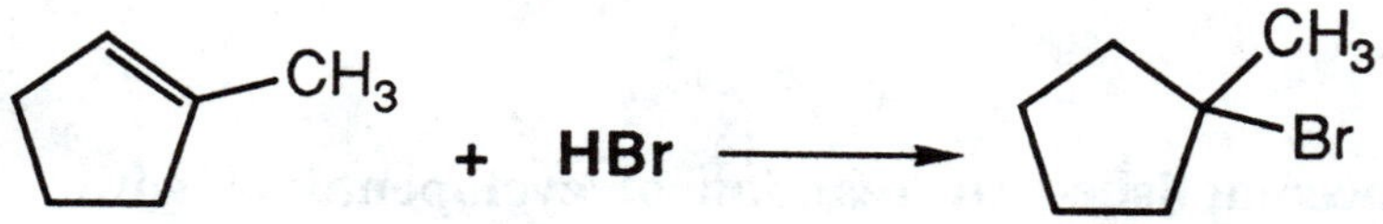 + HBr ⟶

(e) $CH_3\overset{\underset{|}{CH_3}}{\underset{|}{CH_3}}CCH_2CH_2OH$     HCl + ZnCl₂
      or
      SOCl₂     ⟶     $CH_3\overset{\underset{|}{CH_3}}{\underset{|}{CH_3}}CCH_2CH_2Cl$

(f) $CH_3CH_2CH{=}CH_2 \ + \ HI \ \longrightarrow \ CH_3CH_2\overset{\underset{|}{}}{C}HCH_3$ (with I)

## Nucleophilic Aliphatic Substitution

**Problem 8.17** Write structural formulas for these common organic solvents.

(a) Methylene chloride           (b) Acetone                    (c) Ethanol

$$CH_2Cl_2$$

$$CH_3\overset{O}{\overset{\|}{C}}CH_3$$

$$CH_3CH_2OH$$

(d) Diethyl Ether               (e) Dimethyl sulfoxide

$$CH_3CH_2OCH_2CH_3$$

$$CH_3\overset{O}{\overset{\|}{S}}CH_3$$

Problem 8.18 Arrange these protic solvents in order of increasing polarity.
(a) $H_2O$                    (b) $CH_3CH_2OH$                    (c) $CH_3OH$

**Alkyl groups decrease the polarity of a solvent, so the following are ranked in order of least to most polar solvents:**
$$CH_3CH_2OH > CH_3OH > H_2O$$

Problem 8.19 Arrange these aprotic solvents in order of increasing polarity.
(a) Acetone                    (b) Pentane                    (c) Diethylether

**The carbonyl group of acetone is a polar functional group, so acetone is the most polar of the three.  The oxygen atom of diethylether adds polarity to this solvent compared to the hydrocarbon pentane, so the solvents ranked in order of least to most polar are:**
**Pentane > Diethyl ether > Acetone**

Problem 8.20 From each pair, select the stronger nucleophile.

(a)    $H_2O$ or $OH^-$        (b)    $CH_3\overset{\overset{\displaystyle O}{\|}}{C}O^-$ or $OH^-$        (c)    $CH_3SH$ or $CH_3S^-$

$OH^- > H_2O$        $OH^- > CH_3\overset{\overset{\displaystyle O}{\|}}{C}O^-$        $CH_3S^- > CH_3SH$

Problem 8.21 Which statements are true for $S_N2$ reactions of alkyl halides?
(a) Both the alkyl halide and the nucleophile are involved in the transition state.  **True**
(b) Reaction proceeds with inversion of configuration at the site of reaction.  **True**
(c) Reaction proceeds with retention of optical activity.  **True**
(d) The order of reactivity is $3° > 2° > 1° >$ methyl.  **False**
(e) The nucleophile must have an unshared pair of electrons and bear a negative charge.  **False**
(f) The greater the nucleophilicity, the faster the reaction.  **True**

**Statement (d) is false because 3° alkyl halides are the least reactive and methyl halides are the most reactive.  Statement (e) is false because a nucleophile does not have to possess a negative charge.**

Problem 8.22 Complete these $S_N2$ reactions.

(a)    $Na^+I^-$ + $CH_3CH_2CH_2Cl$ $\xrightarrow{\text{acetone}}$ $CH_3CH_2CH_2I$ + $Na^+Cl^-$

(b)    $NH_3$ + —Br $\xrightarrow{\text{ethanol}}$ —$NH_3^+$ $Br^-$

(c) $CH_3CH_2O^- Na^+$ + $CH_2{=}CHCH_2Cl$ $\xrightarrow{\text{ethanol}}$ $CH_2{=}CHCH_2OCH_2CH_3$ + $Na^+Cl^-$

**Problem 8.23** Complete these $S_N2$ reactions.

(a) 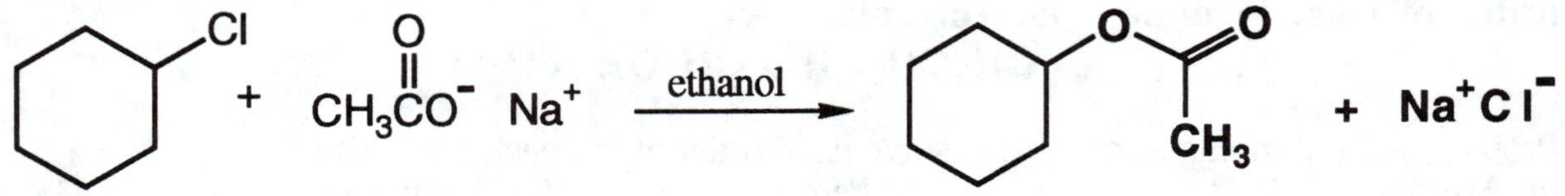 + $CH_3\overset{O}{\overset{\|}{C}}O^- Na^+$ $\xrightarrow{\text{ethanol}}$ + $Na^+Cl^-$

(b) $CH_3\overset{I}{\underset{|}{C}H}CH_2CH_3$ + $CH_3CH_2S^- Na^+$ $\xrightarrow{\text{acetone}}$ $CH_3\overset{SCH_2CH_3}{\underset{|}{C}H}CH_2CH_3$ + $Na^+I^-$

(c) $CH_3\overset{CH_3}{\underset{|}{C}H}CH_2CH_2Br$ + $Na^+I^-$ $\xrightarrow{\text{acetone}}$ $CH_3\overset{CH_3}{\underset{|}{C}H}CH_2CH_2I$ + $Na^+Br^-$

**Problem 8.24** Complete these $S_N2$ reactions.

(a) $CH_3CH_2CH_2Cl$ + $CH_3CH_2O^- Na^+$ $\xrightarrow{\text{ethanol}}$ $CH_3CH_2CH_2OCH_2CH_3$ + $Na^+Cl^-$

(b) $(CH_3)_3N$ + $CH_3I$ $\xrightarrow{\text{acetone}}$ $(CH_3)_4N^+I^-$

(c) 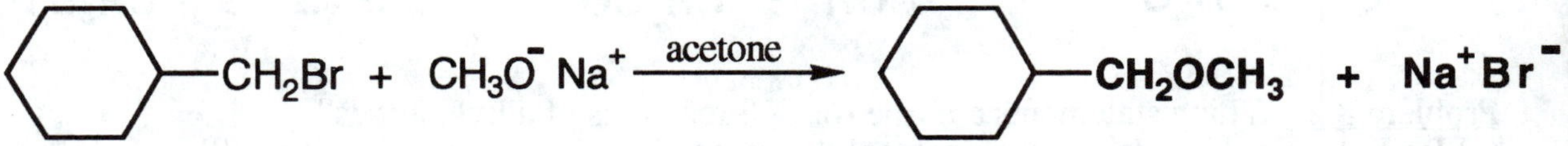 $-CH_2Br$ + $CH_3O^- Na^+$ $\xrightarrow{\text{acetone}}$ $-CH_2OCH_3$ + $Na^+Br^-$

(d) 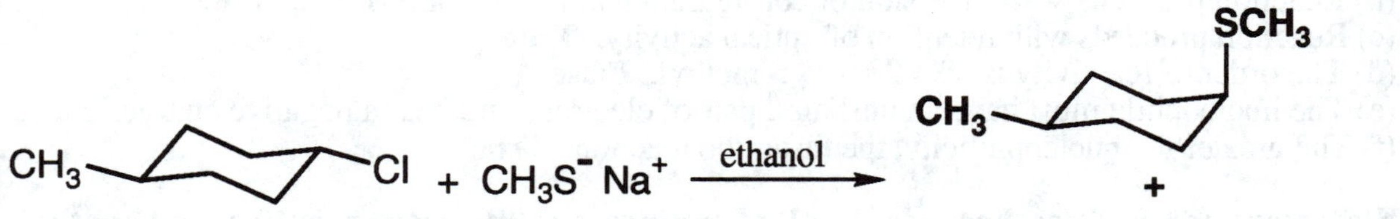 + $CH_3S^- Na^+$ $\xrightarrow{\text{ethanol}}$

+

$Na^+Cl^-$

**Notice the inversion of configuration due to the $S_N2$ mechanism**

(e) 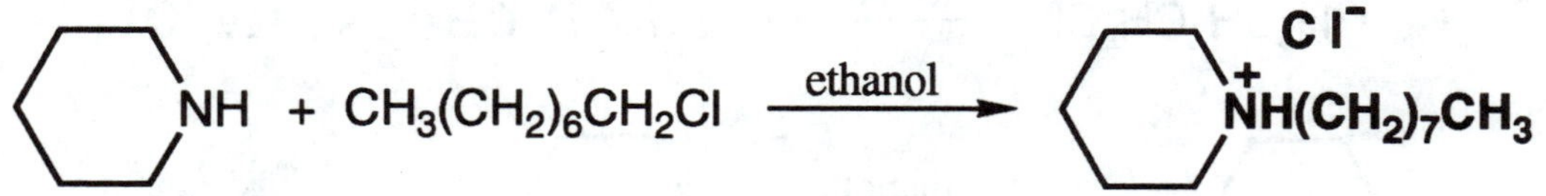 + $CH_3(CH_2)_6CH_2Cl$ $\xrightarrow{\text{ethanol}}$

(f) 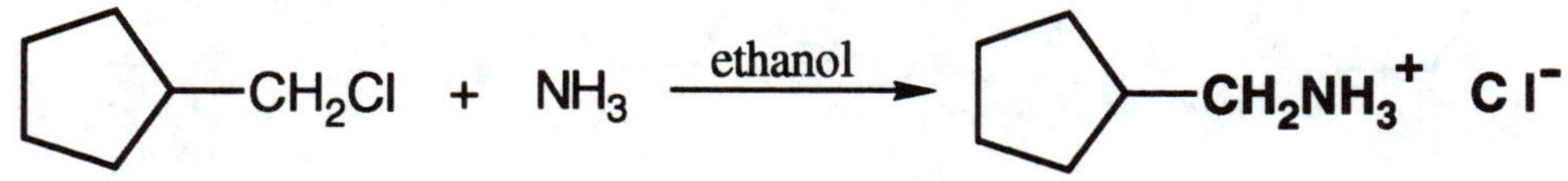 $-CH_2Cl$ + $NH_3$ $\xrightarrow{\text{ethanol}}$ $-CH_2NH_3^+$ $Cl^-$

Problem 8.25  You were told that each reaction in the Problem 8.24 proceeds by an $S_N2$ mechanism.  Suppose you were not told the mechanism.  Describe how you could conclude from the structure of the alkyl halide, the nucleophile, and the solvent that each reaction is in fact an $S_N2$ reaction.

(a)  A primary halide, strong nucleophile/strong base in ethanol, a moderately ionizing solvent all favor $S_N2$.

(b)  Trimethylamine is a moderate nucleophile.  A methyl halide in acetone, a weakly ionizing solvent, all work together to favor $S_N2$.

(c)  A primary halide, strong nucleophile/strong base in ethanol, a moderately ionizing solvent all favor $S_N2$.

(d)  The alkyl chloride is secondary, so either an $S_N1$ or $S_N2$ mechanism is possible.  Ethyl sulfide ion is a strong nucleophile, but weak base.  It, therefore, reacts by an $S_N2$ pathway.

(e)  Piperidine is a moderate nucleophile.  The primary alkyl halide in ethanol, a moderately ionizing solvent, work together to favor $S_N2$.

(f)  Ammonia is a weak base and good nucleophile, and the halide is primary. Therefore, $S_N2$ is favored.

Problem 8.26  Select the member from each pair that shows the faster rate of $S_N2$ reaction with KI in acetone.

The relative rates of $S_N2$ reactions for pairs of molecules in this problem depend on the fact that a primary carbon without β-branching is less hindered and more reactive toward $S_N2$ substitution than a primary carbon with one, two or three branches on the β-carbon.

(a)   $CH_3CH_2CH_2CH_2Cl$ or $CH_3\underset{\underset{CH_3}{|}}{C}HCH_2Cl$

$CH_3CH_2CH_2CH_2Cl$ faster than $CH_3\underset{\underset{CH_3}{|}}{C}HCH_2Cl$

(b)   $CH_3\underset{\underset{CH_3}{|}}{C}HCH_2CH_2Br$ or $CH_3\overset{\overset{CH_3}{|}}{\underset{\underset{CH_3}{|}}{C}}CH_2Br$

$CH_3\underset{\underset{CH_3}{|}}{C}HCH_2CH_2Br$ faster than $CH_3\overset{\overset{CH_3}{|}}{\underset{\underset{CH_3}{|}}{C}}CH_2Br$

(c)    $CH_3CH_2CH_2CHCH_3$   **or**   $CH_3CHCHCH_3$
$\qquad\qquad\qquad\qquad\qquad\qquad\qquad\qquad CH_3$

$CH_3CH_2CH_2CHCH_3$    **faster than**    $CH_3CHCHCH_3$
$\qquad\qquad\qquad\qquad\qquad\qquad\qquad\qquad\qquad\qquad\qquad CH_3$

**Problem 8.27** In these reactions, an alkyl halide is treated with a compound that has two nucleophilic sites. Select the more nucleophilic site in each part and show the product of each $S_N2$ reaction.

(a)   $HOCH_2CH_2NH_2$   +   $CH_3I$   $\xrightarrow{\text{ethanol}}$  
$\qquad\qquad\qquad\qquad\qquad\qquad\qquad\qquad\qquad\qquad CH_3$
$\qquad\qquad\qquad\qquad\qquad\qquad\qquad\qquad\qquad HOCH_2CH_2\overset{+}{N}H_2 \quad I^-$

(b)   [morpholine structure]   +   $CH_3I$   $\xrightarrow{\text{ethanol}}$   [N-methylmorpholinium iodide structure]

(c)   $HOCH_2CH_2SH$   +   $CH_3I$   $\xrightarrow{\text{ethanol}}$   $HOCH_2CH_2SCH_3$   +   $HI$

**Problem 8.28** Which statements are true for $S_N1$ reactions of alkyl halides?
(a) Both the alkyl halide and the nucleophile are involved in the transition state of the rate limiting step. **False**
(b) Reaction proceeds with retention of configuration at the site of reaction. **False**
(c) Reaction at a stereocenter proceeds with loss of optical activity. **True**
(d) The order of reactivity is $3° > 2° > 1° >$ methyl. **True**
(e) The greater the steric crowding around the reactive center, the slower the rate of reaction. **False**
(f) Rate of reaction is greater with good nucleophiles compared with poor nucleophiles. **False**

**Statements (a), (e), and (f) are false because only the alkyl halide is involved with the rate-limiting step. Statement (b) is false because optical activity is lost due to the achiral trigonal planar carbocation intermediate that is formed.**

**Problem 8.29** Draw the structural formula for the product of these $S_N1$ reactions.

$\qquad\qquad\qquad Cl$
(a)   $(S)\text{-}CH_3CHCH_2CH_3 + CH_3CH_2OH \xrightarrow{\text{ethanol}}$

$\qquad\qquad\qquad\qquad\qquad\qquad\qquad\qquad OCH_2CH_3$
$\qquad\qquad\qquad\qquad\qquad (R)+(S)\text{-}CH_3CHCH_2CH_3 + HCl$

(b)

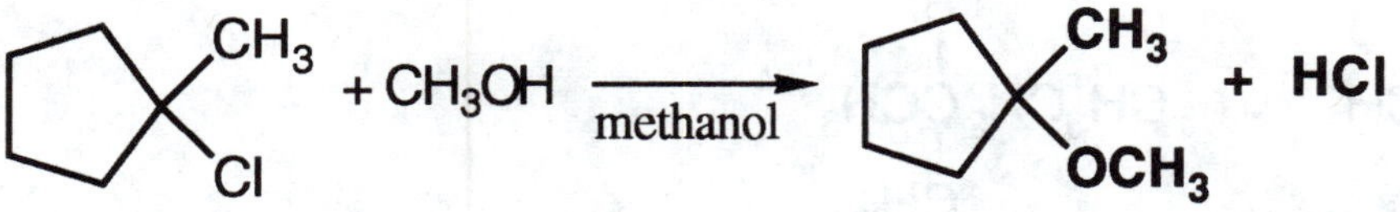

$$\text{(ring with } CH_3 \text{ and } Cl) + CH_3OH \xrightarrow{\text{methanol}} \text{(ring with } CH_3 \text{ and } OCH_3) + HCl$$

(c)

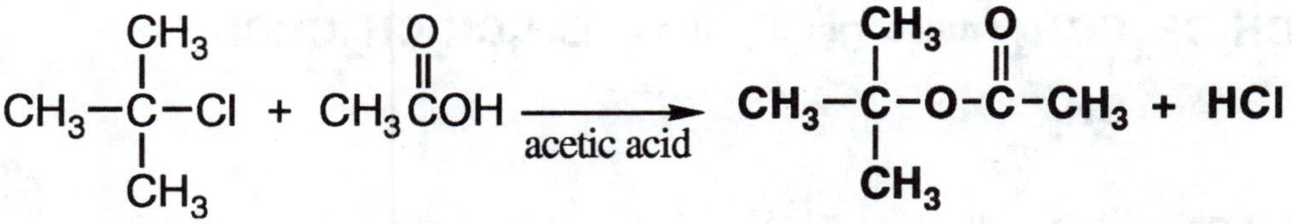

$$CH_3-\overset{\overset{\displaystyle CH_3}{|}}{\underset{\underset{\displaystyle CH_3}{|}}{C}}-Cl + CH_3\overset{\overset{\displaystyle O}{||}}{C}OH \xrightarrow{\text{acetic acid}} CH_3-\overset{\overset{\displaystyle CH_3}{|}}{\underset{\underset{\displaystyle CH_3}{|}}{C}}-O-\overset{\overset{\displaystyle O}{||}}{C}-CH_3 + HCl$$

(d)

$$\text{(cyclohexene ring)}-Br + CH_3OH \xrightarrow{\text{methanol}} \text{(cyclohexene ring)}-OCH_3 + HBr$$

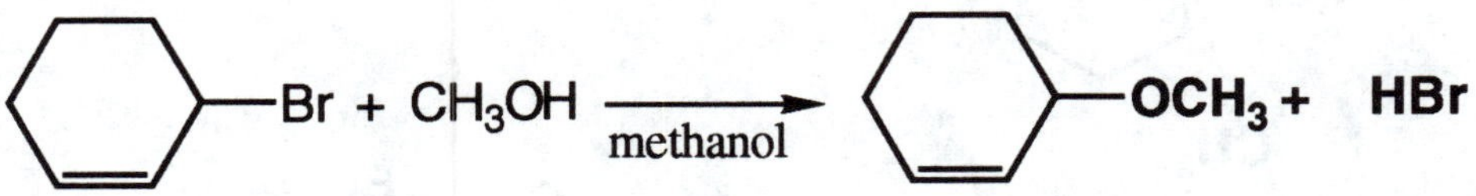

**Problem 8.30** You were told that each reaction in Problem 8.29 proceeds by an $S_N1$ mechanism. Suppose you were not told the mechanism. Describe how you could conclude from the structure of the alkyl halide, the nucleophile, and the solvent that each reaction is in fact an $S_N1$ reaction.

**(a) Chlorine is a good leaving group and the resulting secondary carbocation is a stable carbocation intermediate. Ethanol is a moderately ionizing solvent and a poor nucleophile.**
**(b) Methanol is a moderately ionizing solvent and a poor nucleophile. Chlorine is a good leaving group and the resulting carbocation is tertiary.**
**(c) Acetic acid is a strongly ionizing solvent and a poor nucleophile. Chlorine is a good leaving group and the resulting carbocation is tertiary.**
**(d) Methanol is a moderately ionizing solvent and a poor nucleophile. Bromine is a good leaving group, and the resulting carbocation is both secondary and allylic.**

**Problem 8.31** Select the member of each pair that undergoes nucleophilic substitution in aqueous ethanol more rapidly.

**Water is an ionizing solvent and there are no strong nucleophiles present in the solution, so the following reactions will proceed by an $S_N1$ mechanism. Thus, the alkyl halide of each pair that can form the more stable carbocation (i.e., the carbocation with more alkyl groups attached) will react more rapidly.**

(a)    $CH_3(CH_2)_3CH_2Cl$  or  $CH_3(CH_2)_2\overset{\overset{\displaystyle Cl}{|}}{C}HCH_3$

$CH_3(CH_2)_2\overset{\overset{\displaystyle Cl}{|}}{C}HCH_3$  more rapidly than  $CH_3(CH_2)_3CH_2Cl$

(b) 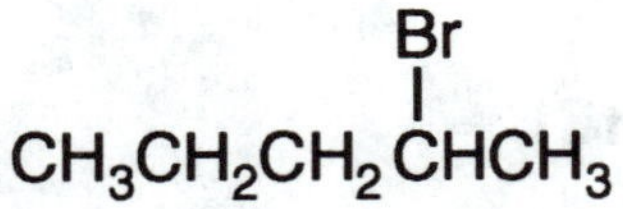 $CH_3CH_2CH_2\overset{\overset{\displaystyle Br}{|}}{C}HCH_3$  or  $CH_3CH_2\overset{\overset{\displaystyle Br}{|}}{\underset{\underset{\displaystyle CH_3}{|}}{C}}CH_3$

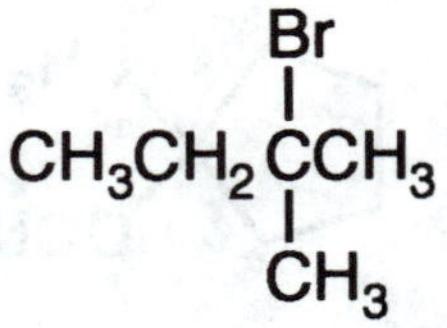

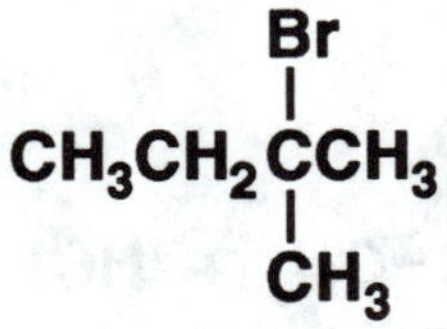

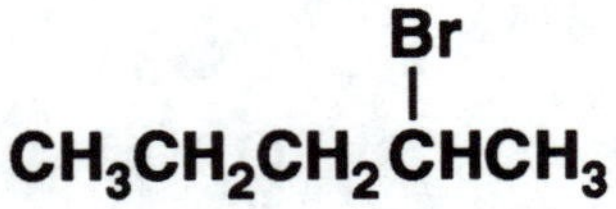

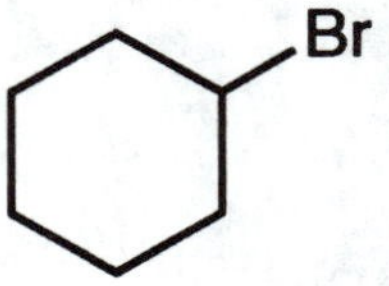

$CH_3CH_2\overset{\overset{\displaystyle Br}{|}}{\underset{\underset{\displaystyle CH_3}{|}}{C}}CH_3$  more rapidly than  $CH_3CH_2CH_2\overset{\overset{\displaystyle Br}{|}}{C}HCH_3$

(c)

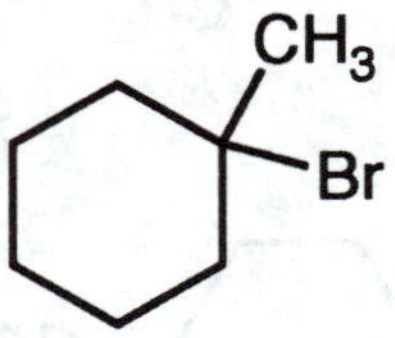

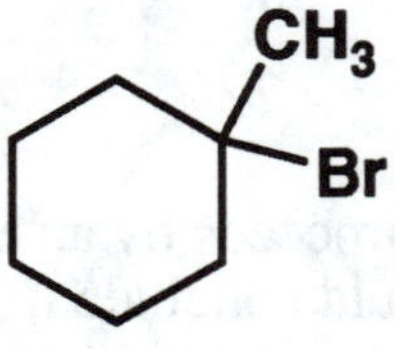

more rapidly than

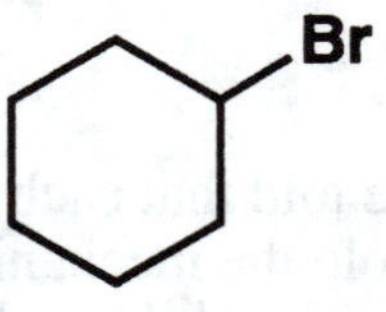

**Problem 8.32**  Propose a mechanism for formation of products (but not their relative percentages) in this reaction.

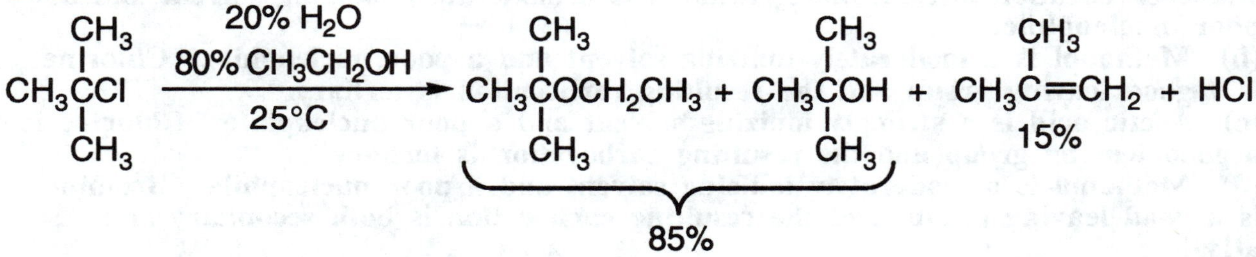

$$CH_3\overset{\overset{\displaystyle CH_3}{|}}{\underset{\underset{\displaystyle CH_3}{|}}{C}}Cl \xrightarrow[\underset{25°C}{}]{\overset{20\% \ H_2O}{80\% \ CH_3CH_2OH}} CH_3\overset{\overset{\displaystyle CH_3}{|}}{\underset{\underset{\displaystyle CH_3}{|}}{C}}OCH_2CH_3 + CH_3\overset{\overset{\displaystyle CH_3}{|}}{\underset{\underset{\displaystyle CH_3}{|}}{C}}OH + CH_3\overset{\overset{\displaystyle CH_3}{|}}{C}=CH_2 + HCl$$

85%

15%

**All of the reaction products shown can be produced from the same intermediate, a carbocation that arises from ionization of the carbon-halogen bond. Notice that the reaction is run in an ionizing solvent, there are no strong nucleophiles around, and the tertiary alkyl halide starting material gives a very stable tertiary carbocation intermediate. All of these factors work in concert to favor $S_N1$ and E1 reaction mechanisms.**

$$CH_3CCl \text{ (with } CH_3 \text{ up and } CH_3 \text{ down)} \longrightarrow CH_3C^+ \text{ (with } CH_3 \text{ up and } CH_3 \text{ down)}$$

$$\xrightarrow[S_N1]{CH_3CH_2OH} CH_3COCH_2CH_3 \text{ (with } CH_3 \text{ up, } CH_3 \text{ down)}$$

$$\xrightarrow[S_N1]{H_2O} CH_3COH \text{ (with } CH_3 \text{ up, } CH_3 \text{ down)}$$

$$\xrightarrow{E1} CH_3C=CH_2 \text{ (with } CH_3 \text{ up)}$$

**Problem 8.33** The rate of reaction in Problem 8.32 increases by 140 times when carried out in 80% water: 20% ethanol compared with 40% water: 60% ethanol. Account for this rate difference.

**The reaction mechanism involves formation of a carbocation intermediate and water is a more ionizing solvent than ethanol. In other words, the carbocation forms easier in water. Thus, the higher the percentage of water in the solvent, the faster the rate-limiting formation of the carbocation and the faster the rate of the reaction.**

**Problem 8.34** Show how you might synthesize these compounds from an alkyl halide and a nucleophile:

(a)  cyclohexane—$NH_2$

**Treatment of a halocyclohexane with two moles of ammonia. The first mole of ammonia is for displacement of bromine. The second mole of ammonia is to neutralize the HBr formed in the substitution reaction.**

$$\text{cyclohexane—Br} + 2NH_3 \longrightarrow \text{cyclohexane—}NH_2 + NH_4^+Br^-$$

(b)  cyclohexane—$CH_2NH_2$

**Treatment of chloromethylcyclohexane with two moles of ammonia. The first mole of ammonia is for displacement of chlorine. The second mole of ammonia is to neutralize the HCl formed in the substitution reaction.**

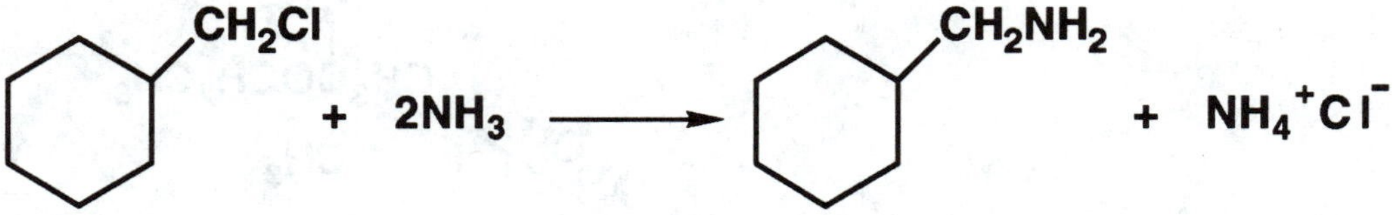

$$CH_2Cl + 2NH_3 \longrightarrow CH_2NH_2 + NH_4^+Cl^-$$

(c)

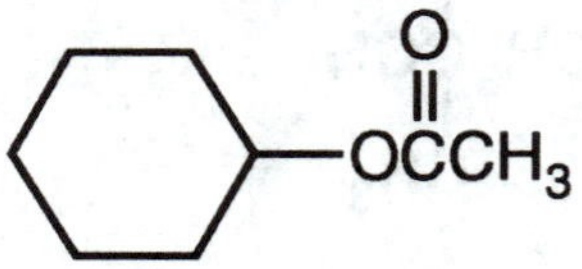

**Treatment of a halocyclohexane with the sodium salt of acetic acid.**

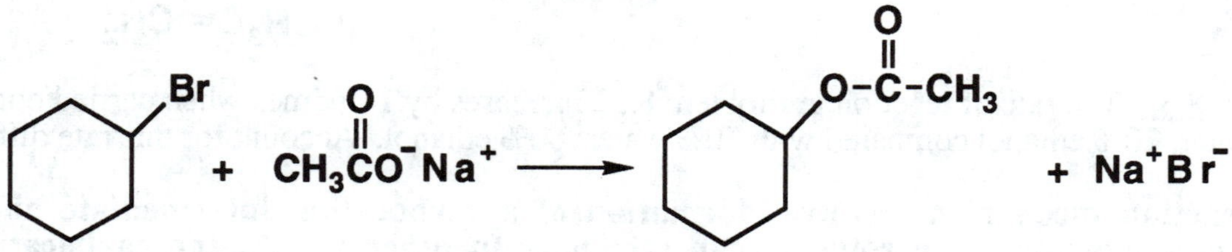

(d)   $CH_3(CH_2)_3CH_2SH$

**Treatment of 1-halopentane with sodium hydrosulfide.**

$$CH_3(CH_2)_3CH_2Br + HS^-Na^+ \longrightarrow CH_3(CH_2)_3CH_2SH + Na^+Br^-$$

(e)

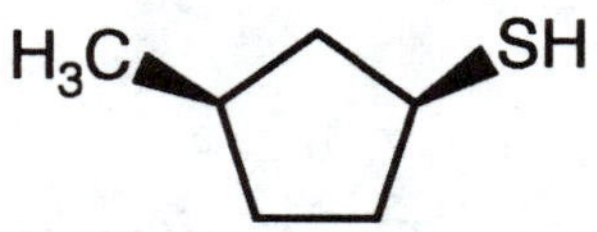

**Treatment of the appropriate halocyclopentane with sodium hydrosulfide.  Note how the stereocenter is inverted upon reaction with the nucleophile.**

(f)   $CH_3CH_2OCH_2CH_3$

**Treatment of a haloethane with sodium or potassium ethoxide in ethanol.**

$$CH_3CH_2O^-Na^+ + CH_3CH_2I \xrightarrow[CH_3CH_2OH]{} CH_3CH_2OCH_2CH_3 + Na^+I^-$$

**Problem 8.35** Draw structural formulas for the alkene(s) formed by treatment of each alkyl halide with sodium ethoxide in ethanol.  Assume that elimination is by an E2 mechanism.  Where two or more alkenes are possible, use Zaitsev's rule to predict which alkene is the major product.

(a)    $\underset{\overset{|}{CH_3}}{\overset{\overset{Br}{|}\ \overset{CH_3}{|}}{CH_3\,CHCCH_3}}$

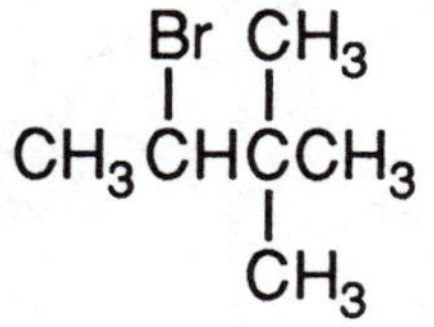

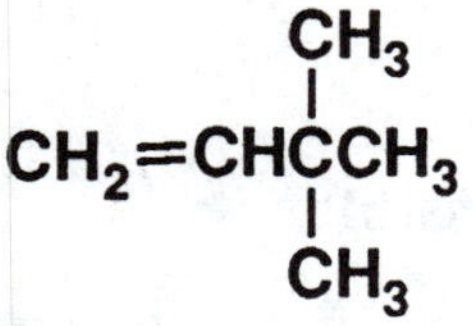

(b)

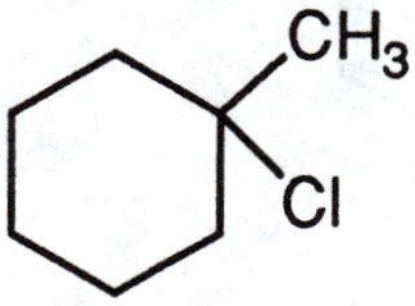

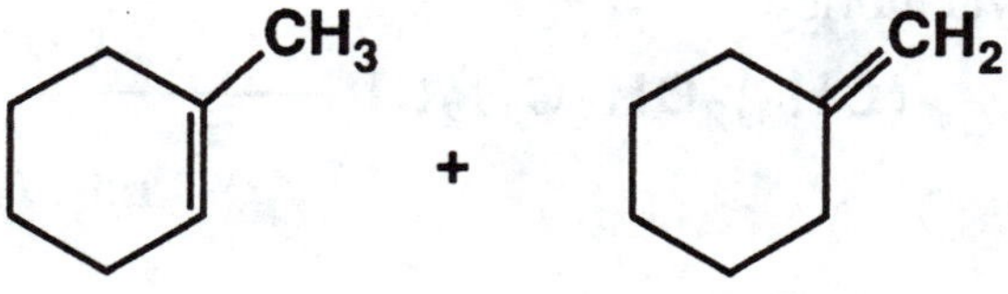

$\underset{\overset{|}{CH_3}}{\overset{\overset{CH_3}{|}}{CH_2{=}CHCCH_3}}$

+

**major**                          **minor**

(c)

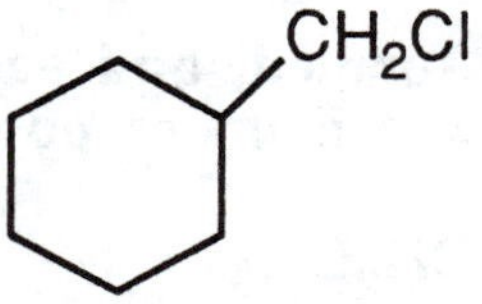

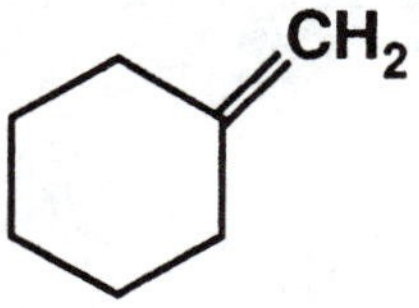

(d)   $\underset{\overset{|}{CH_3}}{\overset{\overset{CH_3}{|}}{H_2C{=}CHCH_2CBr}}$

$H_2C{=}CH{-}CH{=}C\overset{\diagup CH_3}{\diagdown CH_3}$

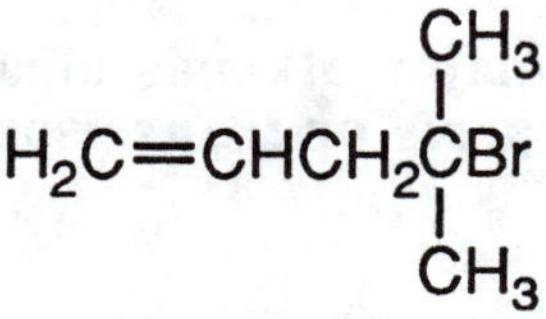

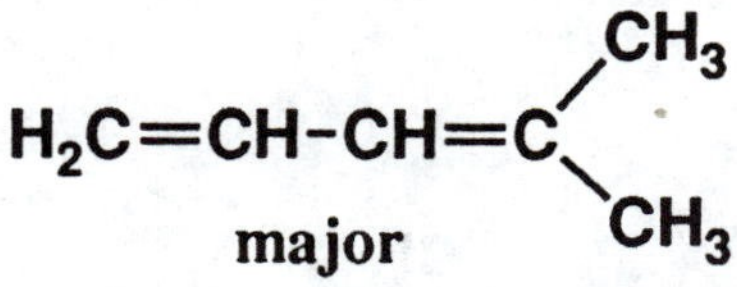

**major**

$H_2C{=}CH{-}CH_2{-}C\overset{\diagup CH_2}{\diagdown CH_3}$

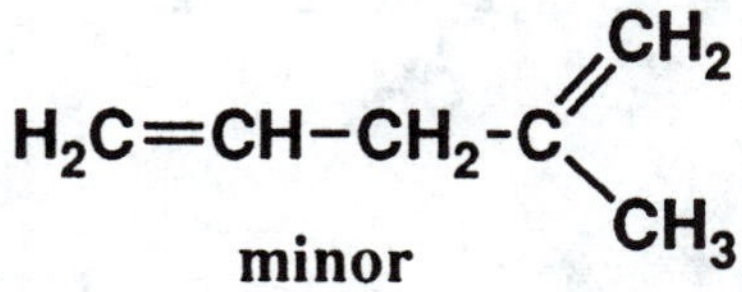

**minor**

**Problem 8.36** Which alkyl halides undergo dehydrohalogenation to give alkenes that do not show *cis-trans* isomerism?
(a) 2-Chloropentane

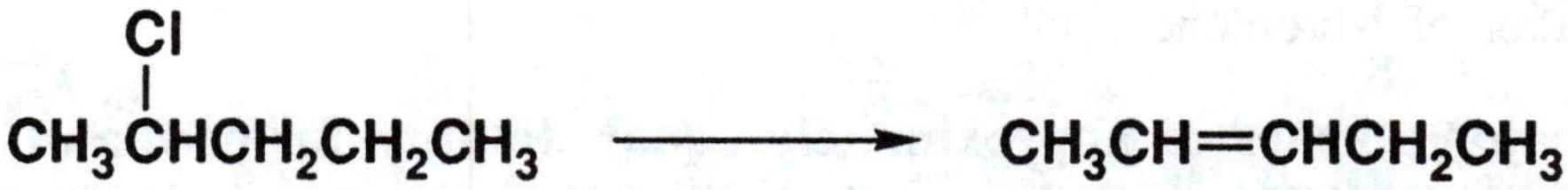

$\underset{}{\overset{\overset{Cl}{|}}{CH_3\,CHCH_2CH_2CH_3}}$   $\longrightarrow$   $CH_3CH{=}CHCH_2CH_3$

**major Zaitsev product**
*cis-trans* **isomers are possible**

(b) 2-Chlorobutane

$\underset{}{\overset{\overset{Cl}{|}}{CH_3\,CHCH_2CH_3}}$   $\longrightarrow$   $CH_3CH{=}CHCH_3$

**major Zaitsev product**
*cis-trans* **isomers are possible**

(c) Chlorocyclohexane

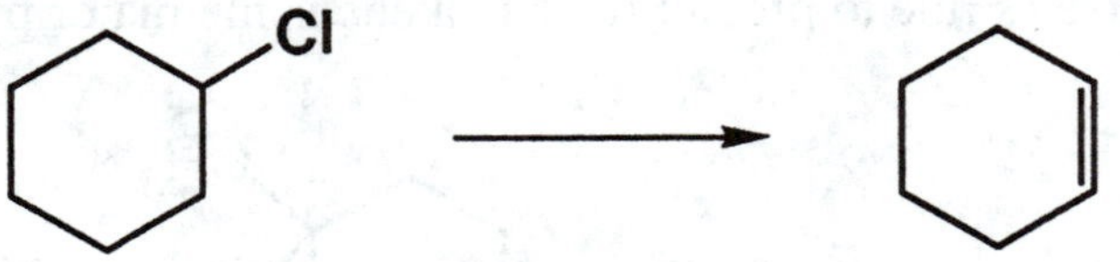

**no *cis-trans* isomers possible**

(d) Isobutyl chloride

$$(CH_3)_2CH_2CH_2Cl \longrightarrow (CH_3)_2CH=CH_2$$

**no *cis-trans* isomers possible**

<u>Problem 8.37</u>  How many isomers, including *cis-trans* isomers, are possible for the major product of dehydrohalogenation of 3-chloro-3-methylhexane?

**There are two major alkenes of approximately equal stability formed, and each of those can have a *cis* or *trans* isomer.  Thus, there are 2 x 2 = 4 isomers possible.**

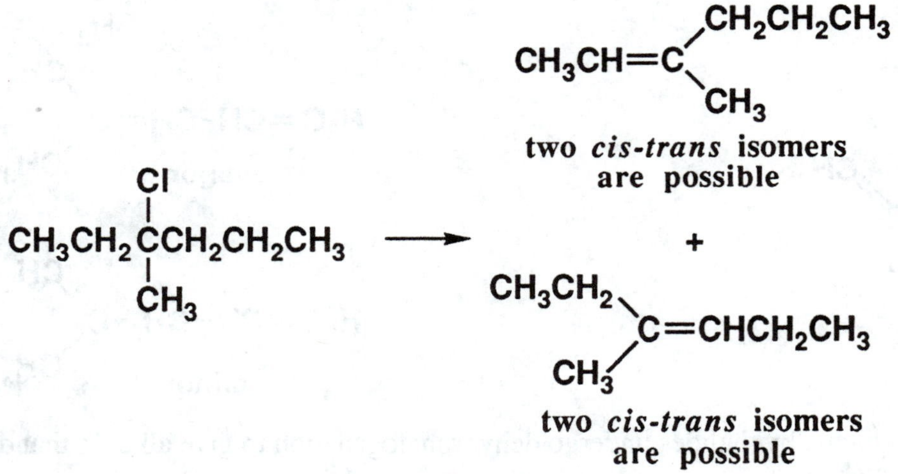

<u>Problem 8.38</u>  How many isomers, including *cis-trans* isomers, are possible for the major product of dehydrohalogenation of 3-bromohexane?

**There are two major alkenes of approximately equal stability formed, and each of those can have a *cis* or *trans* isomer.  Thus, there are 2 x 2 = 4 isomers possible.**

$$CH_3CH=CHCH_2CH_2CH_3$$

two *cis-trans* isomers
are possible

$$\underset{\displaystyle \text{Br}}{CH_3CH_2\,CHCH_2CH_2CH_3} \longrightarrow$$

+

$$CH_3CH_2CH=CHCH_2CH_3$$

two *cis-trans* isomers
are possible

<u>Problem 8.39</u> What alkyl halide might you use as a starting material to produce each alkene in high yield and uncontaminated by isomeric alkenes?

(a)

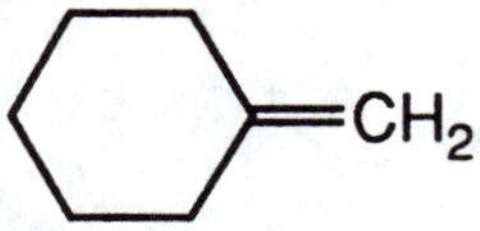

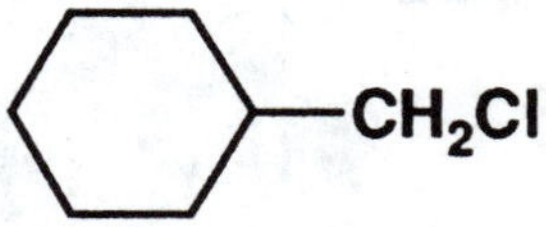

$$\text{(b)}\quad CH_3\underset{\displaystyle CH_3}{CH}CH_2CH=CH_2$$

$$CH_3\underset{\displaystyle CH_3}{CH}CH_2CH_2CH_2Cl$$

## Substitution Versus Elimination

<u>Problem 8.40</u> Consider the following statements in reference to $S_N1$, $S_N2$, E1, and E2 reactions of alkyl halides. To which mechanism(s), if any, does each statement apply?
(a) It involves a carbocation intermediate.
   **$S_N1$ and E1**

(b) It is bimolecular.
   **$S_N2$ and E2**

(c) It involves inversion of configuration at the site of substitution.
   **$S_N2$**

(d) Substitution at a stereocenter gives predominantly a racemic product.
   **$S_N1$**

(e) It is unimolecular
   **$S_N1$ and E1**

(f) The order of reactivity is 3° > 2° > 1° > methyl.
   **$S_N1$, E1, and E2**

(g) The order of reactivity is methyl > 1° > 2° > 3°.
   **$S_N2$**

<u>Problem 8.41</u> Draw a structural formula for the major organic product of each reaction and specify the most likely mechanism for formation of the product you have drawn.

**The substitution and elimination products for the reactions are given in bold.  In each case, the different parameters discussed in the chapter are considered including the type of alkyl halide (primary, secondary, tertiary, etc.) and the relative strength of the nucleophile/base.**

(a)  [cyclohexene ring]—Br + CH$_3$OH $\xrightarrow{\text{methanol}}$

[cyclohexadiene] **(E1)** + [cyclohexenyl]—OCH$_3$ **(S$_N$1)**

(b)  CH$_3$ĊCH$_2$CH$_3$ (with CH$_3$ above, Cl below) + Na$^+$OH$^-$ $\xrightarrow[\text{H}_2\text{O}]{80°}$  [product] **(E2)**

$$\begin{array}{ccc} CH_3 & & H \\ & C=C & \\ CH_3 & & CH_3 \end{array}$$

(c)  CH$_3$ĊHCH$_2$CH$_2$CH$_3$ (Cl above) + CH$_3$C̈O$^-$ Na$^+$ $\xrightarrow{\text{DMSO}}$
    **(R)-Enantiomer**

$$\begin{array}{c} O \\ \| \\ C-CH_3 \\ O \\ | \\ CH_3CHCH_2CH_2CH_3 \end{array}$$   **(S$_N$2)**

    **(S)-Enantiomer**

(d)  CH$_3$ĊHCH$_2$CH$_3$ (Cl above) + HC̈OH $\longrightarrow$
    **(R)-Enantiomer**

$$\begin{array}{c} O \\ \| \\ CH \\ O \\ | \\ CH_3CHCH_2CH_3 \end{array}$$   **(S$_N$1)**    ̸E′

    **(R) + (S)  Enantiomers**

(e)  CH$_3$ĊHCH$_2$CH$_2$CH$_3$ (Cl above) + CH$_3$CH$_2$O$^-$ Na$^+$ $\xrightarrow{\text{ethanol}}$ **CH$_3$CH=CHCH$_2$CH$_3$  (E2)**
    **(R)-Enantiomer**

(f)  CH$_3$CH$_2$O$^-$ Na$^+$ + CH$_2$=CHCH$_2$Cl $\longrightarrow$ **CH$_3$CH$_2$OCH$_2$CH=CH$_2$  (S$_N$2)**

**Problem 8.42** Which set of reagents is more likely to undergo reaction by an E2 mechanism?

(a)  [cyclopentane with O⁻Na⁺ and CH₃ substituents] + CH₃CH₂Cl $\xrightarrow{\text{DMSO}}$

(b)  [cyclopentane with Br and CH₃ substituents] + CH₃CH₂O⁻ Na⁺ $\xrightarrow{\text{DMSO}}$

**The reagents in reaction (b) are more likely to undergo an E2 elimination reaction, because the halide is on a tertiary carbon atom.**

## Williamson Ether Synthesis

**Problem 8.43** Here are two reactions intended to give *tert*-butyl ethyl ether. One reaction gives the ether in good yield; the other reaction does not. Which reaction gives the ether? What is the major product of the other reaction, and how do you account for its formation?

(a)  $\underset{\underset{CH_3}{|}}{\overset{\overset{CH_3}{|}}{CH_3CO^-}} K^+$ + CH₃CH₂Cl $\xrightarrow{\text{2-methyl-2-propanol}}$ $\underset{\underset{CH_3}{|}}{\overset{\overset{CH_3}{|}}{CH_3COCH_2CH_3}}$ + KCl

(b)  CH₃CH₂O⁻ K⁺ + $\underset{\underset{CH_3}{|}}{\overset{\overset{CH_3}{|}}{CH_3CCl}}$ $\xrightarrow{\text{ethanol}}$ $\underset{\underset{CH_3}{|}}{\overset{\overset{CH_3}{|}}{CH_3COCH_2CH_3}}$ + KCl

**The only reaction that will give the desired ether product in good yield is the one shown in (a). In (b), the major product will be the elimination product isobutylene, $CH_2=C(CH_3)_2$, because the halide is on a tertiary carbon atom and ethoxide is a strong base.**

Problem 8.44 Each ether can, in principle, be synthesized by two different combinations of alkyl halide and metal alkoxide. Show one combination of alkyl halide and alkoxide that forms ether bond (1) and another that forms ether bond (2). Which combination gives the higher yield of ether?

(a)

**As the better combination, choose (2) which involves reaction of an alkoxide with a primary halide and will give substitution as the major product. Scheme (1) involves a strong base/strong nucleophile and secondary halide, conditions that will give both substitution and elimination products.**

(1) + $CH_3CH_2O^- Na^+ \longrightarrow$

(2) + $CH_3CH_2Cl \longrightarrow$

(b) $CH_3-O-\underset{\underset{CH_3}{|}}{\overset{\overset{CH_3}{|}}{C}}-CH_3$

**Because of the high degree of branching in the haloalkane in (2), $S_N2$ substitution by this pathway is virtually impossible. Therefore, choose (1) as the only reasonable alternative.**

(1) $CH_3-Cl + CH_3\underset{\underset{CH_3}{|}}{\overset{\overset{CH_3}{|}}{C}}O^- Na^+ \longrightarrow$           (2) $CH_3-O^- Na^+ + CH_3\underset{\underset{CH_3}{|}}{\overset{\overset{CH_3}{|}}{C}}Cl \longrightarrow$

(1) (2)

CH₃

(c) 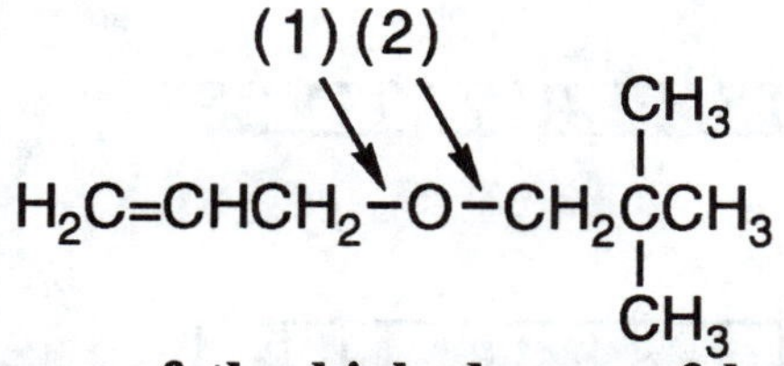 H₂C=CHCH₂–O–CH₂CCH₃

CH₃

**Because of the high degree of branching on the β-carbon in the haloalkane in (2), S_N2 substitution by this pathway is virtually impossible. Therefore, choose (1) as the only reasonable alternative.**

CH₃

**(1)** 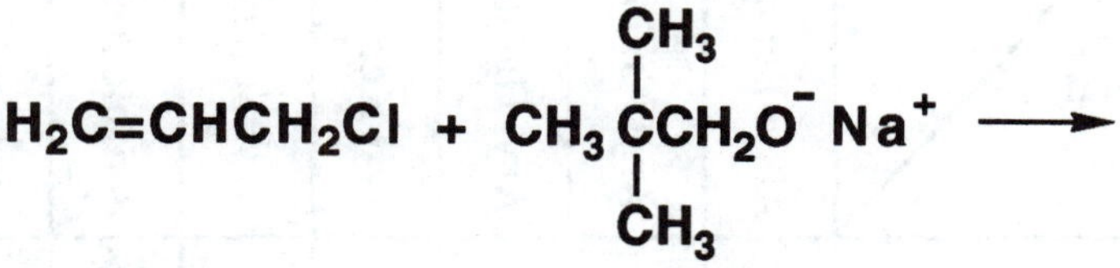 H₂C=CHCH₂Cl + CH₃CCH₂O⁻ Na⁺ ⟶

CH₃

CH₃

**(2)** 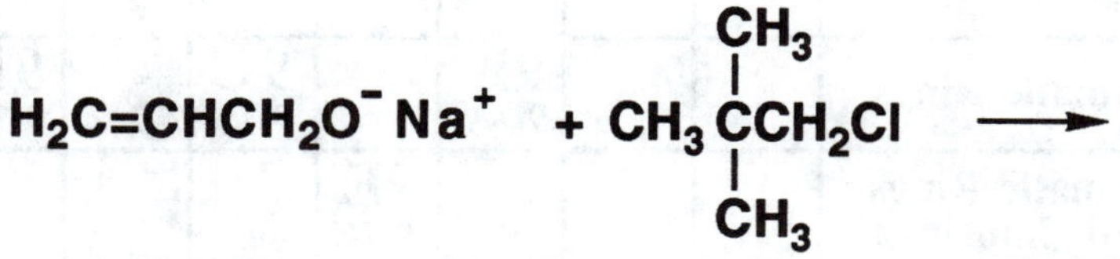 H₂C=CHCH₂O⁻ Na⁺ + CH₃CCH₂Cl ⟶

CH₃

# CHAPTER 9: BENZENE AND ITS DERIVATIVES

## SUMMARY OF REACTIONS

| Starting Material ↓ / Product → | Alkyl Benzenes | Aryl Amines | Aryl Halides | Aryl Ketones | Aryl Sulfonic Acids | Benzoic Acids | Nitro Aromatics | Phenoxides |
|---|---|---|---|---|---|---|---|---|
| Alkyl Benzenes | | | | | | **9A** 9.5* | | |
| Aromatic Rings | | | **9B** 9.7A | | **9C** 9.7B | | **9D** 9.7B | |
| Aromatic Rings Acid Chlorides | | | | **9E** 9.7C | | | | |
| Aromatic Rings Alcohols | **9F** 9.7D | | | | | | | |
| Aromatic Rings Alkenes | **9G** 9.7D | | | | | | | |
| Aromatic Rings Alkyl Halides | **9H** 9.7C | | | | | | | |
| Nitro Aromatics | | **9I** 9.7B | | | | | | |
| Phenols | | | | | | | | **9J** 9.4 |

*Section in book that describes reaction.

## REACTION 9A: OXIDATION AT A BENZYLIC POSITION (Section 9.5)

- Compounds with at least **one benzylic hydrogen** react with $K_2Cr_2O_7$ in aqueous sulfuric acid ($H_2SO_4$) to produce **benzoic acid**. Notice that other groups attached to the benzylic carbon atom are removed in the process. ✳

# REACTION 9B: BROMINATION AND CHLORINATION (Section 9.7A)

- **Aromatic rings** react with $Br_2$ in the presence of the **Lewis acid catalyst FeBr$_3$** to give an **aryl bromide**. This is an example of **electrophilic aromatic substitution**. ✳
- The mechanism involves an initial reaction between $Br_2$ and $FeBr_3$ to generate a molecular complex that can rearrange to give a $Br^+$ $FeBr_4^-$ ion pair. This reacts as a very strong electrophile with the weakly nucleophilic aromatic pi cloud to form a resonance-stabilized cation that loses a proton to give the final product.
- An analogous reaction can be carried out with **$Cl_2$ and $FeCl_3$ or $AlCl_3$** to give an **aryl chloride**.

# REACTION 9C: SULFONATION (Section 9.7B)

- **Aromatic rings** react with **sulfuric acid** upon heating to yield **aryl sulfonic acids** via **electrophilic aromatic substitution**. ✳
- The mechanism involves reaction of $HSO_3^+$ as a very strong electrophile with the weakly nucleophilic aromatic pi cloud to form a resonance-stabilized cation that loses a proton to give the final product.

# REACTION 9D: NITRATION (Section 9.7B)

- **Aromatic rings** react with **nitric acid** in the presence of **sulfuric acid** to yield **nitro aromatic compounds** via **electrophilic aromatic substitution**. ✳
- The mechanism involves an initial reaction between nitric acid and sulfuric acid to yield the nitronium ion $NO_2^+$. This reacts as a very strong electrophile with the weakly nucleophilic aromatic pi cloud to form a resonance-stabilized cation that loses a proton to give the final product.

# REACTION 9E: FRIEDEL-CRAFTS ACYLATION (Section 9.7C)

- **Aromatic rings** react with **acyl chlorides** in the presence of a Lewis acid catalyst like AlCl₃ to produce an aryl ketone via **electrophilic aromatic substitution**. *
- The mechanism involves an initial reaction between the acyl chloride and AlCl₃ to yield the acylium ion R-C⁺=O. This reacts as a very strong electrophile with the weakly nucleophilic aromatic pi cloud to form a resonance stabilized cation that loses a proton to give the final product.
- Rearrangement is not a problem with acylium ions like it is with carbocations.

## REACTION 9F: REACTIONS OF ALCOHOLS WITH AROMATIC RINGS IN THE PRESENCE OF STRONG ACID (Section 9.7D)

- **Aromatic rings** react with **alcohols** in the presence of a strong **acid catalyst** like H₃PO₄ and H₂SO₄ to produce an alkyl benzene via **electrophilic aromatic substitution**. *
- The mechanism involves an initial reaction between the alcohol and strong acid to yield a carbocation. This reacts as a very strong electrophile with the weakly nucleophilic aromatic pi cloud to form a resonance stabilized cation that loses a proton to give the final product.
- Because carbocations are involved in the mechanism, rearrangements can be a problem, especially with primary or secondary alcohols.

## REACTION 9G: REACTIONS OF ALKENES WITH AROMATIC RINGS IN THE PRESENCE OF STRONG ACID OR LEWIS ACID (Section 9.7D)

- **Aromatic rings** react with **alkenes** in the presence of a **strong acid catalyst** like H₃PO₄ to produce an **alkyl benzene** via **electrophilic aromatic substitution**. *
- The mechanism involves an initial reaction between the alkene and strong acid or Lewis acid to yield a positively charged carbocation. This reacts as a very strong electrophile with the weakly nucleophilic aromatic pi cloud to form a resonance stabilized cation that loses a proton to give the final product.

## REACTION 9H: FRIEDEL-CRAFTS ALKYLATION (Section 9.7C)

- **Aromatic rings** react with **alkyl halides** in the presence of a **Lewis acid** like $AlCl_3$ to produce an **alkyl benzene** via **electrophilic aromatic substitution.** ✳
- The mechanism involves an initial reaction between the alkyl halide and Lewis acid to yield an intermediate that can be thought of as a carbocation. This reacts as a very strong electrophile with the weakly nucleophilic aromatic pi cloud to form a resonance stabilized cation that loses a proton to give the final product.
- Because carbocations are involved in the mechanism, rearrangements can be a problem, especially with primary or secondary alkyl halides.

## REACTION 9I: REDUCTION OF A NITROAROMATIC TO PRODUCE A PRIMARY ARYL AMINE (Section 9.7B)

$$\text{NO}_2 \quad \xrightarrow[\substack{\text{or} \\ \text{1) Fe, HCl, H}_2\text{O} \\ \text{2) NaOH}}]{\text{H}_2\text{ / Ni}} \quad \text{NH}_2$$

- **Nitroaromatic compounds** can be **reduced with $H_2$** and a transition metal catalyst such as **Ni, Pd,** or **Pt** to produce a **primary aryl amine.** ✳
- This same reaction can be accomplished using reducing agents such as iron or zinc in HCl.

## REACTION 9J: ACID-BASE REACTIONS OF PHENOLS (Section 9.4)

$$\text{OH} \quad \xrightarrow{\text{NaOH}} \quad \text{O}^-\text{ Na}^+ \quad + \text{ H}_2\text{O}$$

- **Phenols** are **weak acids** that react with strong bases such as NaOH to form water-soluble salts. ✳

## SUMMARY OF IMPORTANT CONCEPTS

### 9.0 OVERVIEW
• Benzene and its derivatives have marked distinctions from other types of molecules, and they are broadly classified as **aromatic.** They have certain similar physical properties as well as a remarkable stability that makes them unreactive toward reagents that attack other species with pi bonds such as alkenes and alkynes. **Aromaticity** is the term used to describe this special stability of benzene and its derivatives. ✳

### 9.1 THE STRUCTURE OF BENZENE
• An important development in the identification of benzene's structure was Kekulé's proposal that benzene is composed of six carbon atoms in a ring, with one hydrogen atom attached to each carbon. ✳
• The six carbon atoms of the ring are equivalent, and the carbon-carbon bond lengths are all intermediate between a single and double bond. Thus, it is not accurate to think of benzene as simply having alternating single and double bonds that are static, because this would predict alternating longer and shorter carbon-carbon bonds.

- **Valence bond description of benzene**.  With the advent of more sophisticated models of chemical bonding, especially the concepts of **hybridization of atomic orbitals** and **resonance**, a more accurate picture of the bonding in benzene was proposed. ✳
  - Each carbon atom of the ring is $sp^2$ hybridized.
  - Each carbon atom of the ring makes sigma bonds by $sp^2$-$sp^2$ overlap with the two adjacent carbon atoms and $sp^2$-1s overlap with a hydrogen atom.
  - Each carbon atom also has a single unhybridized 2p orbital containing one electron.  These six 2p orbitals overlap to form a continuous pi system that extends over all six carbon atoms.  The electron density of this pi system thus lies in two bagel-shaped regions, one above and one below the plane of the ring.
  - Benzene can be represented as a resonance hybrid composed of two resonance forms in which the locations of the double bonds are reversed.  Alternatively, benzene is denoted as a hexagon with a circle drawn on the inside.

  - **Resonance energy** is the difference in energy between a resonance hybrid and the most stable hypothetical contributing structure in which electron density is localized on particular atoms and on particular bonds.  The resonance energy for benzene is large, namely 36.0 kcal/mol.  What this means is that the pi system of benzene is extremely stable, and dramatically less reactive than would be expected for a normal alkene under conditions like catalytic hydrogenation.

## 9.2 HETEROCYCLIC AROMATIC COMPOUNDS
- The German physicist Eric Hückel described a number of criteria to be used when trying to figure out if a molecule is aromatic or not.  For molecules with five or six-member rings, Hückel's criteria are the following:
  - The ring must be planar.
  - Each atom on the ring must have a 2p orbital.
  - There must be six pi electrons in the cyclic arrangement of 2p orbitals.
- A **heterocyclic compound** is one that contains more than one kind of atom in a ring.  Certain heterocycles can be aromatic if the Hückel criteria are met.  Nature is filled with aromatic heterocycles such as indoles, imidazoles, purines, and pyrimidines.
- An important parameter to keep track of in aromatic heterocycles is whether lone pairs of electrons are part of the aromatic pi system or not. ✳
  - For example, in pyridine ($C_5H_5N$) the lone pair of electrons on nitrogen is perpendicular to the six 2p orbitals of the aromatic 6 pi electron system.  Thus, the lone pair of electrons on the nitrogen or pyridine is not part of the aromatic pi system, and is free to take part in interactions with other species.

## 9.3 NOMENCLATURE OF AROMATIC COMPOUNDS
- Monosubstituted alkyl benzenes are named as derivatives of benzene such as ethylbenzene.
- The **IUPAC system retains certain common names** for several of the simpler benzene derivatives including **toluene, cumene, styrene, xylene, phenol, aniline, benzoic acid,** and **anisole**.
- In more complicated molecules, the benzene ring is named as a substituent on a parent chain.
  - The $C_6H_5$ group is given the name **phenyl**.  For example, the IUPAC name for $C_6H_5CH_2CH_2OH$ is 2-phenylethanol.
  - The $C_6H_5CH_2$- group is given the name **benzyl**.  These compounds are derivatives of toluene $C_6H_5CH_3$.  *[The terms phenyl and benzyl are often confused by students.  Make sure you know when each should be used.]*

- For benzene rings with **two substituents**, the three possible constitutional isomers are named as **ortho** (1,2 substitution), **meta** (1,3 substitution), and **para** (1,4 substitution). These are abbreviated as *o*, *m*, and *p*, respectively. It is also acceptable to name these species with numbers as locators. When one of the substituents has a special name (if NH$_2$ is present the molecule is an aniline, etc.) then the molecule is named after that parent molecule. For example, 3-chloroaniline and *m*-chloroaniline are both acceptable names for the same molecule. If neither group imparts a special name, then the substituents are listed in alphabetical order. For example, 1-chloro-4-ethylbenzene and *p*-chloroethylbenzene are acceptable names for the same molecule. ✱

- When three or more substituents are on the same ring, the substituents are given numbers and are listed in alphabetical order. If one of the substituents imparts a special common name, then that substituent is assumed to occupy the position 1, and the molecule is named as a derivative of that parent structure.

- **Polynuclear aromatic hydrocarbons (PAH)** contain more than one benzene ring, each pair of which shares two carbon atoms. For example, naphthalene is two benzene rings fused together and anthracene is three benzene rings fused together in a linear fashion. Other common PAH's include phenanthrene, pyrene, coronene, and benzo[*a*]pyrene. Benzo[*a*]pyrene has been especially well-studied because it is a potent carcinogen. ✱

## 9.4 PHENOLS

- The characteristic feature of **phenols** is a **hydroxyl group attached to a benzene ring**.
- **Phenols are more acidic** than simple alcohols, because the **phenoxide anion is more stable** than an alkoxide anion. This **increased stability** is due to **resonance stabilization of the phenoxide anion**. In other words, upon deprotonation of a phenol, the resulting phenoxide is more stable, because the phenoxide can be considered a resonance hybrid of three contributing resonance structures that delocalize the negative charge onto three different carbon atoms of the ring. This charge delocalization means that no one atom must absorb the entire negative charge, and the delocalized anion is thus more stable. There is no similar resonance stabilization possible for alkoxide anions derived from simple alcohols. ✱

  - **Substituents on the ring** can have a **dramatic influence** on the **acidity** of phenols. The fundamental concept is that **anything** on the ring that **leads to a further stabilization of the phenoxide anion increases acidity** of a phenol, while **anything that destabilizes the phenoxide decreases acidity of a phenol**. Electronegativity effects are important for the stabilization or destabilization of phenoxide anions by ring substituents.

    Atoms or groups that can remove electron density from the ring are said to be **electron-withdrawing**, while atoms or groups that release electron density into **the ring** are said to be **electron-releasing**. Electron-withdrawing groups can have electronegative atoms such as halogens or heteroatoms such as those found in the nitro group, a strongly electron withdrawing group. Electron withdrawing groups stabilize a phenoxide anion by absorbing some of the negative charge, while electron-releasing groups destabilize a phenoxide anion by dumping even more electron density into the ring. The bottom line is that **phenols with electron-withdrawing groups like fluorine atoms are more acidic** than phenol, while **phenols with electron-releasing alkyl groups are less acidic** than phenol. ✱

## 9.7 MECHANISM OF ELECTROPHILIC AROMATIC SUBSTITUTION

- A variety of electrophiles react with aromatic rings via **electrophilic aromatic substitution**. The electrophiles are usually positively charged and examples include reactions 9B-9H. The general **mechanism involves attack on the electrophile** by the weakly nucleophilic aromatic pi cloud to form **a resonance-stabilized cation intermediate** that loses a proton to give the final product. ✱

## 9.8 DISUBSTITUTION AND POLYSUBSTITUTION

• **Substituents** other than hydrogen on an aromatic ring can have a **profound influence** on the **reaction rate** and **substitution pattern** of electrophilic aromatic substitution. In particular, groups can either speed up or slow down the reaction, and can direct new groups meta or ortho-para. ✳

• Substituents can be divided into three broad classes (with a number of subclasses listed in Table 9.2):

  - **Alkyl groups** and all **groups in which the atom bonded to the ring** has an **unshared pair of electrons** are **ortho-para directing**. These groups are **electron-donating** and are thus **activating** toward electrophilic aromatic substitution.

  - Groups in which the **atom bonded to the aromatic ring** bears a **partial or full positive charge** are **meta directing**. These groups often have **multiple bonds such as =O on the atom bonded to the aromatic ring**. These groups are **electron-withdrawing** and are thus **deactivating** toward electrophilic aromatic substitution.

  - **Halogens** are exceptions in that they are **ortho-para directing**, but **electron-withdrawing** and thus **weakly deactivating** toward electrophilic aromatic substitution.

• These effects have a large practical significance, since in synthesizing polysubstituted aromatics, the **order of addition of the substituents must be taken into account**. For example, when making *m*-bromonitrobenzene from benzene, the nitro group (meta directing) must be added before the bromine atom (ortho-para directing). Adding the bromine atom first followed by the nitro group would result in the majority of product being the unwanted ortho and para isomers.

# CHAPTER 9
## *Solutions to the Problems*

<u>Problem 9.1</u>  Write names for these molecules.

(a)

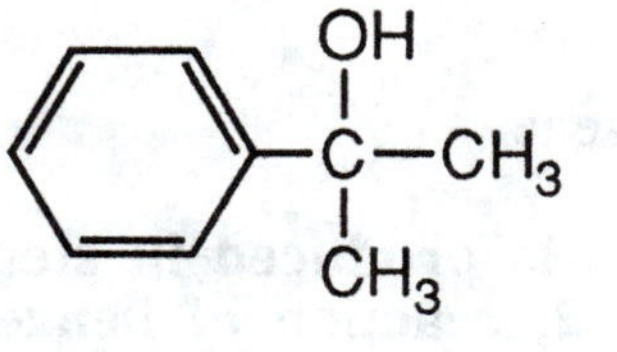

(b)

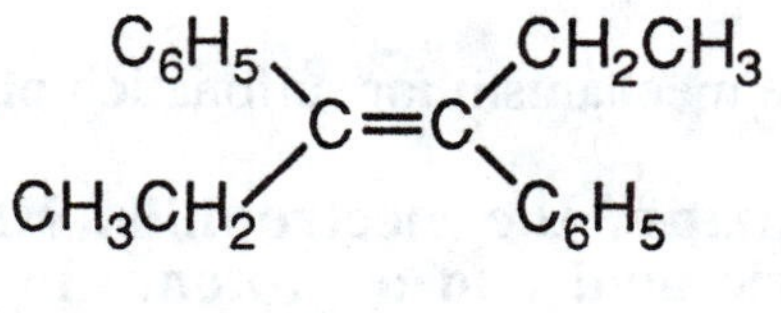

(c)

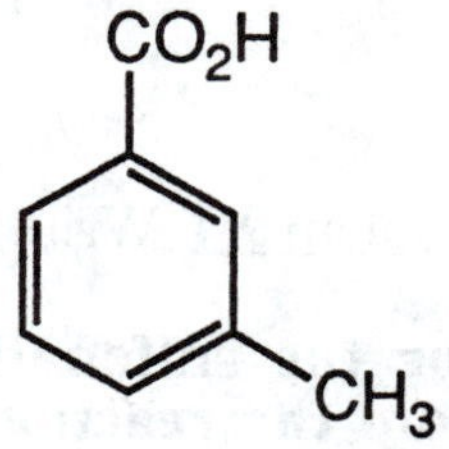

**2-Phenyl-2-propanol**          **(E)-3,4-Diphenyl-3-hexene**          **3-Methylbenzoic acid**

<u>Problem 9.2</u>  Arrange these compounds in order of increasing acidity: 2,4-dichlorophenol, phenol, cyclohexanol.

**The following compounds are ranked from least to most acidic:**

**Cyclohexanol    <    Phenol    <    2,4-Dichlorophenol**

**A good way to predict relative acidities between related compounds is to keep track of the anionic conjugate bases produced upon deprotonation.  In general, the more stable the conjugate base anion, the stronger the acid.  Anions become increasingly stabilized as the negative charge is more delocalized around the molecule.  Thus, phenol is more acidic than an aliphatic alcohol like cyclohexanol, because resonance involving the aromatic ring of phenol leads to increased charge delocalization and thus stabilization of the phenoxide anion compared to the cyclohexylalkoxide anion.  The electronegative chlorine atoms of 2,4-dichlorophenol withdraw electron density from the aromatic ring and thus help to stabilize the 2,4-dichlorophenoxide anion even further than what is seen with phenoxide.**

<u>Problem 9.3</u>  Predict the products resulting from vigorous oxidation of these compounds by $K_2Cr_2O_7$ in aqueous $H_2SO_4$.

(a)

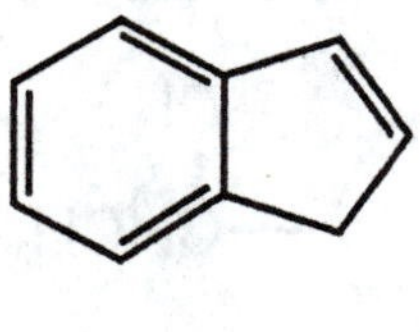

(b)

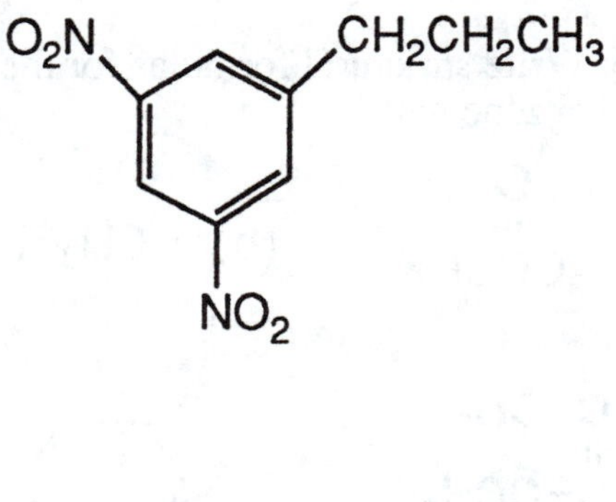

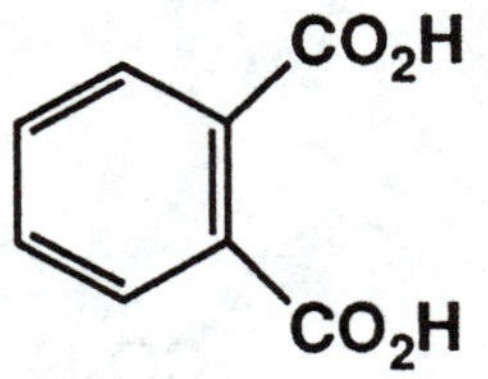

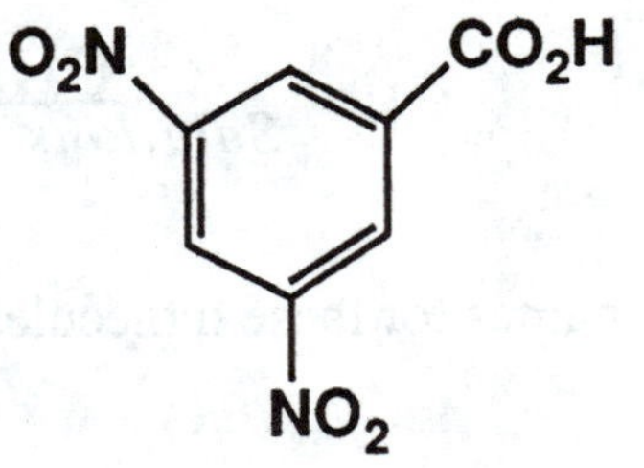

**Problem 9.4** Write the stepwise mechanism for sulfonation of benzene.

For the sulfonation of benzene, the electrophile, $HSO_3^+$, is produced in step 1 from the reaction of sulfuric acid and a proton. In step 2, reaction of benzene with the electrophile yields a resonance-stabilized cation. In step 3, this intermediate loses a proton to complete the reaction.

**Step 1:**

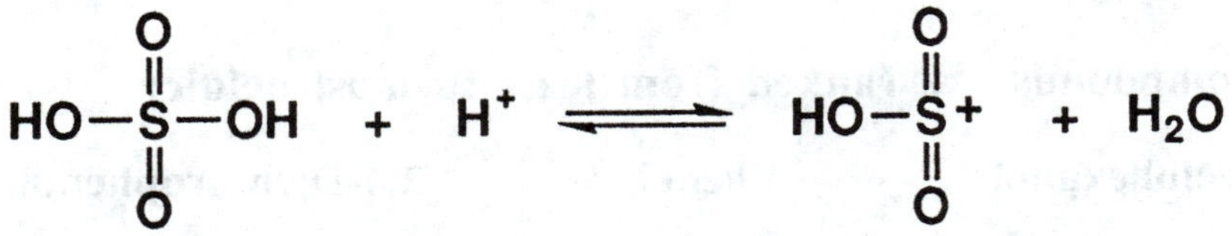

**Step 2:**

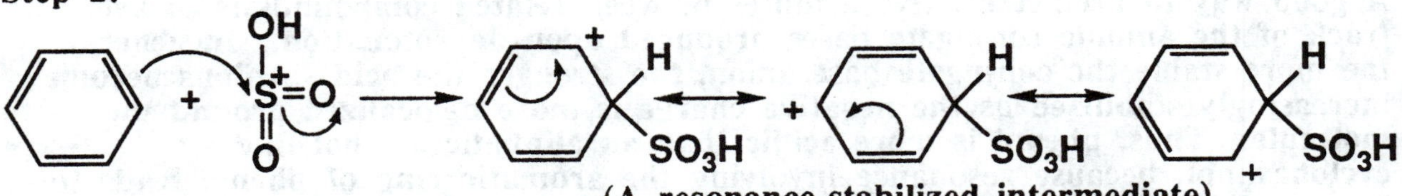

(A resonance-stabilized intermediate)

**Step 3:**

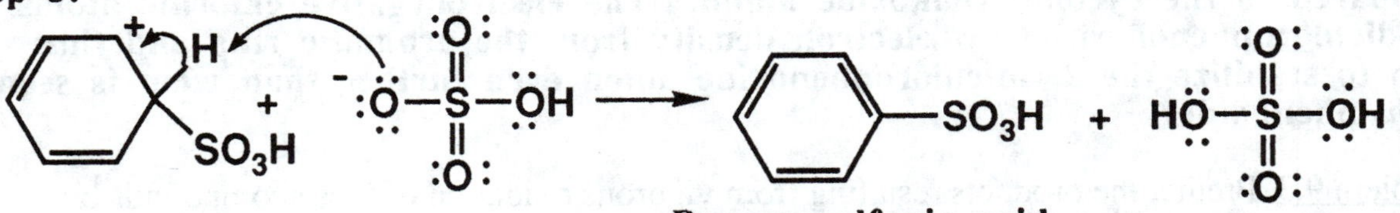

Benzenesulfonic acid

**Problem 9.5** Write structural formulas for the products formed from Friedel-Crafts alkylation or acylation of benzene with:

(a) 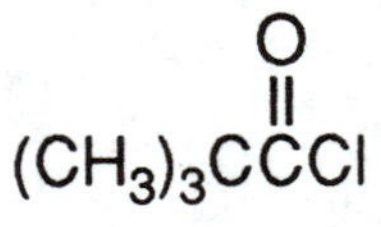  $(CH_3)_3CCCl$

(b)  $CH_3CH_2Cl$

(c)

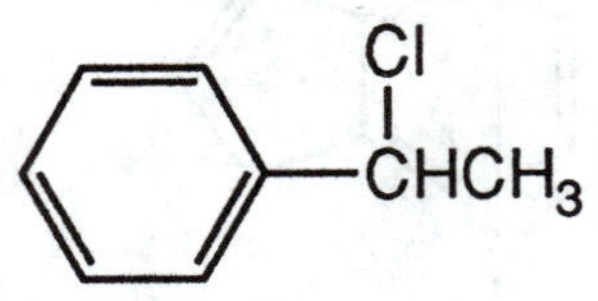

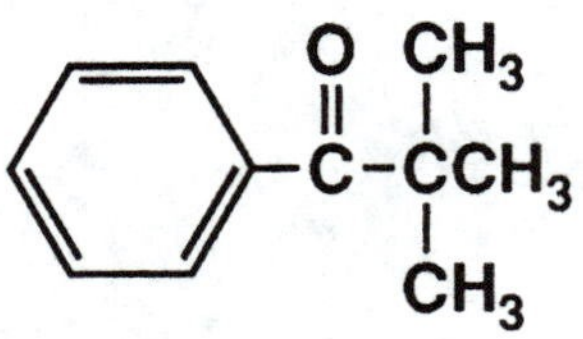

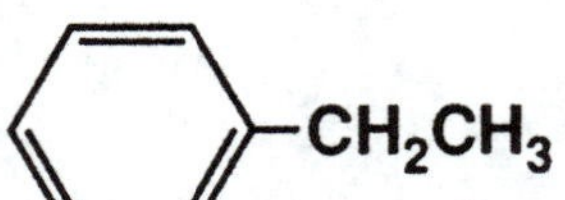

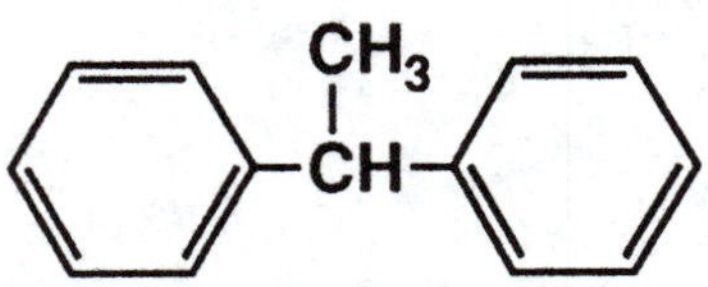

**Problem 9.6** Complete the following electrophilic aromatic substitution reactions. Where you predict meta substitution, show only the meta product. Where you predict ortho-para substitution, show both products.

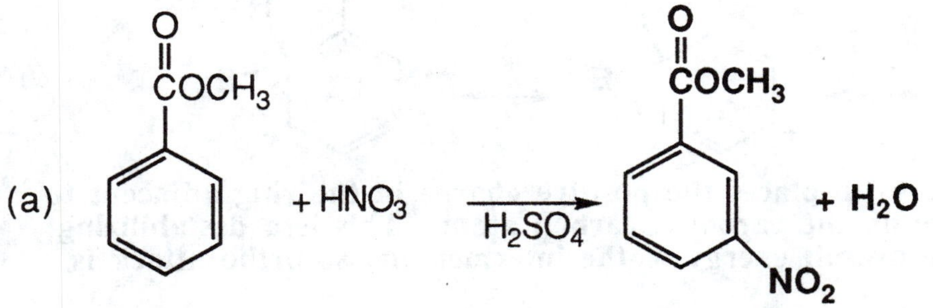

(a)

**The carboxymethyl group is meta directing and deactivating.**

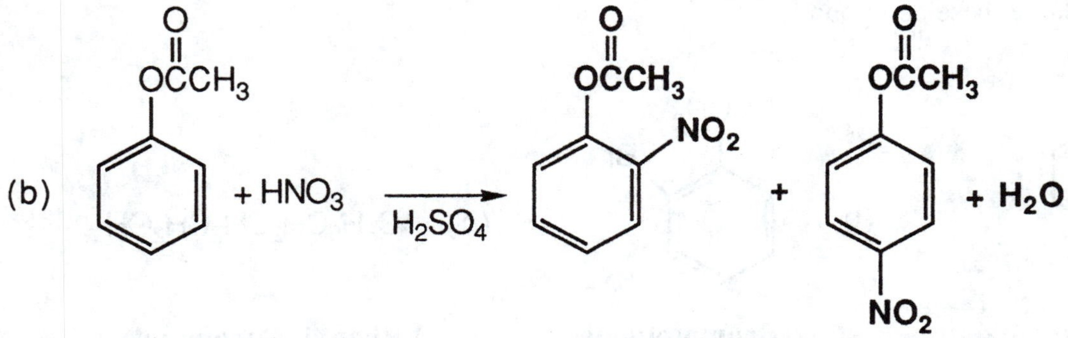

(b)

**The acetoxy group is ortho-para directing and activating.**

**Problem 9.7** Because the electronegativity of oxygen is greater than carbon, the carbon of a carbonyl group bears a partial positive charge, and its oxygen bears a partial negative charge. Using this information, show that a carbonyl group destabilizes the cation intermediate in ortho-para attack and thus acts as a meta directing group.

**Para attack:**

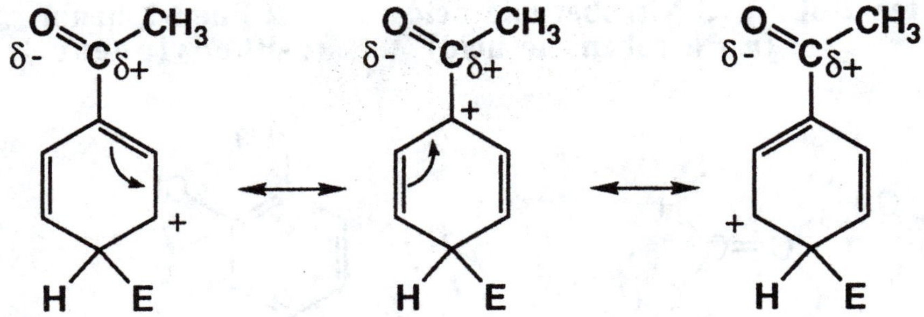

The middle resonance form places the positive charge in the ring adjacent to the partial positive charge on the carbonyl carbon atom. This is a destabilizing interaction that raises the overall energy of the intermediate, so para attack is disfavored.

**Ortho attack:**

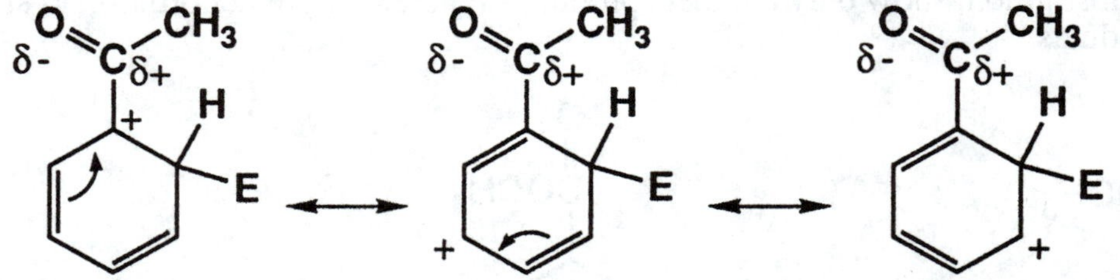

The first resonance form places the positive charge in the ring adjacent to the partial positive charge on the carbonyl carbon atom. This is a destabilizing interaction that raises the overall energy of the intermediate, so ortho attack is disfavored.

Meta attack never places the positive charge on the ring adjacent to the carbonyl carbon atom, so meta products predominate in electrophilic aromatic substitution reactions of benzene rings with attached carbonyl groups.

## Nomenclature and Structural Formulas
<u>Problem 9.8</u> Name these compounds.

(a)

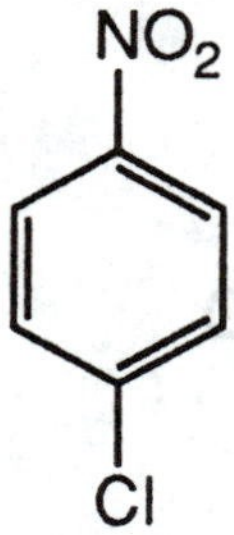

**4-Chloronitrobenzene**

(b)

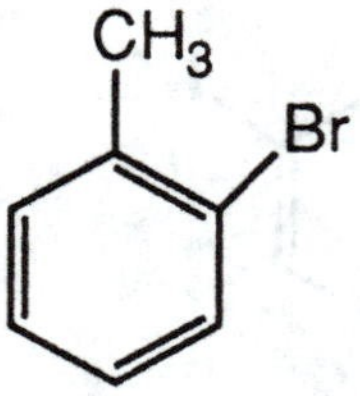

**2-Bromotoluene**
**(*o*-Bromotoluene)**

(c)  $C_6H_5CH_2CH_2CH_2OH$

**3-Phenyl-1-propanol**

(d) 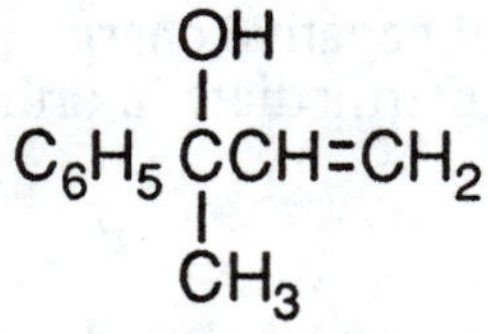  $C_6H_5 \overset{\overset{\displaystyle OH}{|}}{\underset{\underset{\displaystyle CH_3}{|}}{C}} CH=CH_2$

**2-Phenyl-3-buten-2-ol**

(e)

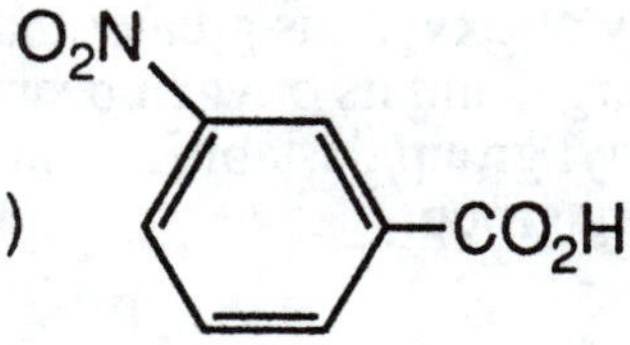

**3-Nitrobenzoic acid**
**(*m*-Nitrobenzoic acid)**

(f)

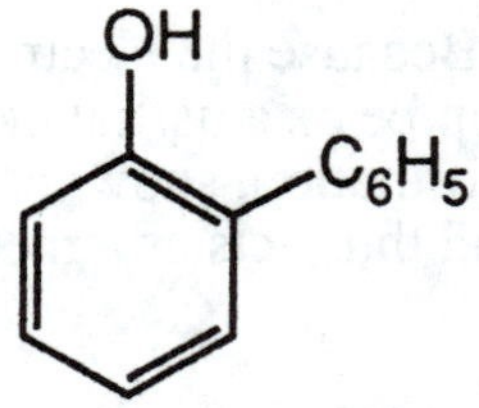

**2-Phenylphenol**
**(*o*-Phenylphenol)**

(g)

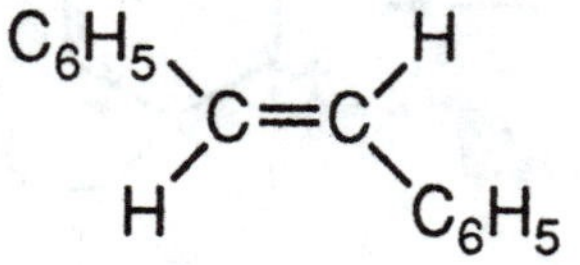

**(E)-1,2-Diphenylethene**
**(*trans*-1,2-Diphenylethylene)**

(h)

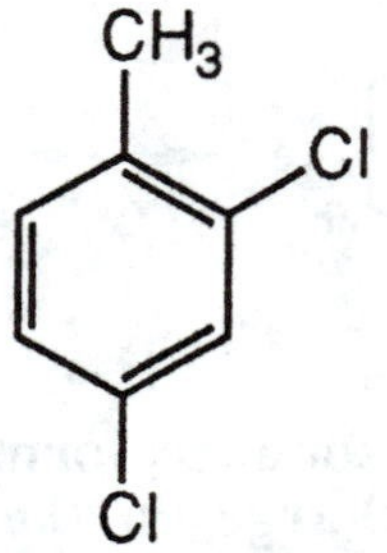

**2,4-Dichlorotoluene**

Problem 9.9 Draw structural formulas for the following molecules.

(a) 1-Bromo-2-chloro-4-ethylbenzene

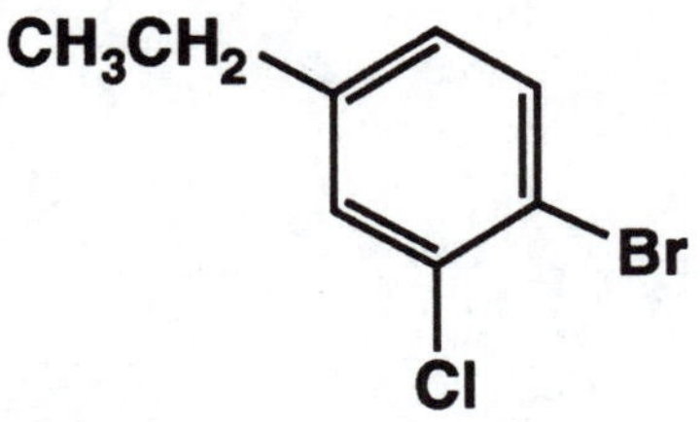

(b) 1,2-Dimethyl-4-iodobenzene

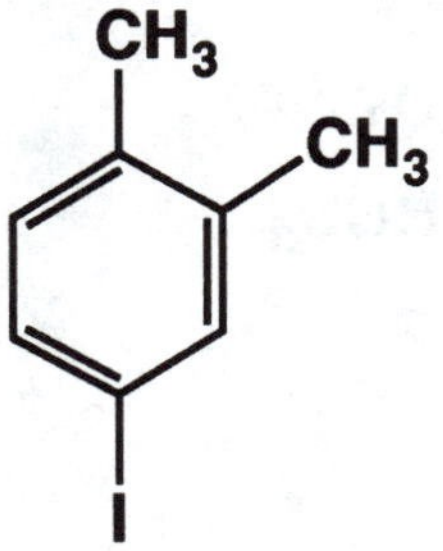

(c) 2,4,6-Trinitrotoluene

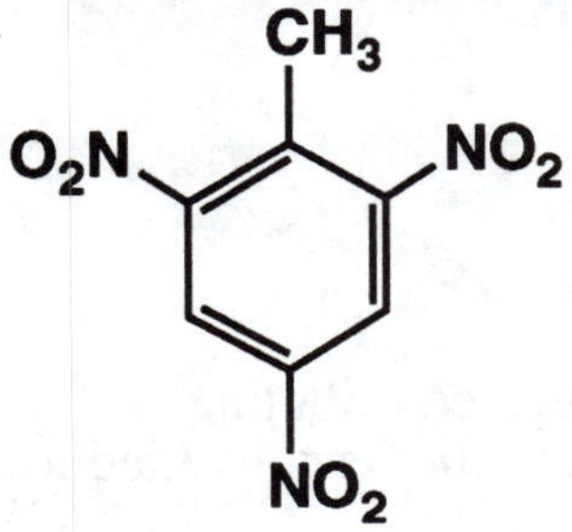

(d) 4-Phenyl-2-pentanol

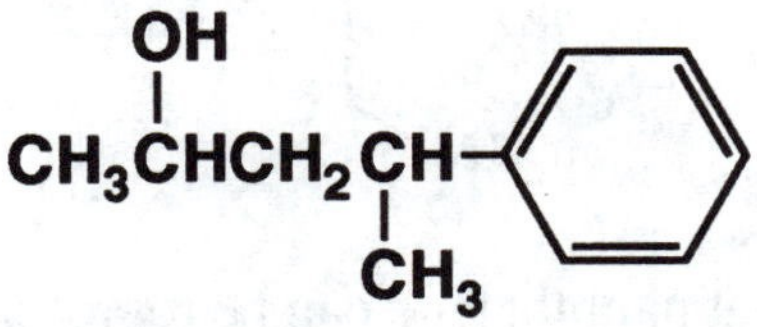

(e) p-Cresol

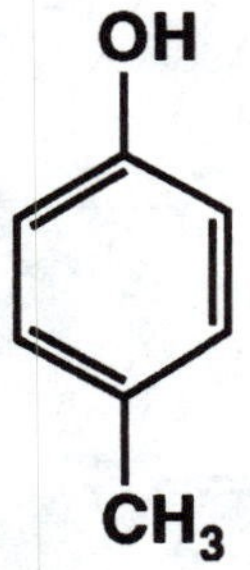

(f) 2,4-Dichlorophenol

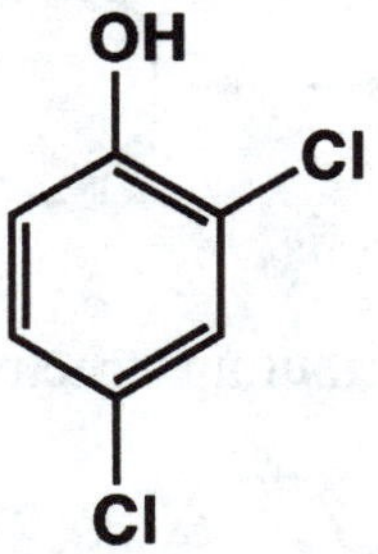

(g) 1-Phenylcyclopropanol

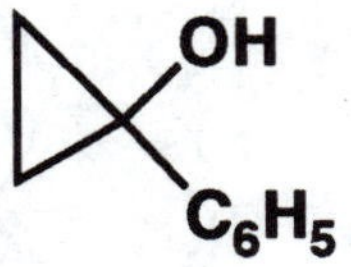

(h) Styrene (phenylethene)

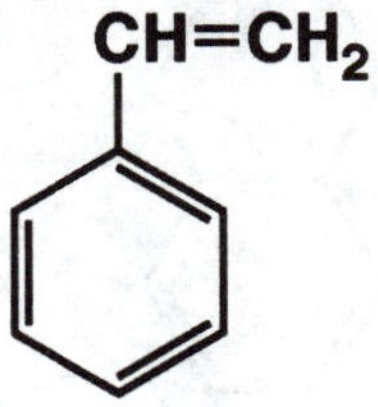

(i) m-Bromophenol

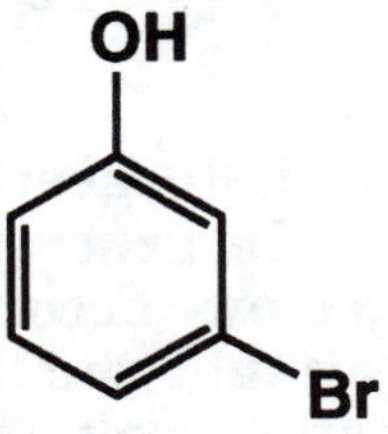

(j) 2,4-Dibromoaniline

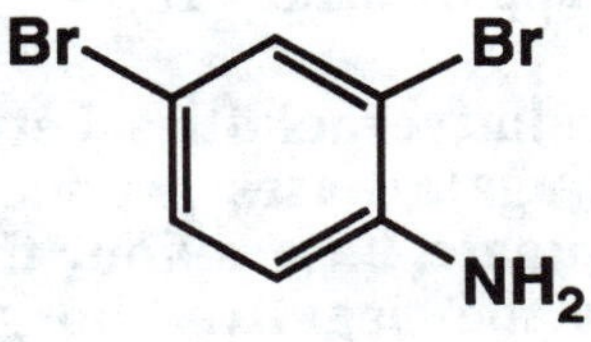

(k)  Isobutylbenzene

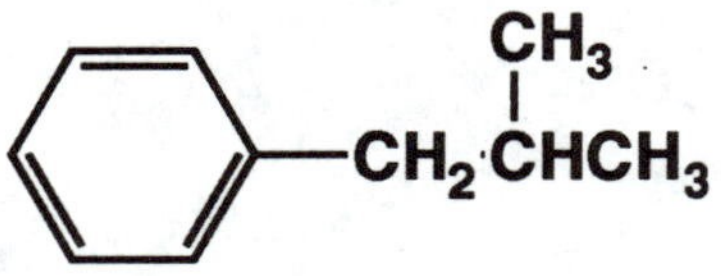

$$CH_3$$
$$-CH_2\cdot CHCH_3$$

(l)  *m*-Xylene

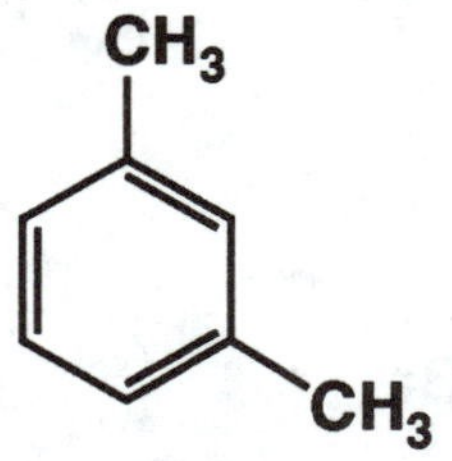

$$CH_3$$

$$CH_3$$

<u>Problem 9.10</u>  Show that pyridine can be represented as a hybrid of two equivalent contributing structures.

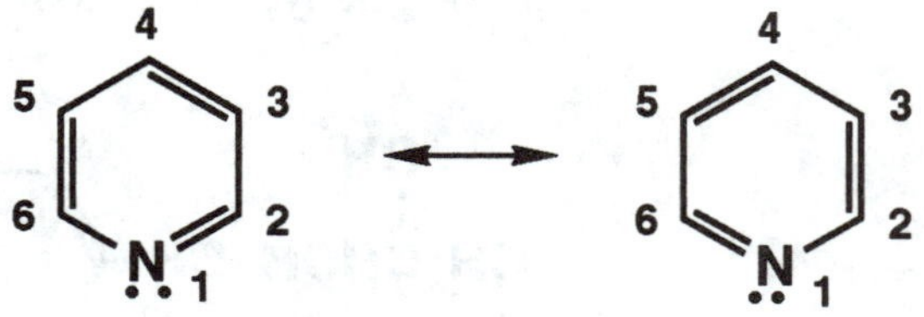

<u>Problem 9.11</u>  Show that naphthalene can be represented as a hybrid of three contributing structures, and show by the use of curved arrows how one contributing structure is converted to the next.

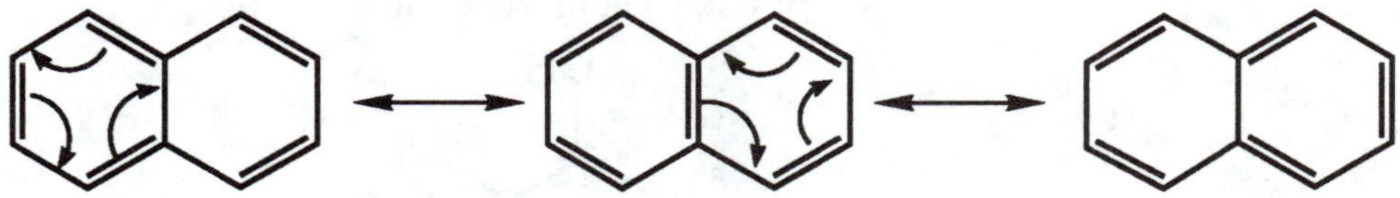

<u>Problem 9.12</u>  Draw four contributing structures for anthracene.

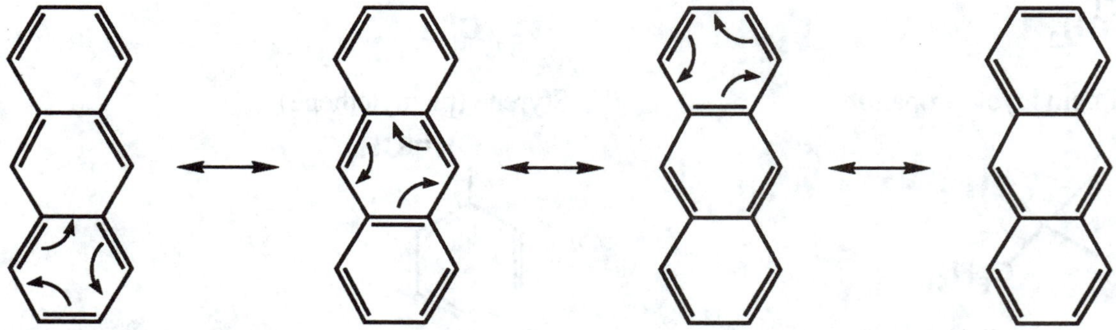

## Acidity of Phenols

<u>Problem 9.13</u> Use resonance theory to account for the fact that phenol (pK$_a$ 9.95) is a stronger acid than cyclohexanol (pK$_a$ approximately 18)

**A good way to predict relative acidities between related compounds is to keep track of the anionic conjugate bases produced upon deprotonation.  In general, the more stable the conjugate base anion, the stronger the acid.  Anions become increasingly stabilized as the negative charge is more delocalized around the molecule.  Thus, phenol is significantly more acidic than an aliphatic alcohol like**

cyclohexanol, because resonance involving the aromatic ring of phenol leads to increased charge delocalization and thus stabilization of the phenoxide anion compared to the cyclohexyloxide anion. Four contributing structures for the phenoxide anion are shown below, illustrating how the negative charge is delocalized into the aromatic ring.

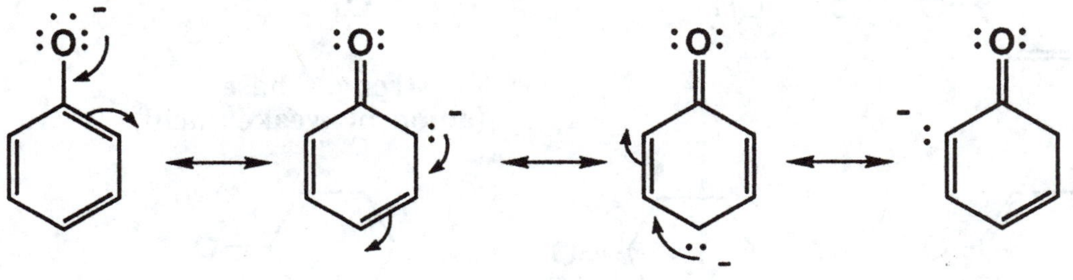

<u>Problem 9.14</u> Arrange the molecules and ions in each set in order of increasing acidity (from least acidic to most acidic).

(a)

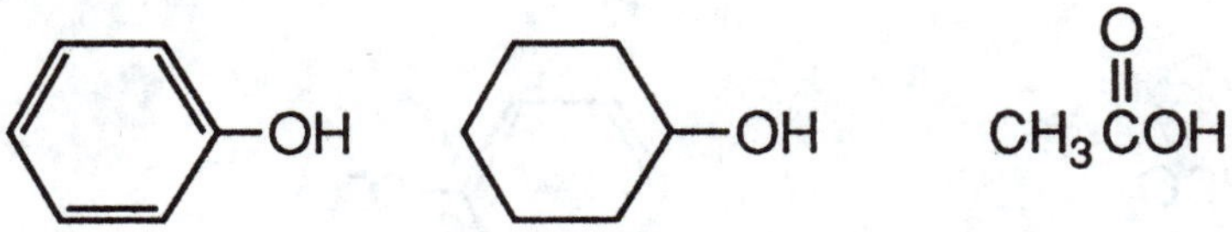

To arrange these in order of increasing acidity, refer to Table 6.1. For those compounds not listed in Table 6.1, estimate pKa using values for compounds that are given in the table.

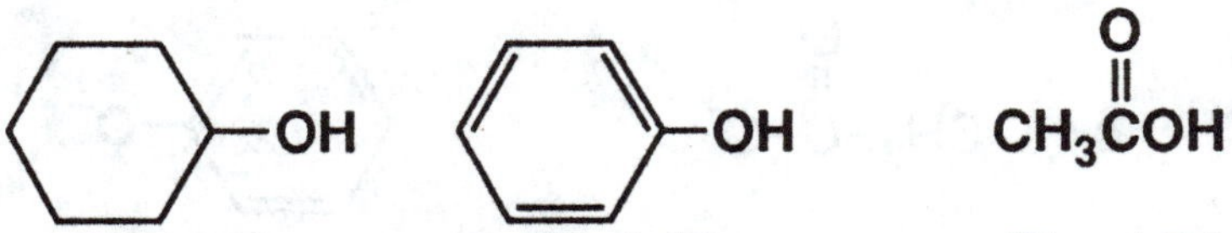

pKa ~ 18          pKa 9.95          pKa 4.76

(b)

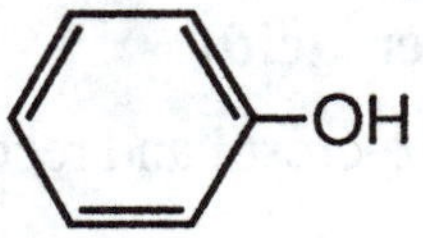

H₂O          HCO₃⁻          OH

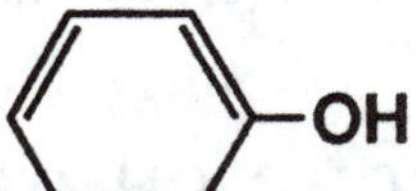

pKa 15.7          pKa 10.33          pKa 9.95

(c) 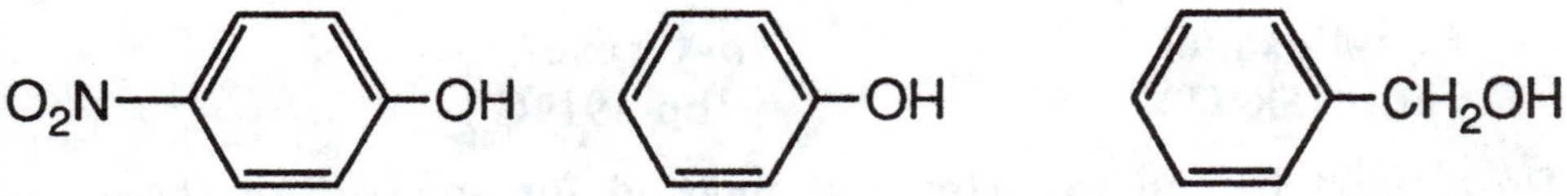  O₂N—⟨ ⟩—OH          ⟨ ⟩—OH          ⟨ ⟩—CH₂OH

⟨ ⟩—CH₂OH          ⟨ ⟩—OH          O₂N—⟨ ⟩—OH

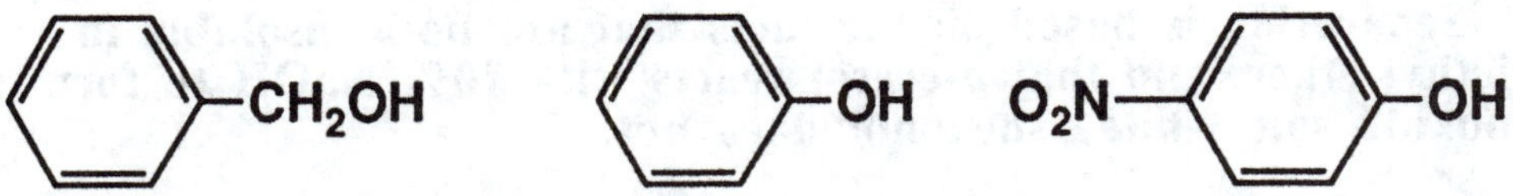

pKa ~ 18          pKa 9.95          pKa 7.15

Problem 9.15 From each pair select the stronger base.

**To estimate which is the stronger base, first determine which conjugate acid is the weaker acid. The weaker the acid, the stronger its conjugate base.**

(a) ⬡—O⁻    or    OH⁻

                              **OH⁻**

                              **stronger base**
                              **(anion of weaker acid)**

(b) ⬡—O⁻    or    ⬡—O⁻

                              ⬡—O⁻

                              **stronger base**
                              **(anion of weaker acid)**

(c) ⬡—O⁻    or    $HCO_3^-$

                              ⬡—O⁻

                              **stronger base**
                              **(anion of weaker acid)**

(d) ⬡—O⁻    or    $CH_3-\overset{\displaystyle O}{\underset{\displaystyle }{C}}-O^-$

                              ⬡—O⁻

                              **stronger base**
                              **(anion of weaker acid)**

Problem 9.16 Describe a procedure to separate a mixture of 1-hexanol and *o*-cresol and recover each in pure form. Each is insoluble in water but soluble in diethyl ether.

$CH_3(CH_2)_4CH_2OH$                    ⬡—CH₃, OH

          **1-Hexanol**                         ***o*-Cresol**
          **(bp 158°C )**                        **(bp 191°C )**

**Following is a flow chart for an experimental method for separating these two compounds. Separation is based on the facts that are both insoluble in water, but soluble in diethyl ether, and that *o*-cresol reacts with 10% NaOH to form a water-soluble phenoxide salt while 1-hexanol does not.**

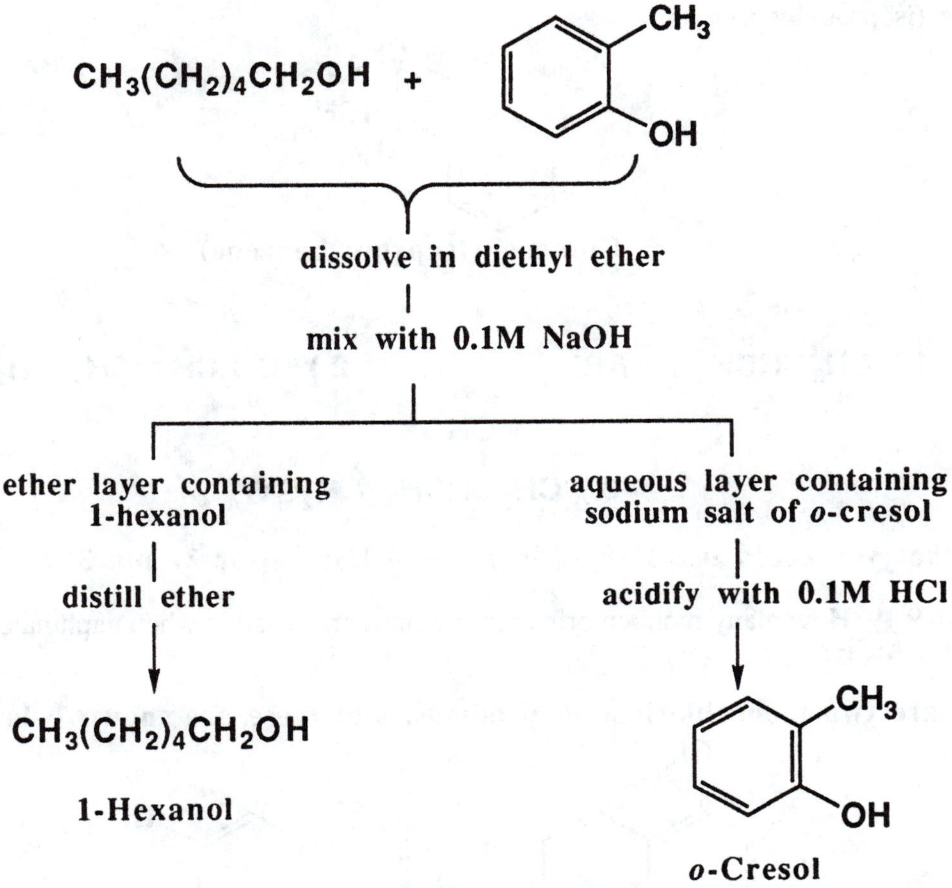

$$CH_3(CH_2)_4CH_2OH \quad + \quad$$

dissolve in diethyl ether

mix with 0.1M NaOH

| ether layer containing 1-hexanol | aqueous layer containing sodium salt of *o*-cresol |
| --- | --- |
| distill ether | acidify with 0.1M HCl |

$$CH_3(CH_2)_4CH_2OH$$

**1-Hexanol**

*o*-Cresol

## Electrophilic Aromatic Substitution: Monosubstitution

Problem 9.17  Draw a structural formula for the compound formed upon treatment of benzene with these combinations of reagents.

(a)  $CH_3CH_2Cl$ / $AlCl_3$          (b) $CH_2{=}CH_2$ / $H_2SO_4$          (c)  $CH_3CH_2OH$ / $H_2SO_4$

**All of these reagents will react with benzene to give the same product, ethyl benzene.**

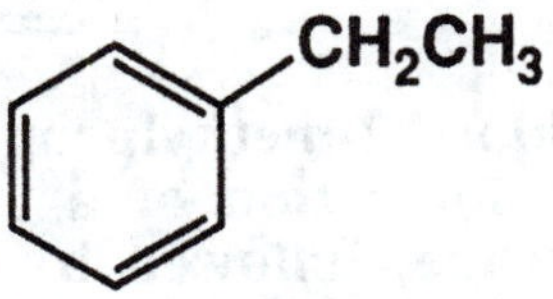

CH₂CH₃

**Ethyl benzene**

Problem 9.18 Show three different combinations of reagents you might use to convert benzene to cumene (isopropylbenzene).

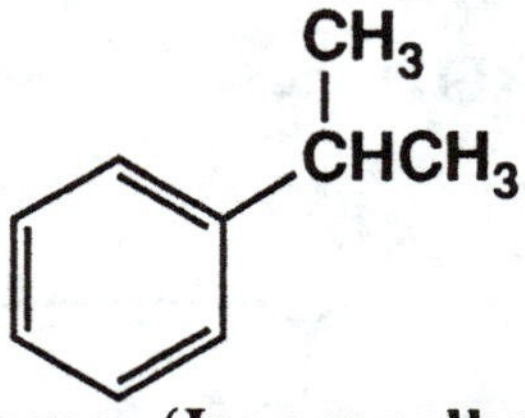

$$\underset{\textbf{Cumene (Isopropylbenzene)}}{\text{(benzene ring with CH(CH}_3)\text{CH}_3\text{ group)}}$$

1 )  $\underset{\underset{\text{Cl}}{|}}{CH_3\,CHCH_3}$  /  $AlCl_3$          2 )  $CH_3CH{=}CH_2$ / $H_2SO_4$

3 )  $\underset{\underset{\text{OH}}{|}}{CH_3\,CHCH_3}$ / $H_2SO_4$

**Note that you could use $H_3PO_4$ in place of $H_2SO_4$ for 2) and 3).**

Problem 9.19 How many monochlorination products are possible when naphthalene is treated with $Cl_2$ / $AlCl_3$?

**There are two monochlorination products, and these are shown below:**

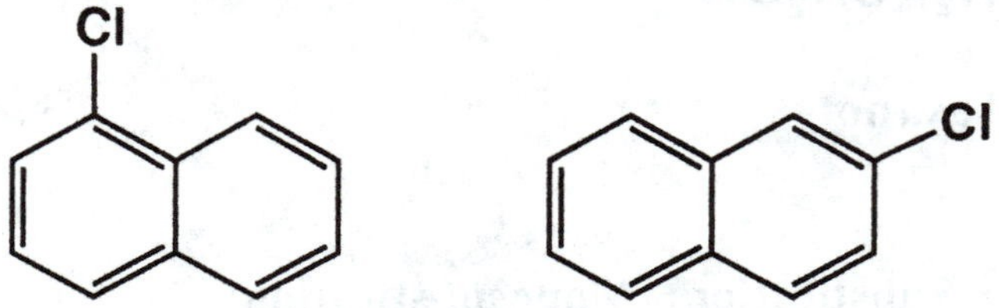

Problem 9.20 Write a stepwise mechanism of this reaction. Use curved arrows to show the flow of electrons in each step.

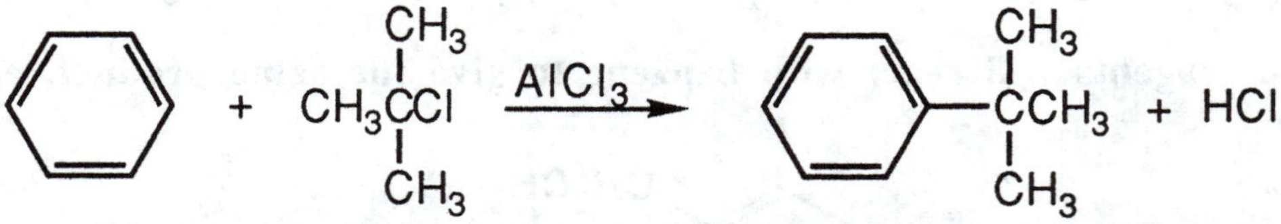

**The reaction of benzene with 2-chloro-2-methylpropane in the presence of aluminum chloride involves initial formation of a complex between 2-chloro-2-methylpropane and aluminum chloride, followed by formation of an alkyl carbocation. This cationic species is the electrophile that undergoes further reaction with benzene.**

**Step 1: Formation of a complex between 2-chloropropane (a Lewis base) and aluminum chloride (a Lewis acid).**

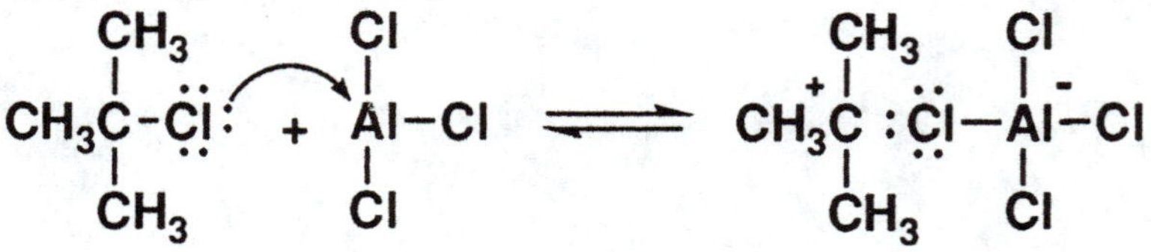

**Step 2:   Formation of *tert*-butyl cation.**

$$CH_3\overset{\overset{CH_3}{|}}{\underset{\underset{CH_3}{|}}{C}}\!\!^+ :\overset{..}{\underset{..}{C}}l\overset{\overset{Cl}{|}}{\underset{\underset{Cl}{|}}{Al}}^-Cl \rightleftharpoons CH_3\overset{\overset{CH_3}{|}}{\underset{\underset{CH_3}{|}}{C}}\!\!^+ \quad AlCl_4^-$$

**Step 3:   Electrophilic attack on the aromatic ring.**

**resonance-stabilized
cation intermediate**

**Step 4:   Proton transfer to regenerate the aromatic ring.**

<u>Problem 9.21</u>  Write a stepwise mechanism for the preparation of diphenylmethane by treating benzene with methylene chloride in the presence of an aluminum chloride catalyst.

**Formation of diphenylmethane involves two successive Friedel-Crafts alkylations.**

**Formation of benzyl chloride completes the first Friedel-Crafts alkylation. This molecule then is a reactant for the second Friedel-Crafts alkylation.**

## Disubstitution and Polysubstitution

<u>Problem 9.22</u> When treated with $Cl_2$ / $AlCl_3$, o-xylene (1,2-dimethylbenzene) gives a mixture of two products. Draw structural formulas for these products.

**For this problem, assume that only one chlorine atom is added to the ring. Because of symmetry considerations, there are two different products that can be produced when o-xylene undergoes chlorination.**

Problem 9.23 How many products are possible when *p*-xylene is treated with Cl$_2$ / AlCl$_3$? When *m*-xylene is treated with Cl$_2$ / AlCl$_3$?

**For this problem, assume that only one chlorine atom is added to the ring. There is only one product formed from *p*-xylene:**

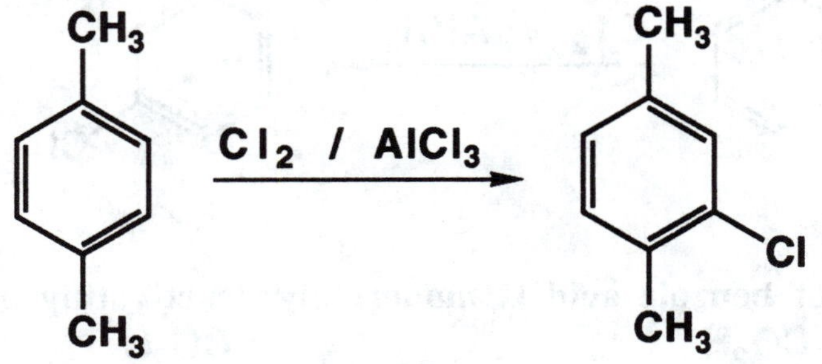

**There are two products formed from *m*-xylene:**

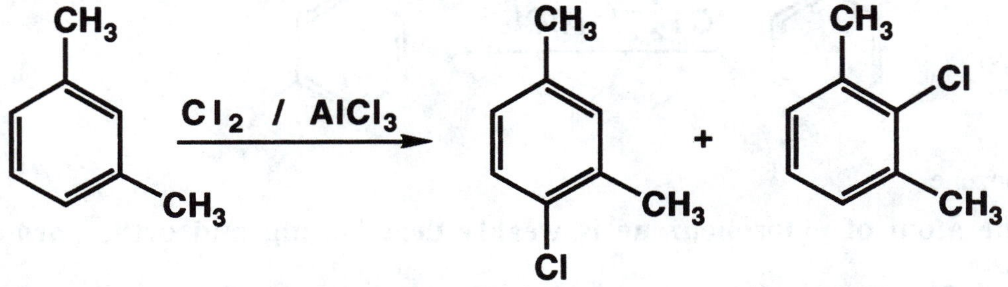

**Note that the last product, the one with the chlorine atom between the methyl groups, will be formed in the smallest amount because of the combined steric effects of the two methyl groups.**

Problem 9.24 Draw the structural formula for the major product formed on treatment of each compound with Cl$_2$ / AlCl$_3$.
(a) Toluene

**The methyl group of toluene is weakly activating and ortho-para directing.**

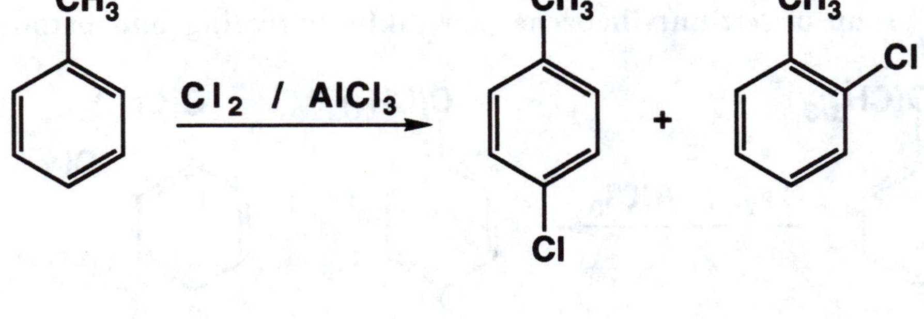

(b) Nitrobenzene

**The nitro group of nitrobenzene is strongly deactivating and meta directing.**

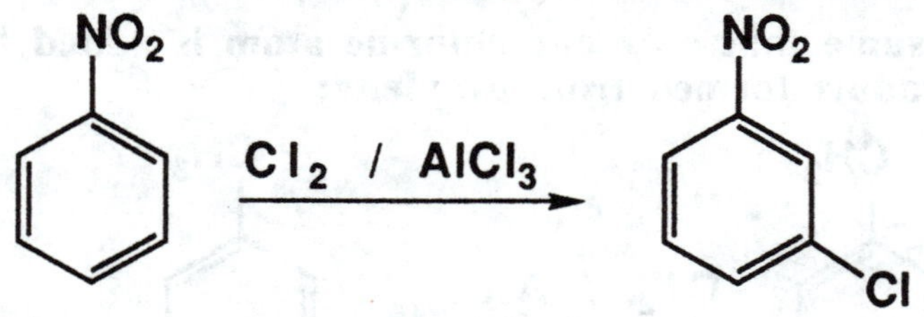

(c) Benzoic acid

**The carboxyl group of benzoic acid is moderately deactivating and meta directing.**

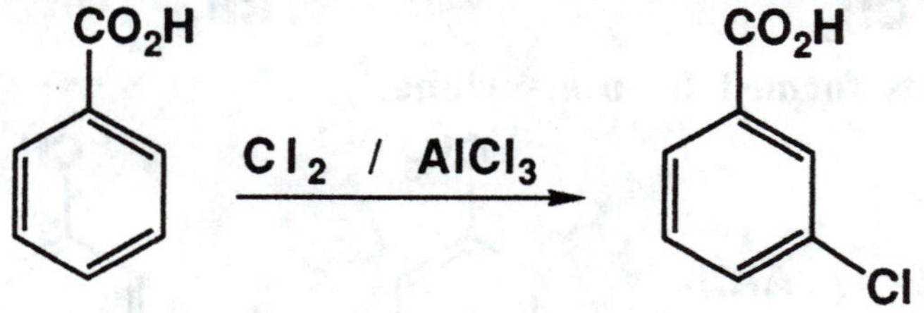

(d) Chlorobenzene

**The chlorine atom of chlorobenzene is weakly deactivating and ortho-para directing.**

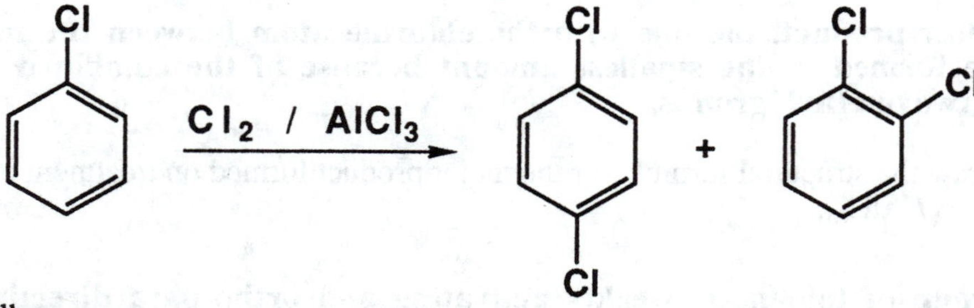

(e) *tert*-Butylbenzene

**The *tert*-butyl group of *tert*-butylbenzene is weakly activating and ortho-para directing.**

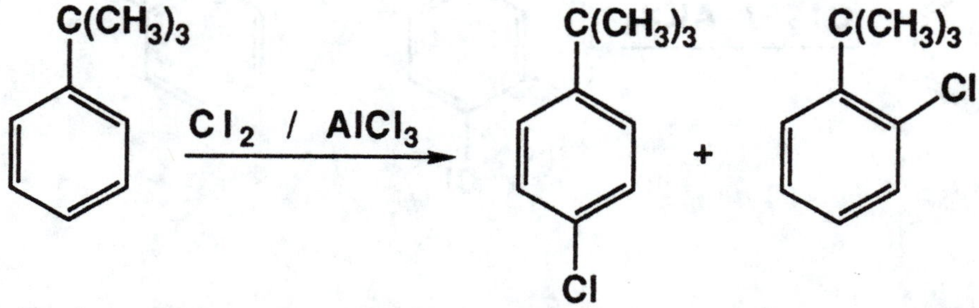

**The ortho product will be formed in smaller amounts because of the steric influence of the *tert*-butyl group.**

(f) Aniline

**The amino group of aniline is strongly activating and ortho-para directing.**

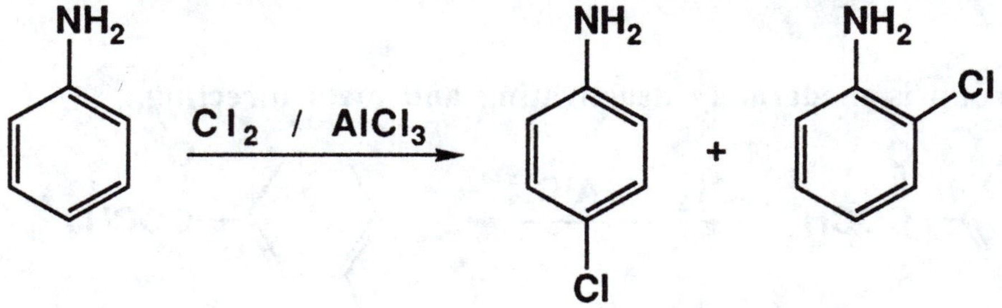

(g)

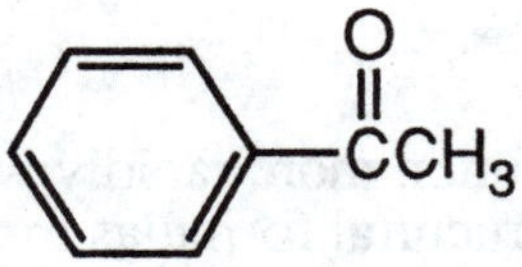

**The carbonyl group of a ketone is moderately deactivating and meta directing.**

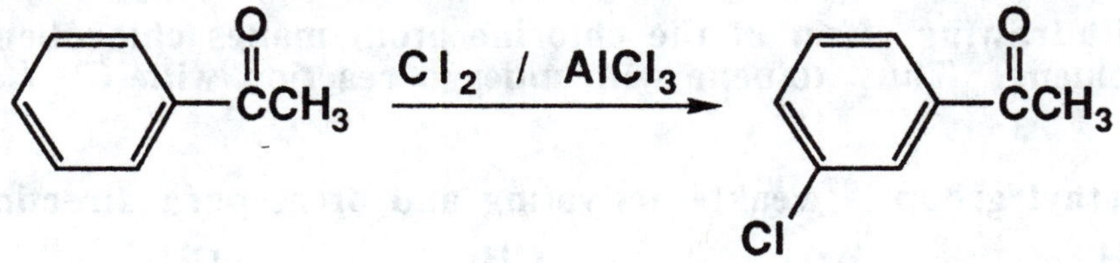

(h)

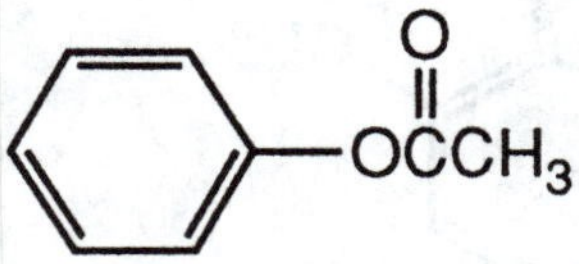

**An ester with the oxygen atom attached to the ring is moderately activating and ortho-para directing.**

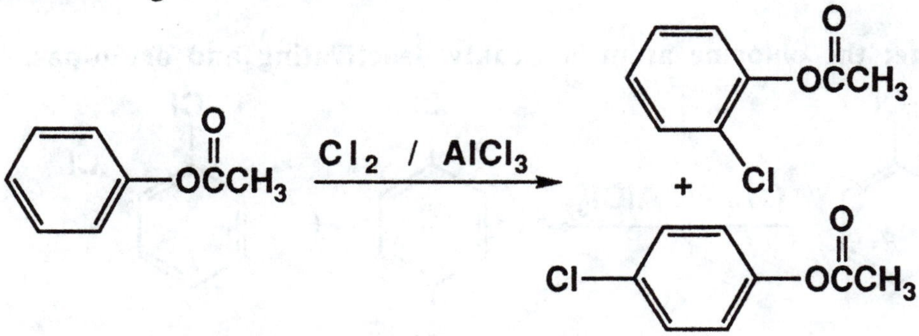

(i)

**A carbonyl group is moderately deactivating and meta directing.**

**Problem 9.25** Which compound undergoes electrophilic aromatic substitution more rapidly when treated with $Cl_2$ / $AlCl_3$, chlorobenzene or toluene? Explain, and draw structural formulas for the major product(s) from each compound.

**The electron-withdrawing effect of the chlorine atom makes chlorobenezene less reactive than toluene. Thus, toluene will undergo reaction with $Cl_2$ / $AlCl_3$ more rapidly.**

**Toluene; the methyl group is weakly activating and ortho-para directing.**

**Chlorobenzene; the chlorine atom is weakly deactivating and ortho-para directing.**

Problem 9.26 Arrange the compounds in each set in order of decreasing reactivity (fastest to slowest) toward electrophilic aromatic substitution.

(a)

(A) 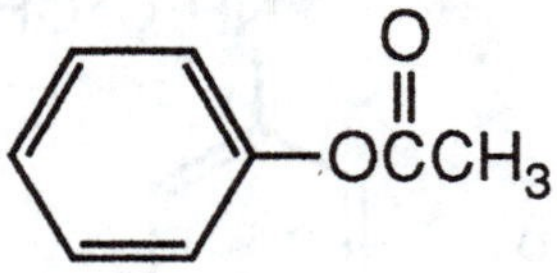          (B) 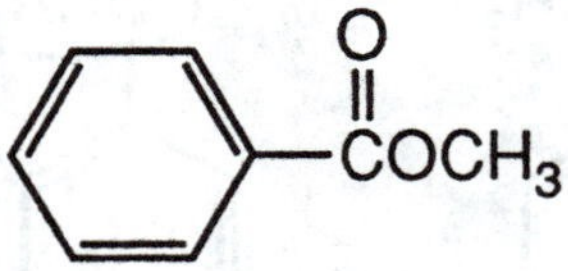          (C)

**B > A > C**

(b)

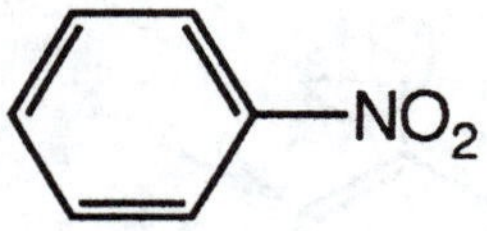

(A) 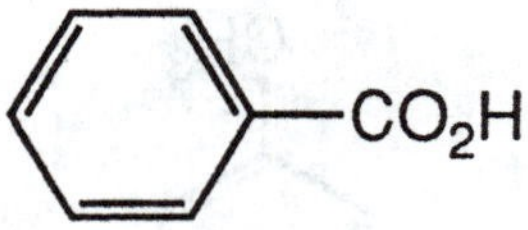          (B)           (C)

**C > B > A**

Problem 9.27 Arrange the compounds in each set in order of decreasing reactivity (fastest to slowest) toward electrophilic aromatic substitution.

(a)

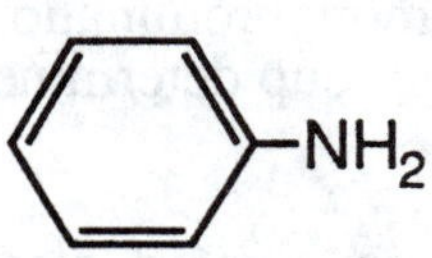

(A) 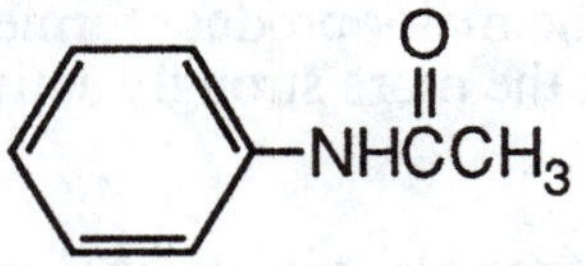          (B) 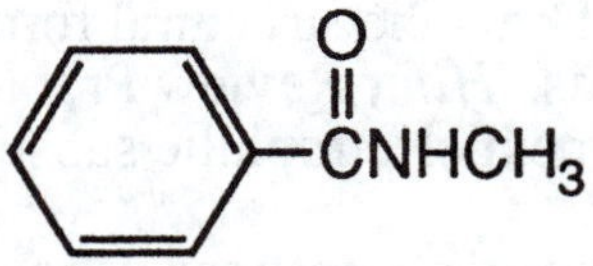          (C)

**A > B > C**

(b)

(A) 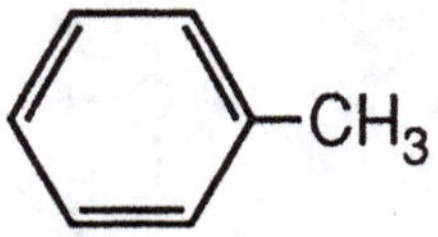          (B) 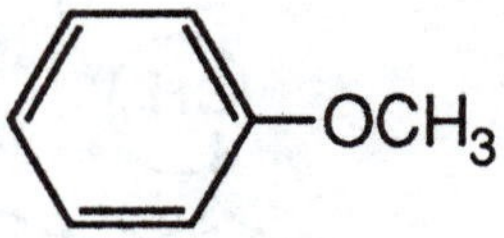          (C)

**C > B > A**

Problem 9.28 In di- and polysubstituted benzene rings, the more strongly activating group exerts the dominant effect. Draw the structural formula of the major product formed by nitration of the ring.

**In the following structures, the more strongly activating group is circled and arrows show the position(s) of nitration. Where both ortho and para nitration are possible, two arrows are shown. A broken arrow shows a product formed in only negligible amounts.**

(a)

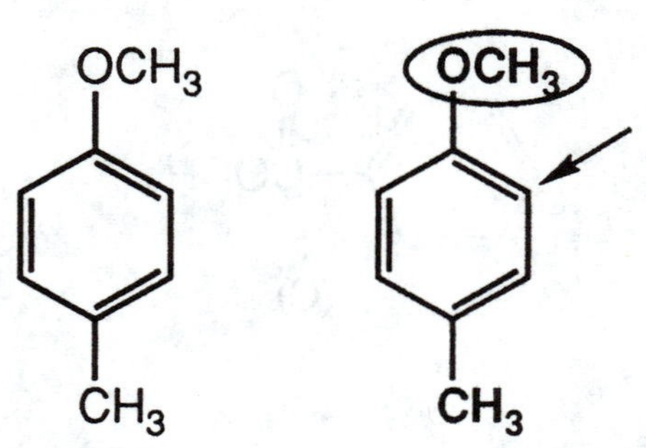

(b)

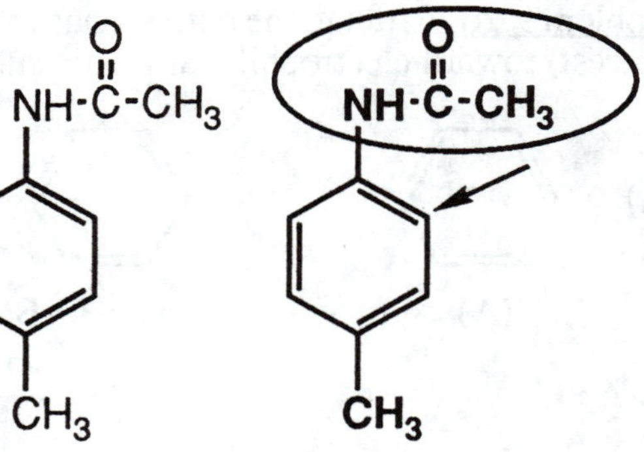

(c)

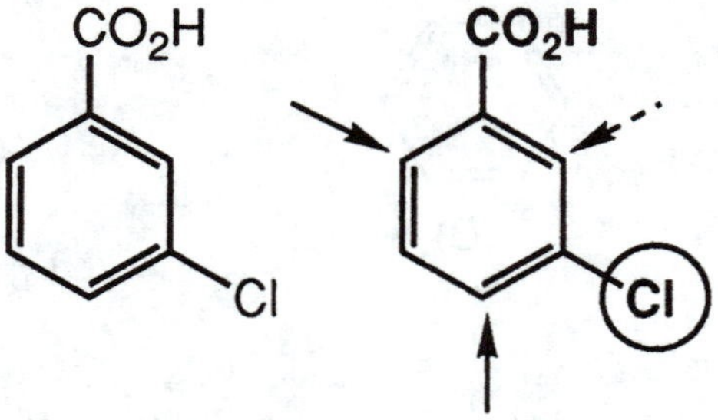

(d)

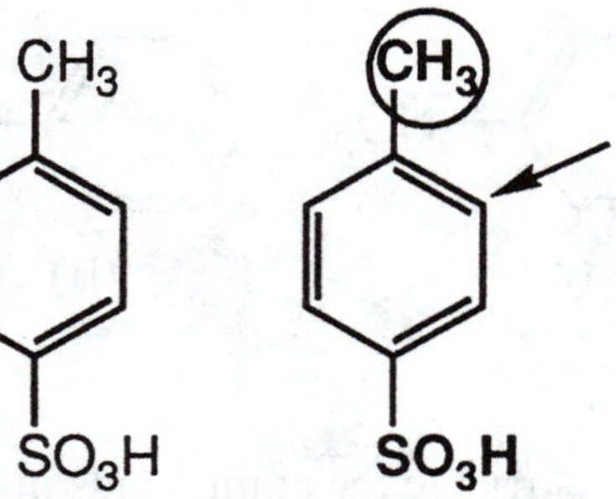

**Problem 9.29** Draw the structural formula of the major product formed by monobromination of each compounds. *Hint;* Review Problem 9.28; the more strongly activating group determines the location of the next electrophilic substitution.

**In the following structures, the more strongly activating group is circled and arrows show the position(s) of monobromination. Where both ortho and para monobromination are possible, two arrows are shown. A broken arrow shows a product formed in only negligible amounts.**

(a)

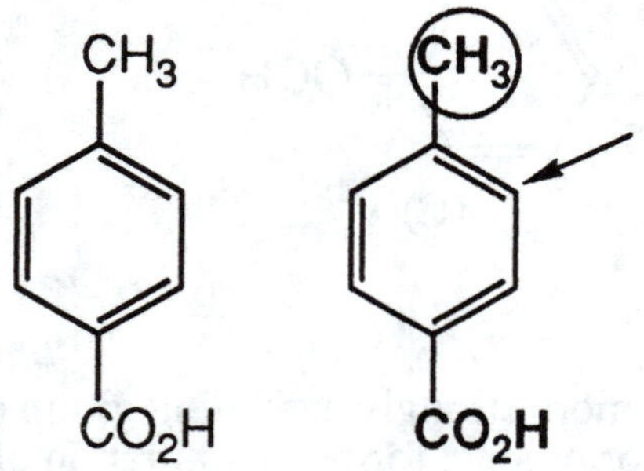

(b)

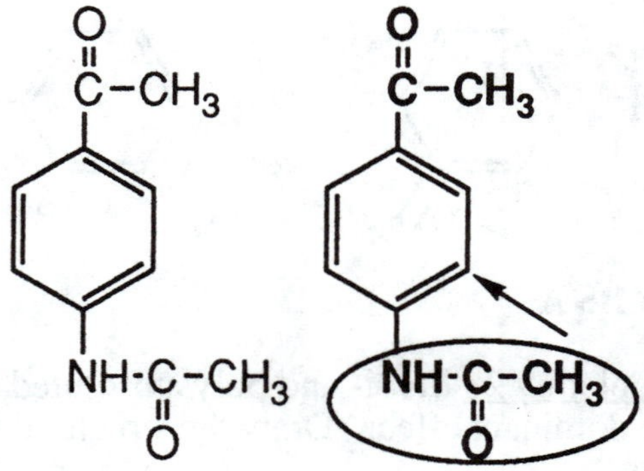

(c)

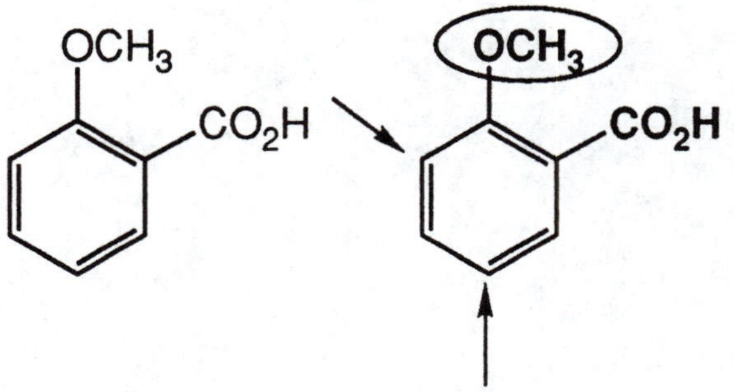

(d)

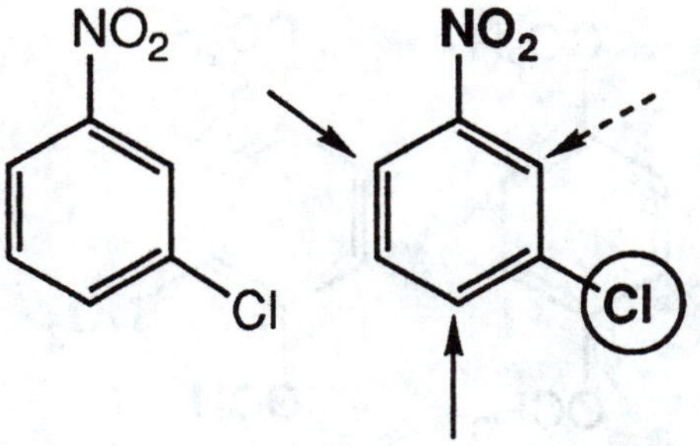

**Problem 9.30** Predict the major product(s) from treatment of each compound with $HNO_3$, $H_2SO_4$.

**When there is more than one substituent on a ring, the predominant product is derived from the orientation preference of the most activating substituent.**

(a)

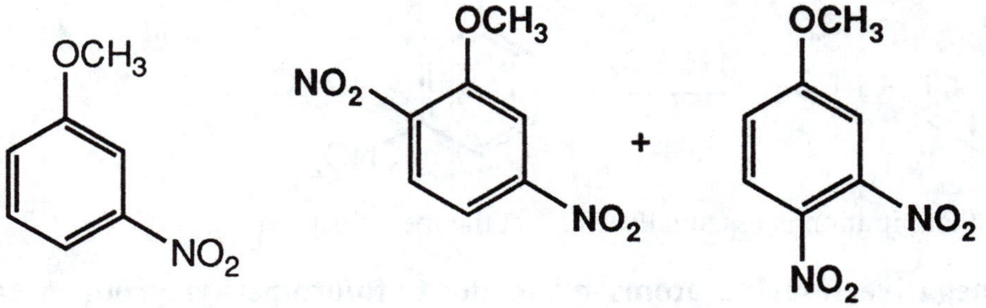

(b)

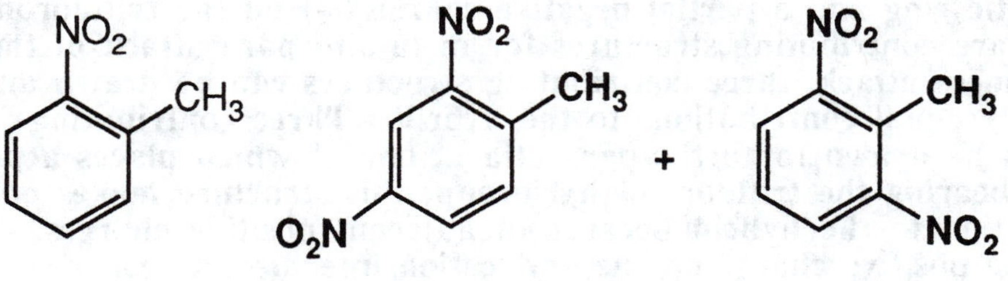

(c)

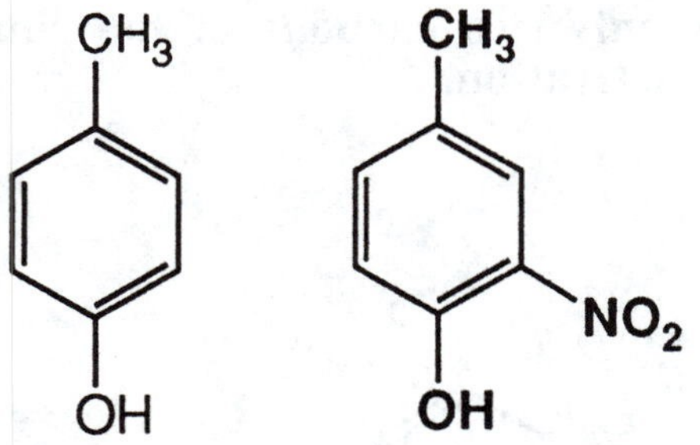

(d)

**Problem 9.31** The trifluoromethyl group is almost exclusively meta directing as shown in the following example:

Account for the fact that nitration is essentially 100% at the meta position.

**The very electronegative flourine atoms make the trifluoromethyl group highly electron withdrawing. Thus, the bond between the trifluoromethyl group and the aromatic ring can be represented with a partial positive charge ($\delta+$) on the carbon atom of the aromatic ring and a partial negative charge ($\delta-$) on the trifluoromethyl group. Following are contributing structures for meta and para attack of the electrophile. For meta attack, three contributing structures can be drawn and all make approximately equal contributions to the hybrid. Three contributing structures can also be drawn for ortho-para attack, one of which places a positive charge on carbon bearing the trifluoromethyl group; this structure makes only a negligible contribution to the hybrid because of adjacent positive charges. Thus, for meta attack, the positive charge on the aryl cation intermediate can be delocalized almost equally over three atoms of the ring giving this cation's formation a lower energy of activation. For ortho-para attack, the positive charge on the aryl cation intermediate is delocalized over only two carbons of the ring, giving this cation's formation a higher energy of activation.**

**meta attack:**

## ortho-para attack:

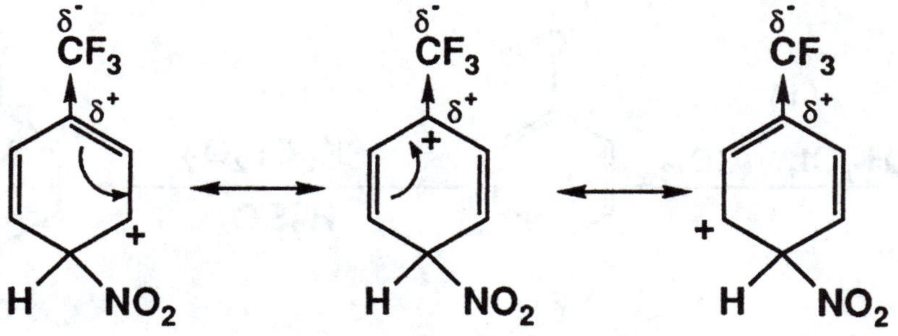

adjacent positive
charges

Problem 9.32  Show how to convert toluene to these carboxylic acids.
(a) 2,4-Dinitrobenzoic acid

**Methyl is ortho-para directing.  Therefore, toluene can be nitrated twice then oxidized with $K_2Cr_2O_7$ in aqueous $H_2SO_4$ to convert the methyl group into the carboxyl group.**

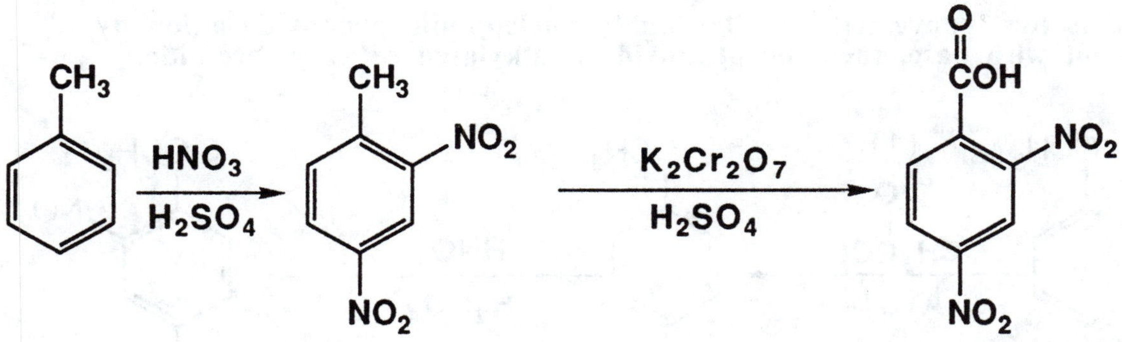

2,4-Dinitrobenzoic acid

(b) 3,5-Dinitrobenzoic acid

**The reaction sequence is very similar to the last one, except now the order of the reactions is reversed because the carboxylic acid group is a meta director.**

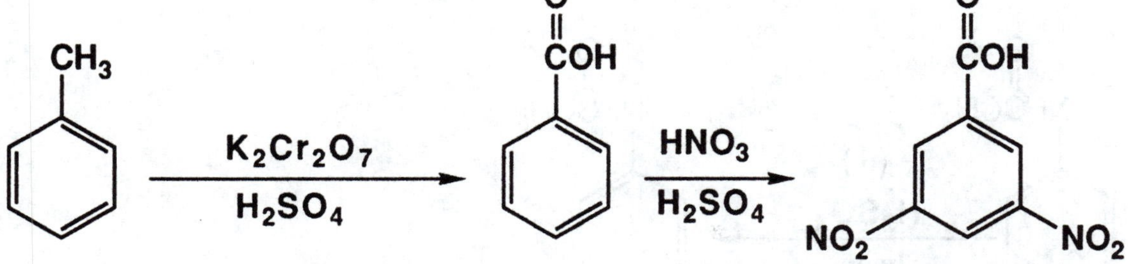

3,5-Dinitrobenzoic acid

Problem 9.33 Show reagents and conditions to bring about these conversions. The numbers over each arrow shows the number of steps required.

(a)

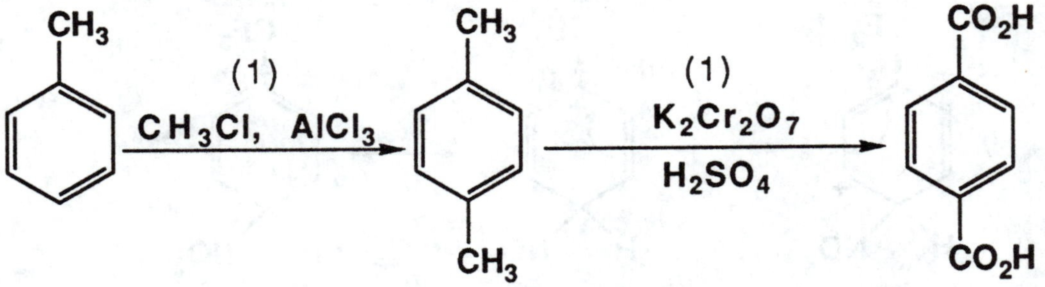

**A Friedel-Crafts alkylation using chloromethane is carried out followed by oxidation of each benzylic carbon to a carboxyl group.**

(b)

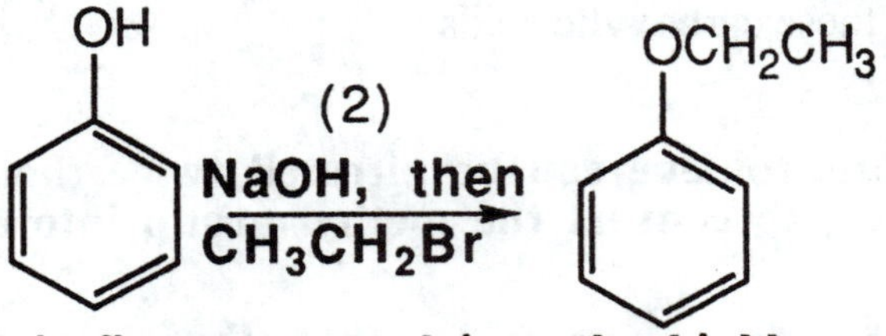

**Phenol is first converted into the highly nucleophilic phenoxide anion by treatment with base, then the phenoxide is alkylated by ethyl bromide.**

(c)

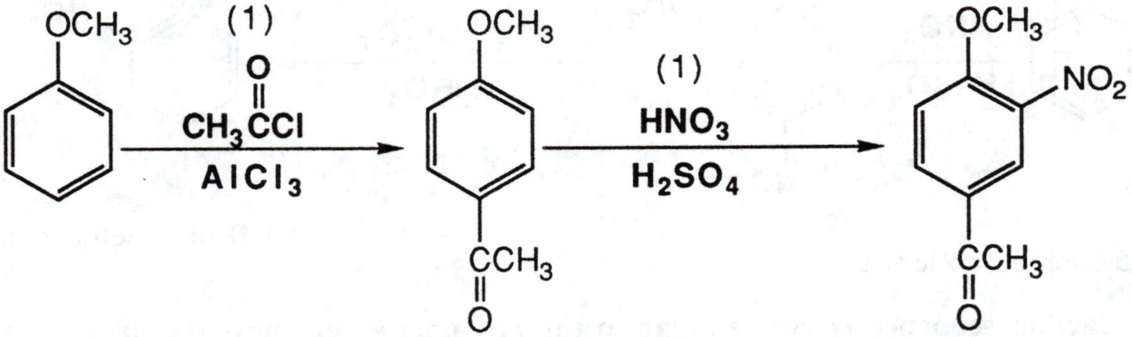

**A Friedel-Crafts acylation of anisole is carried out, followed by nitration, with the orientation directed by the activating methoxy group.**

(d)

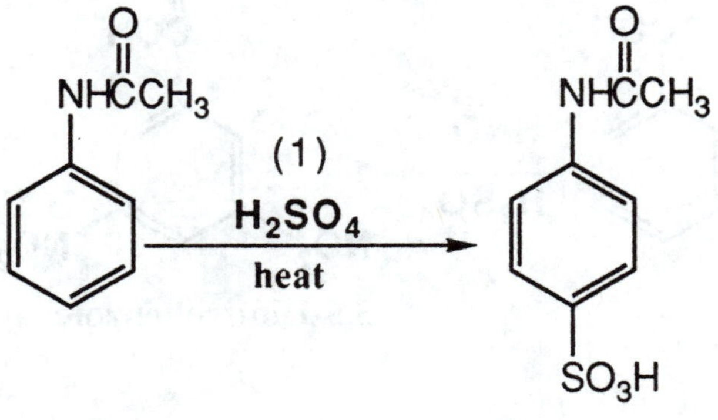

**Problem 9.34** Propose a synthesis of triphenylmethane from benzene, as the only source of aromatic rings, and any other necessary reagents.

**Reaction of three moles of benzene with one mole of trichloromethane (chloroform) will give triphenylmethane.**

$$3 \; \text{C}_6\text{H}_6 + \text{CHCl}_3 \xrightarrow{\text{AlCl}_3} (\text{C}_6\text{H}_5)_3\text{CH} + 3\text{HCl}$$

**Problem 9.35** Reaction of phenol with acetone in the presence of an acid catalyst gives a compound known as bisphenol A. Bisphenol A is used in the production of epoxy resins and polycarbonate resins. Propose a mechanism for the formation of bisphenol A.

$$2 \; \text{C}_6\text{H}_5\text{OH} + \text{CH}_3\text{-CO-CH}_3 \xrightarrow{\text{H}_3\text{PO}_4} \text{HO-C}_6\text{H}_4\text{-C(CH}_3)_2\text{-C}_6\text{H}_4\text{-OH} + \text{H}_2\text{O}$$

Bisphenol A

**The reaction begins with protonation of acetone to form its conjugate acid which may be written as a hybrid of two contributing structures.**

$$\text{CH}_3\text{-CO-CH}_3 + \text{H}^+ \longrightarrow \text{CH}_3\text{-C(}^+\text{OH)-CH}_3 \longleftrightarrow \text{CH}_3\text{-}^+\text{C(OH)-CH}_3$$

**The conjugate acid of acetone is an electrophile and reacts with phenol at the para position by electrophilic aromatic substitution to give 2-(4-hydroxyphenyl)-2-propanol. Protonation of the tertiary alcohol in this molecule and departure of water gives a resonance-stabilized cation that reacts with a second molecule of phenol to give bisphenol A.**

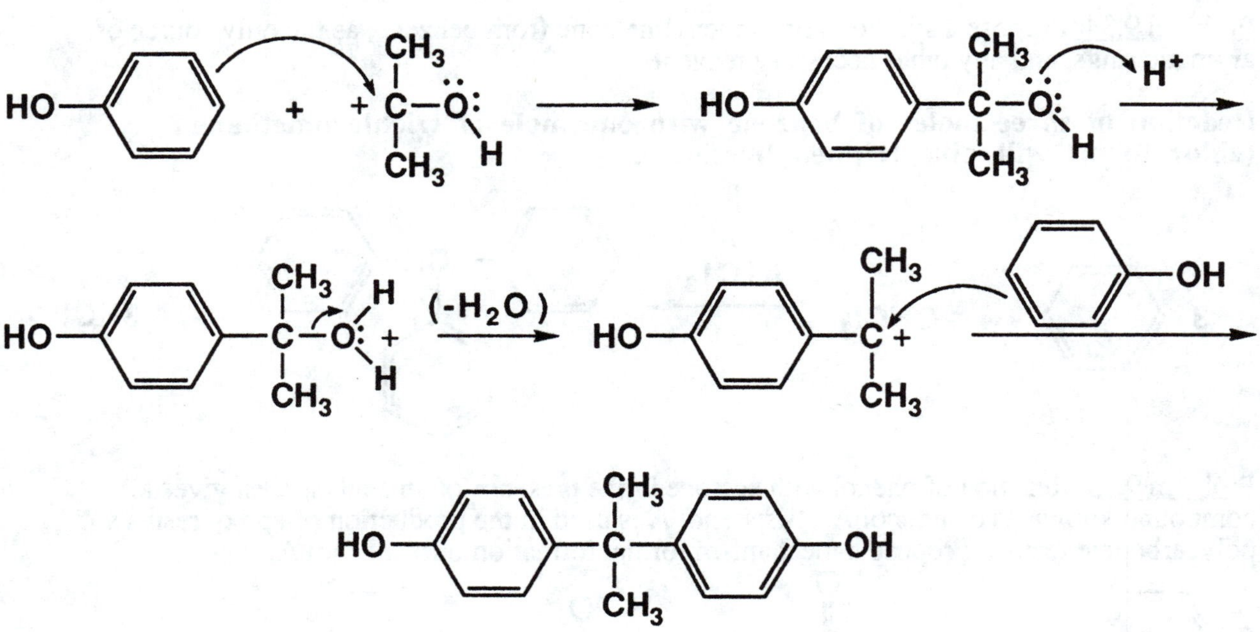

**Bisphenol A**

Problem 9.36 2,6-Di-*tert*-butyl-4-methylphenol, alternatively known as butylated hydroxytoluene (BHT), is used as an antioxidant in foods to "retard spoilage". BHT is synthesized industrially from 4-methylphenol (*p*-cresol) by reaction with 2-methylpropene in the presence of phosphoric acid. Propose a mechanism for this reaction.

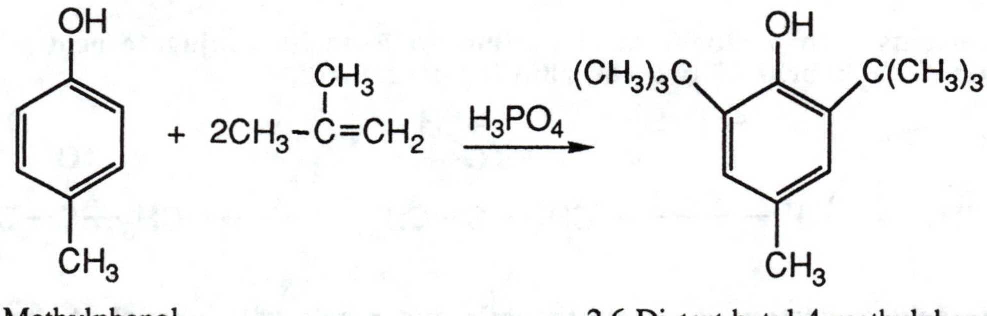

4-Methylphenol
(*p*-cresol)

2,6-Di-*tert*-butyl-4-methylphenol
"butylated hydroxytoluene"
(BHT)

**The reaction involves an initial proton transfer from phosphoric acid to 2-methylpropene to give an electrophilic *tert*-butyl cation that then reacts with the aromatic ring ortho to the strongly activating -OH group to form 2-*tert*-butyl-4-methylphenol. A second electrophilic aromatic substitution gives the final product.**

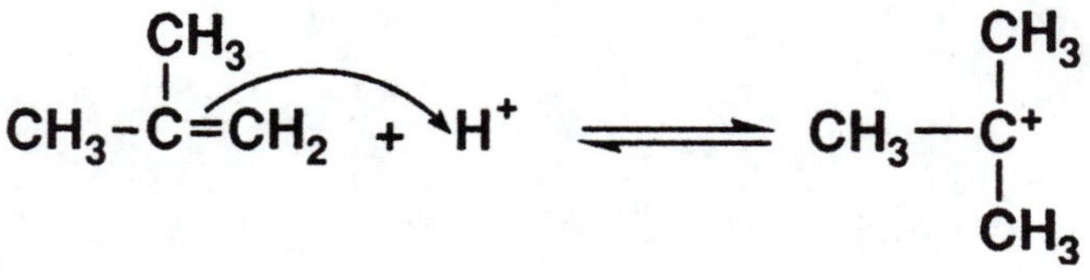

**Problem 9.37** The first widely used herbicide for control of weeds was 2,4-dichlorophenoxyacetic acid (2,4-D). Show how this compound might be synthesized from 2,4-dichlorophenol and 2-chloroacetic acid, ClCH$_2$CO$_2$H.

2,4-Dichlorophenol

2,4-Dichlorophenoxyacetic acid
(2,4-D)

## Syntheses

Problem 9.38  Using styrene, $C_6H_5CH=CH_2$, as the only aromatic starting material, show how to synthesize these compounds.  In addition to styrene, use any other necessary organic or inorganic chemicals.  Any compound synthesized in one part of this problem may be used to make any other compound in the problem.

(a) 
$$\text{C}_6\text{H}_5\text{COOH} \xleftarrow[\text{H}_2\text{SO}_4]{\text{K}_2\text{Cr}_2\text{O}_7} \text{C}_6\text{H}_5\text{CH}=\text{CH}_2$$

(b) 
$$\text{C}_6\text{H}_5\text{CHBrCH}_3 \xleftarrow{\text{HBr}} \text{C}_6\text{H}_5\text{CH}=\text{CH}_2$$

(c) 
$$\text{C}_6\text{H}_5\text{CH(OH)CH}_3 \xleftarrow[\text{H}_2\text{SO}_4]{\text{H}_2\text{O}} \text{C}_6\text{H}_5\text{CH}=\text{CH}_2$$

(d) 
$$\text{C}_6\text{H}_5\text{COCH}_3 \xleftarrow{\text{H}_2\text{CrO}_4} \text{C}_6\text{H}_5\text{CH(OH)CH}_3$$

(e) 
$$\text{C}_6\text{H}_5\text{CH}_2\text{CH}_3 \xleftarrow[\text{Ni}]{\text{H}_2} \text{C}_6\text{H}_5\text{CH}=\text{CH}_2$$

(f) 
$$\text{C}_6\text{H}_5\text{CH(OH)CH}_2\text{OH} \xleftarrow[\text{pH 11.8}]{\text{KMnO}_4} \text{C}_6\text{H}_5\text{CH}=\text{CH}_2$$

Problem 9.39  Starting with benzene, toluene, or phenol as the only sources of aromatic rings, show how to synthesize these compounds.  Assume in all syntheses that ortho-para mixtures can be separated to give the desired isomer in pure form.
(a) *m*-Bromonitrobenzene

**Nitro is meta directing; bromine is ortho-para directing.  Therefore, to have the two substituents meta to each other, carry out nitration first followed by bromination.**

$$\text{benzene} \xrightarrow[\text{H}_2\text{SO}_4]{\text{HNO}_3} \text{Nitrobenzene} \xrightarrow[\text{FeBr}_3]{\text{Br}_2} \text{1-Bromo-3-nitrobenzene}$$

**Nitrobenzene          1-Bromo-3-nitrobenzene
(*m*-Bromonitrobenzene)**

(b) 1-Bromo-4-nitrobenzene

**Reverse the order of steps from part (a). Nitro is meta directing; bromine is ortho-para directing. Therefore, to have the two substituents para to each other, carry out bromination first followed by nitration.**

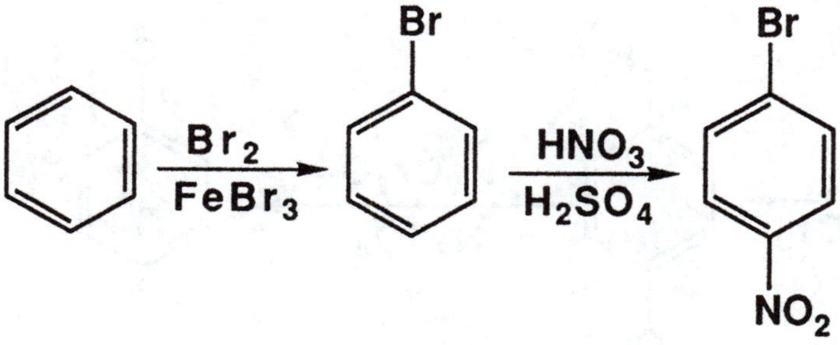

|         | Bromobenzene | 1-Bromo-4-nitrobenzene |
|---------|--------------|------------------------|
|         |              | (*p*-Bromonitrobenzene) |

(c) 2,4,6-Trinitrotoluene (TNT)

**The methyl group is ortho-para directing. Therefore, nitrate toluene three successive times.**

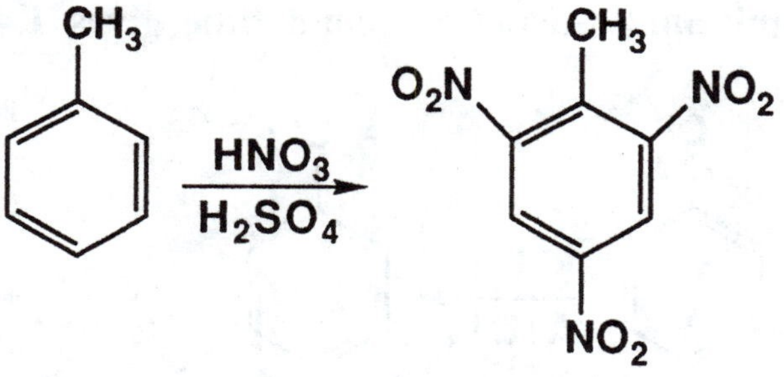

2,4,6-Trinitrotoluene

(d) *m*-Chlorobenzoic acid

**The carboxyl group and chlorine atom are meta to each other, an orientation best accomplished by chlorination of benzoic acid (the carboxyl group is meta-directing). Oxidation of toluene with $K_2Cr_2O_7$ in aqueous $H_2SO_4$ gives benzoic acid. Treatment of benzoic acid with chlorine in the presence of aluminum chloride or ferric chloride gives the desired product.**

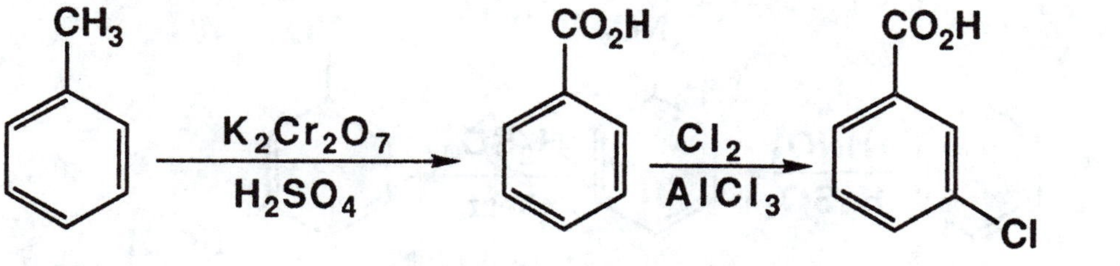

Toluene                    Benzoic acid        3-Chlorobenzoic acid

(e) *p*-Chlorobenzoic acid

**Start with toluene. Methyl is weakly activating and directs chlorination to the ortho-para positions. Separate the desired para isomer and then oxidize the methyl group to a carboxyl group using $K_2Cr_2O_7$ in aqueous $H_2SO_4$ to give 4-chlorobenzoic acid.**

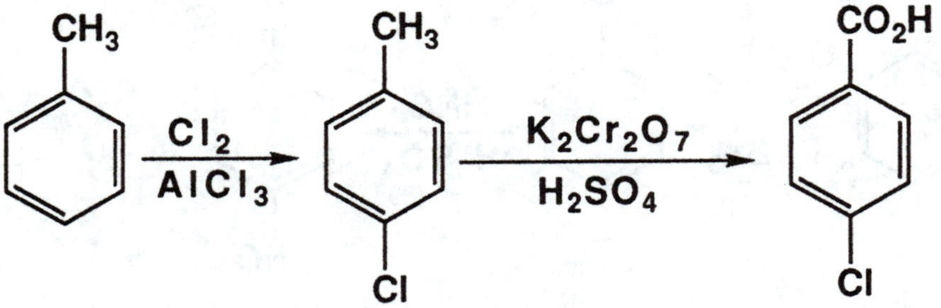

Toluene      4-Chlorotoluene      4-Chlorobenzoic acid

(f) *p*-Dichlorobenzene

**Treatment of benzene with chlorine in the presence of aluminum chloride gives chlorobenzene. The chlorine atom is ortho-para directing. Treatment with chlorine in the presence of aluminum chloride a second time gives 1,4-dichlorobenzene.**

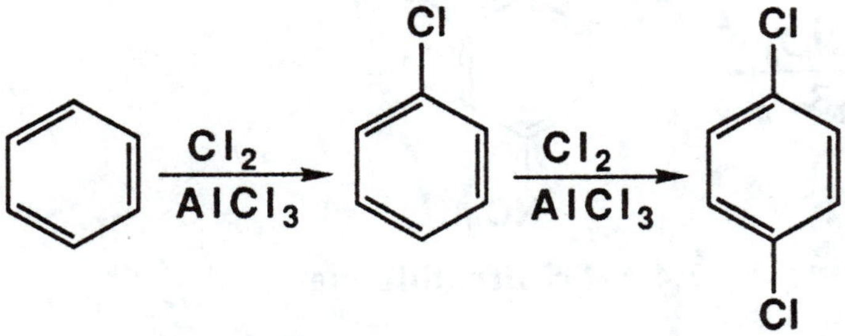

Benzene      Chlorobenzene      *p*-Dichlorobenzene

(g) *m*-Nitrobenzenesulfonic acid

**Both the sulfonic acid group and the nitro group are meta directors. Therefore, the two electrophilic aromatic substitution reactions may be carried out in either order. The sequence shown is nitration followed by sulfonation.**

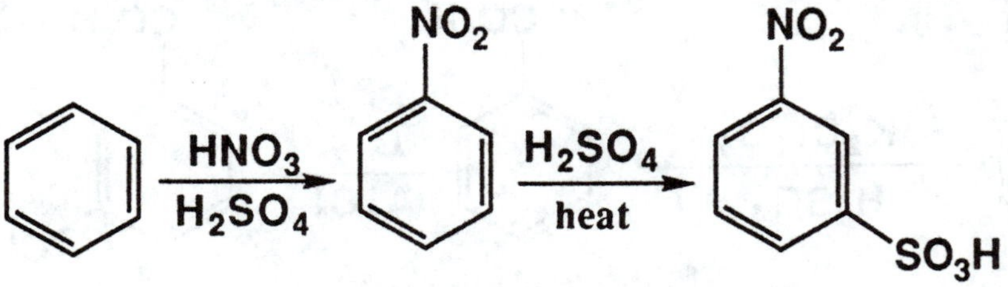

Benzene      Nitrobenzene      *m*-Nitrobenzene sulfonic acid

**(h) 1-Chloro-3-nitrobenzene**

**The nitro group is meta directing, while the chlorine atom is ortho-para directing. Therefore, in order to obtain the desired meta product, the nitration must be carried out prior to the chlorination.**

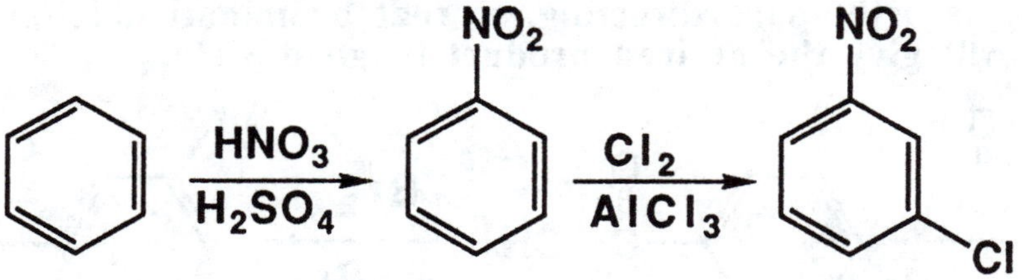

**Benzene**              **Nitrobenzene**        **1-Chloro-3-nitrobenzene**

<u>Problem 9.40</u> Starting with benzene or toluene as the only sources of aromatic rings, show how to synthesize these aromatic ketones. Assume in all syntheses the ortho-para mixtures can be separated to give the desired isomer in pure form.

(a)

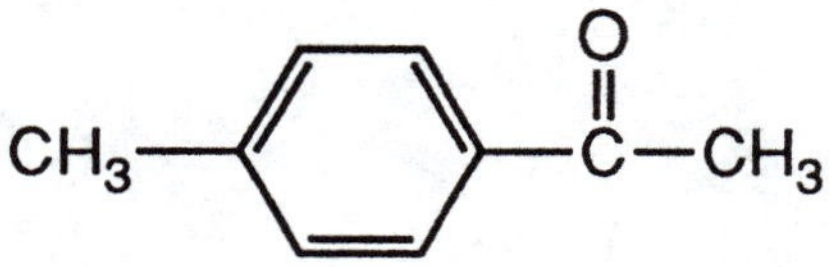

**The methyl group of toluene is ortho-para directing, so a Friedel-Crafts acylation of toluene will give the desired product in good yield.**

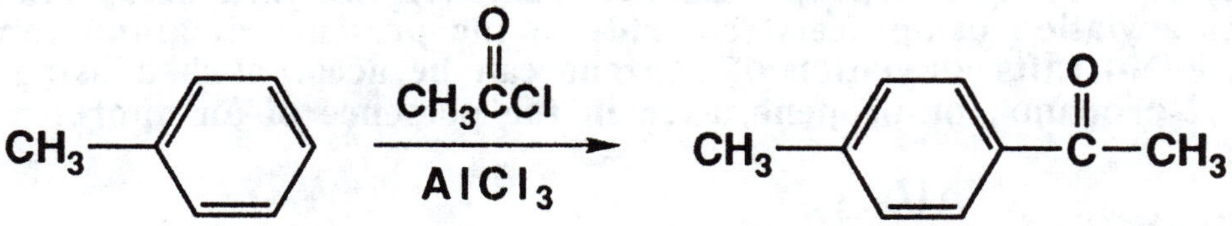

(b)

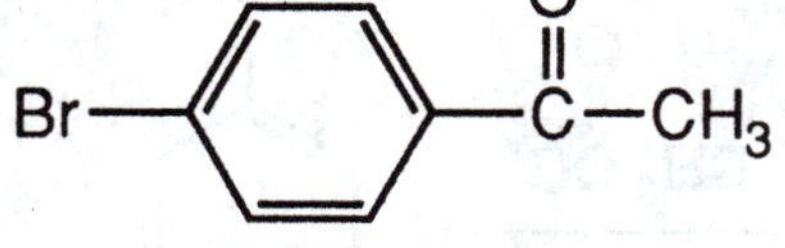

**The bromine atom of bromobenzene is ortho-para directing, so that a Friedel-Crafts acylation following bromination of benzene will give the desired product in good yield.**

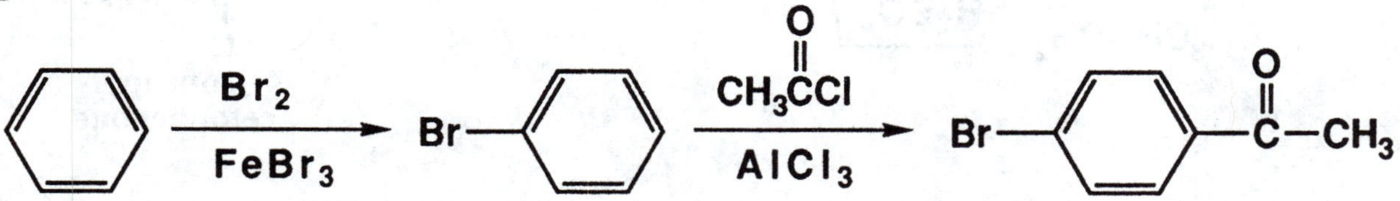

**(c)**

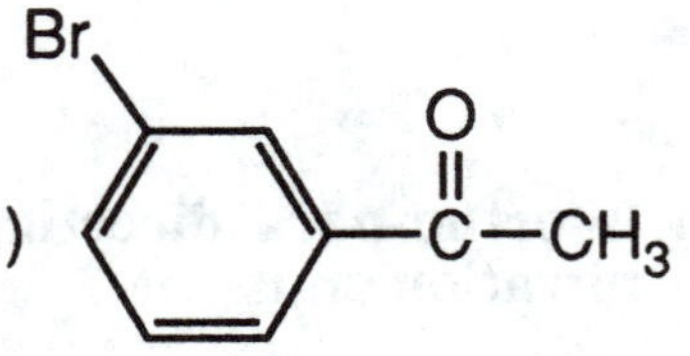

The ketone group is ortho-para directing, so that bromination following a Friedel-Crafts acylation will give the desired product in good yield.

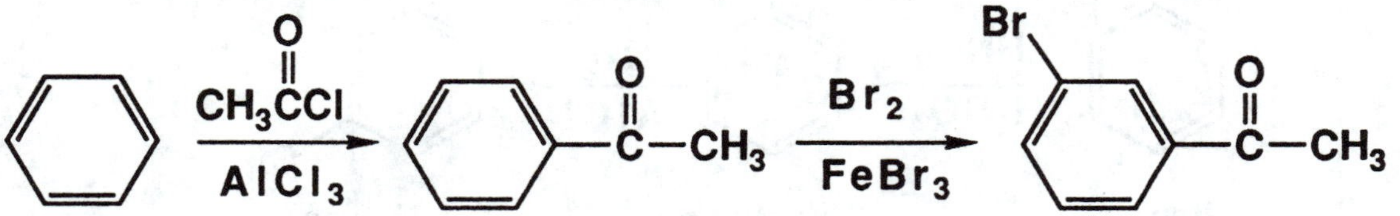

Problem 9.41  The following ketone, isolated from the roots of several members of the iris family, has an odor like that of violets and is used as a fragrance in perfumes.  Describe a synthesis of this ketone from benzene.

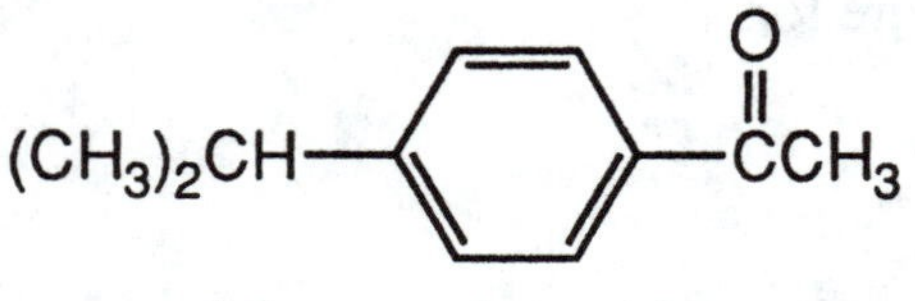

4-Isopropylacetophenone

The isopropyl group is weakly activating and ortho-para directing; the carbonyl group of the acetyl group is deactivating and meta directing.  Therefore, start with benzene, convert it to isopropylbenzene (cumene) and then carry out a Friedel-Crafts acylation using acetyl chloride in the presence of aluminum chloride.  Friedel-Crafts alkylation of benzene can be accomplished using a 2-halopropane, 2-propanol, or propene, each in the presence of an appropriate catalyst.

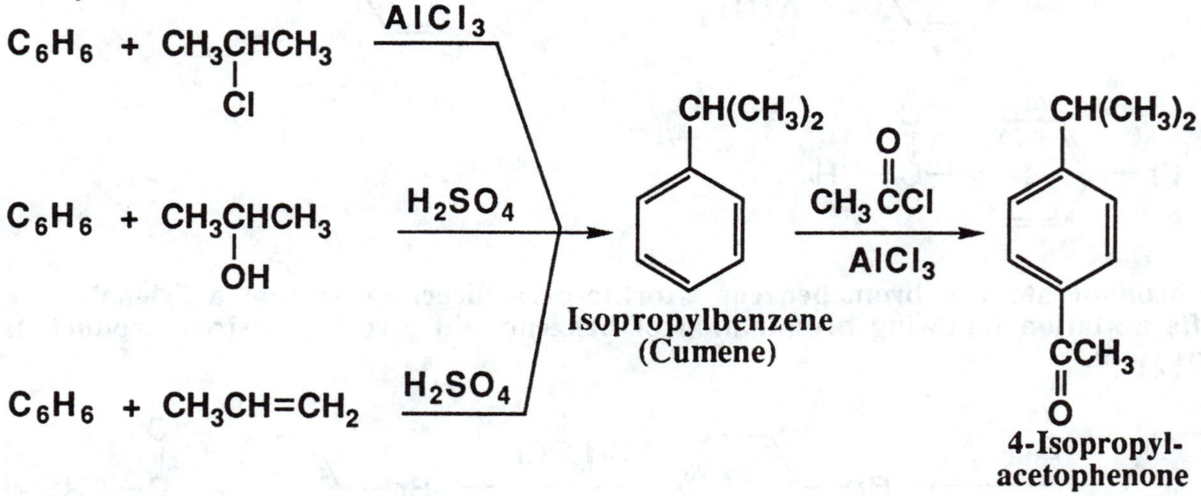

# CHAPTER 10: AMINES

## SUMMARY OF REACTIONS

| Starting Material ↓ / Product → | Ammonium Ions | Arenediazonium Salts | Aryl Amines | Aryl Bromides, Chlorides, Cyanides | Aryl Iodides | Benzene Derivatives | Nitro Aromatics | Phenols |
|---|---|---|---|---|---|---|---|---|
| Amines | **10A** 10.4* | | | | | | | |
| Arenediazonium Salts | | | | **10B** 10.6 | **10C** 10.6 | **10D** 10.6 | | **10E** 10.6 |
| Aromatic Rings | | | | | | | **9D** 9.7B | |
| Nitro Aromatics | | | **9I** 9.7B | | | | | |
| Primary Aromatic Amines | | **10F** 10.6 | | | | | | |

*Section in book that describes reaction.

## REACTION 9D: NITRATION (Section 9.7B)

$$\text{benzene} \xrightarrow[\text{H}_2\text{SO}_4]{\text{HNO}_3} \text{nitrobenzene (NO}_2)$$

## REACTION 9I: REDUCTION OF A NITROAROMATIC TO PRODUCE A PRIMARY ARYL AMINE (Section 9.7B)

$$\text{Ar-NO}_2 \xrightarrow[\begin{array}{l}\text{or}\\ \text{1) Fe, HCl, H}_2\text{O}\\ \text{2) NaOH}\end{array}]{\text{H}_2 \,/\, \text{Ni}} \text{Ar-NH}_2$$

## REACTION 10A: PROTONATION OF AMINES (Section 10.4)

$$—\ddot{N}— \ + \ H\text{-}A \ \longrightarrow \ —\overset{\overset{\displaystyle H \ \ A^-}{|}}{\underset{|}{N^+}}—$$

- **Amines are bases** due to the lone pair of electrons on nitrogen. ✳
- **Aliphatic amines** have $pK_b$'s **near 4**, which is slightly more basic than ammonia itself.
- **Aromatic amines** are much less basic, having $pK_b$'s **near 9**. This is because the nitrogen lone pair is partially delocalized into the adjacent aromatic ring, and thus less available for interacting with protons. Electron-withdrawing groups on the aromatic ring decrease base strength of aromatic amines, and electron-donating groups increase base strength.
- **Aromatic heterocyclic amines** are less basic than non-aromatic heterocyclic amines.

## REACTION 10B: SANDMEYER REACTION (Section 10.6)

- A primary **aromatic amine** can be reacted with **nitrous acid**, followed by **CuBr, CuCl, or CuCN** to produce an **aryl bromide, chloride, or cyanide**, respectively. This reaction is referred to as the **Sandmeyer reaction**. ✳

## REACTION 10C: TREATMENT OF AN ARENEDIAZONIUM ION WITH KI (Section 10.6)

- A **primary aromatic amine** can be reacted with **nitrous acid**, followed by **KI**, to produce an **aryl iodide**. ✳

## REACTION 10D: REDUCTION OF AN ARENEDIAZONIUM ION WITH HYPOPHOSPHOROUS ACID (Section 10.6)

- A **primary aromatic amine** can be reacted with **nitrous acid**, followed by $H_3PO_2$ to produce a product in which the aryl amine function is replaced by hydrogen. ✳
- This reaction is useful for a variety of synthetic applications, especially when the amino group is desired to control orientation or reactivity during electrophilic aromatic substitution.

## REACTION 10E: CONVERSION OF A PRIMARY ARYL AMINE TO A PHENOL
### (Section 10.6)

- **Heating** an **arenediazonium salt** in **water** results in formation of a **phenol**. ✳
- An aryl cation intermediate is formed, that reacts with water to produce the phenol.

## REACTION 10F: FORMATION OF ARENEDIAZONIUM IONS (Section 10.6)

- **Primary aryl amines** are converted to **arenediazonium ions** by treatment with **nitrous acid**. ✳
- Diazonium ions are important, because they can be converted to a large number of other aryl species by treatment with various reagents. For example, see reactions 10B-10E. Thus, primary aromatic amino groups can be replaced by other important functional groups via an aryl diazonium intermediate.

## SUMMARY OF IMPORTANT CONCEPTS

## 10.0 OVERVIEW
• Nitrogen is the **fourth most common element** in organic molecules, and **amines** are the most common functional group containing nitrogen. Amines have a lone pair of electrons on nitrogen, and their two most **important properties** are that they are **nucleophilic and basic**. ✳

## 10.1 STRUCTURE AND CLASSIFICATION
• **Amines** are **derivatives of ammonia** that have one or more of the **hydrogens replaced** with **alkyl** and/or **aryl groups**. ✳
  - **Primary (1°) amines** have **one hydrogen** of ammonia **replaced by a carbon** in the form of an alkyl or aryl group.
  - **Secondary (2°) amines** have **two hydrogens** of ammonia **replaced by a carbon** in the form of alkyl and/or aryl groups.
  - **Tertiary (3°) amines** have **all three hydrogens** of ammonia **replaced by a carbon** in the form of alkyl and/or aryl groups.
  - **Quaternary (4°) ammonium ions** have four alkyl and/or aryl groups attached to nitrogen, resulting in a **positively charged** species.
  - **Aliphatic amines** have **alkyl groups** only bonded to nitrogen, while **aromatic amines** have **at least one aromatic ring** bonded to the nitrogen atom.
  - **Heterocyclic amines** have the **nitrogen atom** as **part of a ring**, and **aromatic heterocyclic amines** have the **nitrogen atom** as **part of an aromatic ring**.

## 10.2 NOMENCLATURE

- **Systematic names** are derived just as they are for alcohols, except the suffix **e** of the alkane is replaced by the suffix **amine**. For secondary and tertiary amines, the largest group is taken as the parent amine, and the smaller alkyl groups are listed as N-substituents. For example, N-methylbutanamine is a systematic name.
- IUPAC uses the common name **aniline** for derivatives of $C_6H_5NH_2$, although certain common names are retained for some substituted anilines such as **toluidine** and **anisidine**.
- Several common heterocycles retain their common names in IUPAC nomenclature as well including **indole, purine, quinoline,** and **isoquinoline**.
- **Common names** are derived by listing the alkyl groups attached to the nitrogen atom in alphabetical order, followed by the suffix **amine**.

## 10.3 PHYSICAL PROPERTIES

- **Amines are polar compounds**, and **primary or secondary amines** can make **intramolecular hydrogen bonds**. As a result, amines have **higher boiling points than analogous hydrocarbons**. Since they can take part in hydrogen bonding, they are also **more soluble in water** than their hydrocarbon counterparts.
- The **N-H---N hydrogen bonds** are not as strong as O-H---O hydrogen bonds, because the difference in electronegativities is not as large for N vs. H as it is for O vs. H.

## 10.4 BASICITY

- **Amines are weak bases**, so aqueous solutions of amines are basic. The **lone pair of electrons on nitrogen** forms a new **bond** with a **hydrogen of water**, producing the **ammonium species and hydroxide**. This equilibrium is often described as $K_b$, that is defined by the following equation:

$$K_b = K_{eq}[H_2O] = \frac{[R_3NH^+][OH^-]}{[R_3N]}$$

where R is an alkyl group, aryl group, or hydrogen. ✳

- The **$pK_b$** of an amine is equal to the **-log $K_b$**.
- The **$pK_a$** of the conjugate acid of an amine is **related to** the **$pK_b$** by the following equation:

$$pK_a + pK_b = 14.00$$

*[This is a particularly useful relationship]*

- In general, any group attached to the nitrogen of an amine that **releases electron density increases basicity**, while any group that **withdraws electron density decreases basicity**.
- The **basicity of amines** and the **water solubility of ammonium salts** in water can be used to **separate amines from water-insoluble**, non-basic compounds.

# CHAPTER 10
## *Solutions to the Problems*

Problem 10.1 Identify all stereocenters in coniine, nicotine, and cocaine.

**The stereocenters are circled in the following structures:**

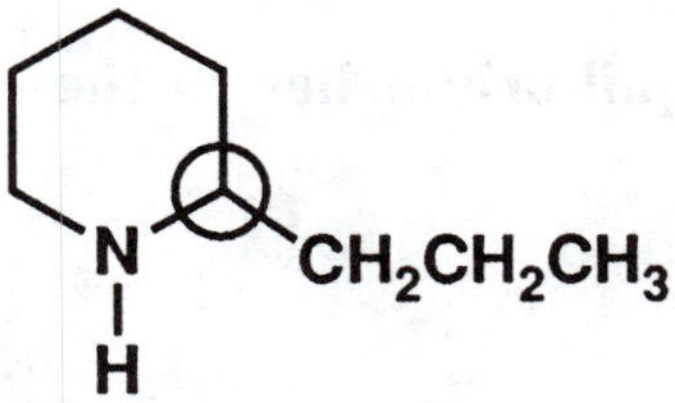

Coniine 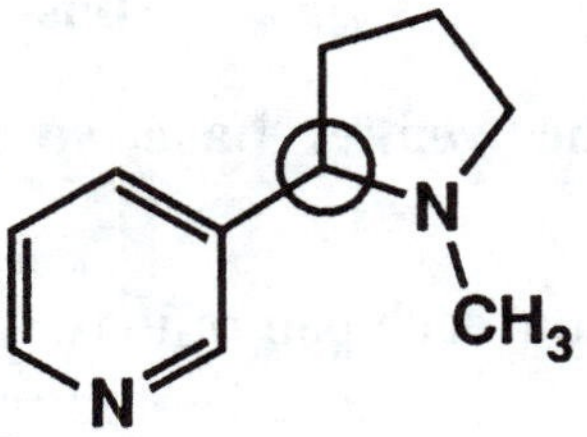                    Nicotine 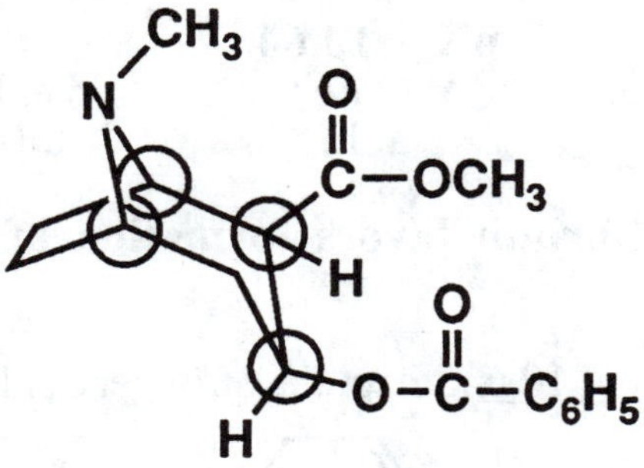                    Cocaine

Problem 10.2 Write structural formulas for these amines.
(a) 2-Methyl-1-propanamine          (b) Cyclohexanamine 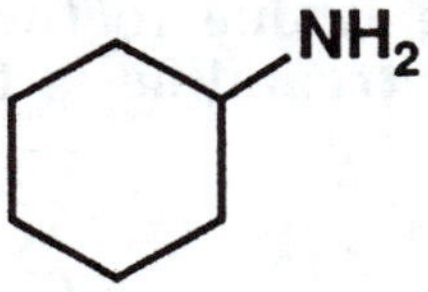          (c) (R)-2-Butanamine

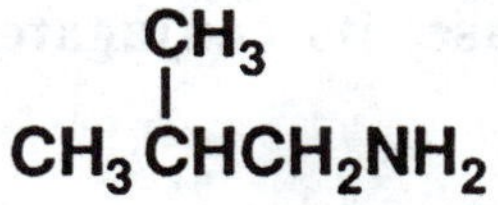

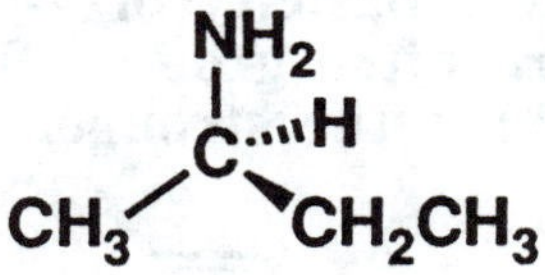

Problem 10.3 Write structural formulas for these amines.
(a) Isobutylamine                    (b) Triphenylamine 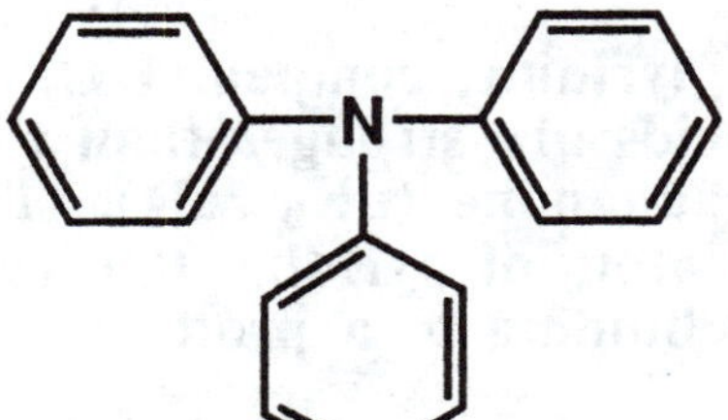                    (c) Diisopropylamine

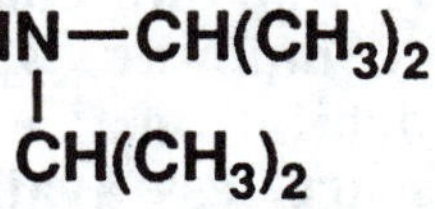

Problem 10.4 Write IUPAC and, where possible, common names for these amines.

(a)

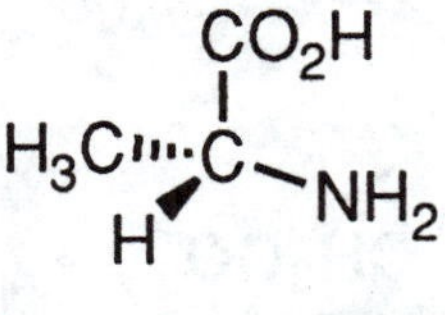

(b)  $H_2NCH_2CH_2CH_2CO_2H$

**(R)-2-Aminopropanoic acid**
**(D-Alanine)**

**4-Aminobutanoic acid**
**(γ-Aminobutyric acid)**

(c)  $(CH_3)_3CCH_2NH_2$

**2,2-Dimethylpropanamine**
   **(Neopentylamine)**

Problem 10.5  Predict the position of equilibrium for this acid-base reaction.

$$CH_3NH_3^+ \quad + \quad H_2O \quad \rightleftharpoons \quad CH_3NH_2 \quad + \quad H_3O^+$$

| pK_a 10.64 | | p Ka -1.74 |
|---|---|---|
| **Weaker** | **Weaker** | **Stronger** **Stronger** |
| **acid** | **base** | **base** **acid** |

**Equilibrium favors formation of the weaker base, so the equilibrium lies to the left.**

Problem 10.6  Select the stronger acid from each pair of ions.

(a)   $O_2N$—⬡—$NH_3^+$    o r    $CH_3$—⬡—$NH_3^+$

   (A)                                         (B)

**4-nitroaniline (pK_b 13.0) is a weaker base than 4-methylaniline (pK_b 8.92). The decreased basicity of 4-nitroaniline is due to the electron-withdrawing effect of the *para* nitro group. Because 4-nitroaniline is the weaker base, its conjugate acid (A) is the stronger acid.**

(b)        ⬡$NH^+$          o r        ⬡—$NH_3^+$

   (C)                                         (D)

**Aromatic heterocycles such as pyridine, compound C, are much weaker bases (pK_b 8.75), so the conjugate acids are stronger than those of aliphatic amines such as compound D, cyclohexanamine (pK_b 3.34). The lone pair of electrons in the sp$^2$ orbital on the nitrogen atom of pyridine (C) has more s character, so these electrons are less available for binding to a proton.**

Problem 10.7  Complete these acid-base reactions and name the salts formed.

(a) $(CH_3CH_2)_3N$  +  HCl  $\longrightarrow$  $(CH_3CH_2)_3NH^+ Cl^-$

                                            **Triethylammonium chloride**

(b)    ⬡N—H  +  $CH_3CO_2H$  $\longrightarrow$  ⬡$N^+$ $\begin{smallmatrix}H\\H\end{smallmatrix}$  $CH_3CO_2^-$

                                            **Piperidinium acetate**

Problem 10.8 In what way(s) might the results of the separation and purification procedure outlined in Example 10.8 be different if
(a) Aqueous $Na_2CO_3$ is used in place of aqueous $NaHCO_3$?

**If $Na_2CO_3$ is used in place of aqueous $NaHCO_3$, then the phenol will be deprotonated along with the carboxylic acid, so they will be isolated together in fraction A.**

(b) The starting mixture contains an aromatic amine, $ArNH_2$, rather than an aliphatic amine, $RNH_2$?

**If the starting mixture contains an aromatic amine, $ArNH_2$, rather than an aliphatic amine, $RNH_2$, then the results will be the same. The aromatic amine will still be protonated by the HCl wash, and deprotonated by the NaOH treatment.**

Problem 10.9 Starting with 3-nitroaniline, show how to prepare these compounds:
(a) 1-Bromo-3-nitrobenzene

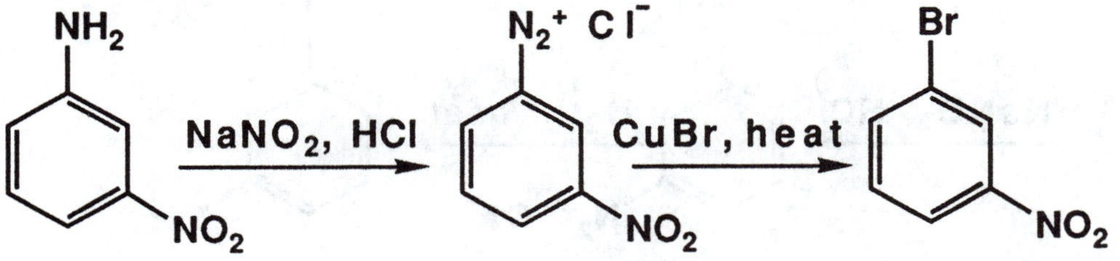

(b) 3-Bromoaniline

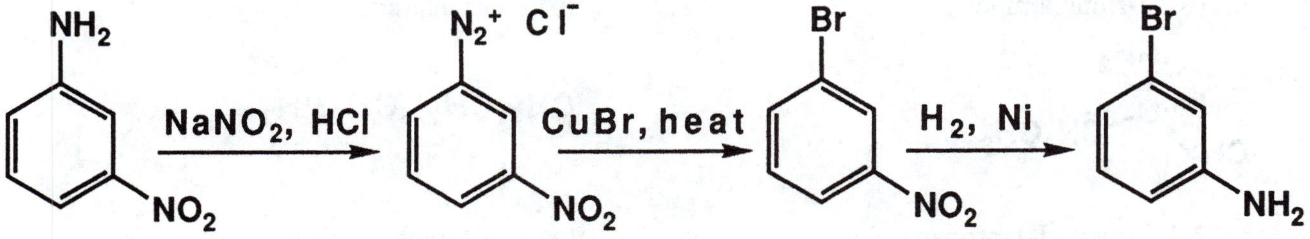

(c) 3-Nitrophenol

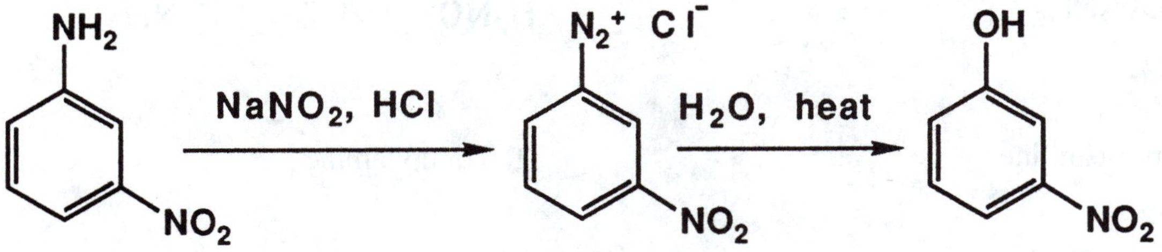

(d) 1,3-Dihydroxybenzene (resorcinol)

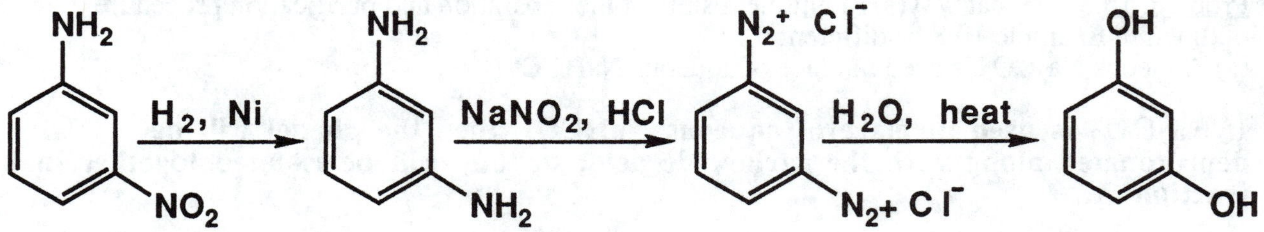

(e) 3-Iodophenol

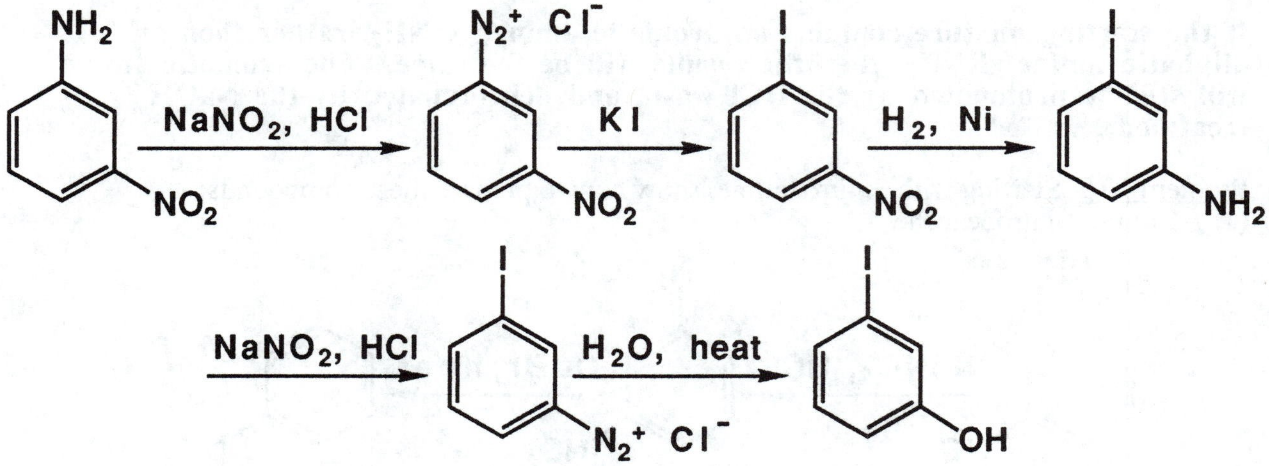

## Structure and Nomenclature

Problem 10.10  Straw structural formulas for these amines.

(a) (R)-2-Butanamine

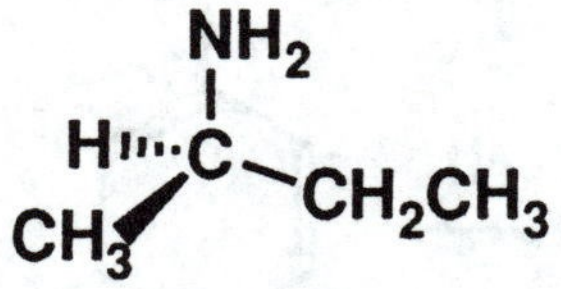

(b) 1-Octanamine

$CH_3(CH_2)_6CH_2NH_2$

(c) 2,2-Dimethyl-1-propanamine

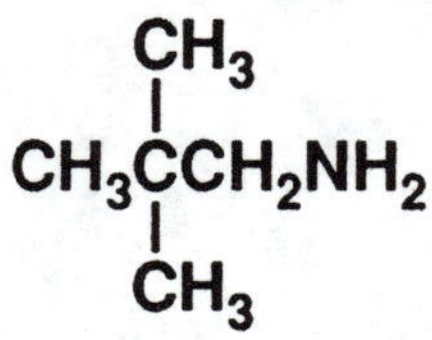

(d) 1,4-Butanediamine

$H_2NCH_2CH_2CH_2CH_2NH_2$

(e) 2-Bromoaniline

(f) Tributylamine

$(CH_3CH_2CH_2CH_2)_3N$

**Problem 10.11** Draw structural formulas for these amines.

(a) N,N-Dimethylaniline

(b) Benzylamine

(c) *tert*-Butylamine

(d) N-Ethylcyclohexylamine

(e) Diphenylamine

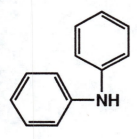

(f) Isobutylamine

$$CH_3CHCH_2NH_2$$
with $CH_3$ on the second carbon

**Problem 10.12** Draw structural formulas for these amines.

(a) 4-Aminobutanoic acid

$$H_2NCH_2CH_2CH_2COH$$ with $=O$ on the carboxyl carbon

(b) 2-Aminoethanol (ethanolamine)

$$H_2NCH_2CH_2OH$$

(c) 2-Aminobenzoic acid

(d) (S)-2-Aminopropanoic acid (alanine)

(e) 4-Aminobutanal

$$H_2NCH_2CH_2CH_2CH$$ with $=O$ on the aldehyde carbon

(f) 4-Amino-2-butanone

$$H_2NCH_2CH_2CCH_3$$ with $=O$ on the ketone carbon

**Problem 10.13** Give an acceptable name for these amines.

(a) 3,4-Dimethoxyaniline

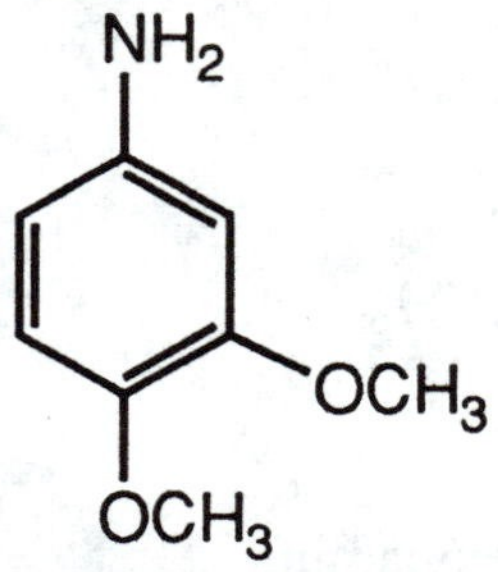

(b) Aniline hydrochloride

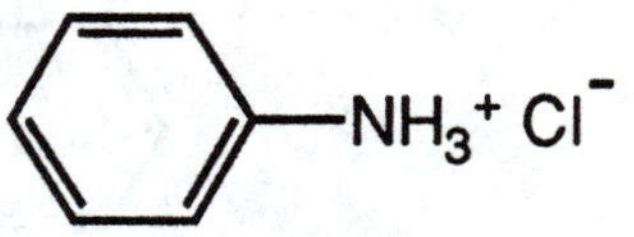

(c) (R)-2-Hexanamine

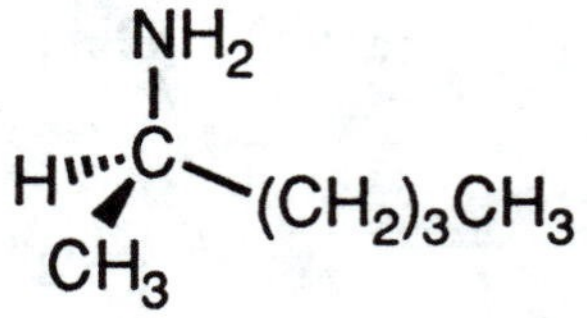

**Problem 10.14** Classify each amine as primary, secondary, or tertiary; as aliphatic or aromatic.

(a) CH₂CH₂NH₂ ← **Primary aliphatic amine**

N—H ← **Heterocyclic aromatic amine**

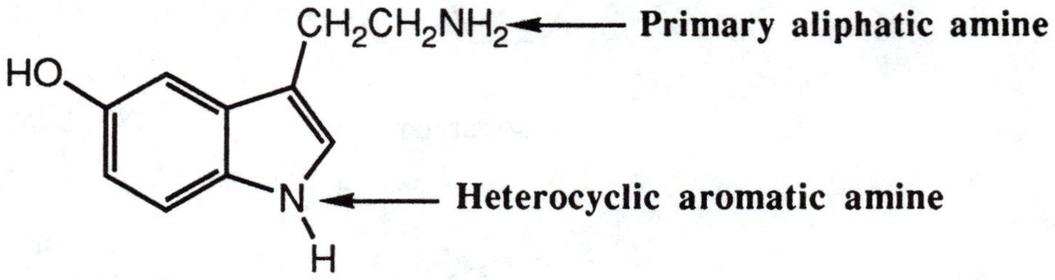

Serotonin
(a neurotransmitter)

**Primary aromatic amine**

(b) H₂N— —COCH₂CH₃

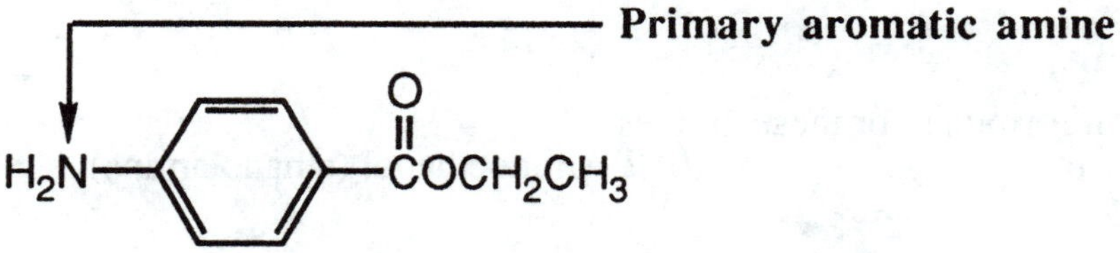

Benzocaine
(a topical anesthetic)

**Tertiary aliphatic amine**

NHCH(CH₂)₃N(C₂H₅)₂

**Secondary aromatic amine**

(c)

Cl

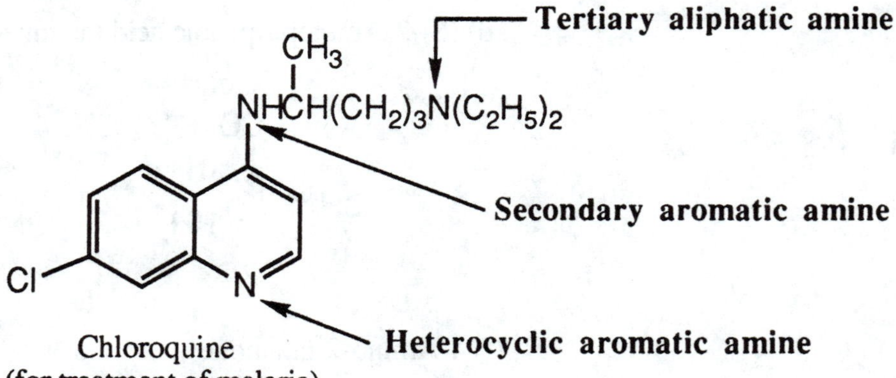

Chloroquine
(for treatment of malaria)

**Heterocyclic aromatic amine**

**Problem 10.15** Epinephrine is a hormone secreted by the adrenal medulla. Among its actions, it is a bronchodilator. Salbutamol, sold as the R enantiomer, is one of the most effective and widely prescribed antiasthma drugs. The R enantiomer is 68 times more effective in the treatment of asthma that the S enantiomer, which is another example of the stereospecificity of many drugs.

Secondary aliphatic amine                    Secondary aliphatic amine

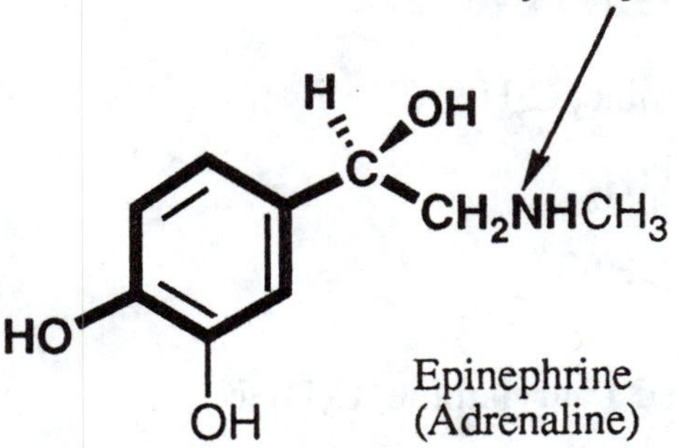

Epinephrine
(Adrenaline)

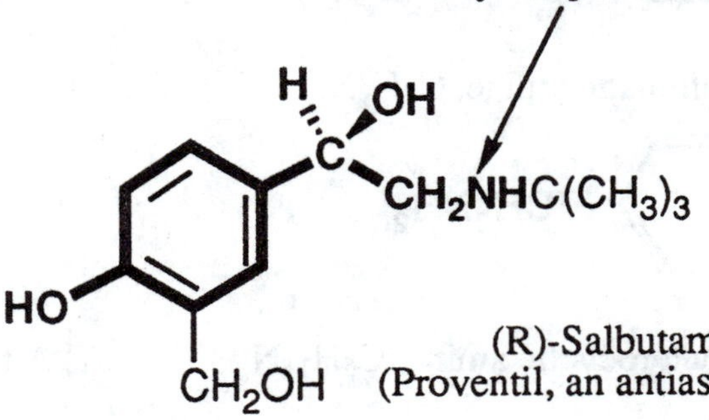

(R)-Salbutamol
(Proventil, an antiasthma drug)

(a) Classify each as a primary, secondary, or tertiary amine.
(b) List the similarities and differences between their structural formulas.

**The parts of the molecules that are identical are indicated in bold on the above structures. As far as differences are concerned, epinephrine posseses a second hydroxyl group on the aromatic ring and a methyl group on the amine, while (R)-salbutamol has a hydroxymethyl on the ring and a *tert*-butyl group on the amine.**

**Problem 10.16** There are eight constitutional isomers of molecular formula $C_4H_{11}N$. Name and draw structural formulas for each, and classify each as a primary, secondary, or tertiary amine

**Primary amines:**

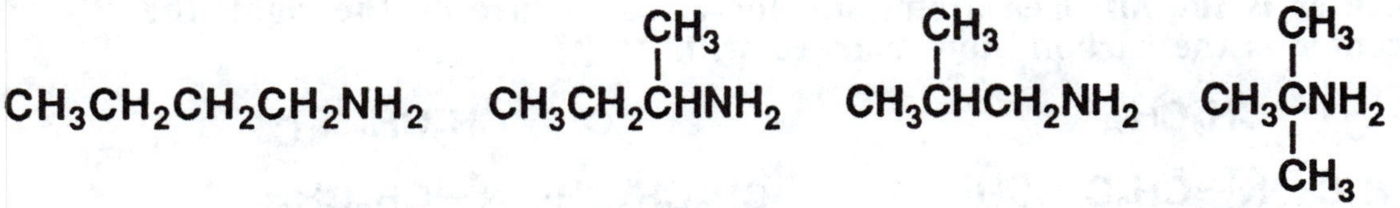

*n*-Butylamine          *sec*-Butylamine          Isobutylamine   *tert*-Butylamine

**Secondary amines:**

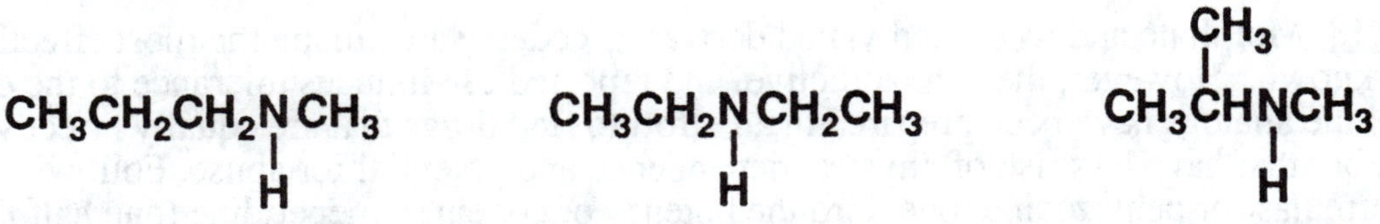

Methylpropylamine          Diethylamine          Isopropylmethylamine

**Tertiary amine:**

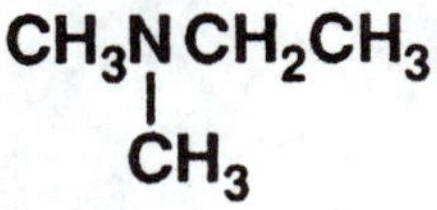

**Dimethylethylamine**

**Problem 10.17** Draw the structural formula for a compound of the given molecular formula.

(a) A 2° arylamine, $C_7H_9N$

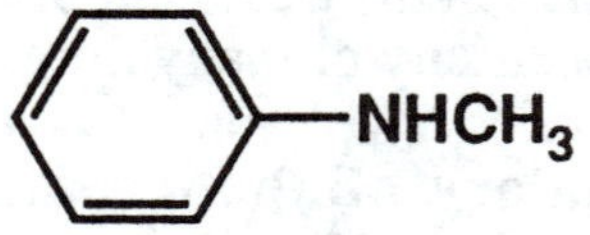

—NHCH$_3$

(b) A 3° arylamine, $C_8H_{11}N$

—N(CH$_3$)$_2$

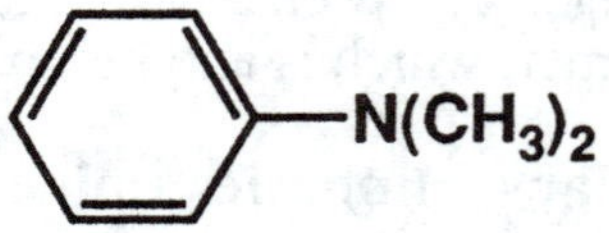

(c) A 1° aliphatic amine, $C_7H_9N$

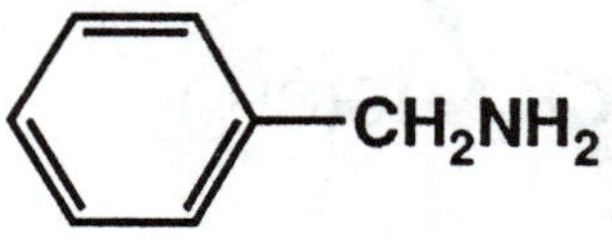

—CH$_2$NH$_2$

(d) A chiral 1° amine, $C_4H_{11}N$

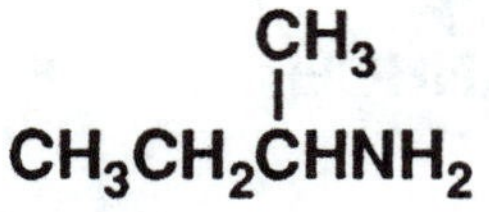

$$\overset{\displaystyle CH_3}{\underset{\displaystyle |}{CH_3CH_2CHNH_2}}$$

(e) A 3° heterocyclic amine, $C_6H_{11}N$

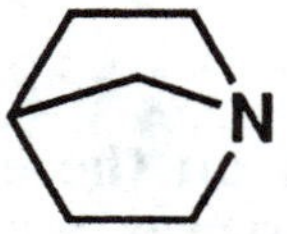

N

(f) A trisubstituted 1° arylamine, $C_9H_{13}N$

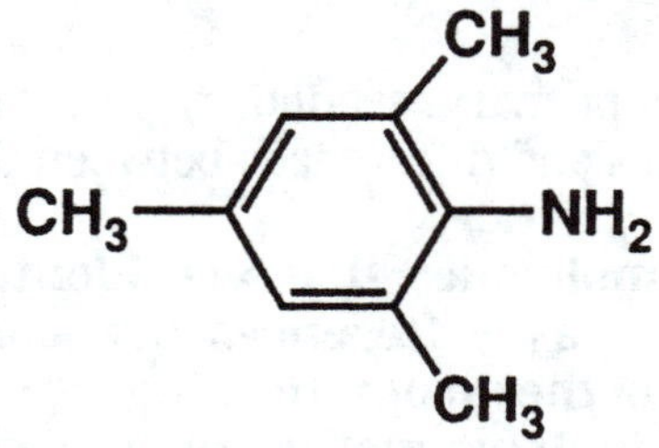

CH$_3$ —

CH$_3$

—NH$_2$

CH$_3$

(Other isomers are possible)

(g) A chiral quaternary ammonium salt, $C_9H_{22}NCl$

**There are two answers to this question. For the structure on the left, the stereocenter is the nitrogen atom, and for the structure on the right, the stereocenter is the carbon atom marked with an "*".**

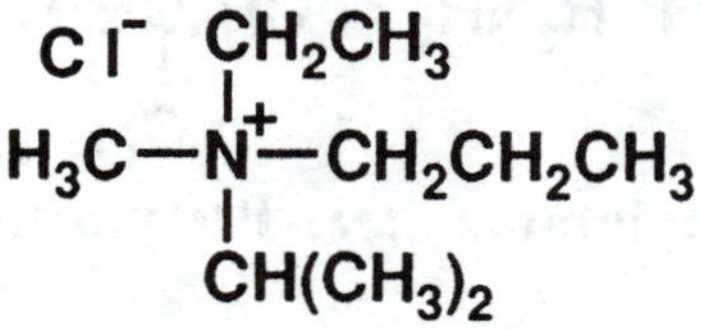

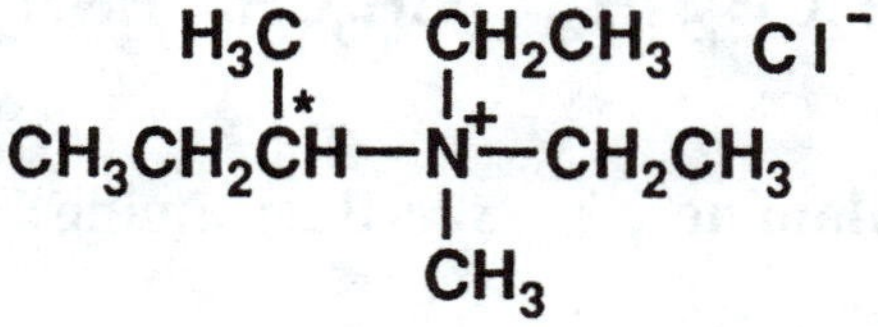

$$Cl^- \quad \overset{\displaystyle CH_2CH_3}{\underset{\displaystyle CH(CH_3)_2}{H_3C-\overset{+}{N}-CH_2CH_2CH_3}}$$

$$\underset{\displaystyle CH_3}{\overset{\displaystyle H_3C}{CH_3CH_2\overset{*}{C}H}-\overset{+}{N}-CH_2CH_3} \quad Cl^-$$

**Problem 10.18** Morphine and its *O*-methylated derivative codeine are among the most effective pain killers known. However, they are addictive, and repeated use induces tolerance to the drug. Many morphine analogs have been prepared in an effort to find drugs that are equally effective as pain killers, but that have less risk of physical dependence and potential for abuse. Following are structural formulas for pentazocine (one third the potency of codeine), meperidine (one half the potency of morphine), and dextropropoxyphene (one half the potency of codeine). Methadone, with a potency equal to that of morphine, is used to treat opiate withdrawal symptoms in heroin abusers.

R = H; Morphine
R = CH₃; Codeine

Pentazocine
(Talwin)

Dextropropoxyphene
(Darvon)

Meperidine
(Demerol)

Methadone

(a) List the structural features common to each of these molecules.

**Each of the above molecules contains a tertiary amine and a phenyl ring that is three sp³ carbons away from the nitrogen atom of the amine.**

(b) The Beckett-Casey rules are a set of empirical rules to predict the structure of molecules that bind to morphine receptors and act as analgesics. According to these rules, to provide effective morphine-like analgesia, a molecule must have (1) an aromatic ring attached to (2) a quaternary carbon and (3) a nitrogen at a distance equal to two carbon-carbon single bond lengths from the quaternary center. Show that these structural requirements are present in the molecules given in this problem.

**By inspection of the structures, it can be seen that all of these three structural requirements are present in the molecules mentioned in this problem.**

## Physical Properties
Problem 10.19 Propylamine, ethylmethylamine, and trimethylamine are constitutional isomers of molecular formula $C_3H_9N$. Account for the fact that trimethylamine has the lowest boiling point of the three isomers.

$CH_3CH_2CH_2NH_2$                $CH_3CH_2NHCH_3$                $(CH_3)_3N$
bp 48°C                          bp 37°C                        bp 3°C

**Trimethylamine does not have any hydrogen atoms attached to the nitrogen atom, so this compound cannot form intermolecular hydrogen bonds. The other two compounds do have hydrogens attached to nitrogen, so they can hydrogen bond. The ability to hydrogen bond increases attraction between molecules and thus raises boiling points.**

<u>Problem 10.20</u>  Account for the fact that 1-butanamine has a lower boiling point than 1-butanol.

$$CH_3CH_2CH_2CH_2OH$$
bp 117°C

$$CH_3CH_2CH_2CH_2NH_2$$
bp 78°C

**An N-H----N hydrogen bond is not as strong as an O-H----O hydrogen bond, because the difference in electronegativity between nitrogen and hydrogen is not as great as the difference in electronegativity between oxygen and hydrogen. Stronger hydrogen bonds between the molecules of 1-butanol lead to increased attraction between the molecules and thus a higher boiling point.**

## Basicity of Amines
<u>Problem 10.21</u>  Account for the fact that amines are more basic than alcohols.

**Nitrogen is less electronegative than oxygen, so a lone pair of electrons on nitrogen is more likely to interact with a proton than a lone pair of electrons on oxygen.  Thus, amines are more basic than alcohols.**

<u>Problem 10.22</u>  Select the stronger base in each pair of compounds.

**In each pair, select the species that has higher electron density on the nitrogen atom.  The more basic amine is circled.**

<u>Problem 10.23</u>  Select the stronger base in each pair of compounds.

**In each pair, select the species that has higher electron density on the nitrogen atom.  The more basic amine is circled.**

(b) or

**Problem 10.24** Account for the fact that substitution of a nitro group makes an aromatic amine a weaker base but makes a phenol a stronger acid. For example, 4- nitroaniline is a weaker base than aniline, but 4-nitrophenol is a stronger acid than phenol.

**The nitro group is acting to withdraw electron density from the aromatic ring. For 4-nitroaniline, this has the effect of increasing the delocalization of the lone pair of electrons on the aryl amine into the aromatic ring. With greater delocalization, the lone pair of electrons is less able to interact with a proton, so 4-nitroaniline is less basic than aniline. In terms of resonance structures, this idea can be understood as the electron withdrawing effect of the nitro group acting to increase the relative contribution of the resonance form on the right.**

In the case of the relative acidities of the phenols, the key interactions involve the deprotonated phenoxide anions. For the 4-nitrophenoxide anion, the electron withdrawing effect of the nitro group helps withdraw more of the negative charge on the oxygen atom into the aromatic ring. Delocalization of a charge leads to stabilization, so this increased delocalization of the negative charge increases the stabilization of the 4-nitrophenoxide anion compared to the phenoxide anion. By stabilizing the 4-nitrophenoxide anion, the equilibrium is moved more toward deprotonation of the 4-nitrophenol; in other words 4-nitrophenol is more acidic. In terms of resonance structures, this idea can be understood as the electron withdrawing effect of the nitro group acting to increase the relative contribution of the resonance form on the right.

**Problem 10.25** Select the stronger base in this pair of compounds.

The stronger base is the molecule on the right, as indicated by the circle around it. This is because the basic site on this molecule is the hydroxide anion, since the quaternary ammonium ion is simply the counterion. For the molecule on the left, the basic site is the tertiary amine. Hydroxide is a stronger base ($pK_b$ = -1.7) than a tertiary amine ($pK_b$ = ~3 - 4).

Problem 10.26 Following are two structural formulas for alanine (2-aminopropanoic acid), one of the building blocks of proteins (Section 17.1) Is alanine better represented by structural formula (A) or structural formula (B)? Explain.

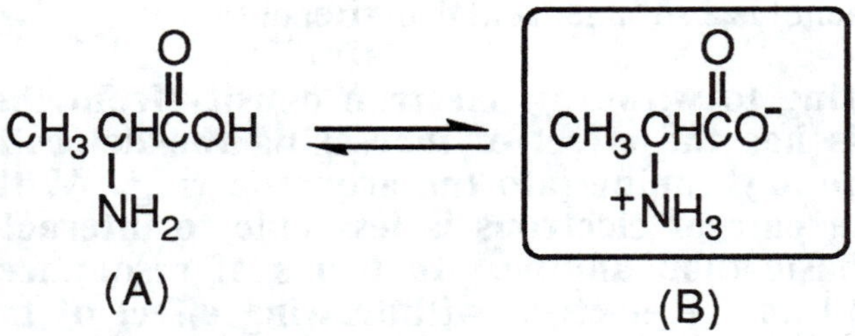

(A)                                              (B)

Carboxylic acids are acidic enough ($pK_a$ = ~3-5) and primary amines are basic enough ($pK_b$ = ~3 - 4) that equilibrium favors the deprotonation of a carboxylic acid by a primary amine to form an ammonium cation and carboxylate anion. The structure on the right, structure B, is the better representation of alanine in neutral solution.

Problem 10.27 Complete the following acid-base reactions and predict the position of equilibrium (to the right or to the left) for each. Justify your predictions by citing values of $pK_a$ for the stronger and weaker acid in each equilibrium. For values of acid ionization constants, consult Table 6.1 (strengths of some inorganic and organic acids), Table 7.4 (acidity of alcohols), Section 9.4B (acidity of phenols), Table 10.2 (base strengths of amines), and Section 13.4 (acidity of carboxylic acids). Where no ionization constants are given, make the best estimate from the information given in the reference tables and sections.

In all cases, the equilibrium favors formation of the weaker acid (higher $pK_a$) and weaker base (higher $pK_b$). Recall that $pK_a + pK_b = 14$ for any conjugate acid-base pair.

(a)

| Acetic acid | Pyridine | | |
|---|---|---|---|
| $pK_a$ 4.76 | $pK_b$ 8.75 | $pK_b$ 9.24 | $pK_a$ 5.25 |

Equilibrium lies to the right, since the acetate anion and pyridinium ion are the weaker acid and base, respectively.

(b)

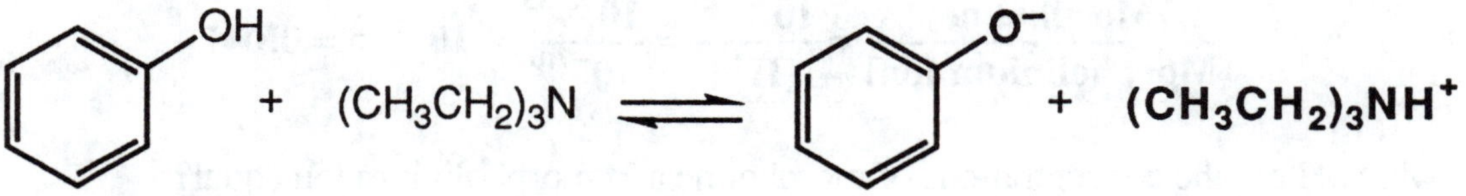

| Phenol | Triethylamine | | |
|---|---|---|---|
| $pK_a$ 9.95 | $pK_b$ 3.25 | $pK_b$ 4.05 | $pK_a$ 10.75 |

**Equilibrium lies to the right, since the phenoxide anion and the triethylammonium species are the weaker base and acid, respectively.**

(c)

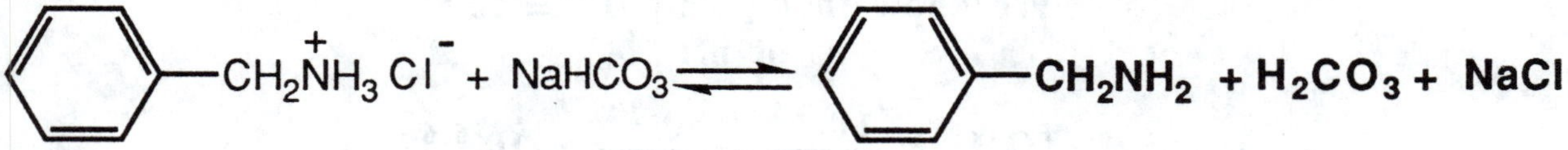

| Benzylammonium chloride | Sodium bicarbonate | | |
|---|---|---|---|
| $pK_a$ ~10-11 | $pK_b$ 7.63 | $pK_b$ ~3-4 | $pK_a$ 6.36 |

**Equilibrium lies to the left, since the ammonium ion and the bicarbonate ion are the weaker acid and base, respectively.**

(d)

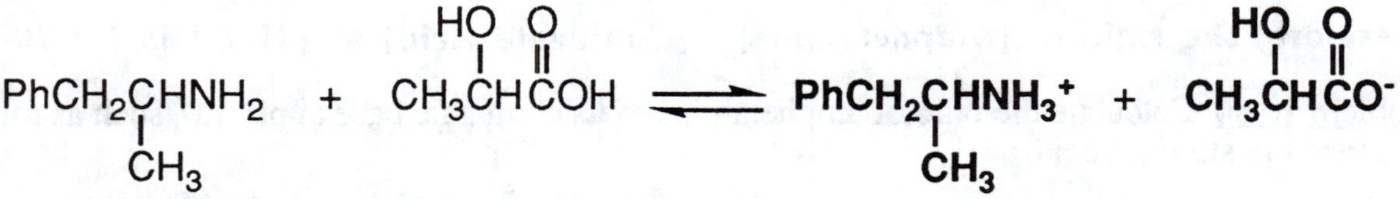

| 1-Phenyl-2-propanamine (Amphetamine) | 2-Hydroxy-propanoic acid (lactic acid) | | |
|---|---|---|---|
| $pK_b$ ~3 | $pK_a$ 3.08 | $pK_a$ ~11 | $pK_b$ 9.92 |

**Equilibrium lies to the right, since the ammonium ion and the carboxylate ion are the weaker acid and base, respectively.**

<u>Problem 10.28</u>  The $pK_a$ of morpholine is 8.33.

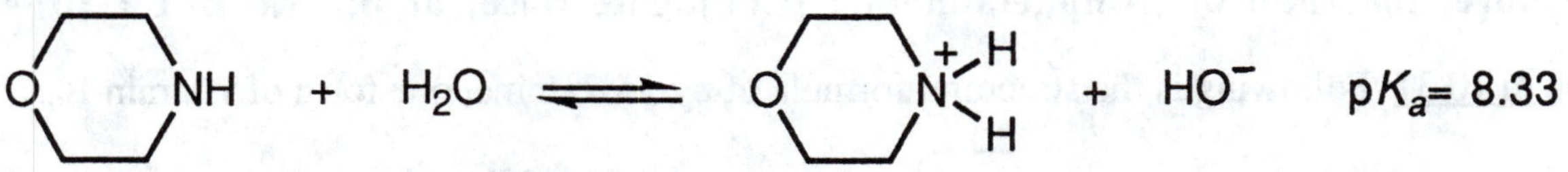

Morpholine                                    Morpholinium ion

(a) Calculate the ratio of morpholine to morpholinium ion in aqueous solution at pH 7.0.

$$K_a = \frac{[\text{Morpholine}][H^+]}{[\text{Morpholinium Ion}]} = 10^{-8.33} \quad \text{At pH 7.0 } [H^+] = 10^{-7}$$

$$\frac{[\text{Morpholine}]}{[\text{Morpholinium Ion}]} = \frac{10^{-8.33}}{[\text{H}^+]} = \frac{10^{-8.33}}{10^{-7.0}} = 10^{-1.33} = 0.047$$

(b) At what pH are the concentrations of morpholine and morpholinium ion equal?

**The concentrations of morpholine and morpholinium ion will be equal when the $pK_a$ is equal to the pH, namely at pH 8.33.**

<u>Problem 10.29</u>  The $pK_b$ of amphetamine (Problem 10.27D) is approximately 3.2.  Calculate the ratio of amphetamine to its conjugate acid at pH 7.4, the pH of blood plasma.

$$\text{We know that } [\text{OH}^-][\text{H}^+] = 10^{-14}$$
$$\text{so at pH 7.4}$$

$$[\text{OH}^-] = \frac{10^{-14}}{[\text{H}^+]} = \frac{10^{-14}}{10^{-7.4}} = 10^{-6.6}$$

$$K_b = \frac{[\text{Conjugate Acid}][\text{OH}^-]}{[\text{Amphetamine}]} = 10^{-3.2}$$

$$\frac{[\text{Conjugate Acid}]}{[\text{Amphetamine}]} = \frac{10^{-3.2}}{[\text{OH}^-]} = \frac{10^{-3.2}}{10^{-6.6}} = 10^{3.4}$$

**Therefore, the ratio of [Amphetamine] : [Conjugate Acid] at pH 7.4 is 1 : $10^{3.4}$**

<u>Problem 10.30</u>  Calculate the ratio of amphetamine to its conjugate base at pH 1.0, such as might be present in stomach acid.

$$\text{We know that } [\text{OH}^-][\text{H}^+] = 10^{-14}$$
$$\text{so at pH 1.0}$$

$$[\text{OH}^-] = \frac{10^{-14}}{[\text{H}^+]} = \frac{10^{-14}}{10^{-1}} = 10^{-13}$$

$$\frac{[\text{Conjugate Acid}]}{[\text{Amphetamine}]} = \frac{10^{-3.2}}{[\text{OH}^-]} = \frac{10^{-3.2}}{10^{-13}} = 10^{9.8}$$

**Therefore, the ratio of [Amphetamine] : [Conjugate Base] at pH 1.0 is 1 : $10^{9.8}$**

<u>Problem 10.31</u>  Following is the structural formula of pyridoxamine, one form of vitamin $B_6$.

Pyridoxamine
(Vitamin $B_6$)

(a) Which nitrogen atom of pyridoxamine is the stronger base?

**The primary amine indicated on the above structure is more basic than the pyridine nitrogen atom. This is because the primary amine nitrogen atom is sp³ hybridized, while the pyridine nitrogen is sp² hybridized. The sp² hybridized nitrogen atom has a greater percentage s character, so the electrons are less available for interactions with protons.**

(b) Draw the structural formula of the hydrochloride salt formed when pyridoxamine is treated with 1 mol of HCl.

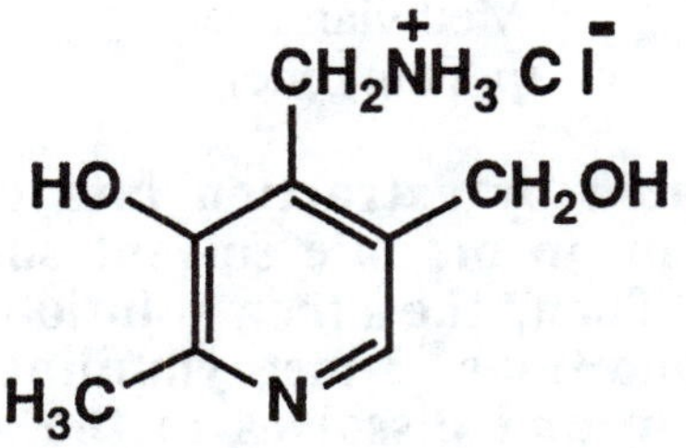

**Problem 10.32** Procaine was one of the first local anesthetics. Its hydrochloride salt is marketed as Novocaine.

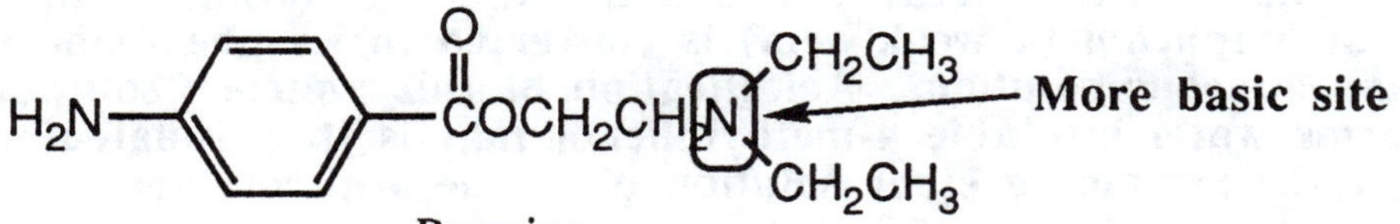

(a) Which nitrogen atom of pyridoxamine is the stronger base?

**The tertiary aliphatic amine is more basic than the primary aryl amine. Aryl amines are less basic because the lone pair of electrons on the nitrogen atom are delocalized into the aromatic ring, thus making them less available for interactions with a proton.**

(b) Draw the formula of the hydrochloride salt formed by treating procaine with 1 mol of HCl.

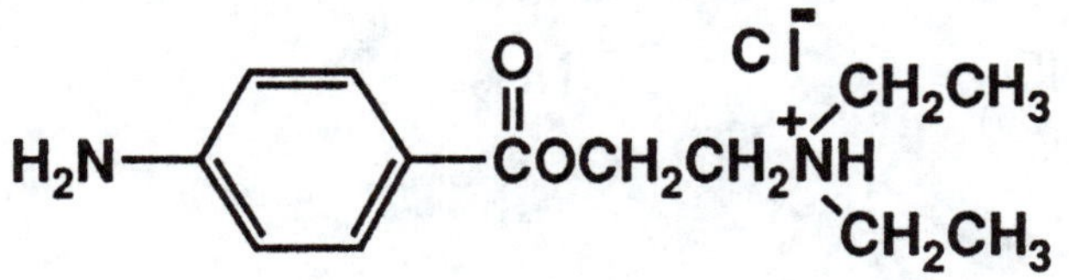

**Problem 10.33** Aniline is prepared by catalytic reduction of nitrobenzene. Devise a chemical procedure based on the basicity of aniline to separate aniline from any unreacted nitrobenzene.

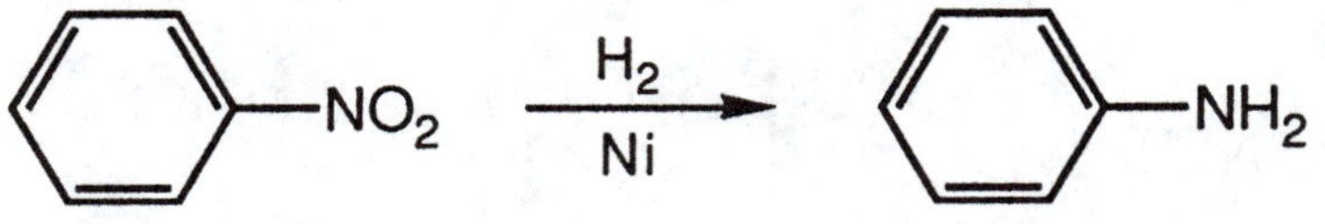

**These compounds can be separated by extraction into an acidic (pH of ~4 or below) aqueous solution. Only the aniline will be protonated, thereby making it**

soluble in the aqueous layer as the anilinium ion.  The hydrophobic nitrobenzene
has no basic sites, so it will not be soluble in the aqueous layer.

Problem 10.34 Suppose you have a mixture of the following three compounds.  Devise a chemical
procedure based on their relative acidity to separate and isolate each in pure form.

4-Nitrotoluene            4-Methylaniline            4-Methylphenol
(*p*--Nitrotoluene)         (*p*-Toluidine)            (*p*-Cresol)

**These molecules can be separated by extraction into different aqueous solutions.
First, the mixture is dissolved in an organic solvent such as ether in which all
three compounds are soluble.  Then, the ether solution is extracted with dilute
aqueous HCl.  Under these conditions, 3-methylaniline (a weak base) is
converted to its protonated form and dissolves in the aqueous solution.  The
aqueous solution is separated, treated with dilute NaOH, and the water-insoluble
4-methylaniline separates, and is recovered.  The ether solution containing the
other two components is then treated with dilute aqueous NaOH.  Under these
conditions, 4-methylphenol (a weak acid) is converted to its phenoxide ion and
dissolves in the aqueous solution.  Acidification of this aqueous solution with
dilute HCl forms water-insoluble 4-methylphenol that is then isolated.
Evaporation of the remaining ether solution gives the 4-nitrotoluene.**

## Arenediazonium Salts
Problem 10.35 Show reagents to bring about the conversion of toluene to *p*-cresol.

Toluene

4-Methylphenol
(*p*-Cresol)

**Notice that an arenediazonium salt is formed after the treatment with NaNO₂ and
HCl.  The arenediazonium salt is then converted to the phenol by heating with
water.**

**Problem 10.36** Show reagents to bring about the conversion of toluene to 3,5-dichlorotoluene.

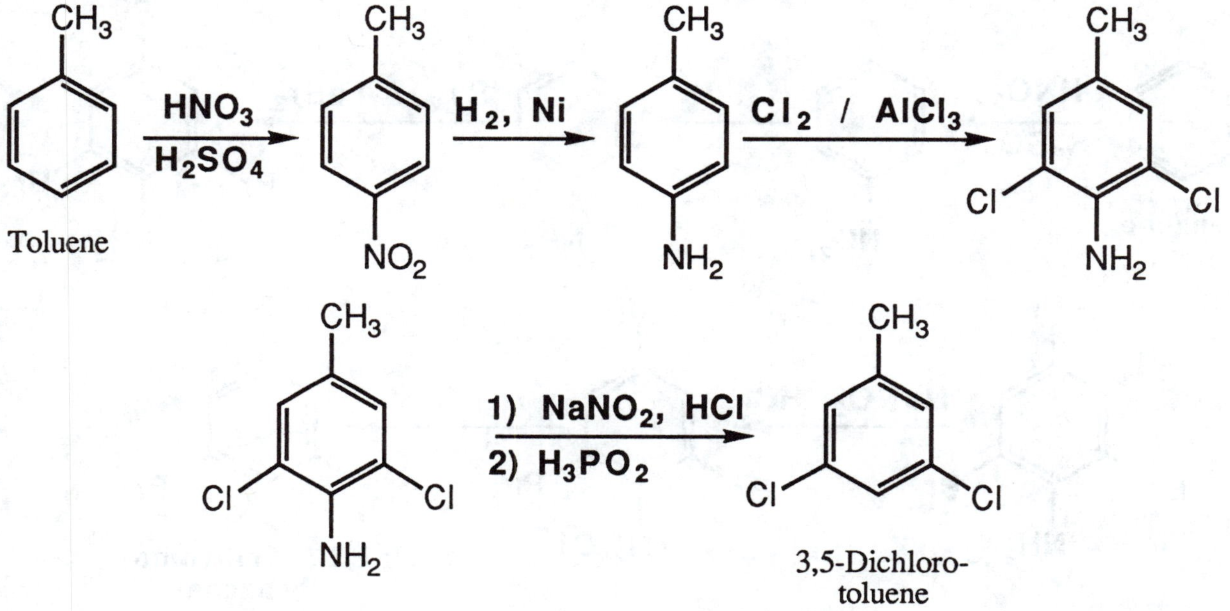

Toluene

3,5-Dichloro-
toluene

**Notice that an arenediazonium salt is formed after the treatment with NaNO$_2$ and HCl. The -N$_2^+$ group of the arenediazonium salt is then converted to -H by treatment with hypophosphorous acid, H$_3$PO$_2$.**

**Problem 10.37** Show how to convert benzene to these compounds.
(a) 2,4,6-Tribromophenol

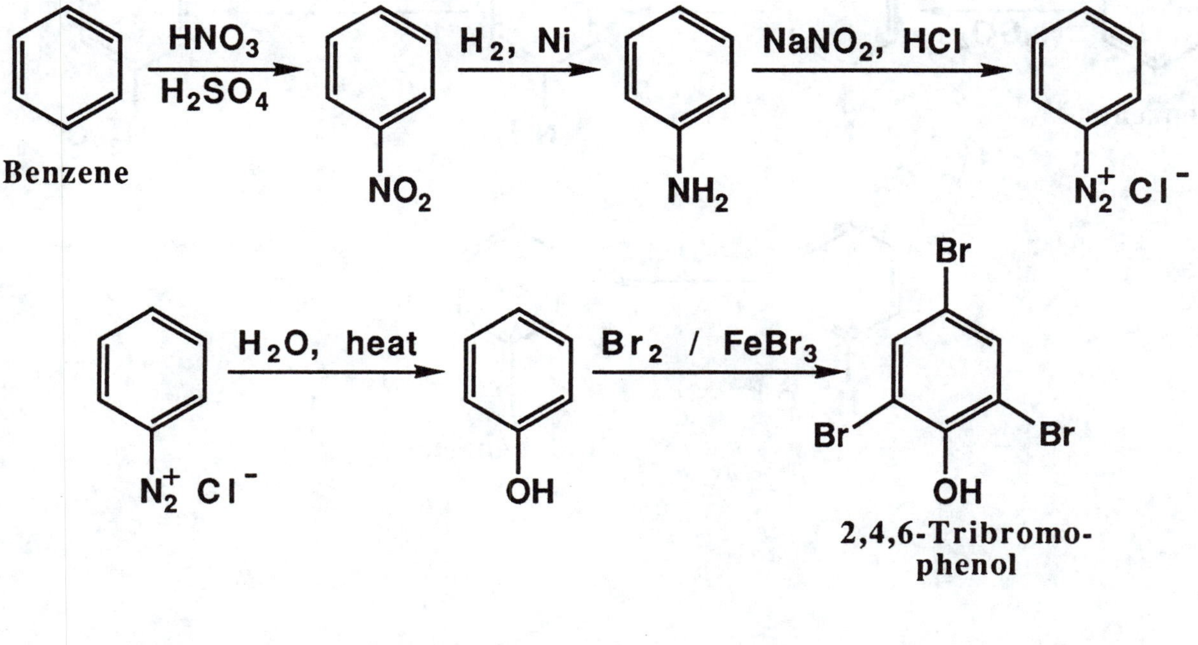

Benzene

2,4,6-Tribromo-
phenol

(b) 1,3,5-Tribromobenzene

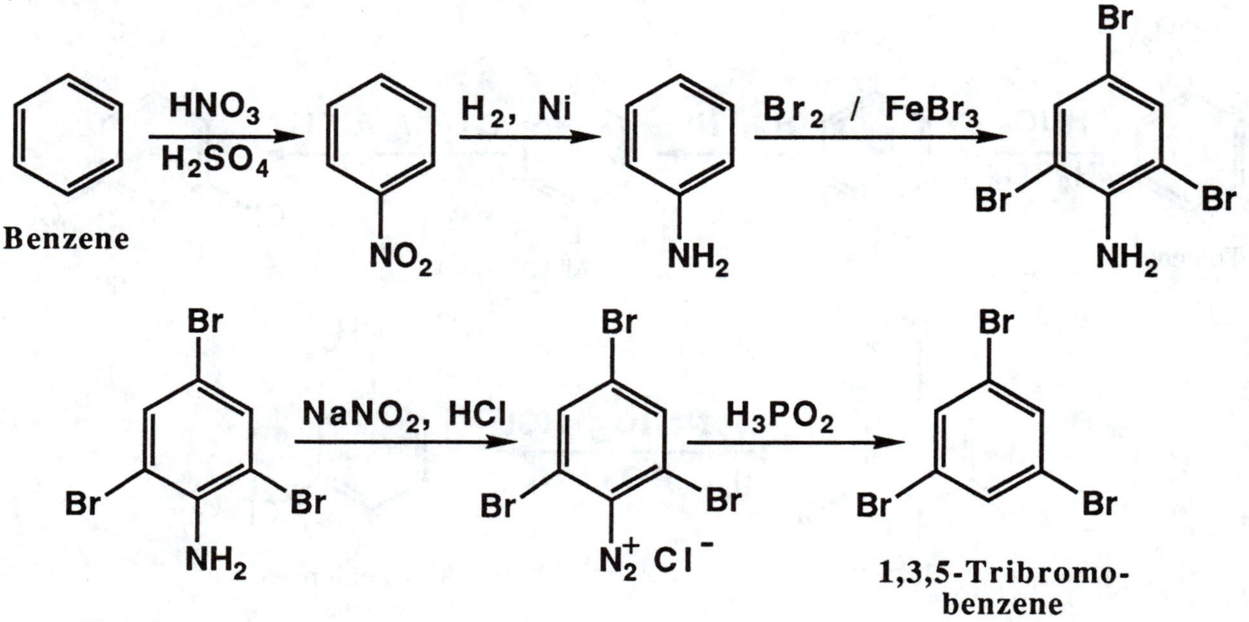

Notice how in the above sequence the -NH₂ group (the most activating group on the benzene ring) directs the three Br atoms ortho/para.

(c) Iodobenzene

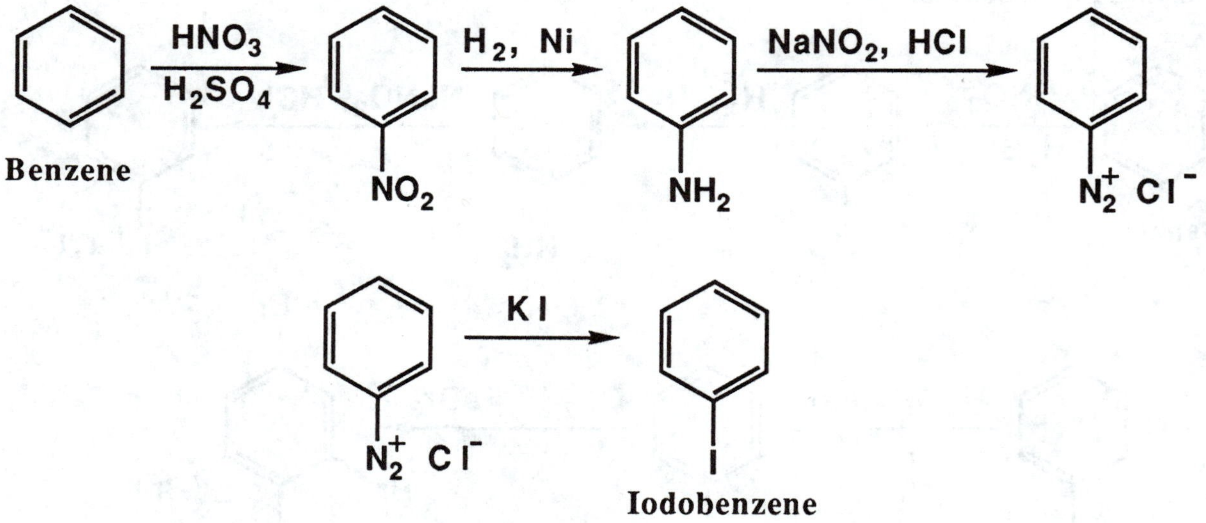

Problem 10.38  Show how to convert toluene to these compounds.
(a) 4-Nitrobenzoic acid

Toluene → (HNO₃ / H₂SO₄) → p-nitrotoluene → (K₂Cr₂O₇ / H₂SO₄) → 4-Nitrobenzoic acid

(b) 4-Aminobenzoic acid

Toluene → (HNO₃ / H₂SO₄) → p-nitrotoluene → (K₂Cr₂O₇ / H₂SO₄) → 4-nitrobenzoic acid → (H₂, Ni) → 4-Aminobenzoic acid

Problem 10.39  Show how to convert chlorobenzene to these compounds.
(a) 2-Chloroaniline

Chlorobenzene → (HNO₃ / H₂SO₄) → (Separate from para isomer) → (H₂, Ni) → 2-Chloroaniline

**(b) 2-Chlorophenol**

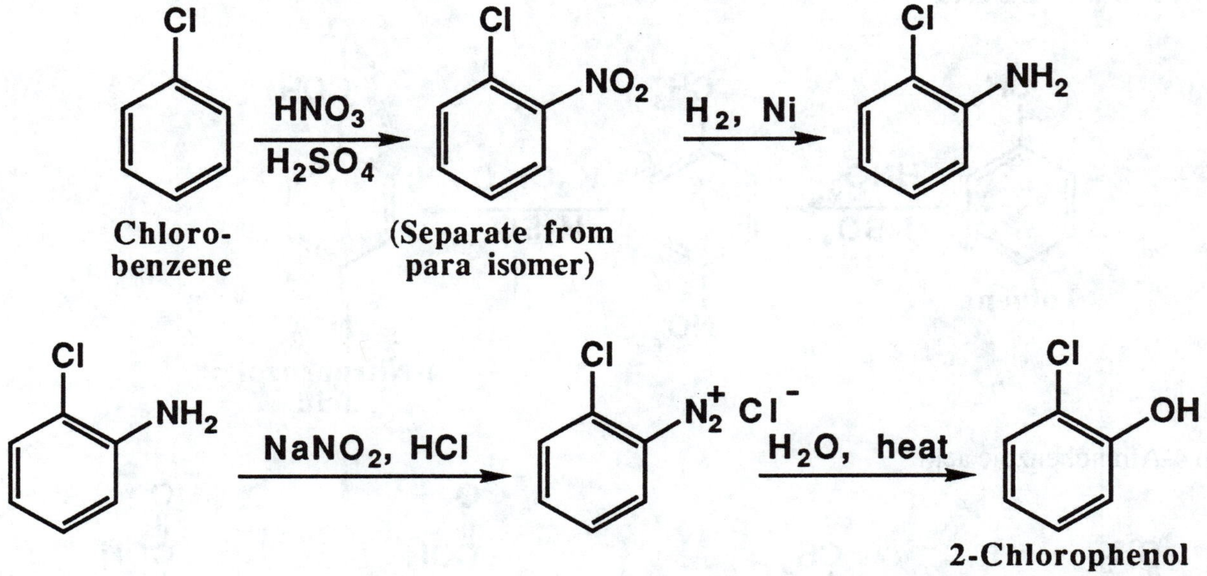

Chloro-
benzene

(Separate from
para isomer)

**2-Chlorophenol**

**(c) 2-Chloro-4-nitroaniline**

Chloro-
benzene

(Separate from
para isomer)

**2-Chloro-4-nitro-
aniline**

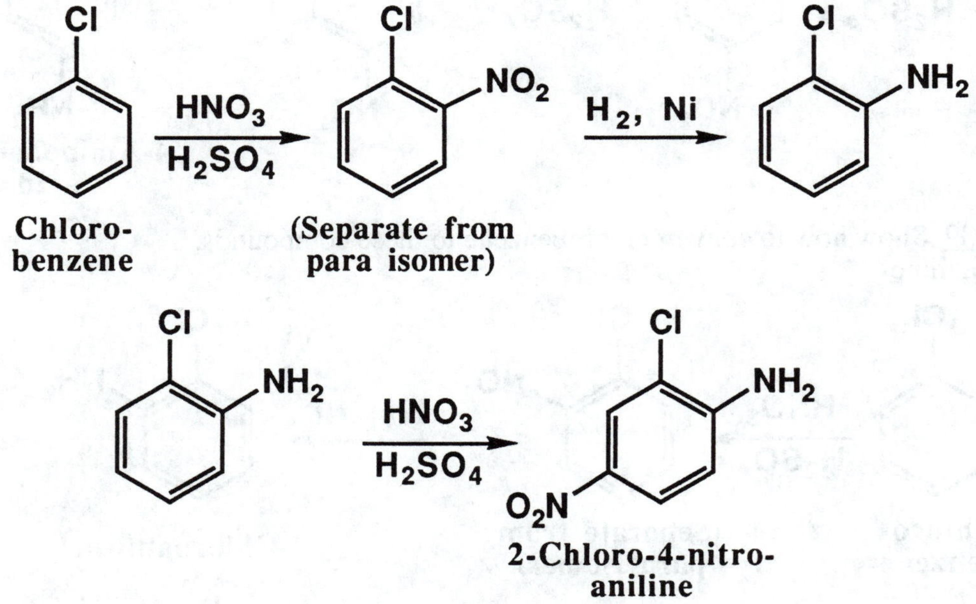

# CHAPTER 11: ALDEHYDES AND KETONES

## SUMMARY OF REACTIONS

| Starting Material ↓ / Product → | Acetals | Alcohols | | Amines | Carboxylic Acids | 1,1 Diols | Enol | Grignard Reagents | Hemiacetals | Imines |
|---|---|---|---|---|---|---|---|---|---|---|
| Aldehydes | | | | | **11A** 11.9A* | | | | | |
| Aldehydes, Ketones | | **11B** 11.10A | **11C** 11.10B | | | **11D** 11.6A | | | | |
| Aldehydes, Ketones Alcohols | **11E** 11.6B | | | | | | | | **11F** 11.6B | |
| Aldehydes, Ketones Amines | | | | **11G** 11.10C | | | | | | **11H** 11.7 |
| Aldehydes, Ketones Grignard Reagents | | **11I** 11.5B | | | | | | | | |
| Alkyl Halides Magnesium | | | | | | | | **11J** 11.5A | | |
| Carbon Dioxide Grignard Reagents | | | | | **11K** 11.5B | | | | | |
| Epoxides Grignard Reagents | | **11L** 11.5B | | | | | | | | |
| Keto | | | | | | | **11M** 11.8 | | | |
| Ketones | | | | | **11N** 11.9B | | | | | |

*Section in book that describes reaction.

## REACTION 11A: OXIDATION OF AN ALDEHYDE TO A CARBOXYLIC ACID (Section 11.9A)

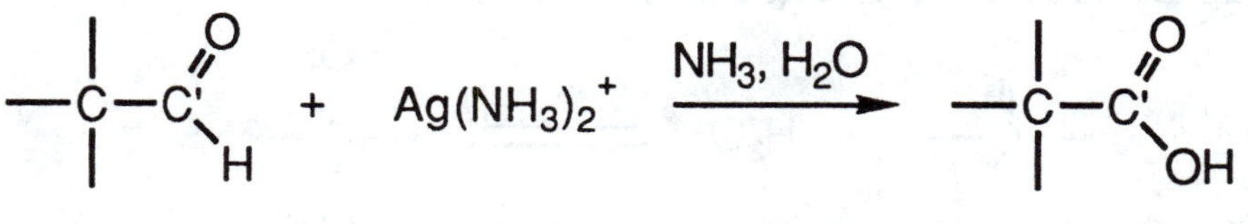

- **Aldehydes** are easily **oxidized to carboxylic acids** by various oxidizing agents including the **Tollens' reagent** shown above, as well as $KMnO_4$, $K_2Cr_2O_7$, $H_2O_2$,

and $O_2$. The Tollens' reagent is especially interesting because a silver mirror can deposit on the side of the flask as the reaction is carried out. ✳
- Note that aldehydes will react with molecular oxygen to give carboxylic acids, so liquid aldehydes are usually stored under an inert atmosphere such as pure nitrogen for long term storage.

## REACTION 11B: CATALYTIC REDUCTION (Section 11.10A)

$$\underset{\overset{\|}{\text{C}}}{\overset{\text{O}}{-}}\!\!- \;+\; H_2 \quad \xrightarrow{\text{Pt}} \quad \underset{\overset{|}{\text{H}}}{\overset{\text{OH}}{-}}\!\!\overset{|}{\text{C}}\!\!-$$

- **Aldehydes and ketones** are easily **reduced to alcohols** in the presence of **hydrogen** and catalysts such as **platinum, palladium, nickel, or rhodium.** ✳
- Unfortunately, other functional groups that may be present in a molecule such as a carbon-carbon double and triple bonds may also be reduced under these conditions. As a result, the metal hydride reductions described below are more common.

## REACTION 11C: METAL HYDRIDE REDUCTIONS (Section 11.10B)

$$\underset{\overset{\|}{\text{C}}}{\overset{\text{O}}{-}}\!\!- \quad \xrightarrow[\text{2) } H_2O]{\text{1) } NaBH_4} \quad \underset{\overset{|}{\text{H}}}{\overset{\text{OH}}{-}}\!\!\overset{|}{\text{C}}\!\!-$$

- **Aldehydes** are **reduced** by **sodium borohydride** ($NaBH_4$) in methanol, ethanol, or mixtures of these solvents with water to form the corresponding **alcohols.** Water is required to convert the initially formed tetraalkoxyborate into the product alcohol and boric acid salts. ✳
- **Lithium aluminum hydride (LiAlH₄)** also reduces aldehydes and ketones to alcohols. $LiAlH_4$ reacts violently with water and other protic solvents, so the first step of the reactions is carried out in aprotic solvents like diethyl ether or tetrahydrofuran. Once the tetraalkoxy aluminate is formed, it is reacted with water in a second step to generate the product alcohol. $LiAlH_4$ also reduces carboxylic acids and their derivatives, whereas $NaBH_4$ does not. Neither reagent will reduce carbon-carbon double bonds.
- The mechanism of the reaction involves an initial transfer of a nucleophilic hydride ion from the metal reagent to the electrophilic carbonyl carbon atom of the carbonyl species. A negatively-charged alkoxide species (negative charge on the oxygen atom) that makes an ionic bond to the metal is the product of this step. This process repeats itself three more times until all of the reducing equivalents are used up. The resulting tetraalkoxy metal species are then hydrolyzed by water to form the product alcohols.

## REACTION 11D: HYDRATION (Section 11.6A)

$$\underset{\overset{\|}{\text{C}}}{\overset{\text{O}}{-}}\!\!- \;+\; H_2O \;\rightleftharpoons\; \underset{\overset{|}{\text{OH}}}{\overset{\text{OH}}{-}}\!\!\overset{|}{\text{C}}\!\!-$$

- **Aldehydes and ketones** react with **water** to form **1,1-diols**, also referred to as geminal diols. ✳

- The position of equilibrium greatly favors the carbonyl form rather than the diol form for all normal ketones, and all but the most simple aldehydes such as formaldehyde.

## REACTION 11E: ADDITION OF ALCOHOLS TO FORM ACETALS (Section 11.6B)

$$O=\overset{|}{\underset{|}{C}} \;+\; 2\; -\overset{|}{\underset{|}{C'}}-OH \;\underset{}{\overset{H^+}{\rightleftharpoons}}\; -\overset{|}{\underset{|}{C'}}-O-\overset{|}{\underset{|}{C}}-O-\overset{|}{\underset{|}{C'}}- \;+\; H_2O$$

- **Aldehydes and ketones** react with **alcohols** in the presence of an **acid catalyst** to generate an **acetal**. $*$
- The reaction is initiated by reversible protonation of the carbonyl oxygen atom. This makes the carbonyl group electrophilic enough to react with the nucleophilic oxygen atom of the alcohol, and following loss of a proton, a hemiacetal is formed. Next, the oxygen atom of the remaining -OH group is protonated to generate a positively-charged oxonium ion that then leaves to produce a highly electrophilic carbocation. Another nucleophilic alcohol molecule reacts with this cation then loses a proton to generate the final acetal product. Acetals are generally more stable than hemiacetals, so acetals are usually the product isolated from the reaction.
- Diols such as ethylene glycol produce cyclic acetals, that can be used as base stable protecting groups for carbonyl groups of aldehydes and ketones.

## REACTION 11F: ADDITION OF ALCOHOLS TO FORM HEMIACETALS (Section 11.6B)

$$O=\overset{|}{\underset{|}{C}} \;+\; -\overset{|}{\underset{|}{C'}}-OH \;\underset{}{\overset{H^+}{\rightleftharpoons}}\; HO-\overset{|}{\underset{|}{C'}}-O-\overset{|}{\underset{|}{C'}}-$$

- **Aldehydes and ketones** react with **alcohols** in the presence of **acid catalysts** to generate **hemiacetals**. Usually, the reaction immediately continues with addition of another molecule of alcohol to form the more stable acetal (Reaction 11E, Section 11.6B), and the hemiacetal cannot be isolated.
- The reaction is initiated by reversible protonation of the carbonyl oxygen atom. This makes the carbonyl group electrophilic enough to react with the nucleophilic oxygen atom of the alcohol, and following loss of a proton, a hemiacetal is formed.
- Hemiacetal formation is highly favored when the -OH and carbonyl group are part of the same molecule and can form a five- or six-member ring cyclic hemiacetal.

## REACTION 11G: REDUCTIVE AMINATION (Section 11.10C)

$$O=\overset{|}{\underset{|}{C}} \;+\; H_2N- \;\underset{}{\overset{(-H_2O)}{\rightleftharpoons}}\; \overset{|}{\underset{}{C}}=N- \;\overset{H_2/Ni}{\longrightarrow}\; \overset{|}{\underset{|}{C}}-NH-$$

(An imine,<br>not isolated)

- **Aldehydes and ketones** react with **ammonia and primary amines** to form **imines** (see Reaction 11H). When this reaction is carried out **in the presence of H$_2$ and a**

**transition metal** catalyst such as Ni, then the **imine is reduced to an amine.** The imine is generally not isolated in these reactions, so the entire process takes place in one laboratory operation. ✳
- Reacting the aldehyde or ketone (in the presence of $H_2$ and Ni) with ammonia, a primary amine or a secondary amine gives primary amine, secondary amine, or tertiary amine products, respectively.

## REACTION 11H: ADDITION OF AMMONIA AND ITS DERIVATIVES: FORMATION OF IMINES (Section 11.7)

$$O=\overset{|}{\underset{|}{C}} \ + \ H_2N- \ \overset{H^+}{\rightleftarrows} \ \overset{|}{C}=N- \ + \ H_2O$$

- **Aldehydes and ketones** react with **ammonia and primary amines** to form **imines,** which are also called **Schiff bases.** ✳
- The mechanism involves an initial attack of the nucleophilic nitrogen atom on the electrophilic carbonyl carbon atom. The proton on the nitrogen is then transferred to the oxygen to create a tetrahedral carbonyl addition intermediate. Acid-catalyzed dehydration involving loss of another proton on the nitrogen atom leads to the imine.
- Note how a total of two hydrogen atoms are removed from the nitrogen atom, explaining why only ammonia or primary amines can react to form imines. Secondary amines, with only one hydrogen atom, react to form another type of compound called enamines.

## REACTION 11I: REACTION WITH GRIGNARD REAGENTS (Section 11.5B)

$$O=\overset{|}{\underset{|}{C}} \ + \ -\overset{|}{\underset{|}{C}}'-MgX \ \xrightarrow{\text{2) } H_2O} \ HO-\overset{|}{\underset{|}{C}}-\overset{|}{\underset{|}{C}}'-$$

- **Formaldehyde, other aldehydes,** and **ketones** react with **Grignard reagents** to produce **primary, secondary, and tertiary alcohols,** respectively. ✳
- The mechanism of the reaction involves an initial attack of the nucleophilic Grignard reagent on the electrophilic carbonyl carbon atom. The magnesium alkoxide is protonated after water is added to the reaction to generate the product alcohol.
- This reaction is important because it involves the formation of carbon-carbon single bonds.

## REACTION 11J: FORMATION OF GRIGNARD REAGENTS FROM THE REACTION BETWEEN ALKYL HALIDES AND MAGNESIUM (Section 11.5A)

$$-\overset{|}{\underset{|}{C}}-X \ + \ Mg \ \xrightarrow{\text{Ether}} \ -\overset{|}{\underset{|}{C}}-MgX$$

Grignard reagent

- **Magnesium metal reacts with alkyl halides**, usually in diethyl ether solvent, to give compounds the have a carbon magnesium bond. These compounds are called Grignard reagents, after their discoverer, Victor Grignard. ✳
- Grignard reagents are special because carbon is so much more electronegative than magnesium, that these reagents react as though they were **carbanions**, that is, as if they have a negative charge on the carbon atom. For this reason, Grignard reagents are strong bases and they are very nucleophilic. For examples of Grignard reagents acting as nucleophiles, see reactions 11I, 11K, and 11L. These reactions are important because **new carbon-carbon bonds are formed**. ✳
- **Grignard reagents** are very **strong bases**. ✳

## REACTION 11K: REACTION OF GRIGNARD REAGENTS WITH CARBON DIOXIDE TO GIVE CARBOXYLIC ACIDS (Section 11.5B)

$$O=C=O \; + \; -\overset{|}{\underset{|}{C'}}-MgX \quad \xrightarrow{\text{2) } H_3O^+} \quad -\overset{|}{\underset{|}{C'}}-C\overset{\displaystyle O}{\underset{\displaystyle OH}{\diagup}}$$

- **Grignard reagents** react with **carbon dioxide** followed by an acidic aqueous workup to give **carboxylic acids**. Note how in these reactions, one carbon atom is added to the carbon skeleton of the Grignard reagent. ✳

## REACTION 11L: REACTION OF GRIGNARD REAGENTS WITH EPOXIDES (Section 11.5B)

$$\overset{O}{\underset{\diagup\;\diagdown}{C-C'}} \; + \; -\overset{|}{\underset{|}{C''}}-MgX \quad \xrightarrow{\text{2) } H_2O} \quad -\overset{HO}{\underset{|}{C}}-\overset{|}{\underset{|}{C'}}-\overset{|}{\underset{|}{C''}}-$$

- **Grignard reagents** react with **epoxides** to give **alcohols**. ✳
- The reactions are regioselective in that the nucleophilic Grignard reagent reacts predominantly with the less sterically hindered epoxide carbon atom, resulting in ring opening. Following protonation with water, an alcohol is produced.

## REACTION 11M: KETO-ENOL TAUTOMERIZATION (Section 11.8)

$$-\overset{\displaystyle O}{\underset{\displaystyle\underset{H}{|}}{\overset{||}{C}}}-\overset{|}{\underset{|}{C'}}- \quad \rightleftharpoons \quad \overset{HO}{\diagdown}C=C'\diagup$$

keto form                                        enol form

- The **keto form of aldehydes and ketones** is in **equilibrium with the enol form**. This interconversion is called **tautomerization**. ✳
- For most simple aldehydes and ketones the equilibrium lies far on side of the keto form.
- Interconversion of these two forms is catalyzed by acids and bases. Remember that catalysis only increases the rate at which the two forms interconvert; even in the presence of a catalyst the keto form still predominates to the same extent.

- The enol form is important to keep in mind because it has unique reactivities that are exploited in some transformations. In addition, a trace of acid or base causes a ketone or aldehyde with an adjacent stereocenter to racemize in solution, *via* an achiral enol intermediate.

## REACTION 11N: REACTION OF KETONES WITH STRONG OXIDIZING AGENTS UNDER FORCING CONDITIONS TO GIVE CARBOXYLIC ACIDS (Section 11.9B)

Enol form

- Ordinarily, ketones do not react with oxidizing agents. However, when **very strong oxidizing agents** such as $K_2Cr_2O_7$, $KMnO_4$ or $HNO_3$ are used **under extremely forcing conditions** such as high heat and/or high concentrations, **oxidative cleavage occurs** to give **carboxylic acids**. ✻
- The reaction actually occurs on the small amount of enol form that is present, so the reaction is most useful on symmetrical ketones such as cyclohexanone in which only one type of enol can be formed.

## SUMMARY OF IMPORTANT CONCEPTS

### 11.0 OVERVIEW
• The **carbonyl group** is one of the most important functional groups in organic chemistry, and it is found in aldehydes, ketones, carboxylic acids, and their functional derivatives. ✻

### 11.1 STRUCTURE
• The **characteristic feature** of **aldehydes** and **ketones** is the **carbonyl group**. In **aldehydes**, the carbonyl carbon atom is bonded to **at least one hydrogen atom**. In **ketones**, the carbonyl carbon atom is bonded to **two other carbon atoms** that are part of alkyl or aryl groups.

### 11.2 NOMENCLATURE
• According to **IUPAC nomenclature**:
  - The **aldehyde group** is named by changing the suffix **e** to **al**. Unsaturated aldehydes are designated with the infixes **en** or **yn** for carbon-carbon double and triple bonds, respectively. The number of the aldehyde carbonyl carbon atom is always 1. For cyclic molecules in which the -CHO group is attached to the ring, the suffix **carbaldehyde** is used.
  - The **ketone group** is named by changing the suffix **e** to **one**. The parent chain is chosen as the longest chain that contains the carbonyl group, and is numbered so that the carbonyl carbon atom has the lowest possible number.
  - The IUPAC nomenclature retains the common names benzaldehyde, acetone, acetophenone, and benzophenone, among others.
  - When a complex molecule contains more than one functional group, an **order of precedence** is used to determine the name. Table 17.1 lists these functional groups in order of precedence that ranges from carboxylic acids at the high end to thiols at the low end. The different functional groups each have a suffix that is used if they are the highest ranking functional group in the molecule, and a prefix if they are not the highest ranking functional

group. For example, aldehydes and ketones use the suffixes **al** and **one** if they are the highest ranking functional groups present, but the prefix of **oxo** in both cases if there are higher ranking functional groups present. The location of the group is indicated by the appropriate number.
• The **common name** for an **aldehyde** is analogous to that for the corresponding carboxylic acid, except the **ic** or **oic** suffix is replaced by the suffix **aldehyde**. The common names for ketones are derived by naming the two alkyl or aryl groups attached to the carbonyl group, followed by the word **ketone**.

## 11.3 PHYSICAL PROPERTIES
• The most **important electronic feature of carbonyl groups** is that they are **polar**, with a **partial negative charge** on the more electronegative **oxygen atom**, and a corresponding **partial positive charge** on the carbonyl **carbon atom**. ✻
  - The **polar nature of carbonyl compounds** means that they **interact via dipole-dipole interactions**, and thus have **higher boiling points** than corresponding hydrocarbons. Similarly, the lower molecular weight aldehydes and ketones are more soluble in water than the corresponding hydrocarbons. Dipole-dipole interactions are not as strong as hydrogen bonds, so **aldehydes and ketones** have **lower boiling points** and lower water solubilities than the **corresponding alcohols**.

## 11.4-11.10 REACTIONS
• **Carbonyl groups** are especially useful from a **synthetic point of view** because they take part in a large number of different types of reactions with **nucleophilic species. Included** within this group are **reactions** that lead to **formation of new carbon-carbon bonds**. Carbon-carbon bond-forming reactions are important because they allow the synthesis of larger, more complex carbon skeletons from smaller ones. *[A secret to learning carbonyl chemistry is to learn the characteristic reaction mechanisms, then figure out which reagents react by which reaction mechanism. In this way, the large number of reactions encountered in this chapter can be effectively considered in groups, thus simplifying greatly the task of learning. The following discussion is intended to help group the reactions, the details of which are described in the "Summary of Reactions" section.]* ✻
  - The **carbonyl carbon atom possesses a partial positive charge**. Therefore, a characteristic reaction of carbonyl groups is that they **react with various nucleophiles** at the carbonyl carbon atom. A list of nucleophiles that react with carbonyl groups in this way includes **carbon nucleophiles** (Grignard reagents), **oxygen nucleophiles** (water and alcohols), and **nitrogen nucleophiles** (amines). ✻
    In **Grignard reagents**, a **carbon atom** is **bonded** to a more **electropositive magnesium atom**. Thus, the bond is highly ionic and polarized with a **partial negative charge on the carbon atom**. Because of this, Grignard reagents **behave as carbon nucleophiles** in the reactions with carbonyl groups. ✻
    **Grignard reagents** are prepared from **alkyl halides** by reaction with free magnesium metal, so they are readily available.
    Because **Grignard reagents** are also **highly basic**, they react with any species that have even moderately acidic protons such as water or alcohols. For this reason, when a Grignard reagent reacts with a carbonyl group, a metal alkoxide is actually the product formed. The more acidic **water is added** to the reaction mixture in a second step to protonate the alkoxide and **generate the product alcohol**.
  - **Oxygen nucleophiles** (alcohols or water) react with aldehydes and ketones to produce molecules that have **two carbon-oxygen single bonds at the same carbon atom**. This species can be **1,1-diols, hemiacetals or acetals**.
  - **Nitrogen nucleophiles** (ammonia or amines) react with aldehydes and ketones to produce molecules that have **carbon-nitrogen double bonds**, called **imines**. When this reaction

is carried out in the presence of $H_2$ and Ni, the carbon-nitrogen double bond is reduced to a single bond, forming an amine.

- A carbonyl group is usually described as a **keto form**, but it is actually **in equilibrium** with a very small amount of the **enol form**. Both **acid and base catalyze** the **interconversion** of the **keto** and **enol forms**. Remember that a catalyst does not change the position of an equilibrium, just the speed at which the species equilibrate. The enol form reacts as an alkene, so certain reactions take place with the small amount of enol that is present at any one time.

- The **hydrogens** on the carbon atoms adjacent to a carbonyl group (the so-called α-**hydrogens**) are **relatively acidic**. The **anion produced** upon **deprotonation** of an α-hydrogen is called an **enolate ion**. The α-hydrogens are relatively acidic for a C-H bond because **enolate ions** are stabilized by **resonance stabilization**. In other words, the negative charge of the enolate ion is shared by the carbonyl oxygen and α-carbon atoms as predicted by the two most important contributing resonance structures. ✳

- **Aldehydes** and **ketones** can also take part in **oxidation reactions**. Aldehydes can be oxidized to carboxylic acids, but ketones are not readily oxidized by the normal reagents. Ketones can only be oxidized under special conditions because ketones do not have any hydrogen atoms attached to the carbonyl carbon atom. Treatment of ketones under harsh conditions such as high temperature in the presence of permanganate and dichromate can lead to oxidative cleavage, generating two carboxylic acids via the enol.

- **Aldehydes** and **ketones** are **reduced** by **hydride reagents** ($LiAlH_4$ or $NaBH_4$) to primary and secondary **alcohols**, respectively.

# CHAPTER 11
## *Solutions to the Problems*

Problem 11.1  Write IUPAC names for these compounds:

(a)  $CH_3CCH_2CHO$ with $CH_3$ groups above and below the central carbon

**3,3-Dimethylbutanal**

(b) cyclohexanone ring with $CH_3$ groups

***trans*-2,5-Dimethyl-
cyclohexanone**

(c)  $CH_3$—benzene ring—$CHCHO$ with $CH_3$

- 4-toluyl

**2-(4-Methylphenyl)propanal**

Problem 11.2  Give IUPAC names for these compounds, each of which is important in intermediary metabolism.  Below each compound is given the name by which it is more commonly known in the biological sciences.

(a)  $CH_3$—$CH$—$CO_2H$ with $OH$

Lactic acid
(product of anaerobic
glycolysis)

**2-Hydroxypropanoic
acid**

(b)  $CH_3$—$C$—$CO_2H$ with $O$

Pyruvic acid
(product of anaerobic
glycolysis)

**2-Oxopropanoic acid**

(c)  $H_2N$-$CH_2$-$CH_2$-$CH_2$-$CO_2H$

$\gamma$–Aminobutyric acid
(a neurotransmitter in the
central nervous system)

**4-Aminobutanoic acid**

Problem 11.3  Explain how these Grignard reagents will react with molecules of their own kind to "self-destruct."

(a)  $HO$—benzene ring—$MgBr$

(b)  $HOCCH_2CH_2CH_2MgBr$ with $O$

**These Grignard reagents cannot be used because they have acidic functions that would be deprotonated by the basic Grignard reagent portion of the molecule.  In (a), the alcohol would be deprotonated to give an alkoxide, and in (b) the carboxylic acid would be deprotonated to give the carboxylate.**

Problem 11.4 Show how these three products can be synthesized from the same Grignard reagent; that is, treating the Grignard reagent with one compound gives (a), with another compound gives (b), and with still another compound gives (c).

(a) [cyclohexene ring]—$CO_2H$   (b) [cyclohexene ring]—$CH_2CH_2OH$   (c) [cyclohexene ring]—$\overset{OH}{\underset{}{CHCH_3}}$

**All of the products can be obtained using 2-cyclohexenylmagnesium bromide and the other reactants shown.**

[cyclohexene ring]—MgBr

1) $CO_2$ → 2) $NH_4Cl$ / $H_2O$ → [cyclohexene ring]—$CO_2H$

1) $H_2C\overset{O}{-\!\!-}CH_2$  2) $NH_4Cl$ / $H_2O$ → [cyclohexene ring]—$CH_2CH_2OH$

1) $CH_3\overset{O}{\overset{\|}{C}}H$  2) $NH_4Cl$ / $H_2O$ → [cyclohexene ring]—$\overset{OH}{\underset{}{CHCH_3}}$

Problem 11.5 Hydrolysis of an acetal forms an aldehyde or ketone and two molecules of alcohol. Following are structural formulas for three acetals. Draw the structural formulas for the products of hydrolysis of each in aqueous acid.

(a)  $CH_3O$—[benzene ring]—$\overset{OCH_3}{\underset{}{CHOCH_3}}$  $\xrightarrow{H_3O^+}$

$CH_3O$—[benzene ring]—$\overset{O}{\overset{\|}{C}}$-H  +  $2CH_3OH$

(b)

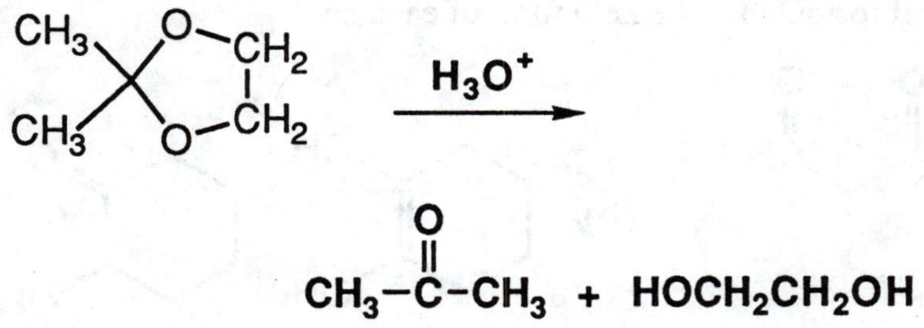

$$CH_3-\overset{\overset{\displaystyle O}{\|}}{C}-CH_3 \; + \; HOCH_2CH_2OH$$

(c)

$$\underset{OH}{\overset{}{CH_3}}\overset{}{C}HCH_2CH_2\overset{\overset{\displaystyle O}{\|}}{C}H \; + \; CH_3OH$$

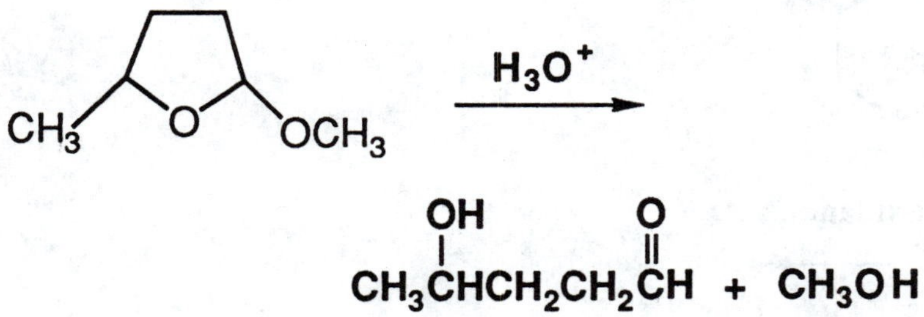

**Problem 11.6** Acid-catalyzed hydrolysis of an imine gives an amine salt and an aldehyde or ketone. Write structural formulas for the products of hydrolysis of these imines.

(a) 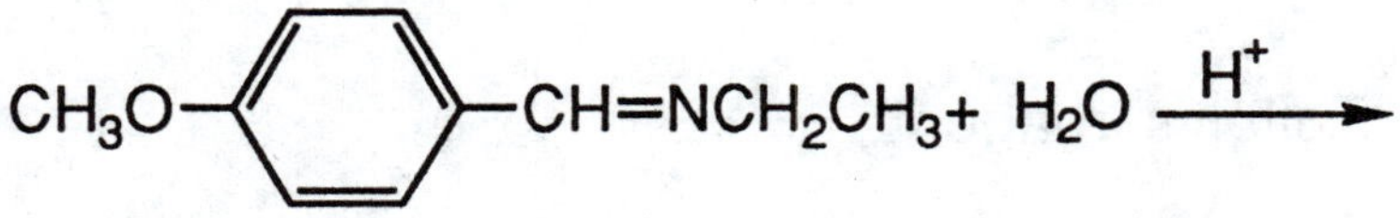   $CH_3O-$ $-CH=NCH_2CH_3 + H_2O \xrightarrow{H^+}$

**Hydrolysis of (a) gives an aldehyde and a primary amine salt.**

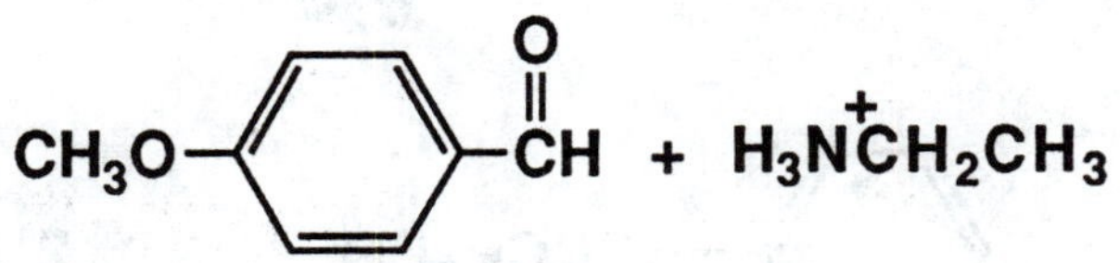

$$CH_3O- \text{⟨aryl⟩} -\overset{\overset{\displaystyle O}{\|}}{C}H \; + \; H_3\overset{+}{N}CH_2CH_3$$

(b) 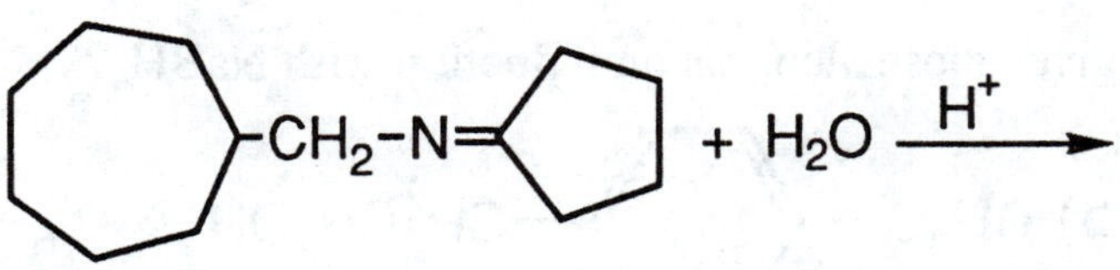   $-CH_2-N= $ $+ H_2O \xrightarrow{H^+}$

**Hydrolysis of (b) gives a ketone and a primary amine salt.**

$$\text{⟨cycloheptyl⟩} -CH_2\overset{+}{N}H_3 \; + \; O= \text{⟨cyclopentyl⟩}$$

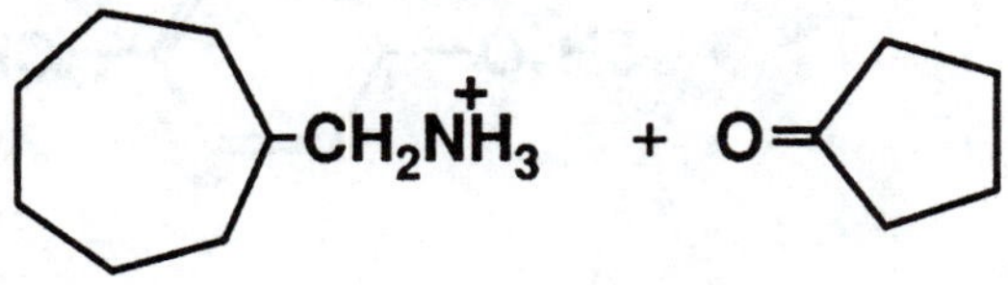

Problem 11.7  Draw the structural formula for the keto form of each enol.

(a)

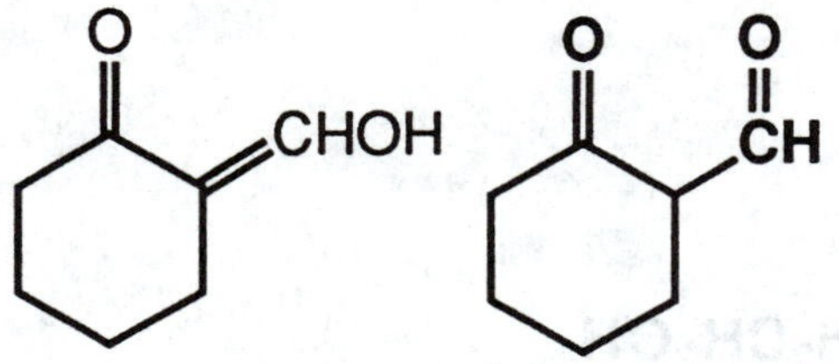

(b)

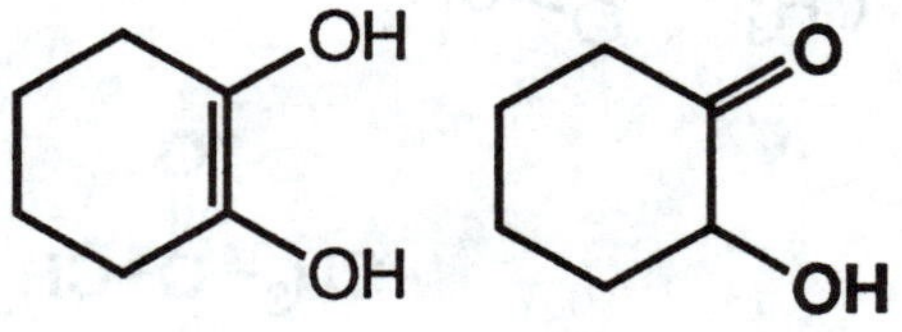

(c)

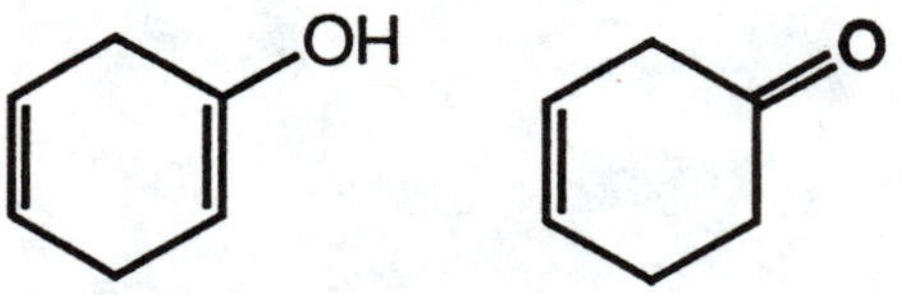

Problem 11.8  Complete these oxidations.

(a)  Hexanedial + H$_2$O$_2$ $\longrightarrow$

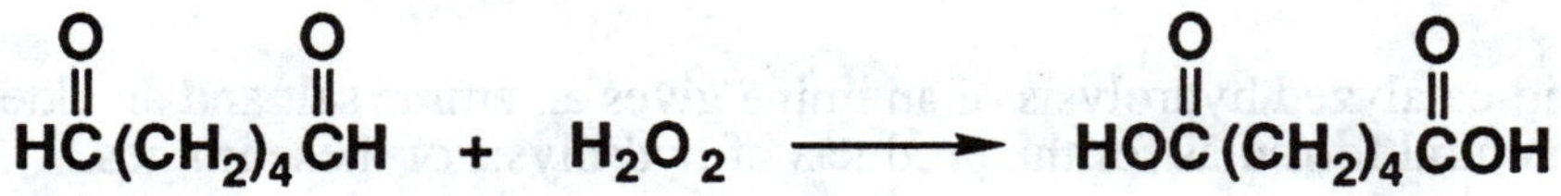

$$\underset{\text{HC}}{\overset{O}{\parallel}}(\text{CH}_2)_4\underset{\text{CH}}{\overset{O}{\parallel}} + H_2O_2 \longrightarrow \underset{\text{HOC}}{\overset{O}{\parallel}}(\text{CH}_2)_4\underset{\text{COH}}{\overset{O}{\parallel}}$$

(b)  3-Phenylpropanal + Tollen's reagent $\longrightarrow$

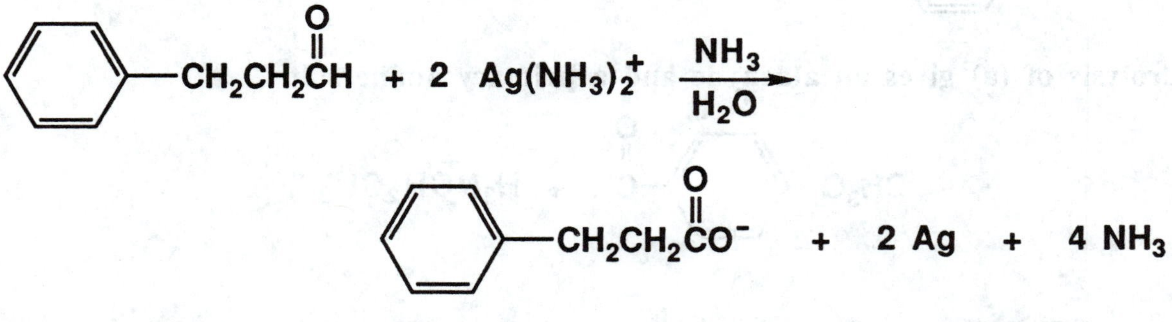

$$-\text{CH}_2\text{CH}_2\overset{O}{\overset{\parallel}{\text{CH}}} + 2\ \text{Ag(NH}_3)_2^+ \xrightarrow[\text{H}_2\text{O}]{\text{NH}_3}$$

$$-\text{CH}_2\text{CH}_2\overset{O}{\overset{\parallel}{\text{CO}}}^- + 2\ \text{Ag} + 4\ \text{NH}_3$$

Problem 11.9  What aldehyde or ketone gives these alcohols on reduction with NaBH$_4$?

(a) ⬡—OH

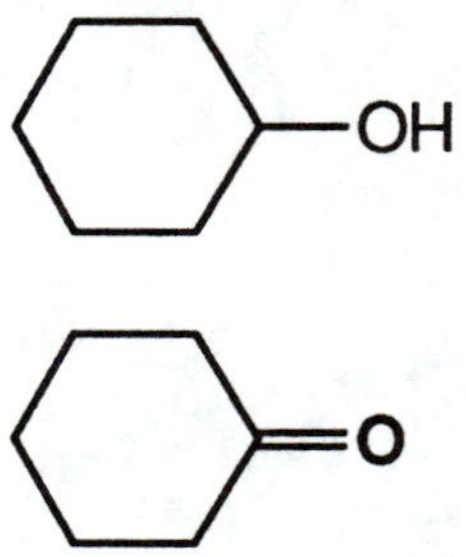

(b) CH$_3$O—⬡—CH$_2$CH$_2$OH

⬡=O

CH$_3$O—⬡—CH$_2$$\overset{O}{\overset{\parallel}{\text{CH}}}$

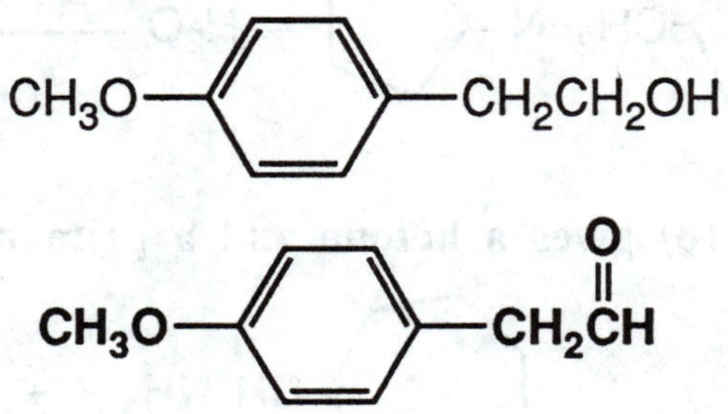

$$\underset{OH}{\underset{|}{}}\qquad \underset{OH}{\underset{|}{}}$$

(c) 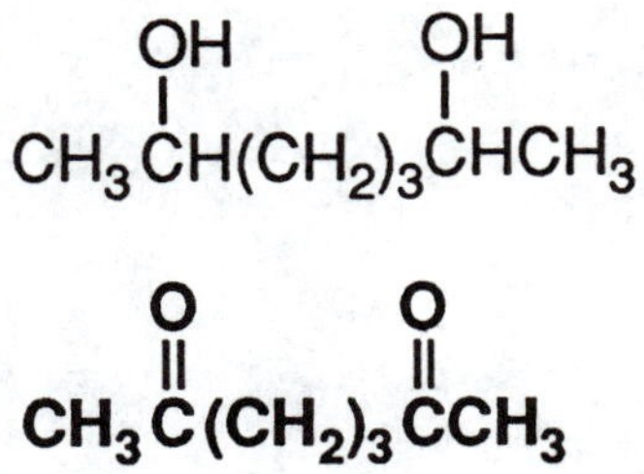 $CH_3CH(CH_2)_3CHCH_3$

$$CH_3\overset{O}{\overset{||}{C}}(CH_2)_3\overset{O}{\overset{||}{C}}CH_3$$

**Problem 11.10** Show how to convert piperidine to these compounds, using any other organic compounds and necessary reagents.

(a) 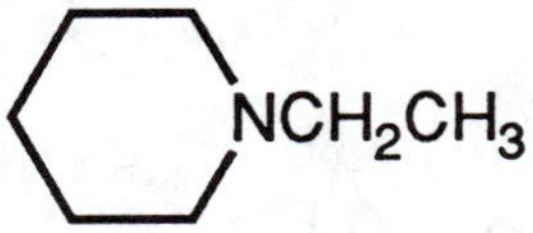  NCH_2CH_3

$$\underset{NH}{} + H\overset{O}{\overset{||}{C}}CH_3 \xrightarrow{H_2, Ni} \underset{NCH_2CH_3}{}$$

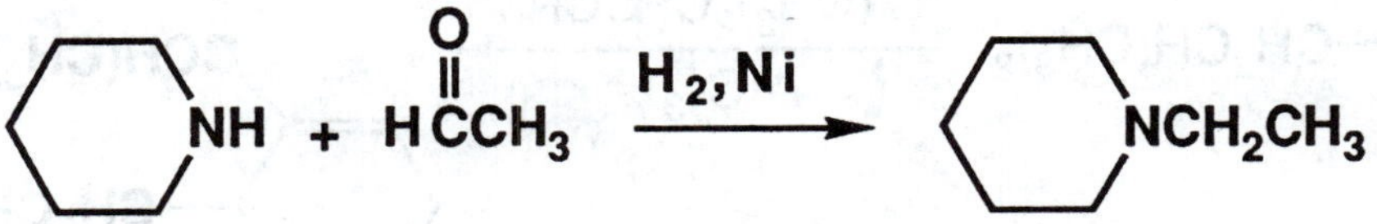

(b) 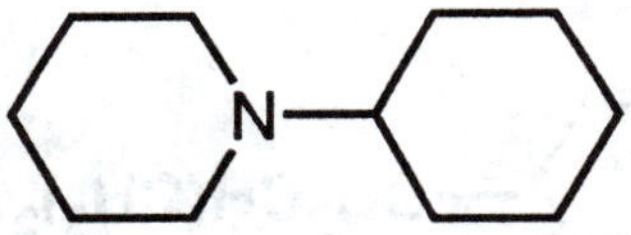  N—

$$\underset{NH}{} + O= \xrightarrow{H_2, Ni} \underset{N—}{}$$

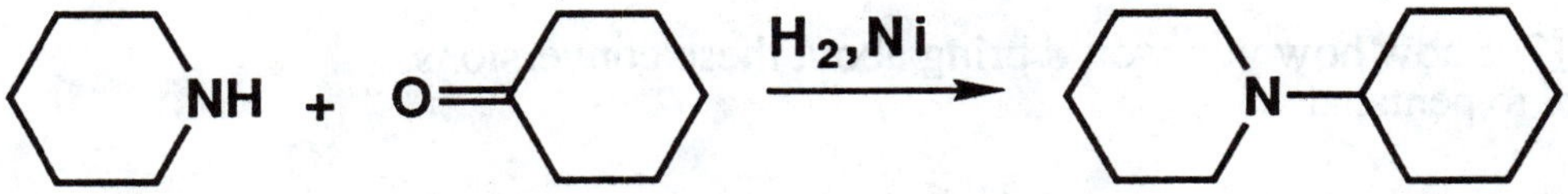

## Preparation of Aldehydes and Ketones
The methods covered to this point for the preparation of aldehydes and ketones are oxidation of primary and secondary alcohols, (Section 7.4F), and Friedel-Crafts acylation of arenes (Section 9.7C).

**Problem 11.11** Complete these reactions.

(a) 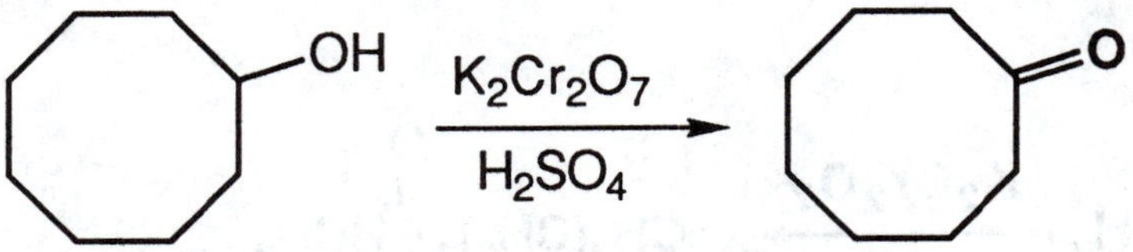  —OH $\xrightarrow[H_2SO_4]{K_2Cr_2O_7}$  =O

(b) 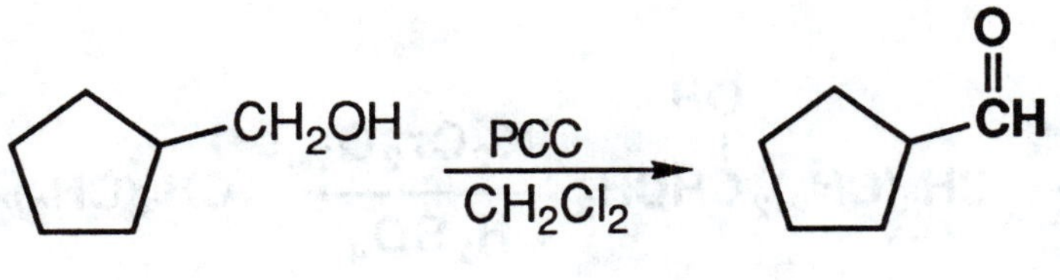  —CH_2OH $\xrightarrow[CH_2Cl_2]{PCC}$  —CH=O

(c)

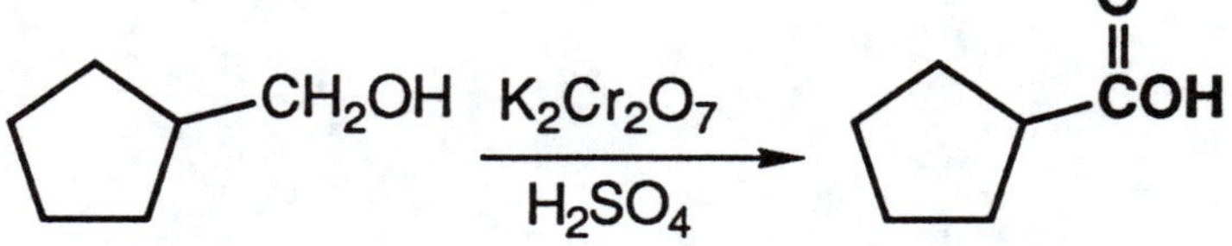

(d)

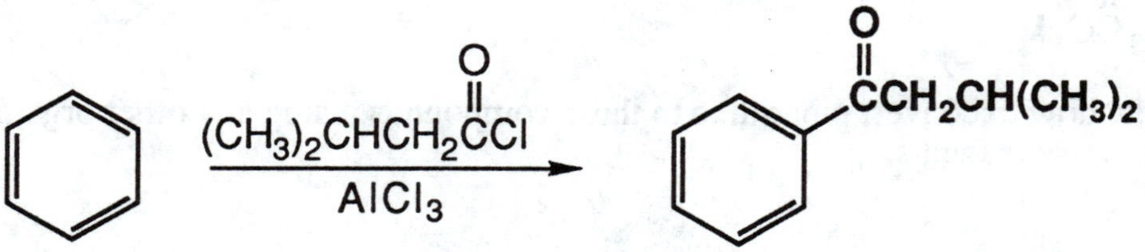

(e)

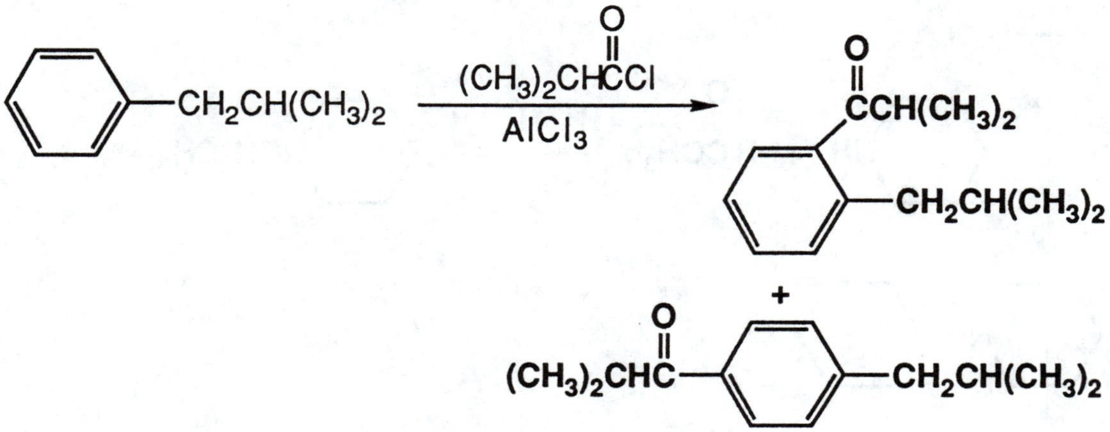

**Problem 11.12** Show how you would bring about these conversions.
(a) 1-Pentanol to pentanal

$$CH_3(CH_2)_3CH_2OH \xrightarrow[CH_2Cl_2]{PCC} CH_3(CH_2)_3\overset{O}{\overset{\|}{C}}H$$

(b) 1-Pentanol to pentanoic acid

$$CH_3(CH_2)_3CH_2OH \xrightarrow[H_2SO_4]{K_2Cr_2O_7} CH_3(CH_2)_3\overset{O}{\overset{\|}{C}}OH$$

(c) 2-Pentanol to 2-pentanone

$$CH_3(CH_2)_2\overset{OH}{\overset{|}{C}}HCH_3 \xrightarrow[H_2SO_4]{K_2Cr_2O_7} CH_3(CH_2)_2\overset{O}{\overset{\|}{C}}HCH_3$$

(d) 1-Pentene to 2-pentanone

$$CH_3(CH_2)_2CH{=}CH_2 \xrightarrow[H_2SO_4]{H_2O} CH_3(CH_2)_2\overset{OH}{\overset{|}{C}}HCH_3 \xrightarrow[H_2SO_4]{K_2Cr_2O_7} CH_3(CH_2)_2\overset{O}{\overset{\|}{C}}HCH_3$$

(e) Benzene to acetophenone

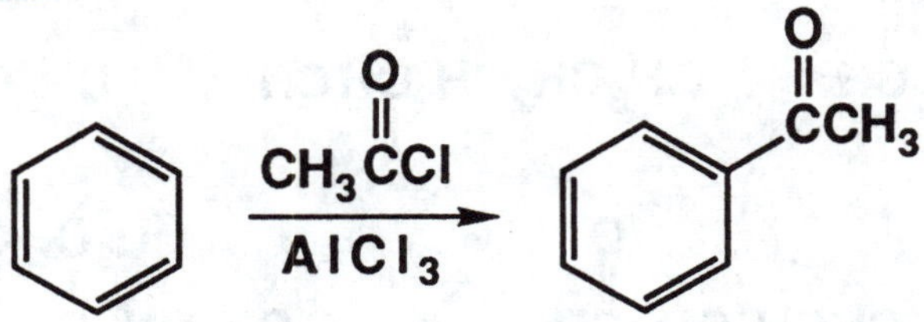

(f) Styrene to acetophenone

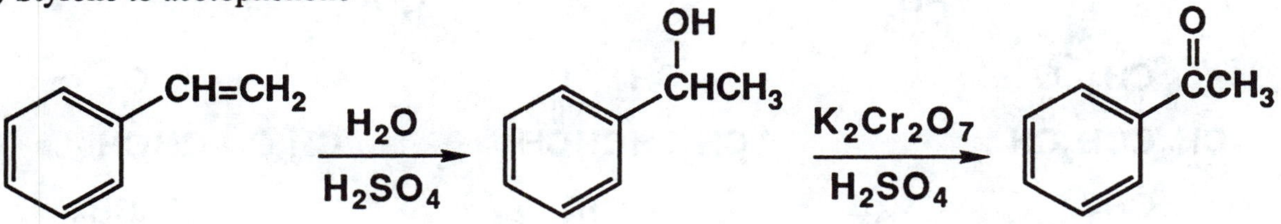

(g) Cyclohexanol to cyclohexanone

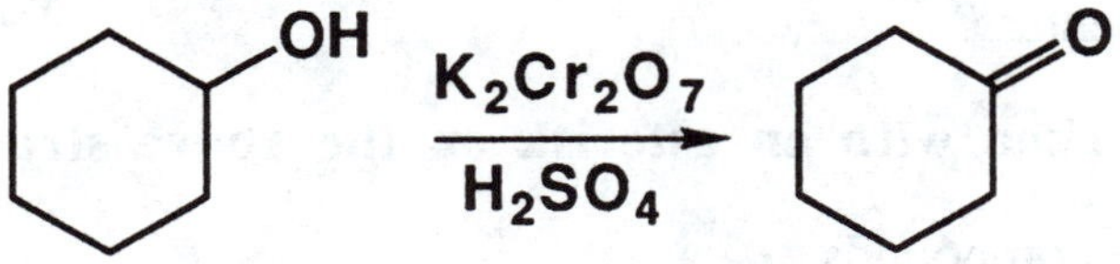

(h) Cyclohexene to cyclohexanone

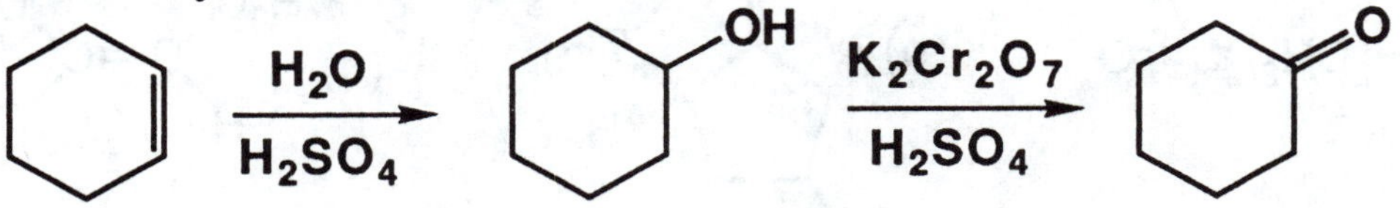

## Structure and Nomenclature

**Problem 11.13** Draw a structural formula for the one ketone of molecular formula $C_4H_8O$ and for the two aldehydes of molecular formula $C_4H_8O$.

$$CH_3CH_2\overset{\overset{O}{\|}}{C}CH_3 \qquad CH_3CH_2CH_2\overset{\overset{O}{\|}}{C}H \qquad (CH_3)_2CH\overset{\overset{O}{\|}}{C}H$$

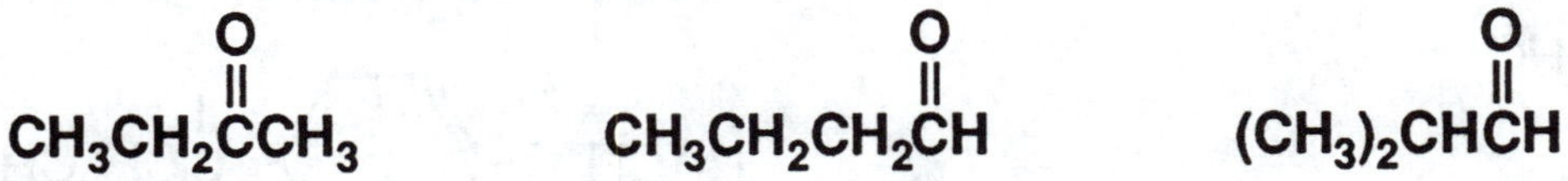

**Problem 11.14** Draw a structural formula for the six ketones of molecular formula $C_6H_{12}O$ and for the eight aldehydes of molecular formula $C_6H_{12}O$.

$$CH_3CH_2CH_2CH_2\overset{\overset{O}{\|}}{C}CH_3 \qquad CH_3CH_2CH_2\overset{\overset{O}{\|}}{C}CH_2CH_3 \qquad \underset{\underset{CH_3}{|}}{CH_3CH}\overset{\overset{O}{\|}}{C}CH_2CH_3$$

$$\underset{\underset{CH_3}{|}}{CH_3CH_2\overset{*}{C}H}\overset{\overset{O}{\|}}{C}CH_3 \qquad \underset{\underset{CH_3}{|}}{CH_3CHCH_2}\overset{\overset{O}{\|}}{C}CH_3 \qquad \underset{\underset{H_3C}{|}}{\overset{\overset{H_3C}{|}}{CH_3C}}-\overset{\overset{O}{\|}}{C}CH_3$$

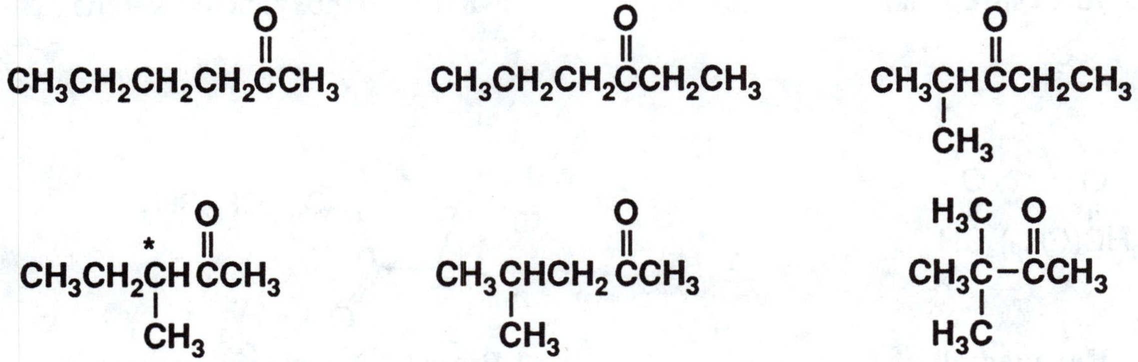

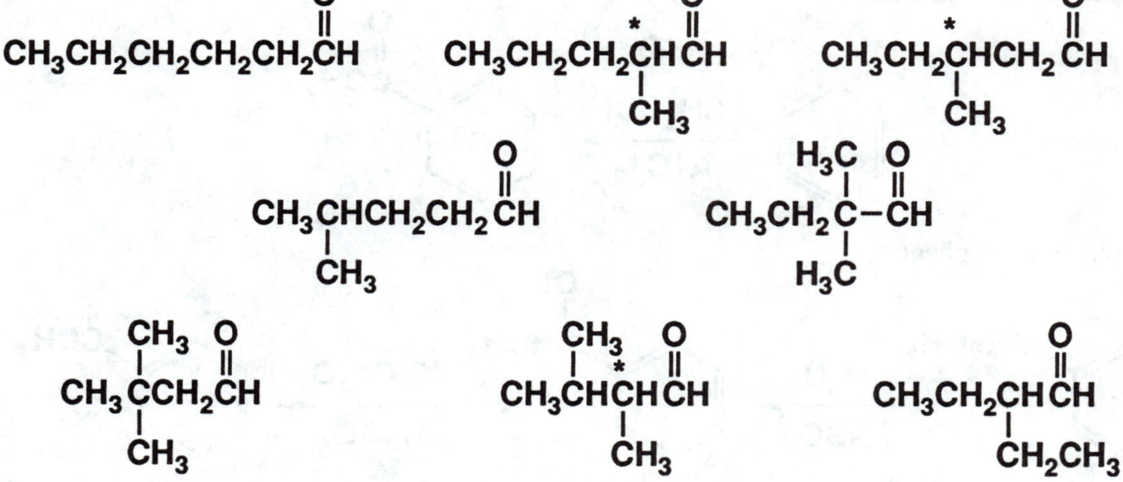

$$CH_3CH_2CH_2CH_2CH_2 \overset{\overset{O}{\|}}{CH}$$

$$CH_3CH_2CH_2 \overset{*}{\underset{\underset{CH_3}{|}}{CH}} \overset{\overset{O}{\|}}{CH}$$

$$CH_3CH_2 \overset{*}{\underset{\underset{CH_3}{|}}{CH}} CH_2 \overset{\overset{O}{\|}}{CH}$$

$$CH_3 \underset{\underset{CH_3}{|}}{CH} CH_2CH_2 \overset{\overset{O}{\|}}{CH}$$

$$CH_3CH_2 \overset{\overset{H_3C}{|}}{\underset{\underset{H_3C}{|}}{C}} {-} \overset{\overset{O}{\|}}{CH}$$

$$CH_3 \overset{\overset{CH_3}{|}}{\underset{\underset{CH_3}{|}}{C}} CH_2 \overset{\overset{O}{\|}}{CH}$$

$$CH_3 \overset{\overset{CH_3}{|}}{CH} \overset{*}{\underset{\underset{CH_3}{|}}{CH}} \overset{\overset{O}{\|}}{CH}$$

$$CH_3CH_2 \underset{\underset{CH_2CH_3}{|}}{CH} \overset{\overset{O}{\|}}{CH}$$

**Problem 11.15** Which of the ketones and aldehydes that answer Problem 11.14 are chiral? Which contain two or more stereocenters?

**Each stereocenter is marked with an asterisk on the above structures.**

**Problem 11.16** Name these compounds.

(a)  $(CH_3CH_2CH_2)_2C{=}O$

**4-Heptanone
(Dipropyl ketone)**

(b)

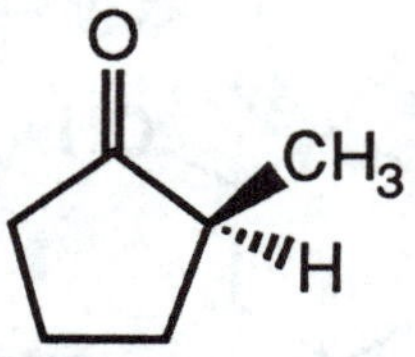

**(S)-2-Methylcyclo-
pentanone**

(c)

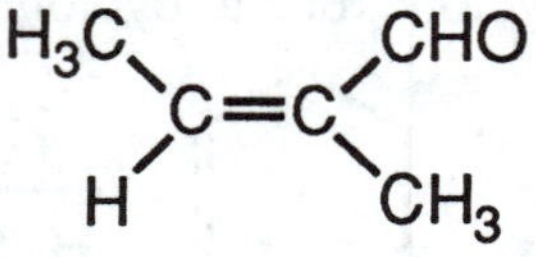

**(Z)-2-Methyl-2-butenal**

(d)

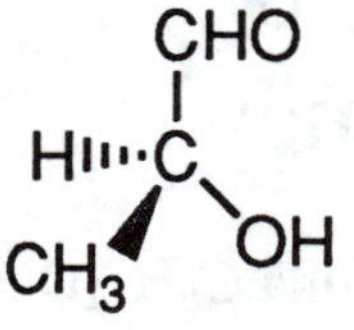

**(S)-2-Hydroxypropanal**

(e) $CH_3O{-}$⟨ring⟩$\overset{\overset{O}{\|}}{C}CH_2CH_3$

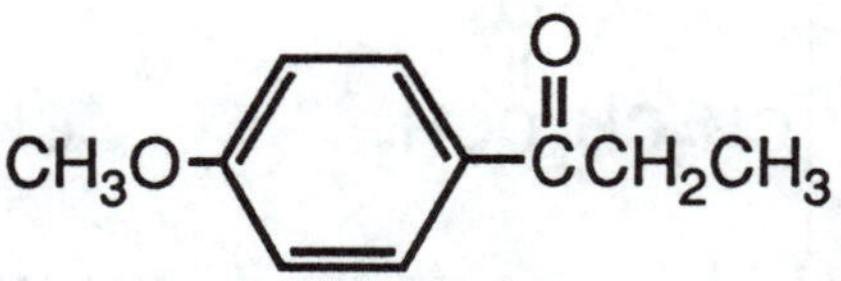

**Ethyl 4-methoxyphenyl ketone**

(f)  $H\overset{\overset{O}{\|}}{C}(CH_2)_4\overset{\overset{O}{\|}}{C}H$

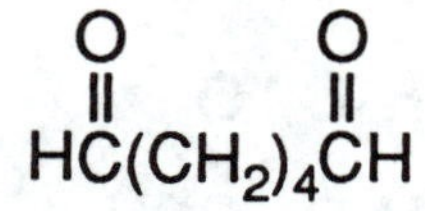

**Hexanedial**

(g)  $CH_2CH_2CH_3$

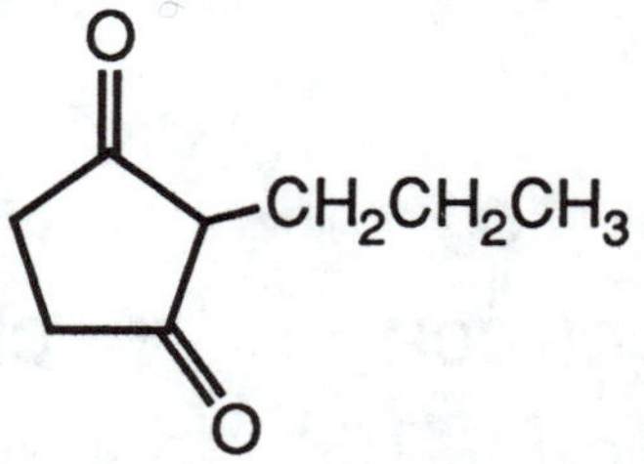

**2-Propyl-1,3-cyclopentanedione**

**Problem 11.17** Draw structural formulas for these compounds.

(a) 1-Chloro-2-propanone

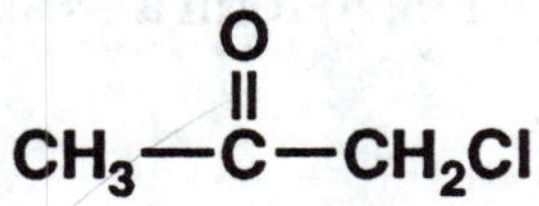

$$CH_3-\overset{\overset{\displaystyle O}{\|}}{C}-CH_2Cl$$

(b) 3-Hydroxybutanal

$$CH_3\overset{\overset{\displaystyle OH}{|}}{CH}CH_2\overset{\overset{\displaystyle O}{\|}}{CH}$$

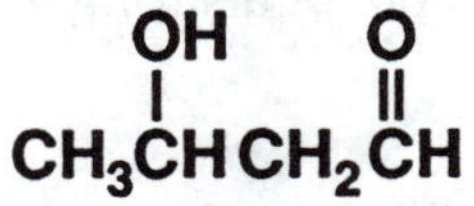

(c) 4-Hydroxy-4-methyl-2-pentanone

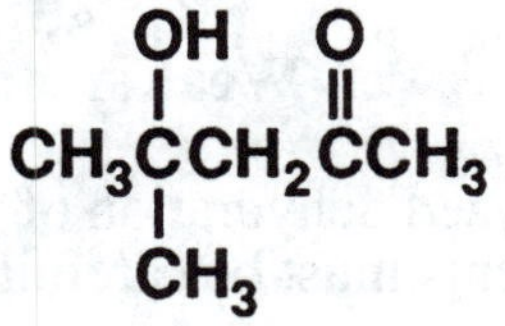

$$CH_3\overset{\overset{\displaystyle OH}{|}}{\underset{\underset{\displaystyle CH_3}{|}}{C}}CH_2\overset{\overset{\displaystyle O}{\|}}{C}CH_3$$

(d) 3-Methyl-3-phenylbutanal

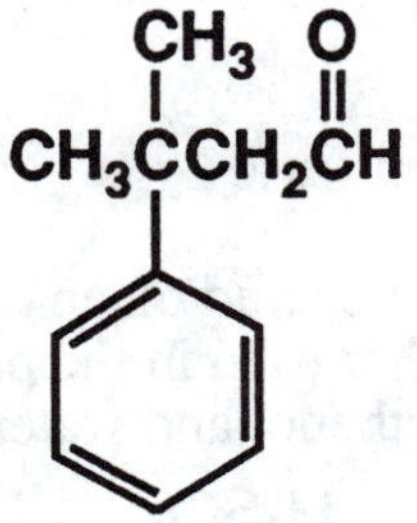

$$CH_3\overset{\overset{\displaystyle CH_3}{|}}{C}CH_2\overset{\overset{\displaystyle O}{\|}}{CH}$$

(e) 1,3-Cyclohexanedione

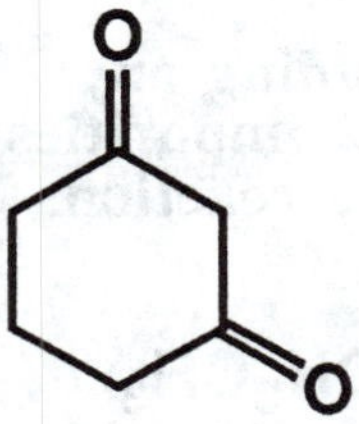

(f) 3-Methyl-3-butene-2-one

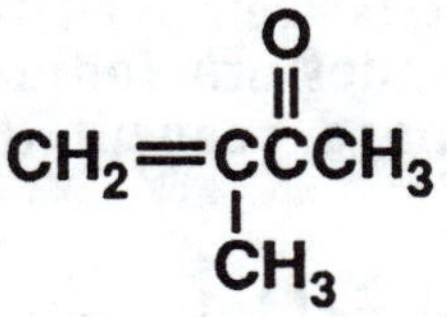

$$CH_2{=}\overset{\overset{\displaystyle O}{\|}}{\underset{\underset{\displaystyle CH_3}{|}}{C}}CH_3$$

(g) 5-Oxohexanal

$$CH_3\overset{\overset{\displaystyle O}{\|}}{C}CH_2CH_2CH_2{-}\overset{\overset{\displaystyle O}{\|}}{CH}$$

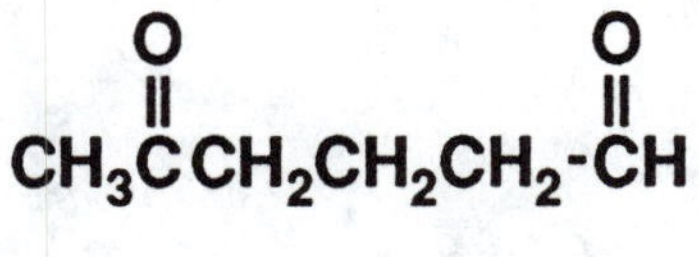

(h) 2,2-Dimethylcyclohexanecarbaldehyde

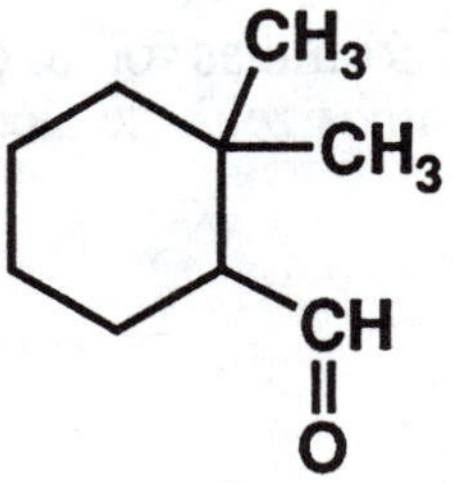

(i) 3-Oxobutanoic acid

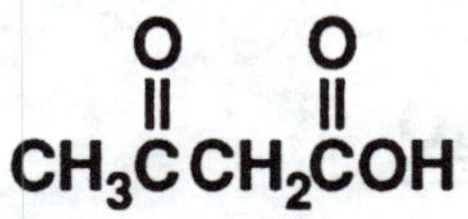

$$CH_3\overset{\overset{\displaystyle O}{\|}}{C}CH_2\overset{\overset{\displaystyle O}{\|}}{C}OH$$

## Addition of Carbon Nucleophiles

**Problem 11.18** Write an equation for the acid-base reaction between phenylmagnesium iodide and a carboxylic acid. Use curved arrows to show the flow of electrons in this reaction. In addition, show that this reaction is an example of a stronger acid and stronger base reacting to form a weaker acid and weaker base.

$$
\underset{\substack{\textbf{pK}_a \ \textbf{4-5} \\ \textbf{Stronger acid}}}{R-\overset{\overset{\displaystyle O}{\|}}{C}-O-H} + \underset{\textbf{Stronger base}}{\text{C}_6\text{H}_5-MgI} \rightleftharpoons \underset{\textbf{Weaker base}}{R-\overset{\overset{\displaystyle O}{\|}}{C}-O^- \ (MgI)^+} + \underset{\substack{\textbf{pK}_a \ \textbf{43} \\ \textbf{Weaker acid}}}{\text{C}_6\text{H}_6}
$$

**Problem 11.19** Diethyl ether is prepared on an industrial scale by acid-catalyzed dehydration of ethanol. Explain why diethyl ether used in the preparation of Grignard reagents must be carefully purified to remove all traces of ethanol and water.

$$2 \ CH_3CH_2OH \xrightarrow[180°C]{H_2SO_4} CH_3CH_2OCH_2CH_3 \ + \ H_2O$$

**All water and alcohol must be removed from diethyl ether prior to adding any Grignard reagent, ethylmagnesium iodide for example, because these impurities would deactivate the Grignard reagent due to the following acid-base reaction.**

$$R-O-H + CH_3CH_2-MgI \longrightarrow R-O^- \ (MgI)^+ \ + \ CH_3CH_3$$

$$R = CH_3CH_2 \ \text{or} \ H$$

**Problem 11.20** Write structural formulas for all combinations of Grignard reagent and aldehyde or ketone that might be used to synthesize these alcohols.

$$
\text{(a)} CH_3CH_2CH_2CH_2\overset{\overset{\displaystyle OH}{|}}{C}HCH_3
$$

$$
CH_3CH_2CH_2CH_2\overset{\overset{\displaystyle O}{\|}}{C}H \ + \ CH_3MgI \xrightarrow{\text{2) } NH_4Cl \ / \ H_2O}
$$

or

$$
CH_3\overset{\overset{\displaystyle O}{\|}}{C}H \ + \ CH_3CH_2CH_2CH_2MgI \xrightarrow{\text{2) } NH_4Cl \ / \ H_2O}
$$

(b)

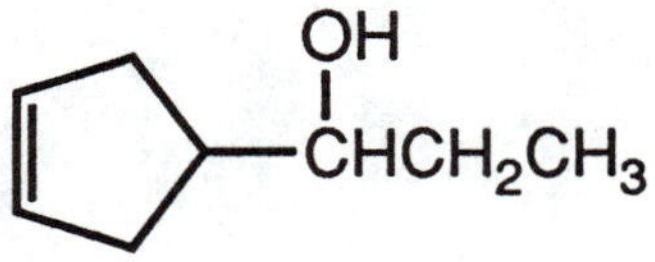

OH
|
$-CHCH_2CH_3$

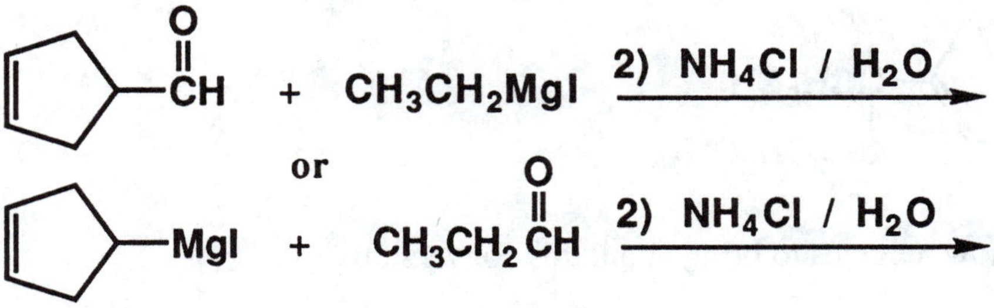

$-CH$  +  $CH_3CH_2MgI$  $\xrightarrow{\text{2) } NH_4Cl \text{ / } H_2O}$

or

$-MgI$  +  $CH_3CH_2\overset{O}{\overset{\|}{C}}H$  $\xrightarrow{\text{2) } NH_4Cl \text{ / } H_2O}$

**Problem 11.21** Draw structural formulas for the product formed by treatment of these compounds with propylmagnesium bromide followed by hydrolysis in aqueous acid.

**The products after acid hydrolysis are given in bold.**

(a)   $CH_2O$

$CH_3CH_2CH_2CH_2OH$

(b) 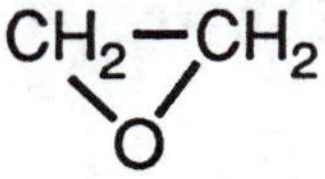   $CH_2-CH_2$ (epoxide, O bridging)

OH
|
$CH_3CH_2CH_2CH_2CH_2$

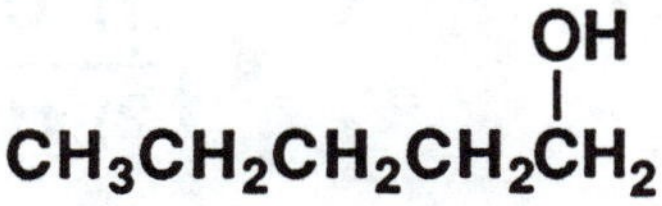

(c) 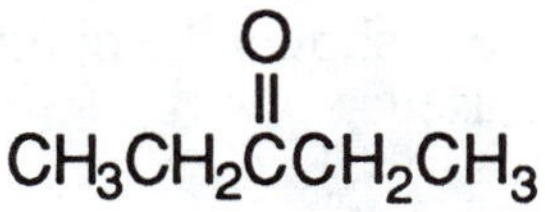   $CH_3CH_2\overset{O}{\overset{\|}{C}}CH_2CH_3$

OH
|
$CH_3CH_2CCH_2CH_3$
|
$CH_2CH_2CH_3$

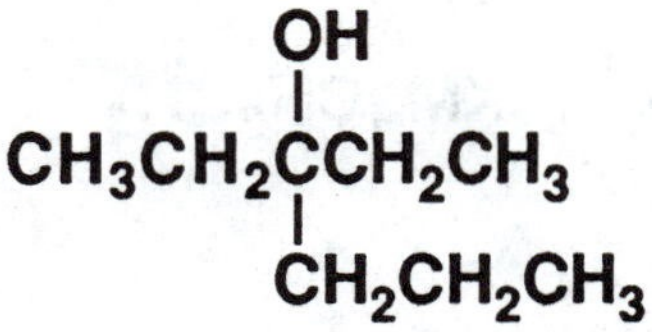

(d)

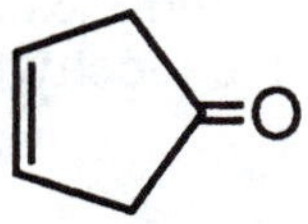

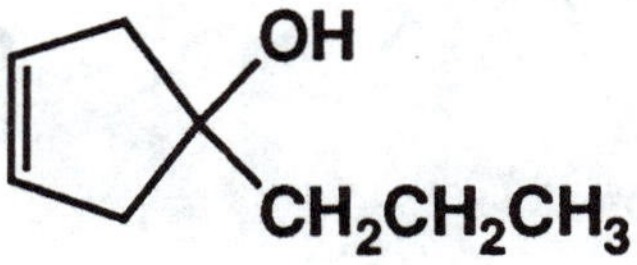

(e)   $CO_2$

$CH_3CH_2CH_2\overset{O}{\overset{\|}{C}}OH$

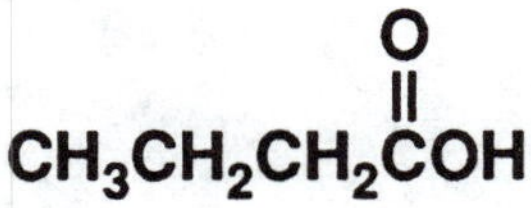

(f)

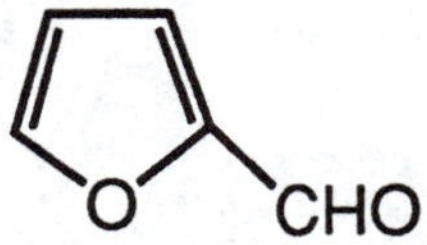

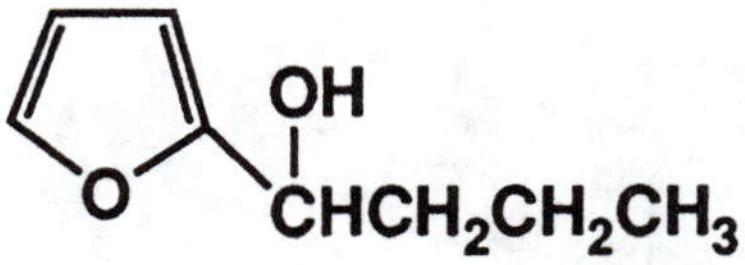

(g) $CH_3O-$⟨benzene ring⟩$-\overset{\overset{O}{\|}}{C}CH_2CH_3$

$CH_3O-$⟨benzene ring⟩$-\overset{\overset{OH}{|}}{\underset{CH_2CH_2CH_3}{C}}CH_2CH_3$

Problem 11.22  Show reagents to bring about this conversion.

$CH_3O-$⟨benzene ring⟩$-Br$ $\xrightarrow{(a)}$ $CH_3O-$⟨benzene ring⟩$-MgBr$ $\xrightarrow{(b)}$ $CH_3O-$⟨benzene ring⟩$-\overset{\overset{O}{\|}}{C}OH$

$CH_3O-$⟨benzene ring⟩$-Br$ $\xrightarrow[\text{Ether}]{\text{Mg}}$ $CH_3O-$⟨benzene ring⟩$-MgBr$

$\xrightarrow[\text{2) } H_3O^+]{\text{1) } CO_2}$ $CH_3O-$⟨benzene ring⟩$-\overset{\overset{O}{\|}}{C}OH$

Problem 11.23  Suggest a synthesis for these alcohols starting from an aldehyde or ketone and an appropriate Grignard reagent.  In parentheses below each target molecule is shown the number of combinations of Grignard reagent and aldehyde or ketone that might be used.

(a)  $CH_3\overset{\overset{OH}{|}}{\underset{CH_2CH_3}{C}}CH_2CH_2CH_3$

(3 combinations)

$CH_3\overset{\overset{O}{\|}}{C}CH_2CH_2CH_3 + XMgCH_2CH_3$

$CH_3MgX + \overset{\overset{O}{\|}}{\underset{CH_2CH_3}{C}}CH_2CH_2CH_3$

$CH_3\overset{\overset{O}{\|}}{\underset{CH_2CH_3}{C}} + XMgCH_2CH_2CH_3$

(b)   CH₃CHCH₂CH=CHCH₃  (with OH on second carbon)

OH

CH₃CHCH₂CH=CHCH₃

(2 combinations)

CH₃CH (=O)  +  XMgCH₂CH=CHCH₃          CH₃CH₂MgX  +  HCCH₂CH=CHCH₃ (with C=O)

(c)   CH₃O—⟨benzene⟩—CH(OH)—⟨benzene⟩

(2 combinations)

CH₃O—⟨benzene⟩—CH(=O)  +  XMg—⟨benzene⟩          CH₃O—⟨benzene⟩—MgX  +  HC(=O)—⟨benzene⟩

---

<u>Problem 11.24</u>  Show how to synthesize 3-ethyl-1-hexanol using 1-bromopropane, propanal and ethylene oxide as the only sources of carbon atoms. It can be done using each compound only once. (*Hint*: Do one Grignard reaction to form an alcohol, convert the alcohol to an alkyl halide, and then do a second Grignard reaction).

$$CH_3CH_2CH_2Br + \overset{\overset{\displaystyle CH_2CH_3}{|}}{CHO} + \underset{\underset{\displaystyle O}{\diagdown \diagup}}{CH_2-CH_2} \xrightarrow{\text{several steps}} CH_3CH_2CH_2\overset{\overset{\displaystyle CH_2CH_3}{|}}{CH}CH_2CH_2OH$$

1-Bromopropane   Propanal   Ethylene oxide          3-Ethyl-1-hexanol

**This synthesis is divided into two stages.  In the first stage, 1-bromopropane is treated with magnesium to form a Grignard reagent and then with propanal followed by hydrolysis to give 3-hexanol.**

$$CH_3CH_2CH_2Br + Mg \xrightarrow{\text{ether}} CH_3CH_2CH_2MgBr \xrightarrow[\substack{1) \overset{\overset{\displaystyle CH_2CH_3}{|}}{CHO}}]{\substack{2)\ NH_4Cl \\ H_2O}} CH_3CH_2CH_2\overset{\overset{\displaystyle CH_2CH_3}{|}}{CH}OH$$

**3-Hexanol**

**In the second stage, 3-hexanol is treated with thionyl chloride followed by magnesium in ether to form a Grignard reagent.  Treatment of this Grignard reagent with ethylene oxide followed by hydrolysis gives 3-ethyl-1-hexanol.**

$$
\underset{\text{CH}_3\text{CH}_2\text{CH}_2\overset{|}{\text{CH}}\text{OH}}{\overset{\overset{\text{CH}_2\text{CH}_3}{|}}{}} \quad \xrightarrow[\text{2)  Mg,  ether}]{\text{1)  SOCl}_2} \quad \underset{\text{CH}_3\text{CH}_2\text{CH}_2\overset{|}{\text{CHMgCl}}}{\overset{\overset{\text{CH}_2\text{CH}_3}{|}}{}} \quad + \quad \underset{O}{\overset{\text{CH}_2\!-\!\text{CH}_2}{}}
$$

$$
\xrightarrow[\text{H}_2\text{O}]{\text{2)  NH}_4\text{Cl}}
$$

$$
\underset{\text{CH}_3\text{CH}_2\text{CH}_2\overset{|}{\text{CH}}\text{CH}_2\text{CH}_2\text{OH}}{\overset{\overset{\text{CH}_2\text{CH}_3}{|}}{}}
$$

**3-Ethyl-1-hexanol**

<u>Problem 11.25</u>  1-Phenyl-2-butanol is used in perfumery.  Show how to synthesize this alcohol from bromobenzene from bromobenzene, 1-butene, and any other needed reagents.

$$
\text{Bromobenzene} \quad \begin{array}{c} \bigcirc\!\!-\!\text{Br} \end{array} \; + \; \text{CH}_3\text{CH}_2\text{CH}=\text{CH}_2 \xrightarrow[\text{steps}]{\text{Several}} \begin{array}{c} \bigcirc\!\!-\!\text{CH}_2\overset{\overset{\text{OH}}{|}}{\text{CH}}\text{CH}_2\text{CH}_3 \end{array}
$$

Bromobenzene                    1-Butene                                      1-Phenyl-2-butanol

**(a)  Bromobenzene is treated with magnesium in diethyl ether to form phenylmagnesium bromide, in preparation for part (c).**

$$
\bigcirc\!\!-\!\text{Br} \; + \; \text{Mg} \xrightarrow{\text{ether}} \bigcirc\!\!-\!\text{MgBr}
$$

**(b)  Treatment of 1-butene with a peroxycarboxylic acid gives 1,2-epoxybutane.**

$$
\text{CH}_3\text{CH}_2\text{CH}=\text{CH}_2 \; + \; \text{R}\!-\!\text{CO}_3\text{H} \longrightarrow \underset{O}{\overset{\text{CH}_3\text{CH}_2\text{CH}\!-\!\text{CH}_2}{}} \; + \; \text{RCO}_2\text{H}
$$

**(c)  Treatment of phenylmagnesium bromide with 1,2-epoxybutane followed by hydrolysis in aqueous acid gives 1-phenyl-2-butanol.**

$$
\bigcirc\!\!-\!\text{MgBr} \; + \; \underset{O}{\overset{\text{CH}_3\text{CH}_2\text{CH}\!-\!\text{CH}_2}{}} \xrightarrow[\text{H}_2\text{O}]{\text{2)  NH}_4\text{Cl}} \bigcirc\!\!-\!\text{CH}_2\overset{\overset{\text{OH}}{|}}{\text{CH}}\text{CH}_2\text{CH}_3
$$

## Addition of Oxygen Nucleophiles
**Problem 11.26** 5-Hydroxyhexanal forms a six-member cyclic hemiacetal, which predominates at equilibrium in aqueous solution.

CH$_3$CHCH$_2$CH$_2$CH$_2$CH $\xrightleftharpoons{H^+}$ A cyclic hemiacetal
|
OH

5-Hydroxyhexanal

(a) Draw a structural formula for this cyclic hemiacetal.

**5-Hydroxyhexanal forms a six-member cyclic hemiacetal.**

(b) How many stereoisomers are possible for 5-hydroxyhexanal?

**Two stereoisomers are possible for 5-hydroxyhexanal**

(c) How many stereoisomers are possible for this cyclic hemiacetal?

**Four stereoisomers are possible for the cyclic hemiacetal; two pairs of enantiomers.**
**Following are planar hexagon formulas for each pair of enantiomers of the cyclic hemiacetal.**

(d) Draw alternative chair conformations for each stereoisomer of the cyclic hemiacetal and label groups axial or equatorial. Also predict which of the alternative chair conformations for each stereoisomer is the more stable.

**Alternative chair conformations are drawn for (A), one of the *cis* enantiomers and for (C), one of the *trans* enantiomers. For (A), the diequatorial chair is the more stable. For (C), the alternative chairs are of approximately equal stability.**

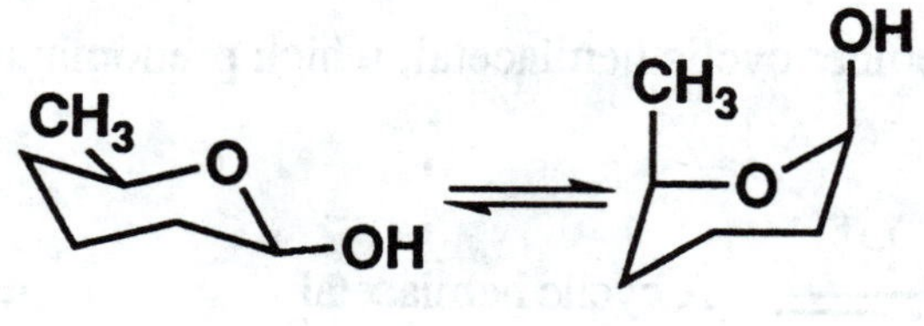

**Alternative chair conformation of (A)** 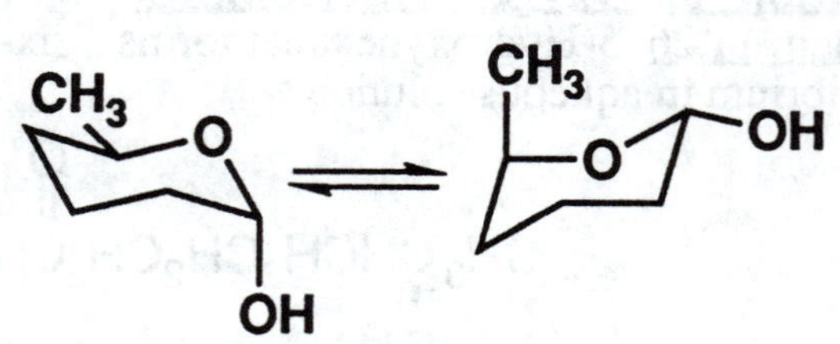        **Alternative chair conformations of (C)**

**Problem 11.27** Draw structural formulas for the hemiacetal and then the acetal formed from each pair of reactants in the presence of an acid catalyst.

(a)

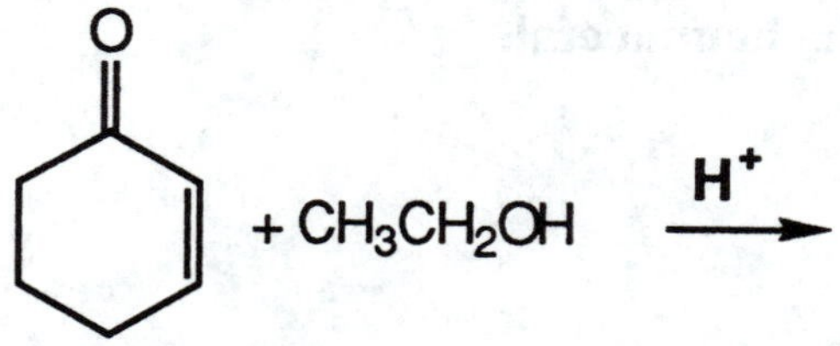

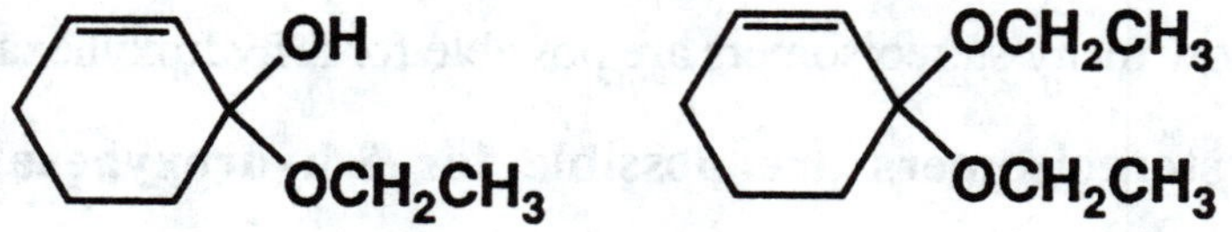

(b)

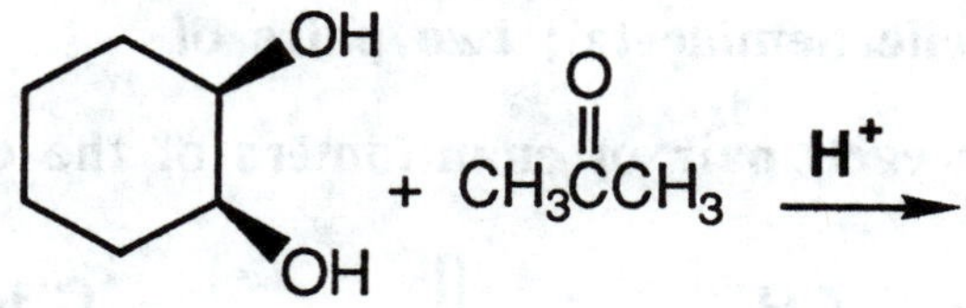

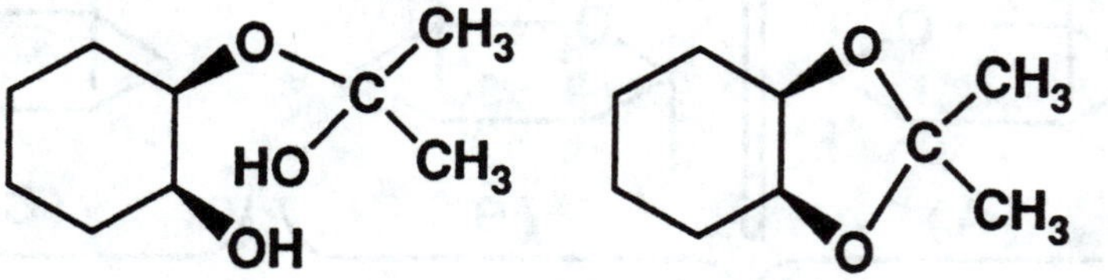

(c) $CH_3CH_2CH_2\overset{O}{\overset{\|}{C}}H$ + $CH_3OH$ $\xrightarrow{H^+}$

$$CH_3CH_2CH_2\overset{OH}{\underset{|}{C}H}OCH_3 \quad CH_3CH_2CH_2\overset{OCH_3}{\underset{|}{C}H}OCH_3$$

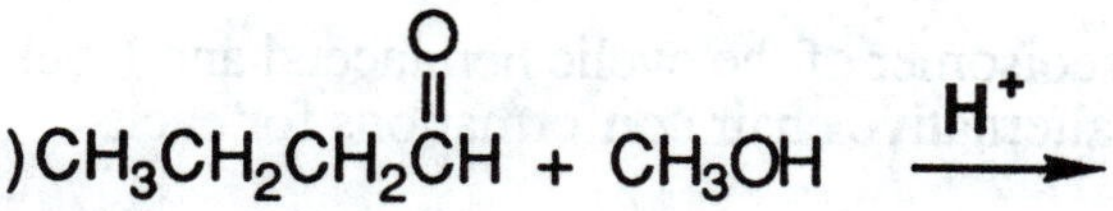

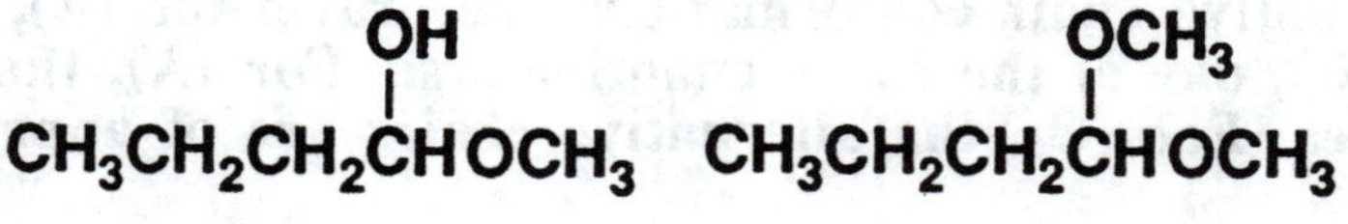

Problem 11.28 Draw structural formulas for the products of hydrolysis of the following acetals:

(a)

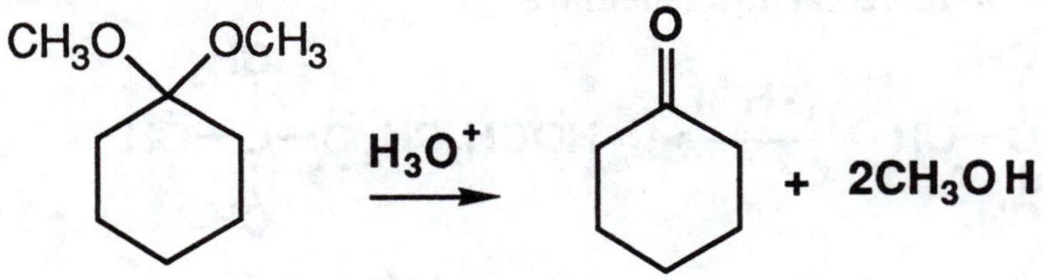

CH3O
OCH3
H3O+
O
+ 2CH3OH


(b)

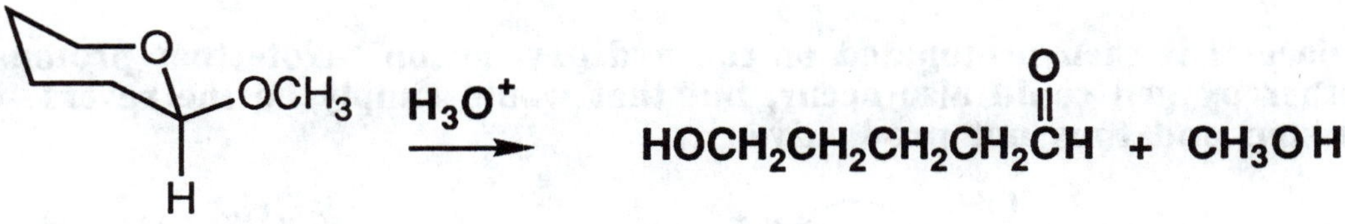

O
OCH3
H
H3O+
O
HOCH2CH2CH2CH2CH + CH3OH


(c)

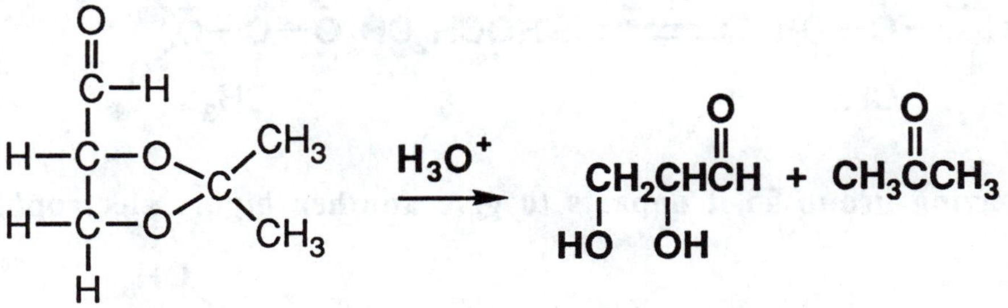

O
C–H
H–C–O
H–C–O
H
C
CH3
CH3
H3O+
O
CH2CHCH
HO OH
+
O
CH3CCH3


Problem 11.29 Propose a mechanism for formation of the cyclic acetal from treatment of acetone with ethylene glycol in the presence of an acid catalyst. Your mechanism must be consistent with the fact that the oxygen atom of the water molecule is derived from the carbonyl oxygen atom of acetone.

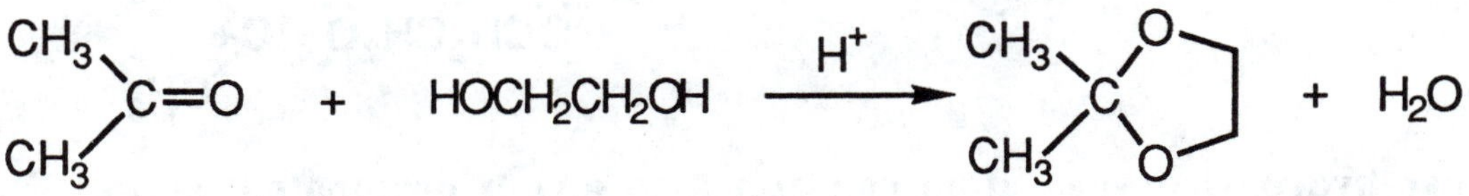

CH3
C=O
CH3
+
HOCH2CH2OH
H+
CH3
C
CH3
O
O
+ H2O


**The first step of the mechanism involves protonation of the carbonyl oxygen atom to form a reactive electrophilic species.**

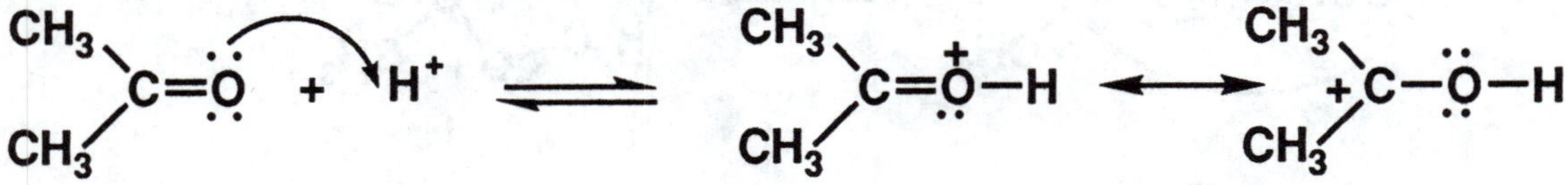

CH3
C=O
CH3
+ H+
CH3
C=O–H
CH3
CH3
+C–O–H
CH3


**The nucleophilic oxygen atoms of ethylene glycol then attack the protonated carbonyl species.**

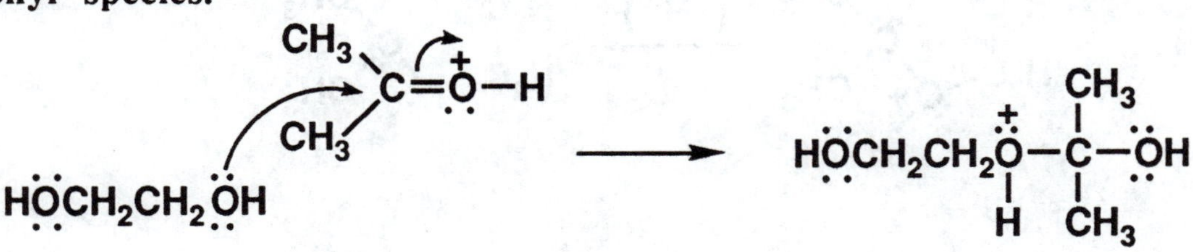

CH3
C=O–H
CH3
HOCH2CH2OH
CH3
HOCH2CH2O–C–OH
H CH3

**A proton is lost to give the hemiacetal intermediate**

$$HOCH_2CH_2\overset{+}{O}-\underset{\underset{CH_3}{|}}{\overset{\overset{CH_3}{|}}{C}}-\ddot{O}H \xrightarrow{(-H^+)} HOCH_2CH_2\ddot{O}-\underset{\underset{CH_3}{|}}{\overset{\overset{CH_3}{|}}{C}}-\ddot{O}H$$

A hemiacetal

**The hemiacetal is then protonated on the hydroxyl group. Note that protonation on the ether oxygen could also occur, but that would simply be the reverse of the previous step and thus not productive.**

$$H\ddot{O}CH_2CH_2\ddot{O}-\underset{\underset{CH_3}{|}}{\overset{\overset{CH_3}{|}}{C}}-\ddot{O}H \overset{H^+}{\underset{}{\rightleftharpoons}} H\ddot{O}CH_2CH_2\ddot{O}-\underset{\underset{CH_3}{|}}{\overset{\overset{CH_3}{|}}{C}}-\overset{+}{\underset{H}{\overset{H}{O}}}$$

**Water is a good leaving group so it departs to give another highly electrophilic species.**

$$H\ddot{O}CH_2CH_2\ddot{O}-\underset{\underset{CH_3}{|}}{\overset{\overset{CH_3}{|}}{C}}-\overset{+}{\underset{H}{\overset{H}{O}}} \xrightarrow{(-H_2O)} \begin{array}{c} H\ddot{O}CH_2CH_2\overset{+}{\ddot{O}}=C\overset{CH_3}{\underset{CH_3}{\big\langle}} \\ \updownarrow \\ H\ddot{O}CH_2CH_2\ddot{O}-\overset{CH_3}{\underset{CH_3}{C+}} \end{array}$$

**The other hydroxyl oxygen atom can attack to give a protonated, cyclic intermediate.**

$$HOCH_2CH_2\overset{+}{\ddot{O}}=C\overset{CH_3}{\underset{CH_3}{\big\langle}} \longrightarrow \underset{\underset{H}{|}}{\overset{}{\begin{array}{c} H_2C-\overset{\ddot{O}:}{}-CH_3 \\ H_2C-\overset{+}{\underset{}{O}}-CH_3 \end{array}}}$$

**Loss of a proton gives the final cyclic acetal product.**

$$\underset{\underset{H}{|}}{\begin{array}{c} H_2C-\overset{\ddot{O}:}{}-CH_3 \\ H_2C-\overset{+}{O}-CH_3 \end{array}} \xrightarrow{(-H^+)} \begin{array}{c} H_2C-\overset{\ddot{O}:}{}-CH_3 \\ H_2C-\ddot{O}:-CH_3 \end{array}$$

**Problem 11.30** Propose a mechanism for the formation of a cyclic acetal from 4-hydroxypentanal and one equivalent of methanol. If the carbonyl oxygen of 4-hydroxypentanal is enriched with oxygen-18, do you predict that the oxygen label appears in the cyclic acetal or in the water?

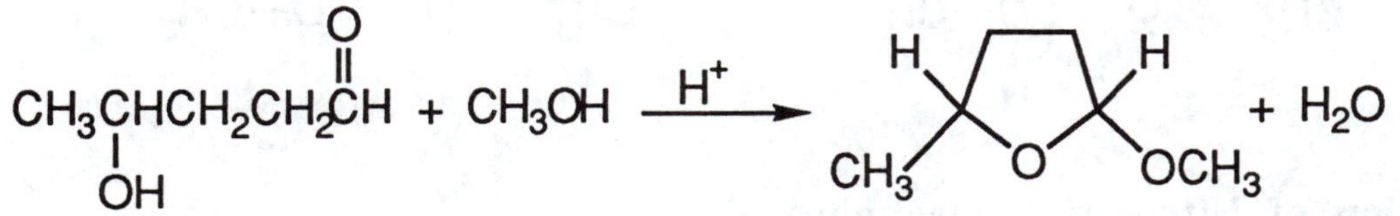

4-Hydroxypentanal                          A cyclic acetal

**Propose formation of a hemiacetal followed by protonation of the hemiacetal-OH and loss of water to form a resonance-stabilized cation. Then propose a Lewis acid-base reaction between this cation and methanol followed by loss of a proton to give the product. If the carbonyl group of 4-hydroxypentanal is enriched with oxygen-18, the oxygen-18 label appears in the water.**

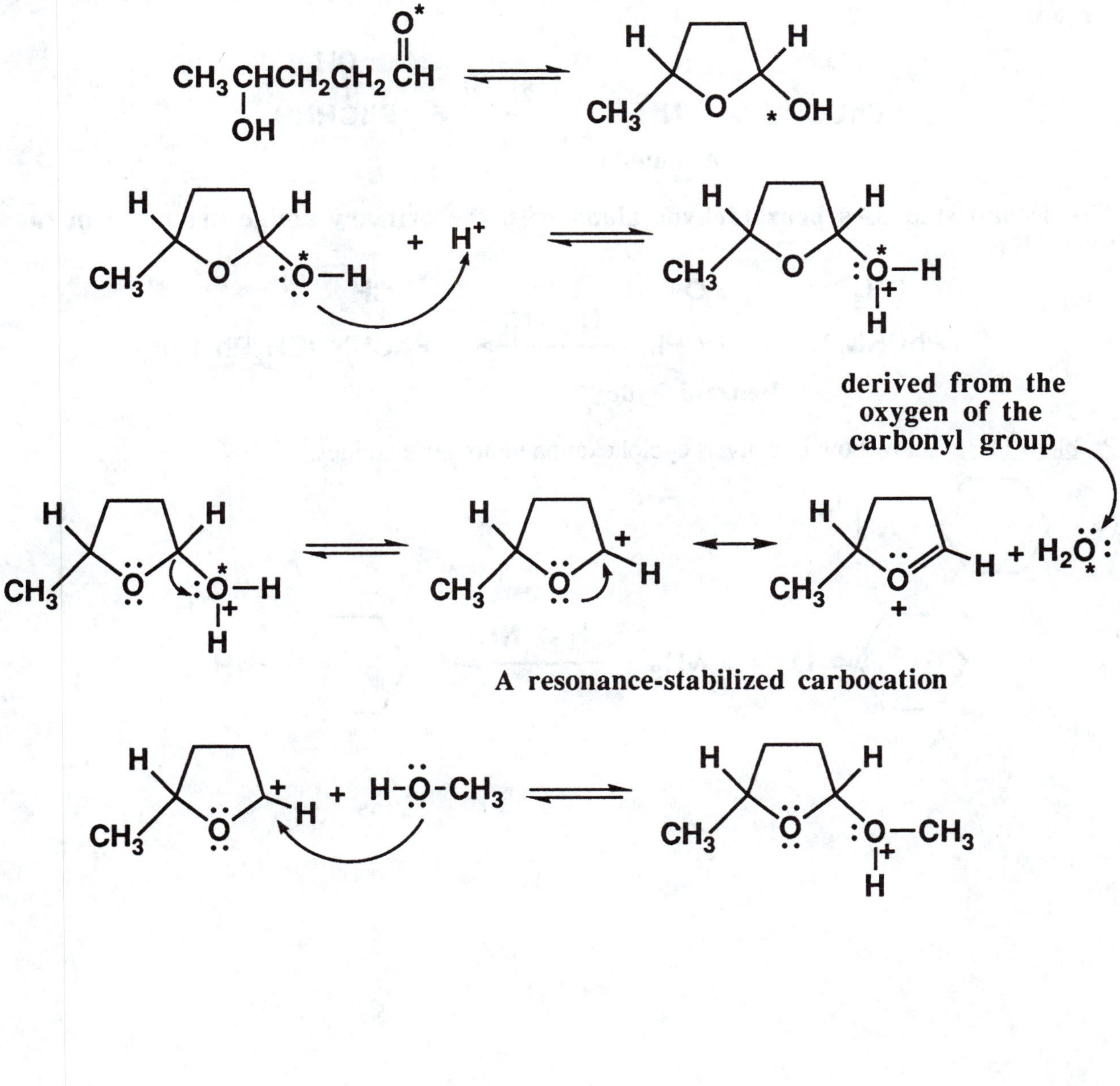

derived from the
oxygen of the
carbonyl group

**A resonance-stabilized carbocation**

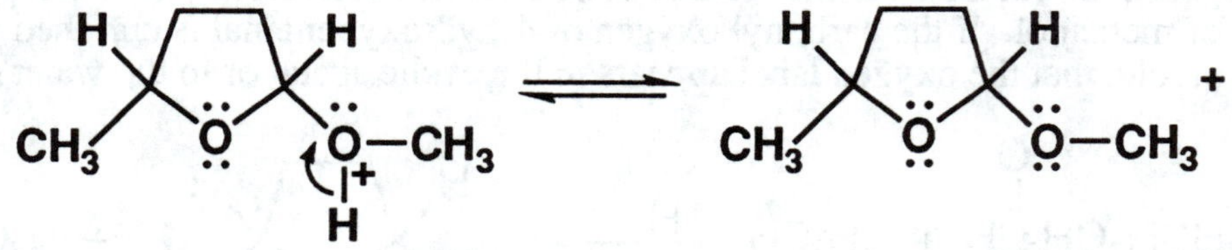

**Addition of Nitrogen Nucleophiles**

Problem 11.31 Show how this secondary amine can be prepared by two successive reductive aminations:

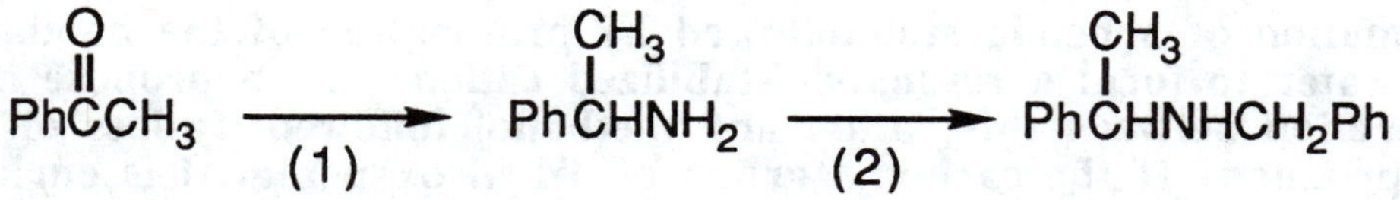

**The first step involves a reductive amination of the ketone shown along with ammonia.**

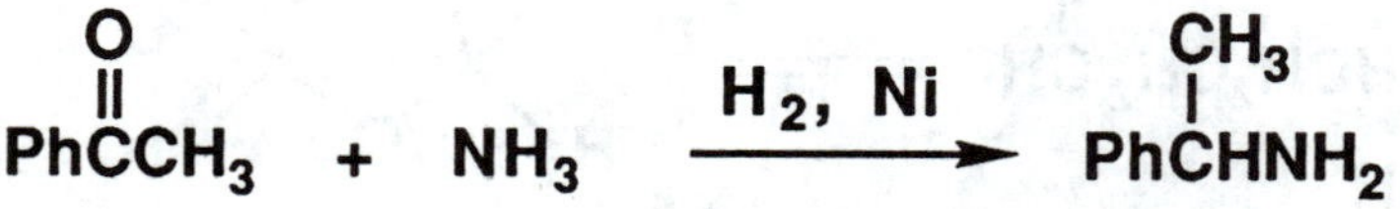

**Ammonia**

**The second step uses benzaldehyde along with the primary amine produced in the first step.**

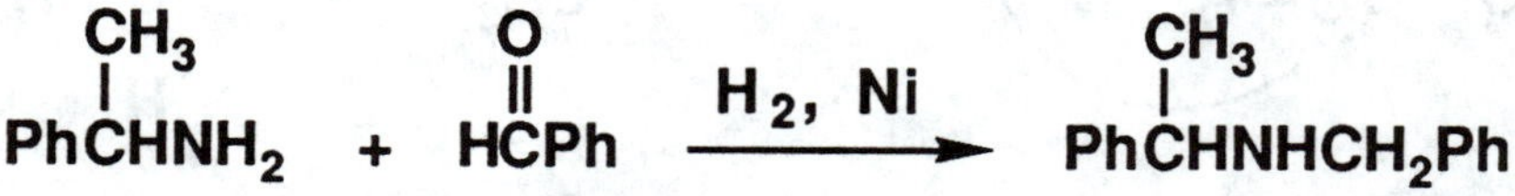

**Benzaldehyde**

Problem 11.32 Show how to convert cyclohexanone into these amines.

(a) 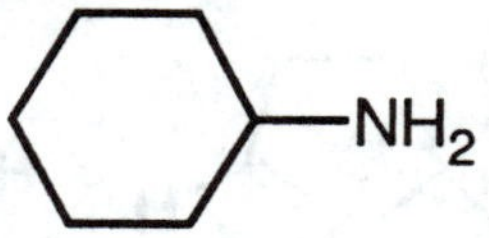

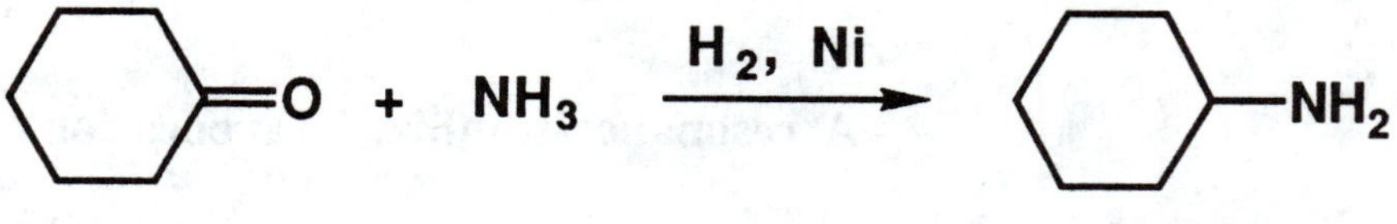

(b)

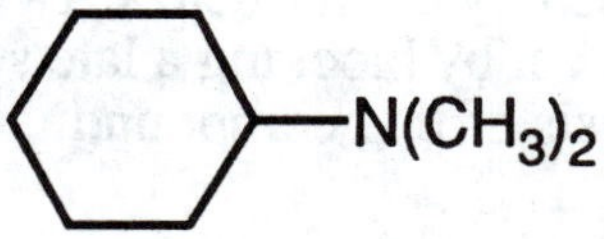

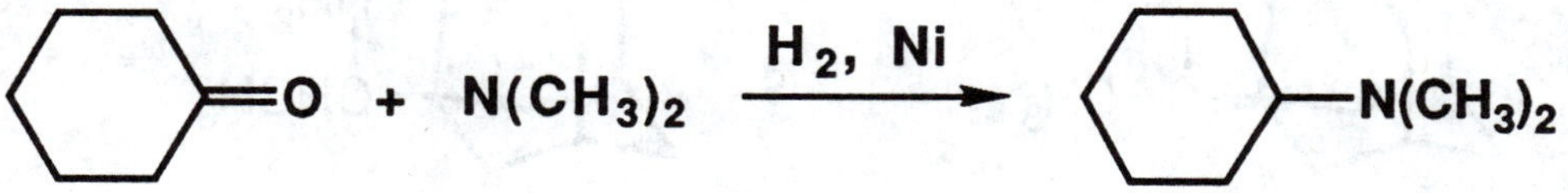

(c)

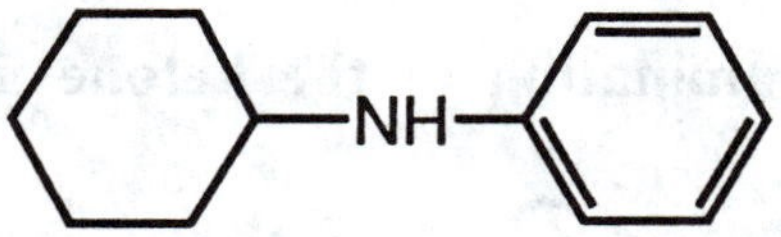

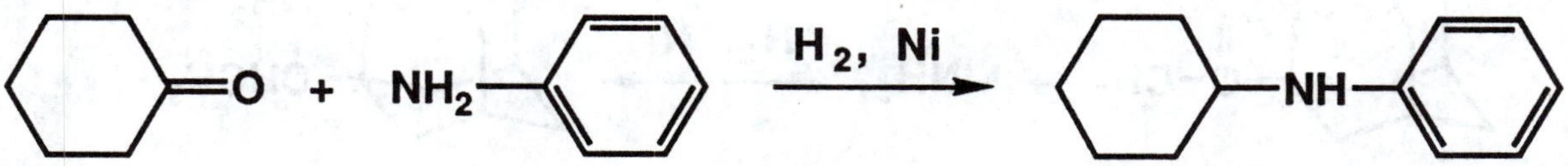

**Problem 11.33** Following are structural formulas for amphetamine, methylamphetamine, and fenfluramine. The major central nervous system effects of amphetamine and amphetamine-like drugs are locomotor stimulation, euphoria and excitement, stereotyped behavior, and anorexia. Show how each of these drugs can be synthesized by reductive amination of an appropriate aldehyde or ketone and amine.

(a)

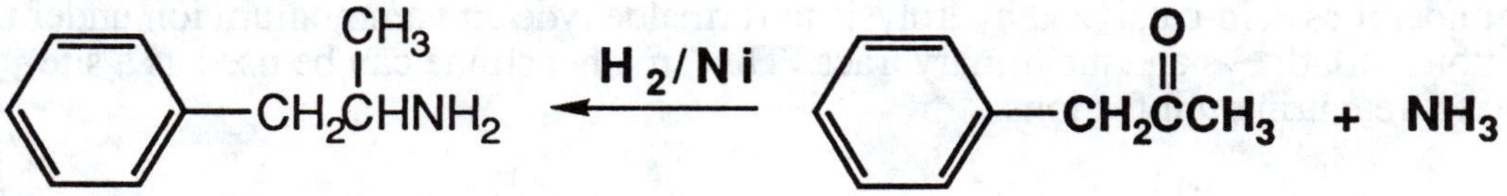

Amphetamine

(b)

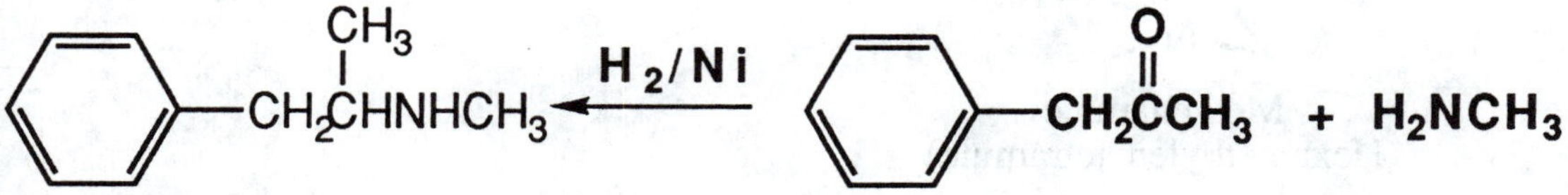

Methylamphetamine

(c)

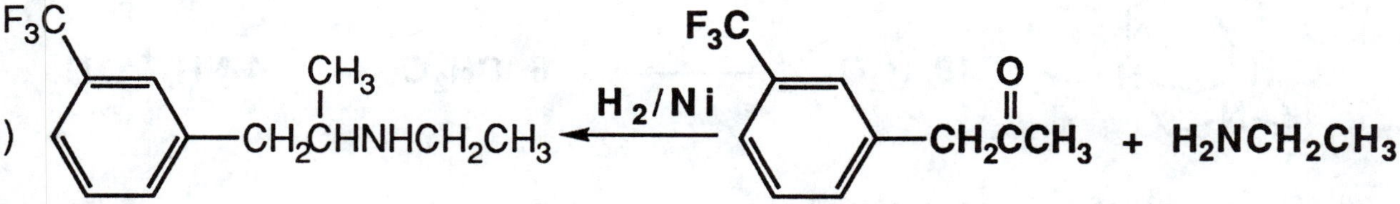

Fenfluramine

Problem 11.34 Rimantadine is effective in preventing infections caused by the influenza A virus and in treating established illness. It is thought to exert its antiviral effect by blocking a late stage in the assembly of the virus. Following is the final step in the synthesis of this compound. Describe experimental conditions to bring about this conversion.

Rimantadine
(an antiviral agent)

**The final step in the synthesis involves a reductive amination of the ketone using ammonia.**

Problem 11.35 Methenamine, a product of the reaction of formaldehyde and ammonia, is an example of a prodrug, a compound that is inactive itself but is converted to an active drug in the body by a biochemical transformation. The strategy behind use of methenamine as a prodrug is that nearly all bacteria are sensitive to formaldehyde at concentrations of 20 mg/mL or higher. Formaldehyde cannot be used directly in medicine, however, because an effective concentration in plasma cannot be achieved with safe doses. Methenamine is stable at pH 7.4 (the pH of blood plasma) but undergoes acid-catalyzed hydrolysis to formaldehyde and ammonium ion under the acidic conditions of kidneys and the urinary tract. Thus, methenamine can be used as a site-specific drug to treat urinary infections.

Methenamine
(Hexamethylenetetramine)

(a) Balance the equation for the hydrolysis of methenamine to formaldehyde and ammonium ion.

(b) Does the pH of an aqueous solution of methenamine increase, remain the same, or decrease as a result of hydrolysis? Explain.

**When methenamine is hydrolyzed, ammonia is released. Ammonia is a base so the pH will increase.**

(c) Explain the meaning of the following statement: The functional group in methenamine is the nitrogen analog of an acetal.

**With an acetal, a single carbon atom is bonded to two oxygen atoms. In the case of methenamine, each carbon atom is bonded to two nitrogen atoms.**

(d) Account for the observation that methenamine is stable in blood plasma, but undergoes hydrolysis in the urinary tract.

**Blood plasma is buffered to the slightly basic pH of 7.4. Methenamine is relatively stable to hydrolysis at this pH, since it is stable to base. Recall that acetals are also stable to base. On the other hand, both methenamine and acetals are readily hydrolyzed at acidic pH. The urinary tract is more acidic, so the methenamine is hydrolyzed more rapidly there.**

### Keto-Enol Tautomerism
<u>Problem 11.36</u> The following molecule belongs to a class of compounds called enediols; each carbon of the double bond carries an -OH group. Draw structural formulas for the α-hydroxyketone and the α-hydroxyaldehyde with which this enediol is in equilibrium.

$$\alpha\text{-Hydroxyaldehyde} \rightleftharpoons \underset{\underset{CH_3}{|}}{\overset{\overset{CHOH}{\|}}{C}}-OH \rightleftharpoons \alpha\text{-Hydroxyketone}$$

An enediol

**Following are formulas for the α-hydroxyaldehyde and α-hydroxyketone in equilibrium by way of the enediol intermediate.**

**α-Hydroxy-          Enediol          α-Hydroxy-**
**aldehyde                            ketone**

<u>Problem 11.37</u> Account for the fact that in dilute aqueous base, (R)-glyceraldehyde is converted into an equilibrium mixture of (R,S)-glyceraldehyde and dihydroxyacetone.

(R)-Glyceraldehyde          (R,S)-Glyceraldehyde          (Dihydroxyacetone)

The key is keto-enol tautomerism. In the presence of base (R)-glyceraldehyde undergoes base-catalyzed keto-enol tautomerism to form an enediol in which carbon-2 is achiral. This enediol is in turn in equilibrium with (S)-glyceraldehyde and dihydroxyacetone.

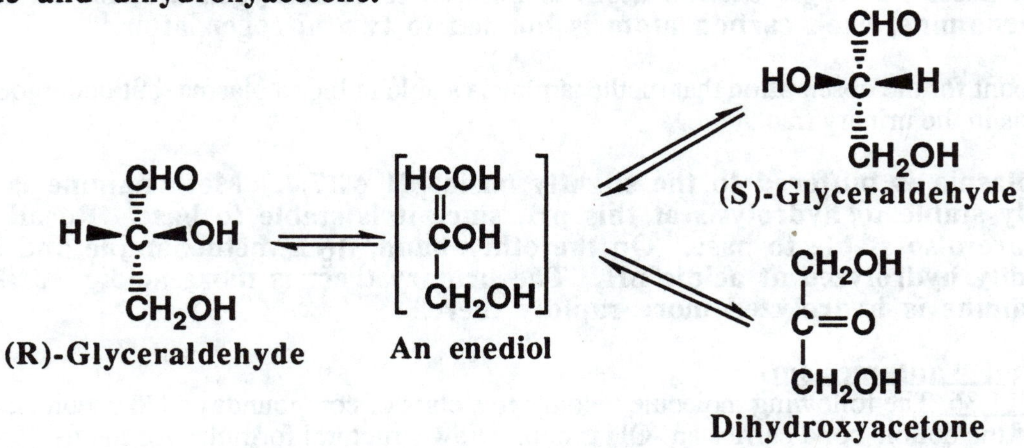

**(R)-Glyceraldehyde**          **An enediol**          **(S)-Glyceraldehyde**

**Dihydroxyacetone**

## Oxidation/Reduction of Aldehydes and Ketones

**Problem 11.38** Draw structural formulas for the products formed by treatment of butanal with the following reagents.

(a)      LiAlH$_4$ followed by H$_2$O

CH$_3$CH$_2$CH$_2$CH$_2$OH

(b)      NaBH$_4$ in CH$_3$OH/H$_2$O

CH$_3$CH$_2$CH$_2$CH$_2$OH

(c)      H$_2$/Pt

CH$_3$CH$_2$CH$_2$CH$_2$OH

(d)      Ag(NH$_3$)$_2$$^+$ in NH$_3$/H$_2$O then HCl/H$_2$O

CH$_3$CH$_2$CH$_2$CO$_2$H

(e)      K$_2$Cr$_2$O$_7$/H$_2$SO$_4$

CH$_3$CH$_2$CH$_2$CO$_2$H

(f) C$_6$H$_5$NH$_2$ in the presence of H$_2$/Ni

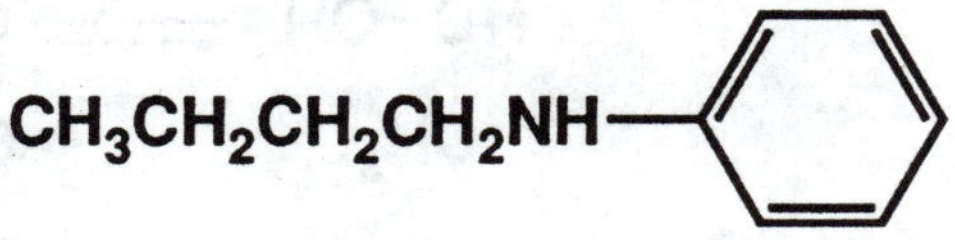

CH$_3$CH$_2$CH$_2$CH$_2$NH—⟨phenyl⟩

**Problem 11.39** Draw structural formulas for the products for the reaction of *p*-bromoacetophenone with each reagent given in Problem 11.38.

**Following are structural formulas for each product.**

(a)

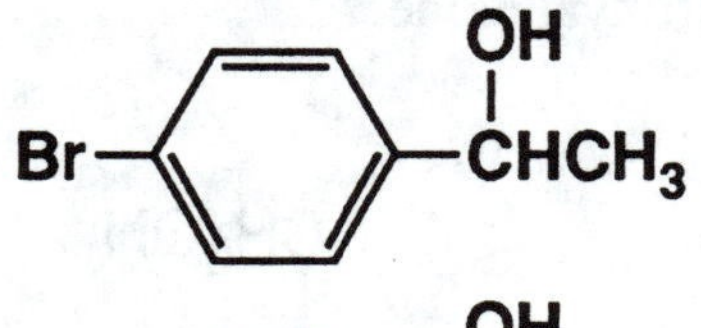

(b)

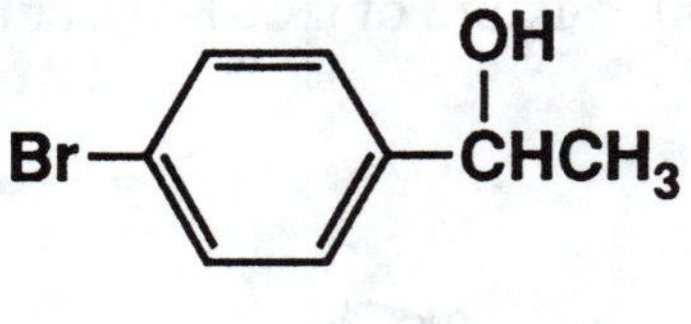

(c)

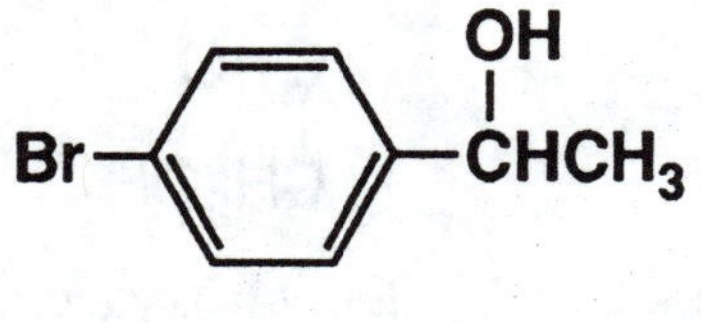

(d)   no reaction

(e) 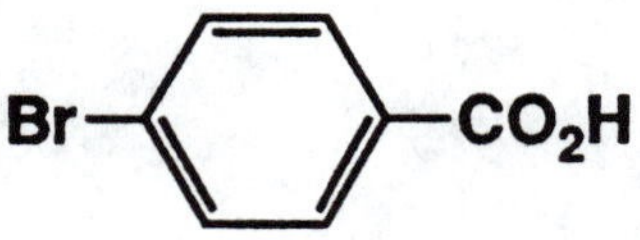  Br—⟨benzene ring⟩—$CO_2H$          (f) 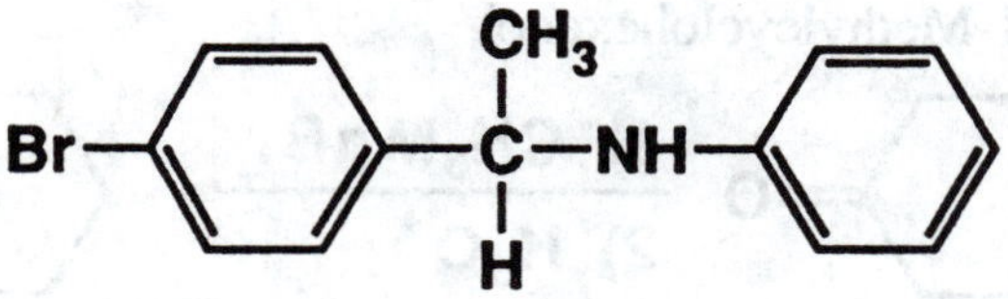  Br—⟨benzene ring⟩—$\overset{CH_3}{\underset{H}{C}}$—NH—⟨benzene ring⟩

## Synthesis

Problem 11.40  Show reagents and conditions to bring about the conversion of cyclohexanol to cyclohexanecarbaldehyde.

**The appropriate reagents are written in the following scheme.**

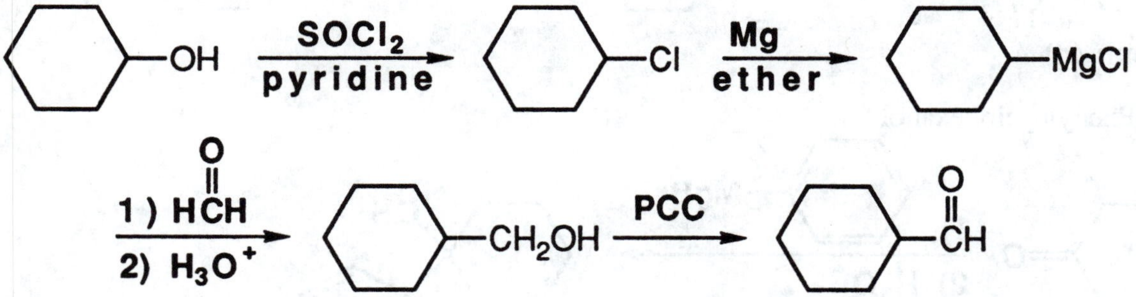

⟨cyclohexane⟩—OH  $\xrightarrow[\textbf{pyridine}]{\textbf{SOCl}_2}$  ⟨cyclohexane⟩—Cl  $\xrightarrow[\textbf{ether}]{\textbf{Mg}}$  ⟨cyclohexane⟩—MgCl

$\xrightarrow[\textbf{2) H}_3\textbf{O}^+]{\textbf{1) H}\overset{O}{\overset{\|}{C}}\textbf{H}}$  ⟨cyclohexane⟩—$CH_2OH$  $\xrightarrow{\textbf{PCC}}$  ⟨cyclohexane⟩—$\overset{O}{\overset{\|}{C}}H$

Problem 11.41  Starting with cyclohexanone, show how to prepare these compounds.  In addition to the given starting material, use any other organic or inorganic reagents as necessary.
(a) Cyclohexanol

**This transformation can be accomplished with any of three different sets of reagents:**

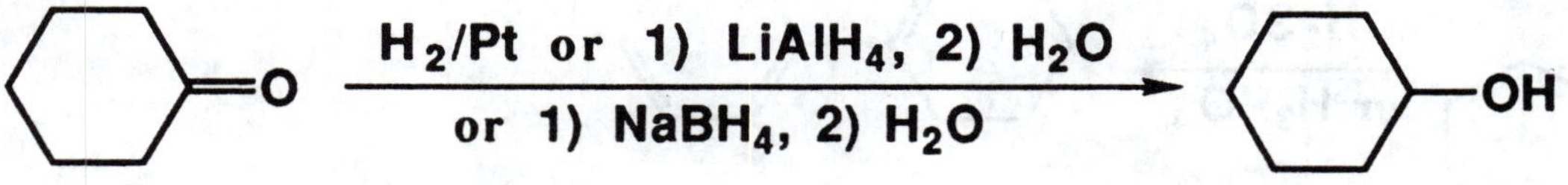

⟨cyclohexanone⟩=O  $\xrightarrow[\textbf{or 1) NaBH}_4\textbf{, 2) H}_2\textbf{O}]{\textbf{H}_2\textbf{/Pt or 1) LiAlH}_4\textbf{, 2) H}_2\textbf{O}}$  ⟨cyclohexane⟩—OH

(b) Cyclohexene

⟨cyclohexane⟩—OH  $\xrightarrow[\textbf{or H}_3\textbf{PO}_4]{\textbf{H}_2\textbf{SO}_4}$  ⟨cyclohexene⟩

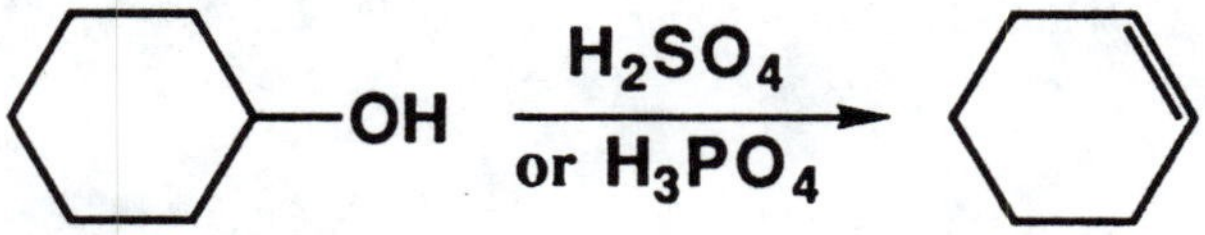

**From (a)**

(c) *cis*-1,2-Cyclohexanediol

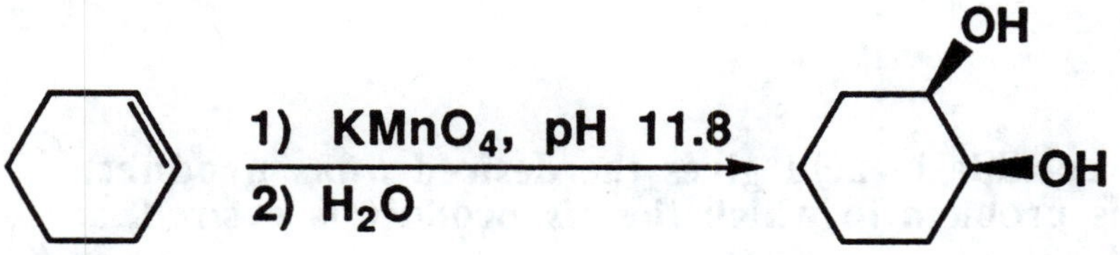

⟨cyclohexene⟩  $\xrightarrow[\textbf{2) H}_2\textbf{O}]{\textbf{1) KMnO}_4\textbf{, pH 11.8}}$  ⟨cyclohexane with OH and OH⟩

**From (b)**

(d) 1-Methylcyclohexanol

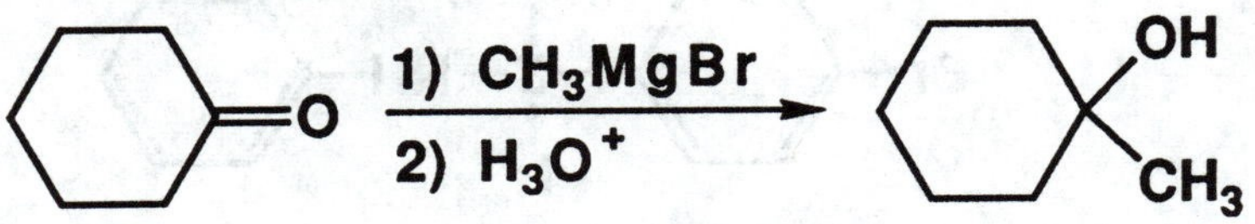

(e) 1-Methylcyclohexane

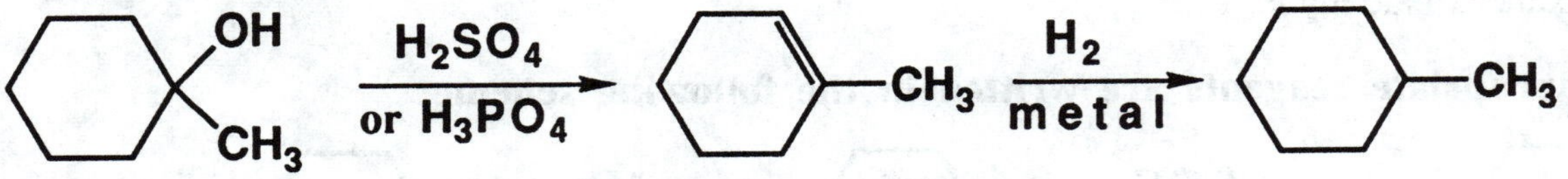

**From (d)**

(f) 1-Phenylcyclohexanol

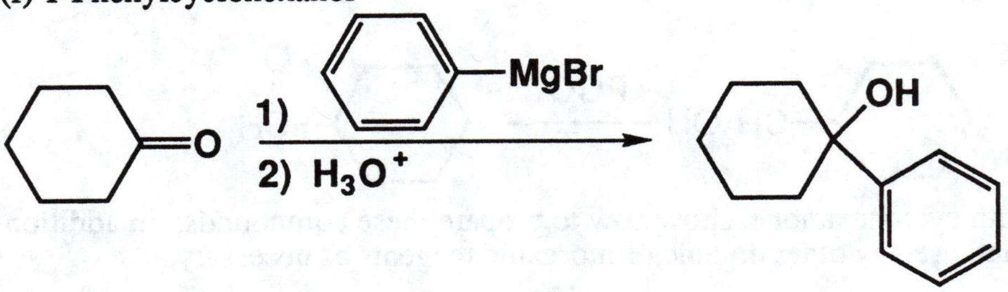

(g) 1-Phenylcyclohexene

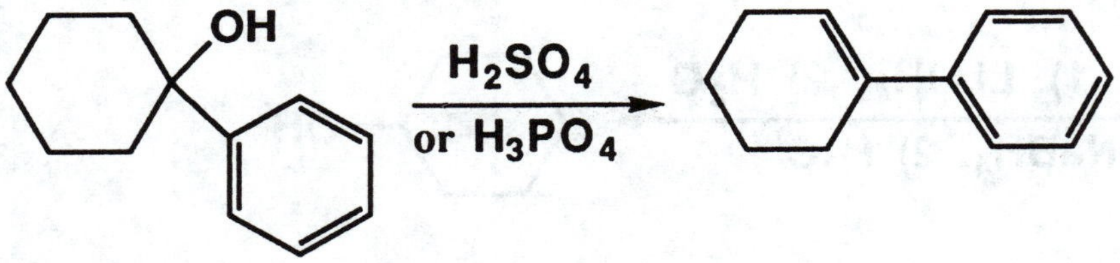

**From (f)**

(h) Cyclohexene oxide

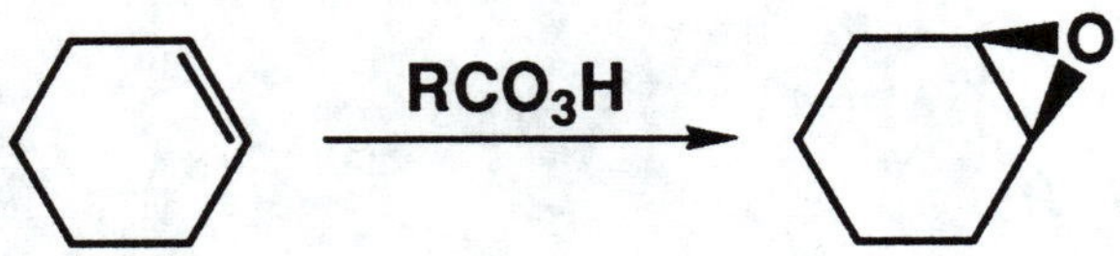

**From (b)**

(i) *trans*-1,2-Cyclohexanediol

**Recall that ring-opening of an epoxide in acid gives the desired *trans* product. Compare this to part (c) of this problem in which the *cis* product is desired, so that KMnO₄ and H₂O are used.**

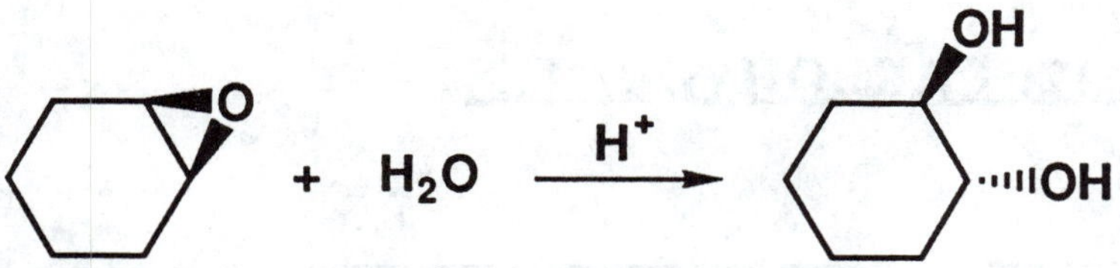

**From (h)**

Problem 11.42 Show how to bring about these conversions. In addition to the given starting material, use any other organic or inorganic reagents as necessary.

(a)

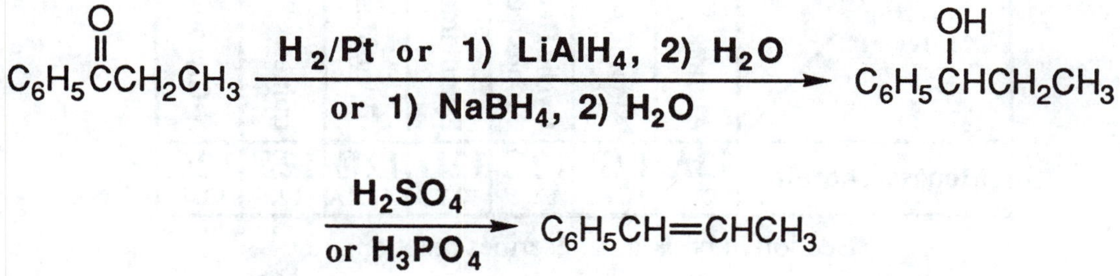

(b)

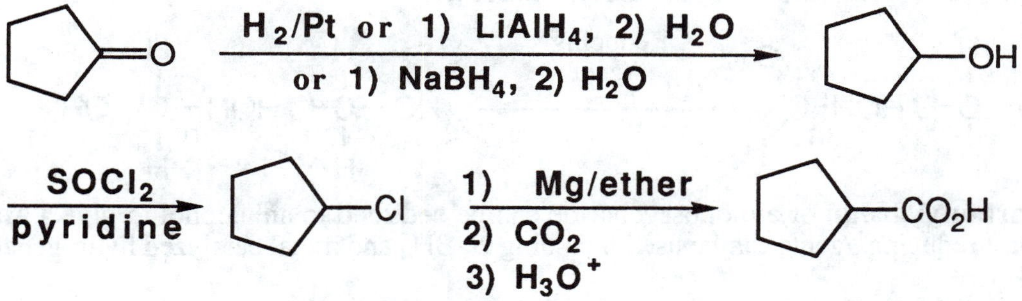

(c)

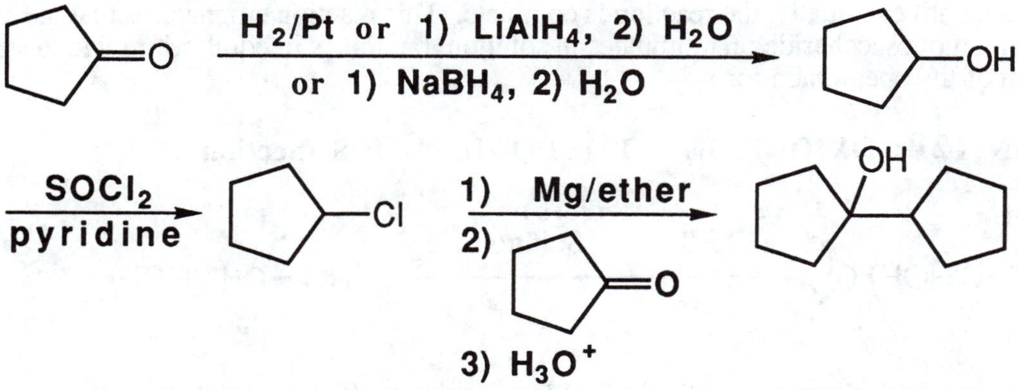

(d)

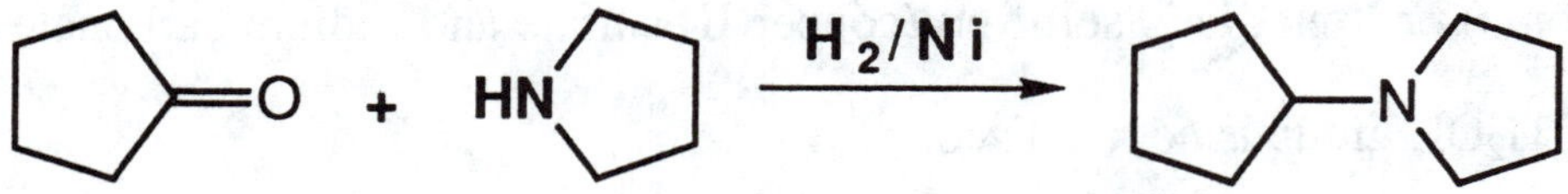

# CHAPTER 12: CARBOHYDRATES

## SUMMARY OF REACTIONS

| Starting Material ↓ / Product → | Alditols | Aldonic Acids | Cyclic Hemiacetal Monosaccharides | Enantiomers | Glycosides | Imine | N-Glycosides |
|---|---|---|---|---|---|---|---|
| Monosaccharides | 12A 12.4B* | 12B 12.4C | 12C 12.2 | 12D 12.2D | 12E 12.4A | 12F 12.5 | 12G 12.4A |

*Section in book that describes reaction.

## REACTION 12A: REDUCTION (Section 12.4B)

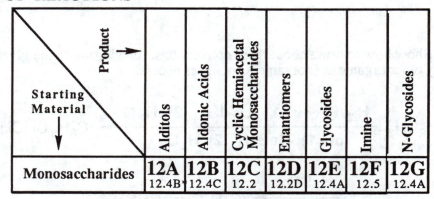

- The **carbonyl group** of a monosaccharide can be **reduced** to an alcohol to give a **alditol**.
- Different reducing agents can be used including $NaBH_4$ and metal catalyzed hydrogenation with $H_2$. ✱
- The reduction actually occurs with the small amount of open-chain sugar that is present at any one time, until eventually, the reaction is complete. This reaction emphasizes that the cyclic form of a monosaccharide predominates in solution, but this is in equilibrium with a small amount of the open-chain form.

## REACTION 12B: OXIDATION TO ALDONIC ACIDS (Section 12.4C)

- **Monosaccharides** can be oxidized to **aldonic** acids by **Tollens' solution** (Silver ion in aqueous ammonia) or **Benedict's solution** (copper(II) sulfate and sodium carbonate in citrate buffer). ✱
- The terminal -$CH_2OH$ group is not oxidized.

## REACTION 12C: FORMATION OF CYCLIC HEMIACETALS (Section 12.2)

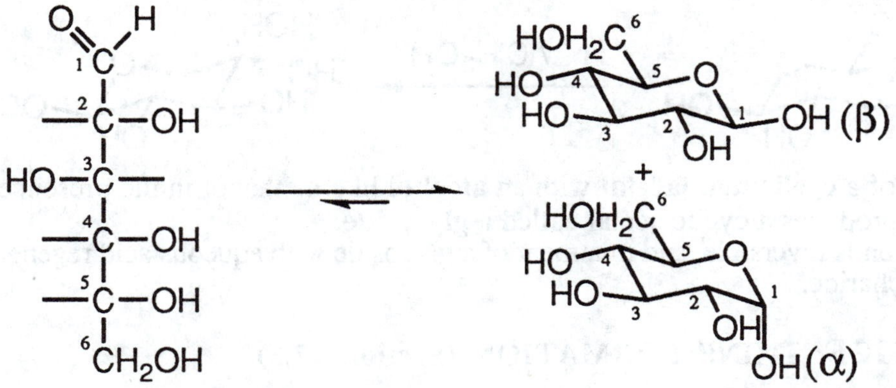

- **Monosaccharides** form **cyclic hemiacetals**, and at equilibrium, the **cyclic forms predominate** to such an extent that only a small amount of the open chain form is present at any one time. Nevertheless, it is important to remember that this equilibration exists, because sometimes the small amount of open chain form can be important in a reaction mechanism. See for example, mutarotation (reaction 12D, section 12.2D). ✳
- Six member rings are designated by the infix **pyran** and five-member rings are designated by the infix **furan**. For example, the cyclic form of glucose is referred to as a D-glucopyranose, and can be either the α or β form, as shown in the figure above.

## REACTION 12D: MUTAROTATION (Section 12.2D)

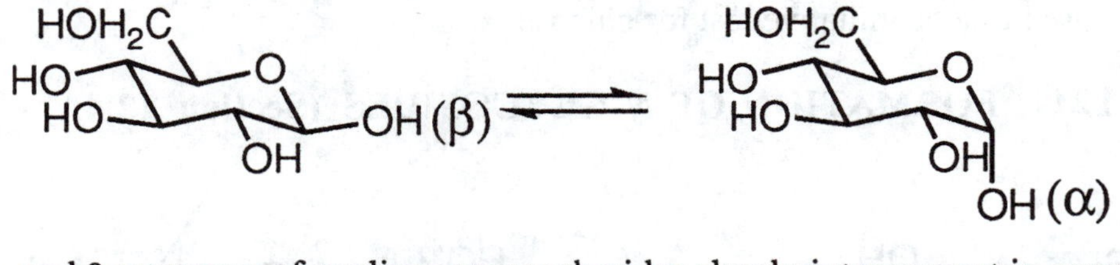

- The α- and β-**anomers** of cyclic monosaccharides slowly interconvert in aqueous solution, and the process is known as **mutarotation**. ✳
- The mechanism of mutarotation involves the open chain form as an intermediate, that quickly recyclizes to the different cyclic forms.
- A number of generalizations can be made about which forms of a given monosaccharide will predominate at equilibrium.
   Little free aldehyde or ketone is present.
   Pyranose forms (six-member rings) predominate over possible furanose forms (five-member rings). Note that here we are referring to the free monosaccharide in solution, but in special biological structures such as nucleic acids, the furanose form of ribose or deoxyribose is found.
   In pyranose forms, the diastereomer that has the larger group equatorial on the anomeric carbon atom predominates.

## REACTION 12E: FORMATION OF GLYCOSIDES (Section 12.4A)

- Reaction of a **cyclic hemiacetal** with an **alcohol** like methanol in the presence of an **acid catalyst** produces a cyclic acetal called a **glycoside**. ✳
- The reaction is reversible, and treatment of a glycoside with aqueous acid regenerates the free monosaccharide.

## REACTION 12F: IMINE FORMATION (Section 12.5)

- **Carbohydrates** such as glucose react with **amines** such as *o*-toluidine to **form imines**. In the case of *o*-toluidine, the imine formed has a blue green color, so this reaction is commonly used as a quantitative test for glucose. ✳

## REACTION 12G: FORMATION OF *N*-GLYCOSIDES (Section 12.4A)

- *N*-glycosides can be formed from a **cyclic monosaccharide** with a compound containing an **N-H bond** to form *N*-glycosides. ✳
- The nucleic acids are based on D-ribose or 2-deoxy-D-ribose, in the furanose form, as the *N*-glycosides with the aromatic bases uracil, cytosine, thymine, adenine, or guanine. In DNA and RNA, these *N*-glycosides are found exclusively as the β-anomers.

## SUMMARY OF IMPORTANT CONCEPTS

### 12.0 OVERVIEW
- **Carbohydrates** are **polyhydroxylated aldehydes** or **ketones**, or compounds that produce polyhydroxylated aldehydes or ketones upon hydrolysis. ✳
- **Carbohydrates** are the most **abundant organic molecules** in the world. They are essential to all forms of life, and perform such functions as energy storage (glucose, starch, glycogen),

structural reinforcement (cellulose), and genetic information storage as components of nucleic acids (DNA, RNA).

## 12.1 MONOSACCHARIDES

• **Monosaccharides** usually have molecular formulas of $C_nH_{2n}O_n$ **(3<n<8)**, and are the **monomers** from which **larger carbohydrates** are constructed. ✱

• Monosaccharides are named by using the suffix **ose**. The prefixes **tri, tetr, pent** are used to indicate 3, 4, or 5 carbon atoms, respectively. An **aldehyde carbohydrate** is called an **aldose**, and is sometimes designated with an **aldo** prefix. A **ketone carbohydrate** is called a **ketose**, and is sometimes designated with a **keto** prefix. For example, glyceraldehyde is an aldotriose and fructose is a ketohexose.

  - The **nomenclature** of monosaccharides is **dominated by common names**. Even though IUPAC names can be derived for each monosaccharide, the common names are much simpler and used almost exclusively. ✱ *[Unfortunately, the system of common names is only slightly systematic, and the names must be simply memorized along with the corresponding structures.]*

• Monosaccharides usually have one or more stereocenters, so **stereochemistry** is of **major importance** to the study of **monosaccharides**. A **Fischer projection** of a monosaccharide is used to show the structure of a monosaccharide and thus keep track of stereochemistry.

  - In a **Fischer projection**, the monosaccharide is usually drawn in the open chain form, and the carbonyl carbon atom is placed at the top of the structure. The **carbon atoms** of **stereocenters** are **not labeled**. They are assumed to be located at the crossing points of bonds. *[Even though it is understood that the cyclic hemiacetal predominates at equilibrium, the open chain Fischer projection is used for clarity]*

  - Since the $sp^3$ carbon atoms are tetrahedral, a monosaccharide that is stretched out analogous to a Fisher projection will have some groups projecting forward, and some projecting backward. In a **Fischer projection**, the **horizontal lines represent the groups** that are **projecting above the plane of the paper**, and **vertical lines** are used to **represent groups** that are **projecting below the plane of the page**. *[Although very difficult to master at first, Fisher projections are useful for understanding and describing the structures of monosaccharides. The only way to get good at using them is to practice.]* ✱

  - The overall **stereochemistry** of monosaccharides is **classified** as **D** or **L**, **based** on a **comparison to glyceraldehyde stereochemistry**. In a monosaccharide, the **point of reference** is the **stereocenter** that is **farthest from** the **carbonyl group**. Since this is a carbon atom that is next to the last carbon atom in the chain (notice that the last carbon atom of the chain has two -H atoms, so it is not a stereocenter) it is referred to as the **penultimate carbon atom**. A monosaccharide that has the **same configuration** about the **penultimate carbon** as **D-glyceraldehyde** is classified as a **D monosaccharide**. In this case the -OH group will be on the right side of the carbon atom in the Fischer projection. Similarly, an **L monosaccharide** has a **configuration** about the **penultimate carbon atom** that is the same as the configuration of **L-glyceraldehyde**, with the -OH group being on the left in a Fischer projection.

  - The enantiomer of a given monosaccharide is not produced by simply changing the configuration of the penultimate carbon atom, but rather by reversing all of the stereocenters. *[This may seem obvious, but it is worth keeping in mind as the different sugars are examined.]*

• Some sugars have an **amino group** (-NH$_2$) **instead of all -OH groups**, and these are called **amino sugars**. Amino sugars are much less common that normal carbohydrates, but important examples include D-glucosamine and D-galactosamine.

## 12.2 THE CYCLIC STRUCTURE OF MONOSACCHARIDES

• The **open chain monosaccharides** are in equilibrium with a **cyclic hemiacetal structure** (Reaction 12C, Section 12.2). The cyclic hemiacetal is greatly favored and thus is found in large excess at equilibrium.

• There are two diastereomers possible, and these are referred to a **anomers**. The two anomers are distinguished by the relative orientation of the **anomeric -OH** group (the -OH group on the so-called **anomeric carbon atom**, the one that was a carbonyl in the open chain form). ✳

• The two anomers are named as α or β. The α–anomer is the one that has the anomeric **-OH group** on the **same side** of the **Fischer projection** as the **-OH group** on the **penultimate carbon** [*Remember that the -OH group on the penultimate carbon atom is the one that determines whether a monosaccharide is named as D or L*]. The β-anomer has the anomeric -OH group and the -OH group of the penultimate carbon atom on opposites sides in the Fischer projection. With D-glucose in the cyclic hemiacetal form, the α-anomer is the one with the anomeric -OH group is axial, while for the β-anomer the anomeric -OH group is equatorial. ✳

• **Cyclic monosaccharide structures** are usually drawn as **Haworth projections**, in which the five-member or six-member **cyclic hemiacetal** is drawn as **planar**, and **perpendicular** to the **plane of the paper**. They are usually drawn with the anomeric carbon to the right, and the hemiacetal oxygen atom in the back. A more accurate "chair" type of structure can be drawn for six-member ring hemiacetals, showing which groups are axial and which are equatorial.

## 12.3 PHYSICAL PROPERTIES

• **Monosaccharides** are all **very soluble** in **water** due to all of the -OH groups that can take part in hydrogen bonding with the water molecules. Monosaccharides also taste sweet to differing degrees.

## 12.5 GLUCOSE ASSAYS: THE SEARCH FOR SPECIFICITY

• For various medical conditions, especially diabetes, an accurate determination of glucose levels in serum and other fluids is required. In order to carry out this determination quantitatively, some chemical and enzymatic tests have been developed.

## 12.6 L-ASCORBIC ACID (VITAMIN C)

• In nature, **L-ascorbic acid** is made through a series of enzyme-catalyzed reactions from D-glucose. Humans, primates and guinea pigs cannot carry out the complete synthesis of ascorbic acid, since not all the required enzymes are present.

• On an industrial level, ascorbic acid is synthesized through a combination of some chemical steps, as well as some microbiological fermentations, starting with inexpensive D-glucose.

## 12.7 DISACCHARIDES AND OLIGOSACCHARIDES

• The monosaccharides can be linked into **disaccharides**, **trisaccharides**, or higher **oligosaccharides**.
  - Important disaccharides include **maltose** (2 D-glucose molecules linked with a β-1,4-glycoside bond), **lactose** (D-glucose and D-galactose linked with a β–1,4-glycoside bond), and **sucrose** (D-glucose and D-fructose linked with an α-1,2-glycoside bond).
  - Larger oligosaccharides are also important in nature, including the **blood group substances** that are on the surfaces of red blood cells. These are used to differentiate blood types.

## 12.8 POLYSACCHARIDES

• Monosaccharides can also be linked together into very large **polysaccharides**.
  - **Starch** is composed entirely of D-glucose units, as a straight chain polymer with α–1,4-glycoside linkages (called amylose) or as a branched polymer with both α-1,4-glycoside linkages and α-1,6-glycoside linkages. Starch is used for energy storage in plants.

- **Glycogen** is the animal equivalent of starch, and is a highly branched structure with both α-1,4-glycoside linkages and α-1,6-glycoside linkages between D-glucose units. Glycogen is found mostly in the liver and muscles, and serves as the carbohydrate reserve.
- **Cellulose** is a linear polymer of D-glucose linked by β-1,4 glycoside bonds. Cellulose is a structural component of plants, comprising almost half of the cell wall material of wood. Cotton is almost entirely composed of cellulose, and the synthetic textile fibers **rayon** and **acetate rayon** are chemically modified forms of cellulose.

# CHAPTER 12
## Solutions to the Problems

Problem 12.1
(a) Draw Fischer projections for all 2-ketopentoses.
(b) Show which are D-ketopentoses, which are L-ketopentoses, and which are enantiomers.
(c) Refer to table 12.2, and write names for the ketopentoses you have drawn.

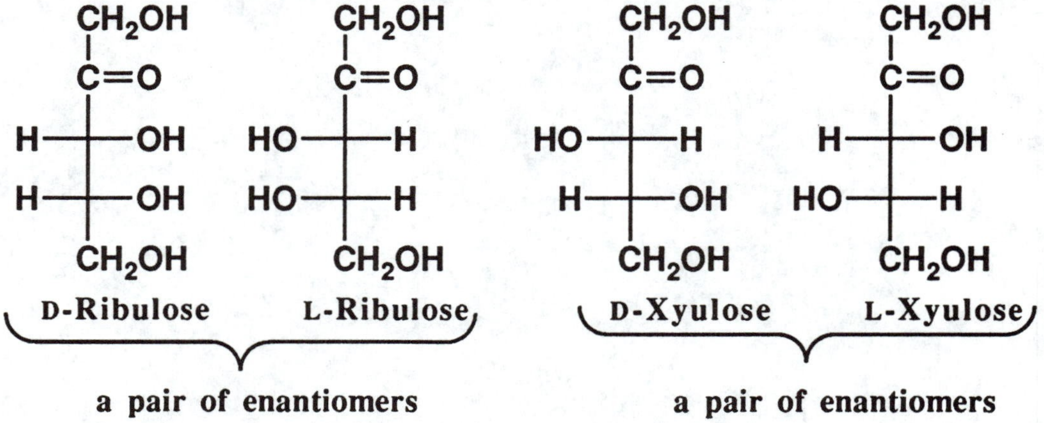

D-Ribulose          L-Ribulose          D-Xylulose          L-Xylulose

a pair of enantiomers              a pair of enantiomers

Problem 12.2  Mannose exists in aqueous solution as a mixture of α-D-mannopyranose and β-D mannopyranose.  Draw Haworth projections for these molecules.

**D-Mannose differs in configuration from D-glucose at carbon-2.  Therefore, the alpha and beta forms of D-mannopyranose differ from those of alpha and beta D-glucopyranoses only in the orientation of the -OH on carbon-2.  Following are Haworth projections for these compounds.**

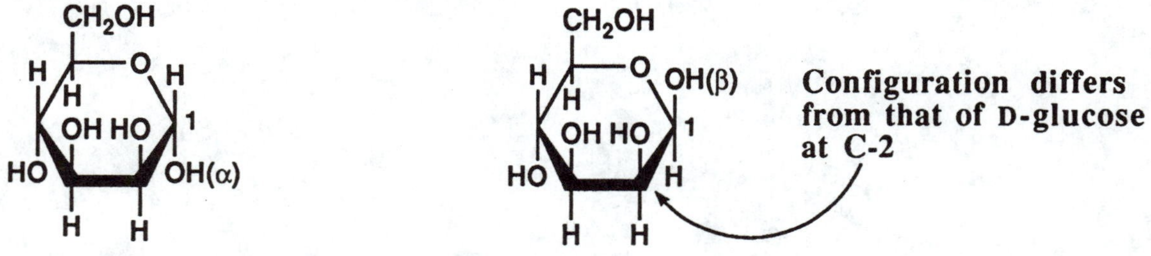

Configuration differs
from that of D-glucose
at C-2

α-D-Mannopyranose          β-D-Mannopyranose
(α-D-Mannose)              (β-D-Mannose)

Problem 12.3  Draw chair conformations for α-D-mannopyranose and β-D-mannopyranose.  Label the anomeric carbon atom in each.

**D-Mannose differs in configuration from D-glucose at carbon-2.  Draw chair conformations for the alpha and beta forms of D-glucopyranose and then invert the configuration of the -OH on carbon-2.  For reference, the open chain form of D-mannose is also drawn.**

β-D-Mannopyranose          D-Mannose          α-D-Mannopyranose
 (β-D-Mannose)                                  (α-D-Mannose)

Problem 12.4 Draw structural formulas for the following glycosides. In each, label the anomeric carbon and the glycoside bond.
(a) Methyl β-D-fructofuranoside (methyl β-D-fructoside).

Methyl β-D-fructofuranoside

(b) Methyl α-D-mannopyranoside (methyl α-D-mannoside).

Methyl α-D-mannopyranoside          Methyl α-D-mannopyranoside
   (Chair conformation)                (Haworth projection)

Problem 12.5 Draw structural formulas for the N-glycoside formed between β-D-ribofuranose (β-D-ribose) and adenine. In each, label the anomeric carbon and the N-glycoside bond.

**Following are structural formulas for adenine, the monosaccharide hemiacetal, and the N-glycoside.**

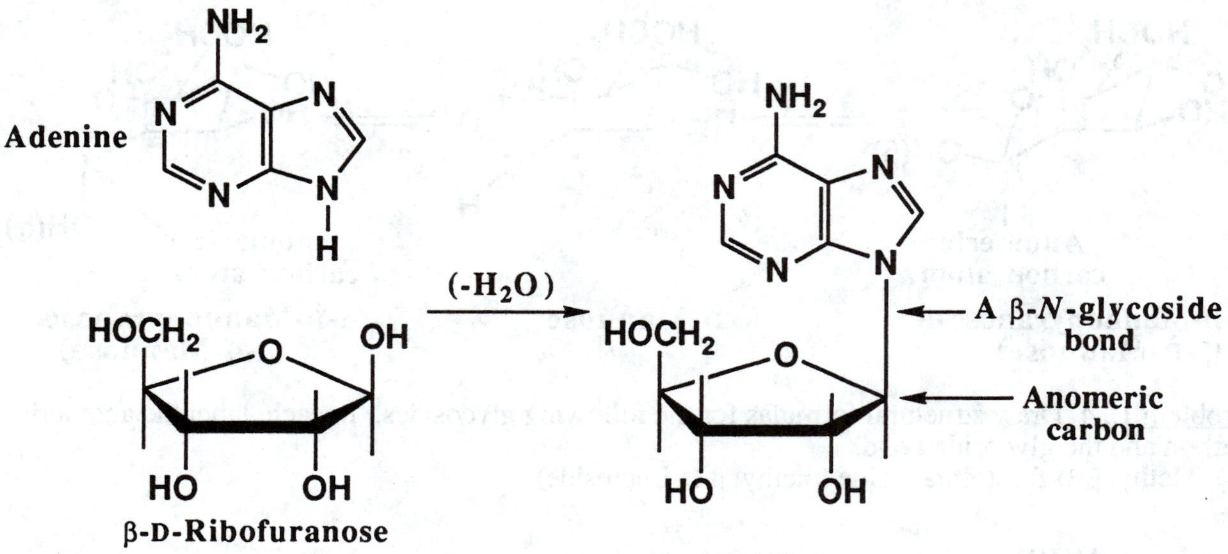

**β-D-Ribofuranose**

A β-*N*-glycoside bond

Anomeric carbon

**Problem 12.6** D-Erythrose is reduced by NaBH$_4$ to erythritol. Do you expect the alditol formed under these conditions to be optically active or optically inactive? Explain.

**Optically inactive. Erythritol is a meso compound and achiral. It is incapable of optical activity.**

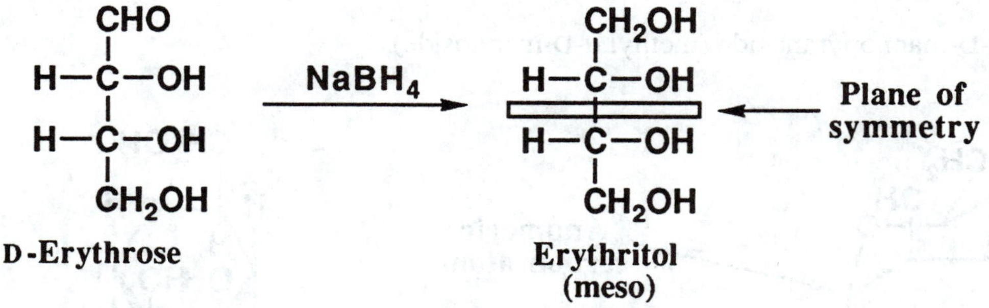

D-Erythrose          Erythritol (meso)          Plane of symmetry

**Problem 12.7** Draw a chair formula for the α form of a disaccharide in which two units of D-glucopyranose are joined by a β-1,3-glycoside bond.

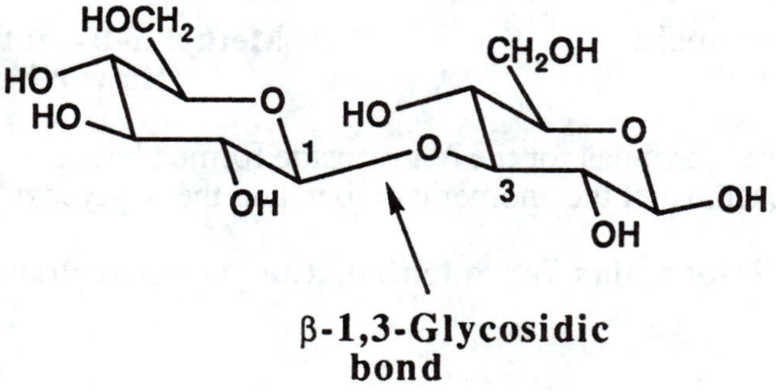

β-1,3-Glycosidic bond

## Fischer Projections

Problem 12.8 List the rules for drawing Fischer projections.

**The carbon chain is drawn vertically with the most highly oxidized carbon toward the top. Horizontal bonds to substituents on the carbon atoms are understood to project forward, out of the plane of the page. Vertical bonds are understood to project backward, behind the plane of the page.**

Problem 12.9 Here are two Fischer projections for 2-hydroxypropanal. Convert each to a tetrahedral representation and assign an R,S configuration to each.

(a)

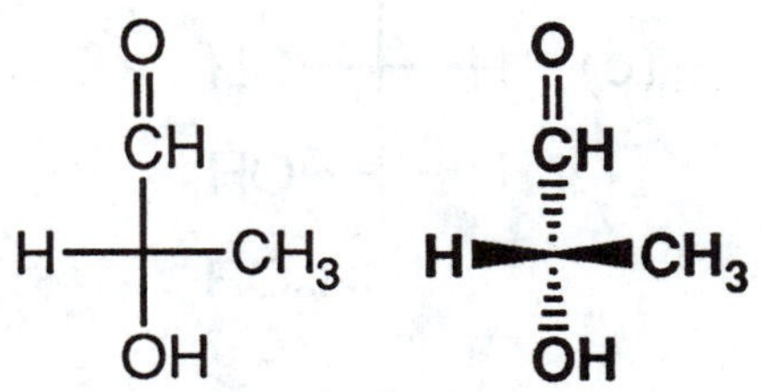

**(S)-2-Hydroxy-
propanal**

(b)

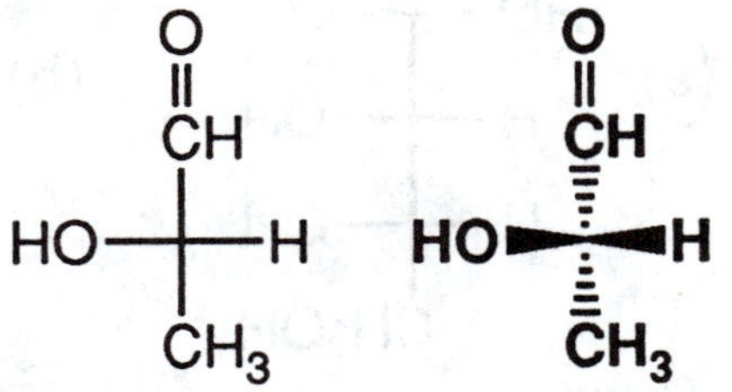

**(S)-2-Hydroxy-
propanal**

Problem 12.10 Draw Fischer projections for the three stereoisomers of tartaric acid (Figure 5.5).

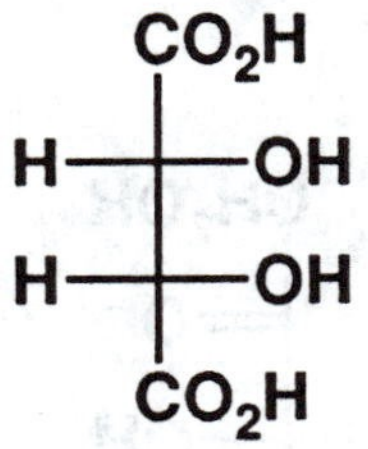

**meso-Tartaric acid** 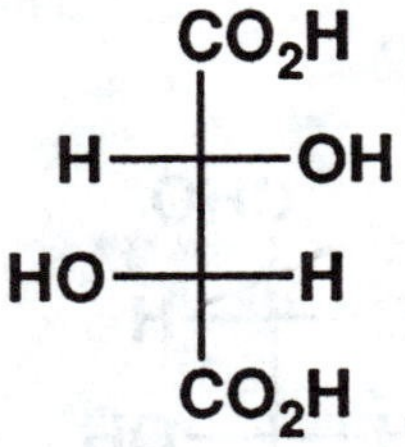　　　　**L-Tartaric acid** 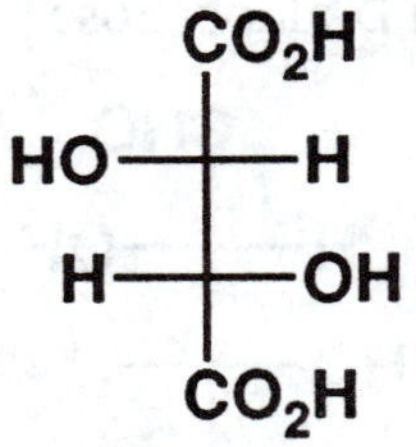　　　　**D-Tartaric acid**

Problem 12.11 Do these Fisher projections represent a pair of enantiomers or do they represent a single meso compound?

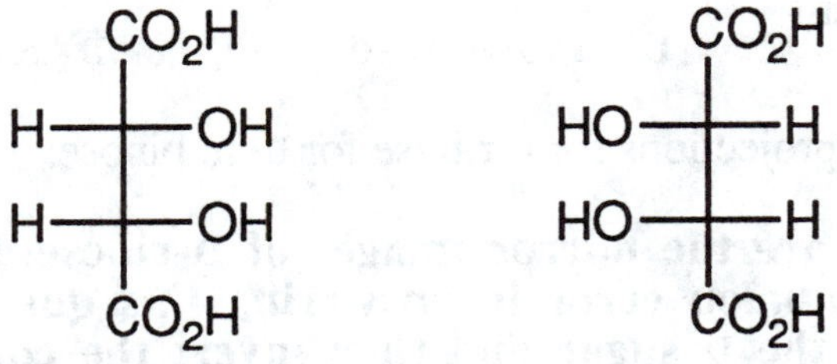

**Both of these structures are the same compound, the meso form of tartaric acid.**

## Monosaccharides

Problem 12.12 Explain the meaning of the designations D and L as used to specify the configuration of carbohydrates.

**The designations D and L refer to the configuration of the chiral center farthest from the carbonyl group of the monosaccharide. When a monosaccharide is**

drawn in a Fischer projection, the reference -OH is on the right in a D-monosaccharide and on the left in an L-monosaccharide.  Note that the conventions D and L specify the configuration at one and only one of however many chiral carbons there are in a particular monosaccharide.

**Problem 12.13** Which compounds are D-monosaccharides and which are L-monosaccharides?

Compounds (a) and (c) are D-monosaccharides, and compound (b) is an L-monosaccharide.

**Problem 12.14** Classify each monosaccharide in Problem 12.13 using the designations D/L and aldose/ketose and according to the number of carbon atoms it contains.  For example, the glucose s classified as a D-aldohexose.

D-Aldohexose             L-Aldopentose             D-Ketopentose

**Problem 12.15** Write Fischer projections for L-ribose for L-arabinose.

L-Ribose and L-arabinose are the mirror images of D-ribose and D-arabinose, respectively.  The most common error in answering this question is to start with the Fischer projection for the D sugar and then invert the configuration of carbon-4 only.  While the monosaccharide thus drawn is an L-sugar, it is not the correct one.  All of the stereocenters must be changed to draw the true enantiomers.

CHO — H—OH — H—OH — H—OH — CH₂OH

**D-Ribose**

CHO — HO—H — HO—H — HO—H — CH₂OH

**L-Ribose**

CHO — HO—H — H—OH — H—OH — CH₂OH

**D-Arabinose**

CHO — H—OH — HO—H — HO—H — CH₂OH

**L-Arabinose**

**Problem 12.16** What is the meaning of the prefix deoxy- as it is used in carbohydrate chemistry?

**The prefix "deoxy-" means without oxygen.**

**Problem 12.17** 2,6-Dideoxy-D-altrose, known alternatively as D-digitoxose, is a monosaccharide obtained on hydrolysis of digitoxin, a natural product extracted from foxglove (*Digitalis purpurea*). Digitoxin has found wide use in cardiology because it reduces pulse rate, regularizes heart rhythm and strengthens heart beat. Draw the structural formula of 2,6-dideoxy-D-altrose.

CHO — H—H — H—OH — H—OH — H—OH — CH₃

**2,6-Dideoxy-D-altose**
**(D-Digitoxose)**

**Problem 12.18** Give L-fucose a name incorporating the prefix deoxy- that shows its relationship to galactose.

**L-Fucose would be 6-deoxy-L-galactose.**

## The Cyclic Structure of Monosaccharides

**Problem 12.19** Define the term anomeric carbon.

**The anomeric carbon is the hemiacetal carbon of the cyclic form of a monosaccharide.**

**Problem 12.20** Explain the conventions for using α and β to designate the configuration of cyclic forms of monosaccharides.

**The designations alpha (α) and beta (β) refer to the configuration of the anomeric hydroxyl in a Fischer projection relative to the configuration of the -OH on the**

reference carbon, that is the carbon by which D or L is determined. If the -OH of the anomeric carbon is on the same side of the ring as the -OH of the reference carbon, then the configuration of the anomeric carbon is alpha. If the -OH on the anomeric carbon is on the opposite side of the ring as the -OH of the reference carbon, then the configuration of the anomeric carbon is beta.

<u>Problem 12.21</u> Draw α-D-glucopyranose (α-D-glucose) in a Haworth projection. Now, using only the information given here, draw Haworth projections for these monosaccharides.
(a) α-D-Mannopyranose (α-D-mannose). The configuration of D-mannose differs from the configuration of D-glucose only at carbon 2.
(b) α-D-Gulopyranose (α-D-gulose). The configuration of D-gulose differs from the configuration of D-glucose only at carbons 3 and 4.

**α-D-Glucopyranose**          **α-D-Mannopyranose**          **α-D-Gulopyranose**
(α-D-Glucose)                   (α-D-Mannose)                   (α-D-Gulose)

<u>Problem 12.22</u> Convert each Haworth projection to an open-chain Fischer projection and name the monosaccharide you have drawn.

(a)                              **D-Allose**

(b)                              **D-Iodose**

**Problem 12.23** Convert each chair conformation to an open-chain Fischer projection and name the monosaccharide you have drawn.

(a)

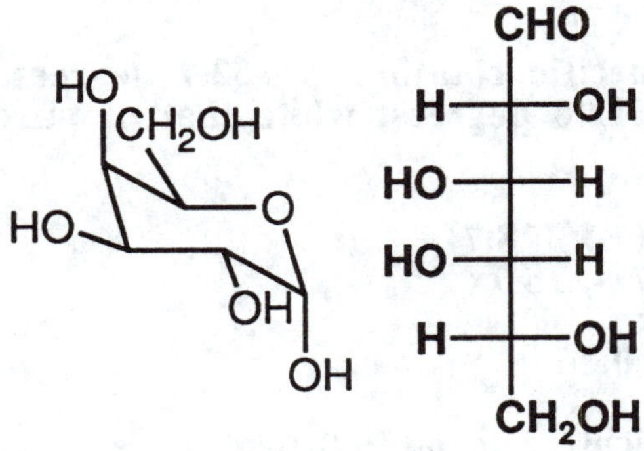

CHO
H——OH
HO——H
HO——H
H——OH
CH$_2$OH

D-Galactose

(b)

CHO
H——OH
H——OH
H——OH
H——OH
CH$_2$OH

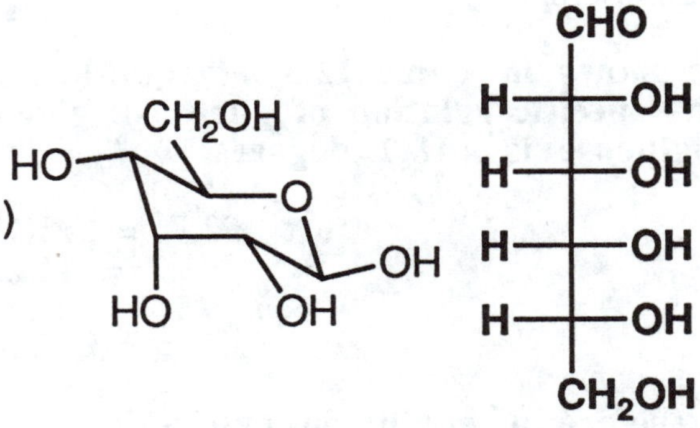

D-Allose

**Problem 12.24** Draw α-D-ribofuranose (α-D-ribose) as a Haworth projection.

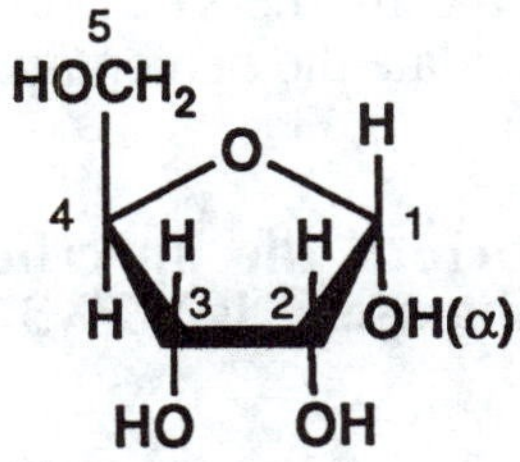

α-D-Ribofuranose (α-D-ribose)

**Problem 12.25** The configuration of D-arabinose differs from the configuration of D-ribose only at carbon 2. Using this information, draw a Haworth projection for α-D-arabinofuranose (α-D-arabinose).

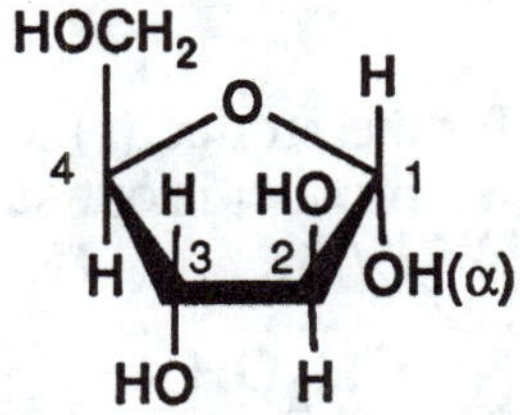

α-D-Arabinofuranose (α-D-arabinose)

**Problem 12.26** Explain the phenomenon of mutarotation with reference to carbohydrates. By what means is it detected?

**Any monosaccharides of four or more carbons can exist in an open-chain form and two or more cyclic hemiacetal (i.e. furanose or pyranose) forms, each having a different specific rotation. The specific rotation of an aqueous solution, measured with a polarimeter, of any one form changes until an equilibrium value is reached, representing an equilibrium concentration of the different forms. Mutarotation is the change in specific rotation toward an equilibrium value.**

<u>Problem 12.27</u>  The specific rotations for the anomers of D-glucose and the value after mutarotation are given in Table 12.3.  Calculate the percentage of each anomer present at equilibrium.

**As shown in Table 12.3, at equilibrium the specific rotation is +52.7 degrees. The specific rotation of pure α-D-glucose is +112.0 degrees, while that of pure β-D-glucose is +18.7 degrees.**

$$52.7 = (x)(112) + (1-x)(18.7)$$
$$52.7 = 112x + 18.7 - 18.7x$$
$$34 = 93.3x$$
$$x = 34/93.3 = 0.36$$

**Therefore, at equilibrium there is 36% α-D-glucose and 64% β-D-glucose.**

<u>Problem 12.28</u>  The specific rotations for the anomers of D-mannose and the value after mutarotation are given in Table 12.3.  Calculate the percentage of each anomer present at equilibrium.

**As shown in Table 12.3, at equilibrium the specific rotation is +14.5 degrees. The specific rotation of pure α-D-mannose is +29.3 degrees, while that of pure β-D-mannose is -16.3 degrees.**

$$14.5 = (x)(29.3) + (1-x)(-16.3)$$
$$14.5 = 29.3x - 16.3 + 16.3x$$
$$30.8 = 45.6x$$
$$x = 30.8/45.6 = 0.68$$

**Therefore, at equilibrium there is 68% α-D-mannose and 32% β-D-mannose.**

## Reactions of Monosaccharides

<u>Problem 12.29</u>  Draw Fischer projections for the product(s) formed by reaction of D-galactose with the following.  In addition state whether each product is optically active or inactive.

(a) NaBH$_4$ in H$_2$O              (b) H$_2$/Pt              (c) AgNO$_3$ in NH$_3$, H$_2$O

|                     |                     |                     |
|---------------------|---------------------|---------------------|
| CH$_2$OH            | CH$_2$OH            | CO$_2$H             |
| H———OH              | H———OH              | H———OH              |
| HO———H              | HO———H              | HO———H              |
| HO———H              | HO———H              | HO———H              |
| H———OH              | H———OH              | H———OH              |
| CH$_2$OH            | CH$_2$OH            | CH$_2$OH            |
| **Galactitol**      | **Galactitol**      | **D-Galactonic acid** |
| **(meso; inactive)** | **(meso; inactive)** | **(chiral; optically active)** |

(d) *o*-Toluidine

CH₃

CH=N—

H——OH

HO——H

HO——H

H——OH

CH₂OH

**A Schiff base of**
**D-Galactose (chiral; optically active)**

Problem 12.30  Repeat problem 12.29 using D-ribose.
(a) NaBH₄ in H₂O                    (b) H₂/Pt                    (c) AgNO₃ in NH₃, H₂O

CH₂OH

H——OH

H——OH

H——OH

CH₂OH

**Ribitol**
**(meso; inactive)**

CH₂OH

H——OH

H——OH

H——OH

CH₂OH

**Ribitol**
**(meso; inactive)**

CO₂H

H——OH

H——OH

H——OH

CH₂OH

**D-Ribonic  acid**
**(chiral; optically active)**

d) *o*-Toluidine

CH₃

CH=N—

H——OH

H——OH

H——OH

CH₂OH

**A Schiff base of**
**D-Ribose (chiral; optically active)**

Problem 12.31 There are four D-aldopentoses (Table 12.1). If each is reduced with NaBH$_4$, which yield optically active alditols? Which yield optically inactive alditols?

**D-Ribose and D-xylose yield different achiral (meso) alditols. D-Arabinose and D-lyxose yield the same chiral alditol.**

D-Ribose  →(NaBH$_4$)→  Ribitol (meso)     D-Xylose  →(NaBH$_4$)→  Xylitol (meso)

D-Arabinose  →(NaBH$_4$)→  D-Arabinitol  ←(NaBH$_4$)←  D-Lyxose

$$[\alpha]_D^{25} = -32°$$

Problem 12.32 Account for the observation that reduction of D-glucose with NaBH$_4$ gives an optically active alditol, whereas reduction of D-galactose with NaBH$_4$ gives an optically inactive alditol.

**Reduction of D-glucose gives an optically active alditol product, while reduction of D-galactose gives an optically inactive meso alditol.**

D-Glucose  →(NaBH$_4$)→  (Optically active)     D-Galactose  →(NaBH$_4$)→  (meso: optically inactive)

**Problem 12.33** Which two D-aldohexoses give optically inactive (meso) alditols on reduction with NaBH₄.

**D-Allose and D-galactose both give optically inactive (meso) alditols.**

D-Allose → (NaBH₄) → (meso: optically inactive)

D-Galactose → (NaBH₄) → (meso: optically inactive)

**Problem 12.34** Name the two alditols formed by NaBH₄ reduction of D-fructose.

D-Fructose → (NaBH₄) → D-Glucitol + D-Mannitol

**Reduction of D-fructose gives D-glucitol and D-mannitol. Each differs in configuration only at carbon-2.**

**Problem 12.35** One pathway for the metabolism of glucose-6-phosphate is its enzyme-catalyzed conversion to fructose-6-phosphate. Show that this transformation can be regarded as two enzyme-catalyzed keto-enol tautomerisms.

D-Glucose-6-phosphate ⇌ (Enzyme catalysis) D-Fructose-6-phosphate

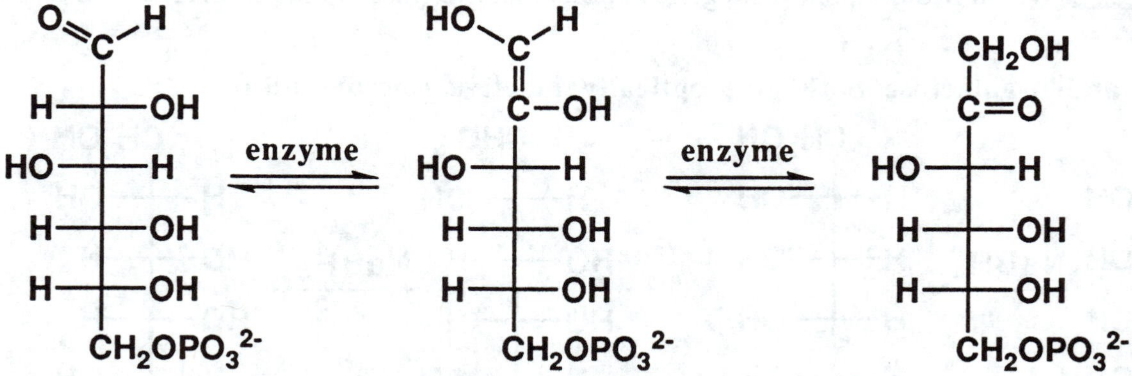

**Enol form**

**Problem 12.36** L-Fucose, one of several monosaccharides commonly found in the surface polysaccharides of animal cells Section 12.7D), is synthesized biochemically from D-mannose in the following eight steps:

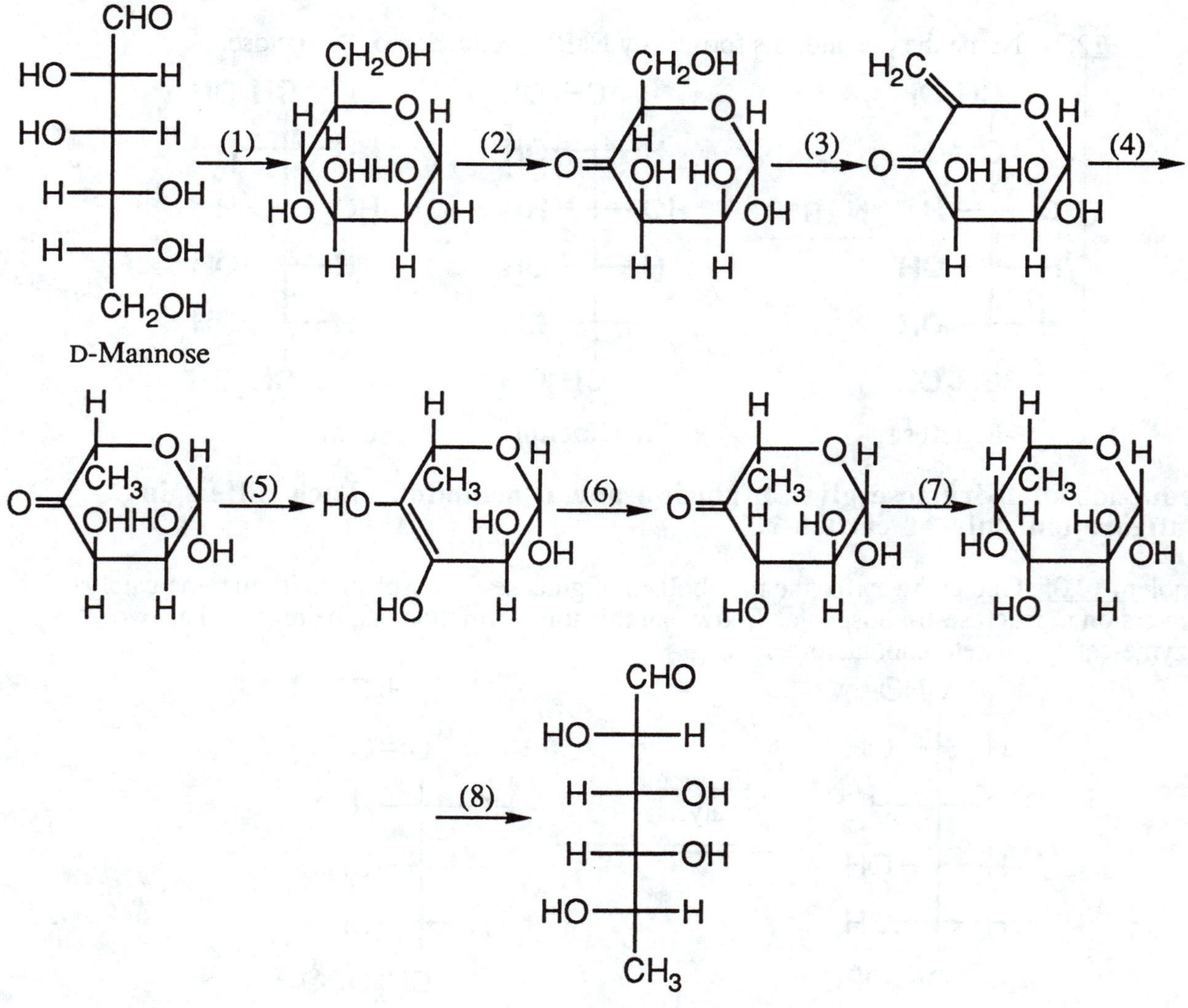

D-Mannose

L-Fucose

(a) Describe the type of reaction ( i.e., oxidation, reduction, hydration, dehydration, etc.) involved in each step.

**Following is the type of reaction in each step.**
**(1)   Formation of a hemiacetal from a carbonyl group and a secondary alcohol.**
**(2)   A two-electron oxidation of a secondary alcohol to a ketone.**
**(3)   Dehydration of a β-hydroxyketone to an α,β-unsaturated ketone.**
**(4)   A two-electron reduction of a carbon-carbon double bond to a carbon-carbon**
       **single bond.**
**(5)   Keto-enol tautomerism of an α-hydroxyketone to form an enediol.**
**(6)   Keto-enol tautomerism of an enediol to form an α-hydroxyketone.**
**(7)   A two-electron reduction of a ketone to a secondary alcohol.**
**(8)   Opening of a cyclic hemiacetal to form an aldehyde and an alcohol.**

(b) Explain why it is that this monosaccharide derived from D-mannose now belongs to the L series.

**It is the configuration at carbon-5 of this aldohexose that determines whether it is of the D-series or of the L-series.   The result of steps 3 and 4 is inversion of configuration at carbon-5 and, therefore, conversion of a D-aldohexose to an L-aldohexose.**

## Ascorbic Acid

Problem 12.37  Write a balanced half-reaction to show that conversion of L-ascorbic acid to L-dehydroascorbic acid is an oxidation.  How many electrons are involved in this oxidation?  Is ascorbic acid a biological oxidizing agent or a biological reducing agent?

**The most direct way to see that this is a two-electron oxidation is to write a balanced half-reaction for the conversion of the enediol to a diketone.**

**Because L-ascorbic acid donates two electrons to another molecule or ion, it is a biological reducing agent.   Conversely, L-dehydroascorbic acid is a biological oxidizing agent.**

Problem 12.38  Ascorbic acid is a diprotic acid with the following acid ionization constants.

$$pK_{a1} = 4.10 \qquad pK_{a2} = 11.79$$

The two acidic hydrogens are those connected with the enediol part of the molecule.  Which hydrogen has which ionization constant? (*Hint:* Draw separately the anion derived by loss of one of these hydrogens and that formed by loss of the other hydrogen.  Which anion has the greater degree of resonance stabilization?)

Following are assignments of the two $pK_a$ values.

The anion derived from ionization of -OH on carbon-3 is stabilized by resonance interaction with the carbonyl oxygen. There is no comparable resonance stabilization of the anion derived from ionization of -OH on carbon-2.

## Disaccharides and Oligosaccharides

Problem 12.39  Define the term glycoside bond.

A glycoside bond is the bond from the anomeric carbon of a glycoside to an -OR group.

Problem 12.40  What is the difference in meaning between the terms glycoside bond and glucoside bond?

A glycoside bond is the bond from the anomeric carbon of a glycoside to an -OR group. A glucoside bond is a glycoside bond that yields glucose upon hydrolysis.

Problem 12.41  In making candy or sugar syrups, sucrose is boiled in water with a little acid, such as lemon juice. Why does the product mixture taste sweeter than the starting sucrose solution?

Sucrose is a disaccharide composed of the monosaccharides glucose and fructose linked through a glycosidic bond. The acid catalyzes hydrolysis of the glycosidic bond, and the monomeric glucose and fructose are more soluble than sucrose itself. Because of this, the syrups and candy have higher concentrations of these sugars than is possible with sucrose. As a result, the candy and syrup taste sweeter. Furthermore, fructose actually tastes sweeter than sucrose, having a relative sweetness of 174, compared with 100 for sucrose. Thus, converting the sucrose into fructose increases the sweetness of the mixture.

Problem 12.42  Which disaccharides are reduced by NaBH₄?
(a) Maltose                        (b) Lactose                        (c) Sucrose

**In order for these disaccharides to react with NaBH₄, they must contain at least one carbonyl group that is equilibrium with the open chain form.  Since all of these molecules have such a carbonyl, they will all be reduced by NaBH₄.**

Problem 12.43' Trehalose is found in young mushrooms and is the chief carbohydrate in the blood of certain insects.  Trehalose is a disaccharide consisting of two units of D-glucose joined by an α-1,1-glycoside bond.

Trehalose

(a)  Is trehalose a reducing sugar?

**Trehalose is not a reducing sugar because each anomeric carbon is involved in formation of the glycoside bond.**

(b)  Does trehalose undergo mutarotation?

**It will not undergo mutarotation since there is no open chain form possible for either monosaccharide.  Both anomeric carbons are involved in the glycoside bond.**

(c)  Name the two monosaccharides units of which trehalose is composed.

**Trehalose is composed of two molecules of D-glucose.**

Problem 12.44  The trisaccharide raffinose occurs principally in cottonseed meal.

Raffinose

(a)  Name the three monosaccharide units in raffinose.

**The three monosaccharide units in raffinose, from top to bottom, are D-galactose, D-glucose, and D-fructose.**

(b)  Describe each glycoside bond in this trisaccharide.

**Reading from left to right, they are D-galactopyranose joined by an α-1,6-glycoside bond to D-glucopyranose and then D-glucopyranose, in turn, joined by an α-1,2-glycoside bond to β-D-fructofuranose.**

(c)  Is raffinose a reducing sugar?

**No, it is not a reducing sugar, since all of the anomeric carbon atoms are involved in glycoside bonds.**

Problem 12.45  Hot water extracts of ground willow bark are an effective pain reliever. Unfortunately, the liquid is so bitter that most persons refuse it.  The pain reliever in these infusions is salicin, a glycoside of glucose and 2-(hydroxymethyl)phenol.
(a)  Given this information, what four structures are possible for salicin?

**Salicin can have the glycoside bond in the α or β conformation of the sugar, linked to the hydroxymethyl or phenol -OH group.**

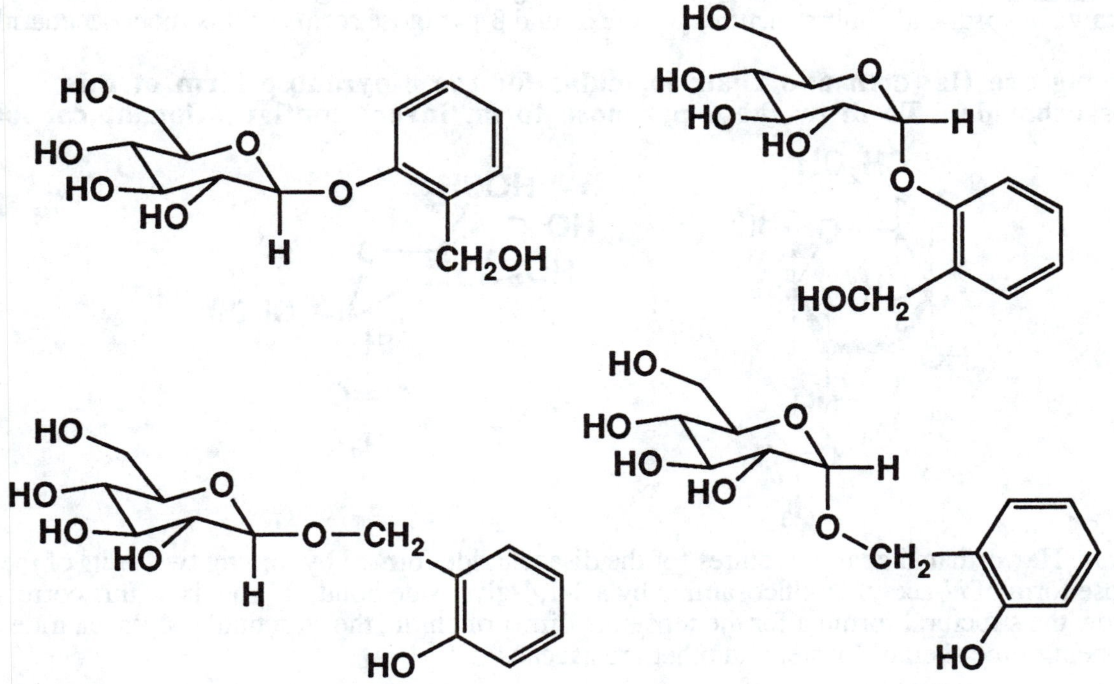

(b) A dilute aqueous solution of salicin is neutral. Which structures are now possible for salicin?

**Dilute aqueous solutions of salicin are neutral indicating that the slightly acidic free phenol function is not present in the molecule. Thus, the glycosidic linkage must involve the phenol group, and salicin must be one of the two upper structures shown above.**

Problem 12.46 Vanillin, the principal component of vanilla, occurs in the vanilla bean and other natural sources as a β-D-glucopyranoside. Draw a structural formula for this glycoside. Draw the D-glucose unit in this glycoside as a chair conformation.

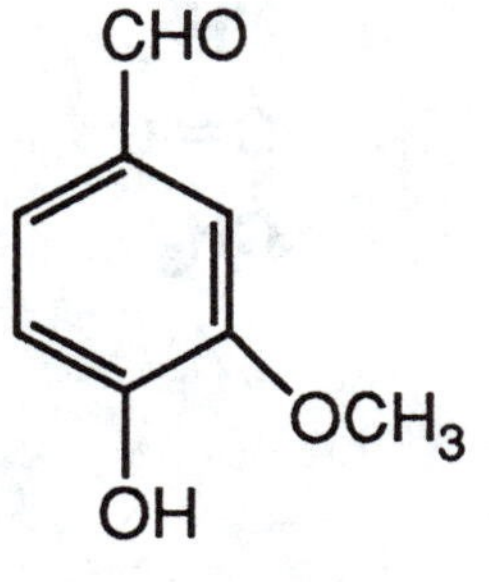

Vanillin

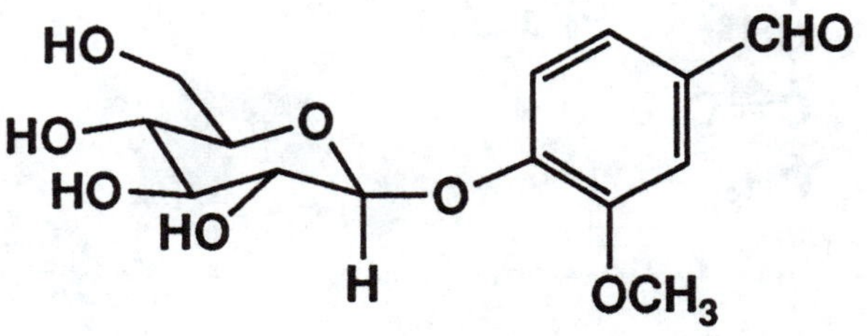

Problem 12.47  A Fischer projection for *N*-acetyl-D-glucosamine is given in Section 12.1E.
(a)  Draw Haworth and chair structures for the α- and β-pyranose forms of this monosaccharide.

**Following are Haworth and chair formulas for the β-pyranose form of this monosaccharide. To draw the α-pyranose form, invert configuration at carbon-1.**

(b)  Draw Haworth and chair structures for the disaccharide formed by joining two units of the pyranose form of *N*-acetyl-D-glucosamine by a β-1,4-glycoside bond.  If you drew this correctly, you drew the structural formula for the repeating dimer of chitin, the structural polysaccharide component of the shell of lobster and other crustaceans.

**Following are Haworth and chair formulas for the β-anomer of this disaccharide.**

Problem 12.48  Propose structural formulas for the following polysaccharides:
(a)  Alginic acid, isolated from seaweed, is used as the thickening agent in ice cream and other foods.  Alginic acid is a polymer of D-mannuronic acid in the pyranose form joined by β-1,4-glycoside bonds.

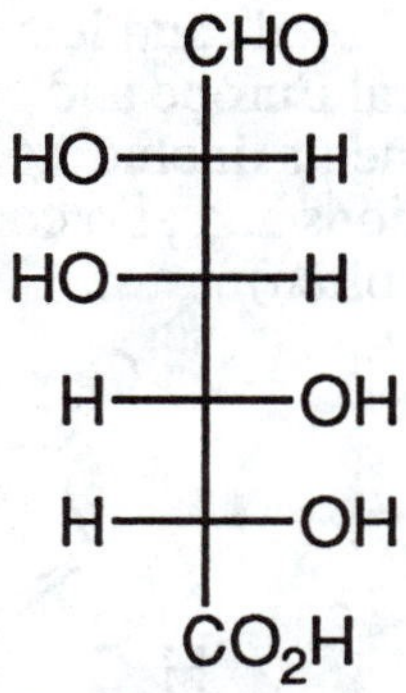
CHO
HO
H
HO
H
H
OH
H
OH
CO2H

D-Mannuronic acid

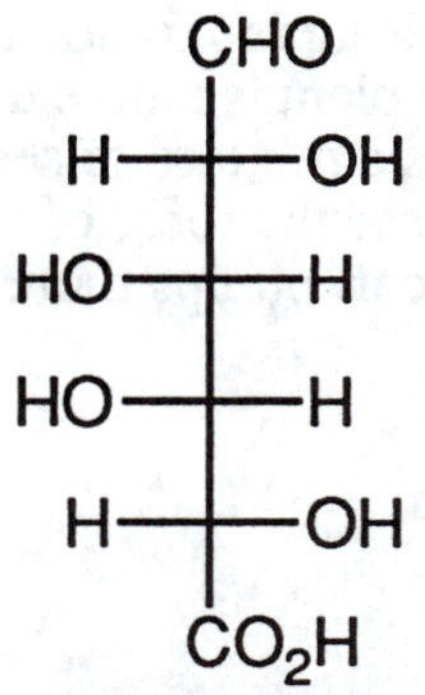
CHO
H
OH
HO
H
HO
H
H
OH
CO2H

D-Galacturonic acid

**Following is the chair conformation for repeating disaccharide units of alginic acid.**

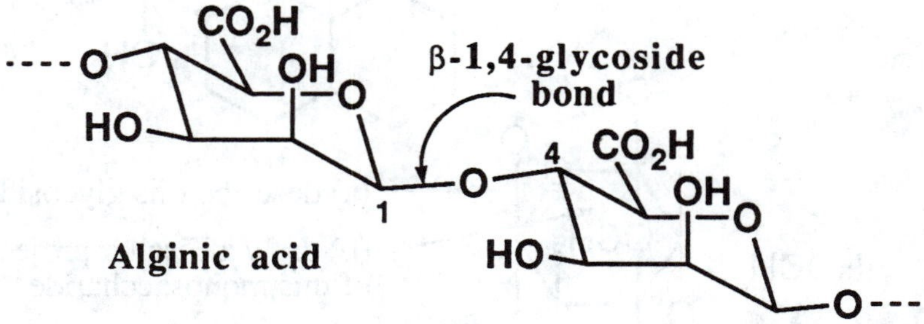
CO2H
OH
O
HO
β-1,4-glycoside bond
1
O
4
CO2H
OH
O
HO
O

Alginic acid

(b) Pectic acid is the main component of pectin, which is responsible for the formation of jellies from fruits and berries. Pectic acid is a polymer of D-galacturonic acid in the pyranose form joined by α-1,4-glycoside bonds.

**Following is the chair conformation for repeating disaccharide units of pectic acid.**

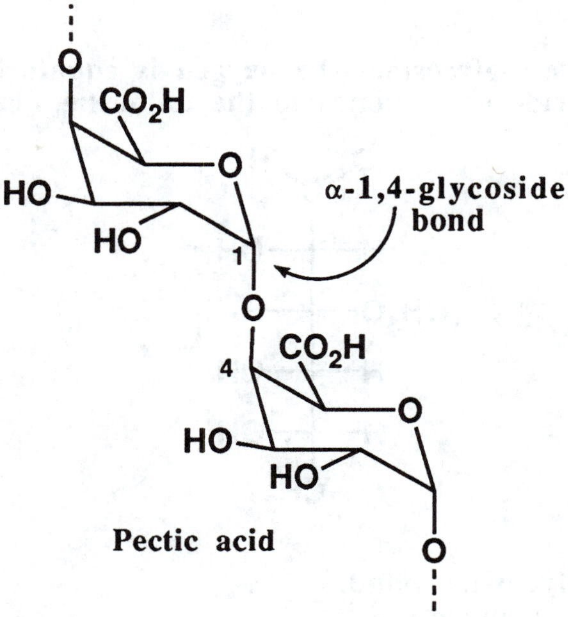
O
CO2H
O
HO
HO
1
α-1,4-glycoside bond
O
4
CO2H
O
HO
HO
O

Pectic acid

**Problem 12.49** Digitalis is a preparation made from the dried seeds and leaves of the foxglove, *Digitalis purpurea*. This plant is native to southern and central Europe and cultivated in the United States. The preparation is a mixture of several active components, including digitalin. Digitalis is used in medicine to increase the force of myocardial contractions and as a conduction depressant to decrease heart rate (the heart pumps more forcefully but less often).

Digitalin

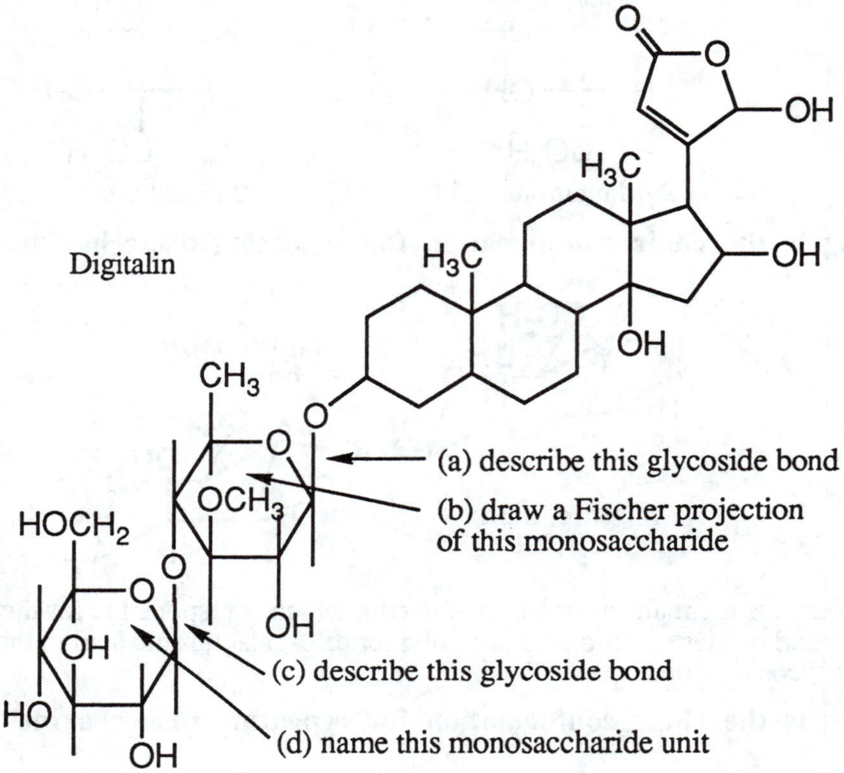

(a) describe this glycoside bond

(b) draw a Fischer projection of this monosaccharide

(c) describe this glycoside bond

(d) name this monosaccharide unit

**a) The indicated bond is a β-glycoside (the oxygen is equatorial).**
**b) The first monosaccharide corresponds to the following Fischer projection.**

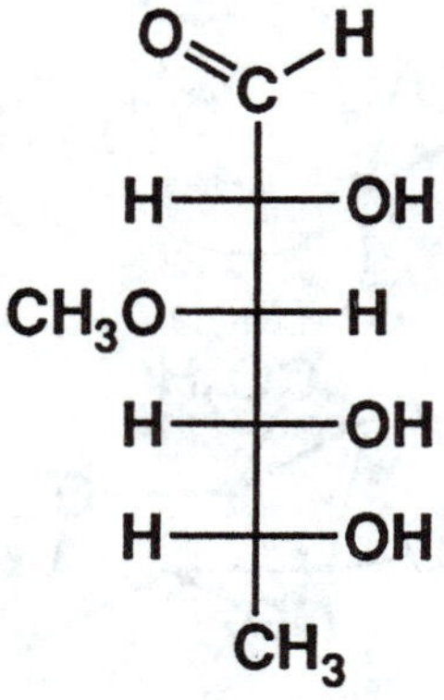

$$
\begin{array}{c}
O{=}C{-}H \\
H{-\!\!|\!\!-}OH \\
CH_3O{-\!\!|\!\!-}H \\
H{-\!\!|\!\!-}OH \\
H{-\!\!|\!\!-}OH \\
CH_3
\end{array}
$$

**c) This bond is a β-1,4-glycoside bond.**
**d) This monosaccharide is glucose.**

# CHAPTER 13: CARBOXYLIC ACIDS

## SUMMARY OF REACTIONS

| Starting Material ↓ / Product → | Acid Chlorides | Alcohols | Carboxyates | Carboxylic Acids | Esters | Ketones |
|---|---|---|---|---|---|---|
| Carboxylic Acids | 13A 13.7* | 13B 13.5A | 13C 13.4B | | | |
| Carboxylic Acids Alcohols | | | | | 13D 13.6 | |
| β-Dicarboxylic Acids | | | | 13E 13.8B | | |
| β-Ketoacids | | | | | | 13F 13.8A |

*Section in book that describes reaction.

## REACTION 13A: CONVERSION TO ACID CHLORIDES (Section 13.7)

- **Carboxylic acids** react with **thionyl chloride** to yield **acid chlorides**. ✳

## REACTION 13B: REDUCTION BY LITHIUM ALUMINUM HYDRIDE (Section 13.5A)

- **Carboxylic acids** react with **lithium aluminum hydride** to give **primary alcohols**.✳
- The reaction involves delivery of two hydride ions from the LiAlH$_4$ to the carbonyl group, while the hydroxyl hydrogen is derived from water in the work-up.
- NaBH$_4$ cannot reduce carboxylic acids to primary alcohols.

## REACTION 13C: REACTION OF CARBOXYLIC ACIDS WITH BASES (Section 13.4B)

$$-\overset{|}{\underset{|}{C}}-\overset{O}{\underset{OH}{C'}} \xrightarrow{\text{Base:}^-} -\overset{|}{\underset{|}{C}}-\overset{O}{\underset{O^-}{C'}} + \text{Base-H}$$

- **Carboxylic acids** are **relatively acidic**, so they react with **bases** like NaOH, KOH, or NH$_3$ to give a **carboxylate anion** and protonated base. ✳
- Carboxylate anions are highly water soluble, and this is often exploited in the isolation of carboxylic acids.

## REACTION 13D: FISCHER ESTERIFICATION (Section 13.6)

$$-\overset{|}{\underset{|}{C}}-\overset{O}{\underset{OH}{C'}} + -\overset{|}{\underset{|}{C''}}-OH \xrightleftharpoons{H_2SO_4} -\overset{|}{\underset{|}{C}}-\overset{O}{\underset{O-\overset{|}{\underset{|}{C''}}-}{C'}} + H_2O$$

- **Carboxylic acids** react with **excess alcohol** in the presence **sulfuric acid** under anhydrous conditions to generate **esters**. ✳
- The mechanism of the reaction involves initial protonation of the carbonyl oxygen atom, followed by nucleophilic attack of the alcohol oxygen atom onto the carbonyl carbon atom, transfer of a proton, and loss of H$_2$O to give the product ester.
- The reaction is reversible, so reaction of an ester with strong aqueous acid can lead to hydrolysis of the ester.

## REACTION 13E: DECARBOXYLATION OF β–DICARBOXYLIC ACIDS (Section 13.8B)

$$\overset{O}{\underset{HO}{C}}-\overset{|}{\underset{|}{C'}}-\overset{O}{\underset{OH}{C''}} \xrightarrow{\text{warm}} \overset{O}{\underset{HO}{C}}-\overset{|}{\underset{|}{C'}}-H + C''O_2$$

- When a β-**dicarboxylic acid** (malonic acid) is **heated**, it can **decarboxylate** to give a carboxylic acid and CO$_2$ as products. ✳
- The mechanism involves a transition state in which six electrons move in a six-member ring to give the enol of a carboxylic acid and CO$_2$. The enol undergoes keto-enol tautomerization to give the carboxylic acid.
- This reaction is important because malonic acid derivatives are easily produced.

## REACTION 13F: DECARBOXYLATION OF β-KETOACIDS (Section 13.8A)

$$-\overset{|}{\underset{|}{C}}-\overset{O}{\underset{}{C'}}-\overset{|}{\underset{|}{C''}}-\overset{O}{\underset{OH}{C'''}} \xrightarrow{\text{warm}} -\overset{|}{\underset{|}{C}}-\overset{O}{\underset{}{C'}}-\overset{|}{\underset{|}{C''}}-H + C'''O_2$$

- β-**Ketoacids** lose CO$_2$ upon **heating** to produce to produce **ketones**. ✳

- The mechanism of the reaction is exactly analogous to that for the decarboxylation of β-dicarboxylic acids, involving a six-member ring transition state that results in formation of an enol and $CO_2$. The enol rapidly undergoes keto-enol tautomerization to give the ketone.

# SUMMARY OF IMPORTANT CONCEPTS

## 13.0 OVERVIEW
• **Carboxylic acids** are acidic, and can be converted to a number of important derivatives such as esters, amides, and anhydrides. ✳

## 13.1 STRUCTURE
• The characteristic feature of carboxylic acids is the **carboxyl group**, which contains a **carbonyl group** that has an **-OH group attached** to the **carbonyl carbon atom**. ✳

## 13.2 NOMENCLATURE
• According to the **IUPAC system**, a carboxylic acid is named by **dropping** the **e** from the name of the longest chain that contains the carboxyl group, and replacing it with the suffix **oic acid**. The carboxyl group takes precedent over most other functional groups.
  - Dicarboxylic acids are named the same as above, except the suffix **dioic acid** is used. Higher carboxylic acids are named with the suffixes **tricarboxylic acid, tetracarboxylic acid**, etc.
  - Aromatic carboxylic acids are named as derivatives of benzoic acid, using the normal aromatic nomenclature to name the other substituents.
• Many **common names** of carboxylic acids are still used quite often, and some of these are listed in table 13.1 of the text. When these common names are used, then the Greek letters α, β, γ, δ, etc. are used to locate substituents.

## 13.3 PHYSICAL PROPERTIES
• In the liquid and solid state, **carboxylic acids exist in dimeric, hydrogen bonded structures**. In these dimeric structures, the hydrogen atom of each carboxyl group is hydrogen bonded to the carbonyl oxygen atom of its partner. ✳
  - Because of these dimeric structures, carboxylic acids have higher boiling points than analogous alcohols or aldehydes.
  - Because carboxylic acids can readily take part in hydrogen bonding with water molecules, they are also more soluble in water than analogous alcohols and aldehydes.

## 13.4 ACIDITY
• Unsubstituted **carboxylic acids** are relatively **acidic**, with **$pK_a$** values in the **range of 4-5.** ✳
  - In general, an anion is generated along with a proton when an acid dissociates, and the more stable the anion, the stronger the acid. The relative **acidity** of carboxylic acids can thus be **attributed to the stability of the carboxylate anion** that is produced upon deprotonation. The **carboxylate anion** is **stabilized** by **delocalization** of the **negative charge** onto both **oxygen atoms**.
  - Alternatively, the acidity of the carboxylic acid can be attributed to a **partial positive charge** on the **carbonyl carbon atom**, that **polarizes** the **electrons** in the **-OH bond away from hydrogen**.
  - **Electron withdrawing groups**, such as **halogen atoms** adjacent to the carboxylic group functions, **increase acid strength**. This effect can be quite large in certain cases.

# CHAPTER 13
## *Solutions to the Problems*

<u>Problem 13.1</u> Each of the following compounds has a well-recognized common name. A derivative of glyceric acid is an intermediate in glycolysis. Maleic acid is an intermediate in the TCA cycle. Mevalonic acid is an intermediate in the biosynthesis of steroids (Section 16.4). Write IUPAC names for these compounds.

(a)

$$CO_2H$$
$$H—C—OH$$
$$CH_2OH$$

Glyceric acid

**(R)-2,3-Dihydroxy-
propanoic acid**

(b)

$$HO_2C \qquad CO_2H$$
$$C=C$$
$$H \qquad H$$

Maleic acid

**(Z)-2-Butenedioic acid**

(c)

$$HO \qquad CH_3$$
$$C$$
$$HOCH_2CH_2 \qquad CH_2CO_2H$$

Mevalonic acid

**(R)-3,5-Dihydroxy-3-
methylpentanoic acid**

<u>Problem 13.2</u> Match each compound with the appropriate $pK_a$ value.

$$CH_3$$
$$CH_3—C—CO_2H \qquad CF_3CO_2H$$
$$CH_3$$

$$OH$$
$$CH_3—CH-CO_2H \qquad pK_a \text{ values} = 5.03, 3.08$$
$$\text{and } 0.22$$

2,2-Dimethyl-
propanoic acid
(Pivalic acid)

Trifluoro-
acetic acid

2-Hydroxy-
propanoic acid
(Lactic acid)

**2,2-Dimethylpropanoic acid has a $pK_a$ comparable to that of an unsubstituted aliphatic carboxylic acid. Lactic acid is a stronger acid due to a combination of the inductive effect of the hydroxyl group on an sp³ carbon atom adjacent to the carboxyl group and the fact that the hydroxyl group can also stabilize the carboxylate anion of the deprotonated acid via an intramolecular hydrogen bond. Trifluoroacetic is an even stronger acid because of the combined inductive effects of the three fluorine atoms. In order of increasing acidity, these acids are:**

$$CH_3$$
$$CH_3—C—CO_2H \qquad CH_3—CHCO_2H \qquad F—C—CO_2H$$
$$CH_3 \qquad\qquad OH \qquad\qquad F$$

$$pK_a \quad 5.03 \qquad\qquad 3.08 \qquad\qquad 0.22$$

$$\longrightarrow$$
**increasing acidity**

<u>Problem 13.3</u> Write equations for the reaction of each acid in Example 13.3 with dimethylamine.

(a) $CH_3(CH_2)_2CO_2H + NH(CH_3)_2 \longrightarrow CH_3(CH_2)_2CO_2^- \; [H_2N(CH_3)_2]^+$

(b) CH₃ĊHCO₂H + NH(CH₃)₂ ⟶ CH₃ĊHCO₂⁻ [H₂N(CH₃)₂]⁺

with OH above each CHCO group

Problem 13.4 Complete these Fischer esterifications.

(a) HOC⬡COH + 2CH₃OH —H⁺→ CH₃OC⬡COCH₃ + 2 H₂O

(b) CH₃CHCOH + HO–⬡ —H⁺→ CH₃CHCO–⬡ + H₂O
         |CH₃                              |CH₃

Problem 13.5 Complete these equations.

(a) ⬡ with CO₂H and OCH₃ + SOCl₂ ⟶ ⬡ with CCl and OCH₃ + SO₂ + HCl

(b) ⬡ with OH + SOCl₂ ⟶ ⬡ with Cl + SO₂ + HCl

Problem 13.6 Draw the structural formula for the indicated starting β-ketoacid.

⬡–C(=O)–CHCH₂CH₃ with CO₂H and CH₃ —-(CO₂), Heat→ ⬡–C(=O)–CHCH₂CH₃ with CH₃

## Structure and Nomenclature

Problem 13.7 Name and draw structural formulas for the four carboxylic acids of molecular formula C₅H₁₀O₂. Which of these carboxylic acids is chiral?

CH₃(CH₂)₃COH          CH₃CHCH₂COH          CH₃—C—COH          CH₃CH₂—C*—COH
                              |CH₃                  |H₃C with H₃C              with H₃C and H

**Pentanoic acid**     **3-Methylbutanoic acid**     **2,2-Dimethyl-propanoic acid**     **2-Methylpropanoic acid (chiral)**

**Problem 13.8** Write the IUPAC names for each carboxylic acid.

(a) 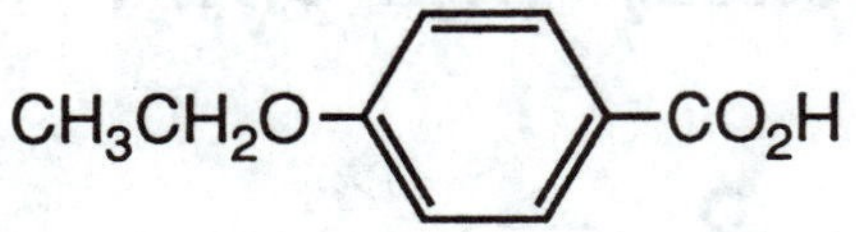  $CH_3CH_2O$—⟨benzene ring⟩—$CO_2H$

**4-Ethoxybenzoic acid**

(b) 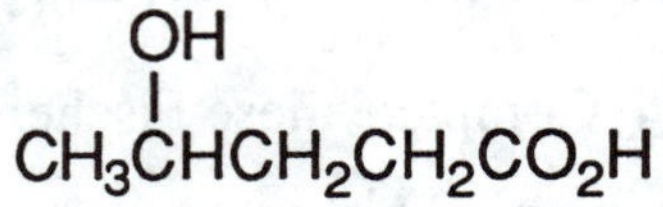  $\overset{OH}{\overset{|}{CH_3CHCH_2CH_2CO_2H}}$

**4-Hydroxypentanoic acid**

(c) 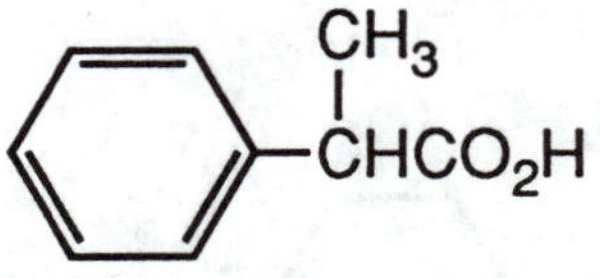  ⟨benzene ring⟩—$\overset{CH_3}{\overset{|}{CHCO_2H}}$

**2-Phenylpropanoic acid**

(d) 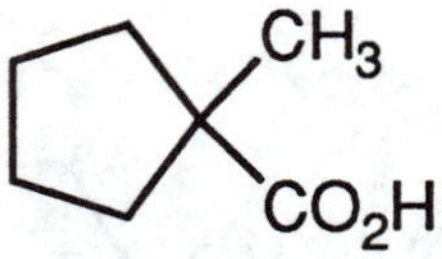  ⟨cyclopentane ring with $CH_3$ and $CO_2H$⟩

**1-Methylcyclopentanecarboxylic acid**

(e)  $CH_3(CH_2)_4CO_2H$

**Hexanoic acid**

(f) 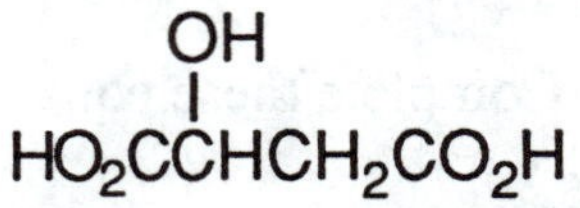  $\overset{OH}{\overset{|}{HO_2CCHCH_2CO_2H}}$

**2-Hydroxybutanedioic acid**

**Problem 13.9** Draw structural formulas for these carboxylic acids.

(a) 4-Nitrophenylacetic acid

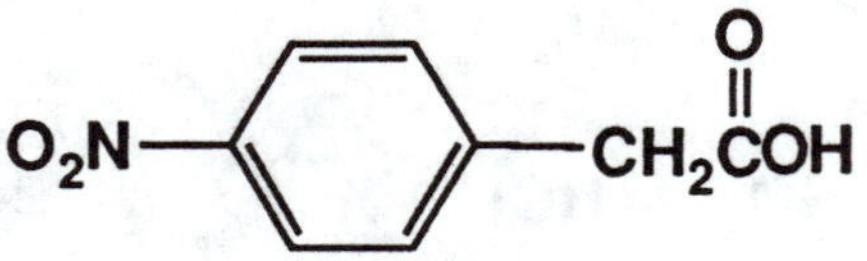

$O_2N$—⟨benzene ring⟩—$\overset{\overset{O}{\|}}{CH_2COH}$

(b) 4-Aminobutanoic acid

$H_2NCH_2CH_2CH_2CO_2H$

(c) 3-Chloro-4-phenylbutanoic acid

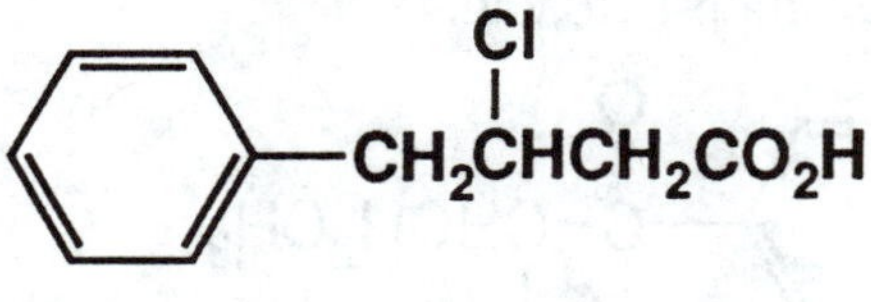

⟨benzene ring⟩—$\overset{Cl}{\overset{|}{CH_2CHCH_2CO_2H}}$

(d) (Z)-3-Hexenedioic acid

$$HO_2CCH_2 \underset{H}{\overset{}{C}}{=}\underset{H}{\overset{}{C}} CH_2CO_2H$$

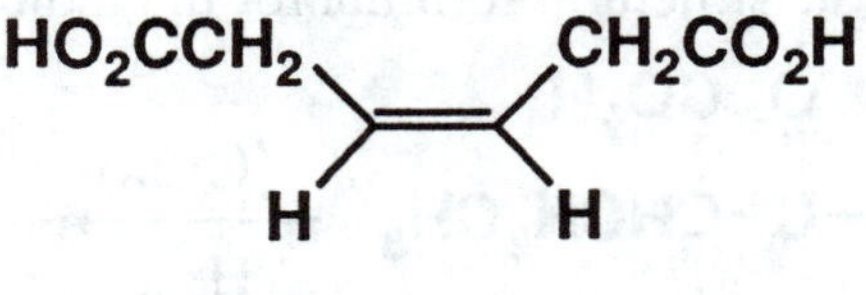

(e) Potassium phenylacetate

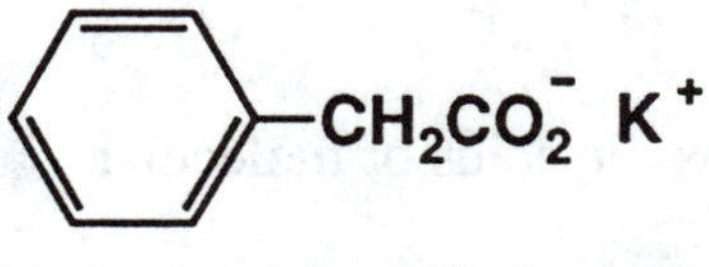

⟨benzene ring⟩—$CH_2CO_2^-\ K^+$

(f) 3-Oxohexanoic acid

$$CH_3CH_2CH_2 \overset{\overset{O}{\|}}{C}CH_2 \overset{\overset{O}{\|}}{C}OH$$

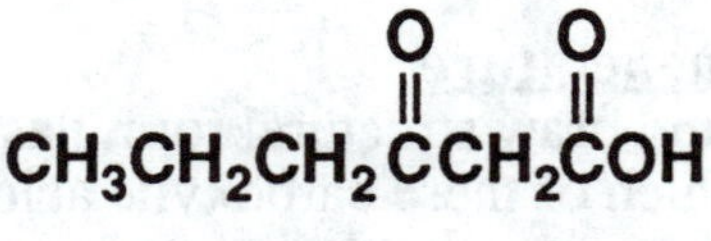

(g) 2-Oxocyclohexanecarboxylic acid

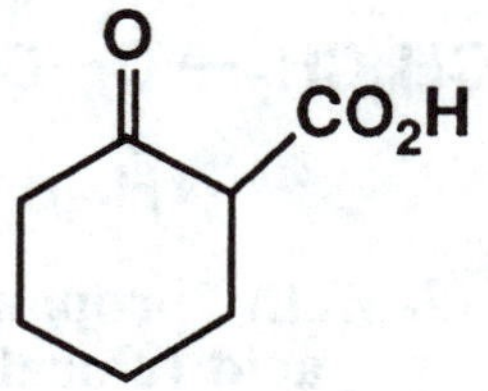

⟨cyclohexanone ring with $CO_2H$⟩

(j) 2,2-Dimethylpropanoic acid

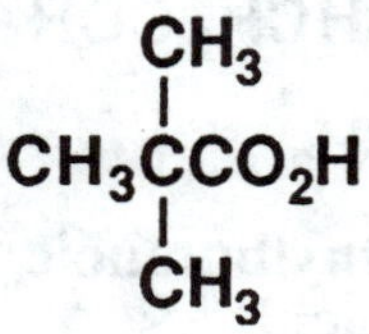

$$\overset{CH_3}{\underset{CH_3}{\overset{|}{CH_3CCO_2H}}}$$

**Problem 13.10** The IUPAC name of ibuprophen is 2-(4-isobutylphenyl)propanoic acid. Draw the structural formula of ibuprophen.

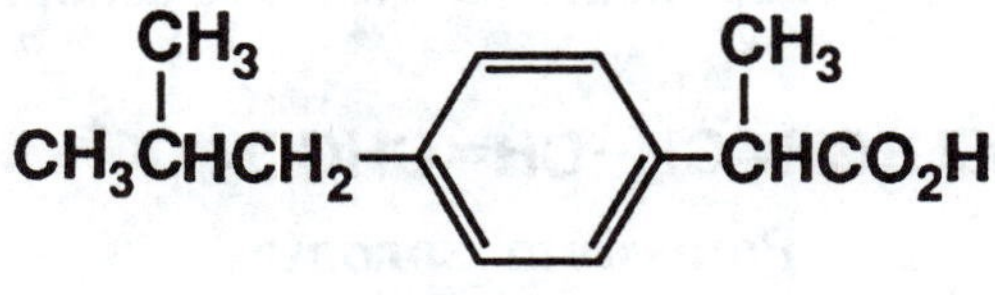

**Ibuprophen**

**Problem 13.11** The IUPAC name of ibuprophen contains the parts isobutyl, phenyl, and propanoic acid. Suggest how these parts might have been combined to arrive at the name ibuprophen.

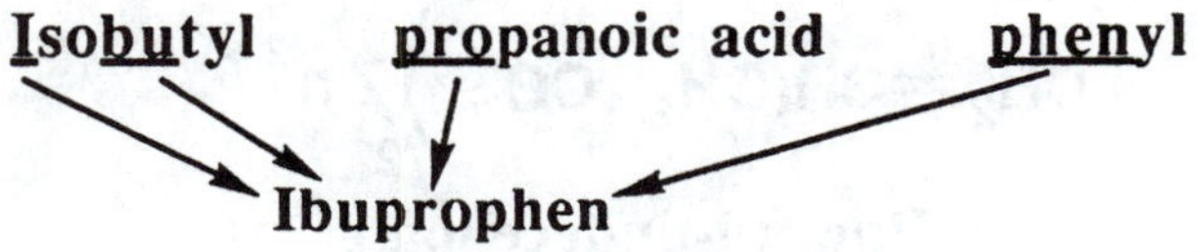

**Problem 13.12** Draw structural formulas for these salts.
(a) Sodium benzoate                    (b) Lithium acetate

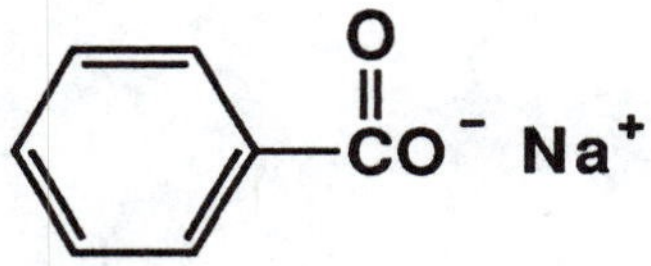

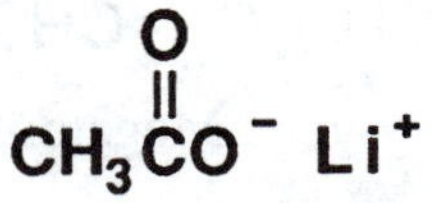

(c) Ammonium acetate 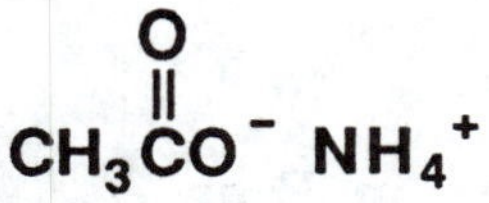                   (d) Disodium adipate

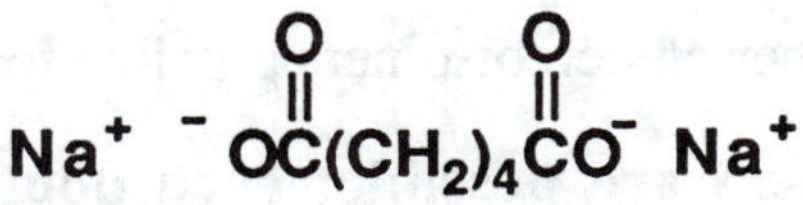

(e) Sodium salicylate 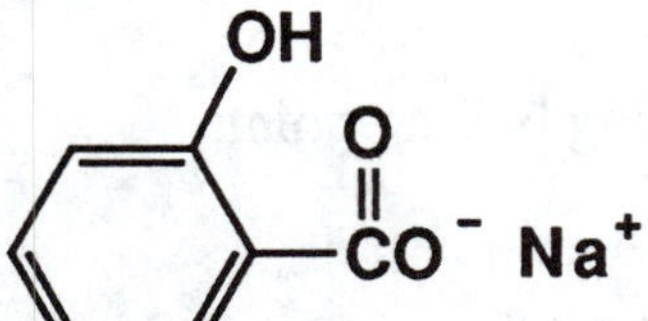                  (f) Calcium butanoate

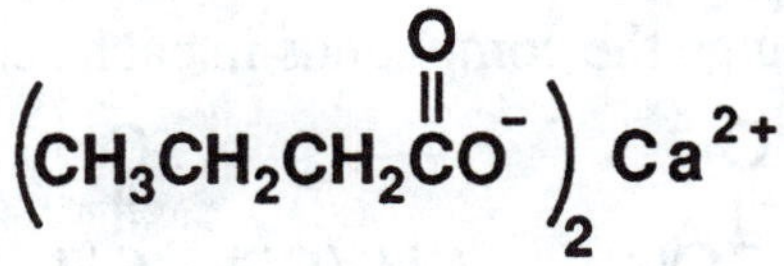

**Problem 13.13** The monosodium salt of oxalic acid is present in certain leafy vegetables, including rhubarb. Both oxalic acid and its salt are poisonous in high concentrations. Draw the structural formula of monopotassium oxalate.

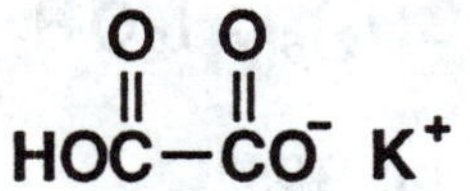

**Monopotassium oxalate**

<u>Problem 13.14</u> Potassium sorbate is added as a preservative to certain foods to prevent bacteria and molds from causing food spoilage and to extend the foods' shelf life. The IUPAC name of potassium sorbate is potassium 2,4-hexadienoate. Draw the structural formula of potassium sorbate.

$$CH_3CH=CH-CH=CHCO_2^- \ K^+$$

**Potassium sorbate**

<u>Problem 13.15</u> Zinc 10-undecenoate, the zinc salt of 10-undecenoic acid, is used to treat certain fungal infections, particularly *tineapidis* (athlete's foot). Draw the structural formula of this zinc salt.

$$\left(CH_2{=}CH(CH_2)_8\overset{\overset{\displaystyle O}{\|}}{C}O^-\right)_2 Zn^{2+}$$

**Zinc 10-undecenoate**

<u>Problem 13.16</u> Megatomoic acid, the sex attractant of the female black carpet beetle, has the structure:

$$CH_3(CH_2)_7CH=CHCH=CHCH_2CO_2H$$

Megatomoic acid

(a) What is its IUPAC name?

**Its IUPAC name is 3,5-tetradecadienoic acid.**

(b) State the number of stereoisomers possible for this compound.

**Four stereoisomers are possible; each double bond can have either an E or Z (*trans* or *cis*) configuration.**

<u>**Physical properties**</u>
<u>Problem 13.17</u> Arrange the compounds in each set in order of increasing boiling point.

(a)    $CH_3(CH_2)_5\overset{\overset{\displaystyle O}{\|}}{C}OH$       $CH_3(CH_2)_6\overset{\overset{\displaystyle O}{\|}}{C}H$       $CH_3(CH_2)_6CH_2OH$

**The better the hydrogen bond capability, the higher the boiling point. In order of increasing boiling point, they are:**

$CH_3(CH_2)_6\overset{\overset{\displaystyle O}{\|}}{C}H$          $CH_3(CH_2)_6CH_2OH$          $CH_3(CH_2)_5\overset{\overset{\displaystyle O}{\|}}{C}OH$

**bp 171°C**                    **bp 195°C**                    **bp 223°C**

(b)

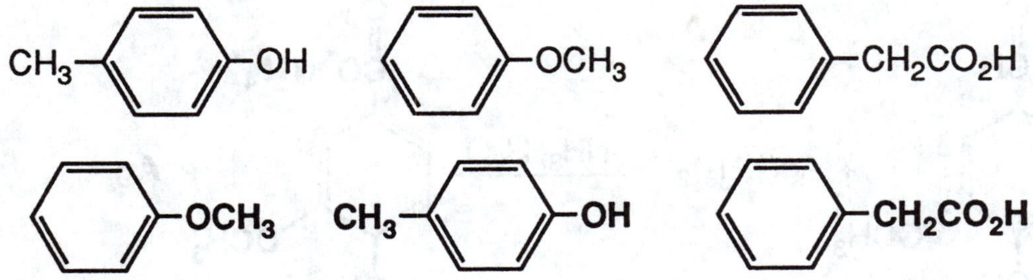

**bp 155°C** **bp 202°C** **bp 266°C**

(c) $CH_3CH_2C(=O)OH$ $CH_3CH_2CH_2CH_2OH$ $CH_3CH_2OCH_2CH_3$

$\mathbf{CH_3CH_2OCH_2CH_3}$ $\mathbf{CH_3CH_2CH_2CH_2OH}$ $\mathbf{CH_3CH_2C(=O)OH}$

**bp 35°C** **bp 117°C** **bp 141°C**

Problem 13.18 Acetic acid has a boiling point of 118°C, whereas its methyl ester has a boiling point of 57°C. Account for the fact that the boiling point of acetic acid is higher than that of its methyl ester, even though acetic acid has a lower molecular weight.

**Acetic acid can make strong hydrogen bonds, but the methyl ester lacks a hydrogen bonding hydrogen atom. Thus, acetic acid will have the much higher boiling point compared to the methyl ester.**

**Preparation of Carboxylic Acids**

We have seen four general methods for the preparation of carboxylic acids:
(1) Oxidation of primary alcohols (Section 7.4F),
(2) Oxidation of arene side chains (Section 9.5),
(3) Oxidation of aldehydes (Section 11.9A),
(4) Carbonation of a Grignard reagent (Section 11.5B).

Problem 13.19 Complete these oxidations.

(a) $CH_3(CH_2)_4CH_2OH + H_2CrO_4 \xrightarrow[\text{heat}]{H^+} \mathbf{CH_3(CH_2)_4CO_2H + Cr^{3+}}$

(b)  Vanillin $+ Ag(NH_3)_2^+ \xrightarrow{NH_3, H_2O}$ ... $+ Ag$

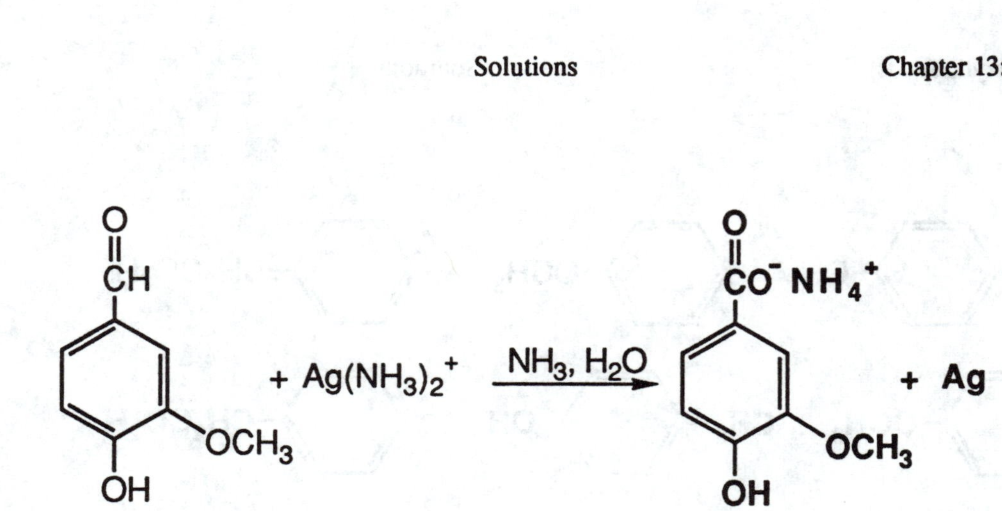

(c) 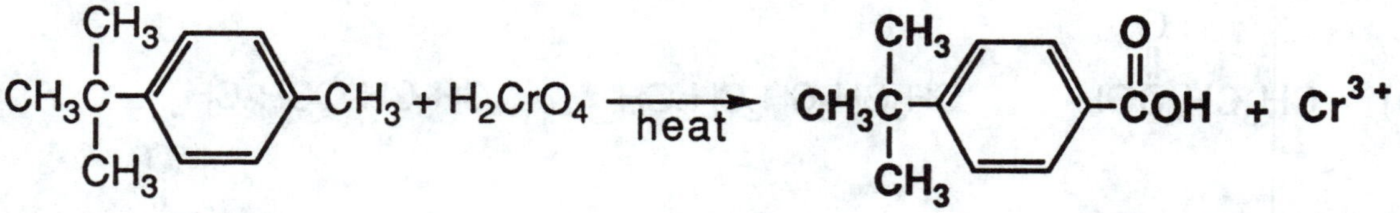  $CH_3C(CH_3)_2$-substituted toluene $+ H_2CrO_4 \xrightarrow{heat}$ ... $+ Cr^{3+}$

(d) 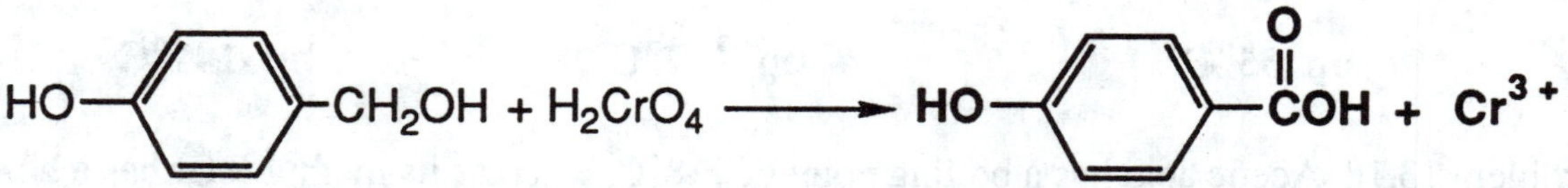  $HO$-benzene-$CH_2OH + H_2CrO_4 \longrightarrow HO$-benzene-$COH + Cr^{3+}$

**Problem 13.20** Show reagents and experimental conditions to complete this synthesis.

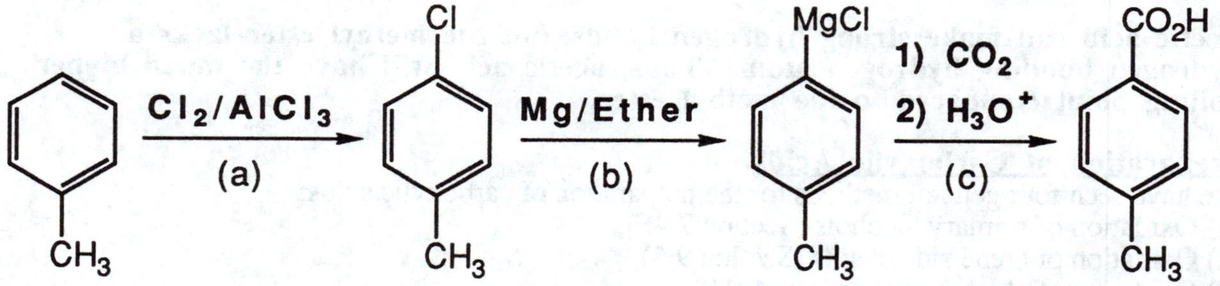

toluene $\xrightarrow[\text{(a)}]{Cl_2/AlCl_3}$ 4-chlorotoluene $\xrightarrow[\text{(b)}]{Mg/Ether}$ 4-methylphenyl-MgCl $\xrightarrow[\text{(c)}]{1) CO_2 \quad 2) H_3O^+}$ 4-methylbenzoic acid ($CO_2H$)

**Problem 13.21** Draw the structural formula of a compound of the given molecular formula that, on oxidation by potassium permanganate in aqueous KOH followed by acidification with aqueous HCl gives the carboxylic acid or dicarboxylic acid shown.

(a)  $C_6H_{14}O \xrightarrow{\text{oxidation}} CH_3(CH_2)_4COH$

$CH_3(CH_2)_4CH_2OH$

(b)  $C_6H_{12}O \xrightarrow{\text{oxidation}} CH_3(CH_2)_4COH$

$CH_3(CH_2)_4CH$

(c)  $C_6H_{14}O_2 \xrightarrow{\text{oxidation}} HOC(CH_2)_4COH$

$HOCH_2(CH_2)_4CH_2OH$

## Acidity of Carboxylic Acids

<u>Problem 13.22</u>  Which is the stronger acid, phenol ($pK_a = 9.95$) or benzoic acid ($pK_a = 4.17$)?

**Recall that $pK_a$ is the negative $\log_{10}$ of $K_a$. The smaller the $pK_a$, the stronger the acid, so benzoic acid is the stronger acid.**

<u>Problem 13.23</u>  Which is the stronger acid, lactic acid ($K_a = 8.4 \times 10^{-4}$) or benzoic acid ($pK_a = 7.9 \times 10^{-5}$)?

**The larger the value of $K_a$, the stronger the acid, so lactic acid is the stronger acid.**

<u>Problem 13.24</u>  Assign the acid in each set its appropriate $pK_a$.

(a)

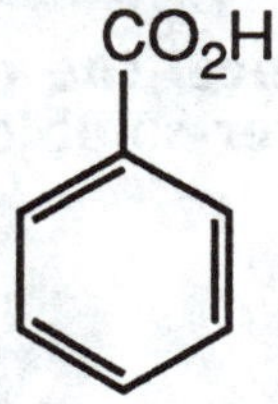

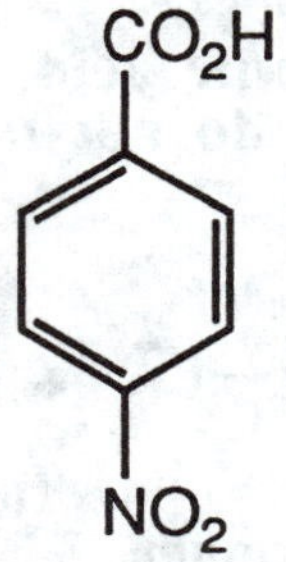

(pK$_a$ 4.19 and 3.14)

4.19          3.14

(b)

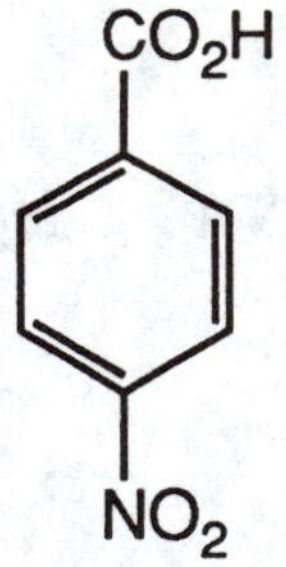

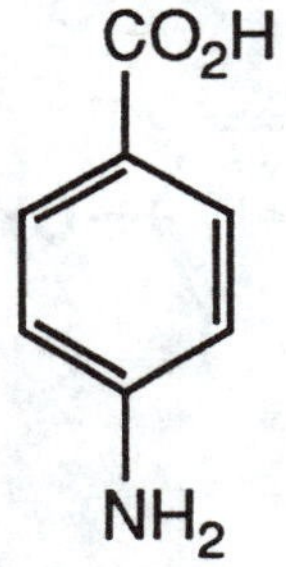

(pK$_a$ 4.92 and 3.14)

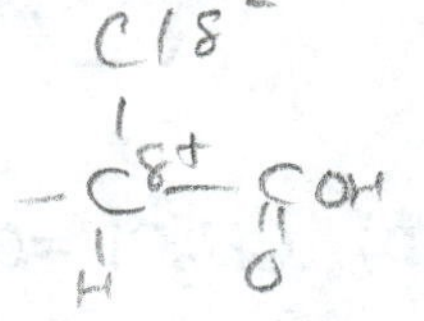

3.14          4.92

(c)  $CH_3CCH_2CO_2H$  and  $CH_3CH_2CCO_2H$   (pK$_a$ 3.58 and 2.49)

3.58                  2.49

(d)   $CH_3CHCO_2H$  and  $CH_3CH_2CO_2H$   (pK$_a$ 4.78 and 3.08)

3.08              4.78

Problem 13.25 Account for the fact that water-insoluble carboxylic acids (pK$_a$ 4-5) dissolve in 10% aqueous sodium bicarbonate (pH 8.5) with the evolution of a gas but water insoluble phenols (pK$_a$ 9.5-10.5) do not dissolve in 10% sodium bicarbonate.

**As shown in the following equilibrium equations, a carboxylic acid is a stronger acid than carbonic acid and the position of equilibrium for this acid-base reaction lies to the right; the carboxylic acid dissolves as a water-soluble sodium salt.**

$$\underset{\substack{(pK_a\ 4.5)\\ \text{Stronger}\\ \text{acid}}}{R-\overset{\overset{\displaystyle O}{\|}}{C}-OH} + \underset{\substack{\text{Stronger}\\ \text{base}}}{HCO_3^-} \quad \rightleftharpoons \quad \underset{\substack{\text{Weaker}\\ \text{base}}}{R-\overset{\overset{\displaystyle O}{\|}}{C}-O^-} + \underset{\substack{(pK_a\ 6.36)\\ \text{Weaker}\\ \text{acid}}}{H_2CO_3}$$

**Phenols are considerably weaker acids than carbonic acid and, hence, the position of this acid-base reaction lies to the left; phenols do not form water-soluble salts in aqueous sodium bicarbonate.**

$$\underset{\substack{(pK_a\ 10.0)\\ \text{Weaker}\\ \text{acid}}}{Ar-OH} + \underset{\substack{\text{Weaker}\\ \text{base}}}{HCO_3^-} \quad \rightleftharpoons \quad \underset{\substack{\text{Stronger}\\ \text{base}}}{Ar-O^-} + \underset{\substack{(pK_a\ 6.36)\\ \text{Stronger}\\ \text{acid}}}{H_2CO_3}$$

Problem 13.26 Complete these acid-base reactions:

(a)  C$_6$H$_5$—CH$_2$CO$_2$H + NaOH $\longrightarrow$ C$_6$H$_5$—CH$_2$CO$_2^-$ Na$^+$ + H$_2$O

(b)  CH$_3$CH=CHCH$_2$CO$_2$H + NaHCO$_3$ $\longrightarrow$

CH$_3$CH=CHCH$_2$CO$_2^-$Na$^+$ + H$_2$O + CO$_2$

(c)  (salicylic acid, CO$_2$H and OH) + NaHCO$_3$ $\longrightarrow$ (CO$_2^-$ Na$^+$ and OH) + H$_2$O + CO$_2$

(d)  $\underset{}{CH_3\overset{\overset{\displaystyle OH}{|}}{C}HCO_2H}$ + H$_2$NCH$_2$CH$_2$OH $\longrightarrow$ $\underset{}{CH_3\overset{\overset{\displaystyle OH}{|}}{C}HCO_2^-}$ + H$_3\overset{+}{N}$CH$_2$CH$_2$OH

(e)   $CH_3CH=CHCH_2CO_2^- \ Na^+ + HCl \longrightarrow CH_3CH=CHCH_2CO_2H + NaCl$

<u>Problem 13.27</u>  The normal pH range for blood plasma is 7.35-7.45.  Under these conditions, would you expect the carboxyl group of lactic acid ($pK_a = 4.07$) to exist primarily as a carboxyl group or as a carboxylic anion?

**Recall from the definition of $K_a$ that:**

$$K_a = \frac{[A^-]\,[H^+]}{[H\text{-}A]} \qquad \text{so dividing both sides by } [H^+] \text{ gives} \qquad \frac{K_a}{[H^+]} = \frac{[A^-]}{[H\text{-}A]}$$

**Here, $[H^+]$ is concentration of $H^+$, $[H\text{-}A]$ is concentration of protonated acid (lactic acid in this case) and $[A^-]$ is the concentration of deprotonated acid (lactic acid carboxylate anion in this case). Therefore, if the ratio of $K_a / [H^+]$ is greater than 1, $[A^-]$ will be the predominant form, and if the ratio of $K_a / [H^+]$ is less than 1, than $[H\text{-}A]$ will be the predominant form.  Recall that $pH = -\log_{10} [H^+]$, so a pH of 7.4 corresponds to a $[H^+]$ of $10^{-(pH)} = 10^{-(7.4)} = 4.0 \times 10^{-8}$.  Similarly, $pK_a = -\log_{10} K_a$, so for lactic acid $K_a = 10^{-(Ka)} = 10^{-(4.07)} = 8.5 \times 10^{-5}$.  Using these numbers:**

$$\frac{[A^-]}{[H\text{-}A]} \;=\; \frac{K_a}{[H^+]} \;=\; \frac{8.5 \times 10^{-5}}{4.0 \times 10^{-8}} = 2.1 \times 10^3$$

**Therefore, lactic acid will exist primarily as the carboxylic anion in blood plasma.**

<u>Problem 13.28</u>  The $K_a$ of ascorbic acid is $7.95 \times 10^{-5}$.  Would you expect ascorbic acid dissolved in blood plasma to exist primarily as ascorbic acid or as ascorbate anion.  Explain.

**Using the same reasoning described in the answer to Problem 13.27:**

$$\frac{[A^-]}{[H\text{-}A]} \;=\; \frac{K_a}{[H^+]} \;=\; \frac{7.9 \times 10^{-5}}{4.0 \times 10^{-8}} = 2.0 \times 10^3$$

**Therefore, ascorbic acid will exist primarily as the ascorbate anion in blood plasma.**

<u>Problem 13.29</u>  Excess ascorbic acid is excreted in the urine, the pH of which is normally in the range 4.8 - 8.4.  What form of ascorbic acid would you expect to be present in urine of pH 8.4, free ascorbic acid or ascorbate anion?  Explain.

**At pH 8.4, $[H^+] = 4.0 \times 10^{-9}$, therefore using the same reasoning as described in the answer to Problem 13.27:**

$$\frac{[A^-]}{[H\text{-}A]} \;=\; \frac{K_a}{[H^+]} \;=\; \frac{7.9 \times 10^{-5}}{4.0 \times 10^{-9}} = 2.0 \times 10^4$$

**Therefore, ascorbic acid will exist primarily as the ascorbate anion in urine of pH 8.4.**

Problem 13.30  The pH of human gastric juice is normally in the range 1.0 - 3.0.  What form of lactic acid would you expect to be present in the stomach, lactic acid or its lactate anion?  Explain.

At pH 3.0, $[H^+] = 1.0 \times 10^{-3}$ and as described in Problem 13.27, the Ka of lactic acid is $8.5 \times 10^{-5}$.  Therefore, using the same reasoning as described in the answer to Problem 13.27:

$$\frac{[A^-]}{[H\text{-}A]} = \frac{K_a}{[H^+]} = \frac{8.5 \times 10^{-5}}{1.0 \times 10^{-3}} = 8.5 \times 10^{-2}$$

Therefore, lactic acid will exist primarily as the lactic acid form in gastric juices of pH <3.0.

## Reactions of Carboxylic Acids
Problem 13.31  Give the expected organic products formed when $PhCH_2CO_2H$, phenylacetic acid, is treated with each of the reagents.

(a) $SOCl_2$

$$PhCH_2\overset{\overset{\displaystyle O}{\|}}{C}OH + SOCl_2 \longrightarrow PhCH_2\overset{\overset{\displaystyle O}{\|}}{C}Cl + SO_2 + HCl$$

(b) $NaHCO_3, H_2O$

$$PhCH_2\overset{\overset{\displaystyle O}{\|}}{C}OH + NaHCO_3 \longrightarrow PhCH_2\overset{\overset{\displaystyle O}{\|}}{C}O^- Na^+ + CO_2 + H_2O$$

(c) $NaOH, H_2O$

$$PhCH_2\overset{\overset{\displaystyle O}{\|}}{C}OH + NaOH \longrightarrow PhCH_2\overset{\overset{\displaystyle O}{\|}}{C}O^- Na^+ + H_2O$$

(d) $NH_3, H_2O$

$$PhCH_2\overset{\overset{\displaystyle O}{\|}}{C}OH + NH_3 \longrightarrow PhCH_2\overset{\overset{\displaystyle O}{\|}}{C}O^- NH_4^+$$

(e) $LiAlH_4$ followed by $H_2O$

$$PhCH_2\overset{\overset{\displaystyle O}{\|}}{C}OH \xrightarrow[\text{2) } H_2O]{\text{1) } LiAlH_4} PhCH_2CH_2OH$$

(f) $CH_3OH + H_2SO_4$ (catalyst)

$$PhCH_2\overset{O}{\overset{\|}{C}}OH + CH_3OH \underset{H_2SO_4}{\overset{H_2SO_4}{\rightleftharpoons}} PhCH_2\overset{O}{\overset{\|}{C}}OCH_3 + H_2O$$

<u>Problem 13.32</u> Show how to convert *trans*-3-phenyl-2-propenoic acid (cinnamic acid) to these compounds.

(a)

$$\begin{array}{c} H \\ \diagdown \\ C_6H_5 \end{array} C=C \begin{array}{c} \overset{O}{\overset{\|}{C}}-OH \\ \diagdown \\ H \end{array} \xrightarrow[\text{2) } H_2O]{\text{1) } LiAlH_4} \begin{array}{c} H \\ \diagdown \\ C_6H_5 \end{array} C=C \begin{array}{c} CH_2OH \\ \diagup \\ H \end{array}$$

(b)

$$\begin{array}{c} H \\ \diagdown \\ C_6H_5 \end{array} C=C \begin{array}{c} \overset{O}{\overset{\|}{C}}-OH \\ \diagdown \\ H \end{array} \xrightarrow[\substack{Pt \quad 25°C \\ 2 \text{ atm}}]{H_2} C_6H_5CH_2CH_2CO_2H$$

(c)

$$\begin{array}{c} H \\ \diagdown \\ C_6H_5 \end{array} C=C \begin{array}{c} \overset{O}{\overset{\|}{C}}-OH \\ \diagdown \\ H \end{array} \xrightarrow[\text{2) } H_2O]{\text{1) } LiAlH_4} \xrightarrow[\substack{Pt \quad 25°C \\ 2 \text{ atm}}]{H_2} C_6H_5CH_2CH_2CH_2OH$$

<u>Problem 13.33</u> Show how to convert 3-oxobutanoic acid (acetoacetic acid) to these compounds.

(a)

$$CH_3\overset{O}{\overset{\|}{C}}CH_2\overset{O}{\overset{\|}{C}}OH \xrightarrow[\text{2) } H_2O]{\text{1) } NaBH_4} CH_3\overset{OH}{\overset{|}{C}}HCH_2\overset{O}{\overset{\|}{C}}OH$$

(b)

$$CH_3\overset{O}{\overset{\|}{C}}CH_2\overset{O}{\overset{\|}{C}}OH \xrightarrow[\text{2) } H_2O]{\text{1) } LiAlH_4} CH_3\overset{OH}{\overset{|}{C}}HCH_2CH_2($$

(c)

$$CH_3\overset{O}{\overset{\|}{C}}CH_2\overset{O}{\overset{\|}{C}}OH \xrightarrow[\text{2) } H_2O]{\text{1) } NaBH_4} CH_3\overset{OH}{\overset{|}{C}}HCH_2\overset{O}{\overset{\|}{C}}OH \xrightarrow{H_2SO_4}$$

$$CH_3CH{=}CHCO_2H$$

Problem 13.34  Complete these examples of Fischer esterification.

(a) $CH_3CO_2H$ + $HOCH_2CH_2CH(CH_3)_2$ $\underset{}{\overset{H^+}{\rightleftharpoons}}$ $CH_3\overset{\overset{\textstyle O}{\|}}{C}OCH_2CH_2CH(CH_3)_2$ + $H_2O$

(b) [benzene ring with $CO_2H$ at top and $CO_2H$ at bottom] + 2 $CH_3OH$ $\overset{H^+}{\rightleftharpoons}$ [benzene ring with $CO_2CH_3$ at top and $CO_2CH_3$ at bottom] + 2 $H_2O$

(c) $HO_2C(CH_2)_2CO_2H$ + 2 $CH_3CH_2OH$ $\overset{H^+}{\rightleftharpoons}$ $CH_3CH_2O\overset{\overset{\textstyle O}{\|}}{C}(CH_2)_2\overset{\overset{\textstyle O}{\|}}{C}OCH_2CH_3$ + 2 $H_2O$

Problem 13.35  Methyl 2-hydroxybenzoate (methyl salicylate) has the odor of oil of wintergreen. This ester is prepared by Fischer esterification of 2-hydroxybenzoic acid (salicylic acid) with methanol. Draw the structural formula of methyl 2-hydroxybenzoic acid.

[benzene ring with $CO_2H$ at top and $OH$ at bottom] + $CH_3OH$ $\overset{H_2SO_4}{\longrightarrow}$ [benzene ring with $\overset{\overset{\textstyle O}{\|}}{C}OCH_3$ at top and $OH$ at bottom] + $H_2O$

**2-Hydroxybenzoic acid**
**(Salicylic acid)**

**Methyl 2-hydroxybenzoate**
**(Oil of wintergreen)**

Problem 13.36  Benzocaine, a topical anesthetic, is prepared by treatment of 4-aminobenzoic acid with ethanol in the presence of an acid catalyst followed by neutralization. Draw the structural formula of benzocaine.

$H_2N$—[benzene ring]—$CO_2H$ + $CH_3CH_2OH$ $\xrightarrow[\substack{\textbf{2) Mild base to}\\\textbf{deprotonate}\\\textbf{amino group}}]{\textbf{1) H}_2\textbf{SO}_4}$

**4-Aminobenzoic acid**

$H_2N$—[benzene ring]—$\overset{\overset{\textstyle O}{\|}}{C}OCH_2CH_3$ + $H_2O$

**Benzocaine**
**(a topical anesthetic)**

**Problem 13.37** From what carboxylic acid and what alcohol is each ester derived?

(a) 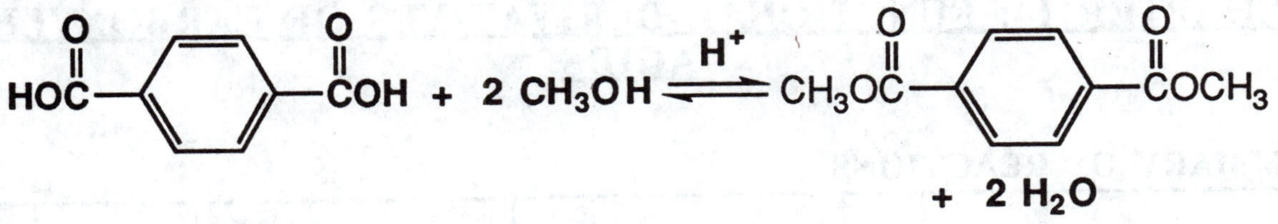 $HO\overset{O}{\overset{\|}{C}}$—⬡—$\overset{O}{\overset{\|}{C}}OH$ + 2 $CH_3OH$ $\overset{H^+}{\rightleftharpoons}$ $CH_3O\overset{O}{\overset{\|}{C}}$—⬡—$\overset{O}{\overset{\|}{C}}OCH_3$

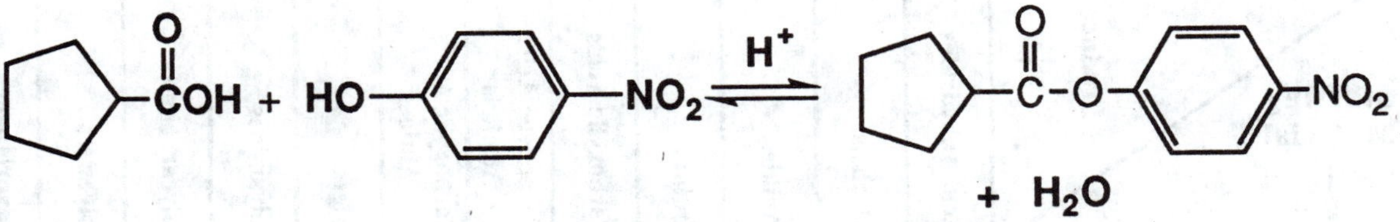

$+ \ 2\ H_2O$

(b) ⬠—$\overset{O}{\overset{\|}{C}}OH$ + $HO$—⬡—$NO_2$ $\overset{H^+}{\rightleftharpoons}$ ⬠—$\overset{O}{\overset{\|}{C}}$-O—⬡—$NO_2$

$+ \ H_2O$

(c) $CH_3\overset{O}{\overset{\|}{C}}OH$ + ⬡—$OH$ $\rightleftharpoons$ ⬡—$O\overset{O}{\overset{\|}{C}}CH_3$ + $H_2O$

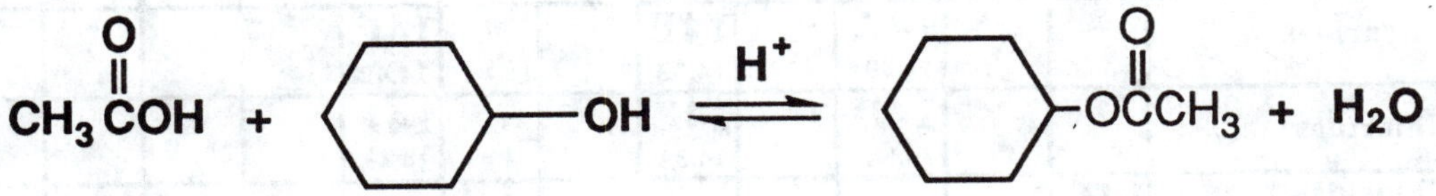

(d) $CH_3CH_2CH{=}CH\overset{O}{\overset{\|}{C}}OH$ + $HOCH(CH_3)_2$ $\overset{H^+}{\rightleftharpoons}$ $CH_3CH_2CH{=}CH\overset{O}{\overset{\|}{C}}OCH(CH_3)_2$

$+ \ H_2O$

**Problem 13.38** Draw the product formed on thermal decarboxylation of these compounds.

(a) $C_6H_5\overset{O}{\overset{\|}{C}}CH_2CO_2H$ $\xrightarrow{Heat}$ $C_6H_5\overset{O}{\overset{\|}{C}}CH_3$ + $CO_2$

(b) $C_6H_5CH_2\overset{CO_2H}{\overset{|}{CH}}CO_2H$ $\xrightarrow{Heat}$ $C_6H_5CH_2CH_2CO_2H$ + $CO_2$

(c) 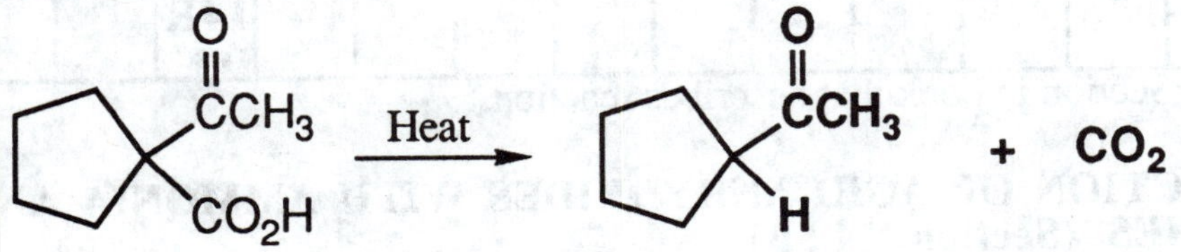 ⬠$\overset{\overset{O}{\overset{\|}{C}CH_3}}{\underset{CO_2H}{}}$ $\xrightarrow{Heat}$ ⬠$\overset{\overset{O}{\overset{\|}{C}CH_3}}{\underset{H}{}}$ + $CO_2$

# CHAPTER 14: FUNCTIONAL DERIVATIVES OF CARBOXYLIC ACIDS

## SUMMARY OF REACTIONS

| Starting Material \ Product → | Acid Anhydrides | Alcohols | Amides | Amines | Carboxylic Acids | Carboxylic Acids Alcohols | Carboxylic Acids Amines | Esters | Polyamides | Polycarbonates | Polyesters | Polyurethanes |
|---|---|---|---|---|---|---|---|---|---|---|---|---|
| Acid Anhydrides | | | 14A 14.5B* | | 14B 14.3B | | | 14C 14.4B | | | | |
| Acid Chlorides | | | 14D 14.5A | | 14E 14.3A | | | 14F 14.4A | | | | |
| Acid Chlorides Carboxylates | 14G 14.6 | | | | | | | | | | | |
| Amides | | | | 14H 14.7B | | | 14I 14.3D | | | | | |
| Dicarboxylic Acids Diamines | | | | | | | | | 14J 14.8A | | | |
| Dicarboxylic Acids Diols | | | | | | | | | | | 14K 14.8B | |
| Diisocyanates Alcohols | | | | | | | | | | | | 14L 14.8D |
| Esters | | 14M 14.7A | 14N 14.5C | | 14O 14.3C | | | | | | | |
| Hydroxy Carboxylic Acids | | | | | | | | | | | 14P 14.8D | |
| Lactams | | | | | | | | | 14Q 14.8A | | | |
| Phosgene Diols | | | | | | | | | | 14R 14.8C | | |

*Section in book that describes reaction.

## REACTION 14A: REACTION OF ACID ANHYDRIDES WITH AMMONIA AND AMINES (Section 14.5B)

- **Acid anhydrides** react with **ammonia, primary amines,** and **secondary amines** to produce **amides.** ✳

- Two moles of amine are usually used in the reaction; one mole to form the amide and one mole to neutralize the carboxylic acid that is also produced.
- For this reaction as well as the others mentioned in this chapter, the acid anhydride will usually be symmetric, that is, derived from two moles of the same carboxylic acid.

# REACTION 14B: HYDROLYSIS OF ACID ANHYDRIDES (Section 14.3B)

- **Lower molecular weight acid anhydrides** react with **water** to from **two moles of carboxylic acid.** *
- Like the other reactions that involve attack by nucleophiles on acid anhydrides, the mechanism of the reaction begins when the nucleophile reacts with the carbonyl carbon atom to form a tetrahedral carbonyl addition intermediate, that converts to product by elimination of a carboxylate leaving group.

# REACTION 14C: REACTION OF ACID ANHYDRIDES WITH ALCOHOLS (Section 14.4B)

- **Acid anhydrides** also react with **alcohols** to produce **one mole** of **ester** and **one mole** of **carboxylic acid.** *

# REACTION 14D: REACTION OF ACID CHLORIDES WITH AMMONIA AND AMINES (Section 14.5A)

- **Acid chlorides** react with **ammonia, primary amines**, and **secondary amines** to produce **amides.** *
- Like the other reactions that involve attack by nucleophiles on acid chlorides, the mechanism of the reaction begins when the nucleophile reacts with the carbonyl carbon atom to form a tetrahedral carbonyl addition intermediate, that converts to product by elimination of a chloride leaving group.

## REACTION 14E: HYDROLYSIS OF ACID CHLORIDES (Section 14.3A)

$$-\overset{|}{\underset{|}{C}}-\overset{O}{\overset{\|}{C}}\diagdown_{Cl} \ + \ H_2O \ \longrightarrow \ -\overset{|}{\underset{|}{C}}-\overset{O}{\overset{\|}{C}}\diagdown_{OH} \ + \ HCl$$

- Lower molecular weight **acid chlorides** are so reactive to nucleophiles that they are **readily hydrolyzed** to a **carboxylic acid** by water. ✱
- The higher molecular weight acid chlorides are less soluble in water, so they react more slowly.

## REACTION 14F: REACTION OF ACID CHLORIDES WITH ALCOHOLS (Section 14.5A)

$$-\overset{|}{\underset{|}{C}}-\overset{O}{\overset{\|}{C}}\diagdown_{Cl} \ + \ -\overset{|}{\underset{|}{C''}}-OH \ \longrightarrow \ -\overset{|}{\underset{|}{C}}-\overset{O}{\overset{\|}{C'}}-O-\overset{|}{\underset{|}{C''}}- \ + \ HCl$$

- **Acid chlorides** react with **alcohols** to produce **esters**. ✱

## REACTION 14G: REACTION OF ACID CHLORIDES WITH CARBOXYLATE SALTS (Section 14.6)

$$-\overset{|}{\underset{|}{C}}-\overset{O}{\overset{\|}{C'}}\diagdown_{Cl} \ + \ -\overset{|}{\underset{|}{C'''}}-\overset{O}{\overset{\|}{C''}}\diagdown_{O^-\ Na^+} \ \longrightarrow \ -\overset{|}{\underset{|}{C}}-\overset{O}{\overset{\|}{C'}}-O-\overset{O}{\overset{\|}{C''}}-\overset{|}{\underset{|}{C'''}}- \ + \ NaCl$$

- **Acid chlorides** react with **carboxylate salts** to produce **acid anhydrides**. ✱

## REACTION 14H: REDUCTION OF AMIDES (Section 14.7B)

$$-\overset{|}{\underset{|}{C}}-\overset{O}{\overset{\|}{C'}}-\overset{|}{\underset{|}{N}}- \ \xrightarrow[\text{2) }H_2O]{\text{1) LiAlH}_4} \ -\overset{|}{\underset{|}{C}}-\overset{H}{\underset{H}{C'}}-\overset{|}{\underset{|}{N}}-$$

- **Amides** are **reduced by LiAlH₄** to produce **primary, secondary**, or **tertiary amines**, depending on the substitution of the original amide. ✱

## REACTION 14I: HYDROLYSIS OF AMIDES (Section 14.3D)

$$-\overset{|}{\underset{|}{C}}-\overset{O}{\overset{\|}{C'}}-\overset{|}{\underset{|}{N}}- \ \xrightarrow[H_2O,\ heat]{H^+} \ -\overset{|}{\underset{|}{C}}-\overset{O}{\overset{\|}{C'}}\diagdown_{OH} \ + \ H-\overset{H}{\underset{|}{N^+}}-$$

$$-\overset{|}{\underset{|}{C}}-\overset{O}{\overset{\|}{C'}}-\overset{|}{\underset{|}{N}}- \ \xrightarrow[H_2O,\ heat]{HO^-} \ -\overset{|}{\underset{|}{C}}-\overset{O}{\overset{\|}{C'}}\diagdown_{O^-} \ + \ \overset{H}{\underset{|}{:N}}-$$

- **Amides** are **hydrolyzed** in the presence of **one equivalent of acid** or **one equivalent of hydroxide** to produce a **carboxylic acid** and a **protonated amine**, or **carboxylate salt** and **amine**, respectively. ✱

## REACTION 14J: REACTION OF DICARBOXYLIC ACIDS AND DIAMINES TO FORM POLYAMIDES (Section 14.8A)

$$\text{HO}-\overset{\overset{\text{O}}{\|}}{\text{C}}-(\text{R})-\overset{\overset{\text{O}}{\|}}{\text{C}'}-\text{OH} + \text{H}_2\text{N}-(\text{R}')-\text{NH}_2 \xrightarrow{\text{Heat}} \left(\overset{\overset{\text{O}}{\|}}{\text{C}}-(\text{R})-\overset{\overset{\text{O}}{\|}}{\text{C}'}-\text{NH}-(\text{R}')-\text{NH}\right)_n$$

- **Dicarboxylic acids** and **diamines** react when heated under pressure to form **polyamides.** ✱
- Important examples of polyamides include **nylon 66**, in which R = -(CH$_2$)$_4$- and R' = -(CH$_2$)$_6$-, or **Kevlar**, in which R and R' = 1,4-linked phenyl.

## REACTION 14K: REACTION OF DICARBOXYLIC ACIDS AND DIOLS TO FORM POLYESTERS (Section 14.8B)

$$\text{HO}-\overset{\overset{\text{O}}{\|}}{\text{C}}-(\text{R})-\overset{\overset{\text{O}}{\|}}{\text{C}'}-\text{OH} + \text{HO}-(\text{R}')-\text{OH} \xrightarrow{\text{Heat}} \left(\overset{\overset{\text{O}}{\|}}{\text{C}}-(\text{R})-\overset{\overset{\text{O}}{\|}}{\text{C}'}-\text{O}-(\text{R}')-\text{O}\right)_n$$

- **Dicarboxylic acids** and **diols** react when heated to form **polyesters.** ✱
- Important examples of polyesters include **Dacron** and **Mylar**, which are different forms of a polymer in which R = 1,4-linked phenyl and R' = -(CH$_2$)$_2$-.

## REACTION 14L: REACTION OF DIISOCYANATES AND DIOLS TO FORM POLYURETHANES (Section 14.8D)

$$\text{O}=\text{C}=\text{N}-(\text{R})-\text{N}=\text{C}'=\text{O} + \text{HO}-(\text{R}')-\text{OH} \longrightarrow$$

Diisocyanate

$$\left(\overset{\overset{\text{O}}{\|}}{\text{C}}-\text{NH}-(\text{R})-\underbrace{\text{NH}-\overset{\overset{\text{O}}{\|}}{\text{C}'}-\text{O}}-(\text{R}')-\text{O}\right)_n$$

Urethane or
Carbamate
Linkage

- **Diisocyanates** and **diols** react to form **polyurethanes.** ✱
- The **urethane linkage, also called a carbamate** linkage, is rigid. Thus, commercial polyurethane fibers are **generally diblock polymers**, consisting of rigid urethane blocks, connected to more flexible polyester or polyether blocks. Such an arrangement can lead to interesting elastic properties and important examples of polyurethane fibers include **Spandex** and **Lycra.**

## REACTION 14M: REDUCTION OF ESTERS (Section 14.7A)

$$
\underset{\text{ester}}{-\overset{O}{\underset{|}{\overset{||}{C}}} - \overset{|}{\underset{|}{C}} ' - O - \overset{|}{\underset{|}{C}} '' -} \quad \xrightarrow[\text{2) } H_2O]{\text{1) } LiAlH_4} \quad -\overset{|}{\underset{|}{C}} - \overset{H}{\underset{H}{\overset{|}{C}}} ' - OH \;+\; HO - \overset{|}{\underset{|}{C}} '' -
$$

- **Esters** are **reduced by LiAlH₄** to produce **two equivalents of alcohols;** one derived from the carboxylic acid portion of the ester, and the other is from the alcohol portion. ✳
- **NaBH₄** reacts much more slowly with esters, so it is possible to use NaBH₄ to reduce an aldehyde or ketone group without reducing an ester in the same molecule.

## REACTION 14N: REACTION OF ESTERS WITH AMMONIA AND AMINES (Section 14.5C)

$$
-\overset{|}{\underset{|}{C}} - \overset{O}{\underset{|}{\overset{||}{C}}} ' - O - \overset{|}{\underset{|}{C}} '' - \;+\; \overset{H}{\underset{|}{\overset{|}{N}}} - \quad \longrightarrow \quad -\overset{|}{\underset{|}{C}} - \overset{O}{\underset{|}{\overset{||}{C}}} ' - N - \;+\; HO - \overset{|}{\underset{|}{C}} '' -
$$

- **Esters** react very slowly with **ammonia, primary amines,** and **secondary amines** to produce **amides.** ✳

## REACTION 14O: HYDROLYSIS OF ESTERS (Section 14.3C)

$$
-\overset{|}{\underset{|}{C}} - \overset{O}{\underset{|}{\overset{||}{C}}} ' - O - \overset{|}{\underset{|}{C}} '' - \quad \xrightarrow[H_2O]{(cat.)H^+} \quad -\overset{|}{\underset{|}{C}} - \overset{O}{\underset{OH}{C}} ' \;+\; HO - \overset{|}{\underset{|}{C}} '' -
$$

$$
-\overset{|}{\underset{|}{C}} - \overset{O}{\underset{|}{\overset{||}{C}}} ' - O - \overset{|}{\underset{|}{C}} '' - \;+\; HO^- \quad \xrightarrow{H_2O} \quad -\overset{|}{\underset{|}{C}} - \overset{O}{\underset{O^-}{C}} ' \;+\; HO - \overset{|}{\underset{|}{C}} '' -
$$

- **Esters** are **hydrolyzed** in the presence of **catalytic acid** or **one equivalent of hydroxide** to produce a **carboxylic acid** and **alcohol,** or **carboxylate salt** and **alcohol,** respectively. ✳
- In the **acid-catalyzed reaction,** the mechanism involves an initial protonation of the carbonyl oxygen atom that facilitates nucleophilic attack of water. A proton transfer gives a tetrahedral carbonyl addition intermediate, that converts to product upon departure of the alkoxide and a proton. The alkoxide is protonated to give an alcohol, thus completing the reaction. Note that **the acid is not consumed** in the reaction, so **it is catalytic.** Furthermore, this reaction is **reversible,** being just the **reverse of the Fischer esterification reaction.**
- In the **base promoted reaction,** the nucleophilic hydroxide attacks the carbonyl carbon atom to produce a tetrahedral carbonyl addition intermediate, that converts to product upon the departure of the alkoxide, that is ultimately protonated by water. This process, sometimes referred to as **saponification,** requires on equivalent of hydroxide, and is **not reversible.**

## REACTION 14P: POLYMERIZATION OF HYDROXY CARBOXYLIC ACIDS TO FORM POLYESTERS (Section 14.8D)

$$HO-(R)-\overset{\overset{\displaystyle O}{\|}}{C}-OH \xrightarrow{\text{Polymerization}} \left(\!\!(R)-\overset{\overset{\displaystyle O}{\|}}{C}-O\!\!\right)_n$$

- **Hydroxy carboxylic acids** can be polymerized to form **polyesters**. ✶
- Important examples of these types of polyesters include the material used to make dissolving stitches, in which R = -CH$_2$-, -CH(CH$_3$)-, or a combination of the two.

## REACTION 14Q: POLYMERIZATION OF LACTAMS TO FORM POLYAMIDES (Section 14.8A)

$$\text{A lactam} \xrightarrow{\text{Polymerization}} \left(\!\!(R)-\overset{\overset{\displaystyle O}{\|}}{C}-NH\!\!\right)_n$$

A lactam

- **Lactams** can be polymerized to form **polyamides**. ✶
- The polymerization is accomplished by partial hydrolysis of the lactam, followed by heating to drive off water and induce polymerization.

## REACTION 14R: REACTION OF PHOSGENE AND DIOLS TO FORM POLYCARBONATES (Section 14.8C)

$$Cl-\overset{\overset{\displaystyle O}{\|}}{C}-Cl \;+\; HO-(R)-OH \longrightarrow \left(\!\!\overset{\overset{\displaystyle O}{\|}}{C}-O-(R)-O\!\!\right)_n$$

Phosgene

- **Phosgene** reacts with diols to form **polycarbonates**. ✶
- An important example is **Lexan**, in which R = ⟨benzene⟩–C(CH$_3$)(CH$_3$)–⟨benzene⟩– . Lexan is used

  for impact resistant applications such as crash helmets, appliances, and automobile parts.


## SUMMARY OF IMPORTANT CONCEPTS

### 14.0 OVERVIEW
• The derivatives of carboxylic acids, namely **acid chlorides, acid anhydrides, esters**, and **amides** all exhibit similar types of chemistry.

### 14.1 STRUCTURE
• The characteristic feature of **acid halides** is a **halogen atom, usually chlorine, attached** to the **carbon atom** of a **carbonyl group**. Acid halides are named the same as a carboxylic acid, but the **ic acid** suffix is replaced by **yl halide**. ✶
• The characteristic feature of **acid anhydrides** is **two acyl groups attached** to an oxygen atom. Acid anhydrides are usually symmetrical, that is both acyl groups are the same, although unsymmetrical acid anhydrides can also be prepared and used. Acid anhydrides are named by

adding the word **anhydride** to the name of the corresponding carboxylic acid. Cyclic anhydrides may be formed from dicarboxylic acids. **Phosphoric anhydrides** are derived from **phosphoric acid ($H_3PO_4$)** and are composed of **two phosphoryl groups bonded** to the **same oxygen atom**. Phosphoric anhydrides are found in a number of important biological molecules. As opposed to acid anhydrides that only have two carboxyl groups bonded together, more than two phosphoryl groups can be bound together into longer phosphoric anhydride structures. ✳

- The characteristic feature of **esters** is an **-OR** group bonded to the **carbon atom** of a **carbonyl group**. The R group may be an alkyl or aryl group. Esters are named by listing the name of the R group first, followed by the name of the carboxylic acid, except the suffix **ic acid** is replaced by the suffix **ate**. **Cyclic esters**, derived from molecules with both an alcohol and carboxylic acid function, are called **lactones**. ✳

- The characteristic feature of **amides** is a **trivalent nitrogen atom attached** to the **carbon atom** of a **carbonyl group**. Primary, secondary, and tertiary amides have zero, one, and two R or Ar groups attached to the nitrogen atom, respectively. Amides are named the same as a carboxylic acid, but the **oic acid** suffix of an IUPAC name or **ic acid** suffix of a common name is replaced by **amide**. If there is an R group on the nitrogen atom, it is listed first and designated with **N-**. Cyclic amides are called **lactams**. **Imides** contain a nitrogen atom attached to two acyl groups. ✳

## 14.2 CHARACTERISTIC REACTIONS

- The different **groups attached** to the **carbonyl groups** of the carboxylic acid derivatives can be **considered** as **leaving groups** of varying ability. Thus, when the **carbonyl group** is **attacked by a nucleophile**, the **tetrahedral carbonyl addition intermediate** that is formed is converted to products by **expulsion** of the **leaving group** to **regenerate** a **carbonyl group**. This process is referred to as **nucleophilic acyl substitution**. ✳ *[Variations of this mechanism operate in a large number of different reactions. It is helpful to distinguish these different reactions by keeping track of 1)the nature of the nucleophile and 2) proton transfers for each of the reactions described throughout the chapter.]*

  - In general, the chloride ion of acid chlorides is the best leaving group, followed by the carboxylate of acid anhydrides, the alkoxy group of esters, and the nitrogen group of amides, respectively. Note that when sufficient acid is present, the leaving group can be protonated as it departs.

  - Because of the relative leaving group abilities listed above, the acid chlorides are the most reactive class of carboxylic acid derivatives; followed by anhydrides, esters, and amides, respectively.

# CHAPTER 14
## *Solutions to the Problems*

<u>Problem 14.1</u> Draw structural formulas for the following compounds.

(a) *N*-Cyclohexylacetamide.

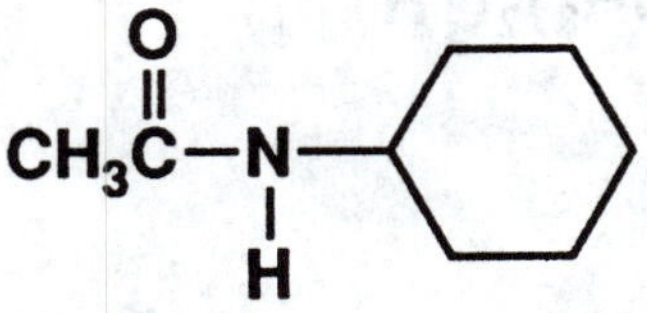

$$CH_3\overset{\overset{\displaystyle O}{\|}}{C}-\underset{\underset{\displaystyle H}{|}}{N}-\bigcirc$$

(b) *sec*-Butyl methanoate

$$H\overset{\overset{\displaystyle O}{\|}}{C}-O-\underset{\overset{\displaystyle |}{CH_3}}{CH}CH_2CH_3$$

(c) Cyclobutyl butanoate

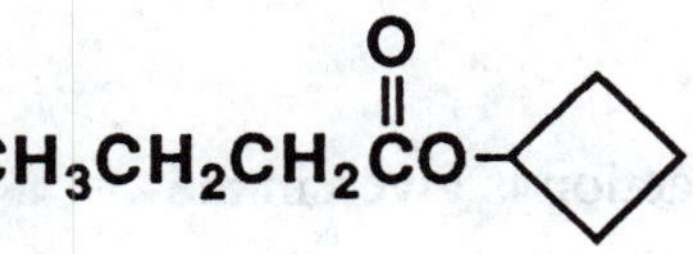

$$CH_3CH_2CH_2\overset{\overset{\displaystyle O}{\|}}{C}O-\square$$

(d) *N*-(2-Octyl)succinimide

$$N-\underset{\underset{\displaystyle CH_3}{|}}{CH}(CH_2)_5CH_3$$

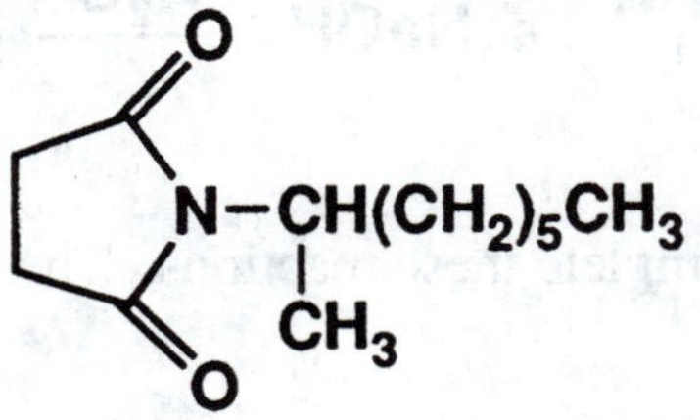

(e) Diethyl adipate

$$C_2H_5O\overset{\overset{\displaystyle O}{\|}}{C}(CH_2)_4\overset{\overset{\displaystyle O}{\|}}{C}OC_2H_5$$

(f) 2-Aminopropanamide

$$CH_3\underset{\underset{\displaystyle NH_2}{|}}{CH}\overset{\overset{\displaystyle O}{\|}}{C}NH_2$$

<u>Problem 14.2</u> Complete and balance equations for hydrolysis of each ester in aqueous acid. Show each product as it is ionized in aqueous acid.

(a)

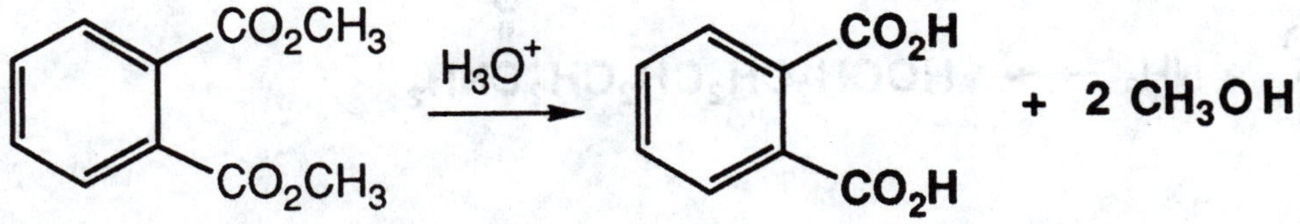

$$\xrightarrow{H_3O^+} \quad + \ 2\ CH_3OH$$

(b)   $CH_3CH_2O\overset{\overset{\displaystyle O}{\|}}{C}OCH_2CH_3 \xrightarrow{H_3O^+} CO_2 + 2\ CH_3CH_2OH$

**Problem 14.3** Complete equations for the hydrolysis of the amides in Example 14.3 in concentrated aqueous NaOH. Show all products as they exist in aqueous NaOH, and show the number of moles of NaOH required for hydrolysis of each amide.

**Each product is shown as it would exist in aqueous NaOH.**

(a)
$$CH_3\overset{O}{\overset{\|}{C}}\text{-N-CH}_3 + NaOH \xrightarrow{H_2O} CH_3\overset{O}{\overset{\|}{C}}O^-Na^+ + (CH_3)_2NH$$
with $CH_3$ on the nitrogen

(b)

piperidinone ring $+ NaOH \xrightarrow{H_2O} H_2NCH_2CH_2CH_2CH_2\overset{O}{\overset{\|}{C}}O^-\ Na^+$

**Problem 14.4** Complete these reactions. The stoichiometry of each reaction is given in the equation.

**Each is an example of aminolysis of an ester.**

(a)
$$CH_3\overset{O}{\overset{\|}{C}}O\text{-}\!\!\!\!\bigcirc\!\!\!\!\text{-}O\overset{O}{\overset{\|}{C}}CH_3 + 2NH_3 \longrightarrow HO\text{-}\!\!\!\!\bigcirc\!\!\!\!\text{-}OH + 2CH_3\overset{O}{\overset{\|}{C}}NH_2$$

(b)
lactone ring $+ NH_3 \longrightarrow HOCH_2CH_2CH_2CH_2\overset{O}{\overset{\|}{C}}NH_2$

**Problem 14.5** Show how to bring about each conversion in good yield. In addition to the given starting material, use any other reagents as necessary.

(a) $C_6H_5CH_2\overset{O}{\overset{\|}{C}}H$ + HN(pyrrolidine) $\xrightarrow{H_2/Ni}$ $C_6H_5CH_2CH_2N$(pyrrolidine)

(b)   $C_6H_5CH_2\overset{\overset{O}{\|}}{C}OH$ + $SOCl_2$ ⟶ $C_6H_5CH_2\overset{\overset{O}{\|}}{C}Cl$ $\xrightarrow{\text{HN} \hspace{-1em} \bigcirc}$

$C_6H_5CH_2\overset{\overset{O}{\|}}{C}-N\hspace{-0.5em}\bigcirc$ $\xrightarrow[\text{2) } H_2O]{\text{1) } LiAlH_4, \text{ ether}}$ $C_6H_5CH_2CH_2N\hspace{-0.5em}\bigcirc$
              3) Mild Base

<u>Problem 14.6</u> Show how to convert (R)-2-phenylpropanoic acid to these compounds.

(a)  **(R)-PhCHCO₂H**   $\xrightarrow[\text{2) } H_2O]{\text{1) } LiAlH_4, \text{ ether}}$   (R)-PhCHCH₂OH
          |                                                    |
          CH₃                                                 CH₃

**(R)-2-Phenyl-1-**                        (R)-2-Phenyl-1-propanol
**propanoic acid**

(b)  **(R)-PhCHCO₂H**   + $SOCl_2$ ⟶   **(R)-PhCHCOCl**   $\xrightarrow{NH_3}$
          |                                      |
          CH₃                                    CH₃

**(R)-2-Phenyl-1-**
**propanoic acid**

**(R)-PhCHCONH₂**   $\xrightarrow[\substack{\text{2) } H_2O \\ \text{3) Mild Base}}]{\text{1) } LiAlH_4, \text{ ether}}$   (R)-PhCHCH₂NH₂
          |                                                            |
          CH₃                                                          CH₃

                                                  (R)-2-Phenyl-1-propanamine

## Structure and Nomenclature
<u>Problem 14.7</u> Draw structural formulas for these compounds.
(a) Dimethyl carbonate                        (b) *p*-Nitrobenzamide

$CH_3O\overset{\overset{O}{\|}}{C}OCH_3$              $O_2N\text{—}\bigcirc\text{—}\overset{\overset{O}{\|}}{C}NH_2$

(c) Octanoyl chloride                         (d) Diethyl oxalate

$CH_3(CH_2)_6\overset{\overset{O}{\|}}{C}Cl$        $CH_3CH_2O\overset{\overset{O}{\|}}{C}\overset{\overset{O}{\|}}{C}OCH_2CH_3$

(e) Ethyl *cis*-2-pentenoate

$$CH_3CH_2 - \overset{\displaystyle H}{\underset{\displaystyle H}{C}} = \overset{\displaystyle H}{\underset{\displaystyle }{C}} - \overset{\displaystyle O}{\overset{\| }{C}}OCH_2CH_3$$

(f) Butanoic anhydride

$$(CH_3CH_2CH_2CO)_2O$$

(g) Dodecanamide

$$CH_3(CH_2)_{10}\overset{O}{\overset{\|}{C}}NH_2$$

(h) Ethyl 3-hydroxybutanoate

$$\overset{OH}{\underset{|}{CH_3CH}}CH_2\overset{O}{\overset{\|}{C}}OCH_2CH_3$$

Problem 14.8  Give each compound an IUPAC name.

(a)

**Benzoic anhydride**

(b) $CH_3(CH_2)_{14}\overset{O}{\overset{\|}{C}}OCH_3$

**Methyl hexadecanoate**

(c) $CH_3(CH_2)_4\overset{O}{\overset{\|}{C}}NHCH_3$

*N*-**Methylhexanamide**

(d) $H_2N-$
$\overset{O}{\overset{\|}{C}}NH_2$

**4-Aminobenzamide**

(e) $CH_2(CO_2CH_2CH_3)_2$

**2,2-Diethylpropanedioate**
**(Diethyl malonate)**

(f) $PhCH_2\overset{O}{\overset{\|}{C}}\overset{O}{\underset{\underset{\displaystyle CH_3}{|}}{C}}H\overset{O}{\overset{\|}{C}}OCH_3$

**Methyl 2-methyl-3-oxo-**
**4-phenylbutanoate**

Problem 14.9  When oil from the head of the sperm whale is cooled, spermaceti, a translucent wax with a white, pearly luster, crystallizes from the mixture.  Spermaceti, which makes up 11% of whale oil, is composed mainly of hexadecyl hexadecanoate (cetyl palmitate).  At one time, spermaceti was widely used in the making of cosmetics, fragrant soaps, and candles.  Draw the structural formula of cetyl palmitate.

$$CH_3(CH_2)_{14}\overset{O}{\overset{\|}{C}}OCH_2(CH_2)_{14}CH_3$$

**Hexadecyl  hexadecanoate**
**(Cetyl  palmitate)**

## Physical Properties

**Problem 14.10** Acetic acid and methyl formate are constitutional isomers. Both are liquids at room temperature; one with a boiling point of 32°C, the other with a boiling point of 118°C. Which of the two has the higher boiling point? Explain your reasoning.

$$\underset{\textbf{Acetic acid}}{CH_3\overset{\displaystyle O}{\overset{\|}{C}}OH} \qquad\qquad \underset{\textbf{Methyl formate}}{HC\overset{\displaystyle O}{\overset{\|}{O}}CH_3}$$

**Because of the polar O-H bond in acetic acid that is not present in methyl formate, only acetic acid has the possibility for intermolecular association by hydrogen bonding. Thus, acetic acid has a higher boiling point than its constitutional isomer methyl formate.**

## Reactions

**Problem 14.11** Arrange these compounds in order of increasing reactivity toward nucleophilic acyl substitution.

$$\underset{A}{CH_3CONH_2} \qquad \underset{B}{CH_3COCl} \qquad \underset{C}{CH_3CO_2CH_2CH_3} \qquad \underset{D}{(CH_3CO)_2O}$$

**In general, the order of reactivity is acid chlorides > acid anhydrides > esters > amides. Therefore, the order for the molecules listed above is:**
**(ranked from least to most reactive) A < C < D < B.**

**Problem 14.12** A common method for preparing acid anhydrides is treatment of an acid chloride with the sodium salt of a carboxylic acid. For example, treatment of benzoyl chloride with sodium acetate gives acetic benzoic anhydride. Write a mechanism for this nucleophilic substitution reaction.

Sodium acetate          Benzoyl chloride          Acetic benzoic anhydride

**Like the other nucleophilic acyl substitution reactions discussed throughout the chapter, the mechanism for this reaction involves attack of the nucleophilic acetate anion on the carbonyl carbon atom of the acid chloride. The resulting tetrahedral carbonyl addition intermediate then decomposes with loss of the good leaving group chloride anion to give the final product.**

**Tetrahedral carbonyl addition intermediate**

**Problem 14.13** Show how to prepare these mixed anhydrides using the method described in Problem 14.12.

**There are two combinations of acid chloride and carboxylate anion that can be used to prepare each mixed anhydride.**

(a) $\quad$ H-C(=O)-O-C(=O)-CH$_3$

(b) $\quad$ CH$_3$C(=O)-O-C(=O)CH$_2$C$_6$H$_5$

**Problem 14.14** A carboxylic acid can be converted to an ester in one reaction by Fischer esterification. Show how to synthesize each ester from a carboxylic acid and an alcohol by Fischer esterification.

(a)

(b)  $(CH_3)_2CHCOCH_2CH_3$
     
     O

$(CH_3)_2CHCOH$  +  $HOCH_2CH_3$  $\xrightarrow[\text{}]{\text{H}_2\text{SO}_4}$  $(CH_3)_2CHCOCH_2CH_3$  +  $H_2O$

**Problem 14.15** A carboxylic acid can be converted to an ester in two steps by first converting the carboxylic acid to its acid chloride and then treating the acid chloride with an alcohol. Show how to prepare each ester in Problem 14.14 from a carboxylic acid and alcohol by this two-step scheme.

(a)  $HOC(CH_2)_4CH_3$  $\xrightarrow{\text{SOCl}_2}$  $Cl-C(CH_2)_4CH_3$  [cyclohexyl]—OH  $\longrightarrow$

[cyclohexyl]—$OC(CH_2)_4CH_3$

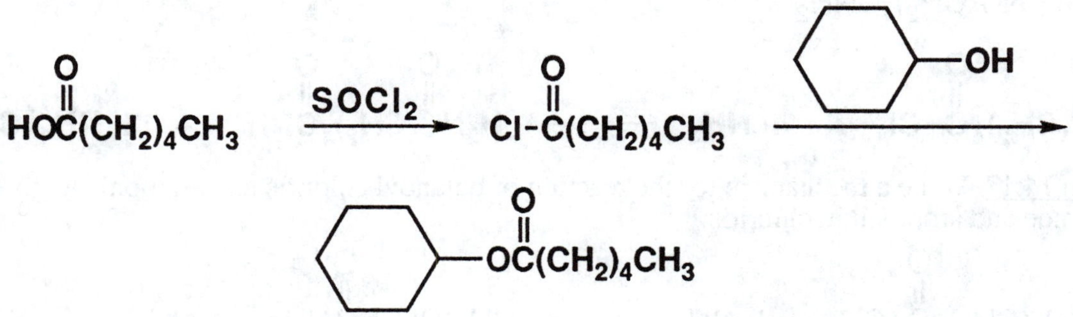

(b)  $(CH_3)_2CHCOH$  $\xrightarrow{\text{SOCl}_2}$  $(CH_3)_2CHC-Cl$  $\xrightarrow{\text{HOCH}_2\text{CH}_3}$

$(CH_3)_2CHCOCH_2CH_3$

**Problem 14.16** Show how to prepare these amides by reaction of an acid chloride with ammonia or an amine.

(a)  [cyclohexyl]—$NHC(CH_2)_4CH_3$

2 [cyclohexyl]—$NH_2$  +  $Cl-C(CH_2)_4CH_3$  $\longrightarrow$  [cyclohexyl]—$NHC(CH_2)_4CH_3$

+

[cyclohexyl]—$NH_3^+$  $Cl^-$

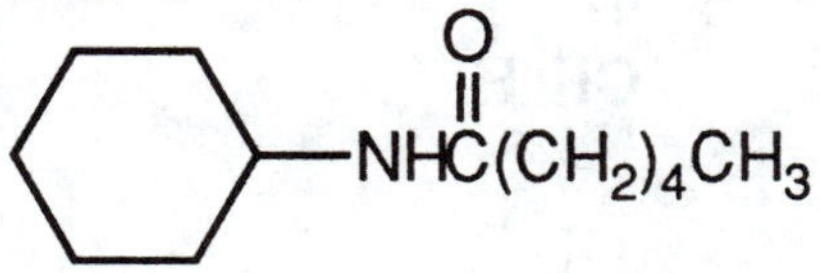

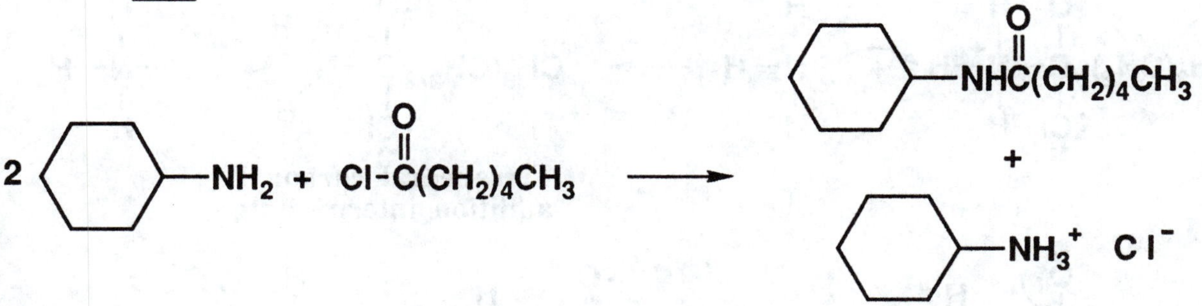

(b)
$$\text{(CH}_3\text{)}_2\text{CHCN(CH}_3\text{)}_2$$
(with C=O)

$$\text{(CH}_3\text{)}_2\text{CHC-Cl} \;+\; 2\ \text{NH(CH}_3\text{)}_2 \;\longrightarrow\;$$
(with C=O)

$$\text{(CH}_3\text{)}_2\text{CHCN(CH}_3\text{)}_2$$
(with C=O)
$$+$$
$$\text{(CH}_3\text{)}_2\text{NH}_2^+ \ \text{Cl}^-$$

(c)
$$\text{H}_2\text{NC(CH}_2\text{)}_4\text{CNH}_2$$
(with two C=O)

$$\text{Cl-C(CH}_2\text{)}_4\text{C-Cl} \;+\; 4\ \text{NH}_3 \;\longrightarrow\; \text{H}_2\text{NC(CH}_2\text{)}_4\text{CNH}_2 \;+\; 2\ \text{NH}_4^+\text{Cl}^-$$
(with C=O groups)

<u>Problem 14.17</u> Write a mechanism for the reaction of butanoyl chloride and ammonia to give butanamide and ammonium chloride.

$$\text{CH}_3\text{(CH}_2\text{)}_2\text{C-Cl} \;+\; 2\ \text{NH}_3 \;\longrightarrow\; \text{CH}_3\text{(CH}_2\text{)}_2\text{CNH}_2 \;+\; \text{NH}_4^+\text{Cl}^-$$
(with C=O groups)

**Like the other reactions that involve attack by nucleophiles on acid chlorides, the mechanism of the reaction begins with the nucleophilic ammonia reacting with the carbonyl carbon atom to form a tetrahedral carbonyl addition intermediate that loses a proton and collapses to product by elimination of a chloride leaving group. Note how the proton ends up on another molecule of ammonia.**

Tetrahedral carbonyl addition intermediate

**Problem 14.18**  What product is formed when benzoyl chloride is treated with these reagents?

(a)  $C_6H_6$, $AlCl_3$

(b)  $CH_3CH_2CH_2CH_2OH$

(structure: diphenyl ketone — two phenyl rings attached to C=O)

(structure: phenyl-C(=O)-OCH₂CH₂CH₂CH₃)

$$\text{C-OCH}_2\text{CH}_2\text{CH}_2\text{CH}_3$$

(c)  $CH_3CH_2CH_2CH_2SH$

(d)  $CH_3CH_2CH_2CH_2NH_2$ (two equivalents)

(structure: phenyl-C(=O)-SCH₂CH₂CH₂CH₃)

$$\text{C-SCH}_2\text{CH}_2\text{CH}_2\text{CH}_3$$

(structure: phenyl-C(=O)-NHCH₂CH₂CH₂CH₃)

$$\text{C-NHCH}_2\text{CH}_2\text{CH}_2\text{CH}_3$$

$$+$$

$$CH_3CH_2CH_2CH_2NH_3^+ \, Cl^-$$

(e)  $H_2O$

(f)  (piperidine N—H) (two equivalents)

(structure: benzoic acid, phenyl-C(=O)-OH)

$$\text{C-OH}$$

(structure: phenyl-C(=O)-N-piperidine)

$$\text{C-N}$$

$$+$$

(piperidinium) $N^+ H_2$  $Cl^-$

**Problem 14.19**  Write the product(s) of treatment of propanoic anhydride with each reagent.

(a) Ethanol (1 equivalent)

$$CH_3CH_2\overset{O}{\underset{\|}{C}}-O-\overset{O}{\underset{\|}{C}}CH_2CH_3 \; + \; CH_3CH_2OH \longrightarrow$$

$$CH_3CH_2\overset{O}{\underset{\|}{C}}-OCH_2CH_3$$

$$+$$

$$CH_3CH_2\overset{O}{\underset{\|}{C}}OH$$

(b) Ammonia (2 equivalents)

$$CH_3CH_2\overset{O}{\underset{\|}{C}}-O-\overset{O}{\underset{\|}{C}}CH_2CH_3 \; + \; 2\ NH_3 \longrightarrow$$

$$CH_3CH_2\overset{O}{\underset{\|}{C}}NH_2$$

$$+$$

$$CH_3CH_2\overset{O}{\underset{\|}{C}}O^-\ NH_4^+$$

**Problem 14.20** Write the product of treatment of succinic anhydride with each reagent.
(a) Ethanol (1 equivalent)

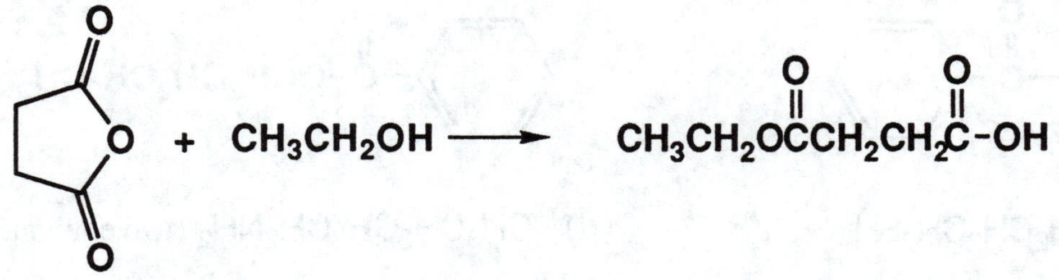

$$\text{+ } CH_3CH_2OH \longrightarrow CH_3CH_2OCCH_2CH_2C\text{-OH}$$

(b) Ammonia (2 equivalents)

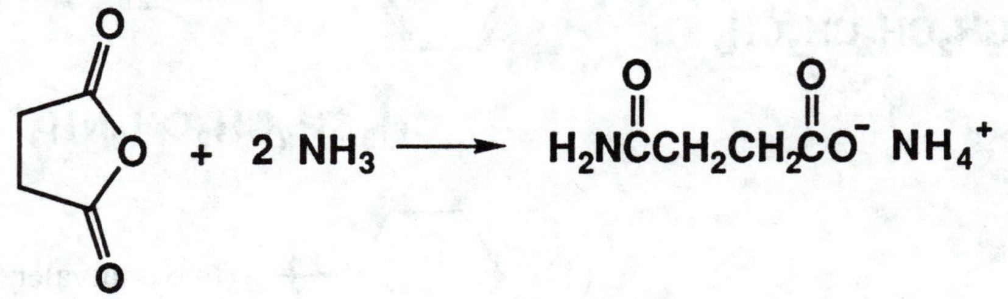

$$\text{+ } 2 \; NH_3 \longrightarrow H_2NCCH_2CH_2CO^- \; NH_4^+$$

**Problem 14.21** The analgesic phenacetin is synthesized by treating 4-ethoxyaniline with acetic anhydride. Write an equation for the formation of phenacetin.

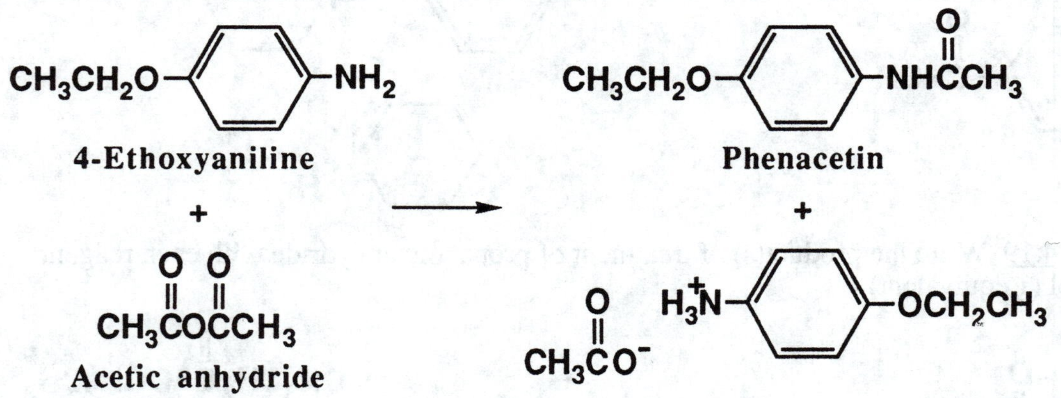

$$CH_3CH_2O-\underset{}{\bigcirc}-NH_2$$

**4-Ethoxyaniline**

+

$$CH_3\overset{O}{\overset{\|}{C}}O\overset{O}{\overset{\|}{C}}CH_3$$

**Acetic anhydride**

$$CH_3CH_2O-\underset{}{\bigcirc}-NH\overset{O}{\overset{\|}{C}}CH_3$$

**Phenacetin**

+

$$H_3\overset{+}{N}-\underset{}{\bigcirc}-OCH_2CH_3$$

$$CH_3\overset{O}{\overset{\|}{C}}O^-$$

**Problem 14.22** The analgesic acetaminophen is synthesized by treating 4-aminophenol with one equivalent of acetic anhydride. Write an equation for the formation of acetaminophen. (*Hint:* An -NH$_2$ group is a better nucleophile than an -OH group.)

**Note how in the following reaction scheme the acylation occurs at the more nucleophilic amino group rather than the less nucleophilic hydroxyl group of 4-aminophenol.**

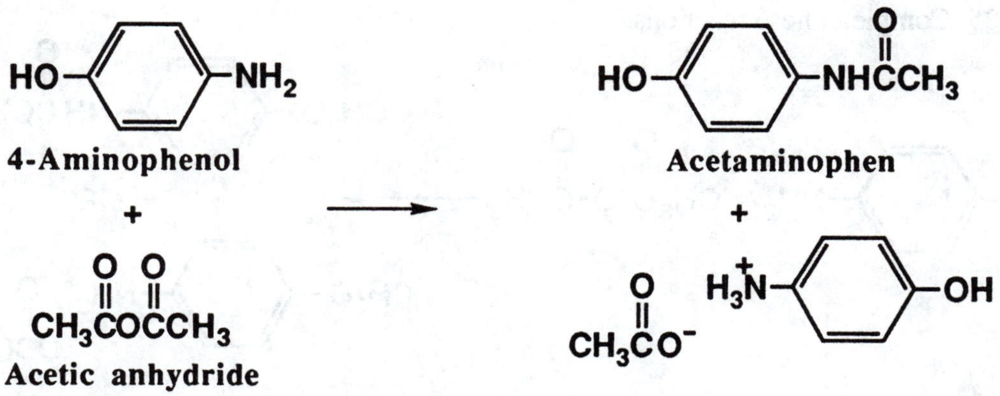

**Problem 14.23** Treatment of choline with acetic anhydride gives acetylcholine, a neurotransmitter. Write an equation for the formation of acetylcholine.

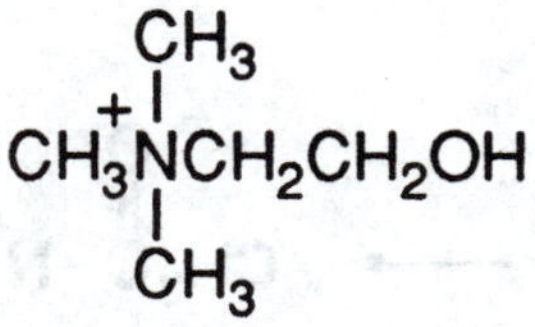

Choline

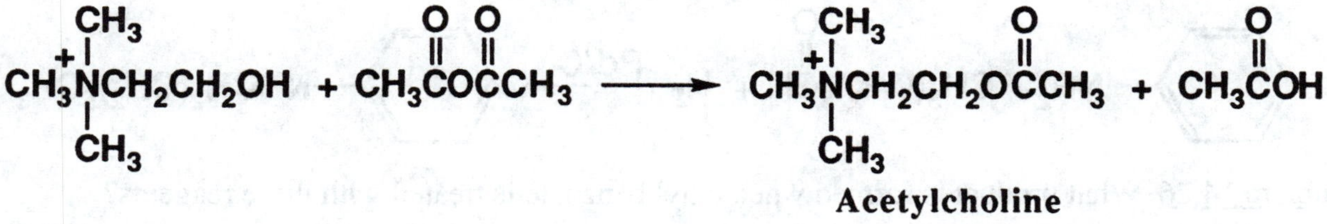

**Acetylcholine**

**Problem 14.24** Nicotinic acid, more commonly named niacin, is one of the B vitamins. Show how nicotinic acid can be converted to (a) ethyl nicotinate and then to (b) nicotinamide.

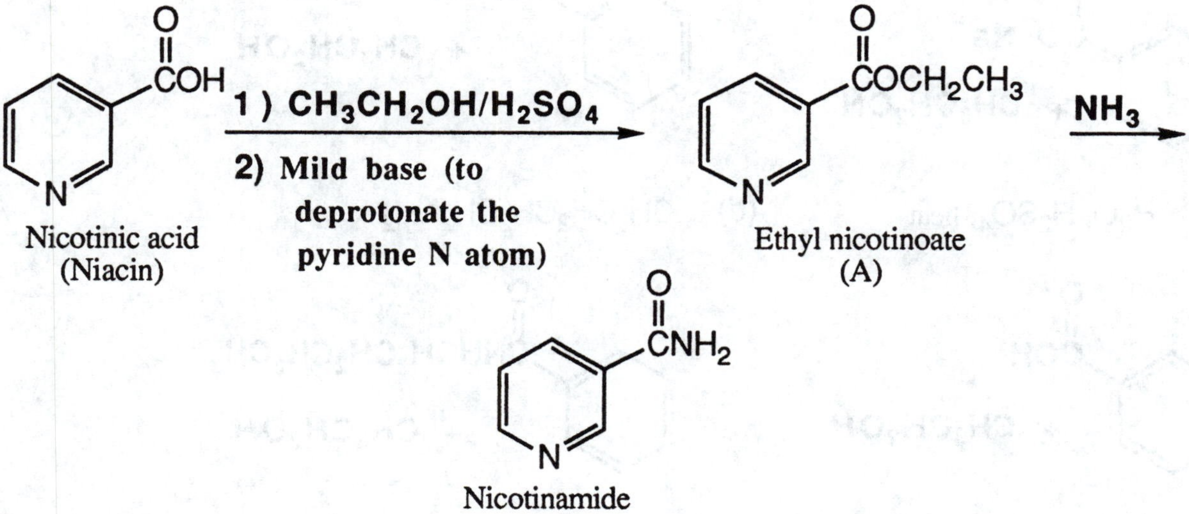

Nicotinic acid
(Niacin)

1) $CH_3CH_2OH/H_2SO_4$

2) **Mild base (to deprotonate the pyridine N atom)**

Ethyl nicotinoate
(A)

$NH_3$

Nicotinamide
(B)

**Problem 14.25** Complete these reactions.

(a) CH₃O—⟨benzene⟩—NH₂ + CH₃C(=O)-O-C(=O)CH₃ ⟶ CH₃O—⟨benzene⟩—NHC(=O)CH₃

+

CH₃O—⟨benzene⟩—NH₃⁺  ⁻OC(=O)CH₃

(b) CH₃C(=O)-Cl + HN⟨piperidine⟩ ⟶ CH₃C(=O)-N⟨piperidine⟩ + Cl⁻ H₂N⁺⟨piperidine⟩

(c) CH₃C(=O)OCH₃ + HN⟨piperidine⟩ ⟶ CH₃C(=O)-N⟨piperidine⟩ + CH₃OH

(d) ⟨benzene⟩—NH₂ + CH₃(CH₂)₅C(=O)H + H₂  —Pd/C→  ⟨benzene⟩—NHCH₂(CH₂)₅CH₃

**Problem 14.26** What product is formed when ethyl benzoate is treated with these reagents?

(a)   H₂O, NaOH, heat                    (b)   LiAlH₄, then H₂O

⟨benzene⟩—C(=O)O⁻ Na⁺                    ⟨benzene⟩—CH₂OH

+ CH₃CH₂OH                               + CH₃CH₂OH

(c)   H₂O, H₂SO₄, heat                   (d)   CH₃CH₂CH₂CH₂NH₂

⟨benzene⟩—C(=O)OH                        ⟨benzene⟩—C(=O)NHCH₂CH₂CH₂CH₃

+ CH₃CH₂OH                               + CH₃CH₂OH

**Problem 14.27** Show how to convert 2-hydroxybenzoic acid (salicylic acid) to these compounds.

(a)

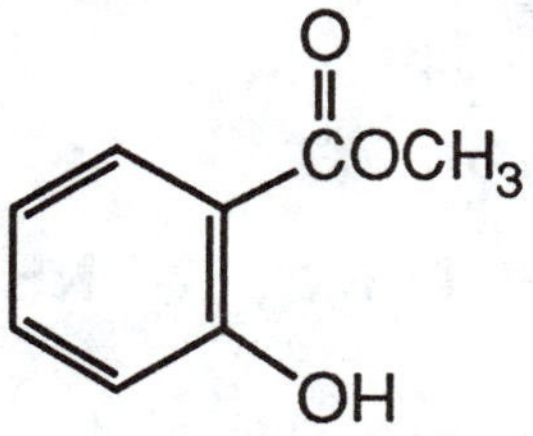

Methyl salicylate
(Oil of wintergreen)

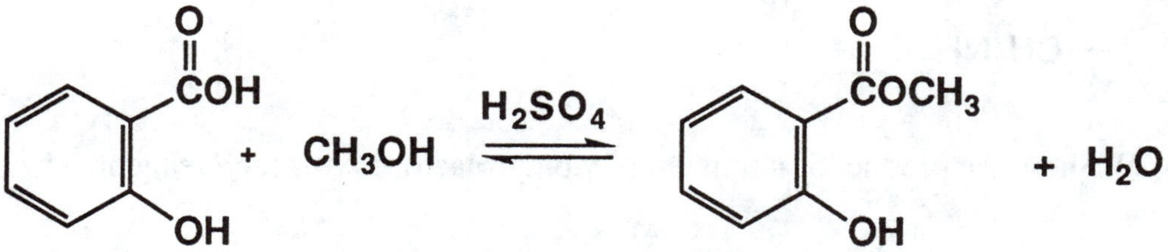

(b)

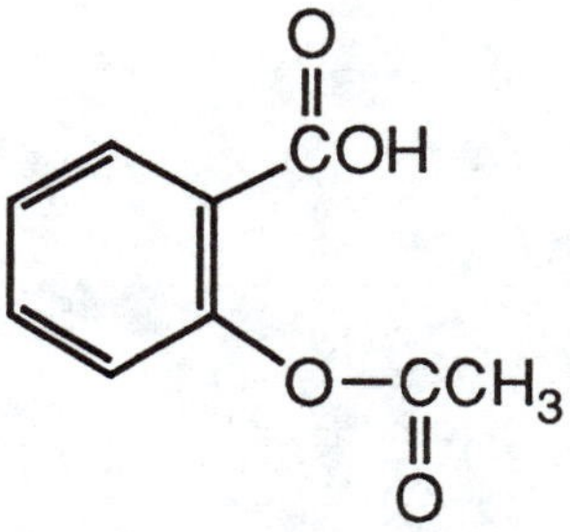

Acetyl salicylic acid
(Aspirin)

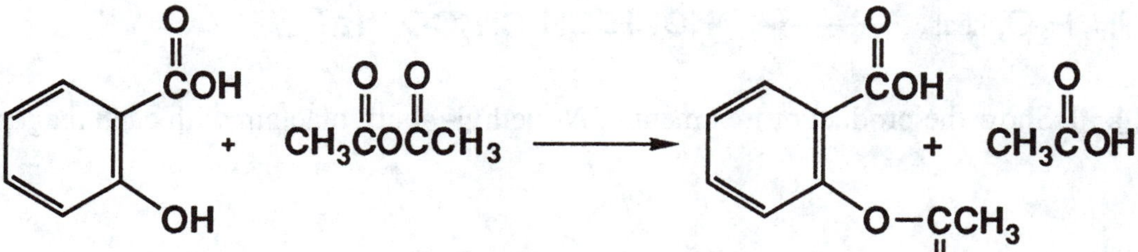

or

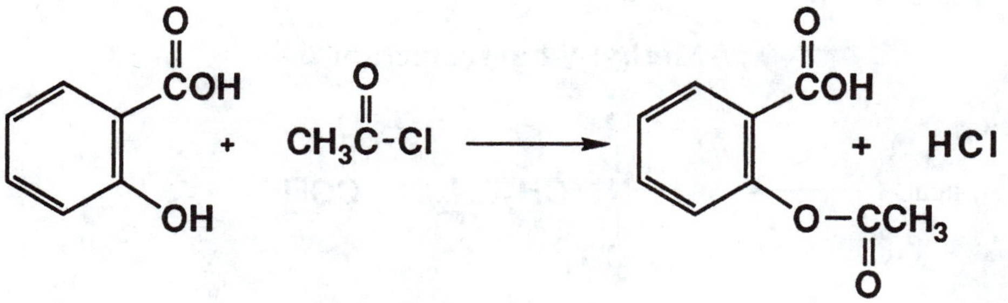

**Problem 14.28** What product is formed when benzamide is treated with these reagents?

(a)  $H_2O$, HCl, heat

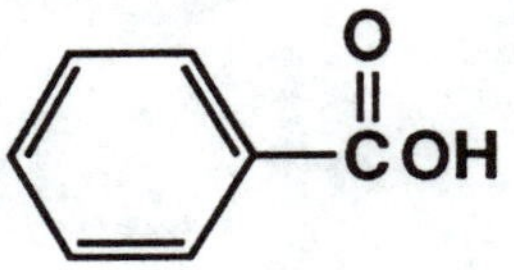

(b)  NaOH, $H_2O$, heat

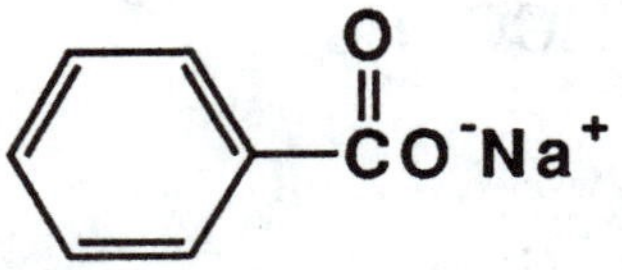

[structure: benzene ring with $-\overset{O}{\underset{||}{C}}OH$] + $NH_4Cl$

[structure: benzene ring with $-\overset{O}{\underset{||}{C}}O^-Na^+$] + $NH_3$

(c)  $LiAlH_4$, then $H_2O$

[structure: benzene ring with $-CH_2NH_2$]

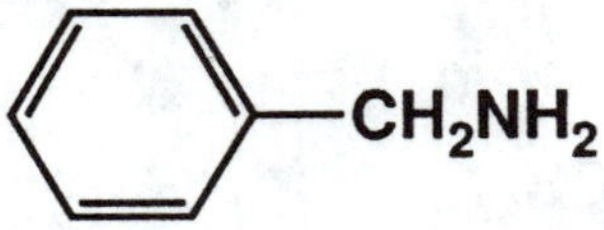

**Problem 14.29** Show the product of treatment of γ-butyrolactone with each reagent.

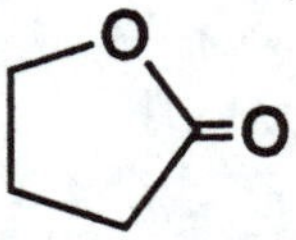

[structure of γ-butyrolactone: five-membered ring with O and =O]

**γ-Butyrolactone**

(a)   $NH_3$  $\longrightarrow$   $HOCH_2CH_2CH_2\overset{O}{\underset{||}{C}}NH_2$

(b)  $LiAlH_4$, then $H_2O$  $\longrightarrow$  $HOCH_2CH_2CH_2CH_2OH$

(c)  NaOH, $H_2O$, heat  $\longrightarrow$  $HOCH_2CH_2CH_2\overset{O}{\underset{||}{C}}O^-Na^+$

**Problem 14.30** Show the product of treatment of *N*-methyl-γ-butyrolactam with each reagent.

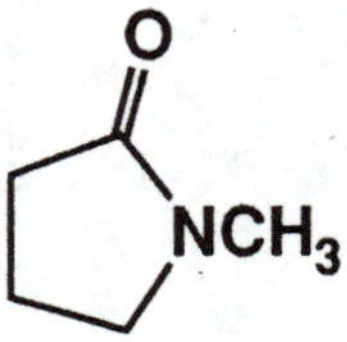

[structure of N-methyl-γ-butyrolactam: five-membered ring with =O and NCH₃]

**N-Methyl-γ-butyrolactam**

(a)  $H_2O$, HCl, heat  $\longrightarrow$  $H-\overset{H}{\underset{\underset{CH_3}{|}}{\overset{|}{N^+}}}-CH_2CH_2CH_2\overset{O}{\underset{||}{C}}OH$   $Cl^-$

(b)   NaOH, $H_2O$, heat  $\longrightarrow$

$$\text{H-N-CH}_2\text{CH}_2\text{CH}_2\overset{\displaystyle O}{\overset{\|}{\text{C}}}\text{O}^-\text{Na}^+$$
$$\underset{\text{CH}_3}{|}$$

(c)   $LiAlH_4$, then $H_2O$  $\longrightarrow$

**Problem 14.31** Reaction of a primary or secondary amine with diethyl carbonate under controlled conditions gives a carbamic ester. Propose a mechanism for this reaction.

$$\underset{\text{Diethyl carbonate}}{\text{EtO-}\overset{\displaystyle O}{\overset{\|}{\text{C}}}\text{-OEt}} + \underset{\text{Butanamine}}{\text{H}_2\text{NCH}_2\text{CH}_2\text{CH}_2\text{CH}_3} \longrightarrow \underset{\text{A carbamic ester}}{\text{EtO-}\overset{\displaystyle O}{\overset{\|}{\text{C}}}\text{-NHCH}_2\text{CH}_2\text{CH}_2\text{CH}_3} + \text{EtOH}$$

**Like the other nucleophilic acyl substitution reactions discussed throughout the chapter, the mechanism for this reaction involves attack of the nucleophilic amine on the carbonyl carbon atom of the diethyl carbonate. The resulting tetrahedral carbonyl addition intermediate loses a proton and decomposes with loss of the ethoxide (which picks up the proton) to give the final products.**

**Tetrahedral carbonyl addition intermediate**

**Problem 14.32** Barbiturates are prepared by treatment of diethyl malonate or a derivative of diethyl malonate with urea in the presence of sodium methoxide as a catalyst. Following is an equation for the preparation of barbital, a long-duration hypnotic and sedative, from diethyl diethylmalonate and urea. Barbital is prescribed under one of a dozen or more trade names.

Diethyl diethylmalonate       Urea                                   5,5-Diethylbarbituric acid
                                                                              (Barbital)

(a)  Propose a mechanism for this reaction.

**Treatment of urea with ethoxide ion gives an anion that then attacks a carbonyl group of the malonic ester to displace ethoxide ion.  This second reaction is an example of nucleophilic displacement at a carbonyl carbon.**

Anion from urea

**This is followed by nucleophilic displacement at the carbonyl group of the remaining ester to complete formation of the six-member ring.**

(b) The pK$_a$ of barbital is 7.4. Which is the most acidic hydrogen in this molecule and how do you account for its acidity?

**The most acidic hydrogen is the imide hydrogen. Acidity results from the inductive effects of the adjacent carbonyl groups and stabilization of the deprotonated anion by resonance interaction with the carbonyl groups. Following are three contributing structures for the barbiturate anion.**

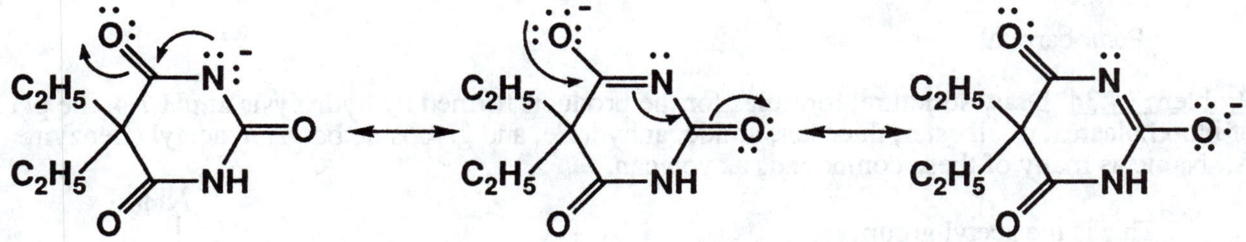

**The two contributing structures that place the negative charge on the more electronegative oxygen atom make the greater contribution to the resonance hybrid.**

Problem 14.33 Draw structural formulas for the products of complete hydrolysis of meprobamate, phenobarbital, and pentobarbital in hot aqueous acid. Meprobamate is a tranquilizer prescribed under one or more of 58 different trade names. Phenobarbital is a long-acting sedative, hypnotic, and anti-convulsant. Luminal is one of over a dozen names under which it is prescribed. Pentobarbital is a short-acting sedative, hypnotic, and anti-convulsant. Nembutal is one of several trade names under which it is prescribed. (*Hint*: Remember that when heated, β-dicarboxylic acids and β-ketoacids undergo decarboxylation.)

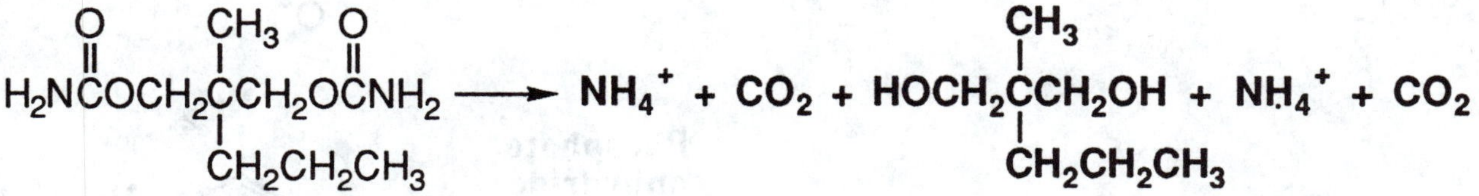

Meprobamate

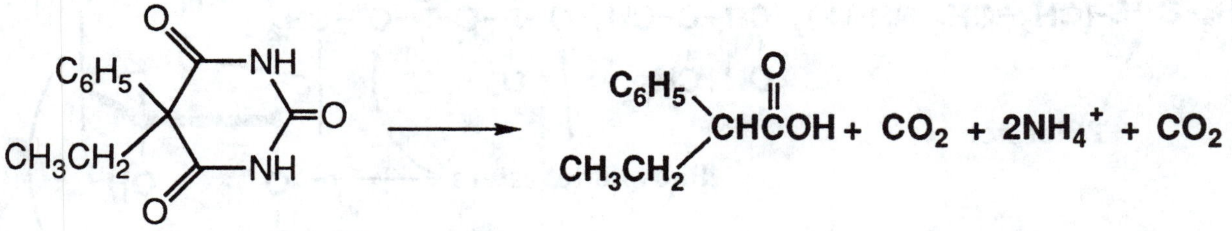

Phenobarbital

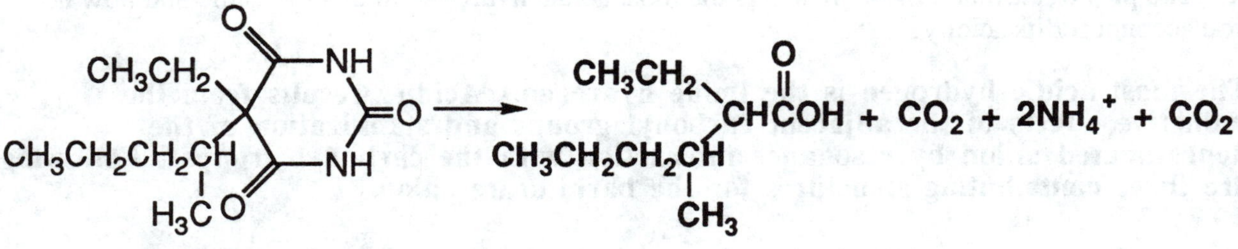

Pentobarbital

**Problem 14.34** Draw structural formulas for the products formed by hydrolysis at pH 7.4, the pH of blood plasma, of all ester, thioester, amide, anhydride, and glycoside bonds in acetyl coenzyme A. Name as many of these compounds as you can.

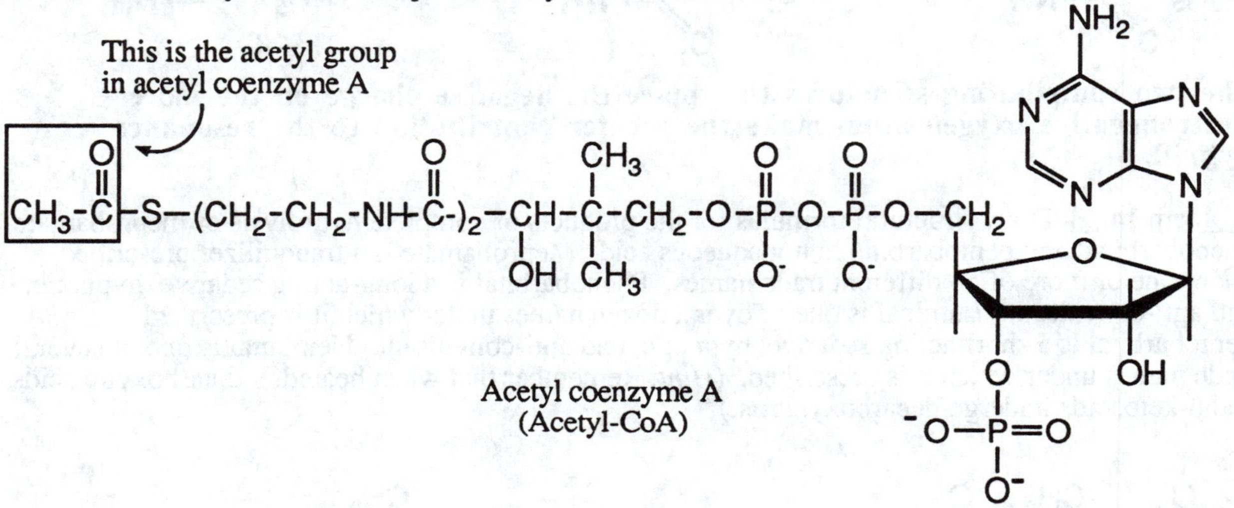

This is the acetyl group
in acetyl coenzyme A

Acetyl coenzyme A
(Acetyl-CoA)

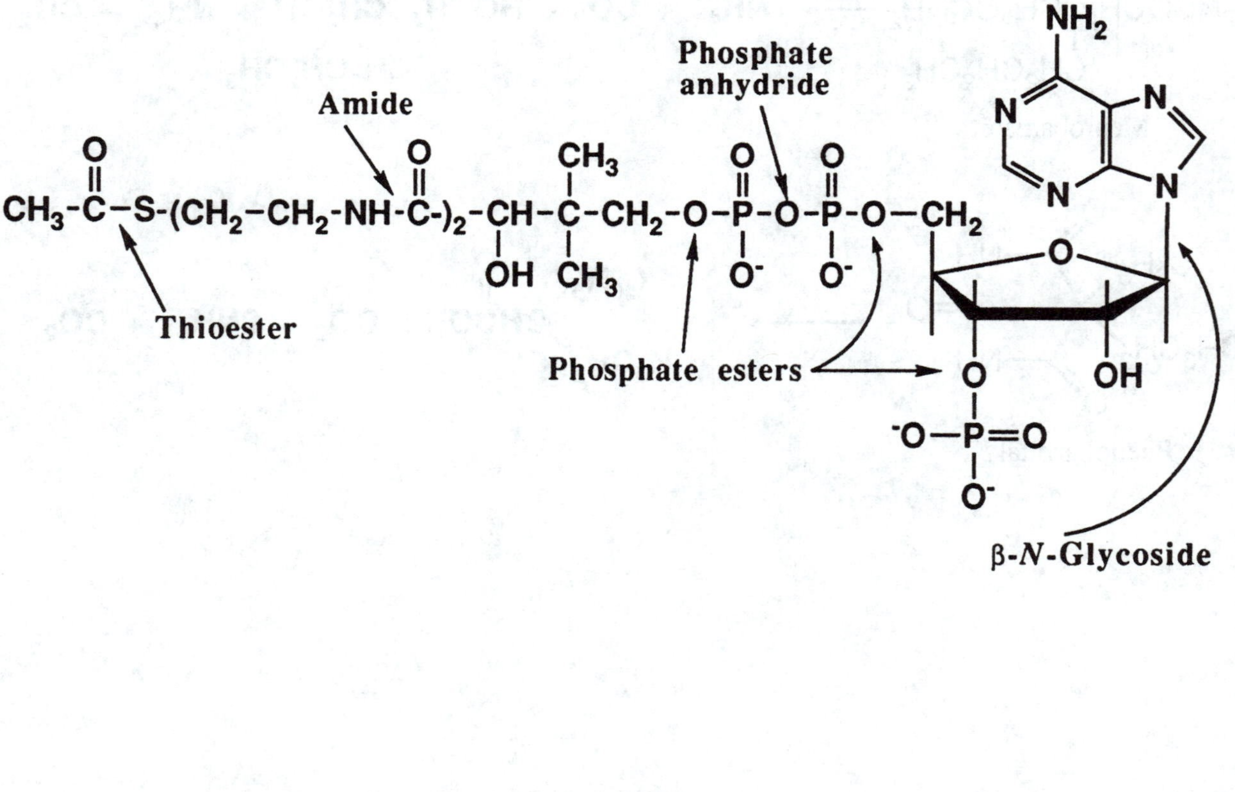

Amide

Phosphate
anhydride

Thioester

Phosphate esters

β-*N*-Glycoside

Following are the smaller molecules formed by hydrolysis of each amide, ester, and anhydride bond. They are arranged to correspond roughly to their location from left to right in acetyl CoA.

$CH_3COH$
**Acetic acid**

$HOCCH_2CH_2NH_2$
**3-Aminopropanoic acid**

$2HPO_4{}^{2-}$
**Phosphate**

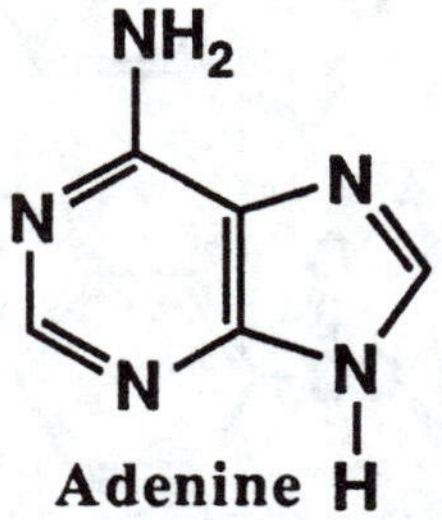

**Adenine**

$HSCH_2CH_2NH_2$
**2-Aminoethanethiol**

$HPO_4{}^{2-}$
**Phosphate**

$HOCCHCHCH_2OH$
**2,4-dihydroxy-3,3-Dimethylbutanoic acid**

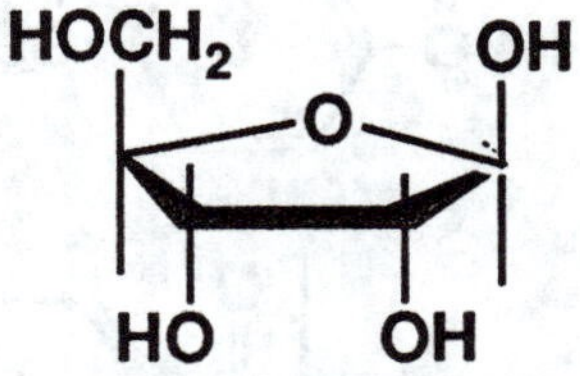

**β-D-Ribofuranose**

The molecule formed by amide formation between 3-aminopropanoic acid and 2,4-dihydroxy-3,3-dimethylbutanoic acid is given the special name pantothenic acid. Pantothenoic acid is a vitamin, most commonly contained in vitamin pills as calcium pantothenate. Its minimum daily requirement (MDR) has not yet been determined.

$HOCCH_2CH_2NHCCHCCH_2OH$
**Pantothenic acid**

## Step-Growth Polymers

**Problem 14.35** Polyethylene terephthalate (PET) can be prepared as shown. Propose a mechanism for the step-growth transesterification reaction of this polymerization.

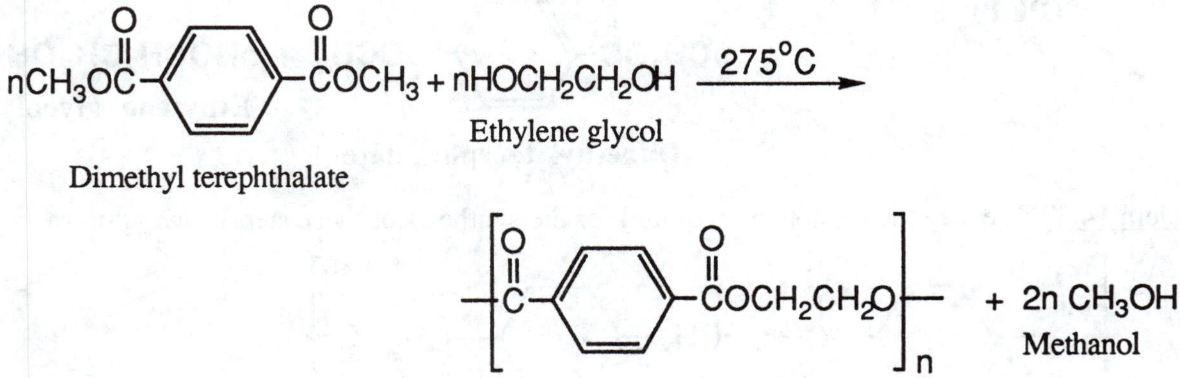

$nCH_3OC \text{—} COCH_3 + nHOCH_2CH_2OH \xrightarrow{275°C}$

**Dimethyl terephthalate**          **Ethylene glycol**

$+ 2n\ CH_3OH$
**Methanol**

**Poly(ethylene terephthalate)**

$$CH_3O\overset{O}{\underset{}{C}}\text{—}\overset{O}{\underset{}{C}}OCH_3 \ + \ HOCH_2CH_2OH \ \xrightarrow[(-CH_3OH)]{275°C}$$

$$CH_3O\overset{O}{\underset{}{C}}\text{—}\overset{O}{\underset{}{C}}OCH_2CH_2OH \ \xrightarrow[(-CH_3OH)]{CH_3O\overset{O}{\underset{}{C}}\text{—}\overset{O}{\underset{}{C}}OCH_3}$$

$$CH_3O\overset{O}{\underset{}{C}}\text{—}\overset{O}{\underset{}{C}}OCH_2CH_2O\overset{O}{\underset{}{C}}\text{—}\overset{O}{\underset{}{C}}OCH_3 \ \longrightarrow \ \longrightarrow \ \longrightarrow$$

$$\left[\overset{O}{\underset{}{C}}\text{—}\overset{O}{\underset{}{C}}OCH_2CH_2O\right]_n \ + \ 2n \ CH_3OH$$

**Problem 14.36** Currently about 30% of PET soft drink bottles are being recycled. In one recycling process, scrap PET is heated with methanol in the presence of a metal catalyst. The methanol reacts with the polymer in a transesterification reaction, liberating ethylene glycol and dimethyl terephthalate. These monomers are then used as feedstock for the production of new PET products. Write an equation for the reaction of PET with methanol to give ethylene glycol and dimethyl terephthalate.

$$\left[\overset{O}{\underset{}{C}}\text{—}\overset{O}{\underset{}{C}}OCH_2CH_2O\right]_n \ + \ 2n \ CH_3OH \ \xrightarrow[\text{Catalyst}]{\text{Heat}}$$

**Poly(ethylene terephthalate)**
**(PET)**

Methanol

$$nCH_3O\overset{O}{\underset{}{C}}\text{—}\overset{O}{\underset{}{C}}OCH_3 \ + \ nHOCH_2CH_2OH$$

**Ethylene glycol**

**Dimethyl terephthalate**

**Problem 14.37** Identify the monomers required for the synthesis of these step-growth polymers.

(a)

$$\left[\overset{O}{\underset{}{C}}\text{—}\overset{O}{\underset{}{C}}\text{—}O\text{—}CH_2\text{—}\bigcirc\text{—}CH_2\text{—}O\right]_n$$

**Kodel (a polyester)**

**The following two monomers can be used to produce Kodel:**

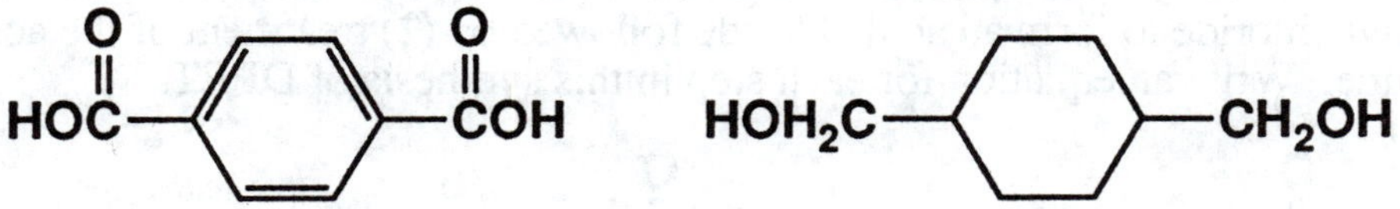

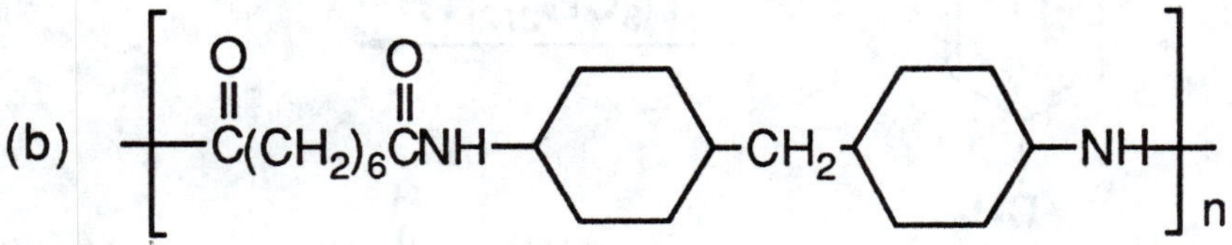

HOC—⬡—COH     HOH₂C—⬡—CH₂OH

(b)

Quiana (a polyamide)

**The following two monomers can be used to produce Quiana:**

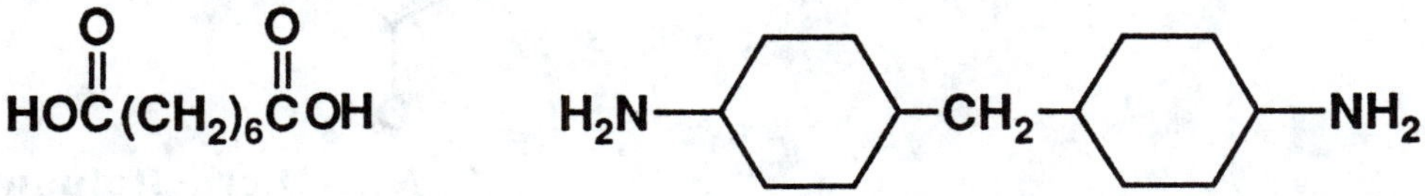

HOC(CH₂)₆COH     H₂N—⬡—CH₂—⬡—NH₂

<u>Problem 14.38</u> Nomex is an aromatic polyamide (aramid) prepared from polymerization of 1,3-benzenediamine and the acid chloride of 1,3-benzenedicarboxylic acid. The physical properties of the polymer make it suitable for high strength, high temperature applications such as parachute chords and jet aircraft tires. Draw a structural formula for the repeating unit of Nomex.

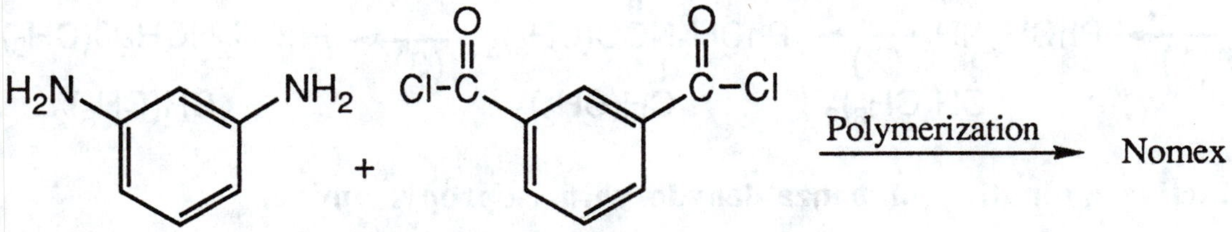

H₂N—⬡—NH₂     Cl-C—⬡—C-Cl

+          $\xrightarrow{\text{Polymerization}}$     Nomex

1,3-Benzenediamine          1,3-Benzenedicarboxylic
acid chloride

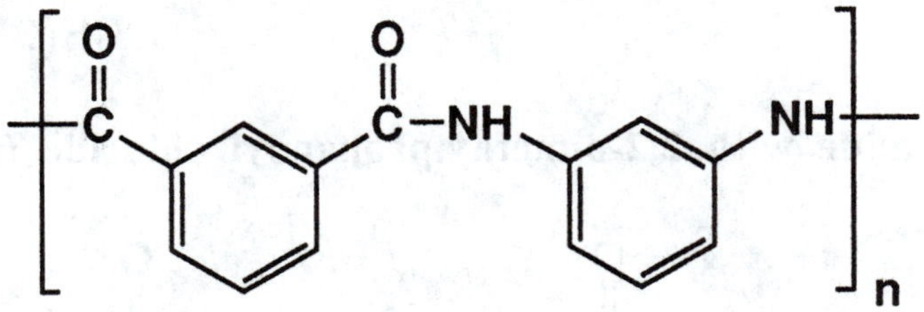

**Nomex**

## Synthesis

**Problem 14.39** *N,N*-Diethyltoluamide (DEET) is the active ingredient in several common insect repellents. DEET can be synthesized in two steps: (1) treatment of 3-methylbenzoic acid (*m*-toluic acid) with thionyl chloride to form an acid chloride followed by (2) treatment of the acid chloride with diethylamine. Write an equation for each step in this synthesis of DEET.

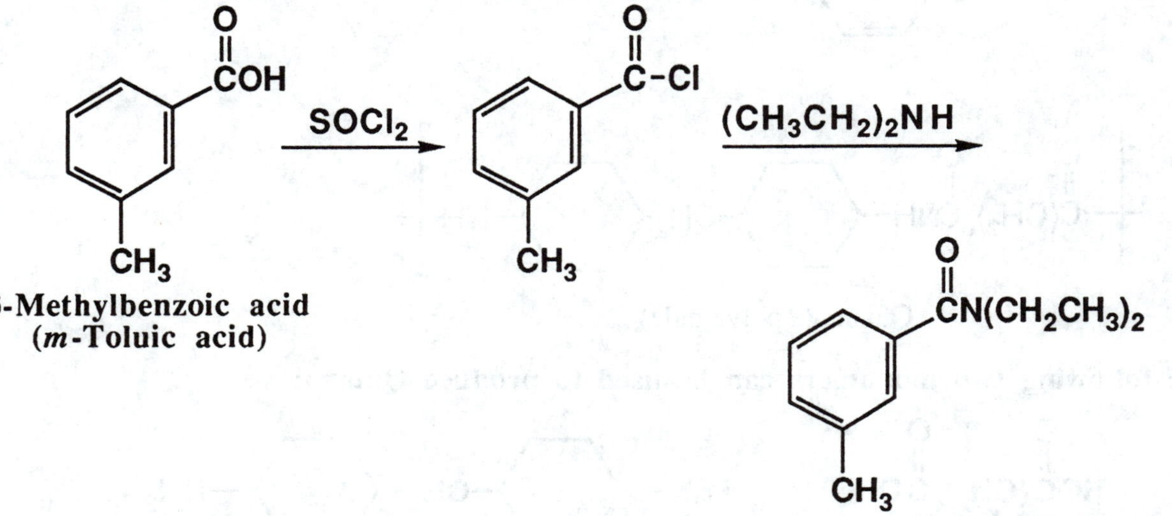

**3-Methylbenzoic acid**
**(*m*-Toluic acid)**

**N,N-Diethyltoluamide**
**(DEET)**

**Problem 14.40** Show reagents for the synthesis of the tertiary amine:

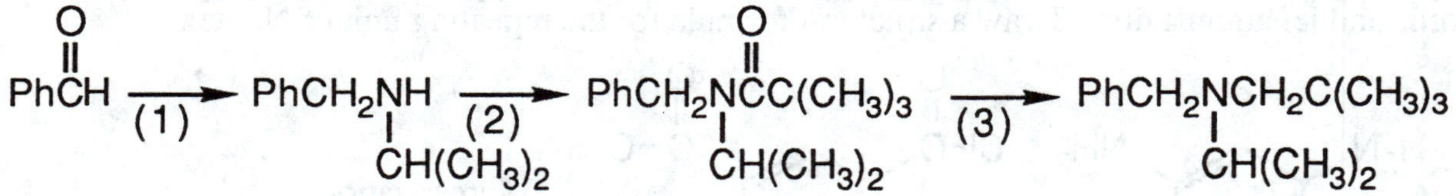

**(1) Reductive amination of benzaldehyde with isopropyl amine.**

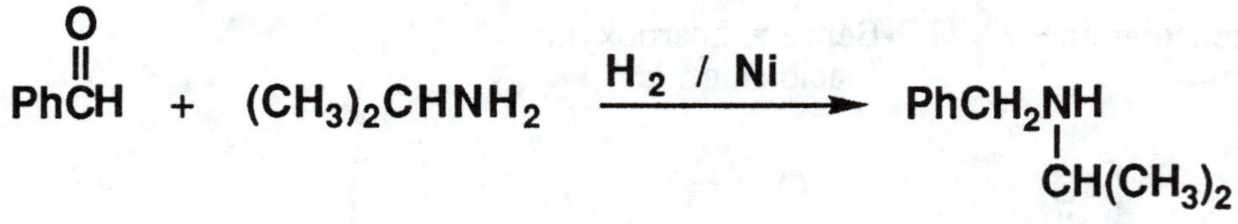

**(2) Formation of an amide with 2,2-dimethylpropanoyl chloride (pivaloyl chloride)**

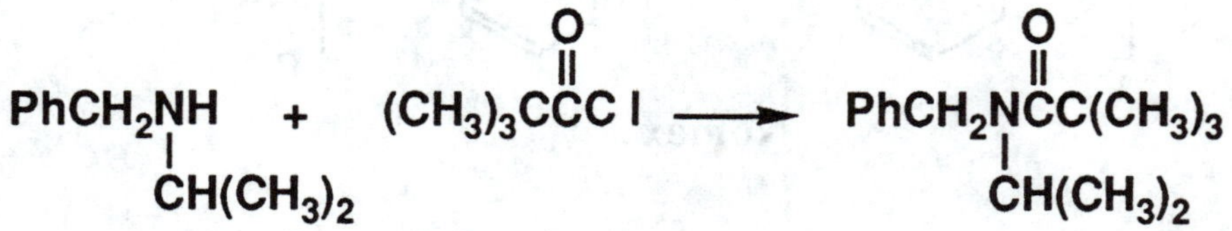

**(3) Reduction of the amide with lithium aluminum hydride or with hydrogen in the presence of a transition metal catalyst.**

$$PhCH_2\underset{\underset{CH(CH_3)_2}{|}}{N}\overset{\overset{O}{\|}}{C}C(CH_3)_3 \quad \xrightarrow[\substack{o\ r \\ H_2,\ Pd}]{\substack{1)\ LiAlH_4 \\ 2)\ H_2O}} \quad PhCH_2\underset{\underset{CH(CH_3)_2}{|}}{N}CH_2C(CH_3)_3$$

Problem 14.41  Show how to convert ethyl 2-pentenoate to these compounds.

$$CH_3CH_2CH\!\!=\!\!CH\overset{\overset{O}{\|}}{C}OCH_2CH_3$$

Ethyl-2-pentenoate

(a)  $CH_3CH_2CH\!\!=\!\!CH\overset{\overset{O}{\|}}{C}OCH_2CH_3 \quad \xrightarrow{H_2/Pd} \quad CH_3CH_2CH_2CH_2\overset{\overset{O}{\|}}{C}OCH_2CH_3$

(b)  $CH_3CH_2CH\!\!=\!\!CH\overset{\overset{O}{\|}}{C}OCH_2CH_3 \quad \xrightarrow[2)\ H_2O]{1)\ LiAlH_4} \quad CH_3CH_2CH\!\!=\!\!CHCH_2OH$

(c)  $CH_3CH_2CH\!\!=\!\!CH\overset{\overset{O}{\|}}{C}OCH_2CH_3 \quad \xrightarrow[\substack{H_2O \\ pH\ 11.8}]{KMnO_4} \quad CH_3CH_2\underset{\underset{OH}{|}}{C}H\!-\!\underset{\underset{OH}{|}}{C}H\!-\!\overset{\overset{O}{\|}}{C}OCH_2CH_3$

Problem 14.42  The World Health Organization estimates that the tropical disease schistosomiasis (bilharziasis) affects between 180 and 200 million persons and, next to malaria, is the world's most serious parasitic infection in humans.  The disease is caused by blood flukes, small flatworms of the family Schistosomatidae, which live in the blood vessels of humans and other mammals.  Female blood flukes release from 300 to 3000 eggs daily into the bloodstream.  Those evacuated in the feces or urine into fresh water hatch to larvae, which find their way to host water snails, and develop further.  The disease is most often contracted by humans from contaminated water populated by snails that carry the worms.  Symptoms of the disease range from cough and fever to liver, lung, and brain damage.  As one attack on this disease, the compound Bayluscid (niclosamide) has been developed to kill infected water snails.

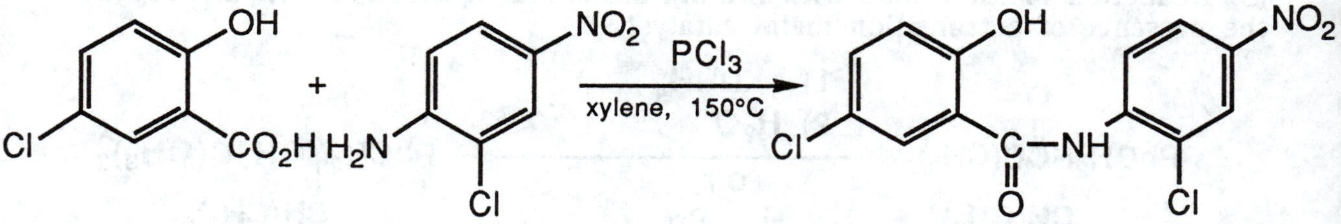

5-Chloro-2-hydroxy-
benzoic acid

2-Chloro-4-
nitroaniline

Bayluscid
(Nilosamide)

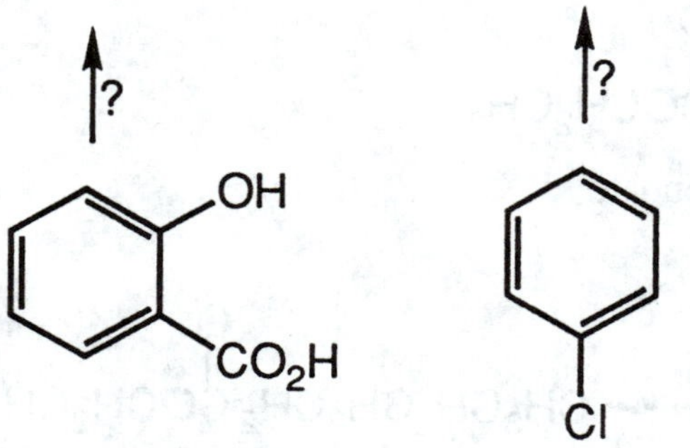

Salicylic acid          Chlorobenzene

(a) Propose a synthesis of 5-chloro-2-hydroxybenzoic acid from salicylic acid.

**The chlorine atom can be introduced via reaction of salicylic acid with $Cl_2$ and $FeCl_3$. Notice how the activating -OH group will direct the Cl atom ortho/para.**

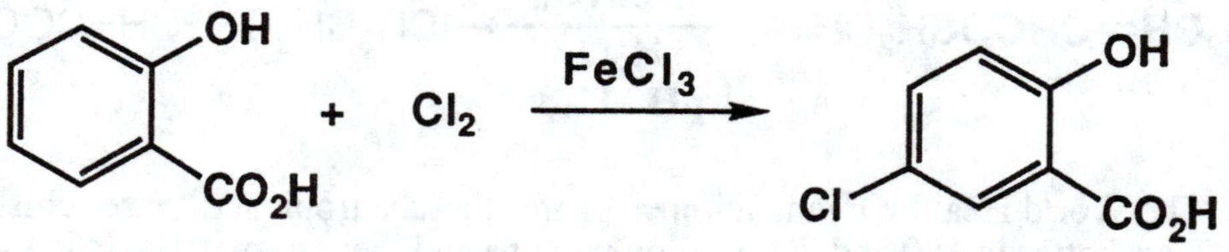

(b) Propose a synthesis of 2-chloro-4-nitroaniline from chlorobenzene.

**A reasonable synthetic plan is shown below:**

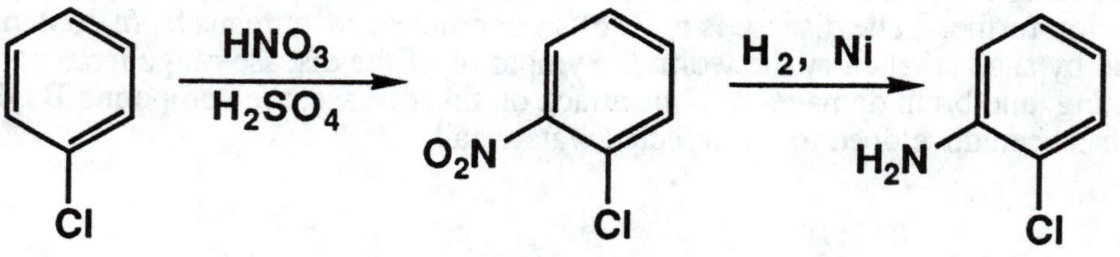

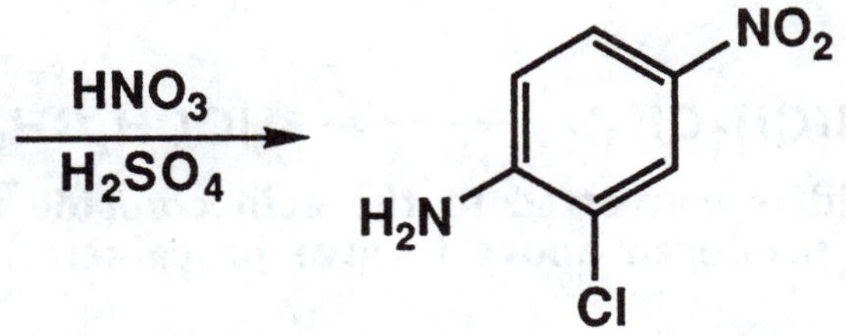

$$\xrightarrow[\text{H}_2\text{SO}_4]{\text{HNO}_3}$$

(structure: benzene ring with NO$_2$, H$_2$N, and Cl substituents)

(c) We have not studied PCl$_3$ as a reagent for organic synthesis. The same transformation can be brought about, however, by treatment of 5-chloro-2-hydroxybenzoic acid with thionyl chloride, SOCl$_2$, followed by treatment of the product of that reaction with 2-chloro-4-nitroaniline. Show how these two steps lead to the synthesis of Bayluscide.

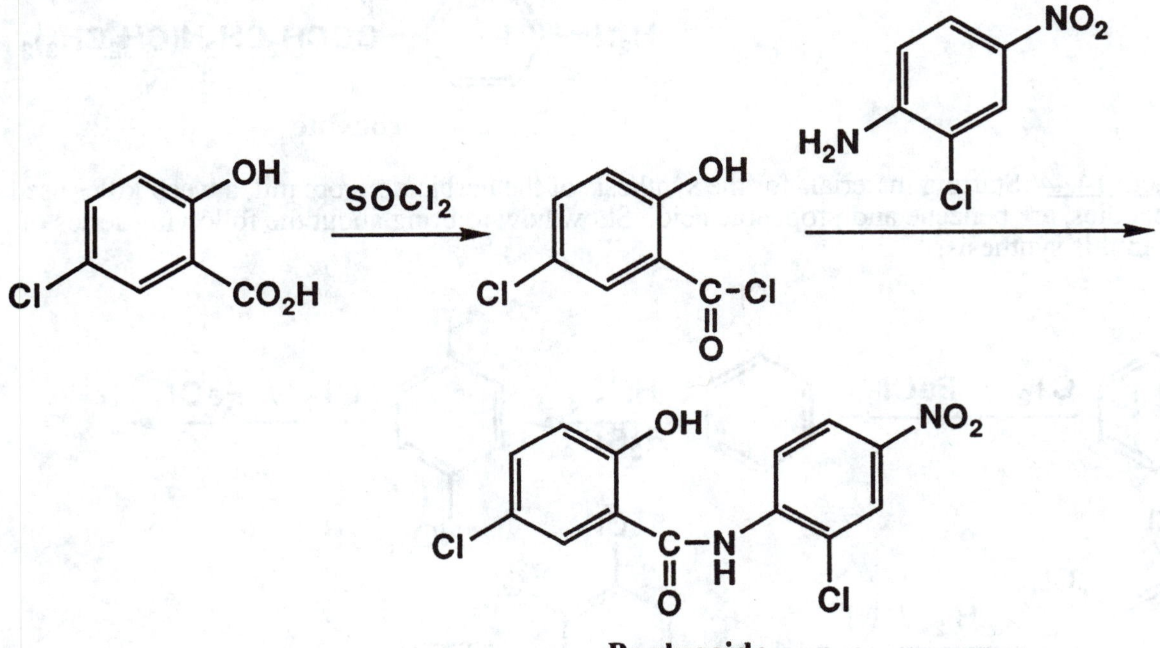

$$\xrightarrow{\text{SOCl}_2}$$

**Bayluscide**

Problem 14.43 Procaine (hydrochloride marketed as Novocaine) was one of the first local anesthetics for infiltration and regional anesthesia. Show how to synthesize this molecule using *p*-aminobenzoic acid, ethylene oxide, and diethylamine as sources of carbon atoms.

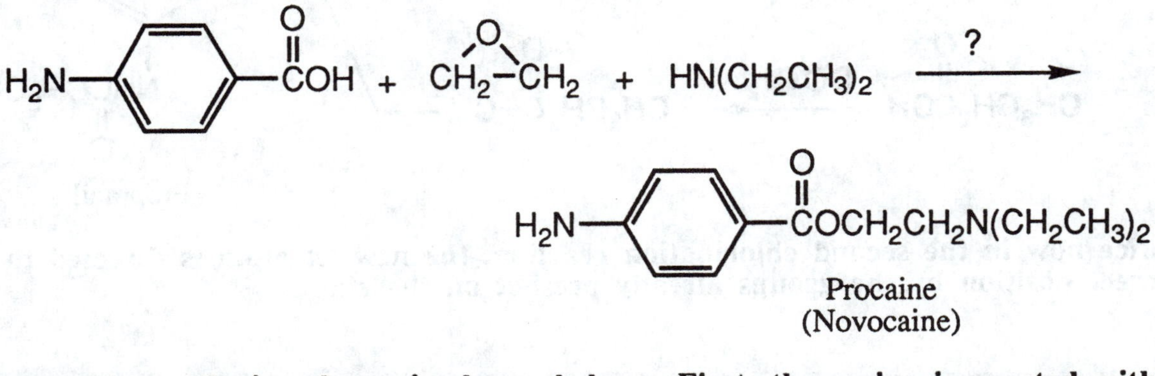

$$\text{H}_2\text{N}-\!\!\!\!\!\bigcirc\!\!\!\!\!-\overset{\overset{\displaystyle O}{\|}}{\text{C}}\text{OH} \;+\; \text{CH}_2-\text{CH}_2 \;+\; \text{HN(CH}_2\text{CH}_3)_2 \;\xrightarrow{\;?\;}$$

$$\text{H}_2\text{N}-\!\!\!\!\!\bigcirc\!\!\!\!\!-\overset{\overset{\displaystyle O}{\|}}{\text{C}}\text{OCH}_2\text{CH}_2\text{N(CH}_2\text{CH}_3)_2$$

Procaine
(Novocaine)

**A reasonable synthetic scheme is shown below. First, the amine is reacted with ethylene oxide to create the corresponding alcohol:**

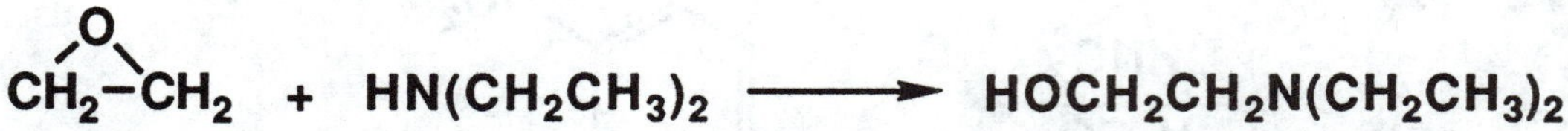
CH2–CH2 + HN(CH2CH3)2
HOCH2CH2N(CH2CH3)2

**Next, *p*-aminobenzoic acid is converted to the acid chloride using $SOCl_2$ and reacted with the alcohol produced above to give procaine.**

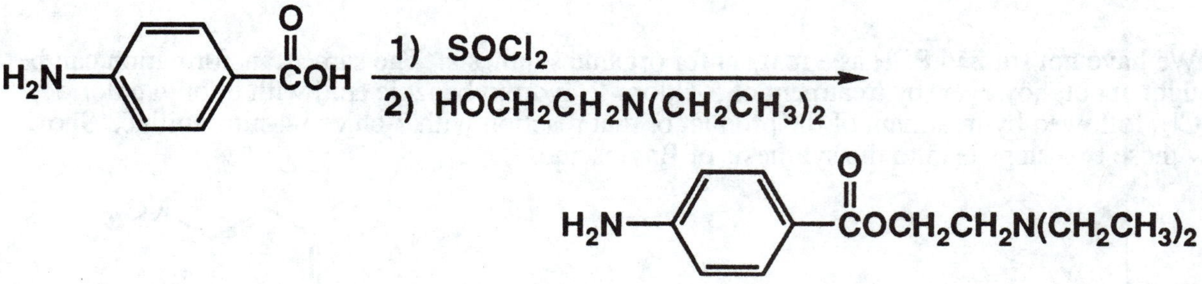
H2N
COH
1) SOCl2
2) HOCH2CH2N(CH2CH3)2
H2N
COCH2CH2N(CH2CH3)2

**Procaine**

Problem 14.44 Starting materials for the synthesis of the herbicide propranil, a weed killer used in rice paddies, are benzene and propanoic acid. Show how to bring about the following series of steps in this synthesis.

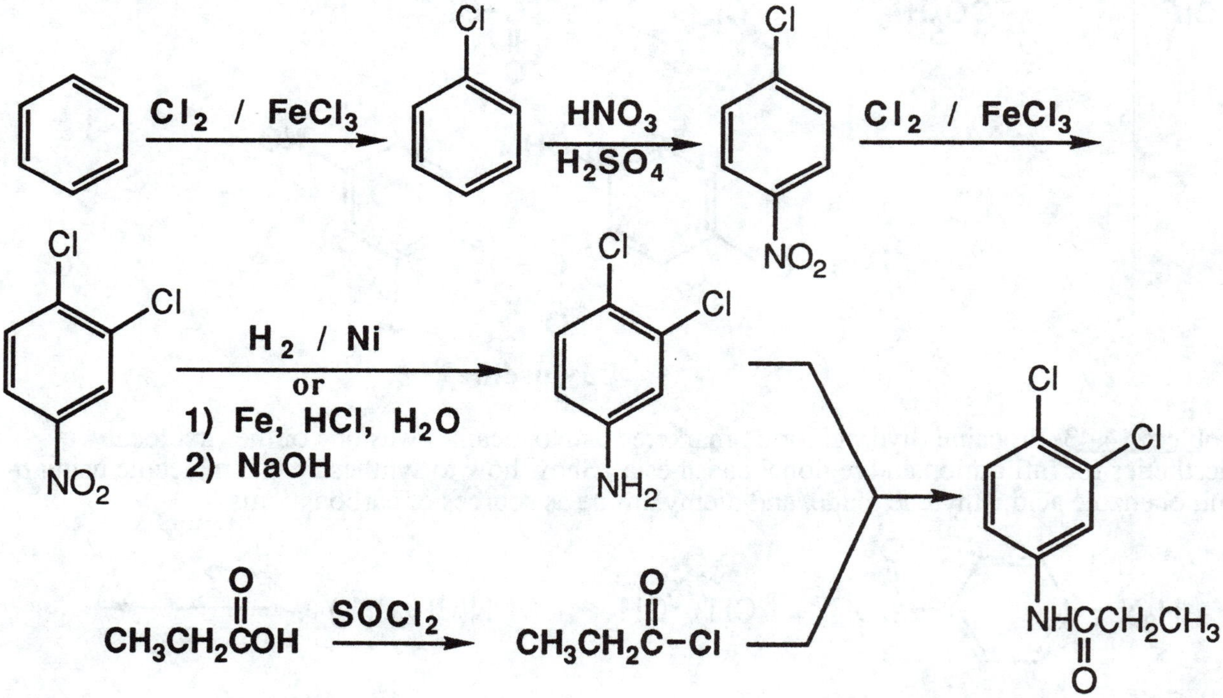
Cl2 / FeCl3
Cl
HNO3
H2SO4
Cl
NO2
Cl2 / FeCl3
Cl
Cl
NO2
H2 / Ni
or
1) Fe, HCl, H2O
2) NaOH
Cl
Cl
NH2
CH3CH2COH
SOCl2
CH3CH2C–Cl
Cl
Cl
NHCCH2CH3

Propranil

**Notice how in the second chlorination reaction, the new Cl atom is directed to the correct position by the groups already present on the ring.**

<u>Problem 14.45</u>  Following are structural formulas for three local anesthetics.  Lidocaine was introduced in 1948 and is now the most widely used local anesthetic for infiltration and regional anesthesia.  Its hydrochloride is marketed under the name Xylocaine.  Etidocaine (hydrochloride marketed as Duranest) is comparable to lidocaine in onset, but its analgesic action lasts two to three times longer.  Anesthetic action from mepivacaine (hydrochloride marketed as Carbocaine) is faster and somewhat longer in duration than lidocaine.

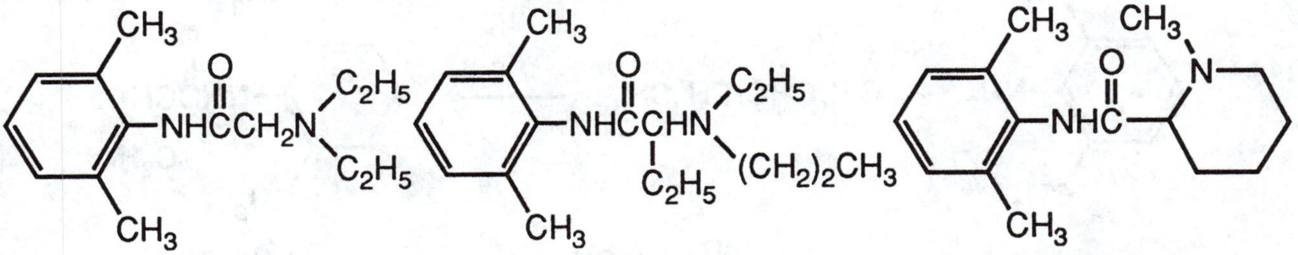

Lidocaine
(Xylocaine)

Etidocaine
(Duranest )

Mepivacaine
(Carbocaine)

(a) Propose a synthesis of lidocaine from 2,6-dimethylaniline, 2-chloroacetyl chloride, and diethylamine.

**The 2-chloroacetyl chloride reacts with 2,6-dimethylaniline to form the amide, since the acid chloride function is more reactive than the chloromethyl moiety. The amide is reacted with diethyl amine to complete the synthesis.**

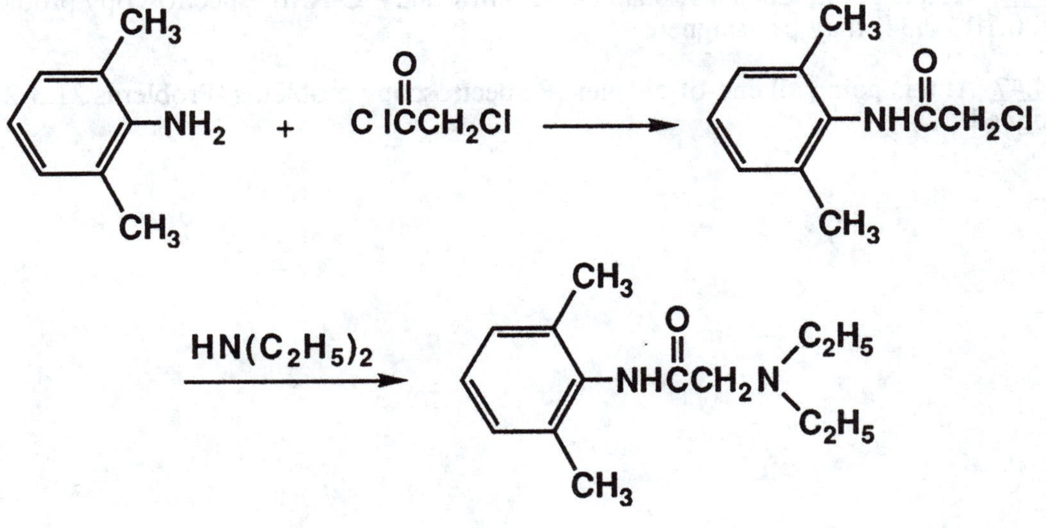

(b) Propose a synthesis of etidocaine form 2,6-dimethylaniline, 2-chlorobutanoyl chloride, and ethylpropylamine.

**This synthesis is the same as above except 2-chlorobutanoyl chloride and N-ethylaminopropane (ethylpropylamine) are used.**

**Problem 14.46** At this point, all end-of-chapter [1]H-NMR and [13]C-NMR spectroscopy problems (Problems 20.10-20.34) may be assigned.

**Problem 14.47** At this point, all end-of-chapter IR spectroscopy problems (Problems 21.3-21.13) may be assigned.

# CHAPTER 15: ENOLATE ANIONS

## SUMMARY OF REACTIONS

|  | Enolates | β-Hydroxyaldehydes β-Hydroxyketones | β-Ketoesters |
|---|---|---|---|
| **Aldehydes, Ketones, Esters** | **15A** 15.1* | | |
| **Enolates Aldehydes, Ketones** | | **15B** 15.2 | |
| **Enolates Esters** | | | **15C** 15.3 |

Starting Material → Product →

*Section in book that describes reaction.

## REACTION 15A: FORMATION OF ENOLATE ANIONS (Section 15.1)

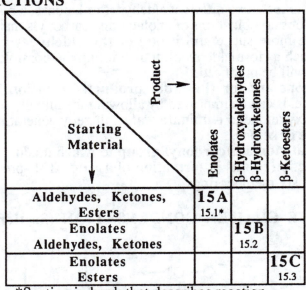

- The α-**hydrogens** of carbonyl **compounds are relatively acidic,** so carbonyl compounds can be converted to **enolate anions** by reaction with **base.** The α hydrogens are so acidic because enolate anions are **stabilized by resonance** involving the carbonyl group. ✻
- Enolate anions are important because they are intermediates in reactions such as the aldol and Claisen condensations, and they also take part as nucleophiles in $S_N2$ reactions.

## REACTION 15B: THE ALDOL REACTION (Section 15.2)

$$2 \ \underset{H}{\overset{O}{\underset{|}{-C}-C'-}} \ \xrightarrow{\text{NaOH}} \ \underset{H}{\overset{OH \ O}{-C-C'-C-C'-}}$$

- In **the aldol reaction, two aldehyde and/or ketone molecules** react to form a β-**hydroxyaldehyde or β-hydroxyketone.** ✻
- The mechanism involves the nucleophilic attack by an enolate anion of one aldehyde or ketone on the electrophilic carbonyl carbon atom of another aldehyde or ketone to produce a tetrahedral carbonyl addition intermediate that is protonated to yield the final product.

- This is an **important reaction** because it represents a relatively **easy way to make a carbon-carbon bond,** and thus create a larger, **more complex molecule from simpler ones.** ✳
- Usually, the two molecules of the same aldehyde or ketone are used to produce the aldol product.
- If attempts are made to react two different aldehydes or ketones, a mixture of aldol products results from the various combinations of enolates and carbonyls that can react. These **mixed aldol reactions** are more successful if **one of the aldehydes or ketones cannot form an enolate** (such as formaldehyde), thus limiting the possible reactions and ensuring that fewer products will be produced.
- The **equilibrium constant for the aldol product formation is usually not large.** However, product yields can be improved by allowing the initially formed β-hydroxy-aldehyde or β-hydroxyketone to **eliminate water,** thereby generating the more stable α,β-unsaturated aldehyde or ketone.
- When both the **enolate ion** and **carbonyl group** to which it adds are in the **same molecule,** aldol reaction results in **formation of a ring.** This process is especially useful for making five- and six-member rings.

## REACTION 15C: THE CLAISEN CONDENSATION (Section 15.3)

- A **Claisen condensation** is carried out by treating **esters** with **base,** often the alkoxide of the alcohol found in the ester, to produce a β-**ketoester.** ✳
- The reaction mechanism involves initial deprotonation of an ester molecule by the base to form an enolate anion that then attacks the carbonyl carbon atom of another ester molecule to produce a tetrahedral carbonyl addition intermediate. This converts to products when the alkoxide departs. The resulting β-keto ester is deprotonated under these basic conditions, but then protonated by the acid work-up to produce the final product.
- A Claisen condensation reaction between **two ester functions on the same molecule** to give a **cyclic product,** usually a five- or six-member ring, is called a **Dieckmann condensation.**
- The reaction usually takes place between two molecules of the same ester. However, **mixed Claisen condensations** are possible if there is **an appreciable difference in the reactivity between the two esters.** For example, crossed Claisen condensations can be carried out if one of the esters cannot form an enolate because it does not have any hydrogens on the α-carbon atom.
- The **Claisen condensation** is important because it allows for the **synthesis of a carbon-carbon bond** under relatively mild conditions.

# CHAPTER 15
*Solutions to the Problems*

<u>Problem 15.1</u>  Identify the acidic hydrogens in these compounds.

**In the following structures, the acidic hydrogens are the ones on carbon atoms that are adjacent to the carbonyl group. These hydrogens are marked by an arrow.**

(a) Cyclohexanone

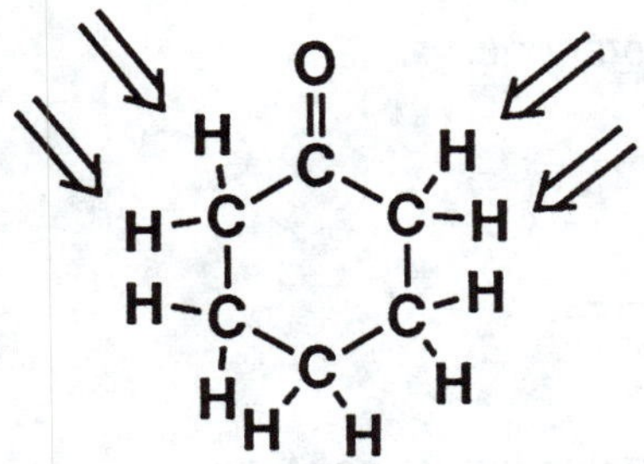

(b) Acetophenone

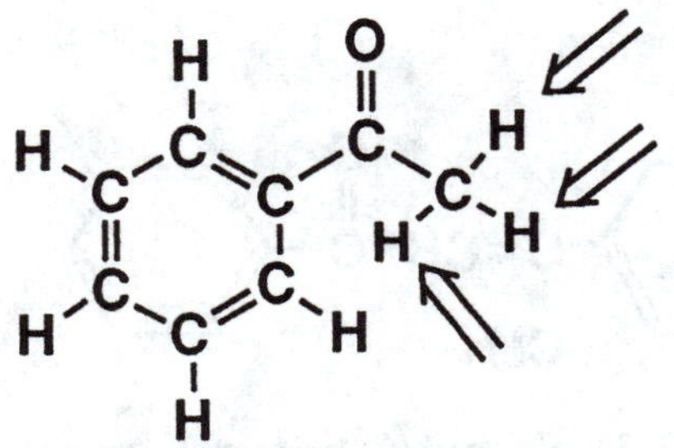

<u>Problem 15.2</u>  Treatment of 2-butanone with base gives two enolate anions. Draw each enolate anion as a hybrid of two contributing structures.

**One enolate is derived from each of the two carbon atoms adjacent to the carbonyl group.**

$$CH_3CH_2\overset{\overset{\displaystyle :O:}{\|}}{C}-\overset{\overset{\displaystyle H}{|}}{\overset{..}{C}}^{-}-H \longleftrightarrow CH_3CH_2\overset{\overset{\displaystyle :\overset{..}{O}:^{-}}{|}}{C}=\overset{\overset{\displaystyle}{}}{\underset{\underset{\displaystyle H}{|}}{C}}-H$$

$$CH_3-\overset{-}{\overset{..}{C}}-\overset{\overset{\displaystyle :O:}{\|}}{C}CH_3 \longleftrightarrow CH_3-\overset{}{C}=\overset{\overset{\displaystyle :\overset{..}{O}:^{-}}{|}}{\underset{\underset{\displaystyle H}{|}}{C}}CH_3$$

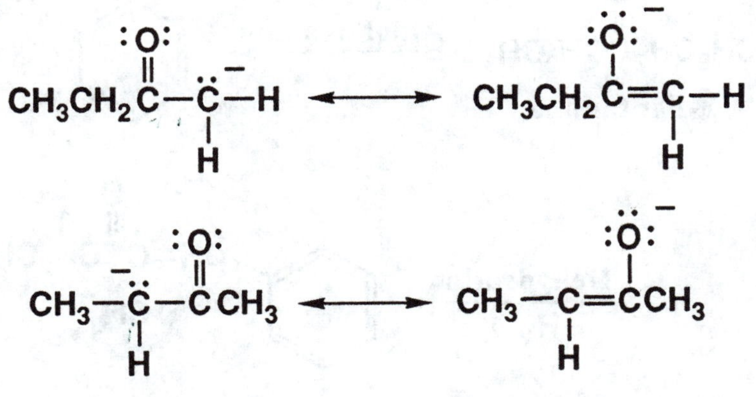

<u>Problem 15.3</u>  Draw the enolate anion formed by treatment of methyl phenylacetate with sodium methoxide.

**In this case, there is only one enolate possible.**

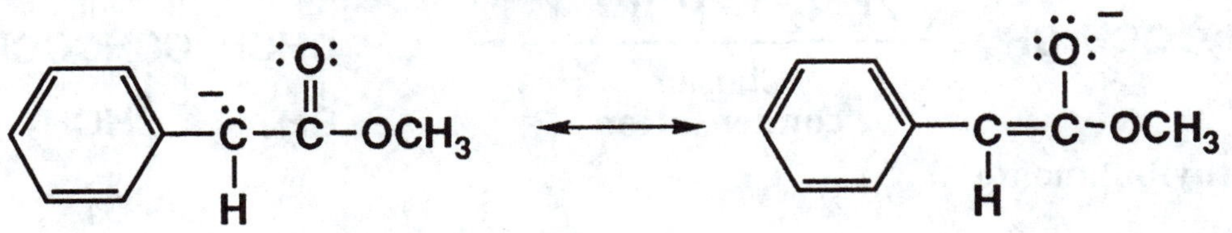

<u>Problem 15.4</u> Draw the product of the aldol reaction of these compounds.
(a) Acetophenone                                            (b) Cyclopentanone

**1,3-Diphenyl-3-hydroxy-**
**3-methyl-1-butanone**

**2-(1-Hydroxycyclopentyl)-**
**cyclopentanone**

<u>Problem 15.5</u> Draw the product of acid catalyzed dehydration of each aldol product from Practice Problem 15.4.
(a) Acetophenone                                            (b) Cyclopentanone

**1,3-Diphenyl-2-buten-1-one**

**2-Cyclopentenylcyclopentane**

<u>Problem 15.6</u> Draw the product of the mixed aldol reaction between benzaldehyde and 3-pentanone and the product formed by its base-catalyzed dehydration.

$$\overset{O}{\overset{\|}{CH}}$$

**Benzaldehyde**

+

$$CH_3CH_2\overset{O}{\overset{\|}{C}}CH_2CH_3$$

**3-Pentanone**

→ **Aldol reaction** →

$$\overset{OH}{\overset{|}{CHCHCCH_2CH_3}}\overset{O}{\overset{\|}{}}$$
$$\overset{|}{CH_3}$$

**Dehydration**
**-H₂O**

→

$$CH=\overset{}{C}\overset{O}{\overset{\|}{C}}CH_2CH_3$$
$$\overset{|}{CH_3}$$

<u>Problem 15.7</u> Show the product of the Claisen condensation of ethyl 3-methylbutanoate in the presence of sodium ethoxide.

$$CH_3\overset{}{C}HCH_2\overset{O}{\overset{\|}{C}}OCH_2CH_3$$
$$\overset{|}{CH_3}$$

**Ethyl 3-methylbutanoate**

1) $CH_3CH_2O^-Na^+$
2) $H_2O$, $HCl$

**Claisen**
**condensation**

→

$$CH_3\overset{}{C}HCH_2\overset{O}{\overset{\|}{C}}\overset{}{C}H\overset{O}{\overset{\|}{C}}OCH_2CH_3$$
$$\overset{|}{CH_3}\qquad\overset{|}{C}HCH_3$$
$$\overset{|}{CH_3}$$

**Problem 15.8** Complete the equation for this mixed Claisen condensation.

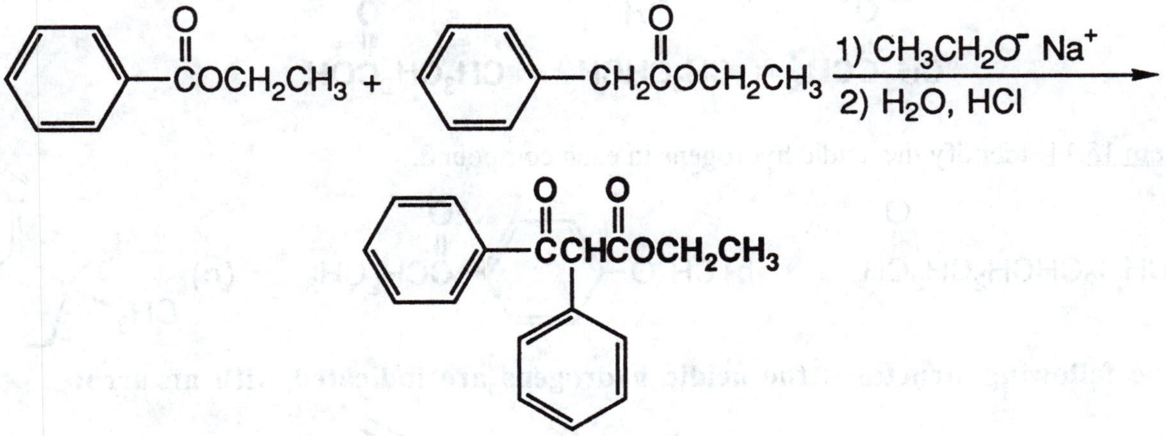

**Problem 15.9** Show how to convert ethyl benzoate to 3-methyl-1-phenyl-1-butanone (isobutyl phenyl ketone) using a Claisen condensation at some stage in the synthesis.

PhCOCH$_2$CH$_3$
Ethyl benzoate

PhCCH$_2$CHCH$_3$ .
3-Methyl-1-phenyl-1-butanone

**A mixed Claisen condensation of ethyl benzoate and ethyl 3-methylbutanoate gives a β-ketoester. Saponification of the ester followed by acidification gives the β-ketoacid. Heating causes decarboxylation and gives the desired product.**

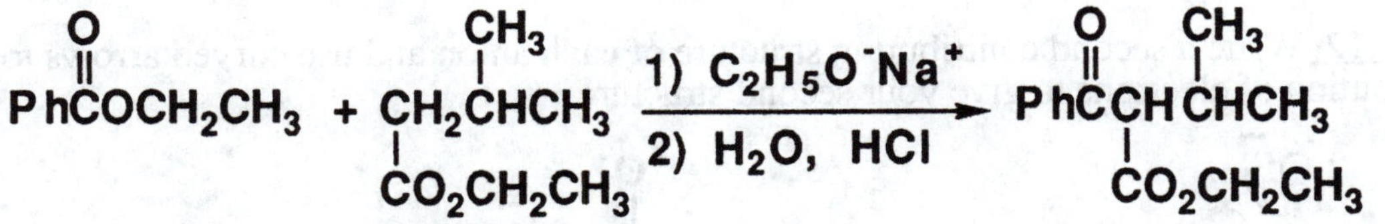

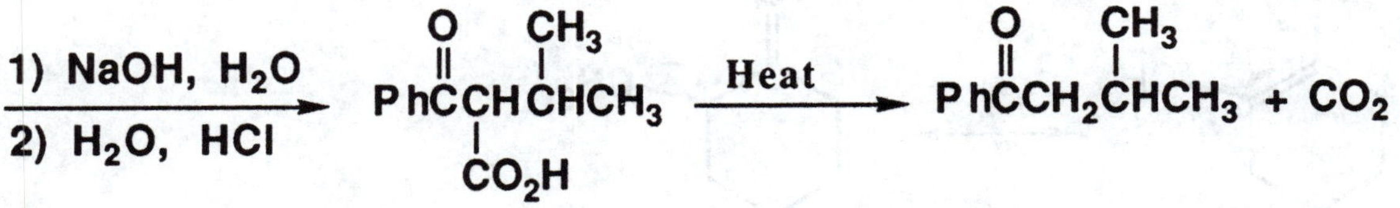

**The Aldol Reaction**
**Problem 15.10** Estimate the p$K_a$ of each compound and then arrange them in order of increasing acidity.

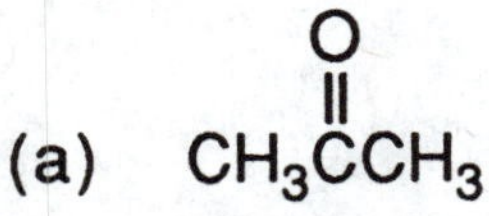

(a)   CH$_3$CCH$_3$

      p$K_a$ 20

(b)   CH$_3$CHCH$_3$

      p$K_a$ 17

(c)   CH$_3$CH$_2$COH

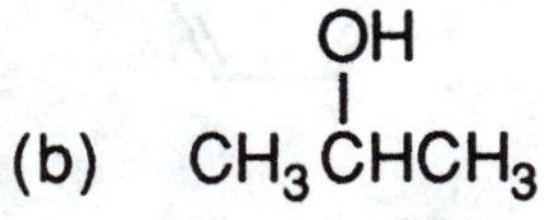

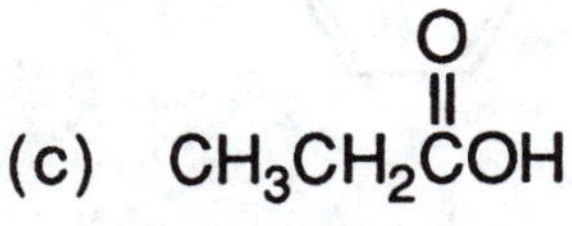

      p$K_a$ 5

**The order of acidity ranked from least to most acidic is:**

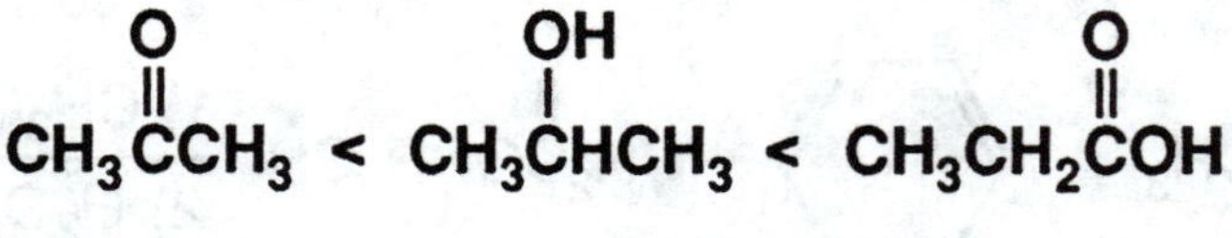

$$CH_3\overset{\overset{\textstyle O}{\|}}{C}CH_3 \quad < \quad CH_3\overset{\overset{\textstyle OH}{|}}{C}HCH_3 \quad < \quad CH_3CH_2\overset{\overset{\textstyle O}{\|}}{C}OH$$

<u>Problem 15.11</u> Identify the acidic hydrogens in each compound.

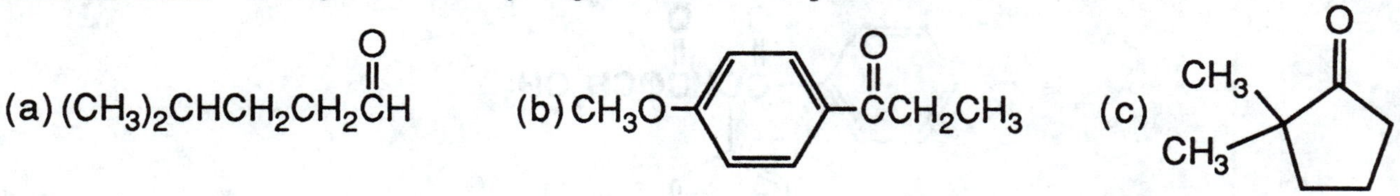

(a) $(CH_3)_2CHCH_2CH_2\overset{\overset{\textstyle O}{\|}}{C}H$     (b) $CH_3O$—⬡—$\overset{\overset{\textstyle O}{\|}}{C}CH_2CH_3$     (c)

**On the following structures the acidic hydrogens are indicated with an arrow,**

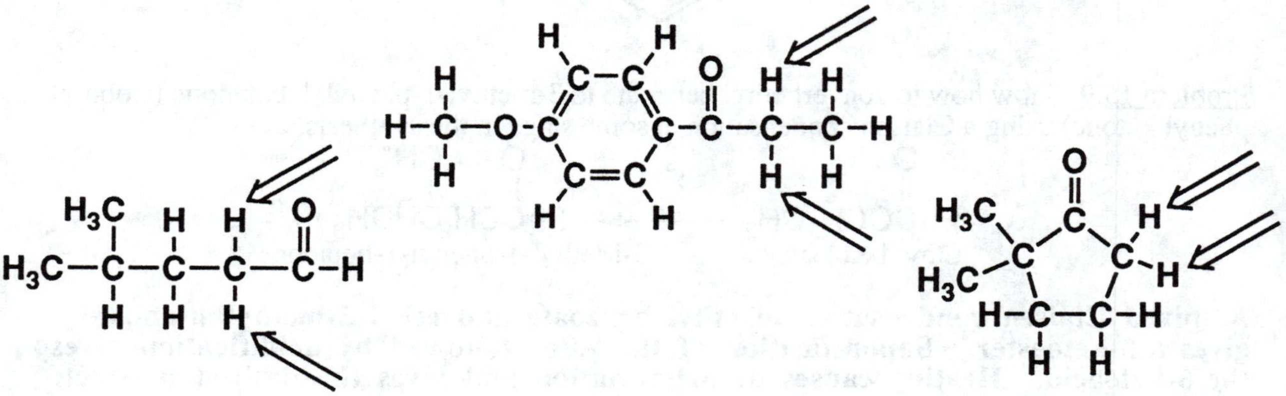

Problem 15.12 Write a second contributing structure of each anion and use curved arrows to show the redistribution of electrons to give your second structure.

(a) 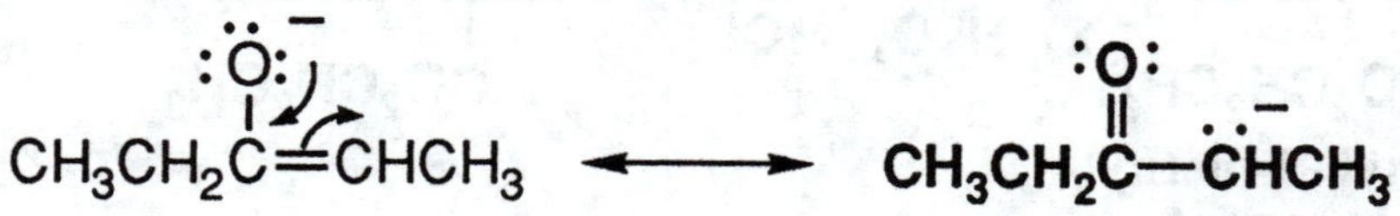   $CH_3CH_2\overset{\overset{\textstyle :\overset{..}{O}:^-}{|}}{C}=CHCH_3 \quad \longleftrightarrow \quad CH_3CH_2\overset{\overset{\textstyle :O:}{\|}}{C}-\overset{..}{C}HCH_3$

(b)

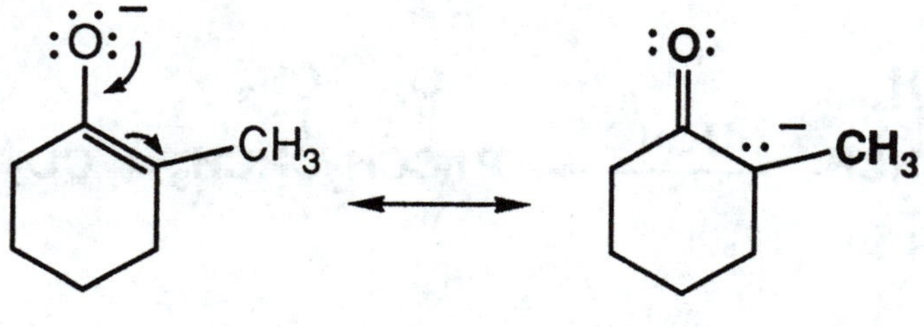

(c)

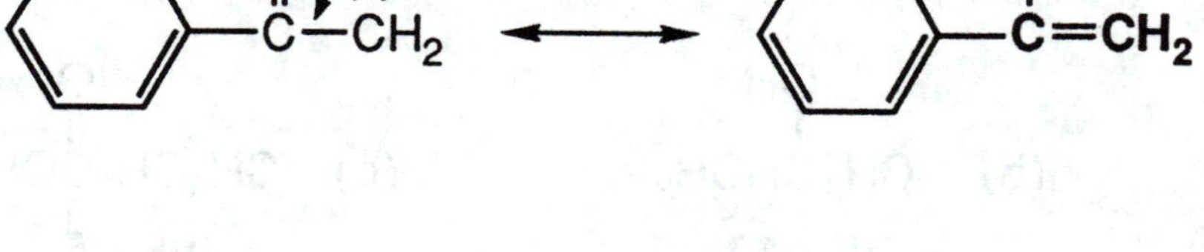

**Problem 15.13** Treatment of 2-methylcyclohexanone with base gives two different enolate anions. Draw the contributing structure for each that places the negative charge on carbon.

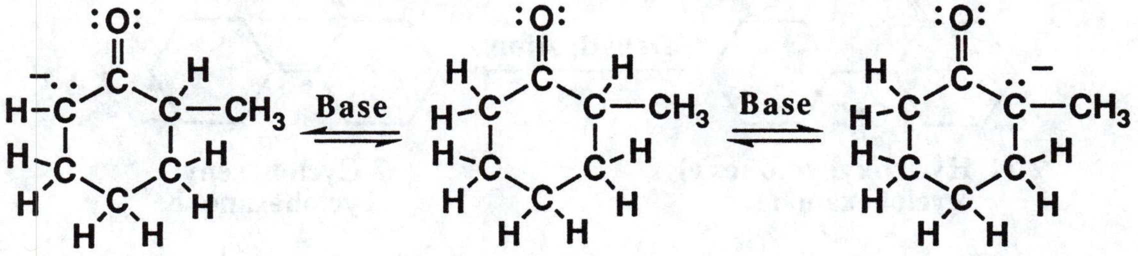

**2-Methyl cyclohexanone**

**Problem 15.14** Draw structural formulas for the products of aldol reactions of the following compounds and for the α,β-unsaturated aldehyde or ketone formed from dehydration of the aldol product.

(a)

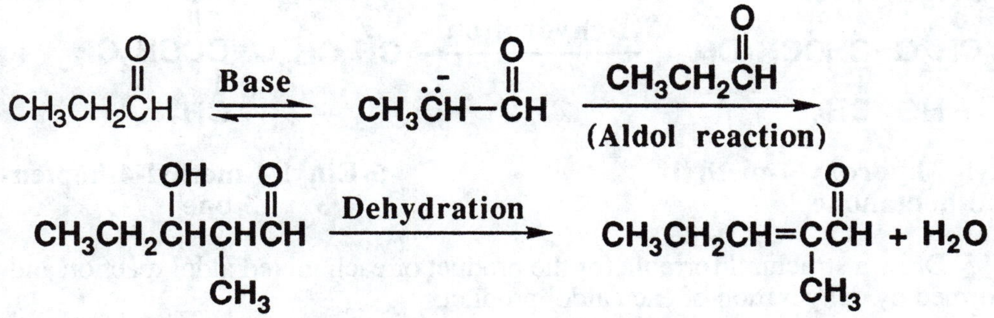

3-Hydroxy-2-methylpentanal                    2-Methyl-2-pentenal

(b)

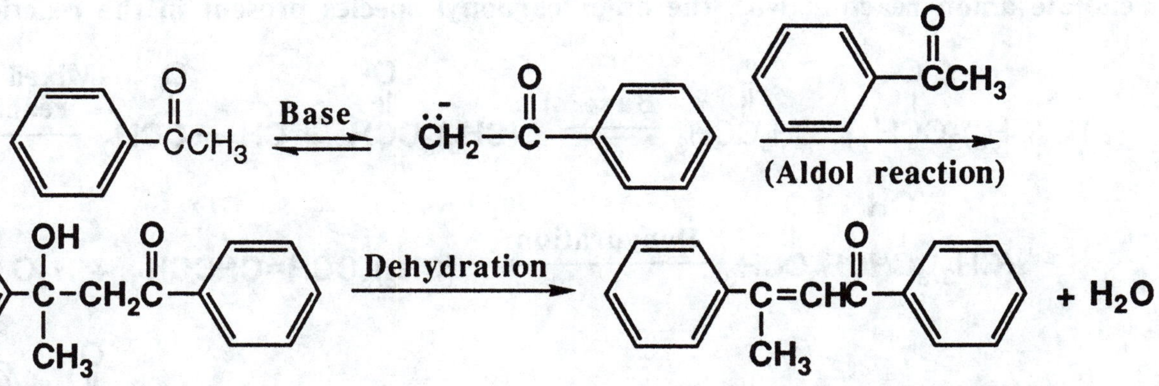

1,3-Diphenyl-3-hydroxy-
3-methyl-1-butanone                            1,3-Diphenyl-2-buten-1-one

(c)

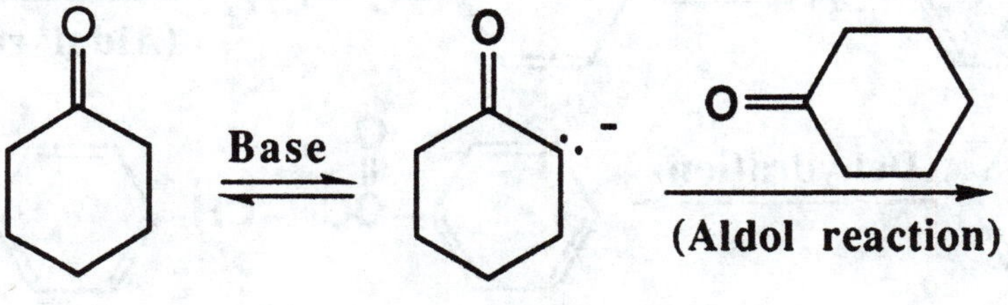

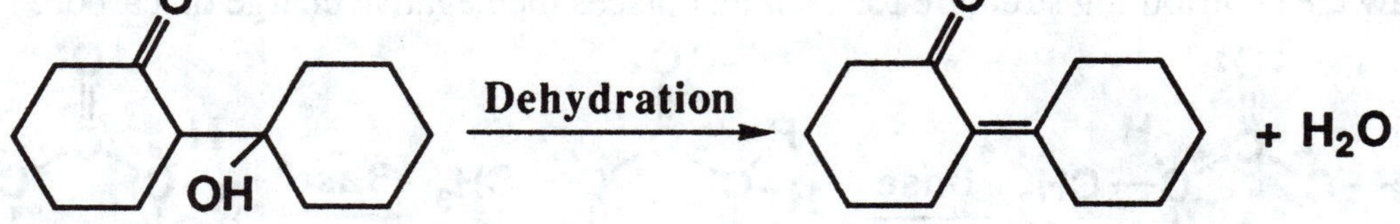

2-(1-Hydroxycyclohexyl)-
cyclohexanone

2-Cyclohexenyl-
cyclohexanone

(d)

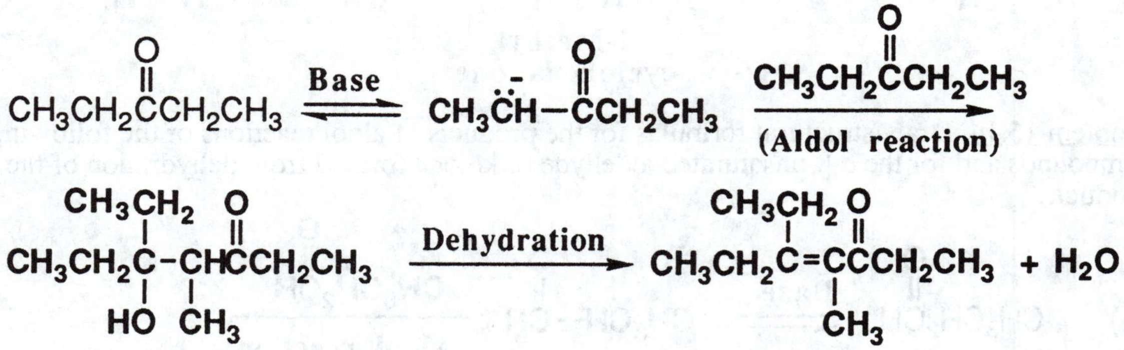

5-Ethyl-5-hydroxy-4-methyl
3-heptanone

5-Ethyl-4-methyl-4-hepten-
3-one

Problem 15.15 Draw a structural formula for the product of each mixed aldol reaction and for the compound formed by dehydration of each aldol product.

**Note that in the following reactions, only one of the carbonyl compounds can form an enolate anion. The aldol product shown is the one derived from that enolate anion reacting with the other carbonyl species present in the reaction.**

(a)

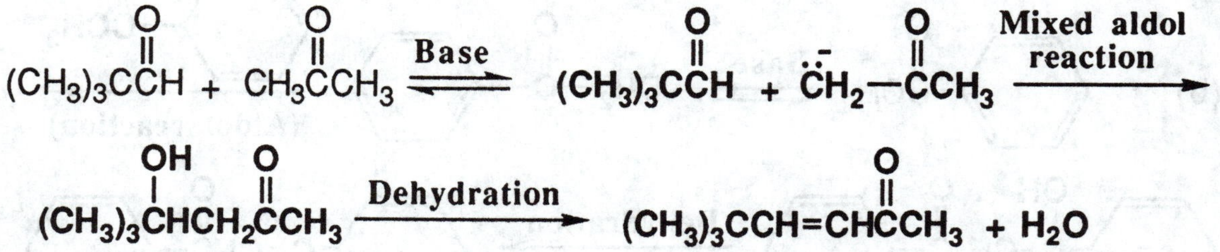

(b)

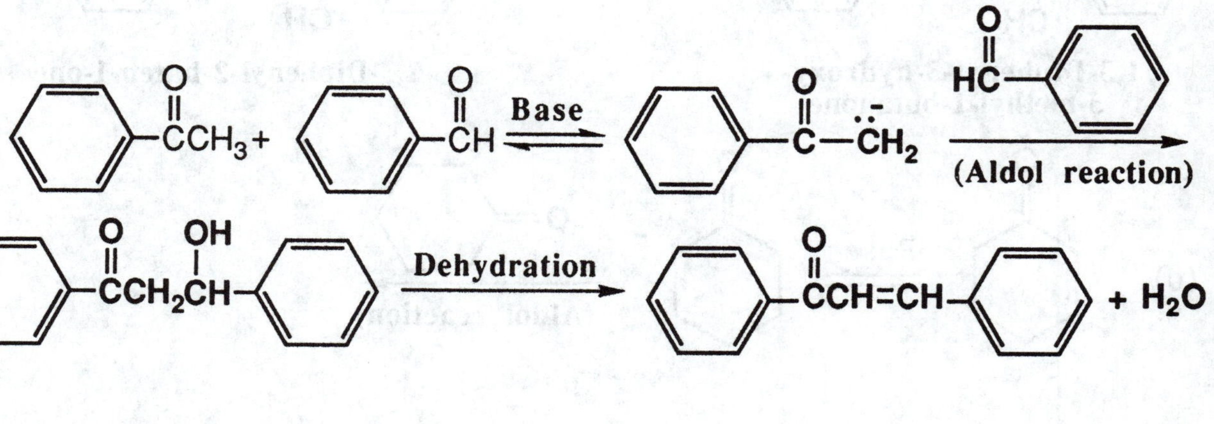

(c)

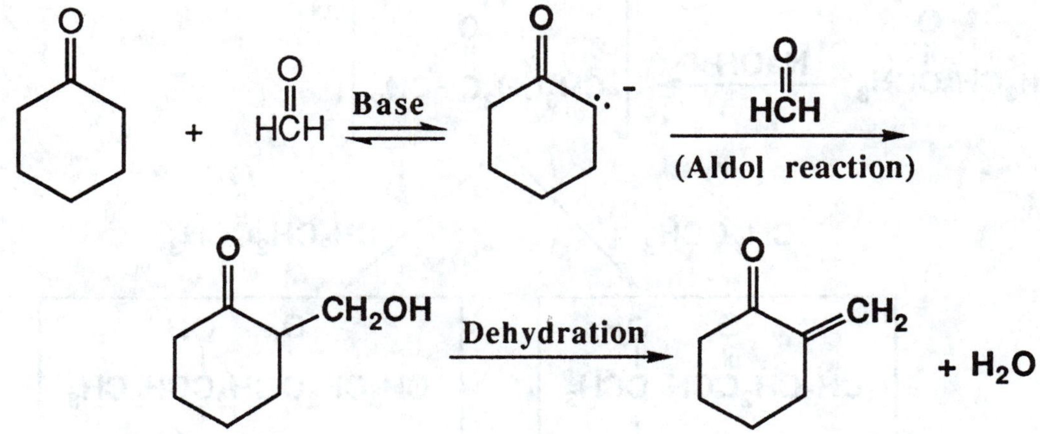

(d)

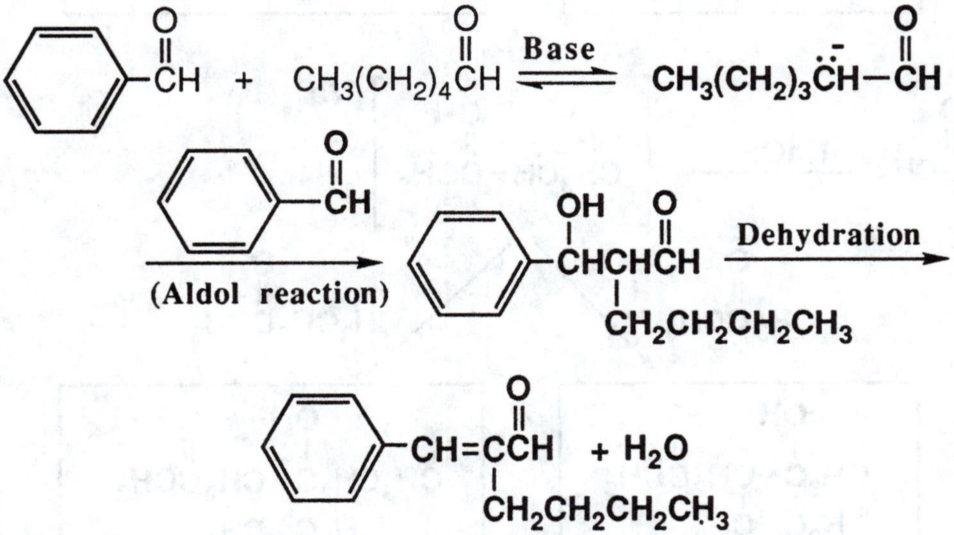

Problem 15.16  When a 1:1 mixture of acetone and 2-butanone is treated with base, a mixture of six aldol products is possible.  Draw structural formulas for these six aldol products.

$$CH_3CH_2\overset{\overset{\displaystyle O}{\|}}{C}CH_3 + CH_3\overset{\overset{\displaystyle O}{\|}}{C}CH_3 \xrightarrow{\text{NaOH}} \text{A mixture of six aldol products}$$

**Three different enolate anions can be formed and each of these can react with either of the two ketones as shown.**

$$CH_3CH_2\overset{\displaystyle O}{\overset{\|}{C}}CH_3 \xrightarrow{\textbf{NaOH}} \left[ CH_3CH_2\overset{\displaystyle O}{\overset{\|}{C}}-\ddot{\overset{\displaystyle -}{C}}H_2 \right]^{Na^+}$$

$$CH_3\overset{\displaystyle O}{\overset{\|}{C}}CH_3$$

$$CH_3CH_2\overset{\displaystyle O}{\overset{\|}{C}}CH_3$$

$$\boxed{CH_3CH_2\overset{\displaystyle O}{\overset{\|}{C}}CH_2\overset{\displaystyle OH}{\overset{|}{\underset{\displaystyle CH_3}{C}}}CH_3}$$

$$\boxed{CH_3CH_2\overset{\displaystyle O}{\overset{\|}{C}}CH_2\overset{\displaystyle OH}{\overset{|}{\underset{\displaystyle CH_3}{C}}}CH_2CH_3}$$

$$CH_3CH_2\overset{\displaystyle O}{\overset{\|}{C}}CH_3 \xrightarrow{\textbf{NaOH}} \left[ CH_3\ddot{\overset{\displaystyle -}{C}}H-\overset{\displaystyle O}{\overset{\|}{C}}CH_3 \right]^{Na^+}$$

$$CH_3\overset{\displaystyle O}{\overset{\|}{C}}CH_3$$

$$CH_3CH_2\overset{\displaystyle O}{\overset{\|}{C}}CH_3$$

$$\boxed{\underset{\displaystyle H_3C\ \ CH_3}{CH_3\overset{\displaystyle OH}{\overset{|}{C}}-CH_2\overset{\displaystyle O}{\overset{\|}{C}}CH_3}}$$

$$\boxed{\underset{\displaystyle H_3C\ \ CH_3}{CH_3CH_2\overset{\displaystyle OH}{\overset{|}{C}}-CH_2\overset{\displaystyle O}{\overset{\|}{C}}CH_3}}$$

$$CH_3\overset{\displaystyle O}{\overset{\|}{C}}CH_3 \xrightarrow{\textbf{NaOH}} \left[ CH_3\overset{\displaystyle O}{\overset{\|}{C}}-\ddot{\overset{\displaystyle -}{C}}H_2 \right]^{Na^+}$$

$$CH_3\overset{\displaystyle O}{\overset{\|}{C}}CH_3$$

$$CH_3CH_2\overset{\displaystyle O}{\overset{\|}{C}}CH_3$$

$$\boxed{CH_3\overset{\displaystyle O}{\overset{\|}{C}}CH_2\overset{\displaystyle OH}{\overset{|}{\underset{\displaystyle CH_3}{C}}}CH_3}$$

$$\boxed{CH_3\overset{\displaystyle O}{\overset{\|}{C}}CH_2\overset{\displaystyle OH}{\overset{|}{\underset{\displaystyle CH_3}{C}}}CH_2CH_3}$$

Problem 15.17 Show how to prepare these α,β-unsaturated ketones by an aldol reaction followed by dehydration of the aldol product.

(a)

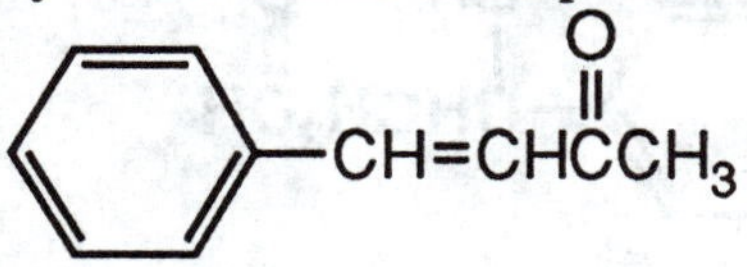

**In the scheme below, there are not a large number of different products produced in this mixed aldol reaction because the benzaldehyde cannot make an enolate ion.**

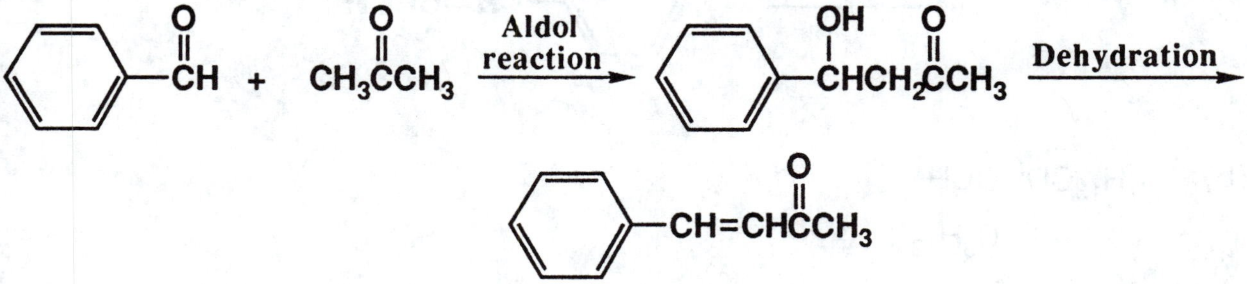

(b)

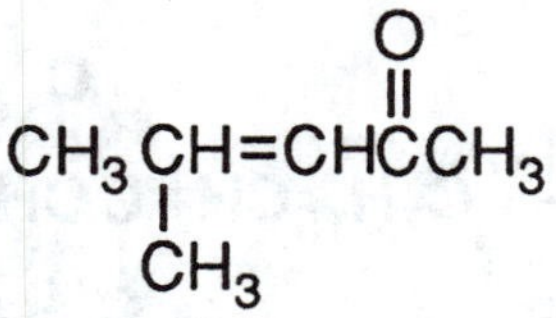

**This compound can be produced from an aldol reaction utilizing only acetone followed by dehydration.**

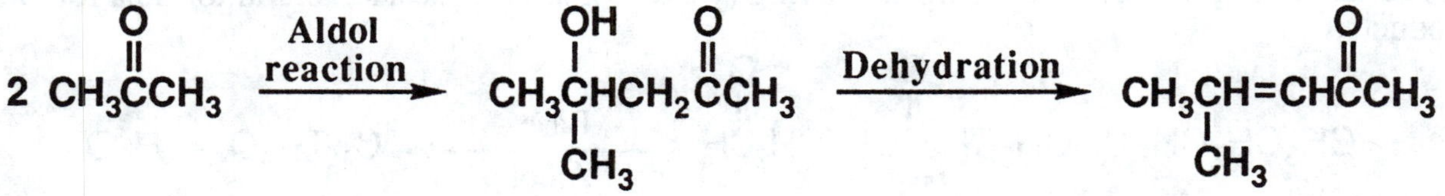

Problem 15.18 Show how to prepare these α,β-unsaturated aldehydes by an aldol reaction followed by dehydration of the aldol product.

(a)

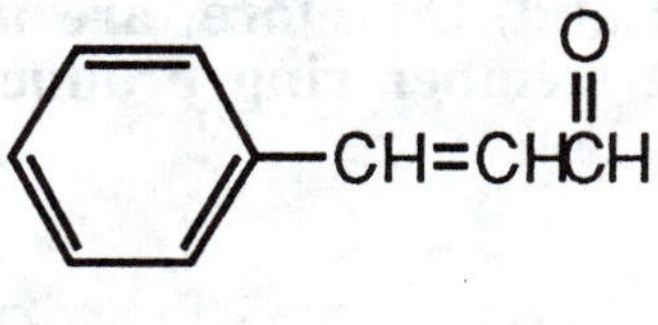

**In the scheme below, there are not a large number of different products produced in this mixed aldol reaction because only the acetaldehyde can make an enolate ion. Benzaldehyde does not have an α hydrogen atom, so it cannot make an enolate.**

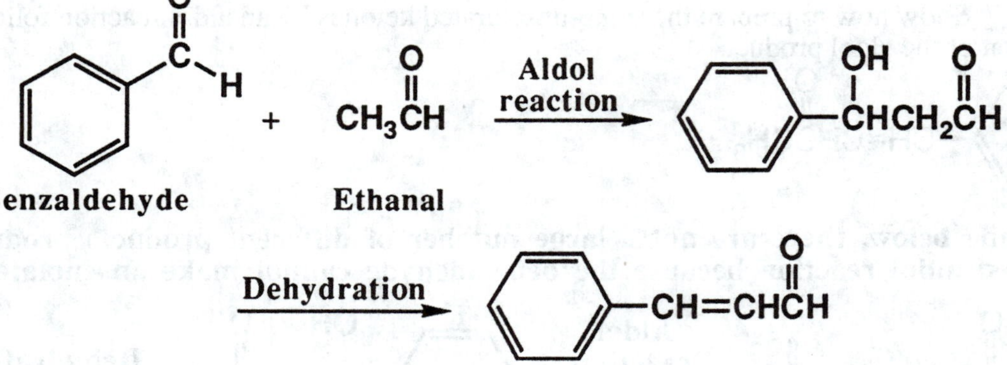

**Benzaldehyde**        **Ethanal**

(b) 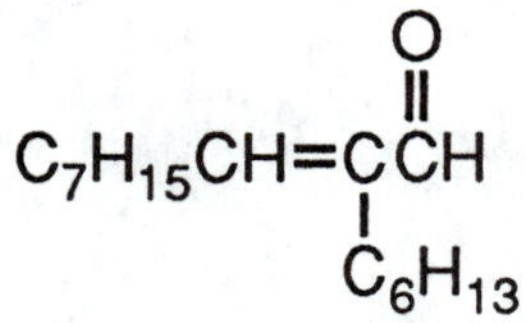    $C_7H_{15}CH=CCH$ with $C_6H_{13}$ and $O$ substituents

**This compound can be produced from an aldol reaction of octanal followed by dehydration.**

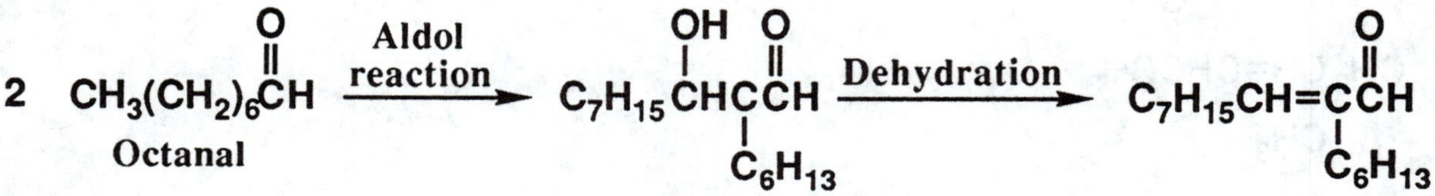

Problem 15.19   When treated with base, the following compound undergoes an intramolecular aldol reaction to give a product containing a ring (yield 78%). Propose a structural formula for this product.

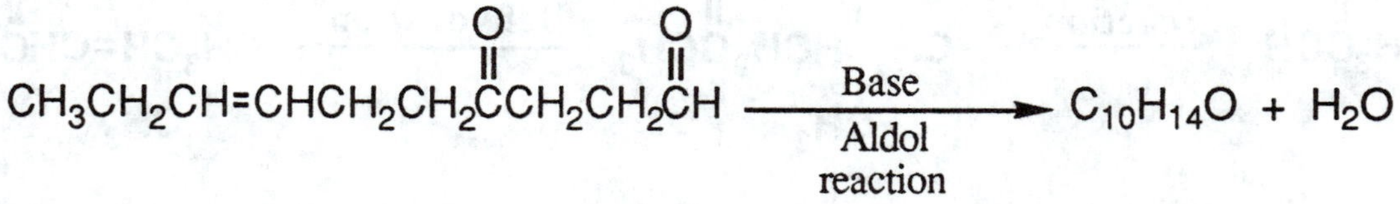

$$CH_3CH_2CH=CHCH_2CH_2CCH_2CH_2CH \xrightarrow[\text{Aldol reaction}]{\text{Base}} C_{10}H_{14}O + H_2O$$

**Analyze this problem in the following way. There are three α-carbons which might from an anion and then condense with either of the other carbonyl groups. Two of these condensations lead to three-member rings and, therefore, are not feasible. The third anion leads to formation of the five-member ring product.**

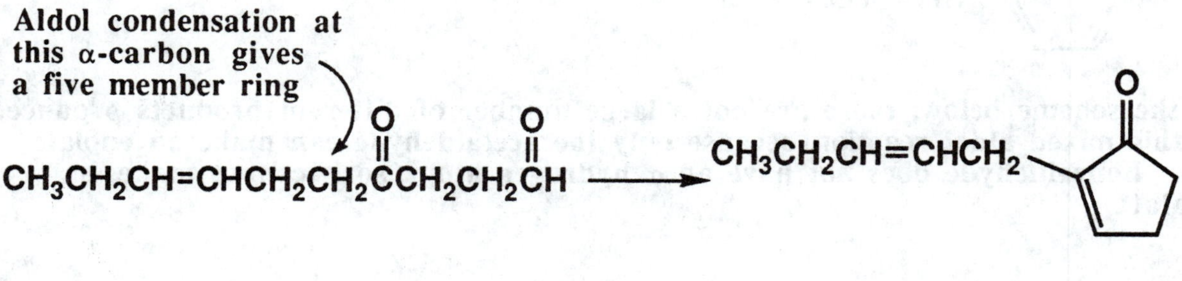

Aldol condensation at
this α-carbon gives
a five member ring

**Problem 15.20** Propose a structural formula for the compound of molecular formula $C_6H_{10}O_2$ that undergoes aldol reaction followed by dehydration to the this α,β unsaturated aldehyde.

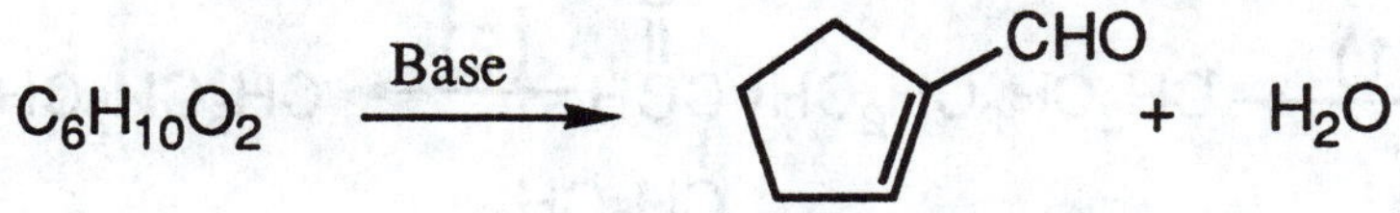

$$C_6H_{10}O_2 \xrightarrow{\text{Base}} \text{1-Cyclopentenecarbaldehyde} + H_2O$$

**The compound in question of molecular formula $C_6H_{10}O_2$ is hexanedial.**

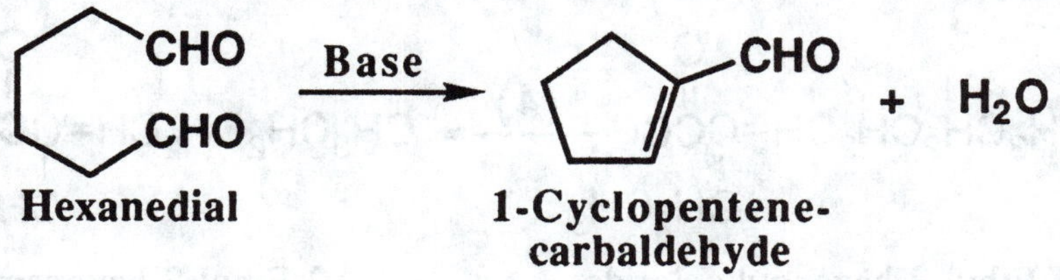

Hexanedial $\xrightarrow{\text{Base}}$ 1-Cyclopentene-carbaldehyde $+$ $H_2O$

**Problem 15.21** Show how to bring about this conversion.

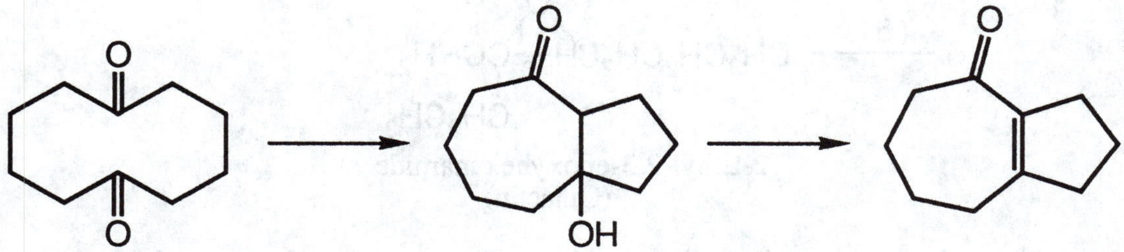

This is actually nothing more than an intramolecular aldol reaction followed by dehydration. Using the numbering shown below, the aldol reaction involves an enolate at carbon 2 reacting with the carbonyl carbon atom 6.

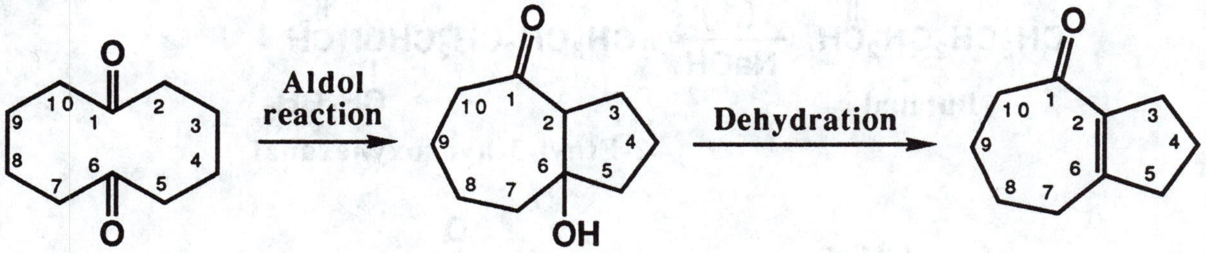

Aldol reaction $\longrightarrow$ Dehydration $\longrightarrow$

<u>Problem 15.22</u> Quiactin, a mild sedative, is synthesized from butanal in these five steps.

$$CH_3CH_2CH_2\overset{\overset{\displaystyle O}{\|}}{C}H \xrightarrow{(1)} CH_3CH_2CH_2CH=\overset{\overset{\displaystyle O}{\|}}{\underset{\underset{\displaystyle CH_2CH_3}{|}}{C}}CH \xrightarrow{(2)} CH_3CH_2CH_2CH=\overset{\overset{\displaystyle O}{\|}}{\underset{\underset{\displaystyle CH_2CH_3}{|}}{C}}COH$$

Butanal                   2-Ethyl-2-hexenal                 2-Ethyl-2-hexenoic acid

$$\xrightarrow{(3)} CH_3CH_2CH_2CH=\overset{\overset{\displaystyle O}{\|}}{\underset{\underset{\displaystyle CH_2CH_3}{|}}{C}}CCl \xrightarrow{(4)} CH_3CH_2CH_2CH=\overset{\overset{\displaystyle O}{\|}}{\underset{\underset{\displaystyle CH_2CH_3}{|}}{C}}CNH_2$$

2-Ethyl-2-hexenoyl chloride              2-Ethyl-2-hexenamide

$$\xrightarrow{(5)} CH_3CH_2CH_2CH\overset{\overset{\displaystyle O}{\diagdown\diagup}}{\underset{\underset{\displaystyle CH_2CH_3}{|}}{C}}\overset{\overset{\displaystyle O}{\|}}{C}NH_2$$

2-Ethyl-2,3-epoxyhexanamide
(Quiactin)

(a) Show reagents and experimental conditions to bring about each step in the synthesis.

**Step 1: Base-catalyzed aldol condensation followed by dehydration to give an α, β-unsaturated ketone.**

$$CH_3CH_2CH_2\overset{\overset{\displaystyle O}{\|}}{C}H \xrightarrow[\text{NaOH}]{(1)} CH_3CH_2CH_2\overset{\overset{\displaystyle OH}{|}}{C}H\underset{\underset{\displaystyle CH_2CH_3}{|}}{C}H\overset{\overset{\displaystyle O}{\|}}{C}H$$

Butanal                 2-Ethyl-3-hydroxyhexanal

$$\xrightarrow{(-H_2O)} CH_3CH_2CH_2CH=\overset{\overset{\displaystyle O}{\|}}{\underset{\underset{\displaystyle CH_2CH_3}{|}}{C}}CH$$

2-Ethyl-2-hexenal

**Step 2: Oxidation of the aldehyde to a carboxylic acid can be accomplished using Tollens' solution or Benedict's solution. In the industrial process, the oxidizing agent is oxygen, $O_2$.**

$$CH_3CH_2CH_2CH=\underset{\underset{CH_2CH_3}{|}}{C}\overset{\overset{O}{||}}{C}H \quad + \quad O_2 \quad \xrightarrow{(2)} \quad CH_3CH_2CH_2CH=\underset{\underset{CH_2CH_3}{|}}{C}\overset{\overset{O}{||}}{C}OH$$

|            2-Ethyl-2-hexenal            |            2-Ethyl-2-hexanoic acid            |

**Step 3: Reaction of the carboxylic acid with thionyl chloride gives the acid chloride.**

$$CH_3CH_2CH_2CH=\underset{\underset{CH_2CH_3}{|}}{C}\overset{\overset{O}{||}}{C}OH \quad + \quad SOCl_2 \quad \xrightarrow{(3)} \quad CH_3CH_2CH_2CH=\underset{\underset{CH_2CH_3}{|}}{C}\overset{\overset{O}{||}}{C}Cl \quad + \quad \begin{matrix} SO_2 \\ HCl \end{matrix}$$

|     2-Ethyl-2-hexenoic acid     |     2-Ethyl-2-hexenoyl chloride     |

**Step 4: Reaction with ammonia to make the amide.**

$$CH_3CH_2CH_2CH=\underset{\underset{CH_2CH_3}{|}}{C}\overset{\overset{O}{||}}{C}Cl \quad + \quad 2\ NH_3 \quad \xrightarrow{(4)} \quad CH_3CH_2CH_2CH=\underset{\underset{CH_2CH_3}{|}}{C}\overset{\overset{O}{||}}{C}NH_2 \quad + \quad NH_4^+\ Cl^-$$

|     2-Ethyl-2-hexenoyl chloride     |     2-Ethyl-2-hexenamide     |

**Step 5: Oxidation of the alkene to an epoxide can be brought about using a peroxy acid, $RCO_3H$.**

$$CH_3CH_2CH_2CH=\underset{\underset{CH_2CH_3}{|}}{C}\overset{\overset{O}{||}}{C}NH_2 \quad + \quad RCO_3H \quad \xrightarrow{(5)} \quad \underset{RCO_2H}{+}\ CH_3CH_2CH_2\overset{\overset{O}{\diagup\diagdown}}{CH-\underset{\underset{CH_2CH_3}{|}}{C}}\overset{\overset{O}{||}}{C}NH_2$$

|     2-Ethyl-2-hexenamide     |     2-Ethyl-2,3-epoxyhexanamide<br>(Quiactin)     |

(b) How many stereocenters are there in Quiactin? How many stereoisomers are possible for this compound?

**There are two stereocenters in Quiactin and these are marked with a "*" in the structure. Four stereoisomers are possible.**

$$CH_3CH_2CH_2\overset{*}{\underset{\underset{H}{|}}{C}}\overset{\overset{O}{\diagup\diagdown}}{\quad}\overset{*}{\underset{\underset{CH_2CH_3}{|}}{C}}\overset{\overset{O}{||}}{C}NH_2$$

**Problem 15.23** This reaction is one of the 10 steps in glycolysis, a series of enzyme-catalyzed reactions by which glucose is oxidized to two molecules of pyruvate. Show that this step is the reverse of an aldol reaction.

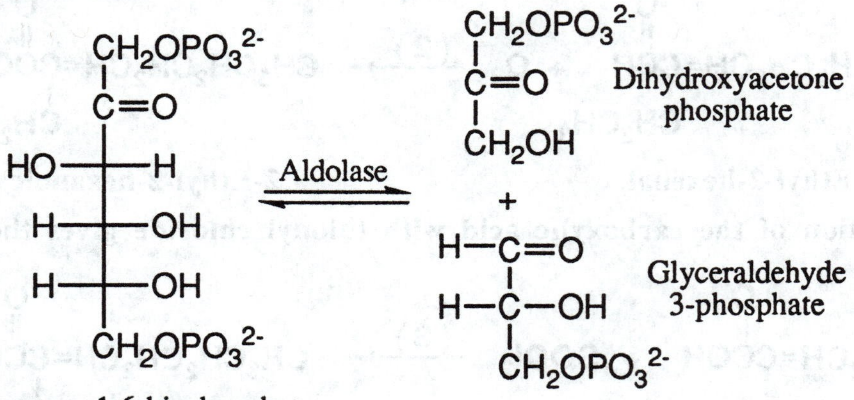

Fructose 1,6-bisphosphate

**As shown in the following scheme, this reaction is functionally the reverse of an aldol reaction between the enolate of dihydroxyacetone phosphate and glyceraldehyde 3-phosphate.**

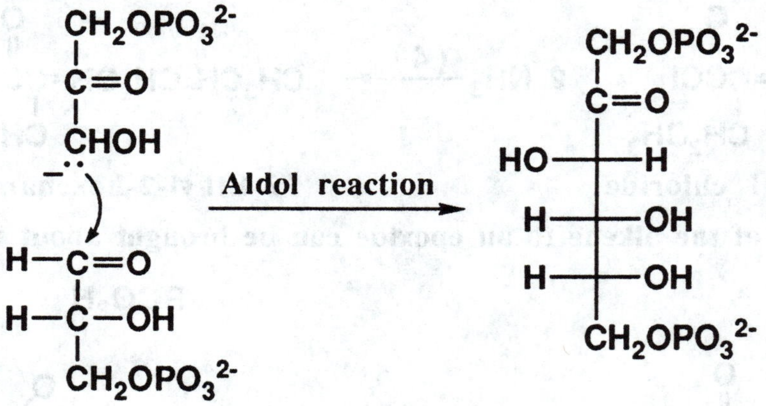

## The Claisen Condensation
**Problem 15.24** Show the product of the Claisen condensation of these esters.
(a) Ethyl phenylacetate in the presence of sodium ethoxide.

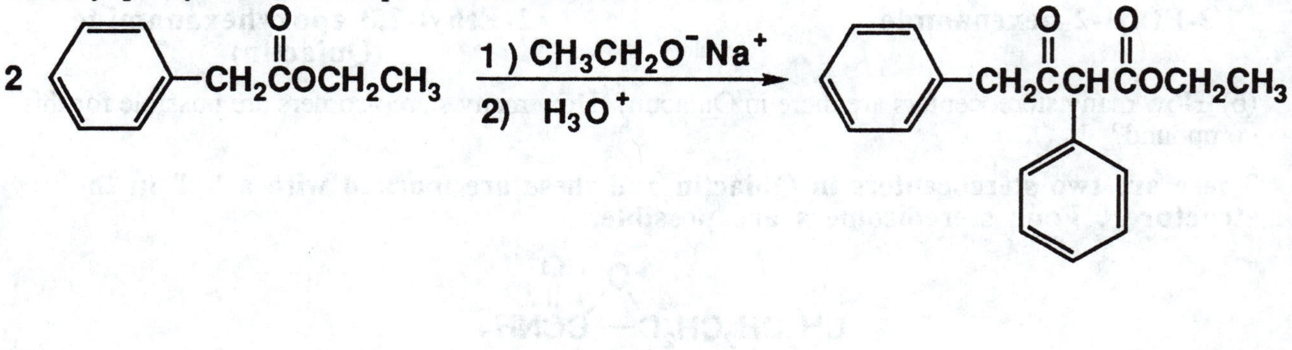

(b) Methyl hexanoate in the presence of sodium methoxide.

$$2 \ CH_3(CH_2)_4\overset{O}{\overset{\|}{C}}OCH_3 \xrightarrow[\text{2) } H_3O^+]{\text{1) } CH_3O^- Na^+} CH_3(CH_2)_4\overset{O}{\overset{\|}{C}}CH\overset{O}{\overset{\|}{C}}OCH_2CH_3$$
$$\underset{(CH_2)_3CH_3}{|}$$

**Problem 15.25** When a 1:1 mixture of ethyl propanoate and ethyl butanoate is treated with sodium ethoxide, a mixture of four Claisen condensation products is obtained. Draw structural formulas for these four products.

**In this case, both esters can form enolates, leading to the four products shown.**

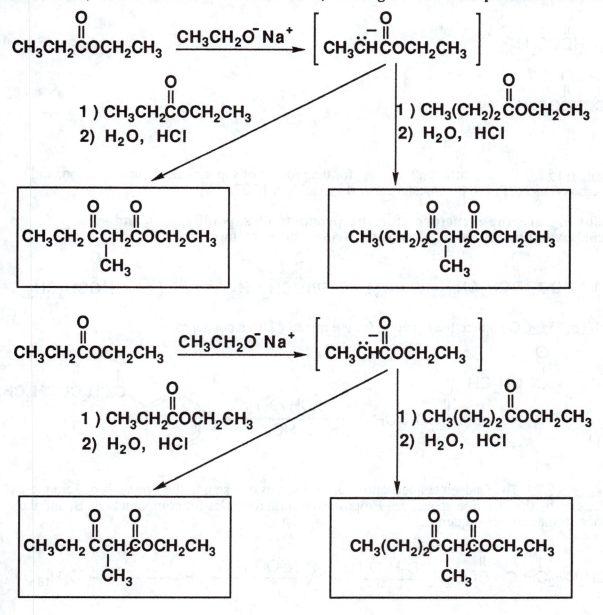

Problem 15.26 Draw structural formulas for the β-ketoesters formed in mixed Claisen condensation of ethyl propanoate with each of the following esters:

(a)   $C_2H_5O\overset{\overset{O}{\|}}{C}-\overset{\overset{O}{\|}}{C}OC_2H_5$

$C_2H_5O\overset{\overset{O}{\|}}{C}-\overset{\overset{O}{\|}}{C}H\overset{\overset{O}{\|}}{C}OC_2H_5$
$\quad\quad\quad\quad\underset{CH_3}{|}$

(b)   $Ph\overset{\overset{O}{\|}}{C}OC_2H_5$

$Ph-\overset{\overset{O}{\|}}{C}\overset{}{C}H\overset{\overset{O}{\|}}{C}OC_2H_5$
$\quad\quad\quad\underset{CH_3}{|}$

(c)   $H\overset{\overset{O}{\|}}{C}OC_2H_5$

$H\overset{\overset{O}{\|}}{C}\overset{}{C}H\overset{\overset{O}{\|}}{C}OC_2H_5$
$\quad\underset{CH_3}{|}$

Problem 15.27 Draw a structural formula for the product of saponification, acidification, and decarboxylation of each β-ketoester formed in Problem 15.27.

**Following are the structures for the products of saponification and decarboxylation of each β-ketoester from the previous problem.**

(a)   $HO\overset{\overset{O}{\|}}{C}-\overset{\overset{O}{\|}}{C}CH_2CH_3$

(b)   $Ph\overset{\overset{O}{\|}}{C}CH_2CH_3$

(c)   $H\overset{\overset{O}{\|}}{C}CH_2CH_3$

Problem 15.28 Complete the equation for the mixed Claisen condensation.

Problem 15.29 The Claisen condensation can be used as one step in the synthesis of ketones, as illustrated by this reaction sequence. Propose structural formulas for compounds A, B, and the ketone formed in this sequence.

$$2\ CH_3CH_2CH_2CH_2\overset{\overset{O}{\|}}{C}OC_2H_5 \xrightarrow[C_2H_5OH]{C_2H_5O^-\ Na^+} A \xrightarrow{NaOH,\ H_2O} B \xrightarrow{HCl,\ H_2O} C_9H_{18}O$$

Compound (A) is a β-ketoester, compound (B) is the sodium salt of a β-ketoacid, and the final ketone is 5-nonanone.

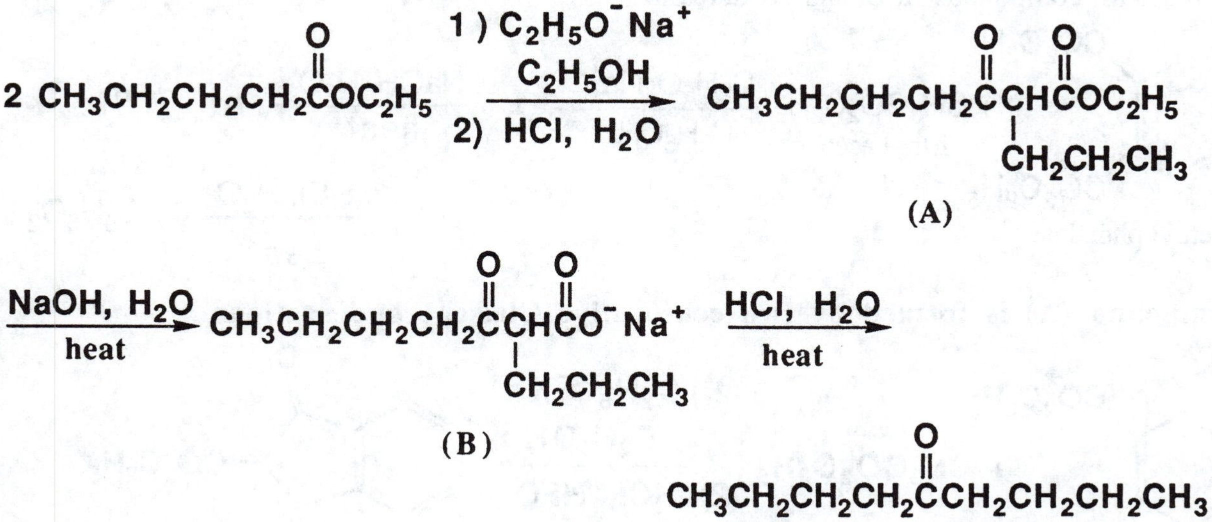

2 CH₃CH₂CH₂CH₂COC₂H₅  →[1) C₂H₅O⁻Na⁺ / C₂H₅OH / 2) HCl, H₂O]  CH₃CH₂CH₂CH₂CCHCOC₂H₅ (with CH₂CH₂CH₃ branch)

(A)

→[NaOH, H₂O / heat]  CH₃CH₂CH₂CH₂CCHCO⁻Na⁺ (with CH₂CH₂CH₃ branch)

(B)

→[HCl, H₂O / heat]

CH₃CH₂CH₂CH₂CCH₂CH₂CH₂CH₃

**5-Nonanone**

Problem 15.30 Draw the product of treating each diester with sodium ethoxide followed by acidification with HCl. (*Hint:* These are Dieckmann condensations.)

(a)

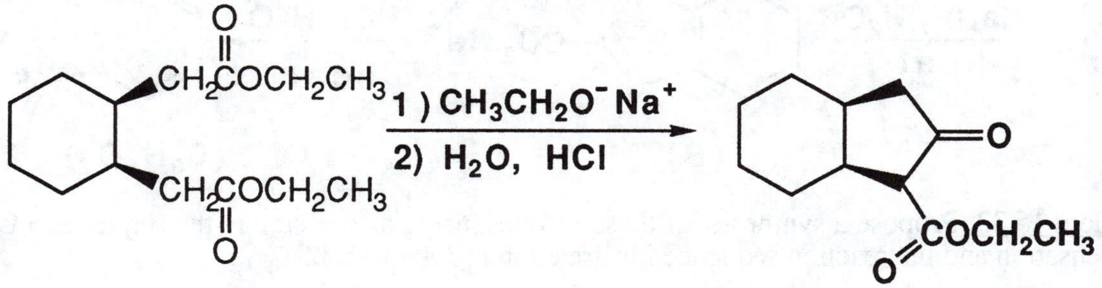

cyclohexane with CH₂COCH₂CH₃ and CH₂COCH₂CH₃ substituents →[1) CH₃CH₂O⁻Na⁺ / 2) H₂O, HCl] bicyclic product with ketone and COCH₂CH₃

(b)

CH₃CH₂OC(CH₂)₅COCH₂CH₃ →[1) CH₃CH₂O⁻Na⁺ / 2) H₂O, HCl] cyclohexanone with COCH₂CH₃ substituent

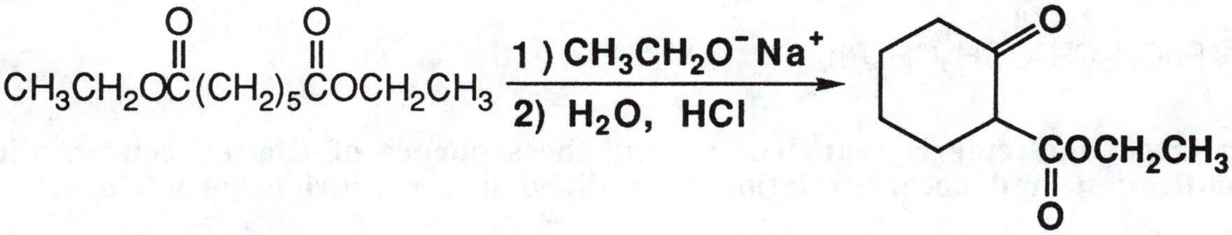

<u>Problem 15.31</u> Claisen condensation between diethyl phthalate and ethyl acetate followed by saponification, acidification, and decarboxylation forms a diketone, $C_9H_6O_2$. Propose structural formulas for compounds A, B, and the diketone.

**Compound (A) is formed by two consecutive Claisen condensations.**

<u>Problem 15.32</u> Propose a synthesis for these ketones, using as one step in the sequence a Claisen condensation and the reaction sequence illustrated in Problem 15.32.

$$\text{(a)} \quad PhCH_2CH_2\overset{\overset{\displaystyle O}{\|}}{C}CH_2CH_2Ph$$

**Each target molecule is synthesized using the sequence of Claisen condensation, saponification, and decarboxylation as outlined in the previous problem.**

**(a) The starting ester is ethyl 3-phenylpropanoate.**

(b)  PhCH₂ÖCCH₂Ph

The starting ester is ethyl phenylacetate.

2 PhCH₂COC₂H₅ $\xrightarrow[\text{2) HCl, H}_2\text{O}]{\substack{\text{1) C}_2\text{H}_5\text{O}^-\text{Na}^+ \\ \text{C}_2\text{H}_5\text{OH}}}$ PhCH₂CCHCOC₂H₅ $\xrightarrow[\text{2) HCl, H}_2\text{O}]{\text{1) NaOH, H}_2\text{O}}$
                                                        |
                                                        Ph

PhCH₂CCH₂Ph

(c)

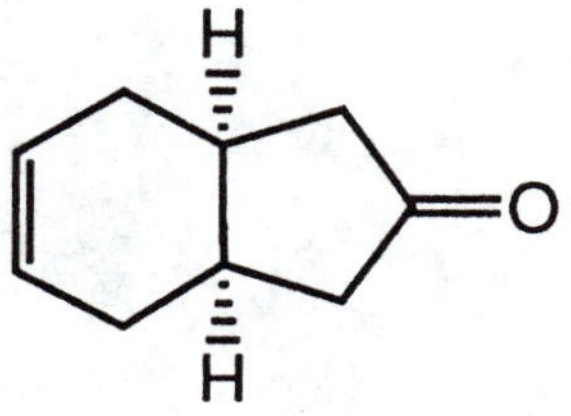

The starting diester is derived from a *cis*-4,5-disubstituted cyclohexene.

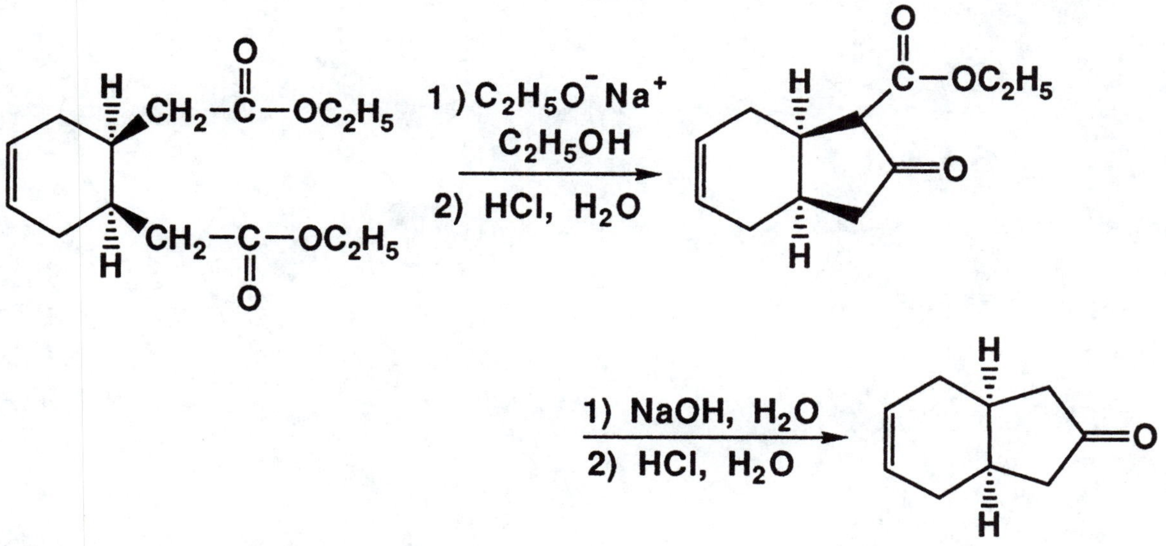

$\xrightarrow[\text{2) HCl, H}_2\text{O}]{\substack{\text{1) C}_2\text{H}_5\text{O}^-\text{Na}^+ \\ \text{C}_2\text{H}_5\text{OH}}}$

$\xrightarrow[\text{2) HCl, H}_2\text{O}]{\text{1) NaOH, H}_2\text{O}}$

<u>Problem 15.33</u>  This reaction is the fourth in the set of four enzyme-catalyzed reactions by which the hydrocarbon chain of a fatty acid (Section 19.3) is oxidized, two carbons at a time, to acetyl-coenzyme A.  Show that this reaction is the reverse of a Claisen condensation.

$$
\underset{\text{β-Ketoacyl-CoA}}{R\text{-}\overset{\displaystyle O}{\overset{\|}{C}}\text{·}CH_2\text{-}\overset{\displaystyle O}{\overset{\|}{C}}\text{-SCoA}} + \underset{\text{Coenzyme A}}{CoA\text{-}SH} \longrightarrow \underset{\text{An acyl-CoA}}{R\text{-}\overset{\displaystyle O}{\overset{\|}{C}}\text{-SCoA}} + \underset{\text{Acetyl-CoA}}{CH_3\text{-}\overset{\displaystyle O}{\overset{\|}{C}}\text{-SCoA}}
$$

**The above enzyme catalyzed process is functionally the reverse of a mixed Claisen condensation that takes place between the enolate of acetyl-CoA and an acyl-CoA.**

$$
R\text{-}\overset{\displaystyle O}{\overset{\|}{C}}\text{-SCoA} + \overset{..}{\underset{}{C}}\overline{H}_2\text{-}\overset{\displaystyle O}{\overset{\|}{C}}\text{-SCoA} \longrightarrow R\text{-}\overset{\displaystyle O}{\overset{\|}{C}}CH_2\overset{\displaystyle O}{\overset{\|}{C}}\text{-SCoA} + CoA\text{-SH}
$$

# CHAPTER 16: LIPIDS
*Outline of Important Concepts*

## 16.0 OVERVIEW
- **Lipids** are a heterogeneous class of biological molecules that are classified together because of their solubility properties. They are **insoluble in water**, but are **soluble in organic solvents** such as diethyl ether, methylene chloride, and acetone. ✳
- This chapter describes **two main classes of lipids**:
  - The first class of lipids has a **large nonpolar hydrophobic region**, and **a polar hydrophilic region**. Members in this class are **triacylglycerols, phospholipids, prostaglandins**, and the **fat-soluble vitamins**.
  - Molecules in the second class, such as **cholesterol and compounds derived from it, contain** a **tetracyclic steroid ring nucleus.**

## 16.1 FATS AND OILS
- **Triacylglycerols**, also called **triglycerides**, are triesters of glycerol and fatty acids.
  - Samples of triacylglycerols rich in unsaturated fatty acids such as oleic and linoleic acids are generally liquids at room temperature, and are referred to as **oils**. Samples of triacylglycerols rich in saturated fatty acids are generally solids or semisolids at room temperature, since the saturated fatty acid chains can pack together well. These triacylglycerols are called **fats**. ✳
- **Fatty acids** are **long-chain monocarboxylic acids** produced by the **hydrolysis of fats and oils**. ✳
  - Nearly all fatty acids have an **even number of carbon atoms**, the most abundant are $C_{16}$ (palmitic acid) and $C_{18}$ (stearic and oleic acids). Different fatty acids can have **different numbers of double bonds**. The number of carbon atoms in the chain and the number of double bonds are separated by a colon when they are named. For example, a 16:2 fatty acid contains sixteen carbon atoms and two double bonds.
  - In most natural **unsaturated fatty acids**, the **Z (*cis*) isomer predominates**, and the **E (*trans*) isomer** is very **rare**. Because they are bent and cannot pack together well, the Z unsaturated fatty acids and molecules that contain them have lower melting points than analogous saturated fatty acids.

## 16.2 SOAPS AND DETERGENTS
- **Natural soaps** are the sodium or potassium salts of fatty acids. They can be prepared from the base-promoted hydrolysis of the ester functions in triacylglycerols, a process called **saponification**. ✳
- **Soaps** act as **cleansing agents** because the long **hydrocarbon chains tend to cluster**, while the polar carboxylate groups remain in contact with the water. The so-called **micelle** structures that are formed "dissolve" nonpolar substances such as dirt and grease in the hydrophobic interior. **Natural soaps** form **insoluble salts** with the ions found in hard water such as **Ca(II), Mg(II),** or **Fe(III)**, leading to **soap scum**.
- **Synthetic detergents** are analogous to natural soaps except the polar group is a sulfonate, not a carboxylate. Synthetic detergents have the advantage that they do not readily form insoluble salts in hard water.

## 16.3 PROSTAGLANDINS
- **Prostaglandins** are a class of compounds that have the **20-carbon skeleton** of **prostanoic acid**. Different prostaglandins have different biological activities, and they are usually very

potent.  As a result, much research has been invested in the understanding of the biological action of natural and synthetic prostaglandins.  Prostaglandins are part of a larger family of biological molecules called **eicosanoids** that include **prostacyclins, thromboxanes,** and **leukotrienes.** ✳

## 16.4 STEROIDS
• **Steroids** are a group of lipids that have a characteristic tetracyclic steroid ring system.

Cholesterol, a component of biological membranes, is a precursor to other important classes of steroids including the **androgens** (male sex hormones), **estrogens** (female sex hormones), **glucocorticoid hormones,** and **mineralocorticoid hormones.** ✳
• Steroids such as cholesterol are built up from two-carbon units derived from **acetyl CoA** via several intermediate steps.  Various intermediate structures are produced along the way including **geranyl pyrophosphate, farnesyl pyrophosphate,** and **squalene.**  The squalene is enzymatically oxidized to give an epoxide, that opens to give a tertiary carbocation.  Formation of this carbocation sets into motion a very remarkable number of probably concerted chemical steps including four concerted cation-initiated cyclizations and four 1,2 shifts.  The product of this sequence, **lanosterol,** is converted to cholesterol though about 25 more enzyme-catalyzed steps.

## 16.5 PHOSPHOLIPIDS
• **Phospholipids** are derivatives of **phosphatidic acid,** having **glycerol esterified** with **two fatty acid molecules** and **one molecule of phosphoric acid**.  Fatty acids such as palmitic, stearic, and oleic acid are the most common.  The phosphoric acid group can be attached to other groups such as ethanolamine, choline, serine, or inositol.  The phosphoric acid group is negatively charged at neutral pH.  Thus, phospholipids have the long hydrophobic chains of the fatty acids, but also the very hydrophilic and charged phosphoric acid group.
• The unique structure of the **phospholipids** allows them to **self-assemble** in water to give a **bilayer** in which the polar headgroups lie on the surface exposed to the water molecules, and the hydrophobic fatty acid alkyl chains are buried within the bilayer.  The hydrophobic interior of the bilayer can vary from rigid to fluid.  Saturated fatty acid chains can pack together more easily, so they make more rigid bilayers than the kinked chains of unsaturated fatty acids.
• According to the **fluid mosaic model** of biological membranes, the membrane is composed of a phospholipid bilayer with membrane proteins associated on the inside and outside surfaces.  Some proteins can also span the distance from the inside to the outside of the membrane bilayer.

## 16.6 FAT-SOLUBLE VITAMINS
• **Vitamins** are classified as either water-soluble or fat-soluble.  The fat-soluble vitamins include **vitamin A** (important in the visual cycle of rod cells), **vitamin D** (important for the regulation of calcium and phosphorus metabolism), **vitamin E** (an antioxidant also important for red blood cell membranes), and **vitamin K** (important for blood clotting).

# CHAPTER 16
### Solutions to the Problems

<u>Problem 16.1</u> How many isomers are possible for a triglyceride containing one molecule each of palmitic, oleic, and stearic acids?

**There are three constitutional isomers possible, the difference being which fatty acid is in the middle of the molecule:**

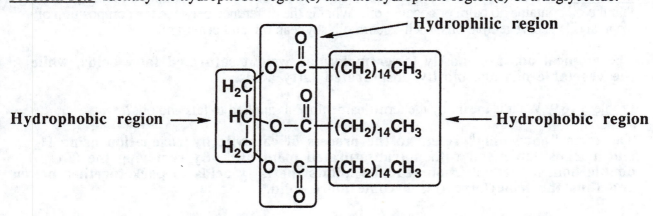

**Each of these molecules has one stereocenter as indicated by the asterisk, so each constitutional isomer shown above can exist as a pair of enantiomers. Thus there are 2 x 3 = 6 total isomers possible. Note that for oleic acid, the carbon-carbon double bond is assumed to have the Z (*cis*) configuration only.**

<u>Problem 16.2</u> Define the term hydrophobic.

**The term literally means "having fear of water." Hydrophobic species will not dissolve in water.**

<u>Problem 16.3</u> Identify the hydrophobic region(s) and the hydrophilic region(s) of a triglyceride.

**Note that the vast majority of the molecule is hydrophobic, thus explaining why triglycerides are so hydrophobic overall.**

Problem 16.4 Explain why the melting points of unsaturated fatty acids are lower than those of saturated fatty acids.

**When fatty acids pack together better, the attractive dispersion forces between molecules are stronger, thereby increasing the melting point. Saturated fatty acids can adopt a much more compact structure compared to unsaturated fatty acids that have a kink induced by the *cis* double bond. The more compact saturated fatty acids can pack together better, so their melting points are higher.**

Problem 16.5 Which would you expect to have the higher melting point, glyceryl trioleate or glyceryl trilinoleate?

**The triglyceride with fewer *cis* double bonds will have the higher melting point. Each oleic acid unit has only one *cis* double bond, while each linoleic acid has two (Please see Table 16.1). Glycerol trioleate will have the higher melting point.**

Problem 16.6 Draw a structural formula for methyl lineolate. Be certain to show the correct stereochemistry about the carbon-carbon double bonds.

**Methyl lineolate**

Problem 16.7 Explain why coconut oil is a liquid triglyceride, even though most of its fatty acid components are saturated.

**Triglycerides having fatty acids with shorter chains have lower melting points. As can be seen in Table 16.2, coconut oil is 45% lauric acid. Lauric acid is only a C12 fatty acid, so coconut oil has a melting point that is low enough to make is a liquid near room temperature.**

Problem 16.8 It is common now to see "contains no tropical oils" on cooking oil labels, meaning that the oil contains no palm or coconut oil. What is the difference between the composition of tropical oils and vegetable oils, such as corn oil, soybean oil, and peanut oil?

**The tropical oils are mostly lower molecular weight saturated fatty acids, while the vegetable oils are mostly unsaturated fatty acids.**

Problem 16.9 What is meant by the term "hardening" as applied to fats and oils?

**The term "hardening" refers to the process of catalytic hydrogenation using $H_2$ and a transition metal with polyunsaturated plant oils. By removing the (Z) double bonds, the reduction reaction allows the fatty acids to pack together better and thus the triacylglycerols become more solid.**

Problem 16.10 How many moles of $H_2$ are used in the catalytic hydrogenation of 1 mole of a triglyceride derived from glycerol, stearic acid, linoleic acid, and arachidonic acid?

**One molecule of $H_2$ is used per double bond in the triglyceride. Stearic acid does not have any double bonds, linoleic acid has 2 and arachidonic acid has 4 double bonds, respectively. Thus, 2 + 4 = 6 moles of $H_2$ will be used per mole of the triglyceride.**

Problem 16.11 Saponification number is defined as the number of milligrams of potassium hydroxide required for saponification of 1.00 g of fat or oil
(a) Write a balanced equation for the saponification of tristearin.

(b) The molecular weight of tristearin is 890 g/mol. Calculate the saponification number of tristearin.

**The molecular weight of potassium hydroxide is 56 g/mol.**

$$(3 \times 56 \text{ g/mol})\left(\frac{1 \text{ g}}{890 \text{ g/mol}}\right) = 0.189 \text{ g}$$

**Therefore, as determined in the above equation, it would take 189 milligrams of KOH to saponify 1 g of tristearin, so it has a saponification number of 189.**

Problem 16.12 The saponification number of butter fat is approximately 230; that of oleomargarine is approximately 195. Calculate the average molecular weight of butter fat and of oleomargarine.

$$\frac{(3 \times 56 \text{ g/mol})(1 \text{ g})}{0.230 \text{ g}} = 730 \text{ g/mol} \qquad \frac{(3 \times 56 \text{ g/mol})(1 \text{ g})}{0.195 \text{ g}} = 862 \text{ g/mol}$$

**As shown in the above equations, the molecular weight of butter fat and oleomargarine are 730 g/mol and 862 g/mol, respectively.**

Problem 16.13 Characterize the structural features necessary to make a good synthetic detergent.

**A good synthetic detergent should have a long hydrocarbon tail and a very polar group at one end. This combination will allow for the production of micelle structures in aqueous solution that will dissolve hydrophobic dirt such as grease**

and oil. The very polar group should not form insoluble salts with the ions normally found in hard water such as Ca(II), Mg(II), and Fe(III).

Problem 16.14 Following are structural formulas for a cationic detergent and a neutral detergent. Account for the detergent properties of each.

$$CH_3(CH_2)_6CH_2\overset{\overset{\displaystyle CH_3}{|}}{\underset{\underset{\displaystyle CH_2C_6H_5}{|}}{\overset{+}{N}}}CH_3 \quad Cl^-$$

Benzyldimethyloctylammonium chloride
(a cationic detergent)

$$HOCH_2\overset{\overset{\displaystyle HOCH_2}{|}}{\underset{\underset{\displaystyle HOCH_2}{|}}{C}}CH_2O\overset{\overset{\displaystyle O}{\|}}{C}(CH_2)_{14}CH_3$$

Pentaerythrityl palmitate
(a neutral detergent)

In each case there is a long hydrocarbon tail attached to a very polar group. This combination will allow for the production of micelle structures in aqueous solution that will dissolve nonpolar, hydrophobic dirt such as grease and oil. In the case of benzyldimethyloctylammonium chloride, the polar group is the positively-charged ammonium group, while for the pentaerythrityl palmitate the polar group is composed of the triol functions.

Problem 16.15 Identify some of the detergents used in shampoos and dish washing products. Are they primarily anionic, neutral, or cationic detergents.

Most detergents in shampoos and dish washing detergents are anionic detergents such as sodium lauryl sulfate.

$$CH_3(CH_2)_{10}CH_2O-\overset{\overset{\displaystyle O}{\|}}{\underset{\underset{\displaystyle O}{\|}}{S}}-O^-Na^+$$

**Sodium Lauryl Sulfate**

Problem 16.16 Show how to convert palmitic acid (hexadecanoic acid) into the following:
(a) Ethyl palmitate

$$CH_3(CH_2)_{14}\overset{\overset{\displaystyle O}{\|}}{C}OH \quad + \quad CH_3CH_2OH \quad \xrightarrow{H^+} \quad CH_3(CH_2)_{14}\overset{\overset{\displaystyle O}{\|}}{C}OCH_2CH_3$$

**Ethyl palmitate**

(b) Palmitoyl chloride

$$CH_3(CH_2)_{14}\overset{\overset{\displaystyle O}{\|}}{C}OH \quad + \quad SOCl_2 \quad \longrightarrow \quad CH_3(CH_2)_{14}\overset{\overset{\displaystyle O}{\|}}{C}Cl$$

**Palmitoyl chloride**

**(c) 1-Hexadecanol (cetyl alcohol)**

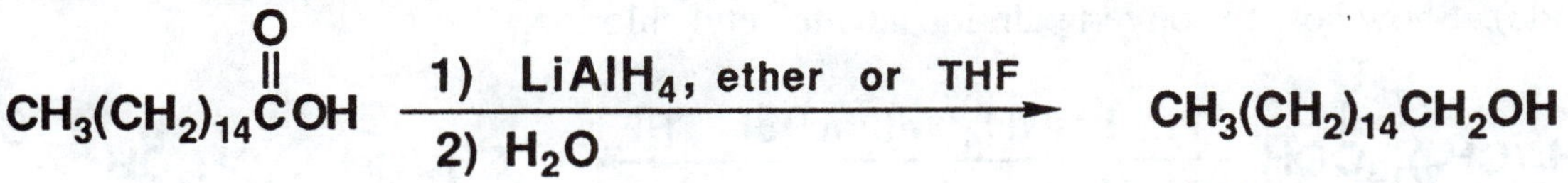

$$CH_3(CH_2)_{14}\overset{O}{\overset{\|}{C}}OH \quad \xrightarrow[\text{2) } H_2O]{\text{1) } \textbf{LiAlH}_4\text{, ether or THF}} \quad CH_3(CH_2)_{14}CH_2OH$$

**1-Hexadecanol**
**(Cetyl alcohol)**

**(d) 1-Aminohexadecane**

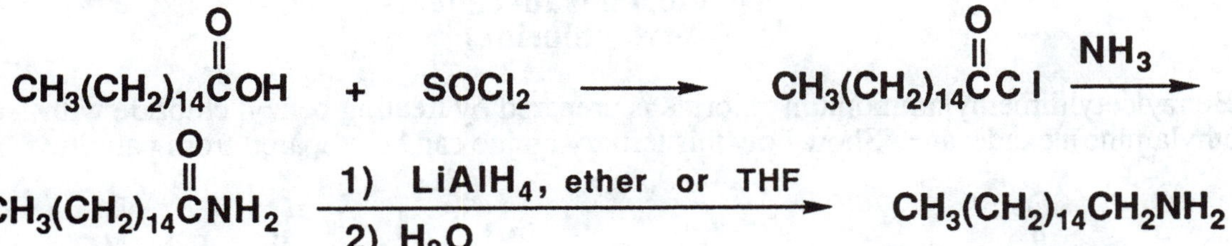

$$CH_3(CH_2)_{14}\overset{O}{\overset{\|}{C}}OH \quad + \quad SOCl_2 \quad \longrightarrow \quad CH_3(CH_2)_{14}\overset{O}{\overset{\|}{C}}Cl \quad \xrightarrow{NH_3}$$

$$CH_3(CH_2)_{14}\overset{O}{\overset{\|}{C}}NH_2 \quad \xrightarrow[\text{2) } H_2O]{\text{1) } \textbf{LiAlH}_4\text{, ether or THF}} \quad CH_3(CH_2)_{14}CH_2NH_2$$

**1-Aminohexadecane**

**(e) *N,N*-Dimethylhexadecanamide**

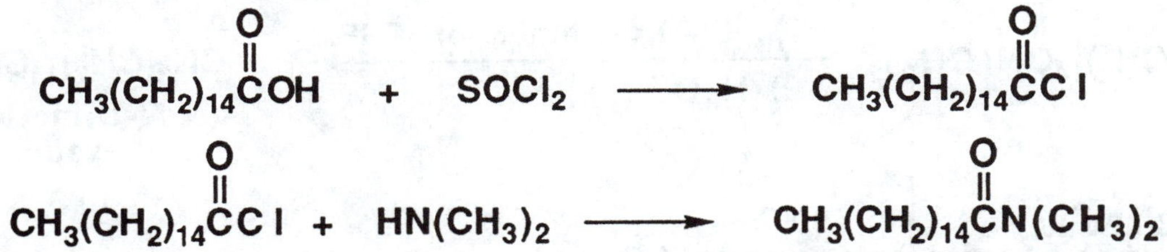

$$CH_3(CH_2)_{14}\overset{O}{\overset{\|}{C}}OH \quad + \quad SOCl_2 \quad \longrightarrow \quad CH_3(CH_2)_{14}\overset{O}{\overset{\|}{C}}Cl$$

$$CH_3(CH_2)_{14}\overset{O}{\overset{\|}{C}}Cl \quad + \quad HN(CH_3)_2 \quad \longrightarrow \quad CH_3(CH_2)_{14}\overset{O}{\overset{\|}{C}}N(CH_3)_2$$

***N,N*-Dimethylhexadecanamide**

<u>Problem 16.17</u> Palmitic acid (hexadecanoic acid) is the source of the hexadecyl (cetyl) group in the following compounds. Each is a mild surface-acting germicide and fungicide and is used as a topical antiseptic and disinfectant.

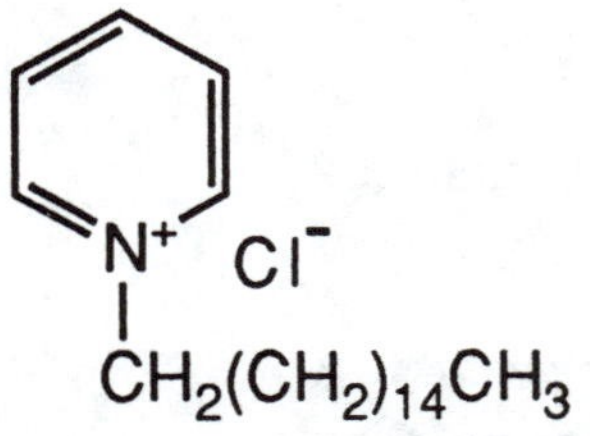

Cetylpyridinium chloride 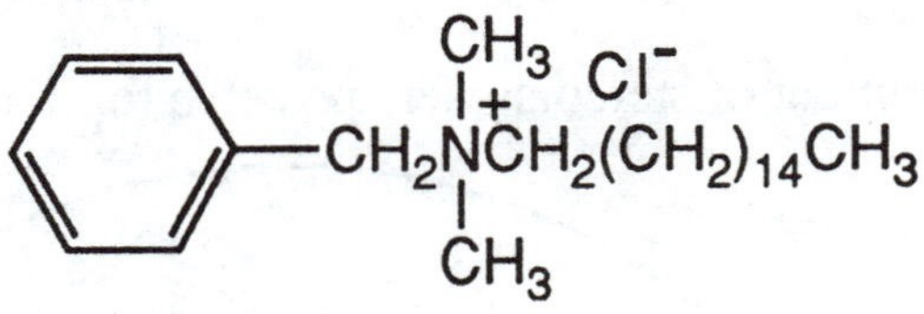          Benzylcetyldimethylammonium chloride

(a)  Cetylpyridinium chloride is prepared by treating pyridine with 1-chlorohexadecane (cetyl chloride).  Show how to convert palmitic acid to cetyl chloride.

$$CH_3(CH_2)_{14}\overset{\overset{\displaystyle O}{\|}}{C}OH \xrightarrow[\text{2) } H_2O]{\text{1)  LiAlH}_4\text{, ether  or  THF}} CH_3(CH_2)_{14}CH_2OH \xrightarrow{\text{SOCl}_2}$$

$$CH_3(CH_2)_{14}CH_2Cl$$

**1-Chlorohexadecane**
**(Cetyl  chloride)**

(b)  Benzylcetyldimethylammonium chloride is prepared by treating benzyl chloride with 1-(*N*,*N*-dimethylamino)hexadecane.  Show how this tertiary amine can be prepared from palmitic acid.

$$CH_3(CH_2)_{14}\overset{\overset{\displaystyle O}{\|}}{C}OH \quad + \quad SOCl_2 \longrightarrow CH_3(CH_2)_{14}\overset{\overset{\displaystyle O}{\|}}{C}Cl \xrightarrow{\text{HN(CH}_3)_2}$$

$$CH_3(CH_2)_{14}\overset{\overset{\displaystyle O}{\|}}{C}N(CH_3)_2 \xrightarrow[\text{2) } H_2O]{\text{1)  LiAlH}_4\text{, ether  or  THF}} CH_3(CH_2)_{14}CH_2N(CH_3)_2$$

**1-(*N*,*N*-Dimethylamino)**
**hexadecane**

## Prostaglandins

Problem 16.18 Examine the structure of PGF$_{2\alpha}$ and
(a) Identify all stereocenters

**The stereocenters are indicated with an asterisk.**

(b) Identify all double bonds about which *cis-trans* isomerism occurs

**These double bonds are indicated by the arrows.**

(c) State the number of stereoisomers possible for a molecule of this structure.

*cis-trans* isomerization
possible

PGF$_{2\alpha}$

**There are $2^5$ stereoisomers possible and $2^2$ *cis-trans* isomers possible for a grand total of 32 x 4 = 128 possible stereoisomers.**

**Problem 16.19** Following is the structure of Doxaprost, an orally active bronchodilator patterned after the natural prostaglandins (Section 16.3). Compare the structural formula of this synthetic prostaglandin with that of $PGF_{2\alpha}$ and note both similarities and differences.

Doxaprost
(an orally active bronchodilator)

**Both Doxaprost have five-member rings connected to alkyl side chains, which are attached *trans* to the ring. In addition, one of these alkyl chains has 7 carbon atoms and terminates in a carboxylic acid group, while the other is 8 carbon atoms long and contains both a *trans* double bond and hydroxyl group. On the other hand, only $PGF_{2\alpha}$ contains two hydroxyl groups attached to the five-member ring and a *trans* double bond within the 7 carbon alkyl chain. Only Doxaprost possesses a ketone on the five-member ring.**

## Steroids
**Problem 16.20** Draw the structural formula for the product formed by treating cholesterol with $H_2/Pd$; with $Br_2$.

Cholesterol

Problem 16.21 List several ways in which cholesterol is necessary for human life. Why do so many people find it necessary to restrict their dietary intake of cholesterol?

**Cholesterol is an important component of biological membranes where it serves to modulate membrane fluidity. In addition, cholesterol is an important precursor to a variety of steroid hormones. Cholesteryl esters are a major component of astherosclerotic plaque, so restricting the dietary intake of cholesterol is helpful for limiting atherosclerosis.**

Problem 16.22 Both low-density lipoproteins (LDL) and high-density lipoproteins (HDL) consist of a core of triacylglycerols and cholesterol esters surrounded by a single phospholipid layer. Draw the structural formula of cholesteryl linoleate, one of the cholesterol esters found in this core.

$$CH_3(CH_2)_4(CH=CHCH_2)_2(CH_2)_6C-O$$

**Cholesteryl linoleate**

Problem 16.23 Examine the structural formulas of testosterone (a male sex hormone) and progesterone (a female sex hormone). What are the similarities in structure between the two? What are the differences?

**Testosterone**                         **Progesterone**

**Overall, these structures are remarkably similar. Both of these two steroids contain the standard four ring steroid structure with the axial methyl groups at C10 and C13. In addition, both structures contain an ene-one group in the A ring. On the other hand, the two structures differ in the nature of the D ring substituent at C17. In testosterone, the substituent is a hydroxy group and in progesterone it is an acetyl group.**

**Problem 16.24** Examine the structural formula of cholic acid and account for the ability of this and other bile acids to emulsify fats and oils and thus aid in their digestion.

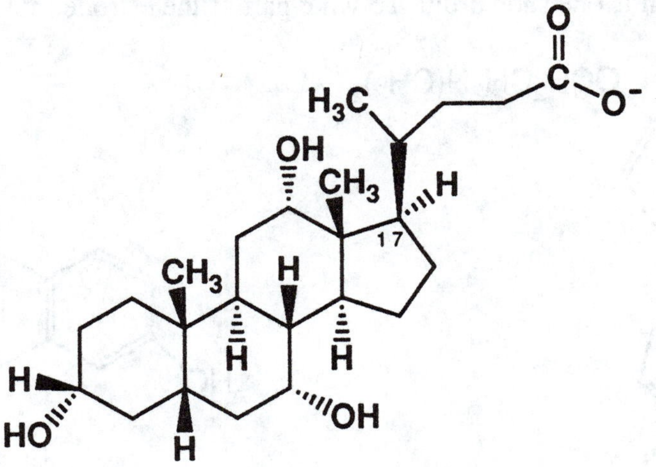

**Cholic acid**

**Cholic acid has the characteristic structure of a soap, so it can emulsify hydrophobic substances. In particular, cholic acid has a large hydrophobic steroid nucleus, and a highly polar carboxylate group.**

**Problem 16.25** Following is a structural formula for cortisol (hydrocortisone). Draw a conformational representation of this molecule.

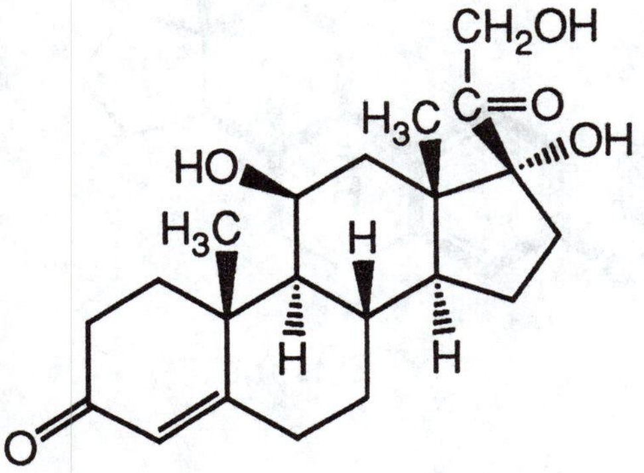

**Cortisol**
**(Hydrocortisone)**
**structural formula**

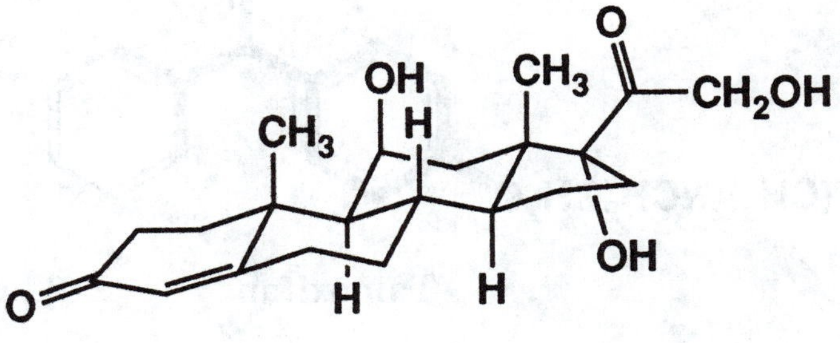

**Cortisol**
**(Hydrocortisone)**
**conformational formula**

**Problem 16.26** Because some types of tumors need an estrogen (a steroid hormone) to survive, compounds that compete with the estrogen receptor on tumor cells are useful anticancer drugs. The compound tamoxifen is one such drug. To what part of the estrone molecule is the shape of tamoxifen similar?

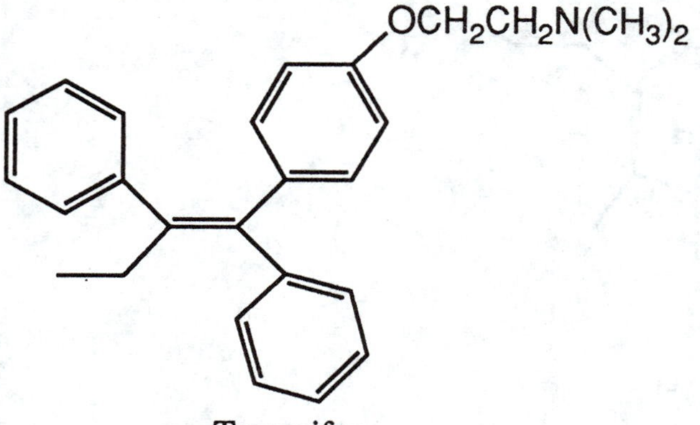

Tamoxifen

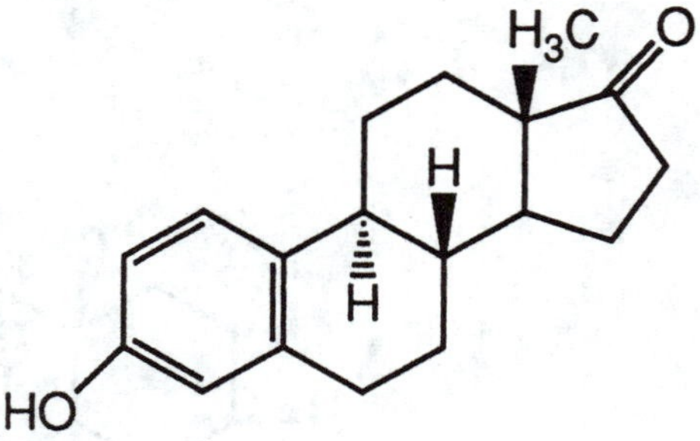

Estrone

**Both tamoxifen and estrone are very hydrophobic. Drawn below are highlighted regions of tamoxifen and estrone that emphasize structural similarity. It should be pointed out that some liberties are taken in the following structures when it comes to some bond angles in tetrahedral and trigonal carbon atoms.**

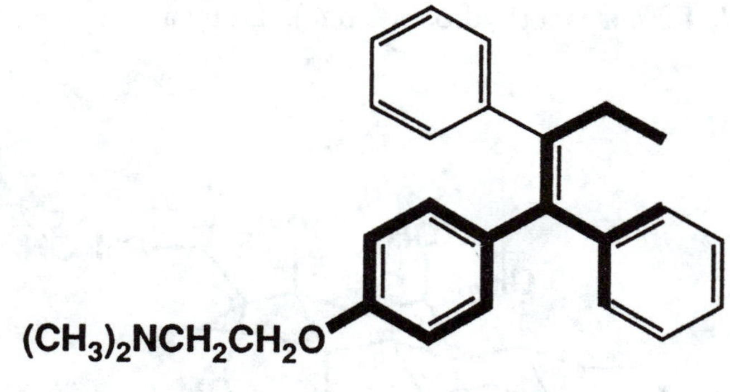

**Tamoxifan**

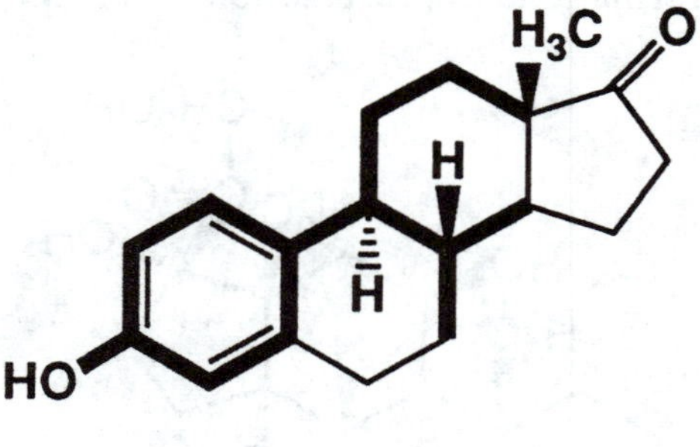

**Estrone**

## Phospholipids

**Problem 16.27** Draw the structural formula of a lecithin containing one molecule each of palmitic acid and linoleic acid.

**Lecithins are phosphoacylglycerols that have choline (HO-(CH$_2$)$_2$-N$^+$(CH$_3$)$_3$) as the alcohol attached to the phosphate group.**

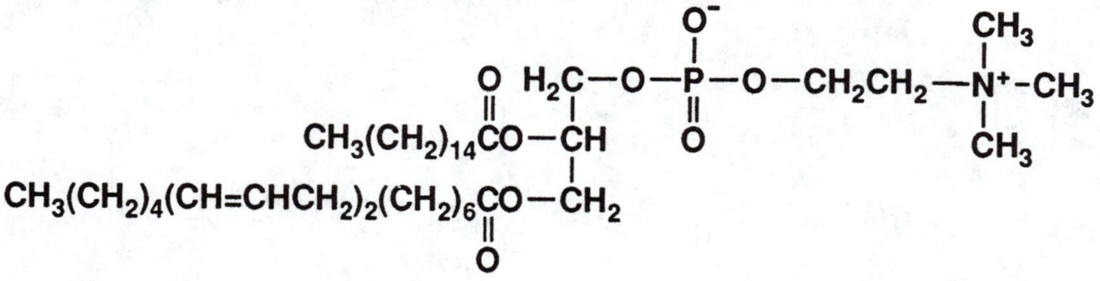

Problem 16.28  Identify the hydrophobic region(s) and the hydrophilic region(s) of a phospholipid.

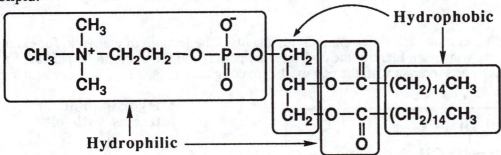

The hydrophobic regions, especially the long hydrocarbon tails are extremely hydrophobic.  As opposed to triacylglycerols, Problem 16.3, the hydrophilic regions of phospholipids, especially the charged head group, are so hydrophilic that they have a tremendous influence over the properties of the molecule.  This "split personality", part hydrophilic and part hydrophobic, is responsible for the ordered structures such as bilayers formed by phospholipids in aqueous solution.

Problem 16.29  The hydrophobic effect is one of the most important noncovalent forces directing the self-assembly of biomolecules in aqueous solution.  The hydrophobic effect arises from tendencies (1) to arrange polar groups so that they interact with the aqueous environment by hydrogen bonding and (2) to arrange nonpolar groups so that they are shielded from the aqueous environment.  Show how the hydrophobic effect is involved in directing:
(a) Formation of micelles by soaps and detergents.

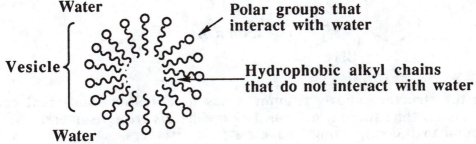

In micelles, the hydrophobic hydrocarbon tails are associated with each other to form the hydrophobic interior, while the polar groups are associated with each other on the outside surface where they interact with water.

(b) Formation of lipid bilayers by phospholipids.

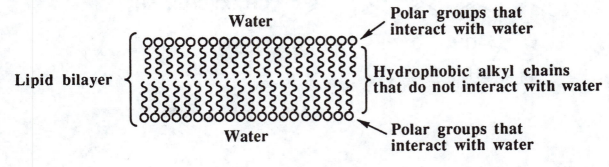

**In lipid bilayers, the hydrophobic hydrocarbon tails are associated with each other to form the hydrophobic inner layer, while the polar head groups are associated with each other on both outside surfaces where they interact with water.**

<u>Problem 16.30</u>  Lecithins can act as emulsifying agents.  The lecithin of egg yolk, for example, is used to make mayonnaise.  Identify the hydrophobic part(s) and the hydrophilic part(s) of a lecithin.  Which parts interact with the oils used in making mayonnaise?  Which parts interact with the water?

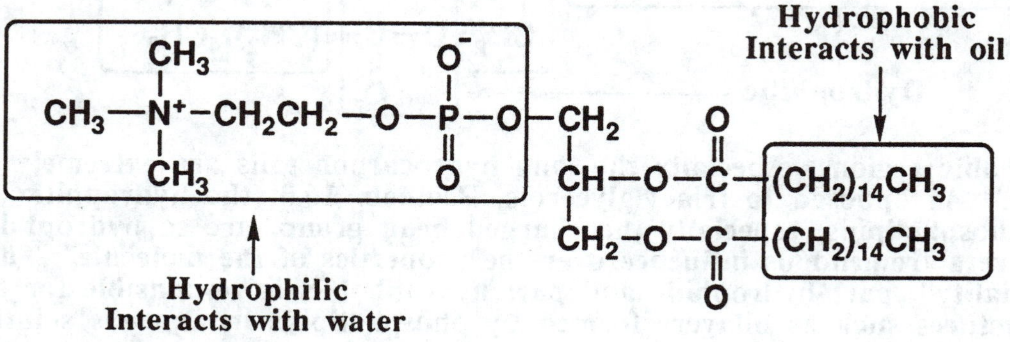

### Fat-soluble Vitamins
<u>Problem 16.31</u>  Examine the structural formula of vitamin A and state the number of *cis-trans* isomers possible for this molecule.

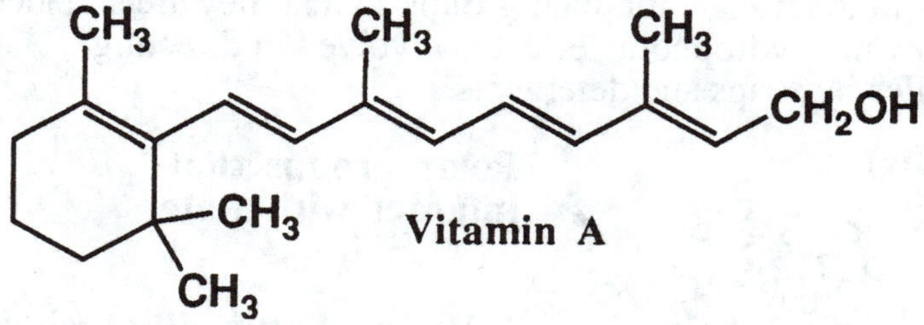

Vitamin A

**As shown in the structure above, vitamin A has four double bonds that can be either *cis* or *trans*, thus there are $2^4$ or 16 possible *cis-trans* isomers.  Note that the double bond in the ring cannot have *cis-trans* isomers.**

<u>Problem 16.32</u>  The form of vitamin A present in many food supplements is vitamin A palmitate.  Draw the structural formula of this molecule.

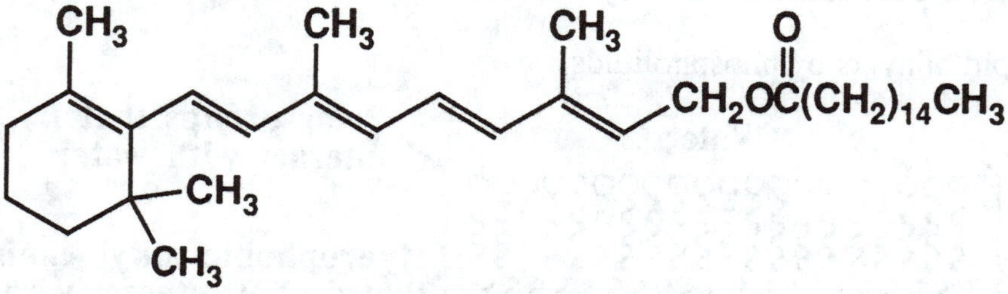

**Problem 16.33** Examine the structural formulas of vitamins A, $D_3$, E, and $K_1$. Do you expect them to be more soluble in water or in dichloromethane? Do you expect them to be soluble in blood plasma?

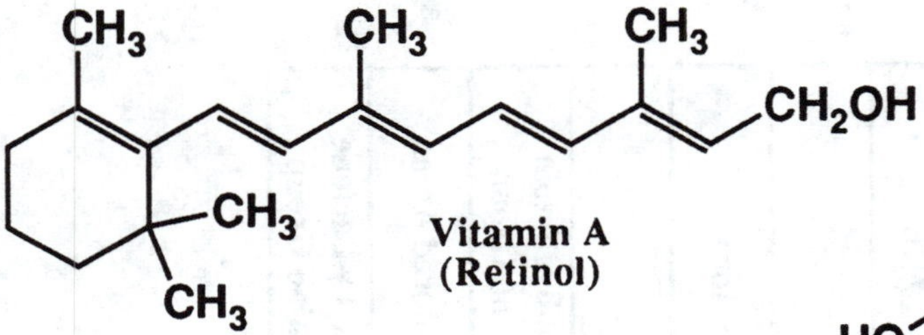

**Vitamin A**
**(Retinol)**

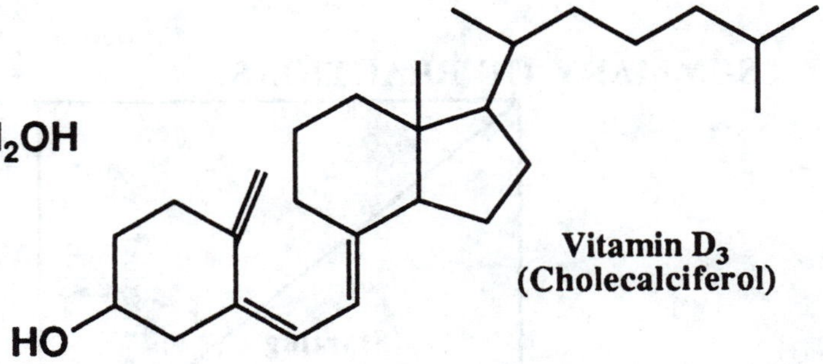

**Vitamin D₃**
**(Cholecalciferol)**

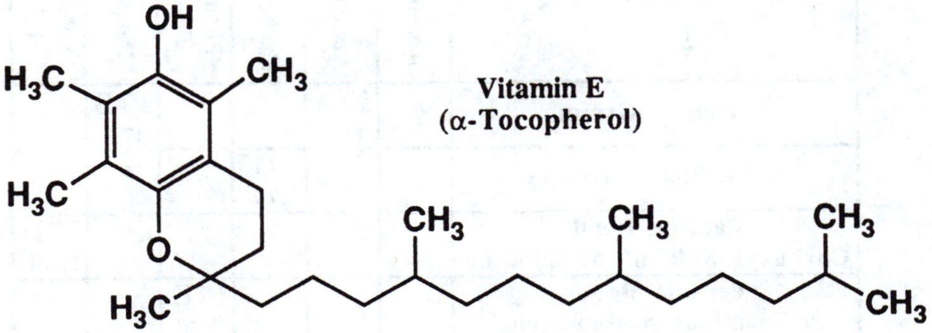

**Vitamin E**
**(α-Tocopherol)**

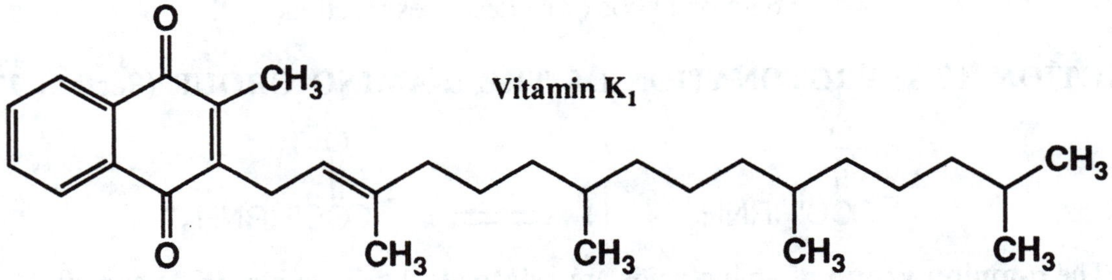

**Vitamin K₁**

All of these structures are extremely hydrophobic, so they will be more soluble in organic solvents such as dichloromethane than polar solvents such as water. Since blood plasma is an aqueous solution, these vitamins will only be sparingly soluble in blood plasma.

# CHAPTER 17: AMINO ACIDS AND PROTEINS

## SUMMARY OF REACTIONS

| Starting Material \ Product → | Amide | Ammonium Ion | Carboxylate | Phenylthiohydantoin Free Amino Group | Purple-Colored Anion | Substituted γ-Lactone Free Amino Group |
|---|---|---|---|---|---|---|
| α-Amino Group | | 17A 17.2* | | | 17B 17.2 | |
| α-Carboxyl Group | | | 17C 17.2 | | | |
| Peptide Bond Carboxyl Side of Methionine | | | | | | 17D 17.4B |
| Peptide Bond N-Terminal Amino Acid | | | | 17E 17.4B | | |

*Section in book that describes reaction.

## REACTION 17A: PROTONATION OF THE α-AMINO GROUP (Section 17.2)

$$^-OCC'HRNH_2 + H^+ \rightleftharpoons \ ^-OCC'HRNH_3{}^+$$

- The **α-amino group** of amino acids are **relatively basic** with a **pK$_a$ near 10.** ✻

## REACTION 17B: THE NINHYDRIN REACTION (Section 17.2)

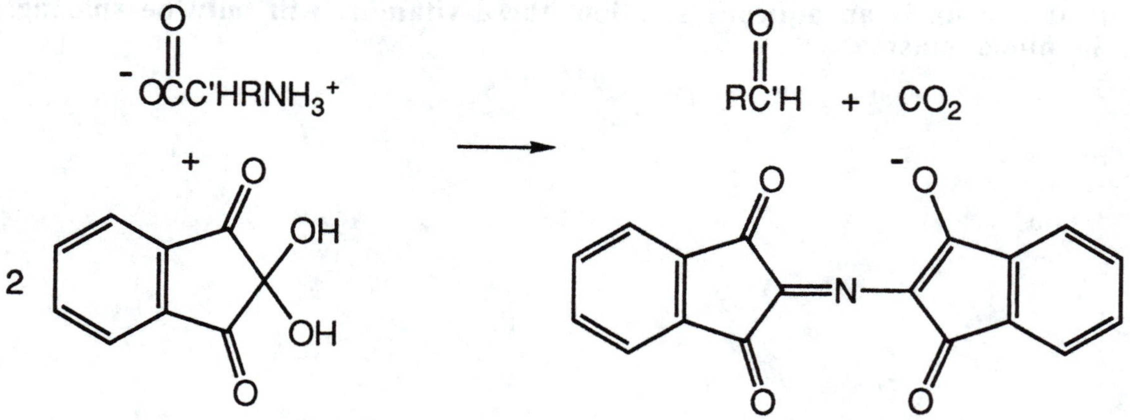

$$^-OCC'HRNH_3{}^+$$

$$\longrightarrow \quad RC'H + CO_2$$

Ninhydrin

Purple-colored anion

- **Ninhydrin** reacts with **primary amine** groups such as the $\alpha$-amino group of amino acids to give a **deep purple-colored anion**. This reaction can be used to detect even very small amounts of primary amine groups. Ninhydrin has been used extensively to detect unreacted $\alpha$–amino groups during solid-phase peptide synthesis. Secondary amines like those in proline react with ninhydrin to form an orange-colored compound. ✳

## REACTION 17C: ACIDITY OF THE $\alpha$-CARBOXYL GROUP (Section 17.2)

$$HOCC'HRNH_3^+ \rightleftharpoons {}^-OCC'HRNH_3^+ + H^+$$

- The **$\alpha$-carboxyl group** is **relatively acidic**, having a **$pK_a$** value near **2.0**, due to the inductive effect of the adjacent electron-withdrawing $-NH_3^+$ group. ✳

## REACTION 17D: THE CYANOGEN BROMIDE CLEAVAGE REACTION OF METHIONINE CONTAINING PEPTIDES AND PROTEINS (Section 17.4B)

- **Cyanogen bromide** reacts with **methionine-containing peptides and proteins** to give products that are **cleaved** at the **amide bond** on the **carboxyl side** of the **methionine residue**. The reaction is used to cleave large proteins into smaller peptides in order to facilitate sequence analysis. A substituted $\gamma$-lactone is the other product of the reaction. ✳

## REACTION 17E: THE EDMAN DEGRADATION (Section 17.4B)

A Phenylthiohydantoin

- Reacting peptides or proteins with **phenyl isothiocyanate** causes **removal of the *N*-terminal amino acid** as a **phenylthiohydantoin** that can be **isolated and identified**. This reaction is referred to as the **Edman degradation** after its inventor Pehr Edman. ✳
- The Edman degradation can be run sequentially on a peptide of unknown sequence, so that the exact amino acid sequence can be determined. Since usually only twenty to thirty amino acids can be sequenced using the Edman degradation, a large protein must first be fragmented into smaller pieces using the CNBr reaction and/or limited hydrolysis with a protease such as trypsin or chymotrypsin. The Edman degradation chemistry has now been automated.

## SUMMARY OF IMPORTANT CONCEPTS

### 17.0 OVERVIEW

• **Proteins** are composed of chains of **amino acids** linked together by amide (**peptide**) bonds. Two triumphs of chemistry are that the sequence of amino acids in proteins can be determined chemically, and that amino acids can be joined together synthetically to produce functional proteins. ✳

### 17.1 AMINO ACIDS

• **Amino acids** are compounds that contain a carboxyl group and an amino group. The $\alpha$-**amino acids** ($H_2N$-CHR-$CO_2H$) are the most important class of amino acids in the biological world. ✳

  - The **amino group** is **basic**, while the **carboxyl** group is relatively **acidic**, thus **amino acids** exist as **internal salts** called a **zwitterions** near neutral pH. ✳
  - **Except for glycine** ($H_2N$-$CH_2$-$CO_2H$), the $\alpha$-**carbon atoms** of amino acids are **stereocenters**. According to the D and L designations used with carbohydrates, the vast majority of amino acids in living systems are of the **L-series**. This corresponds to the **S configuration** in the R-S convention except for cysteine in which the L designation corresponds to the R configuration because of the priorities assigned with the R-S system. Isoleucine and threonine also have a second stereocenter on their side chains.
  - The **20 protein-derived amino acids** are usually grouped according to the chemical properties of the **side chains** as either **nonpolar, polar but unionized, acidic**, and **basic side chains**.
  - Besides these 20 amino acids, small amounts of other amino acids are found in nature. For example, L-ornithine and L-citrulline are components of the urea cycle. In addition, D-amino acids are found as structural components of lower forms of life.

### 17.2 ACID-BASE PROPERTIES OF AMINO ACIDS

• Besides the acid-base properties of $\alpha$-amino group and $\alpha$-carboxyl group, a number of amino acid side chains undergo acid-base reactions. The **guanidine group of arginine**, the **amino group of lysine**, and the **imidazole group of histidine** are **basic**. The **carboxylic acid groups** of **aspartic acid** and **glutamic acid** are acidic. ✳
• The **Henderson-Hasselbalch equation** can be used to calculate the ratio of a conjugate base to weak acid at any pH. This can be used to calculate the amount of protonated carboxylic acid or protonated amine functions in amino acids at a given pH. ✳

$$pH = pK_a + \log \frac{[\text{conjugate base}]}{[\text{weak acid}]}$$

Henderson-Hasselbalch equation

• The **isoelectric point, pI,** for an amino acid is the pH at which the majority of molecules in solution have no charge. At the isoelectric point, the molecules are more likely to aggregate and

thus precipitate. For this reason, amino acids can be precipitated from solution at the isoelectric point, a process known as **isoelectric precipitation**. ✳
• Charged molecules move in an applied electric field toward the electrode carrying the charge opposite their own. This process is called **electrophoresis**, and it can be used to separate amino acids on the basis of charge.

## 17.3 POLYPEPTIDES AND PROTEINS
• **Proteins** are long chains of amino acids **linked** together by **amide bonds** between the α-**amino group** of one amino acid and the α-**carboxyl group** of another. These amides bonds are given the special name of **peptide bonds**. ✳

## 17.4 PRIMARY STRUCTURE OF POLYPEPTIDES AND PROTEINS
• The **primary (1°) structure** of a protein or polypeptide is the **sequence of amino acids** in the polypeptide chain. The sequence of amino acids that make up a protein is determined in several steps. ✳
  - First, the **proportion of different amino acids** is determined by **amino acid analysis**. The **polypeptide is hydrolyzed** into individual amino acids by heating in 6 *M* HCl or 4 *M* NaOH, then techniques such as ion-exchange chromatography are used to **separate, identify, and quantitate** the **different amino acids** present.
  - Next the **polypeptide is selectively cleaved into fragments** using the **CNBr** reaction (Reaction 17D, Section 17.4B) and/or limited proteolysis. The fragments are subjected to sequence analysis using the **Edman degradation** (Reaction 17E, Section 17.4B). The entire sequence of amino acids is reconstructed from the sequences of these fragments. ✳

## 17.5 THREE-DIMENSIONAL SHAPES OF POLYPEPTIDES AND PROTEINS
• The **amide bond is planar**. In other words, the carbonyl carbon atom, the carbonyl oxygen atom, the amide nitrogen atom, the amide hydrogen atom, and both α-carbon atoms are all in the same plane. ✳
  - The planarity is explained by considering that an amide is accurately represented as the **resonance hybrid of two contributing structures,** one with a carbon-oxygen double bond and **one with a carbon-nitrogen double bond.**
  - The partial carbon-nitrogen double bond character of the amide means that **two configurations are possible** for an amide, an *s-trans* or *s-cis* configuration. **Almost all peptide bonds** in proteins are in the *s-trans* **configuration** in which the two α-carbon atoms are *trans* to each other.
• In large part due to the planarity and rigidity of peptide bonds, polypeptide chains can form **secondary (2°) structures** such as α-**helixes** and β-**sheets**. These structures are reinforced by hydrogen bonds between the oxygen atoms and hydrogen atoms of the peptide bonds. ✳
• A polypeptide chain exhibits even higher order structure, referred to as **tertiary (3°) structure**, that describes the way in which the secondary structural units of the chain are oriented in three-dimensions. The 3° structure can be held together by a combination of forces including disulfide bonds between two cysteine residues. ✳
• More than one folded polypeptide chain can come together to form a functional complex, and the association of more than one chain is referred to as **quaternary (4°) structure**. ✳

## CHAPTER 17
### Solutions to the Problems

Problem 17.1  Of the 20 protein-derived amino acids shown in Table 17.1, which contain (a) no stereocenter, (b) two stereocenters.

**The only amino acid with no stereocenters is glycine (Gly, G).  Both isoleucine (Ile, I) and threonine (Thr, T) have two stereocenters as shown with asterisks in the structures below.**

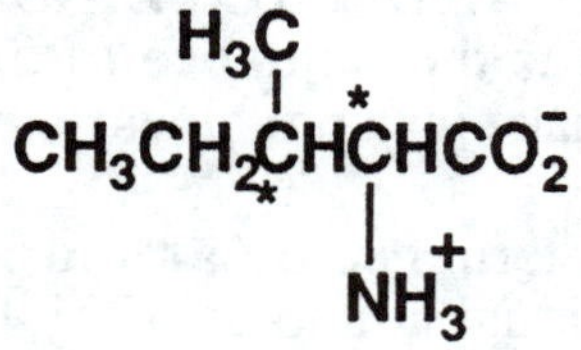

$$CH_3CH_2\overset{*}{C}H\overset{*}{C}HCO_2^-$$

Isoleucine (Ile, I)

$$CH_3\overset{*}{C}H\overset{*}{C}HCO_2^-$$

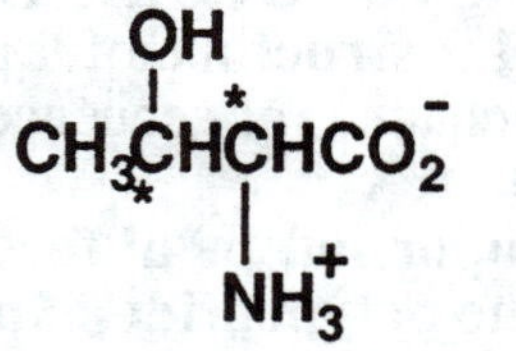

Threonine (Thr, T)

Problem 17.2  Draw a structural formula for lysine, and estimate the net charge on each functional group at pH values of 3.0, 7.0, and 10.0.

**The net charge on the functional groups is calculated as described in example 17.2.  The results of the calculations are shown on the structures at the indicated pH:**

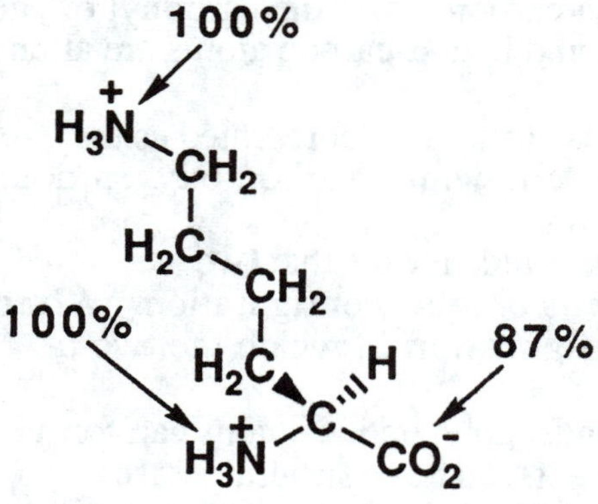

(pH  3.0)

Net charge +

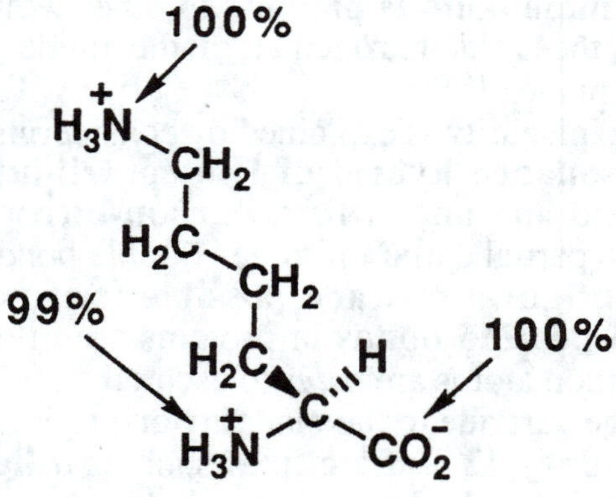

(pH  7.0)

Net charge +

**78%**

H$_3$$\overset{+}{\text{N}}$—CH$_2$

H$_2$C—CH$_2$

**91%**   H$_2$C   H

H$_2$N—C—CO$_2^-$

**100%**

**(pH 10.0)**

**Net charge +**

Problem 17.3  The isoelectric point of histidine is 7.64.  Toward which electrode does histidine migrate on paper electrophoresis at pH 7.0?

**An amino acid will have at least a partial positive charge at any pH that is below its isoelectric point.  A pH of 7.0 is below the isoelectric point of histidine (7.64), so it will have a partial positive charge.  Therefore, at this pH histidine migrates toward the negative electrode.**

Problem 17.4  Describe the behavior of a mixture of glutamic acid, arginine, and valine on paper electrophoresis at pH 6.0.

**The pI's for glutamic acid, arginine, and valine are 3.08, 10.76, and 6.00, respectively.  Therefore, at pH 6.0 glutamic acid is negatively charged, arginine is positively charged, and valine is neutral.  Thus, on paper electrophoresis, glutamic acid will migrate toward the positive electrode, arginine will migrate toward the negative electrode, and valine will not move.**

Problem 17.5  Draw a structural formula for Lys-Phe-Ala.  Label the N-terminal amino acid and the C-terminal amino acid.  What is the net charge on this tripeptide at pH 6.0?

*N*-terminal amino acid

*C*-terminal amino acid

**Due to the presence of the basic lysine residue, this tripeptide will have a net positive charge at pH 6.0**

Problem 17.6 Which of these tripeptides are hydrolyzed by trypsin? by chymotrypsin?
(a) Tyr-Gln-Val          (b) Thr-Phe-Ser          (c) Thr-Ser-Phe

**Based on the substrate specificities listed in Table 17.3, trypsin will not cleave any of these tripeptides because there are no arginine or lysine residues. On the other hand chymotrypsin will cleave peptides (a) and (b) between the Tyr-Gln and Phe-Ser residues, respectively.**

Problem 17.7 Deduce the amino acid sequence of an undecapeptide (11 amino acids) from the experimental results shown in the table.

| Experimental Procedure | Amino Acid Composition |
|---|---|
| Undecapeptide | Ala,Arg,Glu,Lys$_2$,Met,Phe,Ser,Thr,Trp,Val |
| Edman degradation | Ala |
| **Trypsin-Catalyzed Hydrolysis** | |
| Fragment E | Ala,Glu,Arg |
| Fragment F | Thr,Phe,Lys |
| Fragment G | Lys |
| Fragment H | Met,Ser,Trp,Val |
| **Chymotrypsin-Catalyzed Hydrolysis** | |
| Fragment I | Ala,Arg,Glu,Phe,Thr |
| Fragment J | Lys$_2$,Met,Ser,Trp,Val |
| **Reaction with Cyanogen Bromide** | |
| Fragment K | Ala,Arg,Glu,Lys$_2$,Met,Phe,Thr,Val |
| Fragment L | Trp,Ser |

**Based on the Edman degradation result, alanine (Ala) is the N-terminal residue of the peptide.**
**Fragment E must have Arg on the C-terminal end because it is a peptide produced by trypsin cleavage. Since we know Ala is the N-terminal residue, this means fragment E must be of the sequence Ala-Glu-Arg.**
**There must be two lysine residues or an arginine and a lysine residue adjacent to each other based on the appearance of a single lysine residue as Fragment G.**
**Since Fragment J has two lysines and no arginine residues, the two lysine residues must be adjacent to each other.**
**Methionine must be the third to the last residue, because CNBr treatment created fragment L that is only Ser and Trp. In addition, Trp and Ser must be the last two residues. Combining this information with the knowledge that there are two lysine residues adjacent to each other indicates the Fragment J is of the sequence Lys-Lys-Val-Met-Ser-Trp. Note that Val must come after the two Lys residues because of Fragment H. In addition, Trp has to be on the C-terminus or a residue would have been cleaved off by chymotrypsin.**
**Phenylalanine must be on the C terminus of Fragment I since it results from chymotrypsin cleavage. We already know that Fragment I must start with Ala-Glu-Arg, so the entire sequence of Fragment I must be Ala-Glu-Arg-Thr-Phe. Putting Fragments I and J together gives the following sequence for the entire peptide:**

**Ala-Glu-Arg-Thr-Phe-Lys-Lys-Val-Met-Ser-Trp**

<u>Problem 17.8</u>  At pH 7.4, with what amino acid side chains can the side chain of lysine form salt linkages.

**At pH 7.4, the only negatively charged side chains are the carboxylates of glutamic acid and aspartic acid.  Therefore, these are the amino acid side chains with which the side chain of lysine can form a salt linkage.**

<u>Problem 17.9</u>  What amino acids do these abbreviations stand for?
(a) Phe  **Phenylalanine**     (b) Ser  **Serine**        (c) Asp  **Aspartic acid**
(d) Gln  **Glutamine**        (e) His  **Histidine**      (f) Gly  **Glycine**
(g) Tyr  **Tyrosine**

<u>Problem 17.10</u>  Configuration of the stereocenter in $\alpha$-amino acids is most commonly specified using the D, L convention.  It can also be identified using the R,S convention (Section 5.3)  Does the stereocenter in L-serine have the R configuration or the S configuration.

L-Serine

**The configuration of L-serine according to the R,S convention is S.**

<u>Problem 17.11</u>  Assign R or S configuration to these amino acids.
(a) L-Phenylalanine                    (b) L-Glutamic acid              (c) L-Methionine

**The configuration of all of these amino acids according to the R,S convention is S.**

<u>Problem 17.12</u>  The amino acid threonine has two stereocenters.  The stereoisomer found in proteins has the configuration 2S, 3R about the two stereocenters.  Draw a Fischer projection of this stereoisomer and also a three-dimensional representation using solid, wedged, and dashed lines.

L-Threonine

<u>Problem 17.13</u>  Define the term zwitterion.

**A zwitterion is a molecule that has an internal salt linkage between a positively charged group and a negatively charged group.  A zwitterion has no net charge.**

Problem 17.14 Draw zwitterionic forms of these amino acids.
(a) Valine                              (b) Phenylalanine                    (c) Glutamine

$(CH_3)_2CH$—$C$(,,,H)($CO_2^-$), $H_3^+N$

$PhCH_2$—$C$(,,,H)($CO_2^-$), $H_3^+N$

$H_2NCCH_2CH_2$ (O double bond) —$C$(,,,H)($CO_2^-$), $H_3^+N$

Problem 17.15 Why are Glu and Asp often referred to as acidic amino acids.

**The side chains of glutamic acid (Glu) and aspartic acid (Asp) have carboxylic acid functions, so these amino acids are referred to as acidic amino acids.**

Problem 17.16 Why is Arg often referred to as a basic amino acid? Which two other amino acids are also basic amino acids?

**The guanidine function of arginine (Arg) is strongly basic, so this amino acid is referred to as being a basic amino acid. Note that this means arginine is positively charged at neutral pH. Lysine (Lys) and histidine (His) are also referred to as basic amino acids because their side chains contain a basic primary amine and imidazole functions, respectively.**

Problem 17.17 What is the meaning of the alpha as it is used in $\alpha$-amino acid?

**The alpha in $\alpha$-amino acids indicates that the amine group is on the carbon atom that is $\alpha$ to the carboxylic acid group.**

Problem 17.18 Several $\beta$-amino acids exist. There is a unit of $\beta$-alanine, for example, contained within the structure of coenzyme A (Problem 14.34). Write the structural formula of $\beta$-alanine.

$$H_3^+NCH_2CH_2CO_2^-$$
**$\beta$-Alanine**

Problem 17.19 Draw a Fisher projection formula L-serine. Now convert your Fischer projection formula to a three-dimensional representation using solid, wedged, and dashed lines.

$H_3N^+$ —|— $H$ , $CO_2^-$ (top), $CH_2OH$ (bottom)

$H$(,,,,)$C$($CO_2^-$)($^+NH_3$), $HOH_2C$

**L-Serine**

**Problem 17.20** Although only L-amino acids occur in proteins, D-amino acids are often a part of the metabolism of lower organisms. The antibiotic actinomycin D, for example, contains a unit of D-valine, and the antibiotic bacitracin A contains units of D-asparagine and D-glutamic acid. Draw Fischer projections and three-dimensional representations for these three D-amino acids.

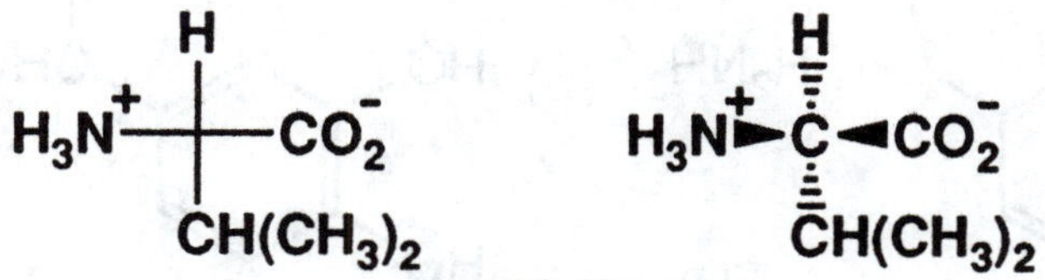

D-Valine

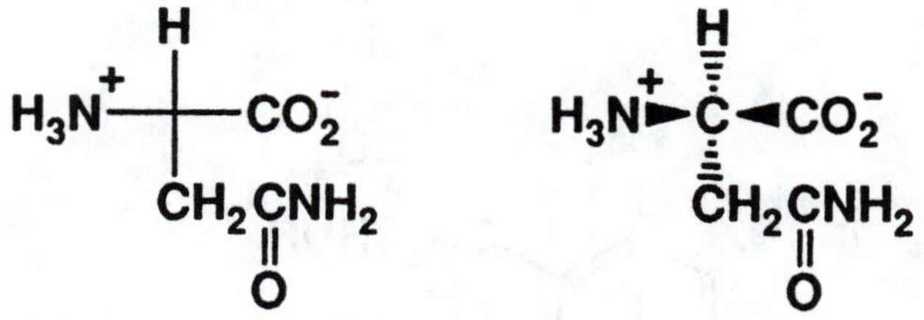

D-Asparagine

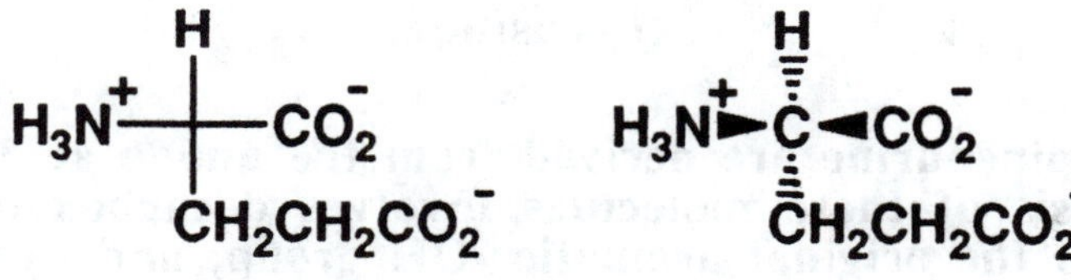

D-Glutamic acid

**Problem 17.21** Histamine is synthesized from one of the 20 protein-derived amino acids. Suggest which amino acid is its biochemical precursor, and the type of organic reaction(s) involved in its biosynthesis (e.g., oxidation, reduction, decarboxylation, nucleophilic substitution).

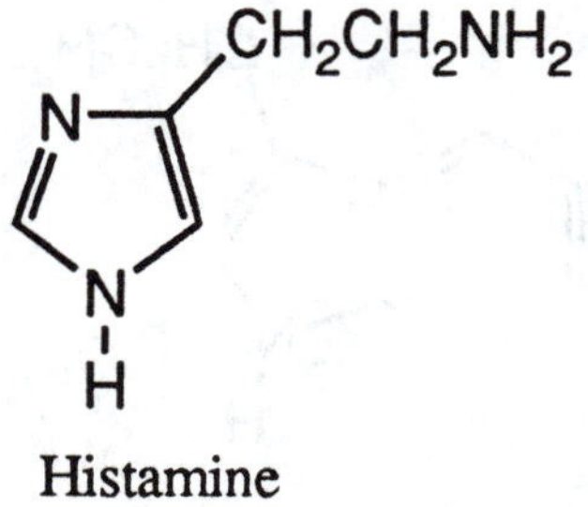

Histamine 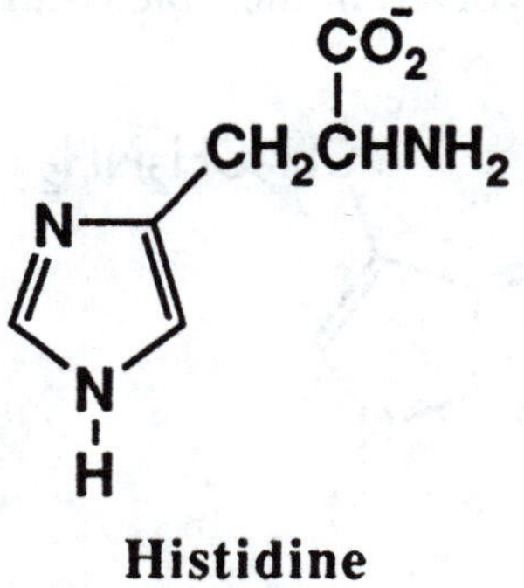                         **Histidine**

**Histamine is derived from the amino acid histidine and is the result of a biosynthetic decarboxylation reaction. Note how both the histamine and histidine are drawn in the form present at basic pH.**

Problem 17.22  Both norepinephrine and epinephrine are synthesized from the same protein-derived amino acid.  From which amino acid are they synthesized and what types of reactions are involved in their biosynthesis?

(a) Norepinephrine

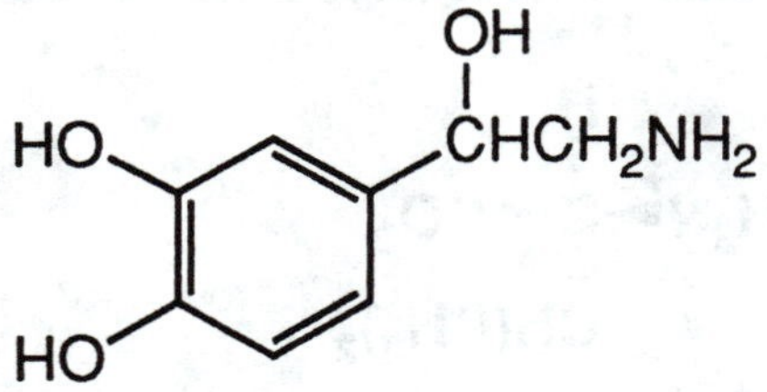

(b) Epinephrine (Adrenaline)

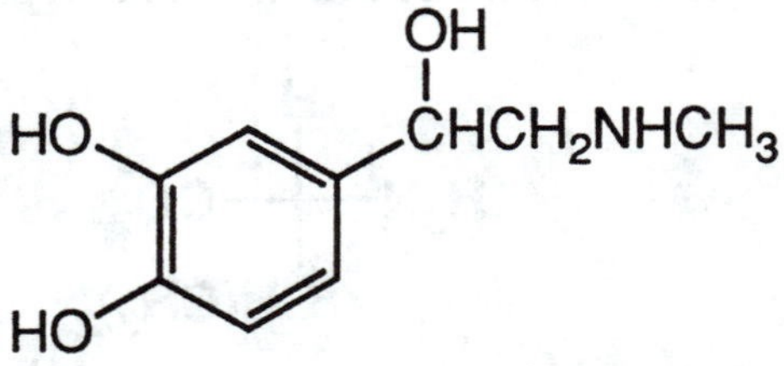

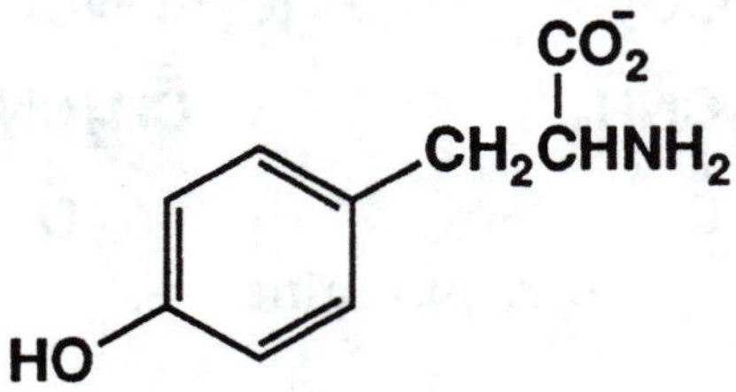

Tyrosine

**Norepinephrine and epinephrine are derived from the amino acid tyrosine.  In both cases, biosynthesis of these molecules involves decarboxylation, aromatic hydroxylation ortho to the original aromatic -OH group, and hydroxylation of the benzylic methylene group.  Epinephrine is also methylated on the $\alpha$-amino group. Note how all of the molecules in the problem are drawn in the form present at basic pH.**

Problem 17.23  From which amino acid are serotonin and melatonin synthesized and what types of reactions are involved in their biosynthesis?

(a) Serotonin

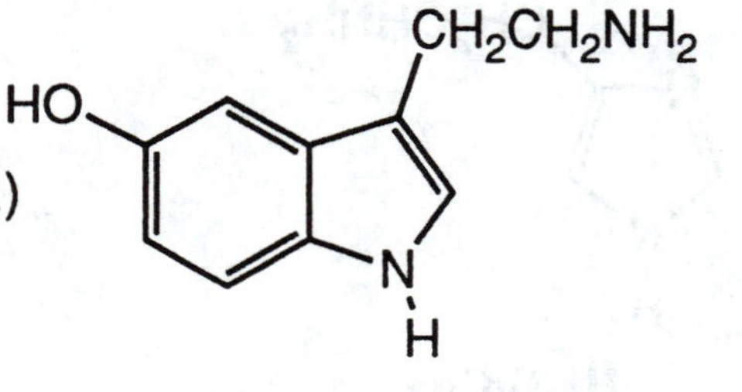

(b) Melatonin

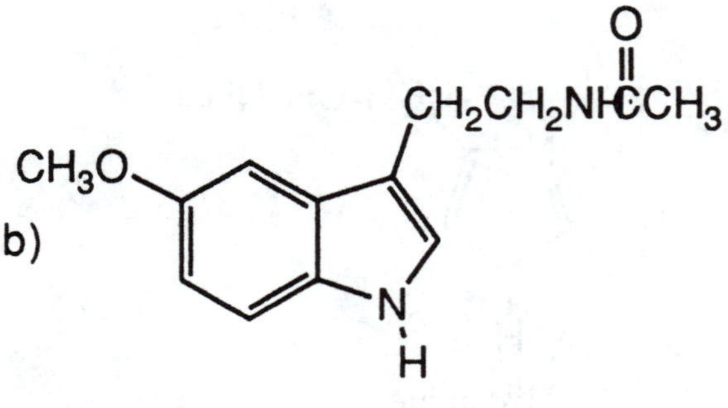

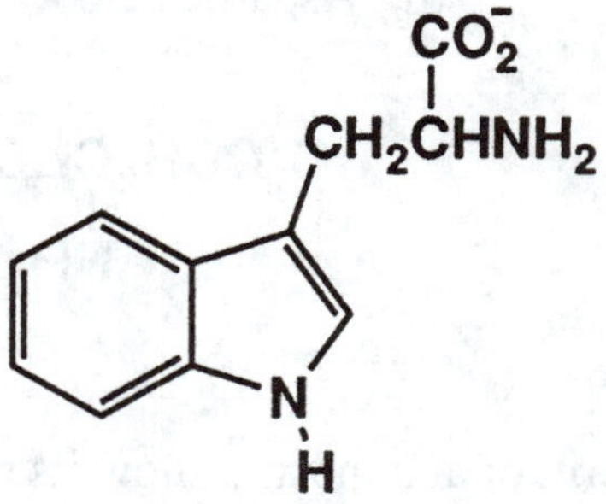

$$CO_2^-$$
$$CH_2CHNH_2$$

**Tryptophan**

Serotonin and melatonin are derived from the amino acid tryptophan. In both cases, biosynthesis of these molecules involves decarboxylation. In the case of serotonin there is also an aromatic hydroxylation. For melatonin there is an aromatic methoxy group added, as well as an acetyl group added to the amine. Note how all of the molecules in the problem are drawn in the form present at basic pH.

## Acid-Base Behavior of Amino Acids

<u>Problem 17.24</u> Draw the structural formula for the form of each amino acid most prevalent at pH 1.0.

(a) Threonine

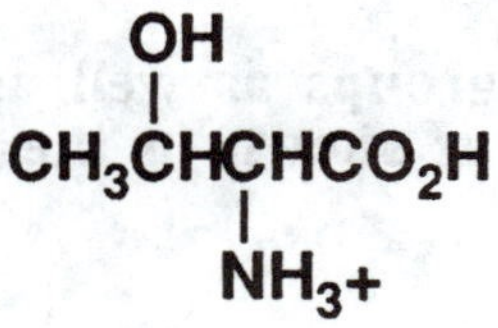

$$OH$$
$$CH_3CHCHCO_2H$$
$$NH_3+$$

(b) Arginine

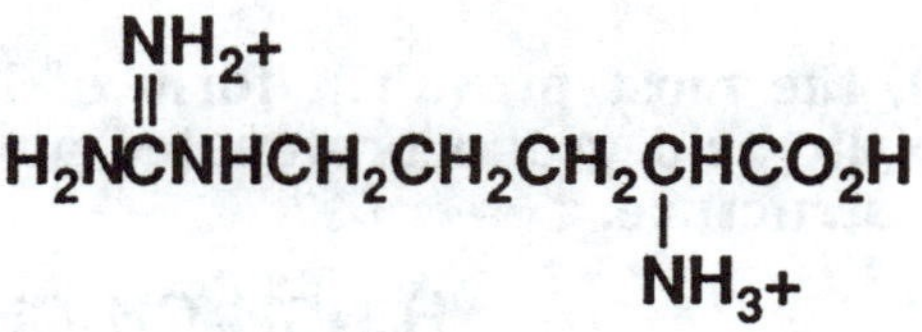

$$NH_2+$$
$$H_2NCNHCH_2CH_2CH_2CHCO_2H$$
$$NH_3+$$

(c) Methionine

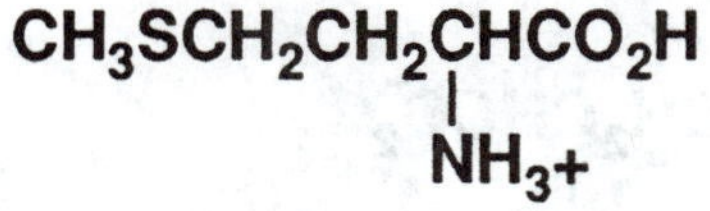

$$CH_3SCH_2CH_2CHCO_2H$$
$$NH_3+$$

(d) Tyrosine

$$HO\!-\!\!\!\bigcirc\!\!\!-CH_2CHCO_2H$$
$$NH_3+$$

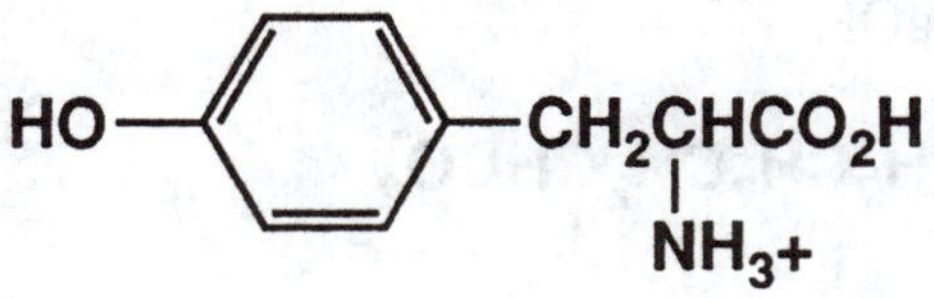

<u>Problem 17.25</u> Draw the structural formula for the form of each amino acid most prevalent at pH 10.0.

(a) Leucine

$$(CH_3)_2CHCH_2CHCO_2^-$$
$$NH_2$$

(b) Valine

$$(CH_3)_2CHCHCO_2^-$$
$$NH_2$$

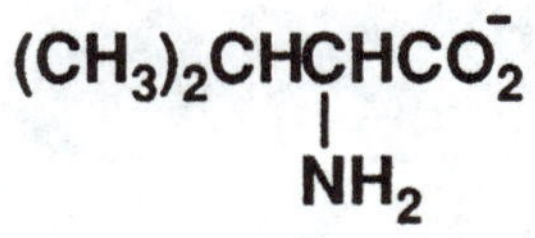

(c)  Proline                                    (d)  Aspartic acid

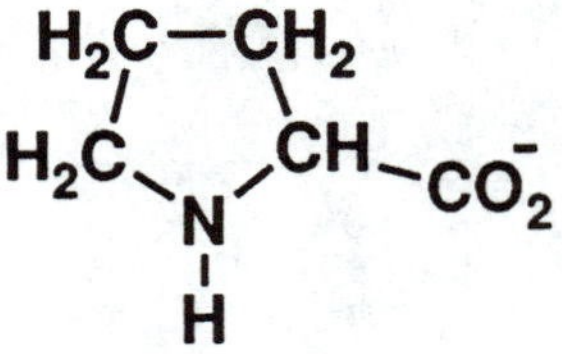

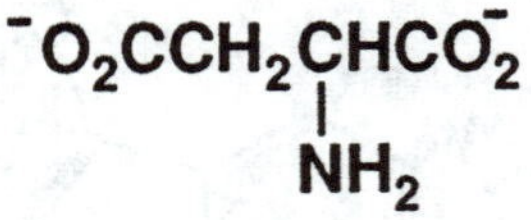

$$H_2C-CH_2$$
$$H_2C \quad CH-CO_2^-$$
$$\underset{H}{N}$$

$$^-O_2CCH_2\underset{NH_2}{CHCO_2^-}$$

<u>Problem 17.26</u>  Write the zwitterion form of alanine and show its reaction with:
(a)  1 mol NaOH

$$CH_3\underset{NH_3+}{CHCO_2^-} \quad + \quad 1 \; mol \; NaOH \longrightarrow CH_3\underset{NH_2}{CHCO_2^-}$$

(b)  1 mol HCl

$$CH_3\underset{NH_3+}{CHCO_2^-} \quad + \quad 1 \; mol \; HCl \longrightarrow CH_3\underset{NH_3+}{CHCO_2H}$$

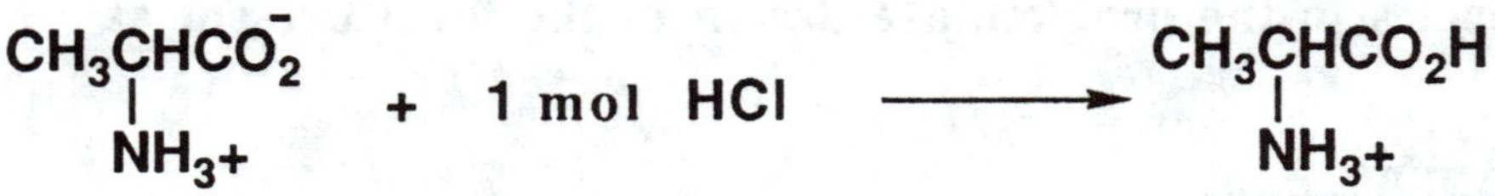

<u>Problem 17.27</u>  Write the form of lysine most prevalent at pH 1.0 and then show its reaction with the following.  Consult Table 17.2 for $pK_a$ values of the ionizable groups in lysine.

**At pH 1.0, the most prevalent form of lysine has both amino groups as well as the carboxylic acid group protonated and a total charge of +2 as shown in the following structure.**

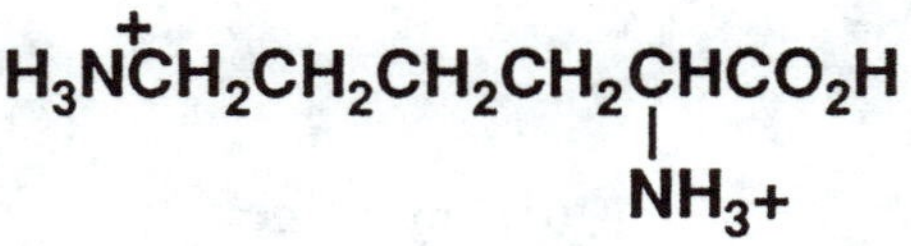

$$H_3\overset{+}{N}CH_2CH_2CH_2CH_2\underset{NH_3+}{CHCO_2H}$$

(a) 1 mol NaOH 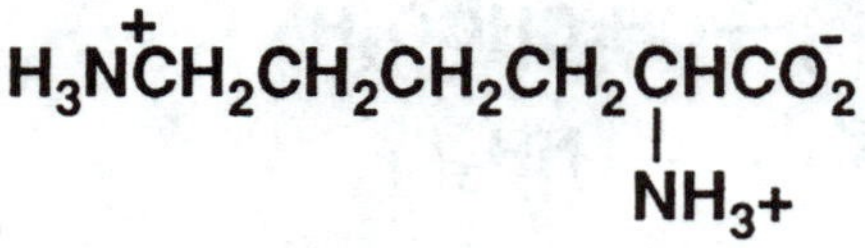                              (b)  2 mol NaOH

$$H_3\overset{+}{N}CH_2CH_2CH_2CH_2\underset{NH_3+}{CHCO_2^-}$$

$$H_3\overset{+}{N}CH_2CH_2CH_2CH_2\underset{NH_2}{CHCO_2^-}$$

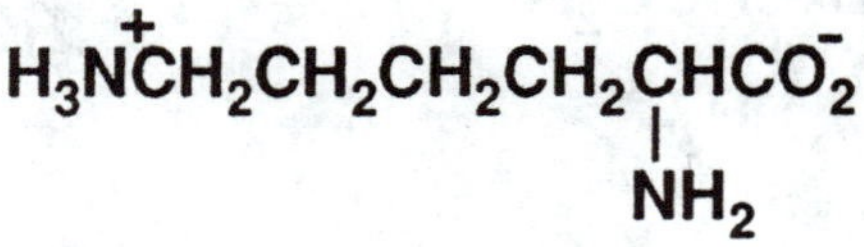

(c)  3 mol NaOH

$$H_2NCH_2CH_2CH_2CH_2\underset{NH_2}{CHCO_2^-}$$

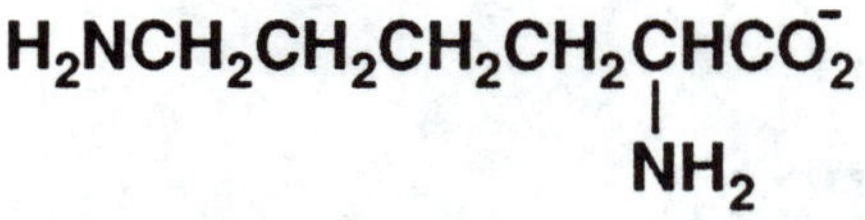

Problem 17.28 Write the form of aspartic acid most prevalent at pH 1.0 and then show its reaction with the following. Consult Table 17.2 for $pK_a$ values of the ionizable groups in aspartic acid.

**At pH 1.0, the most prevalent form of aspartic acid has both carboxylic acid groups as well as the amino group protonated and a total charge of +1 as shown in the following structure.**

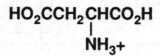

(a) 1 mol NaOH                    (b) 2 mol NaOH                    (c) 3 mol NaOH

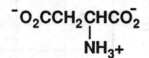   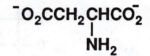

Problem 17.29 Given $pK_a$ values for ionizable groups from Table 17.2, sketch curves for the titration of (a) glutamic acid with NaOH, and (b) histidine with NaOH.

**Glutamic acid has $pK_a$ values of 2.1, 4.07, and 9.47 so the titration curve would look something like the following:**

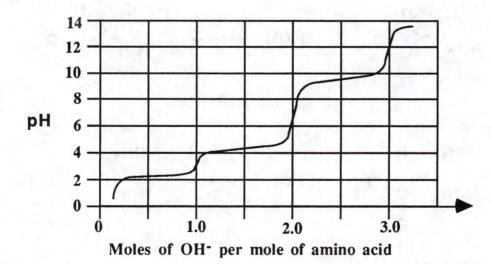

**Histidine has pK$_a$ values of 1.77, 6.10, and 9.18 so the titration curve would look something like the following:**

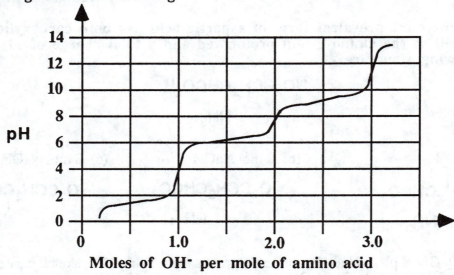

Moles of OH⁻ per mole of amino acid

Problem 17.30 Draw a structural formula for the product formed when alanine is treated with the following reagents:
(a) Aqueous NaOH

(b) Aqueous HCl

(c) CH$_3$CH$_2$OH, H$_2$SO$_4$

(d)  $(CH_3CO)_2O$, $CH_3CO_2Na$

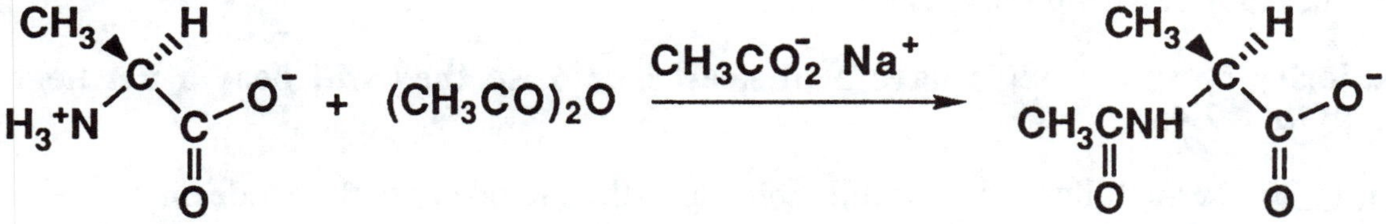

$$CH_3 \overset{H}{\underset{H_3^+N}{\overset{|}{\underset{|}{C}}}}... + (CH_3CO)_2O \xrightarrow{CH_3CO_2^- \ Na^+} ...$$

**Problem 17.31** Account for the fact that the isoelectric point of glutamine (pI 5.65) is higher than the isoelectric point of glutamic acid (pI 3.08).

**Amino acids have no net charge at their pI.  For this to happen with glutamic acid, the net charge on the α-carboxyl and side chain carboxyl groups must be -1 to balance the +1 charge of the α-amino group.  This will occur at a pI = (1/2)(2.10 + 4.07) = 3.08.  The amide side chain of glutamine is already neutral near neutral pH, so the pI of the amino acid is determined by the values for the only ionizable groups, namely the α-carboxyl group and α-amino groups, according to the equation pI = (1/2)(2.17 + 9.03) = 5.6.  This value is near that of the other amino acids with non-ionizable functional groups on their side chains.**

**Problem 17.32** Enzyme-catalyzed decarboxylation of glutamic acid gives 4-aminobutanoic acid (Section 17.1D).  Estimate the pI of 4-aminobutanoic acid.

**There is little if any inductive effect operating between the amino and carboxyl groups of 4-aminobutanoic acid because there are three methylene groups between them.  Thus, the $pK_a$ of the amino group of 4-aminobutanoic acid is like that of a simple amino group, near 10.0.  Similarly, the $pK_a$ of the carboxyl group is like that of a simple carboxyl group, near 4.5.  Given these estimates for the $pK_a$ values, the pI would be:**

$$\text{pI} = \frac{1}{2}(pK_a \ \alpha - CO_2H + pK_a \ \alpha - NH_3^+) = \frac{1}{2}(4.5 + 10.0) = 7.25$$

**Problem 17.33** Guanidine and the guanidino group present in arginine are two of the strongest organic bases known.  Account for this basicity.

**The amino groups are strongly electron-releasing, thus through an inductive effect they are made extremely basic when they are in such close proximity within the same molecule. In addition, the guanidino group is strongly basic because of resonance stabilization of the protonated guanidinium ion as shown below:**

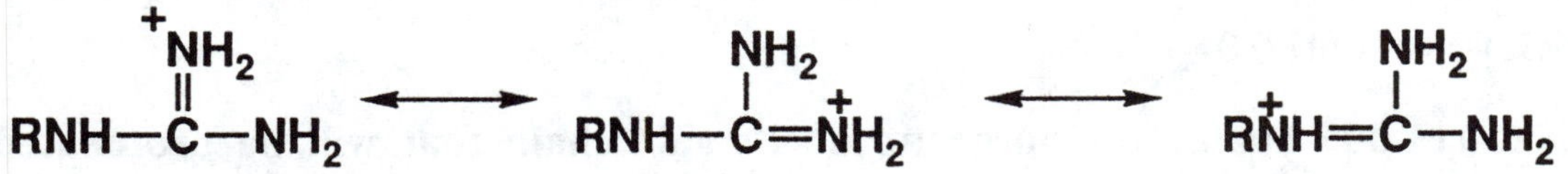

$$\overset{+}{N}H_2 \quad\quad NH_2 \quad\quad NH_2$$
$$RNH-\overset{\|}{C}-NH_2 \longleftrightarrow RNH-\overset{|}{C}=\overset{+}{N}H_2 \longleftrightarrow R\overset{+}{N}H=\overset{|}{C}-NH_2$$

**R = H or alkyl group**

<u>Problem 17.34</u> At pH 7.4, the pH of blood plasma, do the majority of protein-derived amino acids bear a net negative or a net positive charge?

**The majority of amino acids have a pI near 5 or 6, so they will bear a net negative charge at pH 7.4.**

<u>Problem 17.35</u> Do the following molecules migrate to the cathode or to the anode on electrophoresis at the specified pH?

**The key to determining which way the molecules migrate is to estimate the net charge on the molecules at the given pH. Molecules with a net positive charge will migrate toward the negative electrode and molecules with a net negative charge will migrate toward the positive electrode. Molecules at a pH below their isoelectric point (Table 17.2) have a net positive charge, molecules at a pH above their isoelectric point have a net negative charge, and molecules at a pH that equals their isoelectric point have no net charge.**

(a) Histidine at pH 6.8

**pI = 7.64, so at pH 6.8 histidine has a net positive charge and migrates toward the negative electrode (cathode).**

(b) Lysine at pH 6.8

**pI = 9.74, so at pH 6.8 lysine has a net positive charge and migrates toward the negative electrode (cathode).**

(c) Glutamic acid at pH 4.0

**pI = 3.08, so at pH 4.0 glutamic acid has a net negative charge and migrates toward the positive electrode (anode).**

(d) Glutamine at pH 4.0

**pI = 5.65, so at pH 4.0 glutamine has a net positive charge and migrates toward the negative electrode (cathode).**

(e) Glu-Ile-Val at pH 6.0

**The glutamic acid residue has a carboxyl group that will be largely deprotonated at pH 6.0, so the overall molecule will have a net negative charge and will migrate toward the positive electrode (anode).**

(f) Lys-Gln-Tyr at pH 6.0

**The lysine residue has an amino group on its side chain that will be protonated at pH 6.0, so the molecule will have a net positive charge and will migrate toward the negative electrode (cathode).**

Problem 17.36  At what pH would you carry out an electrophoresis to separate the amino acids in each mixture?

**Recall that an amino acid below its isoelectric point will have some degree of positive charge, an amino acid above its isoelectric point will have some degree of negative charge, and an amino acid at its isoelectric point will have no net charge.**

(a)  Ala, His, Lys

**Electrophoresis could be carried out at pH 7.64, the isoelectric point of histidine (His).  At this pH, the histidine is neutral and would not move, the lysine (Lys) will be positively charged and will move toward the negative electrode, and the alanine (Ala) will be slightly negatively charged and will move toward the positive electrode.**

(b) Glu, Gln, Asp

**Electrophoresis could be carried out at pH 3.08, the isoelectric point of glutamic acid (Glu).  At this pH, the glutamic acid is neutral and would not move, the glutamine (Gln) will be positively charged and will move toward the negative electrode, and the aspartic acid (Asp) will be slightly negatively charged and will move toward the positive electrode.**

(c)  Lys, Leu, Tyr

**Electrophoresis could be carried out at pH 6.04, the isoelectric point of leucine (Leu).  At this pH, the leucine is neutral and would not move, the lysine (Lys) will be positively charged and will move toward the negative electrode, and the tyrosine (Tyr) will be slightly negatively charged and will move toward the positive electrode.**

Problem 17.37  Examine the amino acid sequence of human insulin (Figure 17.14) and list each Asp, Glu, His, Lys, and Arg in this molecule.  Do you expect human insulin to have an isoelectric point nearer that of the acidic amino acids (pI 2.0-3.0), the neutral amino acids (pI 5.5-6.5), or the basic amino acids (pI 9.5-11.0)?

**A listing of the amino acids present are shown below:**

|  |  |
|---|---|
| aspartic acid (Asp) | 0 |
| glutamic acid (Glu) | 4 |
| histidine (His) | 2 |
| lysine (Lys) | 1 |
| arginine (Arg) | 1 |

**The charge will only be neutral when there are four positively charged residues to neutralize the four negative charges of the carboxylates from the four Glu residues.  For this to happen, the Lys, Arg, and both His residues must be positively charged.  Since the imidazole of His is not protonated until the pH is below 6 or so, the entire molecule will only be neutral around this pH.  Thus, insulin is expected to have an isoelectric point nearer to that of the neutral amino acids.**

## Primary Structure of Polypeptides and Proteins

Problem 17.38  If a protein contains four different SH groups, how many different disulfide bonds are possible if only a single disulfide bond is formed?  How many different disulfides are possible if two disulfide bonds are formed?

**If only one disulfide bond were to be formed from the four different cysteine residues, then there are a total of 6 different disulfide bonds that can be formed. There are three possibilities if two disulfide bonds are to be formed.**

Problem 17.39  How many different tetrapeptides can be made if
(a)  The tetrapeptide contains one unit each of Asp, Glu, Pro, and Phe?

**There could be any of the four residues in the first position, any of the remaining three amino acids in the second position and so on.  Thus, there are 4 x 3 x 2 x 1 = 24 possible tetrapeptides.**

(b)  All 20 amino acids can be used, but each only once?

**Using the same logic as in (a), there are 20 x 19 x 18 x 17 = 116,280 possible tetrapeptides.**

Problem 17.40  A decapeptide has the following amino acid composition:

$$Ala_2, Arg, Cys, Glu, Gly, Leu, Lys, Phe, Val$$

Partial hydrolysis yields the following tripeptides:

$$Cys\text{-}Glu\text{-}Leu + Gly\text{-}Arg\text{-}Cys + Leu\text{-}Ala\text{-}Ala + Lys\text{-}Val\text{-}Phe + Val\text{-}Phe\text{-}Gly$$

One round of Edman degradation yields a lysine phenylthiohydantoin.  From this information, deduce the primary structure of this decapeptide.

**Due to the Edman degradation result, the Lys residue must be at the N-terminus. Given this information, the rest of the peptide sequence is deduced because of overlap among the tripeptide sequences as shown below.**

**The complete peptide is:**
**Lys-Val-Phe-Gly-Arg-Cys-Glu-Leu-Ala-Ala**

**The peptides fit as follows:**
**Lys-Val-Phe**
**Val-Phe-Gly**
**Gly-Arg-Cys**
**Cys-Glu-Leu**
**Leu-Ala-Ala**

Problem 17.41 Following is the primary structure of glucagon, a polypeptide hormone of 29 amino acids. Glucagon is produced in α-cells of the pancreas and helps maintain blood glucose levels in a normal concentration range.

1                     5                          10                          15
His-Ser-Glu-Gly-Thr-Phe-Thr-Ser-Asp-Tyr-Ser-Lys-Tyr-Leu-Asp-Ser-Arg-Arg-

                    20                          25              29
              Ala-Gln-Asp-Phe-Val-Gln-Trp-Leu-Met-Asn-Thr

                                    Glucagon

Which peptide bonds are hydrolyzed when this polypeptide is treated with
(a) Phenyl isothiocyanate

**This reagent only hydrolyzes the *N*-terminal amino acid, so the His-Ser bond would be hydrolyzed. The site of cleavage is indicated by the ∗.**

**1                         5                              10                              15**
**His]∗[Ser-Glu-Gly-Thr-Phe-Thr-Ser-Asp-Tyr-Ser-Lys-Tyr-Leu-Asp-Ser-Arg-**

**                    20                          25                          29**
**Arg-Ala-Gln-Asp-Phe-Val-Gln-Trp-Leu-Met-Asn-Thr**

(b) Chymotrypsin

**Chymotrypsin catalyzes the hydrolysis of the peptide bonds that are located on the carboxyl side of phenylalanine, tyrosine, and tryptophan residues. The sites of cleavage are indicated by the ∗.**

**1                     5                              10                              15**
**His-Ser-Glu-Gly-Thr-Phe]∗[Thr-Ser-Asp-Tyr]∗[Ser-Lys-Tyr]∗[Leu-Asp-Ser-**

**                    20                          25                          29**
**Arg-Arg-Ala-Gln-Asp-Phe]∗[Val-Gln-Trp]∗[Leu-Met-Asn-Thr**

(c) Trypsin

**Trypsin catalyzes the hydrolysis of the peptide bonds that are located on the carboxyl side of arginine and lysine residues. The sites of cleavage are indicated by the ∗.**

**1                     5                          10                          15**
**His-Ser-Glu-Gly-Thr-Phe-Thr-Ser-Asp-Tyr-Ser-Lys]∗[Tyr-Leu-Asp-Ser-**

**                    20                          25                          29**
**Arg]∗[Arg]∗[Ala-Gln-Asp-Phe-Val-Gln-Trp-Leu-Met-Asn-Thr**

(d)  Br-CN

**Cyanogen bromide cleaves on the *C*-terminal side of methionine residues.  The site of cleavage is indicated by the ✳.**

1                          5                                    10                                      15
His-Ser-Glu-Gly-Thr-Phe-Thr-Ser-Asp-Tyr-Ser-Lys-Tyr-Leu-Asp-Ser-

20                                    25                                    29
Arg-Arg-Ala-Gln-Asp-Phe-Val-Gln-Trp-Leu-Met]✳[Asn-Thr

Problem 17.42  A tetradecapeptide (14 amino acid residues) gives the following peptide fragments on partial hydrolysis.  From this information, deduce the primary structure of this polypeptide. Fragments are grouped according to size.

| Pentapeptide Fragments | Tetrapeptide Fragments |
|---|---|
| Phe-Val-Asn-Gln-His | Gln-His-Leu-Cys |
| His-Leu-Cys-Gly-Ser | His-Leu-Val-Glu |
| Gly-Ser-His-Leu-Val | Leu-Val-Glu-Ala |

**The complete peptide is:**
Phe-Val-Asn-Gln-His-Leu-Cys-Gly-Ser-His-Leu-Val-Glu-Ala

**The peptides fit as follows:**
Phe-Val-Asn-Gln-His
Gln-His-Leu-Cys
His-Leu-Cys-Gly-Ser
Gly-Ser-His-Leu-Val
His-Leu-Val-Glu
Leu-Val-Glu-Ala

Problem 17.43  Write the structural formula of these tripeptides.  Mark each peptide bond, the *N*-terminal amino acid, and the *C*-terminal amino acid.
(a)  Phe-Val-Asn

(b) Leu-Val-Gln

Problem 17.44 Estimate the pI of each tripeptide in Problem 17.43.

**These pI values can be estimated by using the $pK_a$ of the amino group for the $N$-terminal amino acid, and the $pK_a$ of the carboxylic acid group for the $C$-terminal amino acid. Using the values for the appropriate amino groups and carboxylic acid groups listed in table 17.2 leads to the values of pI = 1/2(9.24 + 2.02) = 5.63 and pI = 1/2(9.76 + 2.17) = 5.96 for (a) Phe-Val-Asn and (b) Leu-Val-Gln, respectively.**

Problem 17.45 Glutathione (G-SH), one of the most common tripeptides in animals, plants, and bacteria, is a scavenger of oxidizing agents. In reacting with oxidizing agents, glutathione is converted to G-S-S-G.

Glutathione

(a) Name the amino acids in this tripeptide.

**The amino acids in glutathione are glutamic acid (Glu), cysteine (Cys), and glycine (Gly).**

(b) What is unusual about the peptide bond formed by the $N$-terminal amino acid?

**The $N$-terminal glutamic acid is linked to the next residue by an amide bond between the carboxyl group of the side chain, not the $\alpha$-carboxyl group.**

(c) Write a balanced half-reaction for the reaction of two molecules of glutathione to form a disulfide bond. Is glutathione a biological oxidizing agent or a biological reducing agent?

$$2\text{G-SH} \longrightarrow \text{G-S-S-G} + 2\text{H}^+ + 2e^-$$

**The glutathione is oxidized in this process, so it is a biological reducing agent.**

(d) Write a balanced equation for reaction of glutathione with molecular oxygen, $O_2$, to form G-S-S-G and $H_2O$. Is molecular oxygen oxidized or reduced in this process?

$$2\text{G-SH} \;+\; 1/2\; O_2 \;\longrightarrow\; \text{G-S-S-G} \;+\; H_2O$$

**The molecular oxygen is reduced in this process.**

Problem 17.46 Following is a structural formula for the artificial sweetener aspartame. Each amino acid has the L configuration.

Aspartame

(a) Name the two amino acids in this molecule.

**Aspartame is composed of aspartic acid (Asp) attached via a peptide bond to the methyl ester of phenylalanine (Phe).**

(b) Estimate the isoelectric point of Aspartame?

**Using the values in Table 17.2 for the amino group and side chain carboxylic acid group of aspartic acid leads to pI = 1/2(9.82 + 3.86) = 6.84.**

(c) Write the structural formulas for the products of hydrolysis of aspartame in 1M HCl.

### Three-Dimensional Shapes of Polypeptides and Proteins

Problem 17.47 Examine the α-helix conformation. Are amino acid side chains arranged all inside the helix, all outside the helix, or is their arrangement random?

**All of the amino acid side chains extend outside the helix.**

Problem 17.48 Distinguish between intermolecular and intramolecular hydrogen bonding between the backbone groups on polypeptide chains. In what types of secondary structure do you find intermolecular hydrogen bonds; in what types do you find intramolecular hydrogen bonding?

**Intermolecular hydrogen bonding is possible with β-sheet secondary structures, while only intramolecular hydrogen bonding is possible with α-helix secondary structures.**

<u>Problem 17.49</u>  Many plasma proteins found in aqueous environment are globular in shape. Which amino acid side chains would you expect to find on the surface of a globular protein and in contact with the aqueous environment?  Which would you expect to find inside, shielded from the aqueous environment?  Explain.
(a) Leu                    (b) Arg                    (c) Ser                    (d) Lys          (e) Phe

**In general, charged or hydrophilic amino acids are exposed to the aqueous solution on the surface of a globular protein. Thus, (b) Arg, (c) Ser, and (d) Lys will be on the surface. The hydrophobic amino acids (a) Leu and (e) Phe will generally be inside the protein, shielded from the aqueous environment.**

# CHAPTER 18: NUCLEIC ACIDS

## 18.0 OVERVIEW
• The **genetic information** inside living cells is stored and transmitted in the form of long stretches of deoxyribonucleic acid (DNA) often referred to as **genes**. The genetic information is relayed to the cell in **two stages, transcription of the DNA to ribonucleic acids (RNA)** and then **translation** of the RNA to give **proteins**. ✳

## 18.1 NUCLEOSIDES AND NUCLEOTIDES
• A **nucleoside** is a compound containing a **heterocyclic aromatic amine (base) bonded** to a monosaccharide, **D-ribose or 2-deoxy-D-ribose**, via a **β-$N$-glycoside bond**. ✳
   - The **heterocyclic bases** most common to the nucleic acids are **adenine (A)** and **guanine (G)**, both **purines**, as well as **cytosine (C)**, a **pyrimidine**. DNA also has the pyrimidine **thymidine (T)**, while RNA contains the pyrimidine **uracil (U)**. Uracil has a methyl group at the 5 position on the pyrimidine ring that is not found in thymine, otherwise they are the same.
   - The **β-$N$-glycosidic bond** of nucleosides is between the **anomeric (C-1') carbon atom of the monosaccharide** and the **$N$-1 of the pyrimidine base or $N$-9 of the purine base**, respectively.
   - Nucleosides are named after the heterocyclic base as well as the type of monosaccharide attached. If the monosaccharide is D-ribose, then the common ribonucleosides are named **adenosine, guanosine, cytidine, and uridine**. If the monosaccharide is 2-deoxy-D-ribose then the common deoxyribonucleosides are named **2'-deoxyadenosine, 2'-deoxyguanosine, 2'-deoxycytidine, and 2'-deoxythymidine**.
• A **nucleotide** is a **nucleoside** that has one or more **phosphate groups** attached to one of the hydroxyl groups, usually at the **5' and/or 3' positions**. Nucleotides usually have between one and three phosphate groups attached, and are **named according** to the **nucleoside** present, followed by the position and number of phosphate groups attached. For example, **adenosine 5'-triphosphate (ATP)** is the name given the molecule that has adenosine with three phosphates attached at the 5'-position of the D-ribose ring. The three phosphates groups are held together via phosphate anhydride linkages. The phosphate groups are deprotonated and thus negatively charged at neutral pH. ✳

## 18.2 THE STRUCTURE OF DNA
• There are **three levels of structural complexity of DNA**. ✳
• The **primary structure** of a nucleic acid refers to the **sequence of nucleotides** that are **linked via single phosphate** units between the **5' position of one nucleotide** and the **3' position** of the **adjacent nucleotide**. These chains of nucleotides linked 5' to 3' with phosphates can be extremely long. Because of this structure, the backbone is often referred to as being a **sugar-phosphate backbone**. Because the phosphodiester units have a negative charge, the backbone is polyanionic. ✳
• The **secondary structure of DNA is best thought of as a double helix** as described by Watson and Crick. ✳
   - In the **DNA double helix**, the two strands of nucleic acids are **antiparallel**. In other words, one strand is in the **5' to 3' direction**, while the other is in the **3' to 5' direction**.
   - The **bases project inward toward the axis** of the double helix, and they pair in a specific manner with the bases of the opposite strand. **Guanine makes three specific hydrogen bonds with cytosine, and adenine makes two specific hydrogen**

**bonds with thymine**.  The specific pattern of hydrogen bonding ensures that these are the only two sets of base pairs normally observed.  This means that the two strands are held together by these hydrogen bonds, so the sequences must be complementary for them to be paired in a double helix. ✳
   - Since each of these base pairs is composed of a purine and a pyrimidine, they are the **same general size** and the double helix is relatively regular in structure.
 • There are **several forms** of the **double helix**.
   - The most common type of DNA helix is called the **B-form helix**.  It is a "**right-handed helix**" with so-called minor and major grooves of similar depths but different widths.
   - An **A-form helix** is also known.  It is also a right-handed helix, but it has a **different conformation** of the **2'-deoxy-D-ribose ring** leading to different grooves and a slightly different number of bases per turn of the helix.
   - A so-called **Z-form helix** has also been found that is a **left-handed helix**.
   - The **different types of helices** can be **interconverted** depending on parameters such as **temperature, ionic strength, polarity of the solvent**, and the nature of the **cations** associated with the negatively-charged helix backbone.
 • The **long pieces of DNA** are **flexible** and can exhibit **tertiary structure** called **supercoiling**.  Supercoiling involves a **different number of helical turns** than normal along the DNA helix.  Supercoiling is observed in circular pieces of DNA, as well as in long pieces of linear DNA wound around histone proteins.

## 18.3  RIBONUCLEIC ACIDS (RNA)
 • RNA is different from DNA in three important ways:
   - β-**D-Ribose** is found in **RNA** instead of the β-D-2'-deoxyribose found in DNA.
   - The pyrimidine base **uracil is found in RNA** instead of the thymine found in DNA.
   - **RNA is single-stranded**, rather than double-stranded like DNA.
 • RNA is found in three major forms within the cell, listed in decreasing order of abundance:
   - **Ribosomal RNA (rRNA)** is present in **ribosomes**, the **particles responsible for protein synthesis**.  Ribosomes contain 60% RNA and 40% protein.
   - **Transfer RNA (tRNA)** are relatively small nucleic acid molecules, 73-94 bases in length, that **carry amino acids to the appropriate sites of protein synthesis** on the ribosome.
   - **Messenger RNA (mRNA)** are short-lived, single-stranded pieces of RNA that result from transcription of DNA.  The mRNA serves as the actual template for protein synthesis on the ribosome.

## 18.4  THE GENETIC CODE
 • The **genetic code is a triplet code** since three nucleic acid bases code for a single amino acid or a "stop" signal.  There are $4^3$ **or 64 possible sequences** of three nucleic acid bases, and these code for 20 different amino acids and stop signals.  Therefore, the **genetic code is degenerate** in that **different sequences can code for the same amino acid**, and there are three different "stop" sequences.

## 18.5  SEQUENCING NUCLEIC ACIDS
 • **Sequencing of nucleic acids** is accomplished by **selectively cleaving** a strand of nucleic acids at a given base or bases, then **separating** the cleaved fragments by **electrophoresis on a polyacrylamide gel**.  The individual steps in this process are as follows:
   - Double stranded DNA is cleaved at specific sites using enzymes called **restriction enzymes** that selectively cleave a 4-8 base sequence of DNA.  The resulting fragments, referred to as **restriction fragments**, are purified.  A variety of restriction enzymes are commercially available.

- Both strands of the restriction fragment are **labeled with radioactive $^{32}$P** in the form of phosphate that is added to the **5'-OH group** *via* **an enzyme reaction**. This produces a restriction fragment that has two $^{32}$P labels; one $^{32}$P on the 5' end of each strand.
- In order for sequencing to take place in an unambiguous manner, only one strand can have a $^{32}$P label. This is because trying to carry out sequencing reactions with two different $^{32}$P labels on the same sequence of DNA will lead to double $^{32}$P signals on the polyacrylamide gel.

    In theory, the two labeled single strands could be separated by heating, then isolated as single-stranded pieces of DNA to be sequenced individually. In practice, this is rarely done because of the difficulty associated with the isolation of single-stranded fragments of the same length.

    More often, a new restriction enzyme is used to cut the doubly $^{32}$P-labeled fragment in the middle, thereby creating two new shorter fragments, each with only a single $^{32}$P label at the 5' end of one of the strands. These singly-labeled fragments are easily separated and isolated because they are generally of different lengths. The singly-labeled fragments are sequenced individually.

- The singly-labeled DNA fragments are placed into separate equivalent samples, that are then subjected to **limited base-specific cleavage reactions**. Each sample undergoes a reaction with different base specificity, and conditions are adjusted so that, on average, **each strand is cleaved only once**.

    There are chemical, base-specific cleavage reactions that cleave the DNA at **G, G or A, C, and T residues**. These reactions are run separately.

- The cleaved fragments are subjected to **electrophoresis on a polyacrylamide gel**, where they separate according to size. The **radioactivity from the $^{32}$P label is used to visualize** the different fragments by exposing the finished gel to **photographic film**.
- Shorter pieces of DNA run faster than longer pieces of DNA on the polyacrylamide gels. Therefore, the **lengths of the fragments** in the different lanes correspond to **locations of the different bases**. For example, the G cleavage reaction will generate fragments of DNA corresponding in length to the locations of the G residues, and so on. When the lengths of the fragments for each of the different cleavage reaction are compared, an entire sequence can be determined. Up to 400 or more bases can be sequenced on a single polyacrylamide gel.
- A different strategy based on an enzyme called a polymerase and chain terminating 2',3'-dideoxy nucleotides has been developed to generate labeled sequences of different lengths for use in sequencing. It is this latter method that is generally used in automated DNA sequencing.

# CHAPTER 18
## *Solutions to the Problems*

Problem 18.1 Draw structural formulas for these compounds.
(a) 2'-Deoxythymidine 5'-monophosphate

(b) 2'-Deoxythymidine 3'-monophosphate

<u>Problem 18.2</u> Write the structural formula for the section of DNA that contains the base sequence CTG and is phosphorylated on the 3' end only.

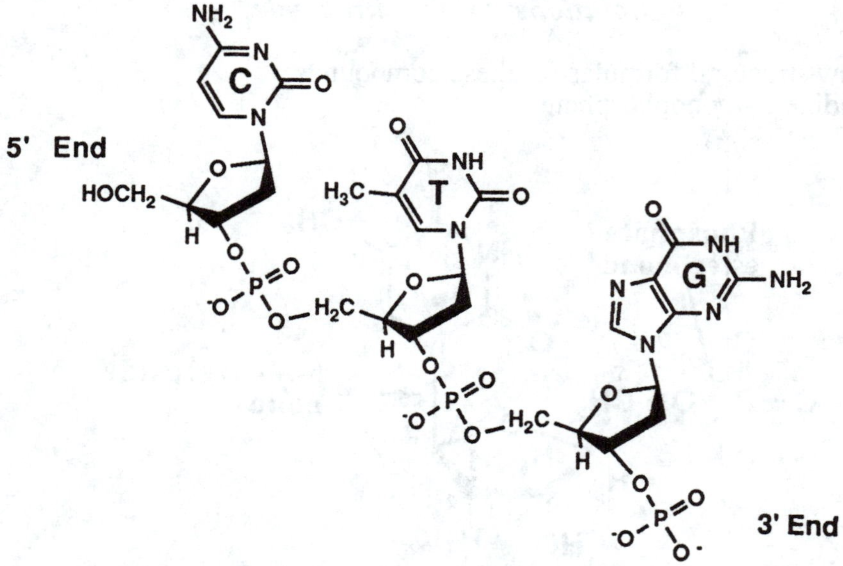

<u>Problem 18.3</u> Write the complementary DNA base sequence for 5'-CCGTACGA-3'.

**The complementary sequence would be 3'-GGCATGCT-5'**

<u>Problem 18.4</u> Here is a portion of the nucleotide sequence in phenylalanine tRNA.

3'-ACCACCUGCUCAGGCCUU-5'

Write the nucleotide sequence of its DNA complement.

**Remember that the base uracil (U) in RNA is complementary to adenine (A) in DNA. The complement DNA sequence of the above RNA sequence would be:**

**5'-TGGTGGACGAGTCCGGAA-3'.**

<u>Problem 18.5</u> The following section of DNA codes for oxytocin, a polypeptide hormone.

3'-ACG-ATA-TAA-GTT-TTA-ACG-GGA-GAA-CCA-ACT-5'

(a) Write the base sequence of the mRNA synthesized from this section of DNA.

**The base sequence of the mRNA synthesized from this section of DNA would be:**

**5'-UGC-UAU-AUU-CAA-AAU-UGC-CCU-CUU-GGU-UGA-3'**

(b) Given the sequence of bases in part (a), write the primary structure of oxytocin.

**The primary sequence of oxytocin would be:**

**Amino terminus- Cys-Tyr-Ile-Gln-Asn-Cys-Pro-Leu-Gly -Carboxyl terminus**

**Note how the last codon, UGA, does not code for an amino acid, but rather is the stop signal.**

<u>Problem 18.6</u> The following is another section of the bovine rhodopsin gene. Which of the endonucleases given in Example 18.6 will catalyze cleavage of this section.

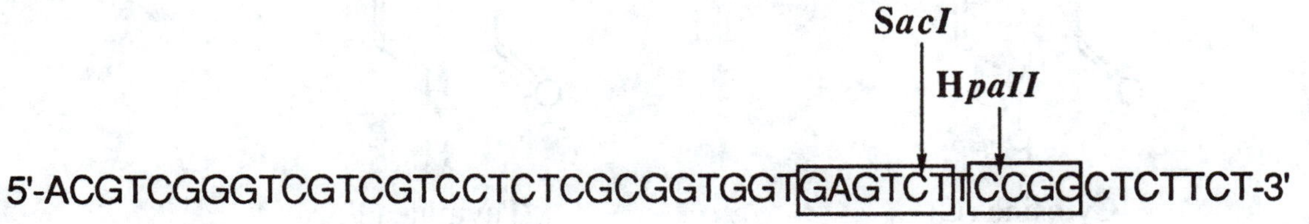

*SacI*

*HpaII*

5'-ACGTCGGGTCGTCGTCCTCTCGCGGTGGT GAGTCTT CCGG CTCTTCT-3'

**The *SacI* and *HpaII* cleavage sites are shown on the sequence above.**

<u>Problem 18.7</u> In what order will the excision fragments in Example 18.7 appear on the developed photographic plate? Remember that only the 5' end of the original restriction fragment is labeled with phosphorus-32.

**On the polyacrylamide gel, shorter fragments migrate faster. Thus, fragment (i) will be closest to the bottom, fragment (ii) will be in the middle, and fragment (iii) will be closest to the top of the gel.**

<u>**Nucleosides and Nucleotides**</u>

<u>Problem 18.8</u> Two important drugs in the treatment of acute leukemia are 6-mercaptopurine and 6-thioguanine. In each of these drugs, the oxygen at carbon 6 of the parent molecule is replaced by divalent sulfur. Draw structural formulas for the enethiol forms of 6-mercaptopurine and 6-thioguanine.

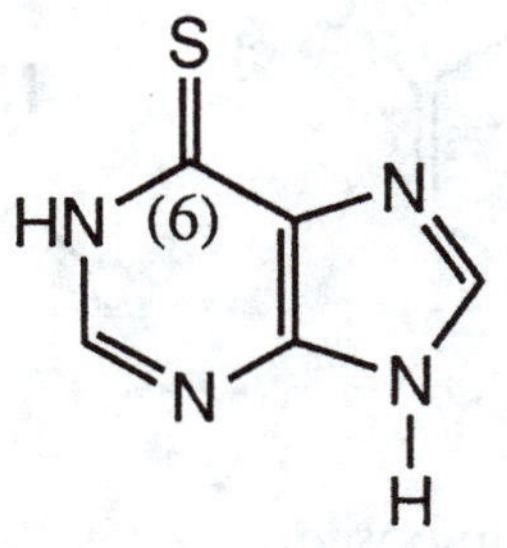

6-Mercaptopurine 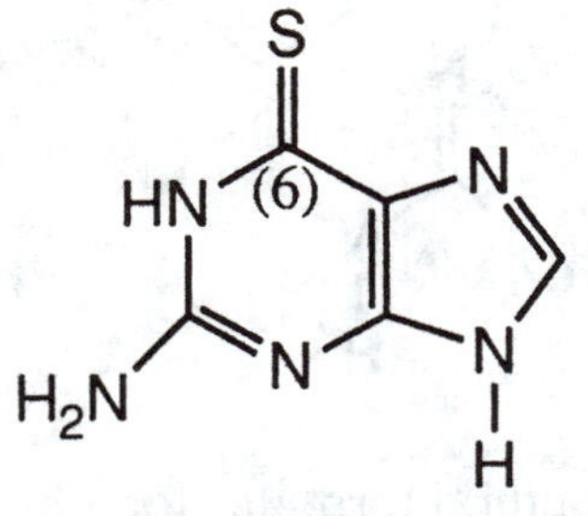          6-Thioguanine

**The enethiol forms are shown below:**

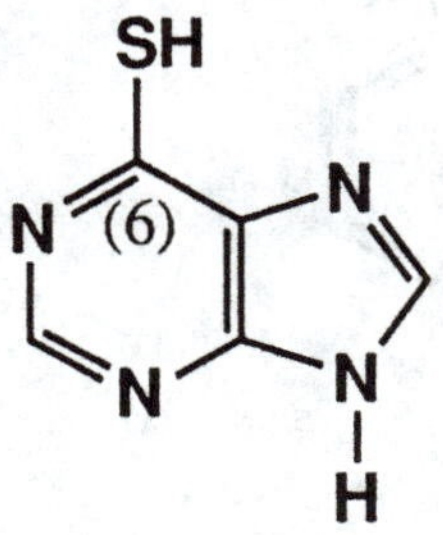

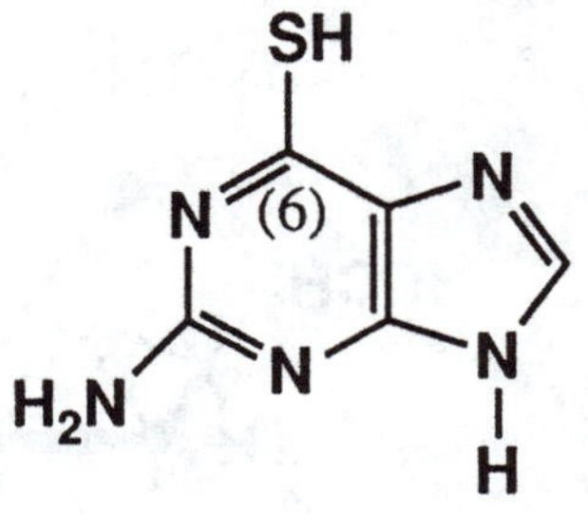

**Problem 18.9** Following are structural formulas for cytosine and thymine. Draw two additional tautomeric forms for cytosine and three additional tautomeric forms for thymine.

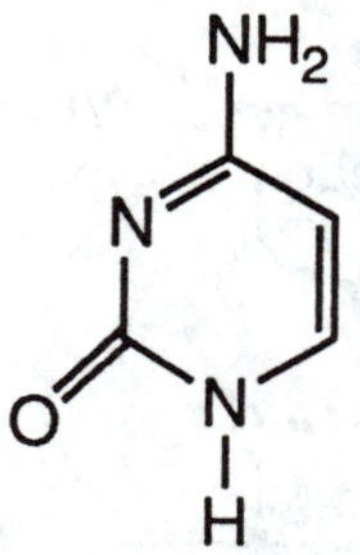

Cytosine (C) 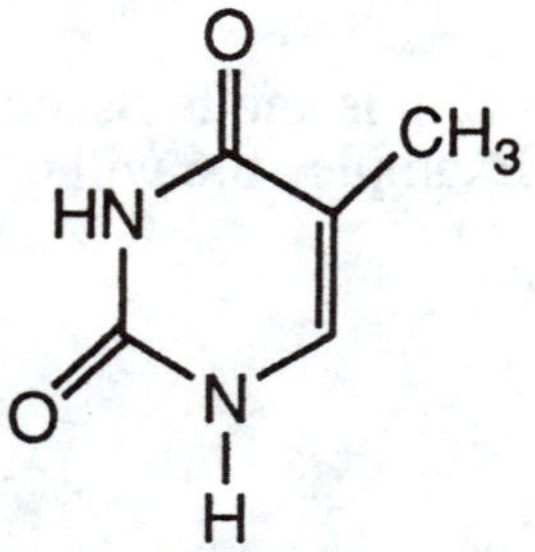                                    Thymine (T)

**Three additional tautomeric forms for cytosine are shown here:**

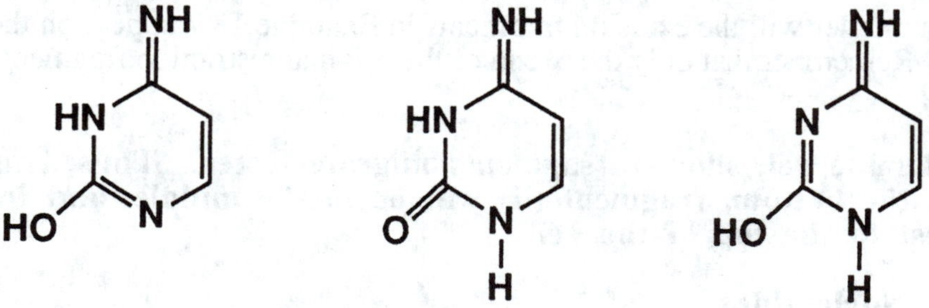

**Four additional tautomeric forms for thymine are shown here:**

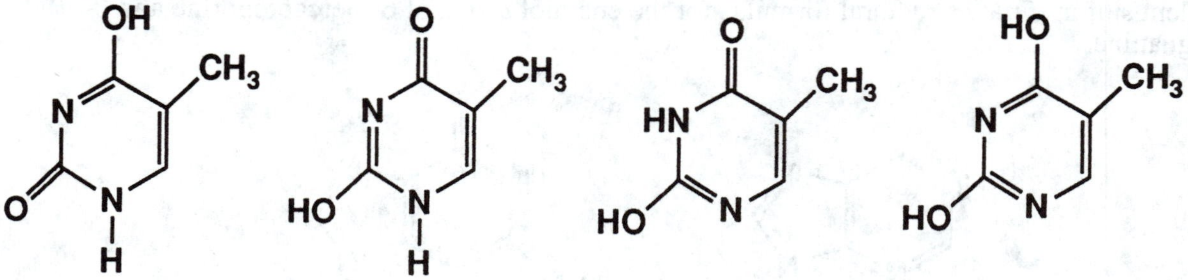

**Problem 18.10** Draw structural formulas for a nucleoside composed of
(a) β-D-Ribose and adenine

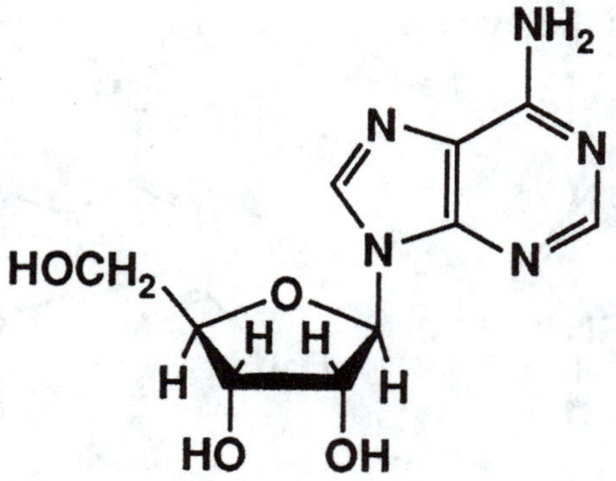

(b) β-D-Deoxyribose and cytosine

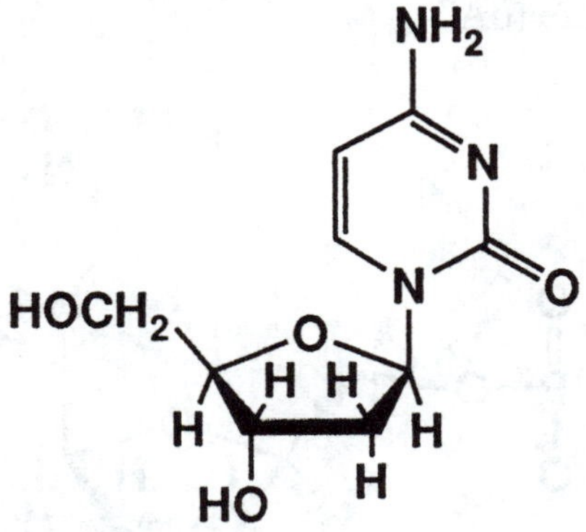

<u>Problem 18.11</u> Nucleosides are stable in water and in dilute base. In dilute acid, however, the glycoside bond of a nucleoside undergoes hydrolysis to give a pentose and a heterocyclic aromatic amine base. Propose a mechanism for this acid-catalyzed hydrolysis.

**Acid-catalyzed glycoside bond hydrolysis is most pronounced for purine nucleosides. A reasonable mechanism involves protonation of the heterocyclic base to create a good leaving group that is displaced by water to produce the product pentose and free base. The reaction of guanosine in acid is shown below.**

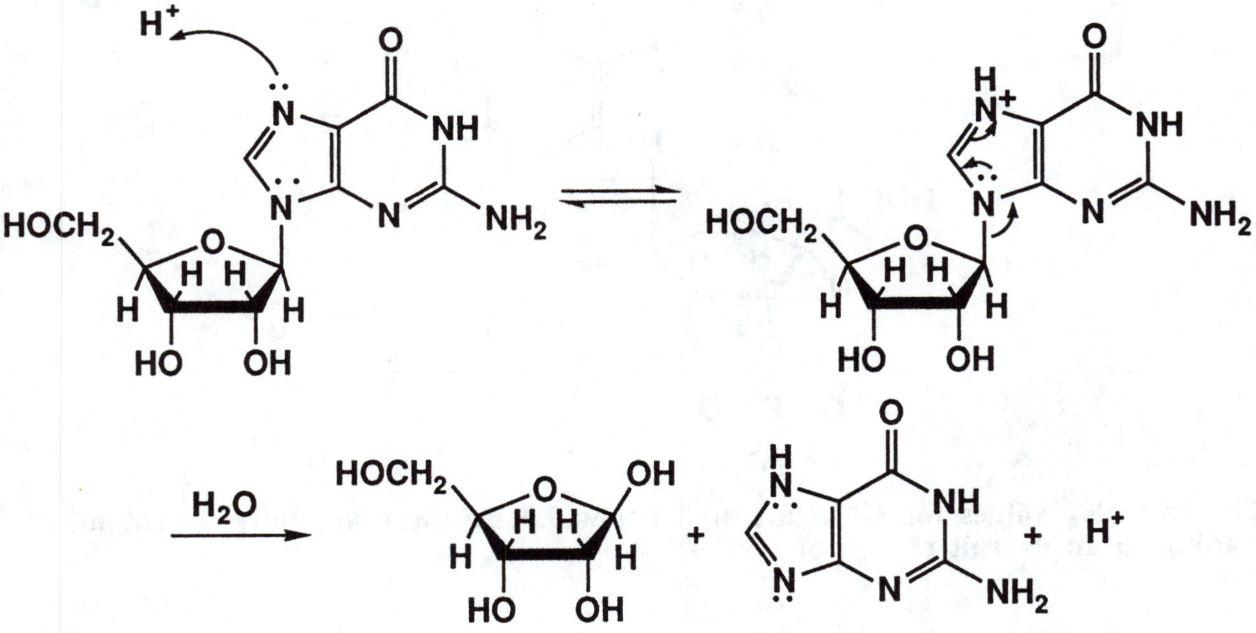

<u>Problem 18.12</u> Explain the difference in structure between a nucleoside and a nucleotide.

**A nucleoside consists of a D-ribose or 2'-deoxy-D-ribose bonded to an heterocylic aromatic base by a β-N-glycoside bond. A nucleotide is a nucleoside that has one or more molecules of phosphoric acid esterified at an -OH group of the monosaccharide, usually at the 3' and/or 5' -OH group.**

**Problem 18.13** Write the structural formula for each nucleotide and estimate its net charge at pH 7.4, the pH of blood plasma.

(a) 2'-Deoxyadenosine 5'-triphosphate (dATP)

The values for the first three $pK_a$'s of dATP are all below 5.0, so these are fully deprotonated at pH 7.4. The fourth $pK_a$ of dATP is 7.0, so that at pH 7.4 there is approximately a 70:30 ratio of species with a net charge of -4 or -3, respectively. This ratio was determined using the Henderson-Hasselbalch equation (Section 17.2).

(b) Guanosine 3'-monophosphate (GMP)

The two $pK_a$ values for GMP are well below 7.4, so these are fully deprotonated, leading to an overall charge of -2.

(c)  2'-Deoxyguanosine 5'-diphosphate (dGDP)

The values for the first two pK$_a$'s of dGDP are all below 5.0, so these are fully deprotonated at pH 7.4.  The third pK$_a$ of dATP is 6.7, so that at pH 7.4 there is approximately a 83:17 ratio of species with a net charge of -3 or -2, respectively. This ratio was determined using the Henderson-Hasselbalch equation (Section 17.2).

Problem 18.14  Cyclic-AMP, first isolated in 1959, is involved in many diverse biological processes as a regulator of metabolic and physiological activity.  In it, a single phosphate group is esterified with both the 3' and 5' hydroxyls of adenosine.  Draw the structural formula of cyclic-AMP.

**Cyclic-AMP**

## The Structure of DNA
Problem 18.15  Why are deoxyribonucleic acids called acids?  What are the acidic groups in their structure?

Deoxyribonucleic acids are called acids because the phosphodiester groups of the backbone are acidic.  At neutral pH, they are fully deprotonated, leading to the anionic nature of DNA.

<u>Problem 18.16</u> Human DNA contains approximately 30.4% A. Estimate the percentages of G, C, and T and compare them with the values presented in Table 18.1.

**The A residues must be paired with T residues, so estimate that there is also 30.4% T. A and T must therefore account for 30.4% + 30.4% = 60.8% of the bases. That leaves (100% - 60.8%) / 2 = 39.2% / 2 = 19.6% each for G and C. In Table 18.1, there is actually slightly less T than expected, so there is also slightly more G and C than expected.**

<u>Problem 18.17</u> Draw the structural formula of the DNA tetranucleotide 5'-A-G-C-T-3'. Estimate the net charge on this tetranucleotide at pH = 7.0. What is the complementary tetranucleotide to this sequence?

**As shown in the preceding structure, there is a net charge of -5 on this tetranucleotide at pH 7.0. This oligonucleotide is self-complementary, that is the complementary oligonucleotide also has the sequence 5'-A-G-C-T-3'.**

<u>Problem 18.18</u> List the postulates of the Watson-Crick model of DNA structure.

**Major postulates of the Watson-Crick model are that:**
1) **A molecule of DNA consists of two antiparallel strands coiled in a right handed manner about the same axis, thereby creating a double helix.**
2) **The bases project inward toward the helix axis.**
3) **The bases are paired through hydrogen bonding, with a purine paired to a pyrimidine so that each base pair is of the same size and shape.**
4) **In particular A pairs with T and G pairs with C.**
5) **The paired bases are stacked one on top of another in the interior of the double helix.**
6) **There is a distance of 0.34 nm between adjacent stacked paired bases.**
7) **There are ten paired bases per turn of the helix, and these are slightly offset from each other. The slight offset provides two grooves of different dimensions along the helix, the so-called major and minor grooves.**

Problem 18.19  The Watson-Crick model is based on certain experimental observations of base composition and molecular dimensions.  Describe these observations and show how the Watson-Crick model accounts for each.

**Chargaff found that in different organisms, the amount of A always equals the amount of T and the amount of G always equals the amount of C, even though different organisms have different ratios of A to G.  The base-pairing postulates of the Watson-Crick model fully explain the observed ratios of bases.  The geometry of the Watson-Crick model also accounts perfectly for the periodicity observed in the X-ray diffraction data.**

Problem 18.20  Compare the α-helix of proteins and the double helix of DNA in these ways.
(a)  The units that repeat in the backbone of the polymer chain.

**The α-helix of a protein is composed of amino acids, so the repeating unit of the backbone is a carboxyl group bonded to a tetrahedral carbon atom and a nitrogen atom.  Of course, the carboxyl group and nitrogen atoms are linked in amide bonds.  The repeating unit of the double helix in DNA is a 2'-deoxy-D-ribose unit linked in 3'-5' phosphodiester bonds.**

(b)  The projection in space of substituents along the backbone (the R groups in the case of amino acids; purine and pyrimidine bases in the case of double-stranded DNA) relative to the axis of the helix.

**The α-helix of a protein has the R groups pointed out away from the helix axis.  The DNA bases of the double helix are pointed inward, toward the helix axis.**

Problem 18.21  Discuss the role of the hydrophobic interactions in stabilizing:
(a)  Double-stranded DNA

**In the DNA double helix, the relatively hydrophobic bases are stacked on the inside, surrounded by the relatively hydrophilic sugar-phosphate backbone that is on the outside of the structure.  The stacking of the hydrophobic bases minimizes contact with water.**

(b)  Lipid bilayers

**In lipid bilayers, the hydrophobic hydrocarbon tails are associated with each other to form the hydrophobic inner layer, while the polar head groups are associated with each other on both outside surfaces.**

(c)  Soap micelles

**In micelles, the hydrophobic hydrocarbon tails are associated with each other to form the hydrophobic interior, while the polar groups are associated with each other on the outside surface.**

Problem 18.22 Name the type of covalent bond(s) joining monomers in these biopolymers.
(a) Polysaccharides    (b) Polypeptides        (c) Nucleic acids

**Polysaccharides have glycosidic linkages, polypeptides have amide linkages and nucleic acids have phosphodiester linkages between the monomers, respectively.**

Problem 18.23 In terms of hydrogen bonding, which is more stable, an A-T base pair or a G-C base pair?

**A G-C base pair is held together by 3 hydrogen bonds, while an A-T base pair is held together by only two hydrogen bonds.  Thus, in terms of hydrogen bonds alone, a G-C base pair is more stable than an A-T base pair.**

Problem 18.24 At elevated temperatures, nucleic acids become denatured, that is, they unwind into single-stranded DNA.  Account for the observation that the higher the G-C content of a nucleic acid, the higher the temperature required for its thermal denaturation.

**G-C base pairs have three hydrogen bonds between them, while A-T base pairs have only two.  Thus, the G-C base pairs are held together with stronger overall attractive forces and require higher temperatures to denature.**

Problem 18.25 Write the DNA complement for 5'-ACCGTTAAT-3'.  Be certain to label which is the 5' end and which is the 3' end of the complement strand.

**The complementary sequence is 3'-TGGCAATTA-5'**

Problem 18.26 Write the DNA complement for 5'-TCAACGAT-3'.

**The complementary sequence is 3'-AGTTGCTA-5'**

## Ribonucleic Acids (RNA)
Problem 18.27 Compare the degree of hydrogen bonding in the base pair A-T found in DNA with that in the base pair A-U found in RNA.

**The only difference between uracil (U) and thymine (T) is the presence of a methyl group at the 5 position of thymine, that is absent in uracil.  As can be seen in the structures, the presence or absence of this methyl group has very little influence on hydrogen bonding.**

<u>Problem 18.28</u>  Compare DNA and RNA in these ways:
(a)  Monosaccharide units

**DNA contains 2'-deoxy-D-ribose units, while RNA contains D-ribose units.**

(b)  Principal purine and pyrimidine bases

| DNA | | RNA | |
|---|---|---|---|
| **Purines** | **Pyrimidines** | **Purines** | **Pyrimidines** |
| Adenine | Thymine | Adenine | Uracil |
| Guanine | Cytosine | Guanine | Cytosine |

(c)  Primary structure

**The monosaccharide unit in DNA is 2'-deoxy-D-ribose, the monosaccharide in RNA is D-ribose. The bases are the same between the two types of nucleic acids, except thymine is found in DNA while uracil is found in RNA. DNA is usually double stranded and RNA is primarily single stranded. In both DNA and RNA, the primary sequence consists of linear chains of the nucleic acids linked by phosphodiester bonds involving the 3' and 5' hydroxyl groups of the monosaccharide units.**

(d)  Location in the cell

**DNA is found in cell nuclei, while the bulk of RNA occurs as ribosome particles in the cytoplasm.**

(e)  Function in the cell

**DNA serves to store and transmit genetic information, and RNA is primarily involved with the transcription and translation of that genetic information during the production of proteins.**

<u>Problem 18.29</u>  What type of RNA has the shortest lifetime in cells?

**Messenger RNA has the shortest lifetime in cells, usually on the order of a few minutes or less. This short lifetime is thought to allow for very tight control over how much protein is produced in the cell at any one time.**

<u>Problem 18.30</u>  Write the mRNA complement for 5'-ACCGTTAAT-3'.  Be certain to label which is the 5' end and which is the 3' end of the mRNA strand.

**The mRNA complement would be 3'-UGGCAAUUA-5'**

<u>Problem 18.31</u>  Write the mRNA complement for 5'-TCAACGAT-3'.

**The mRNA complement would be 3'-AGUUGCUA-5'**

## The Genetic Code

Problem 18.32  What does it mean to say that the genetic code is degenerate?

**The genetic code is referred to as degenerate because more than one codon can code for the same amino acid.  This is because there are 64 different codons, but only twenty amino acids and a stop signal for which coding is needed.**

Problem 18.33  Write the mRNA codons for
(a) Valine  **GUU, GUC, GUA, GUG**   (b) Histidine  **CAU, CAC**
(c) Glycine    **GGU, GGC, GGA, GGG**

Problem 18.34  Aspartic acid and glutamic acid have carboxyl groups on their side chains and are called acidic amino acids.  Compare the codons for these two amino acids.

**All of the codons for these two acidic amino acids begin with GA.  The codons for aspartic acid are GAU and GAC, while the codons for glutamic acid are GAA and GAG.**

Problem 18.35  Compare the structural formulas of the amino acids phenylalanine and tyrosine. Compare also the codons for these two amino acids.

**Phenylalanine**                                    **Tyrosine**

**Phenylalanine has a phenyl group while tyrosine has a phenol group.  The mRNA codons for phenylalanine are UUU and UUC, while the mRNA codons for tyrosine are UAU and UAC.**

Problem 18.36  Glycine, alanine, and valine are classified as nonpolar amino acids.  Compare the codons for these three amino acids.  What similarities do you find?  What differences do you find?

| Glycine | Alanine | Valine |
|---------|---------|--------|
| GGU | GCU | GUU |
| GGC | GCC | GUC |
| GGA | GCA | GUA |
| GGG | GCG | GUG |

**All of these amino acids have four mRNA codons, all codons start with G, and in each case, the first two bases of the codon are identical for a given amino acid. This makes the last base irrelevant.**

Problem 18.37 Codons in the set CUU, CUC, CUA, and CUG all code for the amino acid leucine. In this set, the first and second bases are identical, and the identity of the third base is irrelevant. For what other sets of codons is the third base also irrelevant, and for what amino acid(s) does each set code?

**The third base is also irrelevant for GUX (valine), GCX (alanine), GGX (glycine), ACX (threonine), CCX (proline), CGX (arginine), and UCX (serine). In the preceding codons, X stands for any of the bases.**

Problem 18.38 Compare the codons with a pyrimidine, either U or C, as the second base. Do the majority of the amino acids specified by these codons have hydrophobic or hydrophilic side chains?

**The majority of amino acids with a pyrimidine in the second position of their codons are hydrophobic. This set contains phenylalanine, leucine, isoleucine, methionine, valine, proline, and alanine. Only serine and threonine have a pyrimidine in the second position and also have a somewhat hydrophilic side chain.**

Problem 18.39 Compare the codons with a purine, either A or G, as the second base. Do the majority of the amino acids specified by these codons have hydrophilic or hydrophobic side chains?

**The majority of amino acids with a purine in the second position of their codons are hydrophilic. This set contains histidine, glutamine, asparagine, lysine, aspartic acid, glutamic acid, arginine, cysteine, and serine. Only glycine and tryptophan are not hydrophilic, while tyrosine is a special case that is aromatic with a polar group.**

Problem 18.40 What polypeptide is coded for by this mRNA sequence?

5'-GCU-GAA-GUC-GAG-GUG-UGG-3'

**This mRNA codes for the following polypeptide:**

**Amino terminus- Ala-Glu-Val-Glu-Val-Trp -Carboxyl terminus.**

Problem 18.41 The alpha chain of human hemoglobin has 141 amino acids in a single polypeptide chain. Calculate the minimum number of bases on DNA necessary to code for the alpha chain. Include in your calculation the bases necessary for specifying termination of polypeptide synthesis.

**The minimum number of bases needed for the alpha chain of human hemoglobin must code for the 141 amino acids as well as three extra bases for the stop codon. Therefore, the minimum number of bases that will be required is (3 x 141) + (1 x 3) = 426 bases.**

<u>Problem 18.42</u> In HbS, the human hemoglobin found in individuals with sickle-cell anemia, glutamic acid at position 6 in the beta chain is replaced by valine.
(a) List the two codons for glutamic acid and the four codons for valine.

**The two mRNA codons for glutamic acid are GAA and GAG, while the four mRNA codons for valine are GUU, GUC, GUA, and GUG.**

(b) Show that one of the glutamic acid codons can be converted to a valine codon by a single substitution mutation, that is, by changing one letter in the codon.

**Both of the glutamic acid codons can be converted to valine by replacing the central A with a U residue.**

# CHAPTER 19: THE ORGANIC CHEMISTRY OF METABOLISM

## SUMMARY OF REACTIONS

| Starting Material \ Product → | Acetyl CoA + $CO_2$ + NADH | Acetyl CoA + NADH + $FADH_2$ | Ethanol + $CO_2$ + $NAD^+$ | Lactate + $NAD^+$ | Pyruvate + NADH + ATP |
|---|---|---|---|---|---|
| Fatty Acid + $NAD^+$ + Acetyl CoA + FAD | | **19A** 19.3* | | | |
| Glucose + $NAD^+$ + $HPO_4^{2-}$ + ADP | | | | | **19B** 19.6 |
| Pyruvate + NADH + $H^+$ | | | **19C** 19.7A | **19D** 19.7B | |
| Pyruvate + $NAD^+$ + Acetyl CoA | **19E** 19.7C | | | | |

*Section in book that describes reaction.

## REACTION 19A: β-OXIDATION OF FATTY ACIDS (Section 19.3)

$$CH_3(CH_2)_{14}\overset{O}{\overset{\|}{C}}OH + 8\ CoA\text{-}SH \xrightarrow[\ \ \ \ ]{ATP \quad AMP + P_2O_7^{4-}} 8\ CH_3\overset{O}{\overset{\|}{C}}SCoA + 7\ NADH$$

$$+ 7\ NAD^+ + 7\ FAD \qquad\qquad\qquad + 7\ FADH_2$$

- The overall process of **fatty acid β-oxidation** involves **conversion of a fatty acid** in the presence of coenzyme A (CoA-SH) to molecules of the thioester species **acetyl coenzyme A** along with **reduction of NAD+ and FAD.** ✳
- The first step in the process is the conversion of the fatty acid into an activated form as the thioester derivative of CoA-SH in the cytoplasm. This reaction requires the hydrolysis of ATP to AMP and the pyrophosphate ion. An acyl-AMP mixed anhydride is an intermediate in the process.
- The activated fatty acid is transported into mitochondria where the following four enzyme-catalyzed reactions take place. *[It is very helpful to follow the chemical "logic" of the steps in fatty acid β-oxidation. Notice how the acyl fatty acid is first oxidized to a create a new double bond that is then hydrated and oxidized to set up the reverse Claisen type of cleavage reaction.]*
    1) The alpha- and beta- carbons of the fatty acid chain are oxidized to a double bond while FAD is reduced to $FADH_2$. Enzyme: fatty acyl-CoA dehydrogenase.
    2) The double bond is hydrated to give a β-hydroxyacyl-CoA. The hydroxyl group is added stereoselectively to carbon 3 producing exclusively the R stereoisomer. Enzyme: Enoyl-CoA hydrase.

3) The β-hydroxy group is oxidized to a ketone while NAD$^+$ is reduced to NADH. Enzyme: (R)-β-hydroxyacyl-CoA dehydrogenase.

4) The carbon chain is cleaved between carbons 2 and 3 by an enzymatic reaction that is the functional equivalent of a reverse Claisen condensation using thioesters. The reaction produces the two-carbon fragment acetyl-CoA and an acyl-CoA that is now two carbons shorter than the original fatty acid. Enzyme: thiolase.

- This new shorter acyl-CoA undergoes additional cycles of reactions 1) - 4) until the entire fatty acid is converted to acetyl-CoA.

## REACTION 19B: GLYCOLYSIS (Section 19.6)

$$C_6H_{12}O_6 + 2\ NAD^+ + 2\ HPO_4^{2-} \xrightarrow{\text{Glycolysis}} 2\ CH_3\overset{\overset{\displaystyle O}{\|}}{C}CO_2^- + 2\ NADH + 2\ ATP$$

Glucose          + 2 ADP                                    Pyruvate

- **Glycolysis** is a metabolic pathway for converting the monosaccharide **glucose** into **2 molecules of pyruvate**, along with the **reduction of two molecules of NAD$^+$ to NADH** and the **synthesis of two molecules of ATP** from 2 molecules of ADP and phosphoric acid. It is the synthesis of the **high energy phosphoric anhydride bonds of ATP** that accounts for the energy harvesting of glycolysis. From an evolutionary standpoint, glycolysis is a very old anaerobic metabolic process for producing energy from nutrient molecules and for providing precursors to aerobic pathways such as the tricarboxylic acid cycle. ✳

- The ten enzyme catalyzed reactions of glycolysis are listed below. The first five steps are preparing the molecules for the last five steps, especially steps 7 and 10, which are the energy harvesting steps. *[It is very helpful to follow the chemical "logic" of the steps in glycolysis. Notice how the six carbon sugar is set up for cleavage via a reverse aldol type of reaction. Also notice how high energy bonds are formed then used to drive production of ATP. Finally, notice how keto-enol tautomerization is used several times during the process.]*

1) Transfer of a phosphoric acid group from ATP to the -OH on C6 of glucose. This step is exothermic since a high energy phosphoric anhydride bond is broken and a lower energy phosphoric ester bond to glucose is created. Enzyme: hexokinase.

2) The α-D-glucose-6-phosphate is isomerized to α-D-fructose-6-phosphate via a keto-enol tautomerization to form an enediol intermediate. Enzyme: phosphoglucoisomerase.

3) α-D-Fructose-6-phosphate is phosphorylated at the -OH group on C1 to form α-D-fructose-1,6-diphosphate. Enzyme: phosphofructokinase.

4) α-D-Fructose-1,6-diphosphate is cleaved into dihydroxyacetone phosphate and glyceraldehyde 3-phosphate. There is an imine intermediate in the reaction, which is the functional equivalent of a reverse aldol reaction. Enzyme: aldolase.

5) Dihydroxyacetone phosphate is isomerized to glyceraldehyde 3-phosphate via a keto-enol tautomerization to form an enediol intermediate. Enzyme: triose phosphate isomerase.

6) The aldehyde group of glyceraldehyde 3-phosphate is first oxidized to a carboxylic acid derivative in the form of a thioester with the thiol group of coenzyme A. NAD$^+$ is reduced to NADH in the process. The activated thioester is then converted to a high energy phosphoric anhydride, 1,3-bisphosphoglycerate. Enzyme: glyceraldehyde-3-phosphate dehydrogenase.

7) The high energy acyl-phosphoric mixed anhydride bond of 1,3-bisphosphoglycerate is converted to a phosphoric anhydride bond of ATP to create 3-phosphoglycerate. Enzyme: phosphoglycerate kinase.

8) 3-Phosphoglycerate is isomerized to 2-phosphoglycerate. Enzyme: phosphoglycerate mutase.

9) The 3-OH group of 2-Phosphoglycerate is lost in a dehydration to give a 2-3 double bond in the product phosphoenolpyruvate. Note that the phosphate group keeps the molecule in the relatively high energy enol form. Enzyme: enolase.

10) The phosphate group of phosphoenolpyruvate is transferred to ADP to generate a high energy phosphoric anhydride in ATP and the ketone pyruvate. Notice how pyruvate is just the more stable keto form of enol pyruvate. Enzyme: pyruvate kinase.

## REACTION 19C: REDUCTION OF PYRUVATE TO ETHANOL - ALCOHOL FERMENTATION (Section 19.7B)

$$\underset{\text{Pyruvate}}{CH_3\overset{O}{\overset{\|}{C}}CO_2^-} + 2\,H^+ + NADH \xrightarrow[\text{Fermentation}]{\text{Alcoholic}} \underset{\text{Ethanol}}{CH_3CH_2OH} + CO_2 + NAD^+$$

- **Yeast and other organisms** can **convert pyruvate to ethanol and $CO_2$** in the **absence of oxygen.** That is why fermentation of beer and wine must be carried out in sealed vessels that exclude air. The $CO_2$ is responsible for the natural carbonation of beverages such as beer and champagne. There are two steps involved; a decarboxylation to give $CO_2$ and acetaldehyde followed by reduction of the acetaldehyde to ethanol. Enzymes: pyruvate decarboxylase, alcohol dehydrogenase. ✳

## REACTION 19D: REDUCTION OF PYRUVATE TO LACTATE (Section 19.7A)

$$\underset{\text{Pyruvate}}{CH_3\overset{O}{\overset{\|}{C}}CO_2^-} + H^+ + NADH \underset{}{\overset{\text{Lactate}}{\underset{\text{Dehydrogenase}}{\rightleftharpoons}}} \underset{\text{Lactate}}{CH_3\overset{OH}{\overset{|}{C}}HCO_2^-} + NAD^+$$

- In vertebrates, **pyruvate is converted to lactate,** an **important anaerobic process** for **regenerating NAD+ from NADH.** In the overall process of lactate fermentation, glucose is converted all the way to two molecules of lactic acid, a relatively strong acid that is fully dissociated at the usual pH of blood. Thus, this process generates lactate and protons. The buildup of lactate is associated with muscle fatigue. Enzyme: lactate dehydrogenase. ✳

## REACTION 19E: OXIDATIVE DECARBOXYLATION OF PYRUVATE TO ACETYL-CoA (Section 19.7C)

$$\underset{\text{Pyruvate}}{CH_3\overset{O}{\overset{\|}{C}}CO_2^-} + NAD^+ + CoA\text{-}SH \xrightarrow[\text{Decarboxylation}]{\text{Oxidative}} CH_3\overset{O}{\overset{\|}{C}}SCoA + CO_2 + NADH$$

- **Under aerobic conditions, pyruvate is oxidized and decarboxylated to give Acetyl CoA** and $CO_2$. The acetyl-CoA then becomes fuel for the tricarboxylic acid cycle. Enzyme: pyruvate dehydrogenase complex.

## SUMMARY OF IMPORTANT CONCEPTS

### 19.0 OVERVIEW

- The biochemical pathways of **metabolism** such as **β-oxidation of fatty acids and glycolysis** involve numerous enzyme catalyzed reactions that operate in sequential fashion to effect complex overall processes. **These pathways are actually the biochemical equivalents of organic functional group reactions that have been covered in the previous chapters of the book.** ✳

### 19.1 FIVE KEY PARTICIPANTS IN GLYCOLYSIS AND β-OXIDATION

- **ATP, ADP, and AMP** are **phosphorylated** derivatives of the nucleoside **adenosine**. AMP has a single phosphoric acid group attached to the 5'-OH group of adenosine. ADP has an addition phosphoric acid group attached through a phosphoric anhydride bond. ATP has a third phosphoric acid group attached to the other two through a second phosphoric anhydride bond. ✳
  - **ATP, ADP, and AMP** are involved with **transfer and storage of phosphoric acid groups** within the cell. ✳
- **NAD⁺, nicotinamide adenine dinucleotide**, is composed of a unit of **ADP joined by a phosphoric ester bond** to the terminal **-CH₂OH group of β-D-ribofuranose** that is linked to **nicotinamide via a β-N-glycosyl bond**. **NAD⁺ is a two electron oxidizing agent**, since it is reduced by two electrons and a proton to give NADH. In NADH, the nicotinamide is reduced. Notice that in the following structure, the proton and electrons are shown as being independent of one another. In actual enzyme reactions, they can be considered to be combined in the form of a hydride, H⁻. ✳

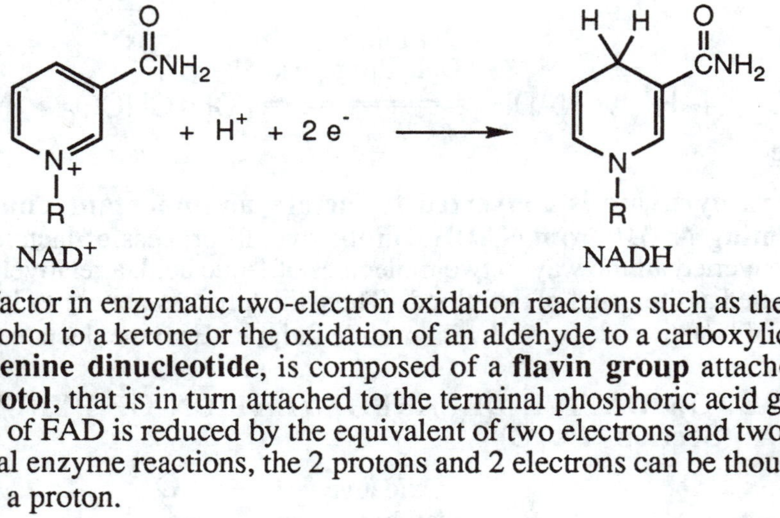

NAD⁺                                          NADH

  - NAD⁺ is a cofactor in enzymatic two-electron oxidation reactions such as the oxidation of a secondary alcohol to a ketone or the oxidation of an aldehyde to a carboxylic acid.
- **FAD, flavin adenine dinucleotide**, is composed of a **flavin group** attached to the five carbon sugar **ribotol**, that is in turn attached to the terminal phosphoric acid group of **ADP**. The flavin group of FAD is reduced by the equivalent of two electrons and two protons to give FADH₂. In actual enzyme reactions, the 2 protons and 2 electrons can be thought of as a hydride, H⁻, and a proton.

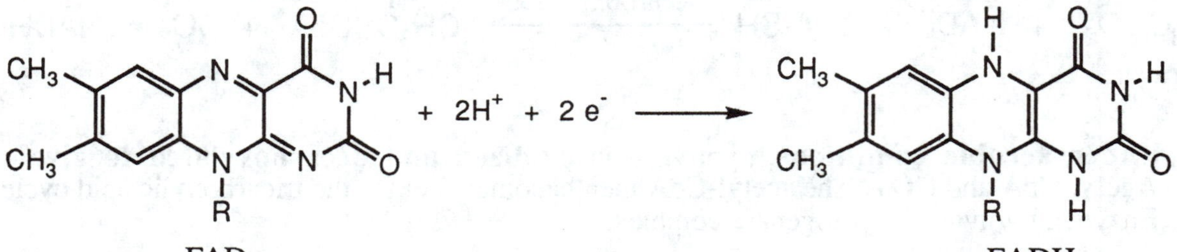

FAD                                          FADH₂

  - FAD is a cofactor in two-electron oxidation reactions such as the oxidation of a carbon-carbon single bond to a carbon-carbon double bond.

• **Every time a substrate is oxidized in the metabolic pathways, either FAD or NAD$^+$ must be reduced, and** *vice versa.* ✳

## 19.2 FATTY ACIDS AS A SOURCE OF ENERGY
• **Fatty acids,** as triglycerides, are the **main storage form of energy** in most organisms. The -CH$_2$- groups of the fatty acid alkyl chains can be oxidized further than oxygenated species such as carbohydrates, so fatty acids are very potent sources of energy.

## 19.4 DIGESTION AND ABSORPTION OF CARBOHYDRATES
• **Carbohydrates** provide about **50 - 60% of daily energy needs**. The **carbohydrates** are consumed in the form of **disaccharides or polysaccharides,** which are first **hydrolyzed** to **monosaccharides** by enzymes called **glycosidases** in the mouth or small intestine. ✳

# CHAPTER 19
## *Solutions to the Problems*

<u>Problem 19.1</u>  Under anaerobic (without oxygen) conditions, glucose is converted to lactate by a metabolic pathway called anaerobic glycolysis or, alternatively, lactate fermentation.  Is anaerobic glycolysis a net oxidation, a net reduction, or neither?

$$C_6H_{12}O_6 \xrightarrow[\text{glycolysis}]{\text{Anaerobic}} 2 \overset{\overset{\displaystyle OH}{|}}{CH_3CHCO_2^-}$$

Glucose                                      Lactate

**The overall process of anaerobic glycolysis that converts glucose to lactate is neither an oxidation or a reduction because the oxidative and reductive steps cancel each other.  Converting one molecule of glucose to two molecules of pyruvate is an oxidation involving a total of four electrons.  The four electrons end up on 2 molecules of NADH.  However, this is balanced exactly by the next step in which conversion of two molecules of pyruvate into two molecules of lactate is a four electron reduction.  The four electrons come from 2 molecules of NADH to regenerate 2 molecules of NAD⁺.**

<u>Problem 19.2</u>  Does lactate fermentation result in an increase or decrease in blood pH?

**Lactate fermentation leads to an increase of the H⁺ concentration in the bloodstream, therefore the bloodstream pH must decrease.**

<u>Problem 19.3</u>  Write structural formulas for palmitic, oleic, and stearic acids, the three most abundant fatty acids.

**Palmitic and stearic acids are fully saturated, having 16 and 18 carbons in their chains, respectively.  Oleic acid has 18 carbons and a single *cis* double bond.**

Palmitic acid

Oleic acid

Stearic acid

<u>Problem 19.4</u>  A fatty acid must be activated before it can be metabolized in cells.  Write a balanced equation for the activation of palmitic acid.

**Activation of a fatty acid involves formation of a thioester with coenzyme A.  The proton is derived from the thiol group of CoA-SH.**

$$CH_3(CH_2)_{14}\overset{\overset{O}{\parallel}}{C}O^- + \ \ CoA\text{-}SH \ + \ ATP \ \longrightarrow \ CH_3(CH_2)_{14}\overset{\overset{O}{\parallel}}{C}SCoA$$

Palmitic acid                                           $+ \ AMP \ + \ P_2O_7{}^{4-} \ + \ H^+$

<u>Problem 19.5</u>  Name three coenzymes necessary for β-oxidation of fatty acids.  From what vitamin is each derived?

**The three coenzymes needed for β-oxidation are:**
  **1) Coenzyme A (CoA-SH) derived from the vitamin pantothenic acid.**
  **2) Nicotine adenine dinucleotide (NAD⁺) derived from the vitamin niacin.**
  **3) Flavin adenine dinucleotide (FAD) derived from the vitamin riboflavin**
     **(vitamin B₂).**
**All three coenzymes contain a molecule of adenosine**

<u>Problem 19.6</u>  We have examined β-oxidation of saturated fatty acids, such as palmitic acid and stearic acid.  Oleic acid, an unsaturated fatty acid, is also a common component of dietary fats and oils.  This unsaturated fatty acid is degraded by β-oxidation but, at one stage in its degradation, requires an additional enzyme named enoyl-CoA isomerase.  Why is this enzyme necessary, and what isomerization does it catalyze? (*Hint:* Consider both the configuration of the carbon-carbon double bond in oleic acid and its position in the carbon chain.)

**If you count the carbon atoms in oleic acid carefully, you will see that after three rounds of β-oxidation you are left with the following fragment that is then isomerized by enoyl-CoA isomerase to the *trans*-enoyl-CoA derivative needed for the next step of β-oxidation.**

Oleic acid

Three rounds of β-oxidation

+  3 AcetylCoA

Enoyl-CoA-isomerase

A *trans*-enoyl-CoA

## GLYCOLYSIS

Problem 19.7 Name two coenzymes required for glycolysis. From what vitamin is each derived?

**The one coenzyme required for glycolysis is NAD+, which is derived from the vitamin niacin.**

Problem 19.8 Number the carbon atoms of glucose 1 through 6 and show from which carbon atom the carboxyl group of each molecule of pyruvate is derived.

**By numbering the carbon atoms of glucose and following the different atoms through the pathway it can be seen that the carboxyl group carbon atoms are derived from carbon atoms 3 and 4 of glucose.**

Problem 19.9 How many moles of lactate are produced from three moles of glucose?

**During anaerobic glycolysis, 2 moles of lactate are produce for each mole of glucose used. 6 moles of lactate will be produced from 3 moles of glucose.**

Problem 19.10 Although glucose is the principal source of carbohydrates for glycolysis, fructose and galactose are also metabolized for energy.
(a) What is the main dietary source of fructose? of galactose?

**The main dietary source of D-fructose is in the disaccharide sucrose, or table sugar, in which D-fructose is combined with D-glucose. The main dietary source of D-galactose is the disaccharide lactose, from milk, in which D-galactose is combined with D-glucose.**

(b) Propose a series of reactions by which fructose might enter glycolysis.

**Fructose could be converted to fructose 6-phosphate, and enter glycolysis at reaction 3, where it will be converted to fructose 1,6-bisphosphate.**

```
        CH2OH                                    CH2OH
         |                                        |
        C=O                                      C=O
         |            Phosphorylation             |
   HO—C—H          ───────────────▶        HO—C—H
         |                                        |
    H—C—OH                                   H—C—OH
         |                                        |
    H—C—OH                                   H—C—OH
         |                                        |
        CH2OH                                    CH2OPO3^{2-}
      D-Fructose                            Fructose 6-phosphate
```

(c) Propose a series of reactions by which galactose might enter glycolysis.

**D-Galactose can be epimerized at C-4 to produce D-glucose and thereby enter glycolysis at the beginning.**

```
        CHO                                      CHO
         |                                        |
    H—C—OH                                   H—C—OH
         |            Epimerization              |
   HO—C—H          ───────────────▶        HO—C—H
         |                                        |
   HO—C—H                                   H—C—OH
         |                                        |
    H—C—OH                                   H—C—OH
         |                                        |
        CH2OH                                    CH2OH
      D-Galactose                            D-Glucose
```

<u>Problem 19.11</u> How many moles of ethanol are produced per mole of sucrose through the reactions of glycolysis and alcoholic fermentation? How many moles of $CO_2$ are produced?

**A total of 4 moles of ethanol and 4 moles of carbon dioxide are produced from 1 mole of sucrose. This can be seen be remembering that 1 mole of the disaccharide sucrose is first hydrolyzed to 1 mole of glucose and 1 mole of fructose. Each of these 6-carbon monosaccharides enter glycolysis to give 2 moles of pyruvate, so a total of 4 moles of pyruvate are produced for each mole of sucrose used. Each mole of pyruvate is converted to 1 mole of ethanol and 1 mole of carbon dioxide, so a total of 4 moles of ethanol and 4 moles of carbon dioxide are produced for each mole of sucrose.**

<u>Problem 19.12</u> Glycerol is derived from hydrolysis of triglycerides and phospholipids. Propose a series of reactions by which the carbon skeleton of glycerol might enter glycolysis and be oxidized to pyruvate.

**Glycerol enters glycolysis through the following enzyme catalyzed steps that lead to glyceraldehyde-3-phosphate, which is converted into pyruvate according to the normal glycolysis pathway.**

$$
\begin{array}{cccc}
\text{CH}_2\text{OH} & \text{CH}_2\text{OH} & \text{CH}_2\text{OH} & \text{CHO} \\
\text{HO}-\text{C}-\text{H} \longrightarrow & \text{HO}-\text{C}-\text{H} \longrightarrow & \text{O}=\text{C} \longrightarrow & \text{H}-\text{C}-\text{OH} \\
\text{CH}_2\text{OH} & \text{CH}_2\text{OPO}_3^{2-} & \text{CH}_2\text{OPO}_3^{2-} & \text{CH}_2\text{OPO}_3^{2-} \\
\textbf{Glycerol} & \textbf{Glycerol-3-} & \textbf{Dihydroxyacetone} & \textbf{Glyceraldehyde-3-} \\
 & \textbf{phosphate} & \textbf{phosphate} & \textbf{phosphate}
\end{array}
$$

<u>Problem 19.13</u> Ethanol is oxidized in the liver to acetate ion by $NAD^+$.
(a) Write a balanced equation for this oxidation.

**The production of acetate ion from ethanol is an overall 4 electron process, so two moles of $NAD^+$ are required for every mole of ethanol. In addition, 2 protons are produced along with the proton that will dissociate from acetic acid to give acetate.**

$$
\text{CH}_3\text{CH}_2\text{OH} + 2\ \text{NAD}^+ \longrightarrow \text{CH}_3\overset{\displaystyle O}{\overset{\|}{\text{C}}}\text{O}^- + 2\ \text{NADH} + 3\ \text{H}^+
$$

(b) Do you expect the pH of blood plasma to increase, decrease, or remain the same as a result of metabolism of a significant amount of ethanol?

**The pH of blood plasma will drop due to the protons produced as the result of metabolism of a significant amount of ethanol.**

<u>Problem 19.14</u> Write a mechanism to show the role of NADH in the reduction of acetaldehyde to ethanol.

**For this reaction, NADH delivers a hydride equivalent, and a group on the enzyme (denoted as A) delivers a proton to the oxygen atom. Note how the lone pair of electrons on the nitrogen in the ring is used as a source of electrons for the reaction.**

Problem 19.15 When pyruvate is reduced to lactate by NADH, two hydrogens are added to pyruvate; one to the carbonyl carbon, the other to the carbonyl oxygen. Which of these hydrogens is derived from NADH?

**As can be seen in the mechanism given in the answer to Problem 19.14, the NADH delivers a hydride equivalent, H-. This species is highly nucleophilic and reacts with the electrophilic carbonyl carbon atom.**

Problem 19.16 Review the oxidation reactions of glycolysis and β-oxidation and compare the types of functional groups oxidized by NAD$^+$ with those oxidized by FAD.

**NAD$^+$ oxidizes a secondary alcohol to a ketone (reaction 3 of β-oxidation) as well as an aldehyde to a carboxylic acid derivative (reaction 6 of glycolysis). FAD oxidizes a carbon-carbon single bond to a carbon-carbon double bond (reaction 1 of β-oxidation).**

Problem 19.17 Why is glycolysis called an anaerobic pathway?

**Glycolysis is called an anaerobic pathway because no oxygen is used. Glycolysis probably first evolved in organisms that appeared before there was oxygen in the environment.**

Problem 19.18 Which carbons of glucose appear as $CO_2$ as a result of alcoholic fermentation?

**As shown in the answer to Problem 19.8, it is carbons 3 and 4 of D-glucose that end up as the carboxylic acid carbons of pyruvate. These same two carbon atoms, carbons 3 and 4, end up as $CO_2$ as a result of alcoholic fermentation.**

Problem 19.19 Which steps in glycolysis require ATP? Which steps produce ATP?

**Reactions 1 and 3 of glycolysis require ATP, while reactions 7 and 10 produce ATP.**

Problem 19.20 The respiratory quotient (RQ) is used in studies of energy metabolism and exercise physiology. It is defined as the ratio of the volume of carbon dioxide produced to the volume of oxygen used:

$$RQ = \frac{\text{Volume } CO_2}{\text{Volume } O_2}$$

(a) Show that RQ for glucose is 1.00. (*Hint:* Look at the balanced equation for complete oxidation of glucose to carbon dioxide and water.)

**In the balanced reaction for the complete oxidation of glucose into $CO_2$ and $H_2O$, 6 moles of $O_2$ are used and 6 moles of $CO_2$ are produced, so the RQ is 6/6 = 1.00.**

$$C_6H_{12}O_6 + 6\ O_2 \longrightarrow 6\ CO_2 + 6\ H_2O$$
**D-Glucose**

(b) Calculate RQ for triolein, a triglyceride of molecular formula $C_{57}H_{104}O_6$.

**In the balanced equation for the complete oxidation of triolein, 80 moles of $O_2$ are used and 57 moles of $CO_2$ are produced for each mole of triolein consumed. The RQ = 57/80 = 0.71**

$$C_{57}H_{104}O_6 + 80\ O_2 \longrightarrow 57\ CO_2 + 52\ H_2O$$
**Triolein**

(c) For an individual on a normal diet, RQ is approximately 0.85. Would this value increase or decrease if ethanol were to supply an appreciable portion of caloric needs?

**In the balanced equation for the complete oxidation of ethanol, $C_2H_6O$, 3 moles of $O_2$ are used and 2 moles of $CO_2$ are produced for each mole of ethanol consumed. The RQ = 2/3 = 0.67, so the individual's RQ would decrease if ethanol were to supply an appreciable portion of caloric needs.**

$$C_2H_6O + 3\ O_2 \longrightarrow 2\ CO_2 + 3\ H_2O$$
**Ethanol**

Problem 19.21 Acetoacetate, β-hydroxybutyrate, and acetone are commonly known within the health sciences as ketone bodies, in spite of the fact that one of them is not a ketone at all. They are products of human metabolism and are always present in blood plasma. Most tissues, with the notable exception of the brain, have the enzyme systems necessary to use them as energy sources. Synthesis of ketone bodies occurs by the following enzyme-catalyzed reactions. Enzyme names are (1) thiolase, (2) β-hydroxy-β-methylglutaryl-CoA synthase, (3) β-hydroxy-β-methylglutaryl-CoA lyase, and (5) β-hydroxybutyrate dehydrogenase. Reaction (4) is spontaneous and uncatalyzed.

Describe the type of reaction involved in each step and the type of mechanism by which each occurs.

**Reaction 1 is a Claisen condensation (Section 15.3) between two molecules of acetyl-CoA.**

**Reaction 2 is an aldol reaction (Section 15.2) that can be thought of as taking place between the enolate of acetyl-CoA and the ketone carbonyl of acetoacetyl-CoA.**

**Reaction 3 is a reverse aldol reaction (Section 15.2) that generates acetyl-CoA and acetoacetate.**

**Reaction 4 is a decarboxylation of a β-ketoacid (Section 13.8A) that generates CO₂ and acetone from acetoacetate.**

**Reaction 5 is a reduction of the ketone group of acetoacetate to a secondary alcohol (Section 11.10).**

Problem 19.22  A connecting point between anaerobic glycolysis and β-oxidation is formation of acetyl-CoA.  Which carbon atoms of glucose appear as methyl groups of acetyl-CoA?  Which carbon atoms of palmitic acid appear as methyl groups of acetyl-CoA?

**As shown in the answer to Problem 19.8, it is carbons 3 and 4 of D-glucose that end up as the carboxylic acid carbons of pyruvate.  These same two carbon atoms, carbons 3 and 4 end up as $CO_2$ as a result of oxidation and decarboxylation to acetyl CoA.  This means that it is carbons 1 and 6 that end up as the methyl groups of acetyl CoA.**

Palmitic acid undergoes β-oxidation to produce acetyl-CoA, so it is the even number carbon atoms (2,4,6,8,10,12,14,16) that end up being the methyl groups.

# CHAPTER 20: NUCLEAR MAGNETIC RESONANCE SPECTROSCOPY

## 20.0 OVERVIEW
• **Nuclear magnetic resonance (NMR)** spectroscopy was developed in the early 1960's, and is now the most important technique for the determination of molecular structure. *[NMR is a very complicated type of spectroscopy, because there are so many difficult concepts involved. The best way to learn about NMR is to read through the concepts, then work through as many spectra as possible.]* ✳

## 20.1 ELECTROMAGNETIC RADIATION
• **Electromagnetic radiation** can be described like a wave, in terms of its wavelength and frequency. ✳
  - **Wavelength** is the distance between any two identical points on a wave and is given the symbol $\lambda$ **(lambda)** and usually expressed in **meters (m)**.
  - **Frequency** is the number of full cycles of a wave that pass a given point in a fixed period of time. Frequency is given the symbol $\nu$ **(nu)** and is usually expressed in **hertz (Hz)** and given the units **sec$^{-1}$**.  For example, 1 Hz corresponds to one cycle per second and 1 MHz corresponds to $10^6$ Hz.
  - Wavelength and frequency are related to each other, and one can be calculated from the other using the expression $\nu = c/\lambda$. Here **c** is the **speed of light**, equal to **$3.00 \times 10^8$ m/sec.**
• **Electromagnetic radiation** can also be described as a particle, and the particle is called a **photon**. ✳
  - The **energy (E)** measured in **kcal** of one mole of photons is related to wavelength and frequency according to the following relationships: **$E=h\nu=hc/\lambda$**.  Here **h** is **Planck's constant** and is equal to **$9.537 \times 10^{-14}$ kcal-sec-mol$^{-1}$**.

## 20.2 MOLECULAR SPECTROSCOPY
• In general, an atom or molecule can be made to **undergo a transition** from energy state $E_1$ to a higher energy state $E_2$ by irradiating it with electromagnetic radiation corresponding to the energy difference between states $E_1$ and $E_2$. During this process, the atom or molecule **absorbs the energy** of the electromagnetic radiation.  When the atom or molecule **returns to the ground state** $E_1$, an equivalent amount of **energy is emitted**. ✳
• **Molecular spectroscopy** involves measuring the frequencies of electromagnetic radiation that are absorbed or emitted by a molecule, then correlating the observed patterns with the details of molecular structure.
  - Specific regions of the electromagnetic spectrum are particularly interesting to the chemist because they represent the energies involved in transition between important types of molecular energy levels. In particular, **radiofrequency electromagnetic radiation** corresponds to transitions between **nuclear spin** energy levels, **infrared electromagnetic radiation** corresponds to transitions between **vibrational levels of chemical bonds** and **ultraviolet-visible electromagnetic radiation** corresponds to **electronic energy levels** of pi and nonbonding electrons. *[It may prove helpful to review how these different regions of electromagnetic radiation fit into the entire spectrum shown in Table 20.2.]*
  - The nuclear spin transitions caused by absorbed radiofrequency electromagnetic radiation form the basis for **nuclear magnetic resonance (NMR) spectroscopy**. ✳

## 20.3 THE ORIGIN OF NUCLEAR MAGNETIC RESONANCE

• Like electrons, certain nuclei have spin. That is, they behave as if they are spinning on an axis and thus have an associated **magnetic moment**. The magnetic moment means that nuclei with spin act like tiny bar magnets. This is only true for nuclei that have **spin.** Both $^1H$ and $^{13}C$ nuclei have spin, so they are the nuclei that are most often studied by NMR spectroscopy. Some common nuclei such as $^{12}C$ and $^{16}O$ do not have spin, so they are not observed by NMR. *[It might help to review the concept of spin quantum numbers in a General Chemistry text.]*

• Ordinarily, nuclear spins are oriented in a completely random fashion. However, in an **applied magnetic field ($B_o$)**, the **nuclear spins of $^1H$ and $^{13}C$ interact with the applied field**. Recall that the nuclear spin produces it own magnetic moment, so this interaction is between the magnetic moments of the nuclei with the applied magnetic field. For $^1H$ and $^{13}C$ nuclei, there are two allowed orientations in the field; alignment with the applied field and alignment against the applied field. **Nuclei aligned with the applied field** are in the **lower energy state**, and **nuclei aligned against the applied field** and are in the **higher energy state.** *

• The **difference** in **energy** between the **nuclear spin states increases** with **increasing applied field strength.** * Nevertheless, these energy differences are small compared to other types of energy levels such as vibrational and electronic energy levels in molecules.

## 20.4 NUCLEAR MAGNETIC "RESONANCE"

• When the nuclei are placed in an applied magnetic field, a majority of spins are aligned with the field and are thus in the lower spin state. **Electromagnetic radiation** can **cause a transition from the lower spin state to the higher spin state.** An NMR spectrum is a plot of how much and of which energies are absorbed by a molecule as its atoms undergo these transitions from the lower to the higher nuclear spin state. *

  - The amount of energy required depends on the strength of the applied field and the type of nuclei being used, but this energy corresponds to electromagnetic radiation somewhere in the radiofrequency range. For example, in an applied magnetic field of strength 7.05 Tesla (T), the energy between the spin states of $^1H$ is around 0.0286 cal/mol, corresponding to electromagnetic energy of approximately 300 MHz (300,000,000 Hz).

• The different types of atomic nuclei in a molecule, for example $^{13}C$ or $^1H$, do not absorb energy (resonate) at the same frequency. If they did, NMR would not be a useful probe of molecular structure. It turns out that there are usually different local chemical environments in a molecule that change the resonance frequencies of the different nuclei. By looking at the **different resonance frequencies measured in the NMR spectrum**, information is obtained about the **different chemical environments in the molecule**, leading to an understanding of **the molecular structure.** *

  - Atomic nuclei from different elements resonate at different frequencies in the same applied field. For example, in the presence of an applied field of 7.05 T, $^1H$ nuclei resonate at about 300 MHz, while $^{13}C$ nuclei resonate at about 75 MHz.

  - The different nuclei in a molecule are surrounded by electron density to varying degrees. NMR can detect these differences in surrounding electron density and thereby provide important structural information. The electrons themselves have spin and thereby create their own **local magnetic field**. These local magnetic fields **shield the nucleus** from the applied magnetic fields. **The greater the electron density around a nucleus**, the **greater the shielding**. This means that at constant magnetic field strength, **the greater the shielding of a nucleus, the lower the frequency of electromagnetic radiation required to bring about a spin flip**. Another way to look at it is from the point of view of constant electromagnetic radiation frequency. At constant frequency of electromagnetic radiation, **the greater the shielding of a nucleus, the higher the magnetic field strength required to bring the nucleus into resonance.** * *[It is*

*important to be able to think about shielding and spin flipping from the point of view of constant frequency of electromagnetic radiation as well as at constant magnetic field strength.]*
- The local magnetic fields are small compared to the applied magnetic fields. The differences in resonance frequencies for different nuclei caused by the local magnetic fields are usually on the order of $1 \times 10^{-6}$ times as large as the original resonance frequencies. In other words, different $^1H$ nuclei in the same molecule have resonance frequencies that differ by an amount on the order of **parts per million (ppm)**. Thus, ppm is a convenient measurement unit for NMR spectroscopy. For example, a difference of 100 Hz is 1 ppm of 100 MHz, and a difference of 300 Hz is 1 ppm of 300 MHz.
- In order to increase the precision of resonance frequency measurements, a reference compound is used. By convention, the $^1H$ resonance in **tetramethylsilane (TMS)** is used as the reference against which the frequency of other $^1H$ resonances are measured. Similarly, the $^{13}C$ resonance in TMS is used as the reference by which the frequency of other $^{13}C$ resonances are measured against.
- A unit called **chemical shift** is used to standardize reporting of NMR data. **Chemical shift ($\delta$)** is the frequency shift from TMS divided by the operating radiofrequency of the spectrometer. **Chemical shift is reported in** the units **ppm**.

$$\delta = \frac{\text{Shift in frequency from TMS (Hz)}}{\text{Frequency of spectrometer (Hz)}} \times 10^6$$

## 20.5 AN NMR SPECTROMETER
• The essential features of an NMR spectrometer are a powerful magnet, a radiofrequency generator, a radiofrequency detector, and a sample tube.

## 20.6 EQUIVALENT HYDROGENS
• All **equivalent hydrogens** have the same chemical environment within a molecule and **have identical chemical shifts**. For example, all of the hydrogens in dimethyl ether ($CH_3$-O-$CH_3$) are equivalent, so there is only one signal in the $^1H$-NMR spectrum of this compound.
   - A good way to see if **two hydrogens** are **equivalent** is to use a **"test atom."** Replace the hydrogens in question one at a time with this test atom, for example, a chlorine atom. If all the hydrogens give the same compound when the test atom is used, then the hydrogens are equivalent. If different compounds are formed, then the hydrogens are not equivalent. *[Understanding how to identify equivalent hydrogens is absolutely essential for the interpretation of $^1H$-NMR spectra, so it is essential that these ideas are understood before going on.]*
• Hydrogens that are not equivalent give rise to different signals with different chemical shifts. For example, there are two different signals in the $^1H$-NMR spectrum of ethyl bromide ($CH_3$-$CH_2$-Br).

## 20.7 SIGNAL AREAS
• **The area under each signal** is **proportional to the number of hydrogens** giving rise to that signal in an $^1H$-NMR spectrum. All modern NMR spectrometers can integrate the area under each signal. Please note that $^{13}C$ signals cannot be integrated accurately (Section 20.13) in $^{13}C$-NMR spectra.

## 20.8 CHEMICAL SHIFT

• Each type of **equivalent hydrogen** within a molecule has only a **limited range of δ values**, and thus the value of the chemical shift for a signal in a $^1$H-NMR spectrum gives valuable information about the type of hydrogen giving rise to that absorption. For example, the three hydrogens on methyl groups bonded to $sp^3$ hybridized carbons resonate near δ 1.0 ppm, while the three hydrogens on methyl groups bonded to an $sp^2$ hybridized carbonyl carbon atom resonate near δ 2.0 ppm. ✴

• A signal is considered **downfield** if it is shifted toward the left (weaker applied field) on the chart paper, and it is **upfield** if it is shifted to the right (stronger applied field).

## 20.9 THE (n+1) RULE

• Signals can be split into several peaks and this phenomenon is called **spin-spin splitting**. In **spin-spin splitting**, the $^1$H-NMR signal from one set of equivalent hydrogens is split by the influence of neighboring nonequivalent hydrogens. ✴

 - If a hydrogen has a set of **n** nonequivalent hydrogens on the same or adjacent atoms, its NMR signal will be split into **(n+1)** peaks. The nuclei of all adjacent hydrogens couple, but it is only between nonequivalent hydrogens that the coupling results in spin-spin splitting. For example, in ethyl bromide ($CH_3$-$CH_2$-Br) the $CH_3$- signal is split into three peaks by the two hydrogens on the adjacent -$CH_2$- group, and the -$CH_2$- signal is split into four peaks by the three hydrogens on the adjacent $CH_3$- group. *[Understanding spin-spin splitting is absolutely essential for the interpretation of $^1$H-NMR spectra, so it is essential that these ideas are understood before going on.]*

## 20.10 $^{13}$C-NMR SPECTROSCOPY

• Carbon-12 ($^{12}$C) is the most abundant natural isotope of carbon (98.89%), but it is not seen in an NMR spectrum because its nucleus has only one allowed spin state. On the other hand, **carbon-13 ($^{13}$C)** (natural abundance of 1.11%) has two allowed nuclear spin states and it **can be detected by NMR.** ✴

• $^1$H nuclei couple to the $^{13}$C nuclei. Unfortunately, this can make the spectra very difficult to interpret. As a result, $^{13}$C spectra are usually measured in the **hydrogen-decoupled mode** in which the sample is irradiated such that all hydrogens are in the same spin state and thus spin-spin splitting is prevented. The $^{13}$C spectra can then be measured without interference from complex signals due to $^1$H-$^{13}$C spin-spin splitting.

• As stated above, $^{13}$C nuclei are very rare, so ordinarily only $^{12}$C carbon atoms are adjacent to a given $^{13}$C nucleus in a molecule. Thus, $^{13}$C-$^{13}$C coupling and/or spin-spin splitting are usually not observed.

• Because of the way $^{13}$C nuclei return to equilibrium states, $^{13}$C signals cannot be integrated accurately in $^{13}$C-NMR spectra.

• $^{13}$C-NMR spectra provide important structural information, because each different carbon atom in a molecule gives rise to a different signal. Thus, by simply looking at a $^{13}$C-NMR spectra, a chemist can tell how many different types of carbon atoms are in a molecule. For example; $sp^3$ alkyl carbon atoms, $sp^2$ atoms in an alkene, or $sp^2$ carbonyl carbon atoms all have characteristic chemical shifts. Please see Table 20.4 for a detailed list of $^{13}$C chemical shifts. ✴

## 20.11 INTERPRETING NMR SPECTRA

• Different types of molecules have **characteristic chemical shifts and splitting patterns** in NMR spectra. The following is a list of these characteristics for a number of important types of molecules.

- **Alkanes**: All hydrogens in alkanes are in similar chemical environments and fall with a narrow range of chemical shifts. $^1$H-NMR chemical shifts for alkanes fall in the range of δ 0.9-1.5. $^{13}$C-NMR chemical shifts for alkanes fall in the range of δ 0 to 60.
- **Alkenes**: The chemical shift of vinylic hydrogens fall in the range δ 4.6-5.7. $^{13}$C-NMR chemical shifts for the sp$^2$ carbon atoms of alkenes fall in the range of δ 100-150.
- **Alcohols**: The hydrogen of the OH group is variable and appears in the range δ 2.0-6.0. This hydrogen is rarely split by spin-spin splitting because of the phenomenon of fast exchange. The hydrogens attached to the carbon adjacent to the OH group are deshielded by the inductive effect of the electron withdrawing nature of the oxygen atom. These hydrogens appear at δ 3.5-4.5.
- **Benzene and its derivatives**: Hydrogens attached to substituted benzenes appear in the range δ 6.5-8.5. Benzene has all equivalent hydrogens so only one $^1$H-NMR signal at δ 7.27. Substituted benzene derivatives can have non-equivalent hydrogens on the ring and exhibit more complex splitting patterns. For example, particularly diagnostic **doublet of doublets splitting pattern** occurs with a benzene ring has non-identical substituents at the **para positions**. $^{13}$C-NMR chemical shifts for the aromatic sp$^2$ carbon atoms fall in the range of δ 110-160. The number of different peaks in the aromatic region of an $^{13}$C-NMR spectrum can be used to distinguish between different constitutional isomers.
- **Amines**: The hydrogens attached to N are variable and appear in the range δ 0.5-5.0. The hydrogens attached to the carbon adjacent to the NH group are deshielded by the inductive effect of the electron withdrawing nature of the nitrogen atom. These hydrogens appear at δ 3.5-4.5. Amine hydrogens are rarely split by spin-spin splitting because of the phenomenon of fast exchange.
- **Aldehydes and ketones**: The hydrogen of an aldehyde comes as a sharp singlet at δ 9.4-9.8. This signal is not split by hydrogens on adjacent carbon atoms, and *vice versa*. $^{13}$C-NMR chemical shifts for the sp$^2$ carbonyl carbon atoms of aldehydes are easy to identify in the range of δ 190-210.
- **Carboxylic acids**: Hydrogens on the α-carbon of a carboxylic acid appear in the range δ 2.0-2.6. The acidic hydrogen of the carboxyl group comes in the range of δ 10-13.
- **Esters**: Hydrogens on the α-carbon to the carbonyl of esters appear in the range δ 2.0-2.5. Hydrogens attached to the carbon on the ester oxygen atom are more strongly deshielded, coming in the range δ 3.6-4.1.

## 20.12 SOLVING NMR PROBLEMS

• Before analyzing the NMR spectrum of a given molecule, it is helpful to analyze the molecular formula. This could be determined by elemental analysis or mass spectrometry. An important piece of information contained in the molecular formula is the **index of hydrogen deficiency**. The **index of hydrogen deficiency** is the **number of rings and/or pi bonds in a molecule**. This is determined by comparing the number of hydrogens in the molecular formula of a compound of unknown structure with the number of hydrogens in a **reference compound**, a compound with the same number of carbon atoms and with no rings or pi bonds. In particular, the **index of hydrogen deficiency** is defined according to the following formula: ✳

$$\text{index of hydrogen deficiency} = \frac{\text{\# of hydrogen atoms}_{(\text{reference})} - \text{\# of hydrogens }_{(\text{molecule of interest})}}{2}$$

- For reference compounds that contain only C and H atoms, the molecular formula is $C_nH_{2n+2}$.
- For each atom of F, Cl, Br, or I subtract one hydrogen.
- No correction is necessary for O, S, or Se.

- For each atom of N or P add one hydrogen.

• **After the index of hydrogen deficiency has been determined**, the following steps should be followed when solving a spectral problem. *[Practice is the best and possibly only way to become good at this.]* ✳

  - **Count the number of signals** to determine how many different types of hydrogens are present.

  - **Analyze the integration of each signal**, to see how many hydrogen atoms of each type are present.

  - **Examine the pattern of chemical shifts** and correlate them with the known characteristic chemical shifts for different types of hydrogen atoms.

  - **Analyze the spin-spin splitting patterns.** This is usually the most difficult task by far, but also the most informative.

  - **Write the formula** that is consistent with all of the above information.

# CHAPTER 20
*Solutions to the Problems*

<u>Problem 20.1</u> Calculate the energy of red light (680 nm) in kilocalories per mole. Which form of radiation carries more energy, infrared radiation of wavelength 2.50 μm or red light of wavelength 680 nm?

**Combining the two equations given in the text gives:**

$$E = h\nu = h\left(\frac{c}{\lambda}\right)$$

**Plugging in the appropriate values gives the desired answer:**

$$E = \frac{(9.537 \times 10^{-14} \text{ kcal-sec-mol}^{-1})(3.00 \times 10^{8} \text{ m-sec}^{-1})}{680 \times 10^{-9} \text{ m}} = \boxed{42.1 \text{ kcal-mol}^{-1}}$$

**Notice how the units canceled to give the final answer in kcal-mol⁻¹. As can be seen from the equations, the longer the wavelength, the lower the energy, thus red light carries more energy.**

<u>Problem 20.2</u> State the number of sets of equivalent hydrogens in each compound and the number of hydrogens in each set.

**Numbers have been added to the carbon atoms of the structures to aid in referring to specific hydrogens. Use the "test atom" approach if you have trouble seeing the answers.**

(a)

$$\overset{4}{C}H_3$$
$$\overset{1}{C}H_3-\overset{2}{C}H_2-\overset{3}{C}H-\overset{5}{C}H_2-\overset{6}{C}H_3$$

**There are four sets of equivalent hydrogens. <u>Set 1:</u> 6 hydrogens from the methyl groups of carbon atoms 1 and 6. <u>Set 2:</u> 4 hydrogens from the -CH₂- groups of carbon atoms 2 and 5. <u>Set 3:</u> 3 hydrogens from the methyl group of carbon atom 4. <u>Set 4:</u> 1 hydrogen from the -CH- group of carbon atom 3.**

(b)

$$\overset{3}{C}H_3 \qquad \overset{6}{C}H_3$$
$$\overset{1}{C}H_3-\overset{2}{C}H-\overset{4}{C}H_2-\overset{5}{C}-\overset{8}{C}H_3$$
$$\overset{7}{C}H_3$$

**There are four sets of equivalent hydrogens. <u>Set 1:</u> 9 hydrogens from the methyl groups of carbon atoms 6, 7, and 8. <u>Set 2:</u> 6 hydrogens from the methyl groups of carbon atoms 1 and 3. <u>Set 3:</u> 2 hydrogens from the -CH₂- group of carbon atom 4. <u>Set 4:</u> 1 hydrogen from the -CH- group of carbon atom 2.**

Problem 20.3 Each compound gives only one signal in its $^1$H-NMR spectrum. Propose a structural formula for each.

**In order for these molecules to give a single signal in their $^1$H-NMR spectrum, each of the hydrogen nuclei must be in an identical environment. This will only occur in symmetrical molecules.**

(a) $C_3H_6O$                    (b) $C_5H_{10}$ 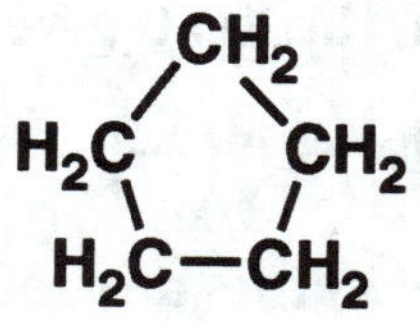                    (c) $C_5H_{12}$

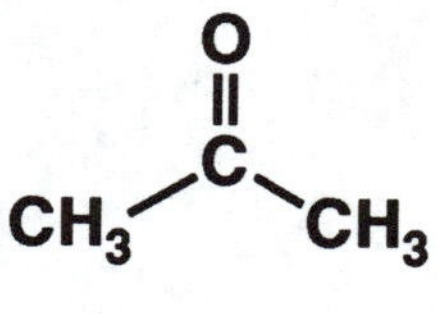

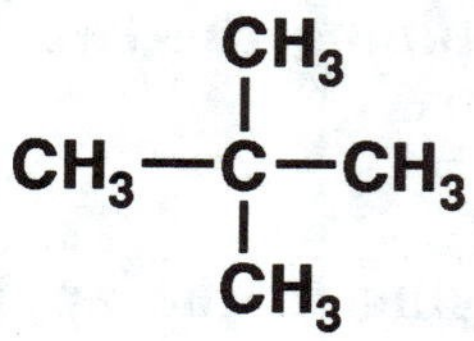

(d) $C_4H_6Cl_4$

**CH$_3$-CCl$_2$-CCl$_2$-CH$_3$**

Problem 20.4 The line of integration of the two signals in the $^1$H-NMR spectrum of a ketone of molecular formula $C_7H_{14}O$ shows a vertical rise of 62 and 10 chart divisions, respectively. Calculate the number of hydrogens giving rise to each signal, and propose a structural formula for this ketone.

**The ratio of signals is approximately 1:6, which corresponds to a 2:12 ratio of hydrogens. Thus, the larger signal represents 12 hydrogens and the smaller signal represents 2 hydrogens. A structure consistent with this assignment is 2,4-dimethyl-3-pentanone as shown below:**

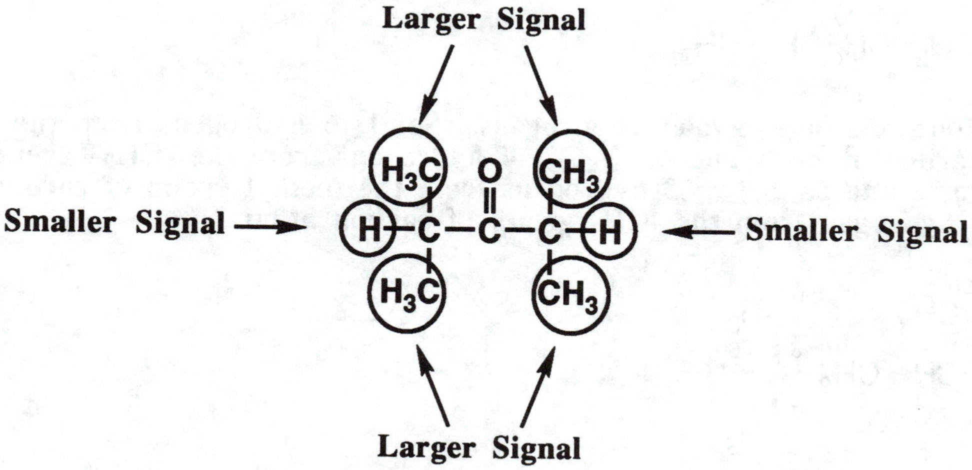

Problem 20.5 Following are two constitutional isomers of molecular formula $C_3H_6O$.

$$\underset{(1)}{CH_3\overset{\overset{\displaystyle O}{\|}}{C}CH_3} \qquad \underset{(2)}{CH_3CH_2\overset{\overset{\displaystyle O}{\|}}{C}H}$$

(a) Predict the number of signals in the $^1$H-NMR spectrum of each isomer.

**Isomer (1) will have 1 signal corresponding to the methyl groups, and isomer (2) will have three signals corresponding to the -CH$_3$, -CH$_2$- and -CH groups, respectively.**

(b) Predict the ratio of areas of the signals in each spectrum.

**Isomer (1) will have only one signal so there is nothing to ratio. Isomer (2) will have the three signals with a ratio of 3:2:1.**

(c) Show how to distinguish between these isomers on the basis of chemical shift.

**These compounds can be easily distinguished using a number of criteria: First, the aldehyde hydrogen of isomer (2) will show up as a singlet at δ 9.4-9.8. Isomer (2) will have three signals while isomer (1) will only have one signal. The methyl groups on isomer (1) are closer to the carbonyl group, so these signals will be deshielded and thus appear farther downfield than the methyl group signal for isomer (2).**

Problem 20.6 Following are pairs of constitutional isomers. Predict the number of signals and the splitting pattern of each signal in the $^1$H-NMR spectrum of each isomer.

(a) $CH_3CH_2\overset{\overset{\displaystyle O}{\|}}{C}H$    and    $CH_3\overset{\overset{\displaystyle O}{\|}}{C}CH_3$

**The aldehyde on the left will have three different signals and the ketone on the right will have one signal with splitting patterns as indicated.**

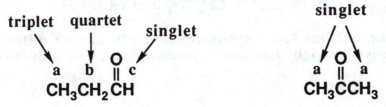

(b) 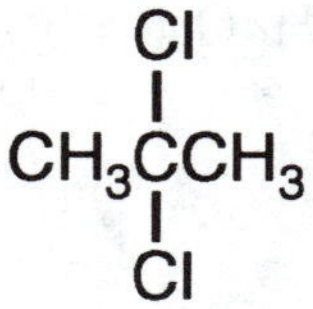   $CH_3\overset{\overset{\displaystyle Cl}{|}}{\underset{\underset{\displaystyle Cl}{|}}{C}}CH_3$   and   $ClCH_2CH_2CH_2Cl$

**The molecule on the left will have one signal and the molecule on the right will have two signals with splitting patterns as indicated.**

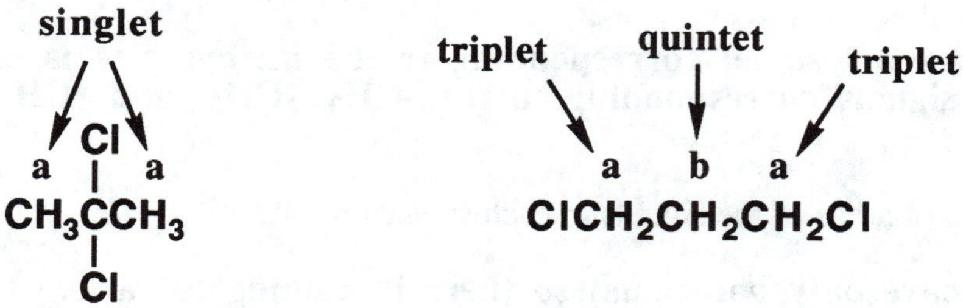

singlet

a $\quad$ Cl $\quad$ a

$CH_3\overset{\overset{\displaystyle Cl}{|}}{\underset{\underset{\displaystyle Cl}{|}}{C}}CH_3$

triplet $\qquad$ quintet $\qquad$ triplet

a $\qquad$ b $\qquad$ a

$ClCH_2CH_2CH_2Cl$

<u>Problem 20.7</u> Explain how to distinguish between members of each pair of constitutional isomers based on the number of signals in the $^{13}$C-NMR spectrum of each member.

(a)

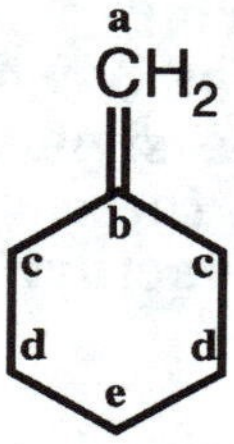

a
$CH_2$

b

c $\qquad$ c

d $\qquad$ d

e

and

a
$CH_3$

g $\qquad$ b

f $\qquad$ c

$\qquad$ d

e

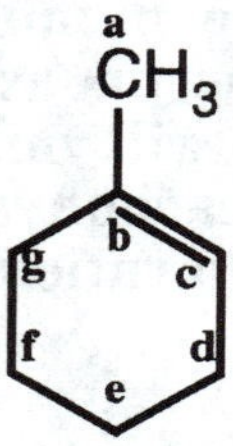

**These molecules can be distinguished because they have different numbers of nonequivalent carbon nuclei and thus will have different numbers of $^{13}$C-NMR signals.   Different signals are indicated by different letters on the above structures. The molecule on the left has higher symmetry and will have 5 different signals, while the molecule on the right has less symmetry and will have 7 different signals.**

(b)    $\overset{a}{C}H_3\overset{b}{C}H=\overset{c}{C}H\overset{d}{C}H_2\overset{e}{C}H_2\overset{f}{C}H_3$   and   $\overset{c}{C}H_3\overset{b}{C}H_2\overset{a}{C}H=\overset{a}{C}H\overset{b}{C}H_2\overset{c}{C}H_3$

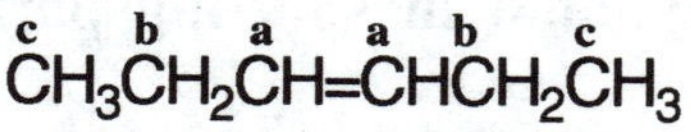

**The molecule on the left has lower symmetry and will have 6 different signals, while the molecule on the right has more symmetry and will only have 3 different signals.**

<u>Problem 20.8</u> Calculate the index of hydrogen deficiency of cyclohexene, and account for this deficiency by reference to its structural formula.

**The molecular formula for cyclohexene is $C_6H_{10}$. The molecular formula for the reference compound with 6 carbon atoms is $C_6H_{14}$. Thus the index of hydrogen deficiency is (14-10)/2 or 2. This makes sense since cyclohexene has one ring and one pi bond.**

**Cyclohexene**

<u>Problem 20.9</u> The index of hydrogen deficiency of niacin is 5. Account for this index of hydrogen deficiency by reference to the structural formula of niacin.

Nicotinamide
(Niacin)

**The index of hydrogen deficiency of niacin is 5 because there are four pi bonds and one ring in the structure.**

## Index of Hydrogen Deficiency
<u>Problem 20.10</u> Complete the following table.

| Class of compound | Molecular formula | Index of hydrogen deficiency | Reason for hydrogen deficiency |
|---|---|---|---|
| alkane | $C_nH_{2n+2}$ | 0 | (reference hydrocarbon) |
| alkene | $C_nH_{2n}$ | 1 | one pi bond |
| alkyne | $C_nH_{2n-2}$ | 2 | **two pi bonds** |
| alkadiene | $C_nH_{2n-2}$ | 2 | **two pi bonds** |
| cycloalkane | $C_nH_{2n}$ | 1 | **one ring** |
| cycloalkene | $C_nH_{2n-2}$ | 2 | **one ring and one pi bond** |
| bicycloalkane | $C_nH_{2n-2}$ | 2 | **two rings** |

<u>Problem 20.11</u> Calculate the index of hydrogen deficiency of the following compounds:
(a) Aspirin, $C_9H_8O_4$                    (b) Ascorbic acid (vitamin C), $C_6H_8O_6$

**(20-8)/2 = 6**                    **(14-8)/2 = 3**

(c) Pyridine, $C_5H_5N$

(13-5)/2 = 4  (nitrogen correction)

(d) Urea, $CH_4N_2O$

(6-4)/2 = 1 (nitrogen correction)

(e) Cholesterol, $C_{27}H_{46}O$

(56-46)/2 = 5

(f) Trichloroacetic acid, $C_2HCl_3O_2$

(3-1)/2 = 1  (halogen correction)

## Interpretation of $^1$H-NMR and $^{13}$C-NMR Spectra

**Problem 20.12** Following are structural formulas for the *cis* isomers of 1,2-, 1,3-, and 1,4-dimethylcyclohexanes and three sets of $^{13}$C-NMR spectral data. Assign each constitutional isomer its correct spectral data.

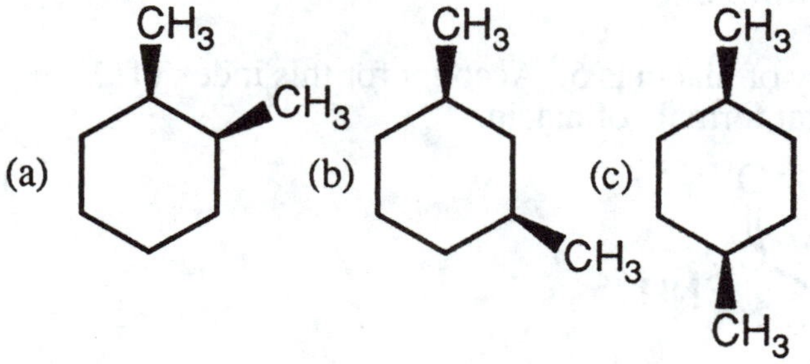

|  | Spectrum 1: | Spectrum 2: | Spectrum 3: |
|---|---|---|---|
|  | 31.35 | 34.20 | 44.60 |
|  | 30.67 | 31.30 | 35.14 |
|  | 20.85 | 23.56 | 32.88 |
|  |  | 15.97 | 26.54 |
|  |  |  | 23.01 |

(a)        (b)        (c)

These constitutional isomers are most readily distinguished by the number of sets of nonequivalent carbon atoms and thus different $^{13}$C signals. Using the analysis shown below, it can be seen that compound (a) has 4 sets of nonequivalent carbon atoms corresponding to spectrum 2, compound (b) has 5 sets of nonequivalent carbon atoms corresponding to spectrum 3, and compound (c) has 3 sets of nonequivalent carbon atoms corresponding to spectrum 1. The different sets of equivalent carbon atoms are indicated by the letters.

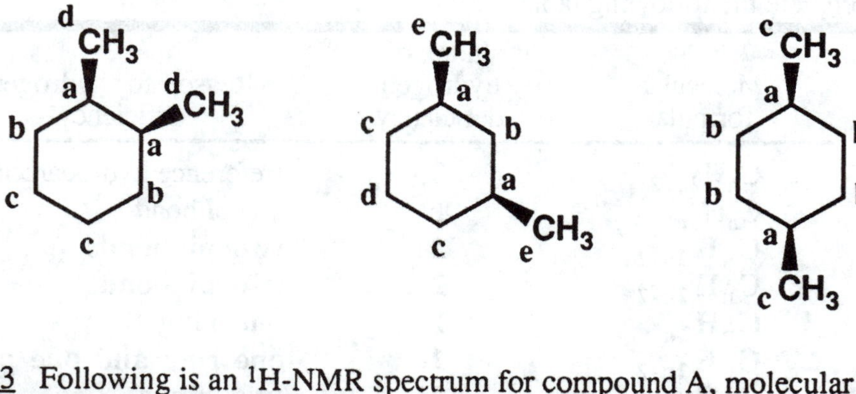

**Problem 20.13** Following is an $^1$H-NMR spectrum for compound A, molecular formula $C_6H_{12}$. Compound A decolorizes aqueous $KMnO_4$ with formation of a brown precipitate of $MnO_2$. Propose a structural formula for compound A.

Compound A has an index of hydrogen deficiency of 1, in the form of a double bond as evidenced by the reaction with $KMnO_4$. The rest of the detailed structures can be deduced from the spectra.

For the spectral interpretations in this problem and the rest of the chapter, the chemical shift ($\delta$) is given followed by the relative integration, the multiplicity of the peak (singlet, doublet, triplet, etc.) and finally the identity of the hydrogens giving rise to the signal are shown in bold.

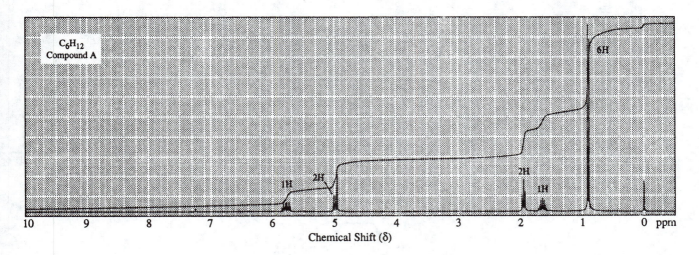

$$CH_3$$
$$\underset{0.9}{CH_3}-\underset{1.6}{CH}-\underset{1.9}{CH_2}-\underset{5.8}{CH}=\underset{5.0}{CH_2}$$

**Compound A**

**¹H-NMR δ 5.8 (1H, multiplet; this is more complex than expected because the adjacent vinylic hydrogens are not equivalent, -CH=), 5.0 (2H, multiplet; this is asymmetric because these two vinylic hydrogens are not equivalent, =CH₂), 1.9 (2H, multiplet; doublet of doublets, -CH₂-), 1.6 (1H, multiplet; a triplet of septets, -CH-), 0.9 (6H, one doublet, -CH₃). The chemical shifts associated with each set of hydrogens is indicated on the structure.**

<u>Problem 20.14</u>  Following is the ¹H-NMR spectrum of compound B, $C_7H_{12}$. Compound B decolorizes both aqueous $KMnO_4$ with formation of a brown precipitate, and a solution of $Br_2$ in $CCl_4$.  Propose a structure for compound B.

**The molecular formula indicates that there is an index of hydrogen deficiency of 2, so there are two rings and/or pi bonds.**

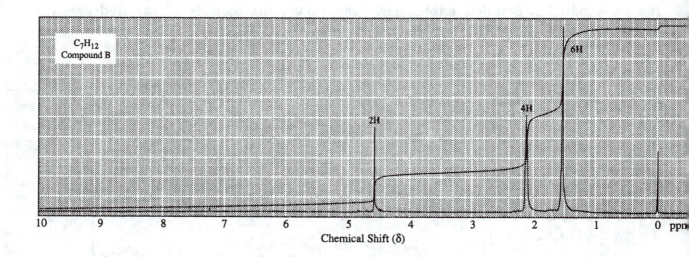

**Compound B**

$^1$**H-NMR** $\delta$ **4.6 (2H, singlet, $=CH_2$), 2.1 (4H, broad peak, the $-CH_2-$ groups adjacent to the sp$^2$ carbon atom on the ring labeled as "a" on the structure above), 1.6 (6H, broad peak, the three $-CH_2-$ groups labeled as "b" on the structure above)**

<u>Problem 20.15</u>  Following are $^1$H-NMR spectra for compounds C and D, each of molecular formula $C_5H_{12}O$.  Each is a liquid at room temperature, slightly soluble in water, and reacts with sodium metal with the evolution of a gas.  Propose structural formulas for compounds C and D..

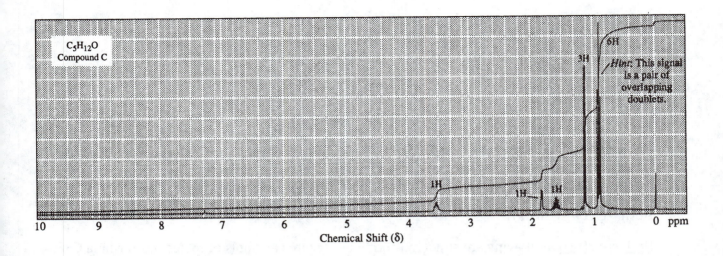

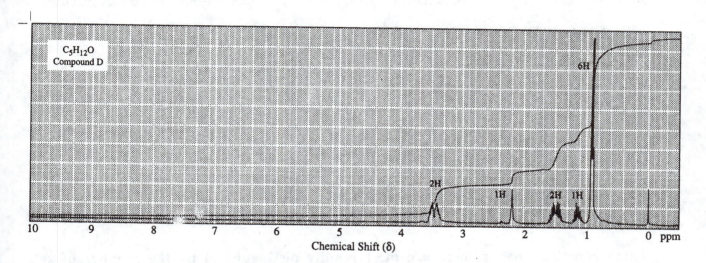

The index of hydrogen deficiency is 0 for these molecules, so there are no rings or double bonds. The fact that the compounds are slightly soluble in water and react with sodium metal indicate that each molecule has an -OH group.

$$CH_3-CH-CH-CH_3$$

**Compound C**

¹H-NMR δ 3.5 (1H, multiplet, -CH-OH-), 1.85 (1H, doublet, -OH), 1.6 (1H, multiplet, -CH-(CH₃)₂), 1.15 (3H, doublet, -C(OH)-CH₃), 0.9 (6H, overlapping doublets, -CH-(CH₃)₂).

$$\underset{\underset{1.1}{}}{\overset{0.8-0.9}{CH_3}}$$

$$\underset{1.1}{\overset{0.8-0.9 \quad 1.4-1.6 \quad | \quad 3.4-3.5}{CH_3CH_2CHCH_2OH}}$$

**Compound D**

[1]H-NMR δ 3.4-3.5 (2H, multiplet; this is more complex than expected because it is adjacent to a stereocenter, -CH$_2$-OH), 2.2 (1H, broad triplet, -OH), 1.4-1.6 (2H, multiplet; this is more complex than expected because it is adjacent to a stereocenter, CH$_3$-CH$_2$-), 1.1 (1H, multiplet, -CH-), 0.8-0.9 (6H, broad multiplet, both -CH$_3$ groups).

<u>Problem 20.16</u> Following are structural formulas for three alcohols of molecular formula C$_7$H$_{16}$O and three sets of [13]C-NMR spectral data. Assign each constitutional isomer its correct spectrum.

(a)  CH$_3$CH$_2$CH$_2$CH$_2$CH$_2$CH$_2$CH$_2$OH

(b)  CH$_3$ C̈CH$_2$CH$_2$CH$_2$CH$_3$ with OH above and CH$_3$ below

(c)  CH$_3$CH$_2$C̈CH$_2$CH$_3$ with OH above and CH$_2$CH$_3$ below

| Spectrum 1: | Spectrum 2: | Spectrum 3: |
|---|---|---|
| 74.66 | 70.97 | 62.93 |
| 30.54 | 43.74 | 32.79 |
| 7.73 | 29.21 | 31.86 |
| | 26.60 | 29.14 |
| | 23.27 | 25.75 |
| | 14.09 | 22.63 |
| | | 14.08 |

These constitutional isomers are most readily distinguished by the number of sets of nonequivalent carbon atoms and thus different [13]C signals. Using the following analysis, it can be seen that compound (a) has 7 sets of nonequivalent carbon atoms corresponding to spectrum 3, compound (b) has 6 sets of nonequivalent carbon atoms corresponding to spectrum 2, and compound (c) has 3 sets of nonequivalent carbon atoms corresponding to spectrum 1.

$$\underset{}{\overset{g \quad f \quad e \quad d \quad c \quad b \quad a}{CH_3CH_2CH_2CH_2CH_2CH_2CH_2OH}}$$

$$\underset{e}{\overset{OH}{\underset{}{\overset{e \quad a \, | \quad b \quad c \quad d \quad f}{CH_3CCH_2CH_2CH_2CH_3}}}}$$  with CH$_3$ (e) below

$$\underset{b \quad c}{\overset{OH}{\underset{}{\overset{c \quad b \quad a \, | \, b \quad c}{CH_3CH_2CCH_2CH_3}}}}$$  with CH$_2$CH$_3$ (b c) below

**Problem 20.17** Alcohol E, molecular formula $C_6H_{14}O$, undergoes acid-catalyzed dehydration when warmed with phosphoric acid to give compound F, molecular formula $C_6H_{12}$, as the major product. The $^1H$-NMR spectrum of compound E shows peaks at δ 0.89 (t, 6H), 1.12 (s, 3H), 1.38 (s, 1H), and 1.48 (q, 4H). The $^{13}C$-NMR spectrum of compound E shows peaks at δ 72.98, 33.72, 25.85, and 8.16. Propose structural formulas for compounds E and F.

**From the molecular formula, there is a hydrogen deficiency index of 0, so there are no rings or pi bonds in compound E. From the $^{13}C$-NMR peak at 72.98 there is a carbon bonded to an -OH group. The rest of the structure can be deduced from the $^1H$-NMR spectrum. The chemical shifts associated with each set of hydrogens are indicated on the structure.**

**Compound E**

$^1H$-NMR δ 1.48 (4H, quartet, $-CH_2$), 1.38 (1H, singlet, $-OH$), 1.12 (3H, singlet, $-C(OH)-CH_3$), 0.89 (6H, triplet, both $-CH_2-CH_3$ groups)

**Dehydration of compound E gives the following alkene.**

**Compound F**

**Problem 20.18** Compound G, $C_6H_{14}O$, does not react with sodium metal and does not discharge the color of $Br_2$ in $CCl_4$. Its $^1H$-NMR spectrum consists of only two signals, a 12H doublet at δ 1.1 and a 2H septet at δ 3.6. Propose a structural formula for compound G.

**From the molecular formula, there is a hydrogen deficiency index of 0, so there are no rings or pi bonds in compound G. Since it does not react with sodium metal there cannot be an -OH group. The oxygen atom must therefore be contained within an ether group. The simplicity of the $^1H$-NMR spectrum indicates a highly level of symmetry in the molecule, with each methyl group being attached to a carbon with a single hydrogen atom. The only structure consistent with all of this information is the following ether. The chemical shifts associated with each set of hydrogens are indicated on the structure.**

**Compound G**

<u>Problem 20.19</u>  Propose a structural formulas for each haloalkane:.
(a) $C_2H_4Br_2$: δ 2.5 (d, 3H) and 5.9 (q, 1H)

$$\overset{2.5}{CH_3}-\overset{5.9}{CH}Br_2$$

(b) $C_4H_8Cl_2$: δ 1.60 (d, 3H), 2.15 (q, 2H), 3.72 (t, 2H), and 4.27 (sextet, 1H)

$$\overset{1.6}{CH_3}-\overset{4.27}{CH}Cl-\overset{2.15}{CH_2}-\overset{3.72}{CH_2}Cl$$

(c) $C_5H_8Br_4$: δ 3.6 (s, 8H)

$$\overset{3.6}{CH_2}Br-\underset{\underset{CH_2Br}{\overset{3.6}{|}}}{\overset{\overset{3.6}{CH_2Br}}{\overset{|}{C}}}-\overset{3.6}{CH_2}Br$$

(d) $C_4H_9Br$: δ 1.1 (d, 6H), 1.9 (m, 1H), and 3.4 (d, 2H)

$$\overset{1.1}{CH_3}-\underset{1.9}{\overset{\overset{1.1}{CH_3}}{\overset{|}{CH}}}-\overset{3.4}{CH_2}Br$$

(e) $C_5H_{11}Br$: δ 1.1 (s, 9H) and 3.2 (s, 2H)

$$\overset{1.1}{CH_3}-\underset{\underset{CH_3}{\overset{1.1}{|}}}{\overset{\overset{1.1}{CH_3}}{\overset{|}{C}}}-\overset{3.2}{CH_2}Br$$

(f) $C_7H_{15}Cl$: δ 1.1 (s, 9H) and 1.6 (s, 6H)

$$\overset{1.1}{CH_3}-\underset{\underset{CH_3}{\overset{1.1}{|}}}{\overset{\overset{1.1}{CH_3}}{\overset{|}{C}}}-\underset{\underset{CH_3}{\overset{1.6}{|}}}{\overset{\overset{1.6}{CH_3}}{\overset{|}{C}}}-Cl$$

<u>Problem 20.20</u>  Following are structural formulas for esters (1), (2), and (3) and three ¹H-NMR spectra.  Assign each compound its correct spectrum and assign all signals to their corresponding hydrogens.

$$\overset{O}{\overset{||}{HCO}}CH_2CH_2CH_3$$
(1)

$$CH_3\overset{O}{\overset{||}{CO}}CH_2CH_3$$
(2)

$$CH_3O\overset{O}{\overset{||}{C}}CH_2CH_3$$
(3)

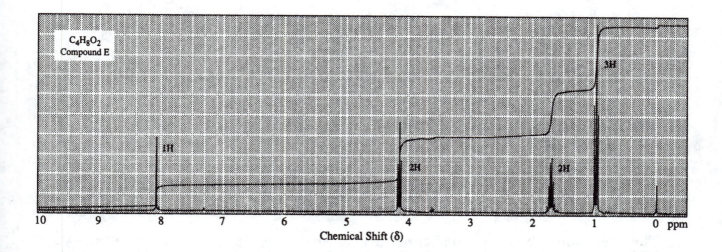

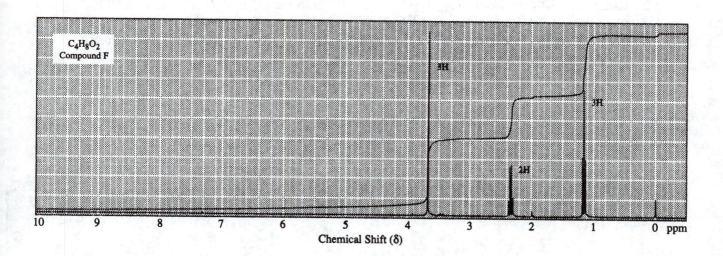

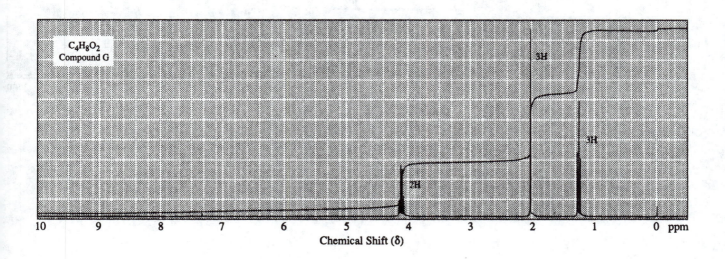

$$\overset{O}{\overset{\|}{\underset{8.1}{\text{HC}}\underset{}{\text{OCH}_2}\underset{4.1}{}\underset{1.7}{\text{CH}_2}\underset{1.0}{\text{CH}_3}}}$$

( 1 )

**Spectrum labeled
Compound E**

$$\overset{O}{\overset{\|}{\underset{2.0}{\text{CH}_3}\underset{}{\text{C}}\underset{4.1}{\text{OCH}_2}\underset{1.25}{\text{CH}_3}}}$$

( 2 )

**Spectrum labeled
Compound G**

$$\overset{O}{\overset{\|}{\underset{3.7}{\text{CH}_3}\underset{}{\text{O}}\underset{}{\text{C}}\underset{2.3}{\text{CH}_2}\underset{1.2}{\text{CH}_3}}}$$

( 3 )

**Spectrum labeled
Compound F**

**The spectrum labeled as compound E corresponds to the ester (1): $^1$H-NMR δ 8.1 (1H, singlet, H-C(O)-), 4.1 (2H, triplet, -O-CH$_2$-), 1.7 (2H, multiplet; a doublet of triplets, -CH$_2$-), 1.0 (3H, triplet, -CH$_3$).**

**The spectrum labeled as compound F corresponds to the ester (3): $^1$H-NMR δ 3.7 (3H, singlet, CH$_3$-O-), 2.3 (2H, quartet, -C(O)-CH$_2$-), 1.2 (3H, triplet, -CH$_3$).**

**The spectrum labeled as compound G corresponds to the ester (2): $^1$H-NMR δ 4.1 (2H, quartet, -C(O)-CH$_2$-), 2.0 (3H, singlet, CH$_3$-C(O)-), 1.25 (3H, triplet, -CH$_3$).**

Problem 20.21  Compound H, $C_{10}H_{12}O_2$, is insoluble in water, 10% NaOH, and 10% HCl. Given are the $^1$H-NMR spectrum and $^{13}$C-NMR spectral data of compound H.  Compound H is reduced by sodium borohydride to compound I, $C_{10}H_{14}O_2$.  Propose structural formulas for compounds H and I

$^{13}$C-NMR

| | |
|---|---|
| 206.51 | 114.17 |
| 158.67 | 55.21 |
| 130.33 | 50.07 |
| 126.31 | 29.03 |

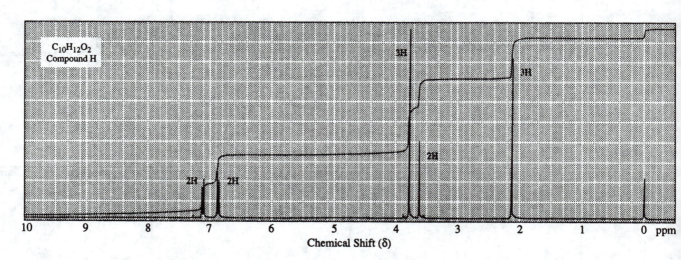

The $^{13}$C-NMR signal at δ 206.51 indicates the presence of a carbonyl group of an aldehyde or ketone, and the absence of any aldehyde signal in the $^1$H-NMR means this must be a ketone.  The $^{13}$C-NMR signals between δ 114 and δ 158 indicate the presence of a phenyl ring. The symmetric doublets at δ 6.88 and δ 7.12, integrating to 2H each, indicate that the ring is 1,4 disubstituted.  The other three signals are singlets; two of which represent methyl groups since they integrate to 3H ($\delta$ 2.10 and $\delta$ 3.76) and the third represents a -$CH_2$- group ($\delta$ 3.61). The only structure that is consistent with the molecular formula and the spectral information is 1-(4-methoxyphenyl)-2-propanone (4-methoxy-phenylacetone).

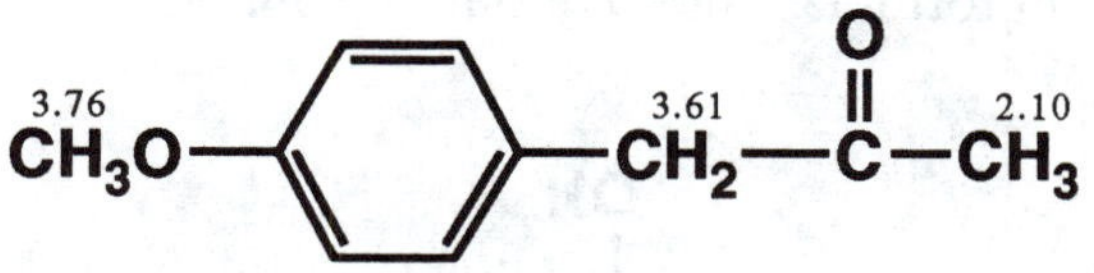

**Compound H**

Compound H will be reduced by sodium borohydride to the secondary alcohol shown below.

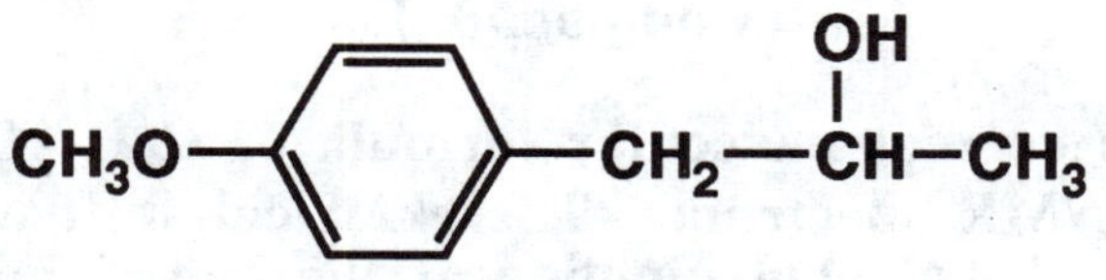

**Compound I**

Problem 20.22 Propose a structural formula for each compound. Each contains an aromatic ring.
(a) $C_9H_{10}O$; $\delta$ 1.2 (t, 3H), 3.0 (quartet, 2H), 7.4-8.0 (m, 5H)

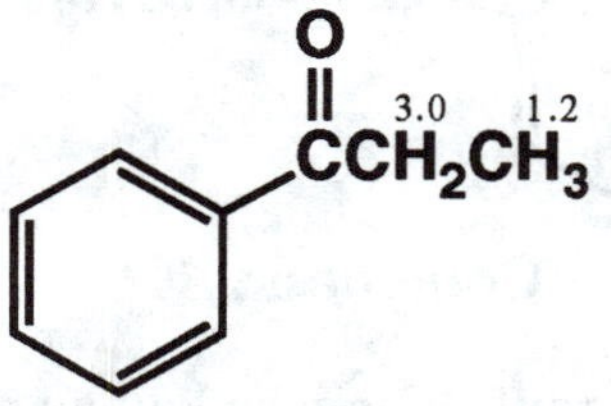

(b) $C_{10}H_{12}O_2$; $\delta$ 2.2 (s, 3H), 2.9 (t, 2H), 4.3 (t, 2H), and 7.3 (s, 5H)

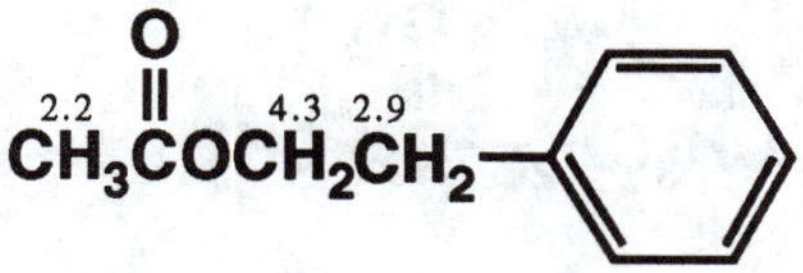

(c) $C_{10}H_{14}$; $\delta$ 1.2 (d, 6H), 2.3 (s, 3H), 2.9 (septet, 1H), and 7.0 (s, 4H)

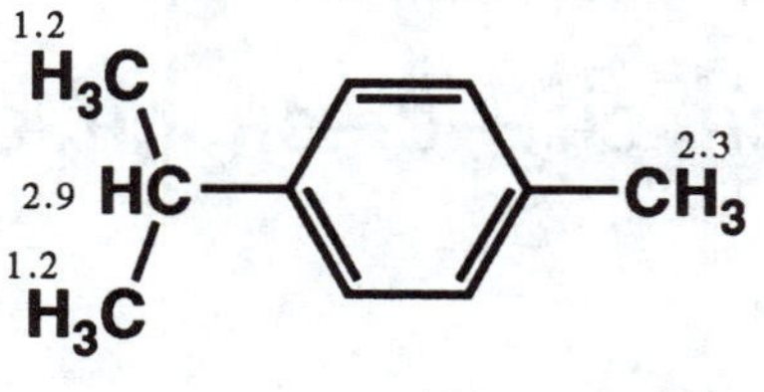

(d) $C_8H_9Br$; $\delta$ 1.8 (d, 3H), 5.0 (quartet, 1H), 7.3 (s, 5H)

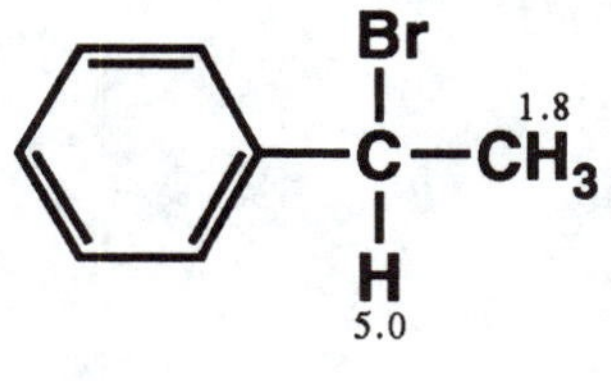

Problem 20.23 Compound J, molecular formula $C_9H_{12}O$, readily undergoes acid-catalyzed dehydration to give compound K, $C_9H_{10}$. The $^1H$-NMR spectrum of compound J shows signals at δ 0.91 (t, 3H), 1.78 (m, 2H), 2.26 (d, 1H), 4.55 (m, 1H), and 7.31 (m 5H). From this information, propose structural formulas for compounds J and K.

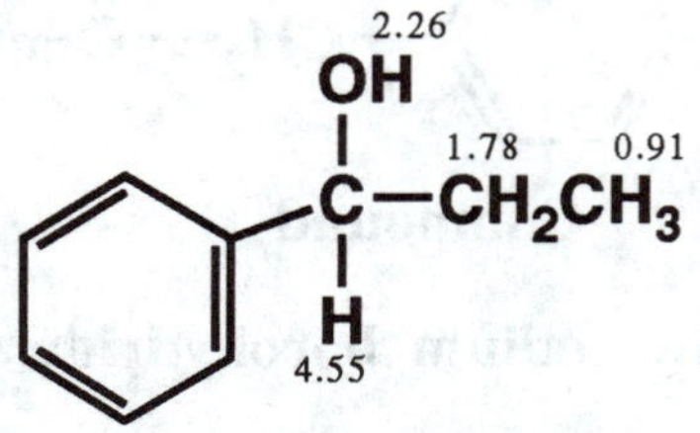

**Compound J**

**This compound has the correct molecular formula of $C_9H_{12}O$ and is fully consistent with the $^1H$-NMR spectrum. The chemical shift of each hydrogen is given on the structure. An alcohol function at the benzyl position explains the acid-catalyzed dehydration that gives the alkene Compound K.**

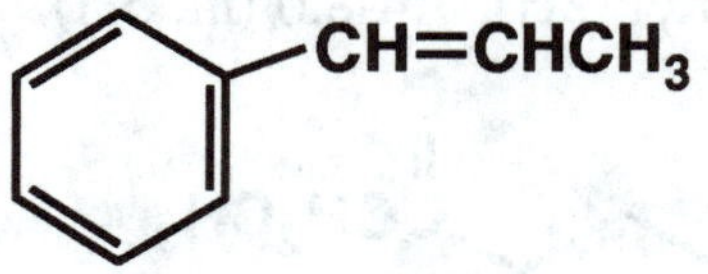

**Compound K**

Problem 20.24 Propose a structural formulas for each ketone or aldehyde.
(a) $C_4H_8O$: δ 1.0 (t, 3H), 2.1 (s, 3H), and 2.4 (q, 2H)

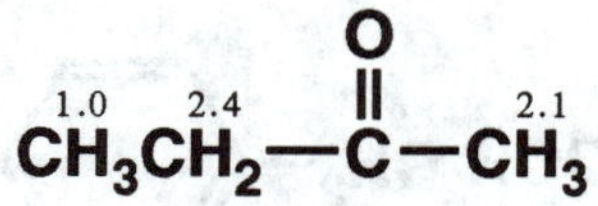

(b) $C_7H_{14}O$: δ 0.9 (t, 6H), 1.6 (sextet, 4H), and 2.4 (t, 4H)

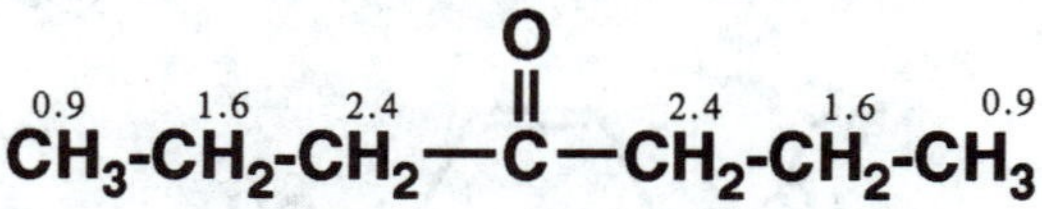

<u>Problem 20.25</u>  Propose a structural formula for compound L, a ketone of molecular formula $C_{10}H_{12}O$.

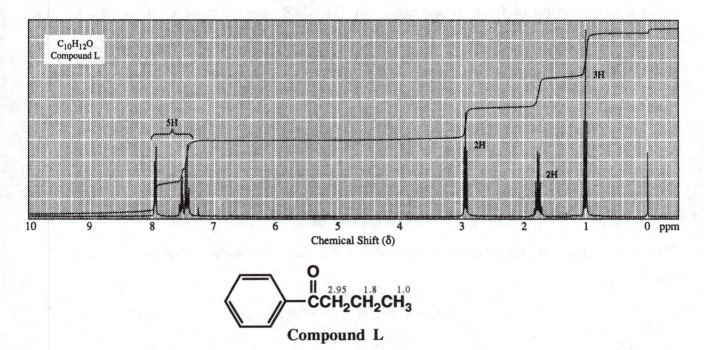

<u>Compound L</u>

This compound $C_{10}H_{12}O$ has the correct molecular formula of $C_{10}H_{12}O$. This compound also has the ketone described in the question. The $^1$H-NMR can be assigned as follows: δ 7.4-8 (5H, multiplet, aromatic hydrogens), 2.95 (2H, triplet, -C(O)CH$_2$-), 1.8 (2H, multiplet, -CH$_2$-CH$_3$), 1.0 (3H, triplet, -CH$_3$).

<u>Problem 20.26</u>  Following is a $^1$H-NMR spectra for compounds M, $C_6H_{12}O_2$. Compound M undergoes acid-catalyzed dehydration to give compound N, $C_6H_{10}O$. Propose structural formulas for compounds M and N.

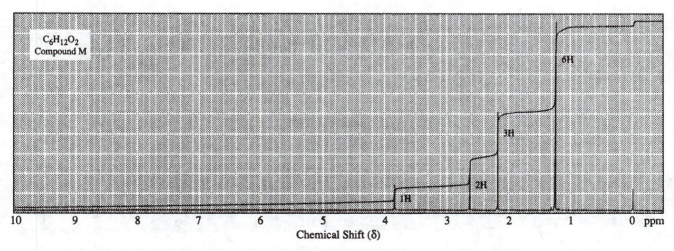

From the molecular formulas it is clear that compound M undergoes an acid-catalyzed dehydration reaction to create compound N. Thus, compound M must have an -OH group. Furthermore, because its index of hydrogen deficiency is 1, compound B must have one pi bond or ring. Compound N has an index of hydrogen deficiency of 2, so it must have two pi bonds and/or rings, consistent with a dehydration to form an alkene. The ¹H-NMR spectrum of compound M shows all singlets. Especially helpful are the methyl group resonances; the singlet integrating to 6H at δ 1.22 and the singlet integrating to 3H at δ 2.18. This latter signal is assigned as a methyl ketone. The other two methyl groups are equivalent. There is the -OH hydrogen at δ 3.85 and a -CH₂- resonance at δ 2.62. The only structure consistent with these signals is 4-hydroxy-4-methyl-2-pentanone.

$$CH_3\!-\!\underset{\underset{\underset{OH}{|}}{\overset{\overset{CH_3}{|}}{C}}}{}\!-\!CH_2\!-\!\underset{\overset{O}{\|}}{C}\!-\!CH_3$$

**Compound M**

Upon dehydration, Compound M would be turned into 4-methyl-3-pentene-2-one.

$$CH_3\!-\!\underset{\overset{CH_3}{|}}{C}\!=\!CH\!-\!\underset{\overset{O}{\|}}{C}\!-\!CH_3$$

**Compound N**

<u>Problem 20.27</u> Propose a structural formula for compound O, C₁₂H₁₆O. Following is its ¹H-NMR spectrum and the position of signals in its ¹³C-NMR spectrum.

### ¹³C-NMR

| | |
|---|---|
| 207.82 | 50.88 |
| 134.24 | 50.57 |
| 129.36 | 24.43 |
| 128.60 | 22.48 |
| 126.86 | |

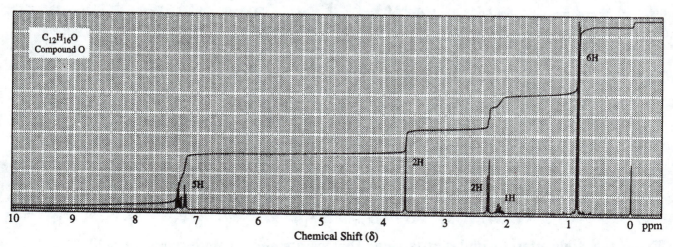

The signal in the $^{13}$C-NMR spectrum at δ 207.82 indicates the presence of a carbonyl group. The signals around δ 130 indicate there is an aromatic ring. The doublet in the $^1$H-NMR at δ 0.84 that integrates to 6H indicates two methyl groups adjacent to a -CH- group. There are also two -CH$_2$- groups; one that is not adjacent to other hydrogens (the singlet at δ 3.62) and one next to a -CH- group (the doublet at δ 2.30). The multiplet at δ 2.12 must be this -CH- group that is also adjacent to the two methyl groups. Five aromatic hydrogens are found in the complex set of signals at δ 7.3. The only structure that is consistent with all of these facts is 4-methyl-1-phenyl-2-pentanone.

$$\underset{}{\text{C}_6\text{H}_5}\underset{3.62}{-\text{CH}_2}\underset{\underset{\text{O}}{\|}}{-\overset{}{\text{C}}}\underset{2.30}{-\text{CH}_2}\underset{\underset{\underset{0.84}{\text{CH}_3}}{|}}{-\underset{2.12}{\text{CH}}}\underset{0.84}{-\text{CH}_3}$$

**Compound O**

Problem 20.28  Propose a structural formula for each carboxylic acid.

(a) $C_5H_{10}O_2$

| $^1$H-NMR | $^{13}$C-NMR |
|---|---|
| 0.94 (t, 3H) | 180.71 |
| 1.39 (m, 2H) | 33.89 |
| 1.62 (m, 2H) | 26.76 |
| 2.35 (t, 2H) | 22.21 |
| 12.0 (s, 1H) | 13.69 |

$$\underset{0.94}{\text{CH}_3}\underset{1.39}{\text{CH}_2}\underset{1.62}{\text{CH}_2}\underset{2.35}{\text{CH}_2}\underset{12.0}{\text{CO}_2\text{H}}$$

(b) $C_6H_{12}O_2$

| $^1$H-NMR | $^{13}$C-NMR |
|---|---|
| 1.08 (s, 9H) | 179.29 |
| 2.23 (s, 2H) | 47.82 |
| 12.1 (s, 1H) | 30.62 |
|  | 29.57 |

$$\underset{1.08}{\text{CH}_3}\underset{\underset{\underset{1.08}{\text{CH}_3}}{|}}{\overset{\overset{\overset{1.08}{\text{CH}_3}}{|}}{\text{C}}}\underset{2.23}{\text{CH}_2}\underset{12.1}{\text{CO}_2\text{H}}$$

(c) $C_5H_8O_4$

| $^1$H-NMR | $^{13}$C-NMR |
|---|---|
| 0.93 (t, 3H) | 170.94 |
| 1.80 (m, 2H) | 53.28 |
| 3.10 (t, 1H) | 21.90 |
| 12.7 (s, 2H) | 11.81 |

$$\underset{12.7}{\text{HO}_2\text{C}}\underset{3.10}{\text{CH}}\underset{12.7}{\text{CO}_2\text{H}}$$
$$\underset{\underset{\underset{1.8 \quad 0.93}{\text{CH}_2\text{CH}_3}}{|}}{}$$

20.29  Following is the $^1$H-NMR spectrum of compound P. Reduction of compound P with lithium aluminum hydride gives two alcohols, molecular formulas $C_4H_{10}O$ and $C_3H_8O$. Propose structural formulas for compound P and the two alcohols.

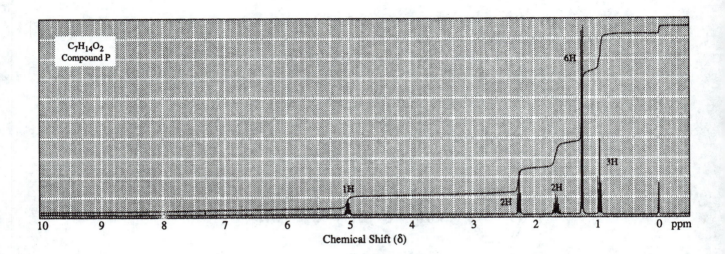

The reduction of compound P with lithium aluminum hydride into two alcohols indicates that compound P is an ester. The multiplet at $\delta$ 5.1 integrating to 1H indicates the carbon atom bound to the ester oxygen atom has a single hydrogen. The triplet at $\delta$ 2.25 integrating to 2H indicates the $-CH_2-$ group adjacent to the carbonyl of the ester is also next to a $-CH_2-$ group, presumably the multiplet at $\delta$ 1.65 that is the only other signal that integrates to 2H. The doublet at $\delta$ 1.3 integrating to 6H must be two methyl groups attached to a -CH- group, and the remaining triplet at $\delta$ 0.95 integrating to 3H must be another methyl group that is attached to a $-CH_2-$ group. The only structure that is consistent with all of this information is isopropyl butyrate.

$$\underset{0.95}{CH_3}\underset{1.65}{CH_2}\underset{2.25}{CH_2}\overset{\overset{\displaystyle O}{\|}}{C}-O\underset{\overset{|}{\underset{1.3}{CH_3}}}{\underset{5.1}{CH}}\overset{1.3}{CH_3}$$

**Compound P**

Compound P will be reduced by lithium aluminum hydride to the two alcohols *n*-butanol and 2-propanol.

$$CH_3CH_2CH_2CH_2OH$$

*n*-Butanol
($C_4H_{10}O$)

$$\underset{}{CH_3}\overset{\overset{\displaystyle OH}{|}}{CH}CH_3$$

2-Propanol
($C_3H_8O$)

<u>Problem 20.30</u>  Propose a structural formula for each ester.

(a) $C_6H_{12}O_2$

| $^1$H-NMR | $^{13}$C-NMR |
|---|---|
| 1.18 (d, 6H) | 177.16 |
| 1.26 (t, 3H) | 60.17 |
| 2.51 (m, 1H) | 34.04 |
| 4.13 (q, 2H) | 19.0 |
| | 14.25 |

$$\underset{1.18}{(CH_3)_2}\underset{2.51}{CH}\overset{\displaystyle O}{\overset{\displaystyle \|}{C}}\underset{4.13}{OCH_2}\underset{1.26}{CH_3}$$

(b) $C_7H_{12}O_4$

| $^1$H-NMR | $^{13}$C-NMR |
|---|---|
| 1.28 (t, 6H) | 166.52 |
| 3.36 (s, 2H) | 61.43 |
| 4.21 (q, 4H) | 41.69 |
| | 14.07 |

$$\underset{1.28}{CH_3}\underset{4.21}{CH_2}O\overset{\displaystyle O}{\overset{\displaystyle \|}{C}}\underset{3.36}{CH_2}\overset{\displaystyle O}{\overset{\displaystyle \|}{C}}\underset{4.21}{OCH_2}\underset{1.28}{CH_3}$$

(c) $C_7H_{14}O_2$

| $^1$H-NMR | $^{13}$C-NMR |
|---|---|
| 0.92 (d, 6H) | 171.15 |
| 1.52 (m, 2H) | 63.12 |
| 1.70 (m, 1H) | 37.31 |
| 2.09 (s, 3H) | 25.05 |
| 4.10 (t, 2H) | 22.45 |
| | 21.06 |

$$\underset{2.09}{CH_3}\overset{\displaystyle O}{\overset{\displaystyle \|}{C}}O\underset{4.10}{CH_2}\underset{1.52}{CH_2}\underset{1.7}{CH}\underset{0.92}{(CH_3)_2}$$

<u>Problem 20.31</u>  Following is the $^1$H-NMR spectrum of compound Q, $C_{11}H_{14}O_3$.  Propose a structural formula for this compound.  (*Hint:* The signal at δ 1.4 is actually two signals, each a closely spaced triplet.

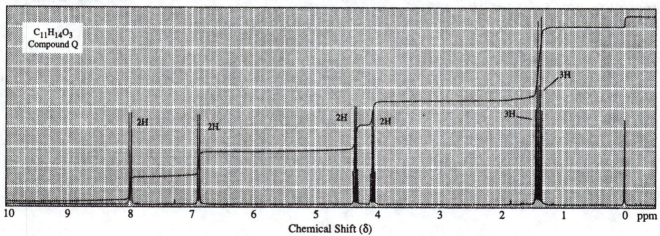

The two sets of doublets integrating to 2H each between δ 6.9 and 8.0 indicate the presence of a 1,4 disubstituted aromatic ring.  The two quartets integrating to 2H each at δ 4.35 and δ 4.1 indicate there are two -CH$_2$- groups attached to oxygen atoms, we know one of which is part of an ester, and each attached to methyl groups.  The two triplets integrating to a total of 6H at δ 1.4 are the signals from the two methyl groups. The only structure that is consistent with the molecular formula $C_{11}H_{14}O_3$ and his spectrum is ethyl 4-ethoxybenzoate.

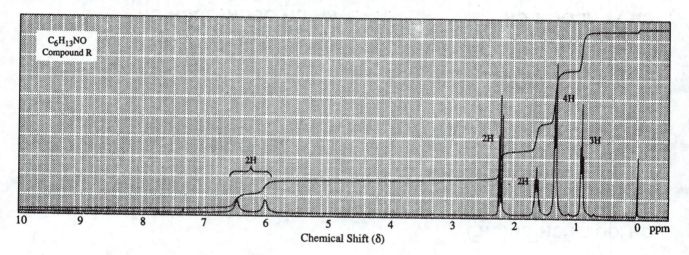

**Compound Q**

Problem 20.32  Propose a structural formula for amide R, molecular formula $C_6H_{13}NO$.

The signals at δ 0.9-2.2 are consistent with a $CH_3CH_2CH_2CH_2CH_2-$ structure, with the last $-CH_2-$ being adjacent to a carbonyl group indicated by a chemical shift of δ 2.2. The two signals integrating to 1H each at δ 6.0 and δ 6.55 are from two different amide N-H's, indicating a primary amide. The only structure of molecular formula $C_6H_{13}NO$ that is consistent with this spectrum is *n*-hexanamide.

$$\underset{0.9}{CH_3}\underset{1.4-22}{(CH_2)_4}\overset{\overset{O}{\|}}{C}\underset{6.0,\ 6.55}{NH_2}$$

**Compound R**

Problem 20.33  Propose a structural formula for the analgesic phenacetin, molecular formula $C_{10}H_{13}NO_2$, based on its $^1H$-NMR spectrum.

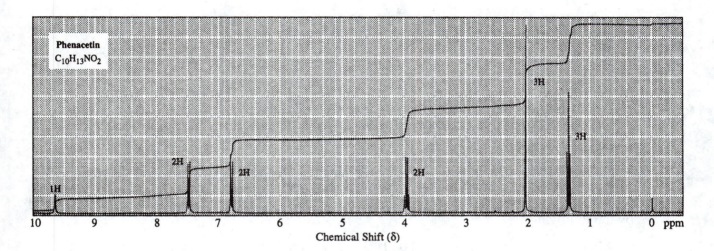

**Phenacetin**

This structure is not only consistent with the molecular formula, but also with the $^1$H-NMR spectrum. The characteristic two doublets centered at δ 7.2 indicates the presence of a 1,4-disubstituted phenyl ring. The singlet at δ 9.65 integrating to 1H indicates a primary amide, and the singlet integrating to 3H at δ 2.02 indicates this is an acetamide. Finally, the typical ethyl splitting pattern for the signals at δ 1.32 and δ 3.96 indicates the presence of an ethyl group. These signals are shifted so far downfield that they must be part of an ethoxy group.

Problem 20.34 Propose a structural formula for compound S, an oily liquid of molecular formula $C_8H_9NO_2$. Compound S is insoluble in water and aqueous NaOH, but dissolves in 10% HCl. When its solution in HCl is neutralized with NaOH, compound S is recovered unchanged. The $^1$H-NMR spectrum of compound S shows signals at δ 3.84 (s, 3H), 4.18 (s, 2H), 7.60 (d, 2H), and 8.70 (d, 2H).

Compound S is an amine based on its solubility in dilute HCl. The two doublets integrating to 2H each at δ 7.60 and δ 8.70 in the $^1$H-NMR spectrum indicate there is a 1,4-disubstituted benzene ring. The singlet integrating to 3H at δ 3.84 indicates the presence of a methyl ester, and the signal integrating to 2H at δ 4.18 indicated the presence of a primary amine. The only structure consistent with the molecular formula of $C_8H_9NO_2$ and the $^1$H-NMR spectrum is methyl 4-aminobenzoate.

**Compound S**

# CHAPTER 21: INFRARED SPECTROSCOPY

## 21.0 OVERVIEW
- **Infrared spectroscopy provides information about molecular vibrations.** ✻
  - The main use of infrared spectroscopy is to determine the presence or absence of certain functional groups in a molecule that give rise to the characteristic molecular vibrations.

## 21.1 INFRARED SPECTROSCOPY
- The **vibrational infrared** region of electromagnetic radiation has wavelengths that extend from 2.5 μm to 25 μm. For the purposes of **infrared spectroscopy (IR spectroscopy)** it is useful to describe infrared radiation in terms of a unit called the **wavenumber** that is defined by the following:

$$\text{wavenumber} = \frac{10,000}{\lambda} \qquad \text{where } \lambda \text{ is wavelength measured in micrometers}$$

**Wavenumbers** are reported in the units of **$cm^{-1}$**.
- **Atoms** joined by covalent bonds can **vibrate in a quantized fashion**, that is only specific vibrational energy levels are allowed. The **energy of these vibrations** corresponds to that of the **vibrational infrared region**. Therefore, absorption of infrared radiation of the appropriate wavelength results in a vibrationally excited state. ✻
- **Infrared radiation is absorbed** if the **frequency** of the infrared radiation **matches** that of an **allowed vibrational transition.** Furthermore, the **bond(s)** undergoing the vibrational transition **must have a dipole moment**, and **absorption** of radiation must result in a **change of** that **dipole moment.** The **greater** this **change in dipole moment**, the **more intense** the **absorption. Infrared active** vibrations meet the above criteria. Note that symmetrical bonds such as those in homonuclear diatomics ($Br_2$, $O_2$, etc.) do not absorb infrared radiation, because they do not have a dipole moment. ✻
- **Nonlinear molecules** (molecules with branches, etc.) that have **n** atoms will have **3n-6 allowed fundamental vibrations.** The simplest vibrations are **stretching** and **bending.** Stretching vibrations can be symmetric or asymmetric. Bending vibrations can be relatively complicated motions such as scissoring, rocking, wagging, or twisting. ✻
- An **infrared spectrophotometer** is the instrument that measures which frequencies of infrared radiation are absorbed by a given sample.
  - IR spectra are recorded from samples that are usually neat if they are liquids, or compacted wafers mixed with KBr if they are solids.
  - **Characteristic absorptions** for different functional groups are recorded in **correlation tables.** The **intensity** of a particular absorption is referred to as being **strong (s)**, **medium (m)**, or **weak (w).**
  - In general, most attention is paid to the region from 4000 $cm^{-1}$ to 1600 $cm^{-1}$ of an infrared spectrum, because most functional groups have characteristic absorptions in this area. Absorptions in the 1000 $cm^{-1}$ to 400 $cm^{-1}$ are much more complex and difficult to analyze, so this area is referred to as the **fingerprint region.**

## 21.2 INTERPRETING INFRARED SPECTRA
- **Alkanes** have **C-H stretching** vibrations near **2900 $cm^{-1}$** and **methylene bending** at **1450 $cm^{-1}$**.
- **Alkenes** have a vinylic **C-H stretch** near **3000 $cm^{-1}$**. The **C=C stretching** near **1600-1660 $cm^{-1}$** is often **weak** and may not be visible.

- **Alkynes** with a **hydrogen on the triple bonded carbon** have a **C-H stretching** vibration near **3300 cm$^{-1}$** that is usually sharp and strong. The **C≡C stretching** occurs at **2150 cm$^{-1}$**.
- **Alcohols** have an **OH stretch** that **depends on the amount of hydrogen bonding** present in the sample. With **no hydrogen bonding** such as when a dilute solution of the alcohol is used, the **OH stretch** is a **sharp peak** of weak intensity at **3625 cm$^{-1}$**. Normally, alcohol samples show **extensive hydrogen bonding** and a **broad absorption** near **3350 cm$^{-1}$**. The **C-O stretch** occurs at **1050 cm$^{-1}$, 1100 cm$^{-1}$, and 1150 cm$^{-1}$** for **primary, secondary, and tertiary alcohols,** respectively.
- **Ethers** have a **strong C-O stretch** near **1000 to 1300 cm$^{-1}$**.
- **Carbonyl**-containing species such as aldehydes, ketones, carboxylic acids, and carboxylic acid derivatives also have **very strong** and **characteristic IR absorptions** in **1630 cm$^{-1}$** to **1800 cm$^{-1}$** region. As a result, infrared spectroscopy is especially useful for characterizing these species.

# CHAPTER 21
## *Solutions to the Problems*

<u>Problem 21.1</u>  A compound shows strong broad IR absorption at 2860 cm$^{-1}$ and strong absorption at 1705 cm$^{-1}$.  What functional group accounts for both of these absorptions?

**The broad IR absorption at 2860 cm$^{-1}$ and strong sharp absorption at 1705 cm$^{-1}$ are characteristic of -C-H and -C=O bonds, respectively.**

<u>Problem 21.2</u>  Propanoic acid and methyl ethanoate are constitutional isomers.  Show how to distinguish between them by IR spectroscopy.

$$CH_3CH_2\overset{\overset{\displaystyle O}{\|}}{C}OH$$
Propanoic acid

$$CH_3\overset{\overset{\displaystyle O}{\|}}{C}OCH_3$$
Methyl ethanoate
(Methyl acetate)

**The propanoic acid will have a strong broad absorption at 3200 to 3550 cm$^{-1}$ due to the presence of the -OH bond of the carboxylic acid group.  The IR spectrum of methyl ethanoate will not have an absorption in this region.**

<u>Problem 21.3</u>  Compound A, molecular formula, $C_6H_{10}$, reacts with $H_2$/Ni to give compound B, $C_6H_{12}$.  See also the IR spectrum of compound A.  From this information about compound A, tell:
(a)  Its index of hydrogen deficiency.

**The index of hydrogen deficiency for compound A is two.**

(b)  The number of its rings and/or pi bonds.

**Compound A must have two rings and/or pi bonds.**

(c)  What structural feature(s) would account for its index of hydrogen deficiency.

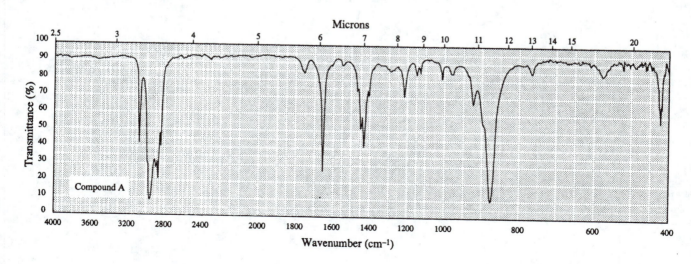

Compound A can only have one pi bond, because reaction with $H_2$/Ni adds only 2 more hydrogen atoms.  Thus, there must be one pi bond and one ring present in the structure.  In addition, the double bond must be highly unsymmetrical, having a permanent dipole, to explain the prominent C=C stretching band at 1654 cm$^{-1}$ seen in the IR spectrum.

It would probably take more information than just the IR spectrum to deduce unambiguously the detailed structure of compound A.  For example, $^1$H-NMR and $^{13}$C-NMR spectra would be helpful.  Nevertheless, for the sake of completeness, the structure of compound A is methylenecyclopentane:

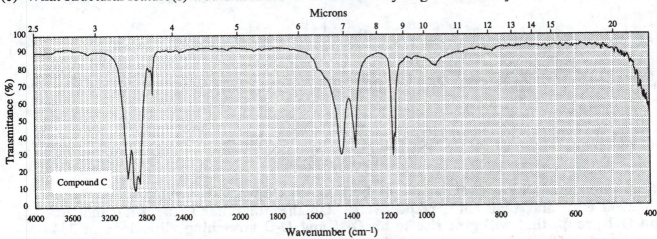

**Methylenecyclopentane**

<u>Problem 21.4</u>  Compound C, molecular formula, $C_6H_{12}$, reacts with $H_2$/Ni to give compound D, $C_6H_{14}$.  See also the IR spectrum of compound C.  From this information about compound C, tell:

(a)  Its index of hydrogen deficiency.

**The index of hydrogen deficiency of compound C is one.**

(b)  The number of its rings and/or pi bonds.

**Compound C must have one ring or pi bond.**

(c)  What structural feature(s) would account for its index of hydrogen deficiency.

Compound C must have one pi bond, since it can react with $H_2$/Ni and add two hydrogen atoms.  However, the absence of any C=C stretching bands near 1650 cm$^{-1}$ in the IR spectrum indicates that this must be an entirely symmetrical double bond.

It would probably take more information than just the IR spectrum to deduce unambiguously the detailed structure of compound C.  For example, $^1$H-NMR and $^{13}$C-NMR spectra would be helpful.  Nevertheless, for the sake of completeness, the structure of compound C is 2,3-dimethyl-2-butene:

$$(CH_3)_2C{=}C(CH_3)_2$$

## 2,3-Dimethyl-2-butene

<u>Problem 21.5</u>  Following are infrared spectra of compounds E and F.  One spectrum is of 1-hexanol, the other of nonane.  Assign each compound its correct infrared spectrum.

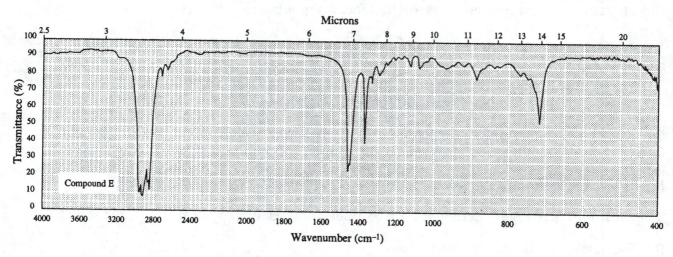

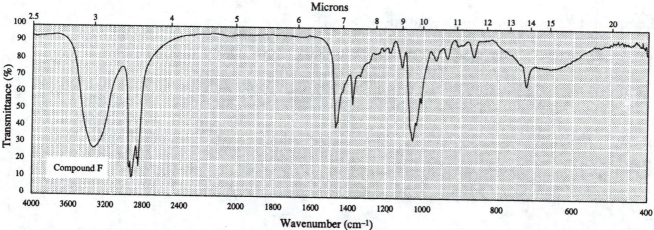

**Both compounds have C-H bonds, so both spectra have C-H stretches and bends at 2900 cm⁻¹ and 1450 cm⁻¹, respectively.  On the other hand, the 1-hexanol has an OH group, that will give rise to an O-H and C-O stretching vibrations at 3340 cm⁻¹ and 1050 cm⁻¹, respectively.  These two features are in the second spectra, so the second spectra must correspond to the 1-hexanol and the first spectra must correspond to nonane.**

<u>Problem 21.6</u>  2-Methyl-1-butanol and *tert*-butyl methyl ether are constitutional isomers of molecular formula $C_5H_{12}O$.  Assign each compound its correct infrared spectrum, G or H.

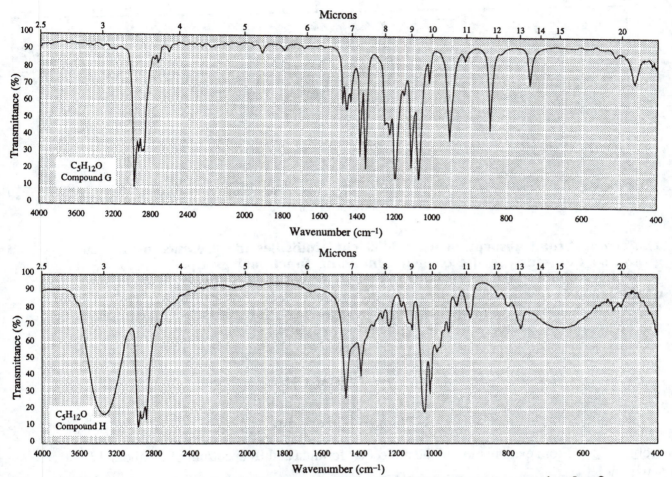

The molecules are extremely similar except for the -OH group present in the 2-methyl-1-butanol. Since the characteristic O-H stretch is present at 3625 cm$^{-1}$ in the second spectrum, this verifies that the second spectrum corresponds to 2-methyl-1-butanol. Therefore, the first spectrum corresponds to *tert*-butyl methyl ether.

Problem 21.7 From examination of the molecular formula and IR spectrum of compound I $C_9H_{12}O$, tell:
(a) Its index of hydrogen deficiency.

**The index of hydrogen deficiency of compound I is four.**

(b) The number of its rings and/or pi bonds.

**Compound I has four rings and/or pi bonds.**

(c) What one structural feature would account for this index of hydrogen deficiency.

**A single benzene ring could account for an index of hydrogen deficiency of four.**

(d) The oxygen-containing functional group.

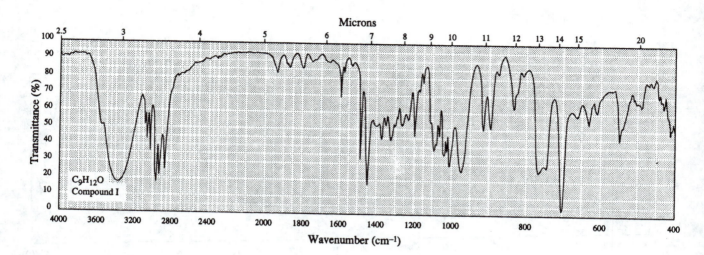

The strong broad absorption near 3400 cm$^{-1}$ indicates the presence of an -OH group, so this must be the oxygen containing functional group.

It would probably take more information than just the IR spectrum to deduce unambiguously the detailed structure of compound I. For example, $^1$H-NMR and $^{13}$C-NMR spectra would be helpful. Nevertheless, for the sake of completeness, the structure of compound I is 1-phenyl-1-propanol:

1-Phenyl-1-propanol

Problem 21.8 From examination of the molecular formula and IR spectrum of compound J $C_5H_{13}N$, tell:

(a) Its index of hydrogen deficiency.

**The index of hydrogen deficiency of compound J is 0.**

(b) The number of its rings and/or pi bonds.

**Compound J has no rings and/or pi bonds.**

(c) The nitrogen-containing functional group(s) it might contain.

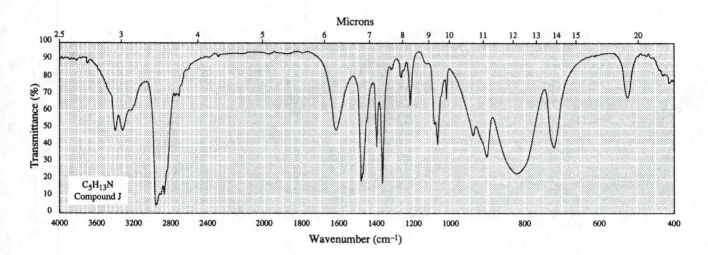

Compound J must contain an amine functional group, a prediction that is confirmed by the presence of the two broad absorptions at 3300 and 3400 cm$^{-1}$. The presence of the two bands indicates compound J is a primary amine. It would probably take more information than just the IR spectrum to deduce unambiguously the detailed structure of compound J. For example, $^1$H-NMR and $^{13}$C-NMR spectra would be helpful. Nevertheless, for the sake of completeness, the structure of compound J is 3,3-dimethylbutanamine:

$$CH_3CCH_2NH_2$$

with CH$_3$ above and CH$_3$ below the central carbon.

**3,3-Dimethylbutanamine**

Problem 21.9  From examination of the molecular formula and IR spectrum of compound K, $C_{10}H_{12}O$, tell:
(a)  Its index of hydrogen deficiency.

**The index of hydrogen deficiency of compound K is five.**

(b)  The number of its rings and/or pi bonds.

**Compound K has five rings and/or pi bonds.**

(c)  What structural features would account for this index of hydrogen deficiency.

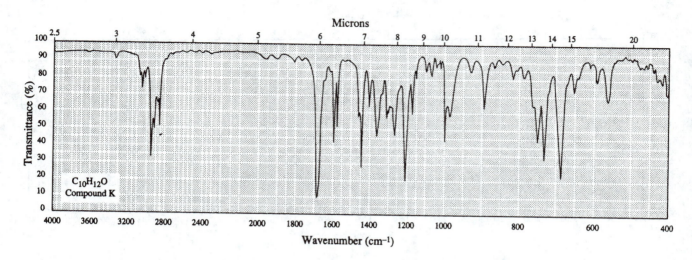

A benzene ring could account for part of this index of hydrogen deficiency, a prediction that is confirmed by the sharp absorptions at 1475 and 1600 cm$^{-1}$. In addition, compound J must contain a carbonyl group because of the strong absorption at 1690 cm$^{-1}$.

It would probably take more information than just the IR spectrum to deduce unambiguously the detailed structure of compound K. For example, $^1$H-NMR and $^{13}$C-NMR spectra would be helpful. Nevertheless, for the sake of completeness, the structure of compound K is 1-phenyl-1-butanone:

**1-Phenyl-1-butanone**

Problem 21.10 From examination of the molecular formula and IR spectrum of compound L, $C_7H_{14}O_2$, tell:

(a) Its index of hydrogen deficiency.

**The index of hydrogen deficiency of compound L is one.**

(b) The number of its rings and/or pi bonds.

**Compound L has one ring and/or pi bonds.**

(c) The oxygen-containing functional group(s) it might contain.

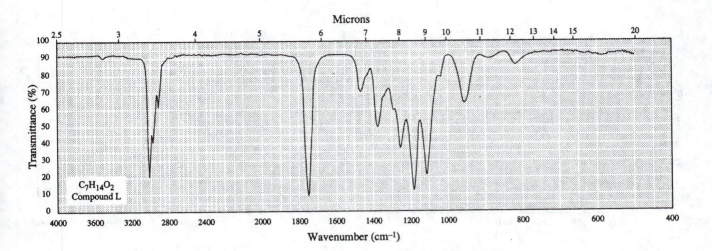

The presence of the strong C=O absorption band at 1750 cm$^{-1}$ indicates the presence of a carbonyl group in compound L. There is no broad -OH absorption between 2500 and 3300 cm$^{-1}$ to indicate the presence of a carboxylic acid. Thus, it is likely that the oxygen-containing functional group of compound L is an aldehyde, a ketone, or an ester. The strong C-O absorptions between 1100 and 1200 cm$^{-1}$ indicate that compound L is an ester.

It would probably take more information than just the IR spectrum to deduce unambiguously the detailed structure of compound L. For example, $^1$H-NMR and $^{13}$C-NMR spectra would be helpful. Nevertheless, for the sake of completeness, the structure of compound L is 2-propyl butanoate:

$$CH_3CH_2CH_2\overset{\overset{\displaystyle O}{\|}}{C}OCH(CH_3)_2$$

**2-Propyl butanoate**

Problem 21.11 From examination of the molecular formula and IR spectrum of compound M C$_6$H$_{13}$NO, tell:
(a) Its index of hydrogen deficiency.

**The index of hydrogen deficiency of compound M is one.**

(b) The number of its rings and/or pi bonds.

**Compound M has one ring and/or pi bonds.**

(c) The oxygen and nitrogen-containing functional group(s).

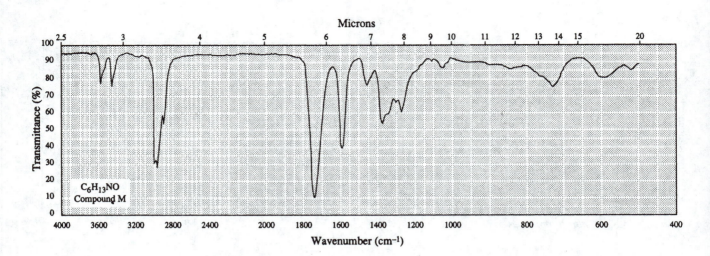

The presence of the carbonyl absorption at 1740 cm$^{-1}$ and the two broad
absorptions at 3460 and 3590 cm$^{-1}$ indicate that the oxygen and nitrogen
containing functional group in compound M is a primary amide.
It would probably take more information than just the IR spectrum to deduce
unambiguously the detailed structure of compound M. For example, $^{1}$H-NMR
and $^{13}$C-NMR spectra would be helpful. Nevertheless, for the sake of
completeness, the structure of compound M is hexanamide:

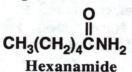

**Hexanamide**

<u>Problem 21.12</u> Show how IR spectroscopy can be used to distinguish between the compounds in
each set.
(a) 1-Butanol and diethyl ether

The 1-butanol will have a strong, broad O-H absorption between 3200 and 3550
cm$^{-1}$, the diethyl ether will not.

(b) Butanoic acid and 1-butanol

The butanoic acid will have a strong C=O absorption between 1710 and 1780
cm$^{-1}$, the 1-butanol will not.

(c) Butanoic acid and 2-butanone

The butanoic acid will have a strong, broad OH absorption between 2500 and
3000 cm$^{-1}$, the 2-butanone will not.

(d) Butanal and 1-butene

The butanal will have a strong C=O absorption between 1690 and 1740 cm$^{-1}$, and
the 1-butene will have a medium C=C absorption between 1620 and 1680 cm$^{-1}$
along with a strong vinylic C-H absorption near 3010-3095 cm$^{-1}$.

(e) 2-Butanone and 2-butanol

**2-Butanone will have a strong C=O absorption between 1680 and 1750 cm⁻¹, and the 2-butanol will have a strong, broad OH absorption between 3200 and 3550 cm⁻¹ along with a medium C-O absorption between 1050 and 1150 cm⁻¹.**

(f) Butane and 2-butene

**2-Butene will have a strong vinylic C-H absorption near 3030 cm⁻¹.**

<u>Problem 21.13</u> For each set of compounds, list one major feature that appears in the IR spectrum of one compound but not the other. Your answer should state what type of bond vibration is responsible for the spectral feature you list and its approximate position in the IR spectrum.

(a) 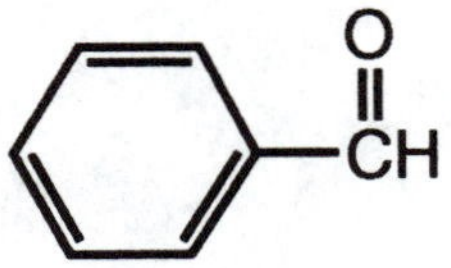  and

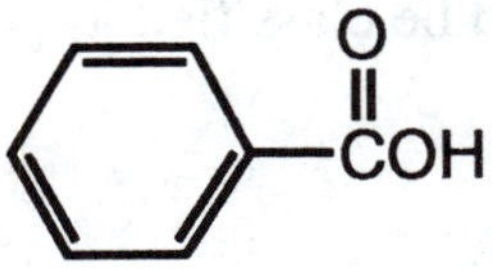

**The benzoic acid on the right will have a strong, broad OH absorption between 2500 and 3000 cm⁻¹, the benzaldehyde will not.**

(b)

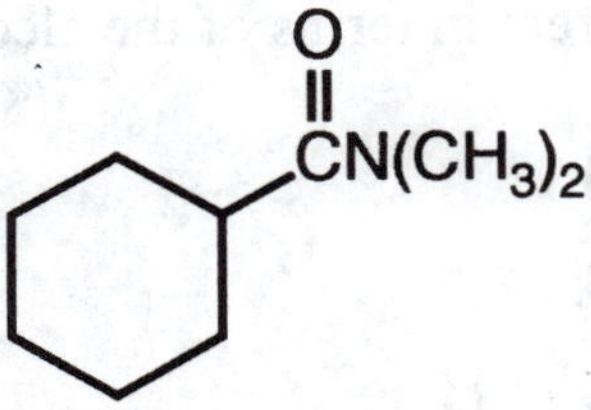

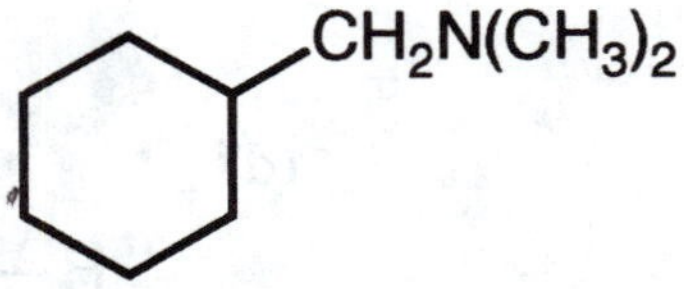

**The amide on the left will have a strong C=O absorption between 1630 and 1690 cm⁻¹, the amine on the right will not.**

(c) 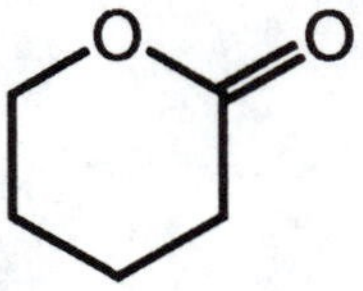  and  $HO(CH_2)_4COH$

**The 5-hydroxypentanoic acid on the right will have a strong, broad OH absorption between 2500 and 3000 cm⁻¹ from the acid OH and a strong, broad OH absorption between 3200 and 3550 cm⁻¹ from the alcohol OH. The lactone on the left will have neither of these absorptions.**

(d) 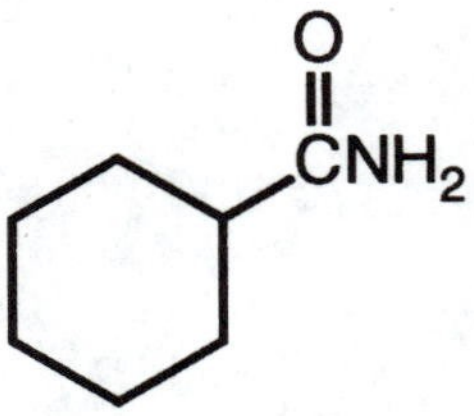  and

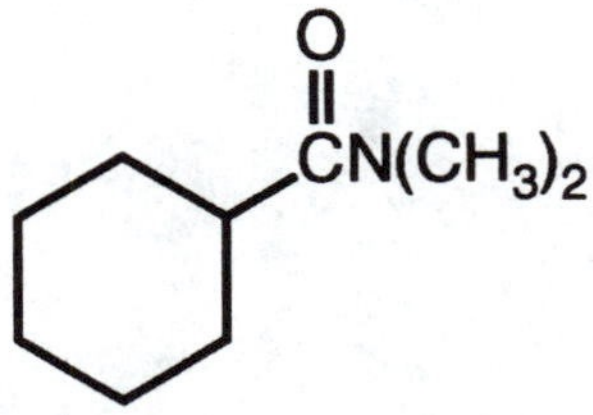

**The primary amide on the left will have two broad N-H absorptions between 3200 and 3400 cm⁻¹, the N,N-dimethyl amide on the right will not.**

# CHAPTER 1: QUIZ

*Determine all of the correct answers for each question. Each question may have more than one correct response.*

1. Which of the electronic configurations is correct for the ground state energy of an atom or ion of the element specified?

    (a) Carbon atom: $1s^2 2s^2 2p_x^1 2p_y^0 2p_z^1$

    (b) Oxygen atom: $1s^2 2s^2 2p_x^1 2p_y^2 2p_z^1$

    (c) Nitrogen atom: $1s^2 2s^2 2p_x^2 2p_y^1 2p_z^0$

    (d) Sodium ion ($Na^+$): $1s^1 2s^2 2p_x^2 2p_y^2 2p_z^2$

    (e) Chloride ion ($Cl^-$): $1s^2 2s^2 2p_x^2 2p_y^2 2p_z^2 3s^2 3p_x^2 3p_y^2 3p_z^1$

2. Which of the following bonds would be classified as polar covalent?

    (a) C—Br

    (b) H—Br

    (c) C—O

    (d) Na—O

    (e) C—N

3. Which of the following Lewis diagrams are incomplete or incorrect in terms of the electron distribution?

    (a)

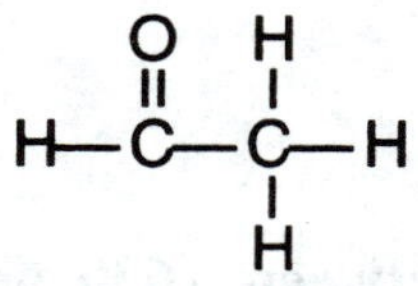

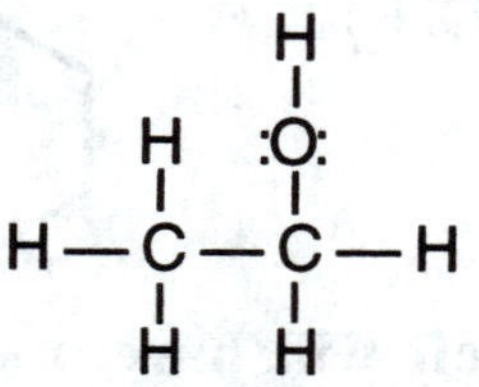

    (b)

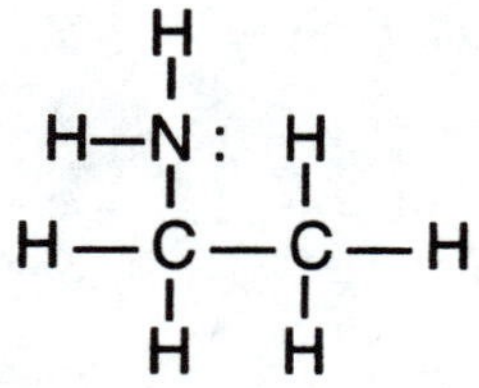

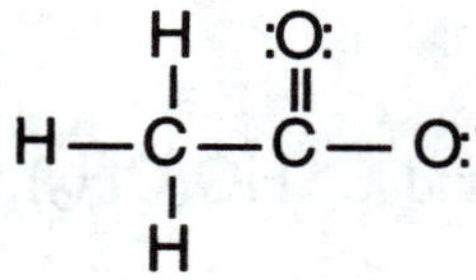

    (c)

    (d)

    (e)

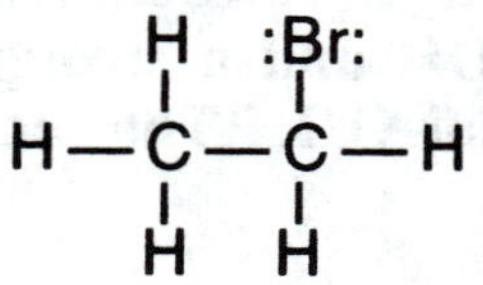

**Use the following compounds to answer questions 4-7.**

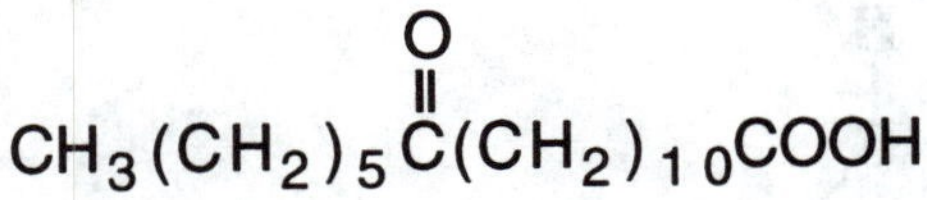

$$CH_3(CH_2)_5 \overset{\overset{O}{\|}}{C}(CH_2)_{10}COOH$$

**a**

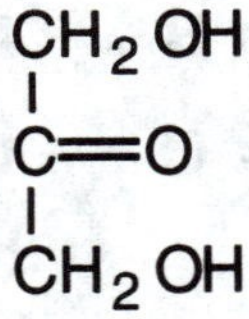

$$\begin{array}{c} CH_2OH \\ | \\ C=O \\ | \\ CH_2OH \end{array}$$

**b**

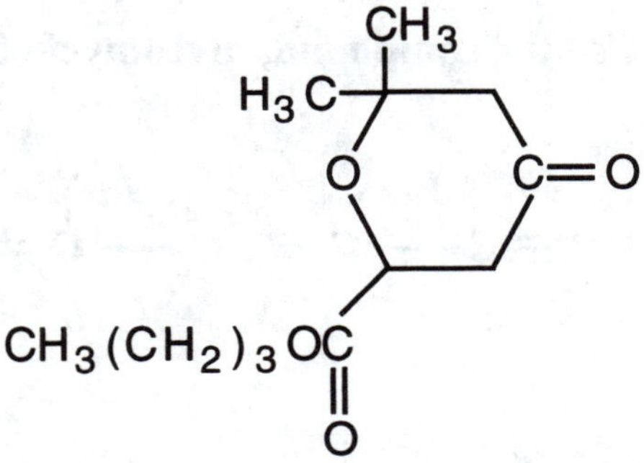

**c**

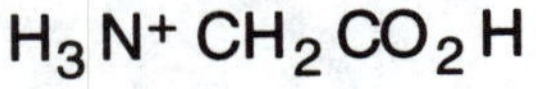

$$H_3N^+CH_2CO_2H$$

**d**

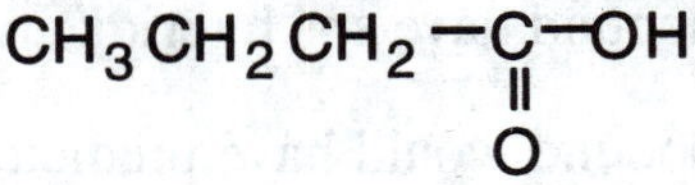

$$CH_3CH_2CH_2 \overset{\overset{}{\underset{\underset{O}{\|}}{C}}}{-}OH$$

**e**

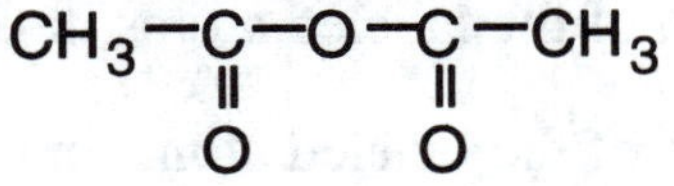

$$CH_3 \overset{\overset{}{\underset{\underset{O}{\|}}{C}}}{-}O \overset{\overset{}{\underset{\underset{O}{\|}}{C}}}{-}CH_3$$

**f**

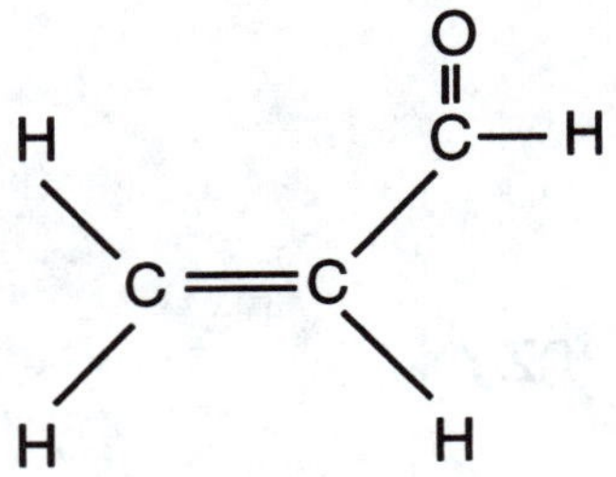

**g**

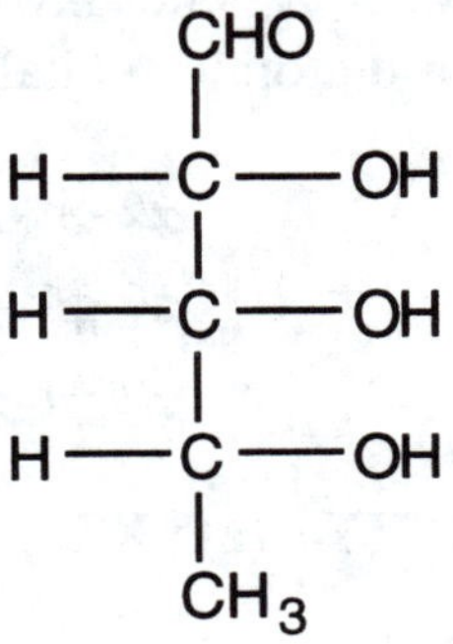

$$\begin{array}{c} CHO \\ | \\ H-C-OH \\ | \\ H-C-OH \\ | \\ H-C-OH \\ | \\ CH_3 \end{array}$$

**h**

$$CH_3 \overset{\overset{}{\underset{\underset{O}{\|}}{C}}}{-}O-CH_2CH_3$$

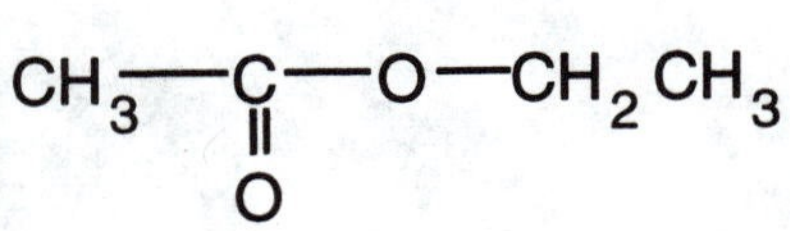

**i**

$$\overset{\overset{O}{\|}}{HCOH}$$

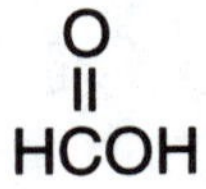

**j**

4. Which compounds contain a carbonyl group?

5. Which compounds are aldehydes?

6. Which compounds are carboxylic acids?

7. Which compounds are alcohols?

**Use the compound, mycomycin isolated from a fungus, to answer questions 8 and 9.**

8. Which of the labeled atoms in the compound have $sp^2$ hybridization?

9. Which of the labeled atoms in the compound would have predicted bond angles of 120°?

10. The symbol $\mathcal{H}$ will represent a single hybrid orbital in the following systems. The 1s atomic orbital is not included since it is not involved in bonding or hybrid atomic orbital formation. Place the following hybrid atomic orbital systems for nitrogen in the order $sp^3$, $sp^2$, sp.

X)  $2s^2\ 2p_x^{\ 1}\ 2p_y^{\ 1}\ 2p_z^{\ 1}$          $\mathcal{H}^2\ \mathcal{H}^1\ \mathcal{H}^1 + 2p_z^{\ 1}$

Y)  $2s^2\ 2p_x^{\ 1}\ 2p_y^{\ 1}\ 2p_z^{\ 1}$          $\mathcal{H}^2\ \mathcal{H}^1 + 2p_y^{\ 1} + 2p_z^{\ 1}$

Z)  $2s^2\ 2p_x^{\ 1}\ 2p_y^{\ 1}\ 2p_z^{\ 1}$          $\mathcal{H}^2\ \mathcal{H}^1\ \mathcal{H}^1\ \mathcal{H}^1$

The order is: (a) XYZ  (b)XZY  (c)YXZ  (d)YZX  (e)ZXY  (f)ZYX

**Chapter 1 Quiz Answers**
(1) **a, b** (2) **a, b, c, e** (3) **a, c, e** (4) **a, b, c, g, h** (5) **g, h** (6) **a, d, e, j** (7) **b, h** (8) **b, d, e, g, h**
(9) **b, d, e, g** (10) **e**

# CHAPTER 2: QUIZ

*Determine all of the correct answers for each question. Each question may have more than one correct response.*

**Use these compounds to answer questions 1-3.**

Reference
compound

$$H-\overset{\overset{\displaystyle H}{|}}{\underset{\underset{\displaystyle H}{|}}{C}}-\overset{\overset{\displaystyle CH_3}{|}}{\underset{\underset{\displaystyle H}{|}}{C}}-\overset{\overset{\displaystyle H}{|}}{\underset{\underset{\displaystyle H}{|}}{C}}-\overset{\overset{\displaystyle CH_3}{|}}{\underset{\underset{\displaystyle CH_3}{|}}{C}}-\overset{\overset{\displaystyle H}{|}}{\underset{\underset{\displaystyle H}{|}}{C}}-\overset{\overset{\displaystyle H}{|}}{\underset{\underset{\displaystyle H}{|}}{C}}-H$$

(a)    $CH_3CH_2CH_2CH_2CH_2CH_2CH_2CH_2CH_3$

(b)    $CH_3-CH_2-CH_2-\underset{\underset{\displaystyle CH_3}{|}}{CH}-\underset{\underset{\displaystyle CH_3}{|}}{CH}-CH_2-CH_3$

(c)    $CH_3-CH_2-CH_2-\underset{\underset{\displaystyle CH_3}{|}}{CH}-\underset{\underset{\displaystyle CH_3}{|}}{CH}-CH_3$

(d)    $CH_3-CH_2-\underset{\underset{\displaystyle CH_3}{|}}{\overset{\overset{\displaystyle CH_3}{|}}{C}}-CH_2-\underset{\underset{\displaystyle CH_3}{|}}{CH}-CH_3$

(e)    $CH_3-\underset{\underset{\displaystyle CH_2}{\underset{\underset{\displaystyle CH_3}{|}}{|}}}{\overset{\overset{\displaystyle CH_3}{|}}{C}}-CH_2-\underset{\underset{\displaystyle CH_3}{|}}{CH}-CH_3$

(f)    $CH_3-CH_2-\underset{\underset{\displaystyle CH_3}{|}}{CH}-CH_2-CH_2-\underset{\underset{\displaystyle CH_3}{|}}{CH}-CH_3$

(g)    $CH_3-\underset{\underset{\displaystyle CH_3}{|}}{\overset{\overset{\displaystyle CH_3}{|}}{C}}-(CH_2)_4CH_3$

1. Which compounds are constitutional isomers of the reference compound?

2. Which structures are identical to the reference compound?

3. The following list of names corresponds to each of the above compounds. Which names are correct as written for the specified compounds a-g?
   (a) nonane
   (b) 4,5-dimethylheptane
   (c) 2,3-dimethylhexane
   (d) 3,3,5-trimethylhexane
   (e) 2-ethyl-2,4-dimethylpentane
   (f) 2,5-dimethylheptane
   (g) 2,2-dimethylbutane

4. Which of the attached methyl groups in 1, 2, 3, 4, 5, 6-hexamethylcyclohexane are drawn in an axial position?

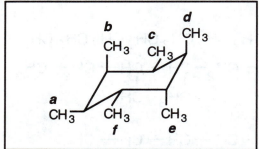

5. List the following conformations for butane drawn using Newman projections in the order, conformation exhibiting most steric strain first; conformation exhibiting least steric strain last.

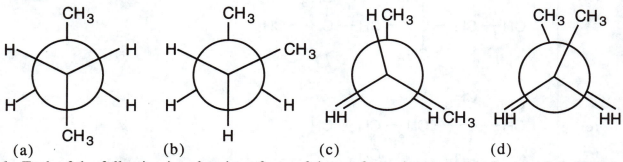

(a)          (b)          (c)          (d)

6. Each of the following is a drawing of one of the conformations possible for menthol. Which is the best conformation for menthol?

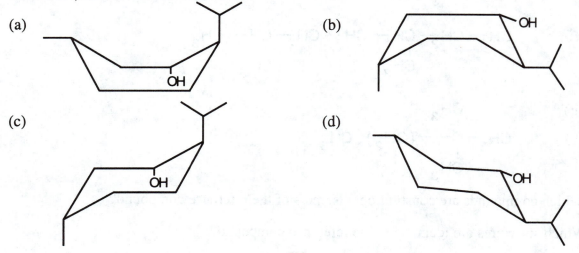

(a)                                    (b)

(c)                                    (d)

7. Which of the following cycloalkanes exhibit *trans* geometry?

(a)

$CH_3$

$CH_3$

(d)

$CH_3$

$CH_3$

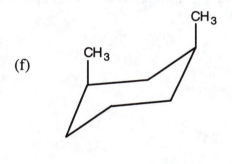

(b)

$CH_3$

$CH_3$

(e)

$CH_3$

$CH_3$

(c)

$CH_3$

$CH_3$

(f)

$CH_3$

$CH_3$

8. The following reaction is used to make the commercial solvent called Cellosolve. Which of the following are true for the manufacture of Cellosolve?

$$CH_3CHCH_2CH_3 \quad + \quad Cl_2 \longrightarrow$$

with $CH_3$ on the second carbon

(a) Initiation of the reaction will require light energy.
(b) Four separate monochlorinated products could be formed in the reaction.
(c) The following compound is the most probable monohalogenated product formed in the course of the reaction.

$$CH_3CHCH_2CH_3$$

with $CH_2Cl$ on the second carbon

(d) If Cellosolve is mainly composed of monochlorinated products, excess 2-methylbutane must be used in the reaction.
(e) If the reaction was run using excess chlorine the eventual product should be a single compound containing no hydrogens.

9. Which of the following is no longer an allowed use of fluorocarbons or chlorofluorocarbons in the United States?
   (a) refrigerant
   (b) synthetic blood substitute
   (c) propellants for aerosol sprays
   (d) non-stick surfaces for cooking ware
   (e) inhalation anesthetic

10. Which of the following are true concerning gasoline?
    (a) The primary products when burned in an efficient engine are carbon dioxide, water, and heat energy.
    (b) Gasoline with higher octane ratings have the burning characteristics that do not cause engine knocking.
    (c) Gasoline is obtained by distillation of petroleum.
    (d) Highly branched chain hydrocarbons are desirable as components of gasoline.
    (e) Eventually gasoline or a replacement product for use in our vehicles may have to be made from coal.

**Chapter 2 Quiz Answers**
(1) **a, b, f, g** (2) **d, e** (3) **a, c, f** (4) **b, d, e** (5) **DCBA** (6) **d** (7) **b, c, d** (8) **a, b, d, e** (9) **c** (10) **a, b, c, d, e**

# CHAPTER 3: QUIZ

*Determine all of the correct answers for each question. Each question may have more than one correct response.*

1. Which of the following are true concerning a carbon-carbon double bond in an alkene?
   (a) The two bonds of the double bond are identical.
   (b) The bond angles around the carbon-carbon double bond are $120°$.
   (c) Each carbon in the carbon-carbon double bond exhibits $sp^2$ hybridization.
   (d) The four atoms bonded to the two carbons of a double bond must all lie in the same plane.
   (e) There is free rotation around the carbon-carbon double bond.

2. Which of the following compounds have a double bond exhibiting *cis* geometry?

515

**Use the following compounds to answer questions 3-4.**

(a)

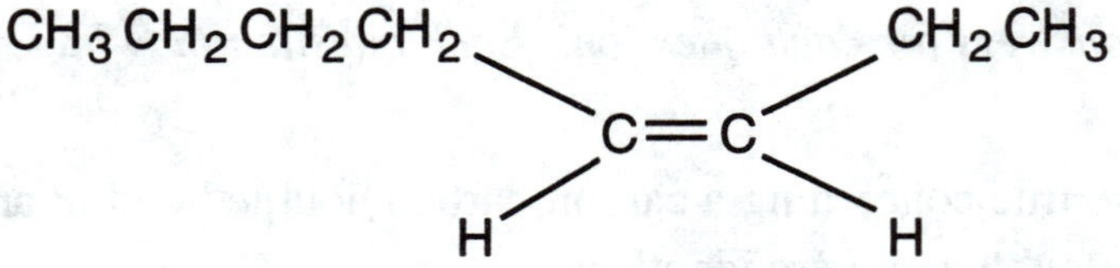

$$CH_3CH_2CH_2CH_2 \quad CH_2CH_3$$
$$C=C$$
$$H \qquad H$$

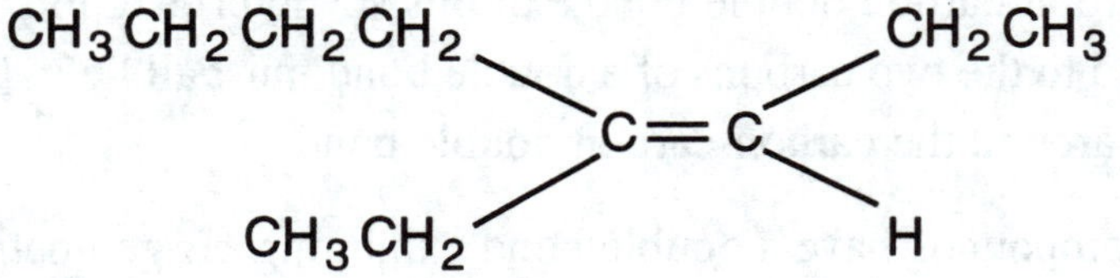

(b)

$$CH_3CH_2CH_2CH_2 \quad CH_2CH_3$$
$$C=C$$
$$CH_3CH_2 \qquad H$$

(c)

$$CH_3$$
$$|$$
$$CHCH_2CH_3$$
$$CH_3CH_2CH_2 \qquad$$
$$C=C$$
$$CH_3CH_2CH_2 \qquad H$$

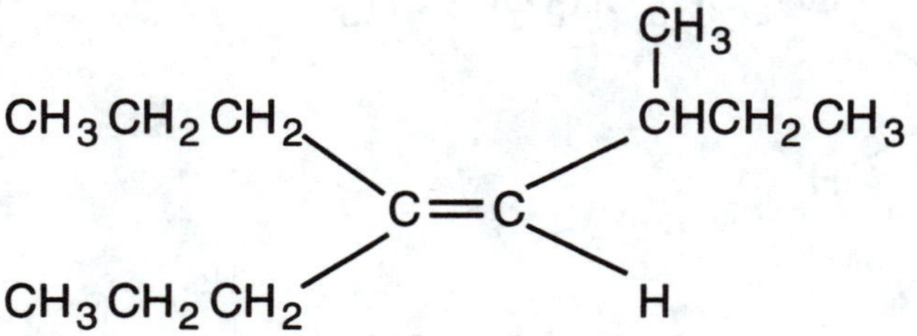

(d)

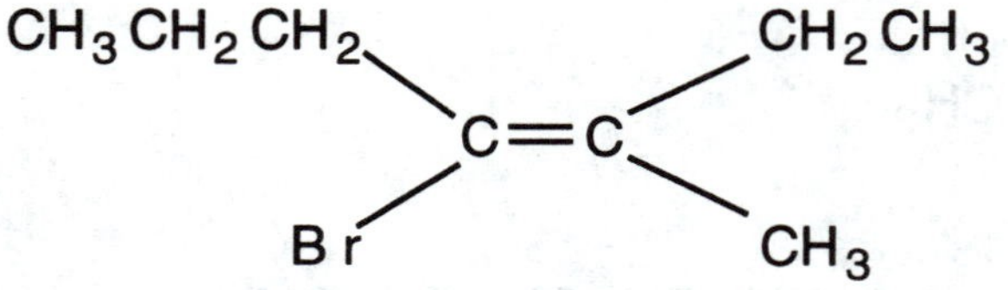

$$CH_3CH_2CH_2 \quad CH_2CH_3$$
$$C=C$$
$$Br \qquad CH_3$$

(e)

$$CH_3CH_2 \quad CH_2CH_3$$
$$C=C$$
$$CH_3CHCH_2 \qquad H$$
$$|$$
$$CH_3$$

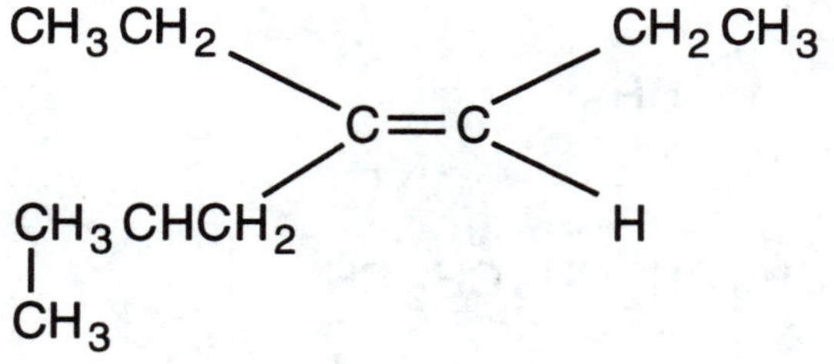

3. Which of the following are correct names for the compounds listed above? Write correct names in place of the incorrect ones using the *cis-trans* system.
   (a) *cis*-5-octene
   (b) *cis*-4-ethyl-3-heptene
   (c) *trans*-3-methyl-5-propyl-4-octene
   (d) *cis*-4-bromo-3-methyl-3-heptene
   (e) *cis*-4-ethyl-6-methyl-3-heptene

4. Which of the following are the correct names for the compounds listed above? Write the correct names in place of the incorrect ones using the E-Z system.
   (a) (Z)-3-octene
   (b) (E)-4-ethyl-3-heptene
   (c) (Z)-3-methyl-5-propyl-4-octene
   (d) (E)-4-bromo-3-methyl-3-heptene
   (e) (Z)-4-ethyl-6-methyl-3-heptene

5. Which of the following compounds have the double bond labeled Z.

(a)

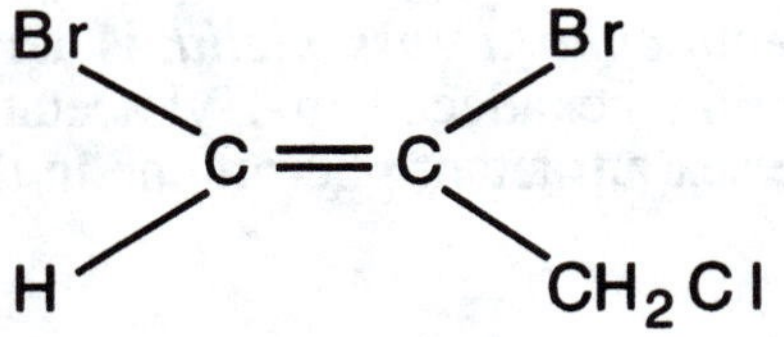

$Br$ $\quad$ $Br$

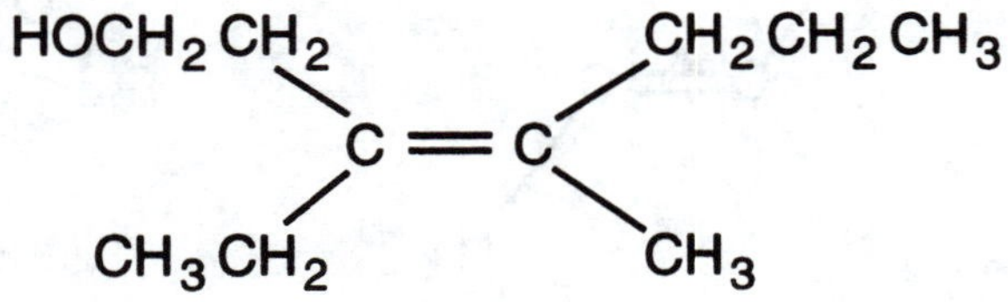

$C = C$

$H$ $\quad$ $CH_2Cl$

(b) $HOCH_2CH_2$ $\quad$ $CH_2CH_2CH_3$

$C = C$

$CH_3CH_2$ $\quad$ $CH_3$

(c) $\quad CH_3$

$CH_3$ $\quad CH$ $\qquad CHO$

$C = C$

$CH_3$ $\quad CH_2$ $\qquad CH_3$

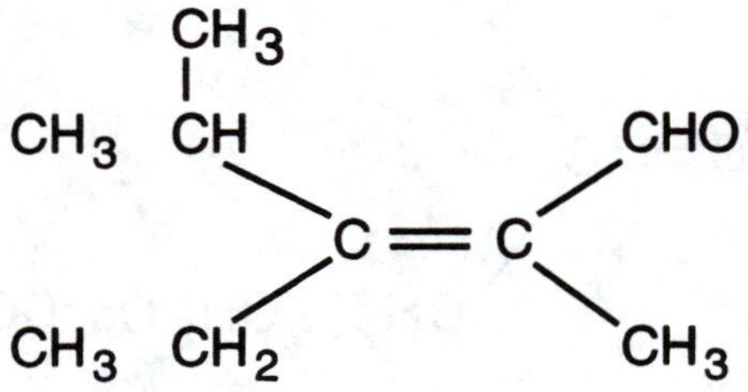

(d) $\quad Br$ $\qquad CH_2OH$

$C = C$

$CH_3CH_2CH_2$ $\qquad CH_3$

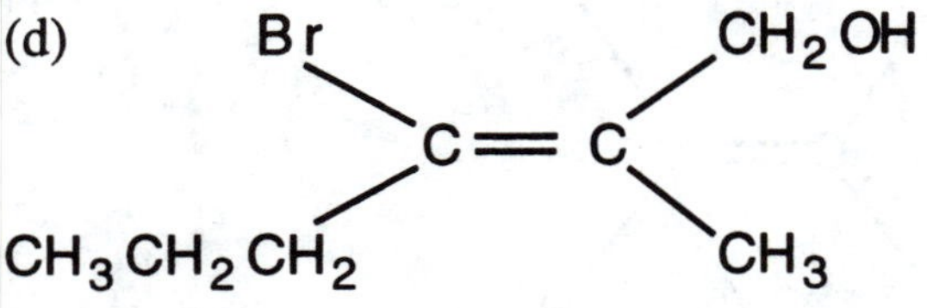

(e) $\quad Br$ $\qquad CHO$

$C = C$

$CH_3CH_2CH_2$ $\qquad CH_2Br$

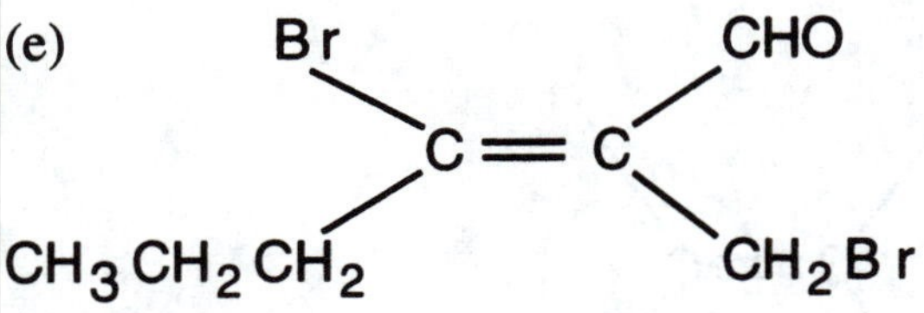

6. Which of the indicated double bonds in this terpene have a *cis* geometry?

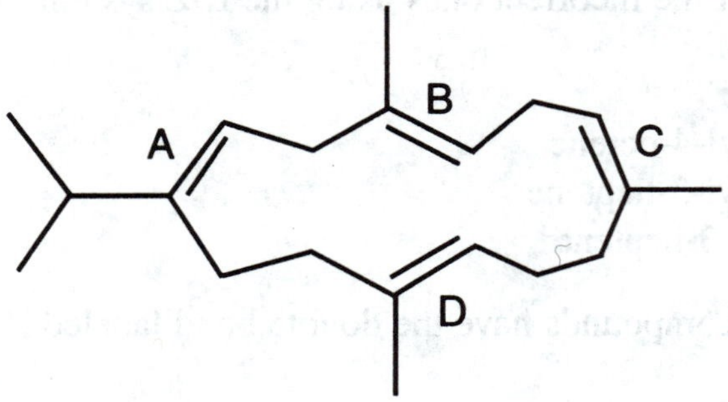

7. The sex attractant pheromone for the pink bollworm, *Pectinophora gossypiella*, is a 1:1 ratio of *cis*-7-*cis*-11-hexadecadien-1-yl acetate and *cis*-7-*trans*-11-hexadecadien-1-yl acetate. Which of the following compounds will be found as the sex attractant pheromone in the pink bollworm?

(a)

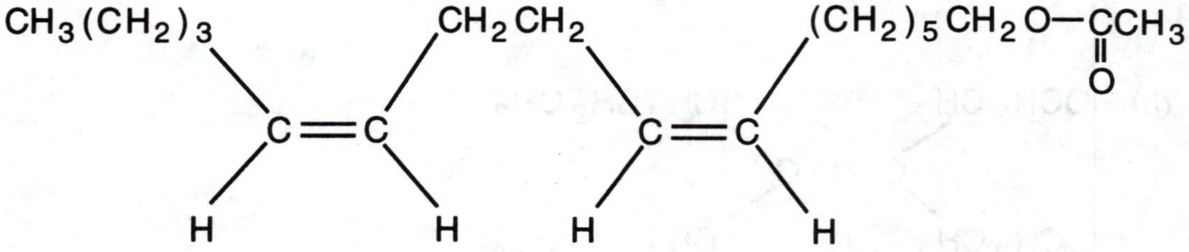

(b)

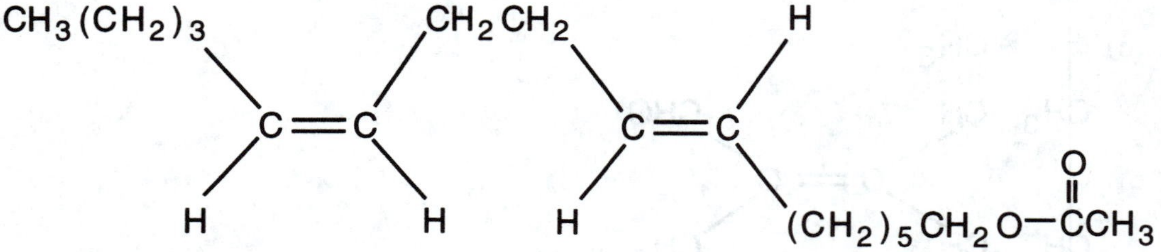

(c)

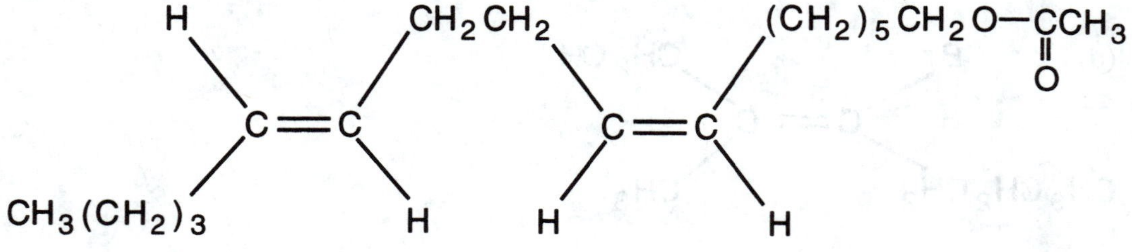

(d)

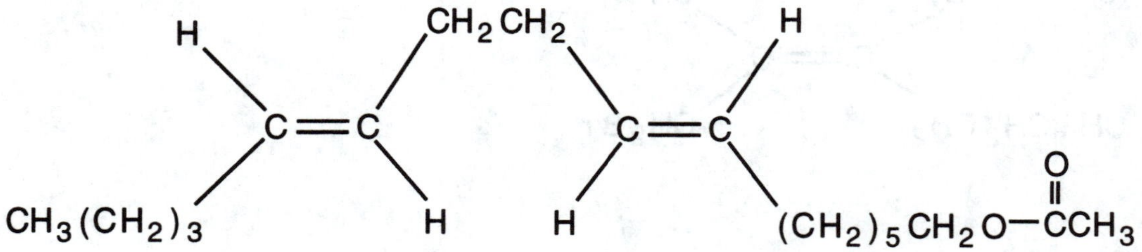

518

8. Based on the information given in question 7 which of the following compounds are not found in the pink bollworm?
   (a) (7Z, 11Z)-7,11-hexadecadien-1-yl acetate
   (b) (7E, 11Z)-7,11-hexadecadien-1-yl acetate
   (c) (7Z, 11E)-7,11-hexadecadien-1-yl acetate
   (d) (7E, 11E)-7,11-hexadecadien-1-yl acetate

9. Which of the following compounds isolated from plants would be classified as terpenes?

(a)

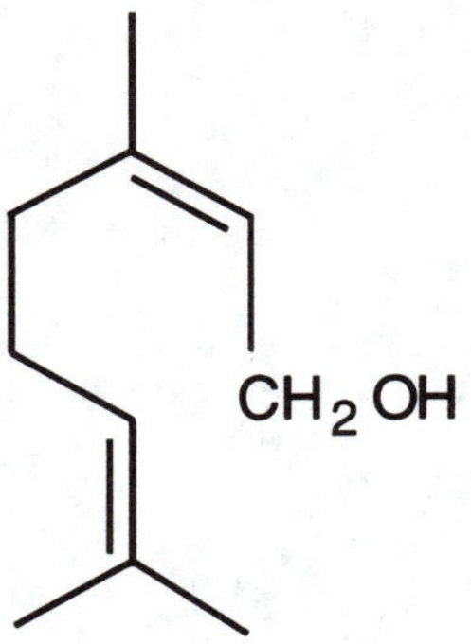

(d)

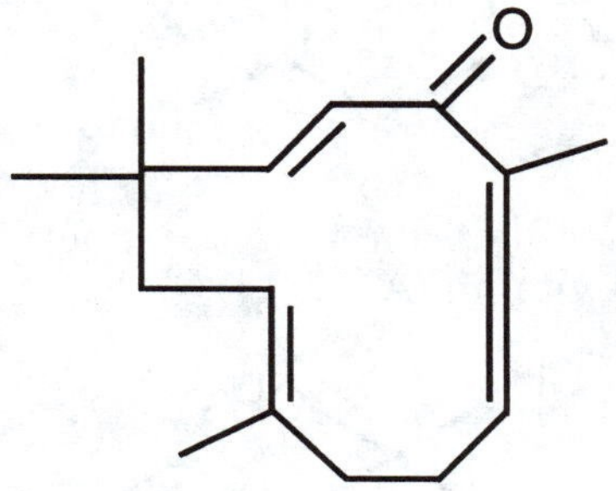

(b)

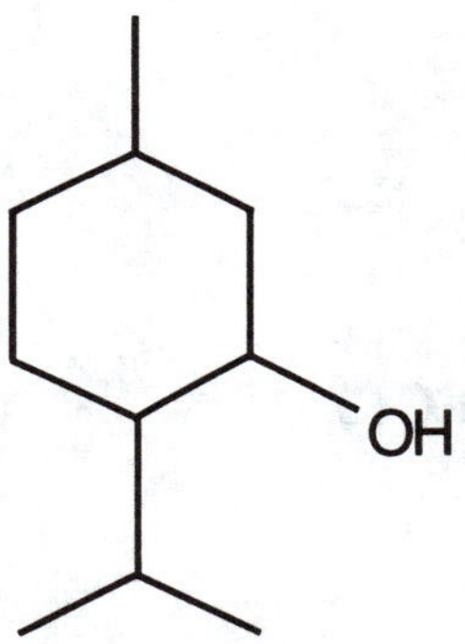

(e)

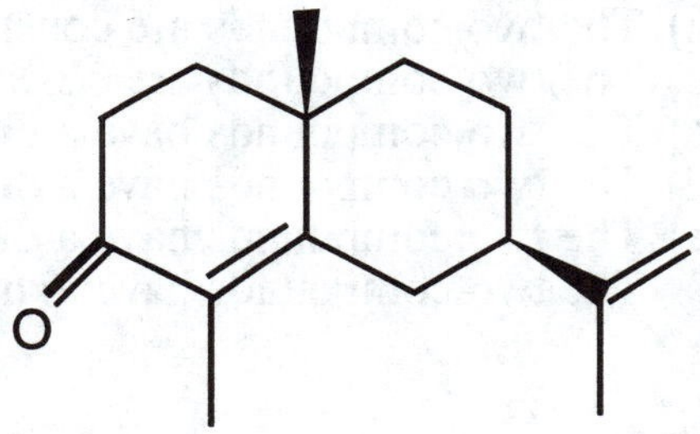

(c)

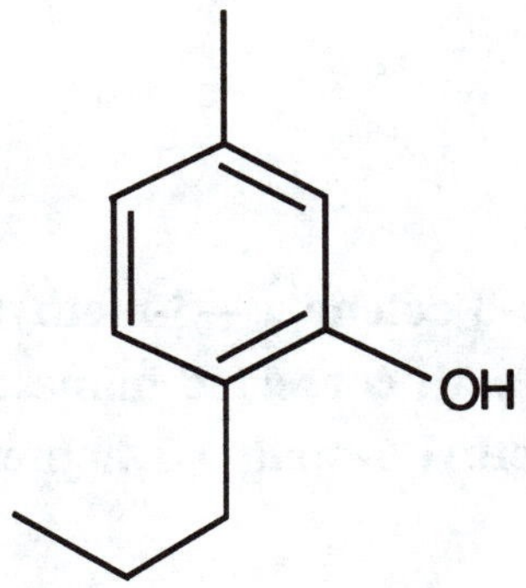

(f)

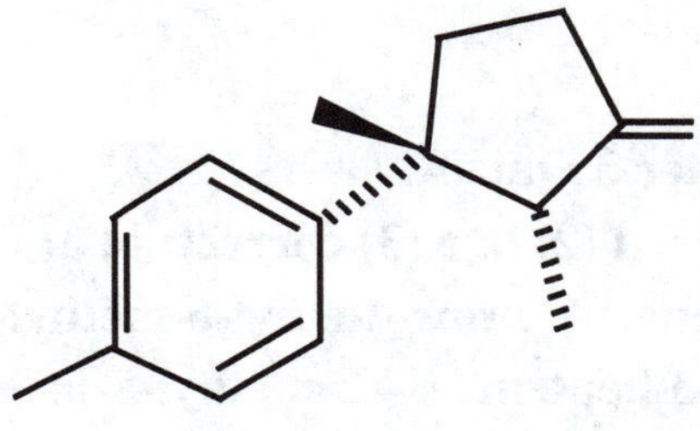

10. Which of the following are correct statements that can be made in describing the differences between vitamin A and neo-B vitamin A?

vitamin A

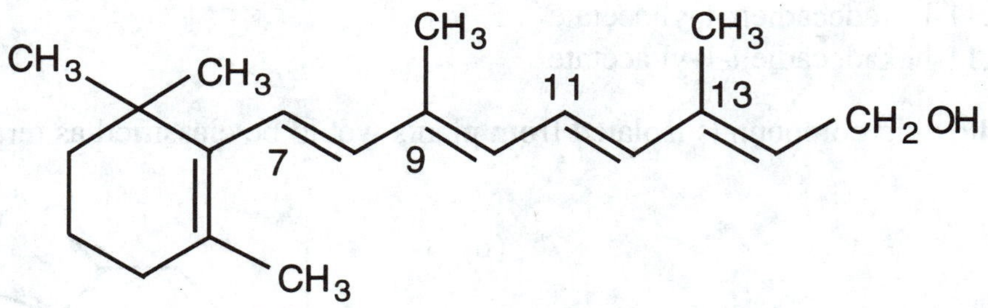

neo-B vitamin A

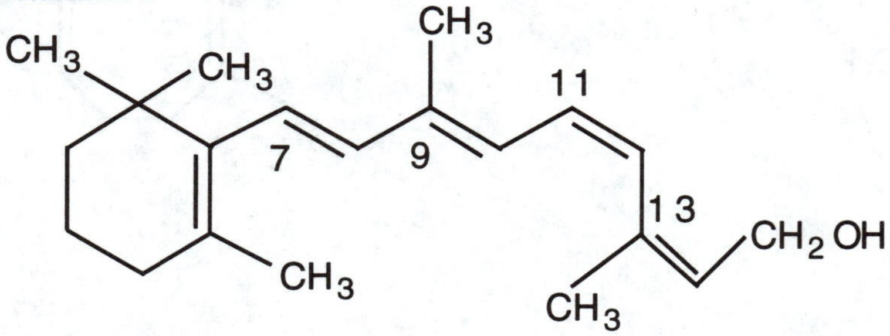

(a) The two compounds are constitutional isomers.
(b) The two compounds are *cis-trans* isomers.
(c) The two compounds have a difference in *cis* versus *trans* at C7.
(d) The two compounds have a difference in *cis* versus *trans* at C9.
(e) The two compounds have a difference in *cis* versus *trans* at C11.
(f) The two compounds have a difference in *cis* versus *trans* at C13.

**Chapter 3 Quiz Answers**

(1) **b, c, d** (2) **b, e** (3) **correct: b, d; corrected names: a—*cis*-3-octene, c—3-methyl-5-propyl-4-octene, e—*trans*-4-ethyl-6-methyl-3-heptene** (4) **correct: a, d; corrected names: b—(Z)-4-ethyl-3-heptene, c—3-methyl-5-propyl-4-octene, e—(E)-4-ethyl-6-methyl-3-heptene** (5) **a, b, c, d** (6) **a, c** (7) **a, c** (8) **b, d** (9) **a, b, d, e, f** (10) **b, e, f**

# CHAPTER 4: QUIZ

*Determine all of the correct answers for each question. Each question may have more than one correct response.*

**Use these reactions to answer questions 1-6.**

(a)

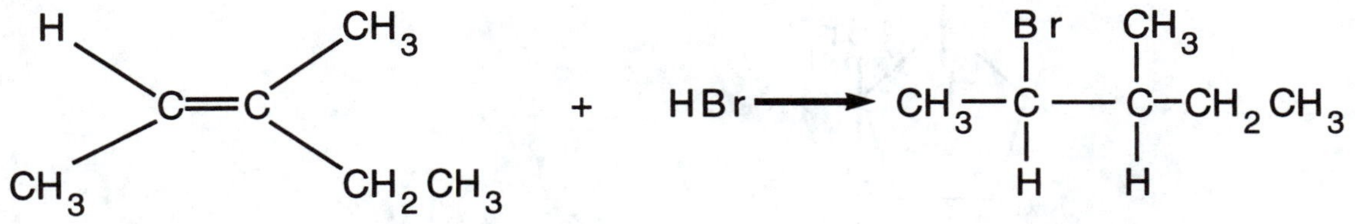

(b)

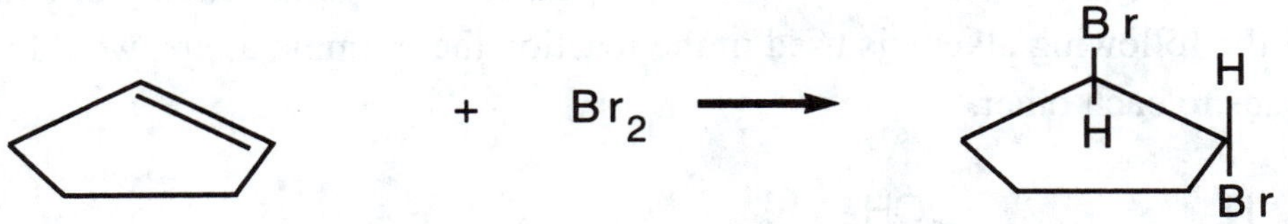

(c)

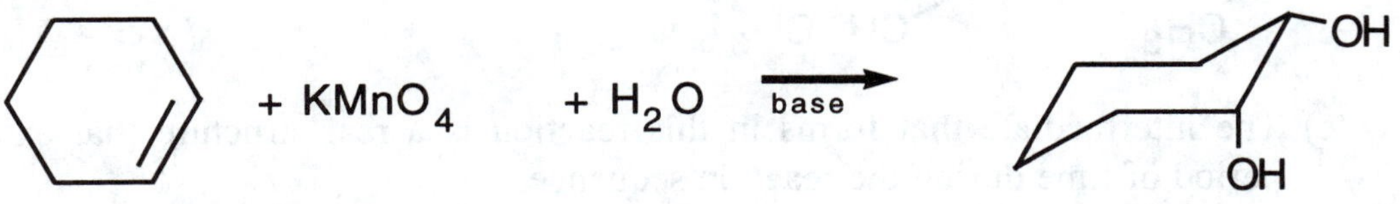

(d)

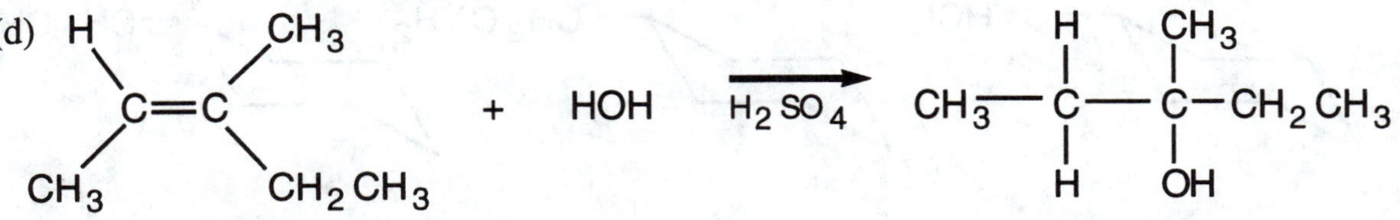

(e)

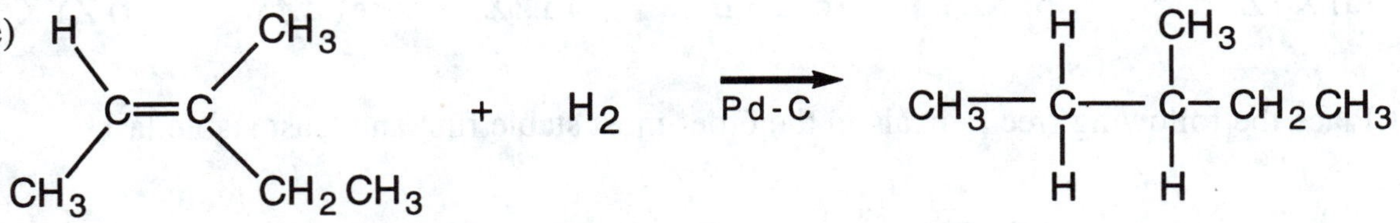

1. Which reactions are alkene addition reactions?

2. Which reactions will have a carbocation intermediate?

3. In which of the reactions would we expect the product shown to be at most a minor component?

4. In which reaction does oxidation of the alkene take place?

5. In which reaction does reduction of the alkene take place?

6. For reaction B which of the following are true?
   (a) It has a bridged halonium ion intermediate.
   (b) The following product is also formed in the reaction.

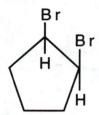

   (c) The energy diagram for the reaction will have two separate energy of activations.
   (d) If the following alkene is used in the reaction the bromine atoms would not have to add trans to each other.

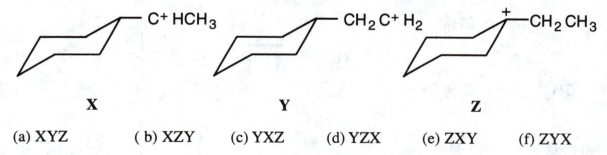

   (e) The intermediate that forms in this reaction is a real structure that exists for a finite period of time during the reaction sequence.

7. Place the following carbocations in the order most stable first and least stable last.

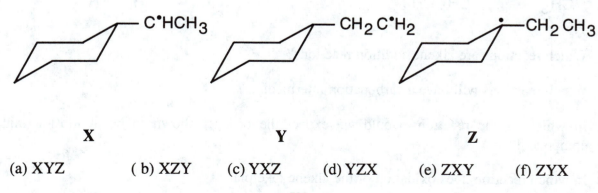

   (a) XYZ      ( b) XZY      (c) YXZ      (d) YZX      (e) ZXY      (f) ZYX

8. Place the following free radicals in the order most stable first and least stable last.

   X      Y      Z

   (a) XYZ      ( b) XZY      (c) YXZ      (d) YZX      (e) ZXY      (f) ZYX

9. Place the following three steps in the proper order for the acid-catalyzed hydration of an alkene.

(X) 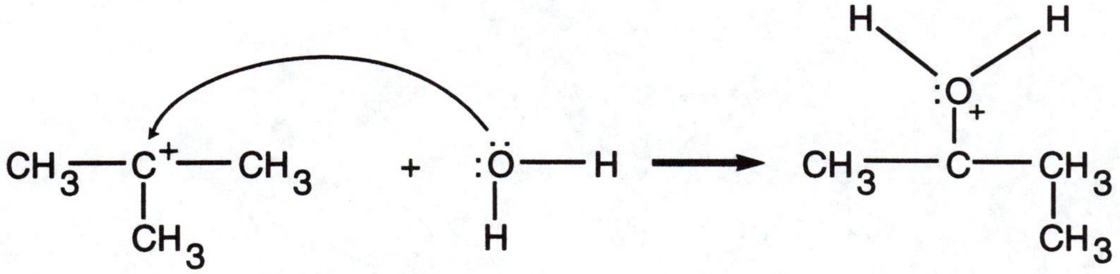 $CH_3-\overset{+}{\underset{\underset{CH_3}{|}}{C}}-CH_3$ + $:\overset{..}{\underset{\underset{H}{|}}{O}}-H$ → $CH_3-\overset{\overset{\overset{\displaystyle H\,\,\,\,\,\,\,\,\,\,H}{\searrow\,\,\,\swarrow}}{\underset{+}{\overset{..}{O}}}}{\underset{\underset{CH_3}{|}}{C}}-CH_3$

(Y) 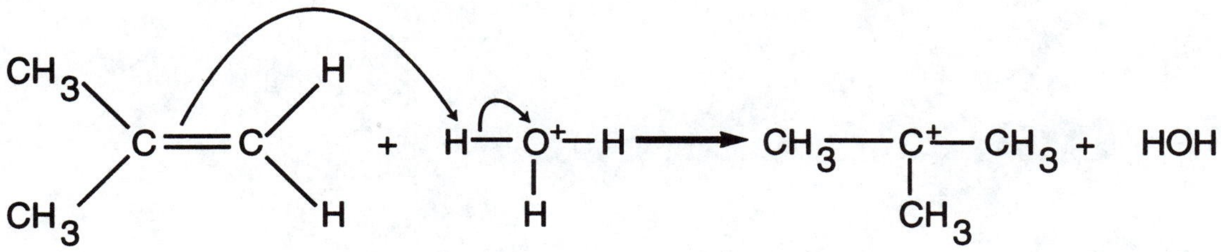 $\underset{CH_3}{\overset{CH_3}{>}}C=C\overset{H}{\underset{H}{<}}$ + $H-\overset{+}{\underset{\underset{H}{|}}{O}}-H$ → $CH_3-\overset{+}{\underset{\underset{CH_3}{|}}{C}}-CH_3$ + HOH

(Z) 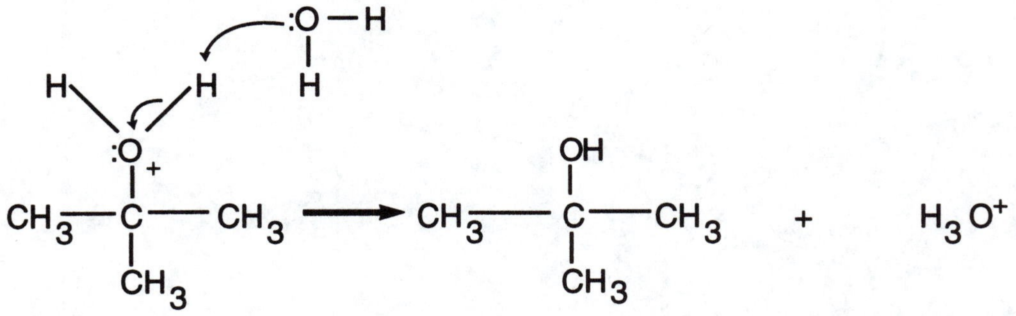 $\overset{:O-H}{\underset{|}{H}}$   $\underset{CH_3}{\overset{H\searrow\,\swarrow H}{\underset{+}{:O}}}$

$CH_3-\overset{\overset{..}{\underset{+}{O}}}{\underset{\underset{CH_3}{|}}{C}}-CH_3$ → $CH_3-\overset{OH}{\underset{\underset{CH_3}{|}}{C}}-CH_3$ + $H_3O^+$

(a) XYZ (b) XZY (c) YXZ (d) YZX (e) ZXY (f) ZYX

10. Which of the following are true concerning polystyrene which is synthesized using benzoyl peroxide as a catalyst?
(a) The following would be formed as the initial intermediate.

$In\,CH_2\,C^\bullet HC_6\,H_5$

(b) Polystyrene has the following structure.

$-CH_2\,CH_2\,CHC_6\,H_5\,CHC_6\,H_5\,CH_2\,CH_2\,CHC_6\,H_5\,CHC_6\,H_5\,CH_2\,CH_2\,CHC_6\,H_5\,CHC_6\,H_5-$

(c) Polystyrene synthesized by use of a Ziegler catalyst is essentially identical to the one made using a free radical catalyst and thus could be used for the same purposes.
(d) The polystyrene formed by use of a free radical catalyst has a termination step equivalent to that found in the formation of polyethylene.
(e) Since atmospheric oxygen ($O_2$) is a free radical it could be expected to cause the polymerization of polystyrene over a period of time.

**Chapter 4 Quiz Answers**
(1) **a, b, d** (2) **a, d** (3) **a** (4) **c** (5) **e** (6) **a, c, e** (7) **e** (8) **e** (9) **c** (10) **a, d, e**

# CHAPTER 5: QUIZ

*Determine all of the correct answers for each question. Each question may have more than one correct response.*

**Use these compounds to answer questions 1-5.**

(a)

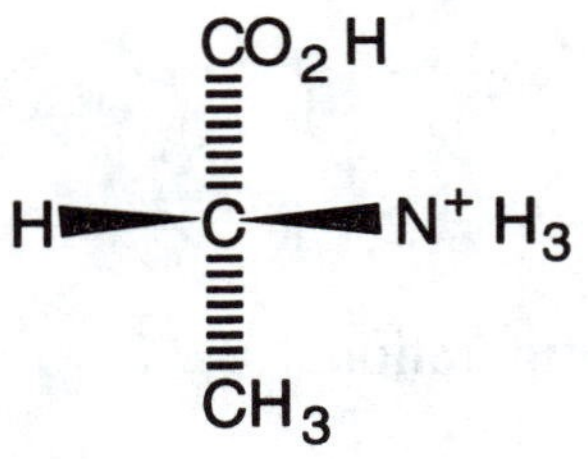

$$CO_2H$$
$$H-C-N^+H_3$$
$$CH_3$$

(b)

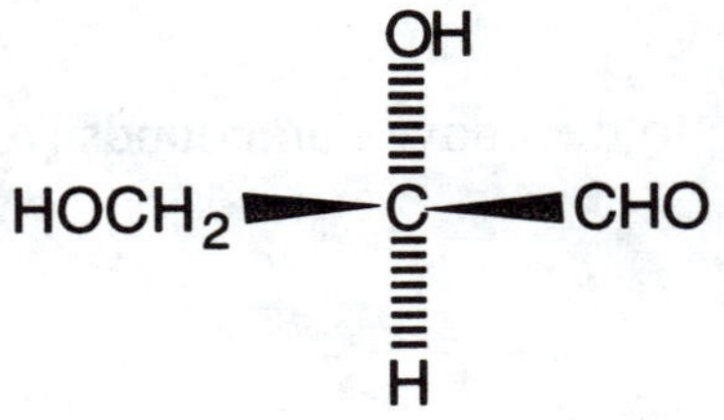

$$OH$$
$$HOCH_2-C-CHO$$
$$H$$

(c)

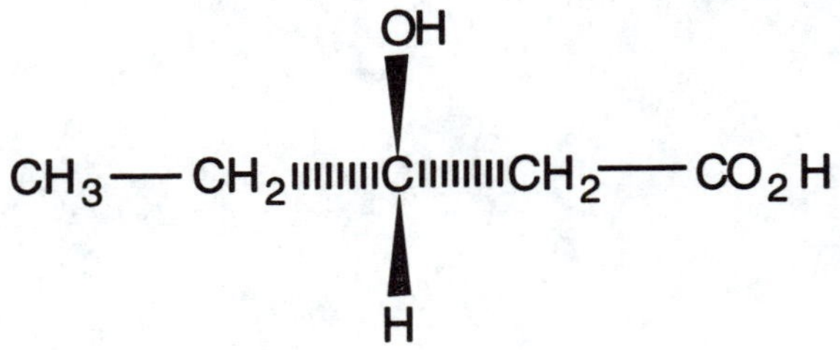

$$OH$$
$$CH_3-CH_2\text{——}C\text{——}CH_2-CO_2H$$
$$H$$

(d)

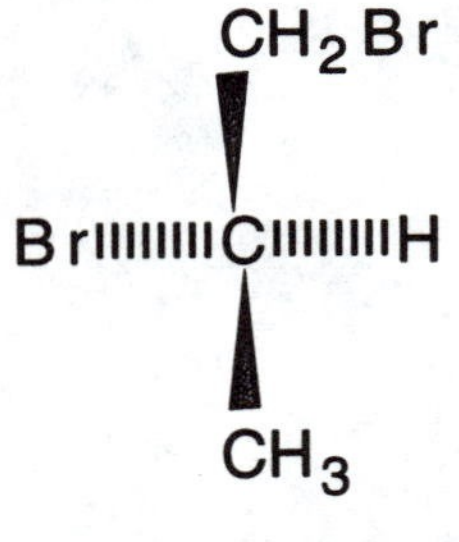

$$CH_2Br$$
$$Br\text{——}C\text{——}H$$
$$CH_3$$

(e)

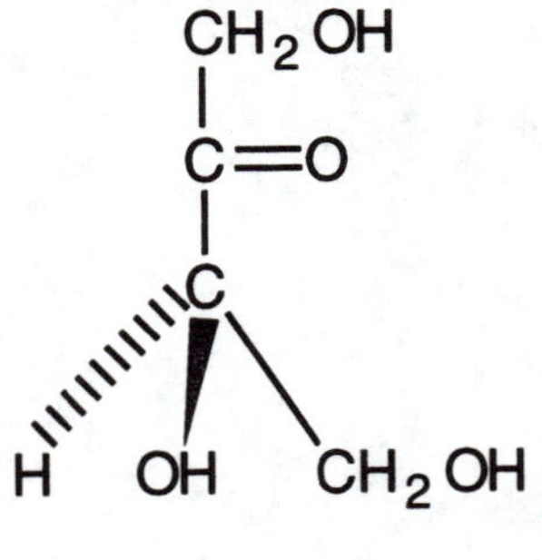

$$CH_2OH$$
$$C=O$$
$$C$$
$$H \quad OH \quad CH_2OH$$

525

(f)

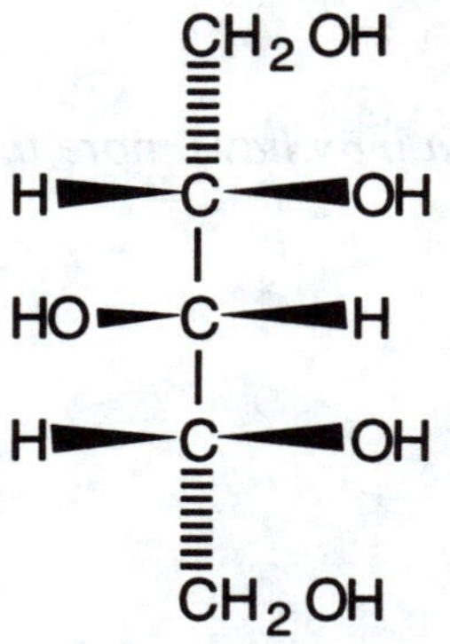

$$CH_2OH$$
$$H\!-\!\!\!-C\!-\!\!\!-OH$$
$$HO\!-\!C\!-\!\!\!-H$$
$$H\!-\!\!\!-C\!-\!\!\!-OH$$
$$CH_2OH$$

1. Which compounds have only one chiral center?

2. Which compounds have at least one chiral center with an S configuration?

3. Which compounds will rotate the plane of polartized light?

4. Which compounds would be labeled meso?

5. Which of the following Fischer projections corresponding to the above compounds (A-F) are correct?

(a)

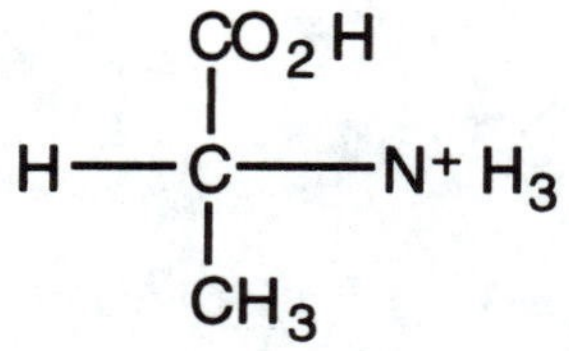

$$CO_2H$$
$$H\!-\!C\!-\!N^+H_3$$
$$CH_3$$

(b)

$$OH$$
$$HOCH_2\!-\!C\!-\!CHO$$
$$H$$

(c)

$$OH$$
$$CH_3\!-\!CH_2\!-\!C\!-\!CH_2\!-\!CO_2H$$
$$H$$

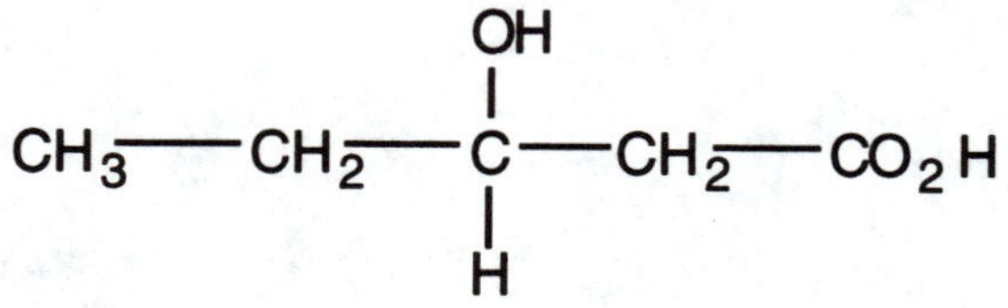

(d)

$$CH_2Br$$
$$H\!-\!C\!-\!Br$$
$$CH_3$$

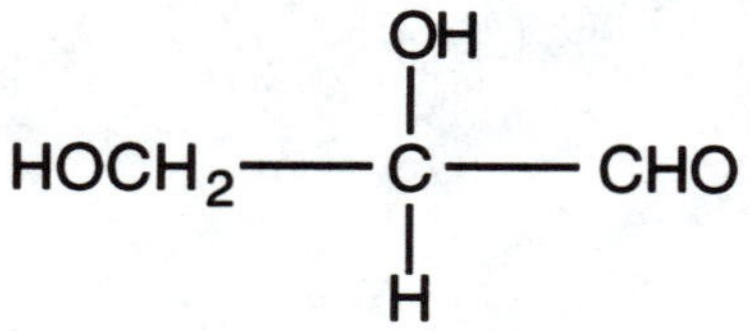

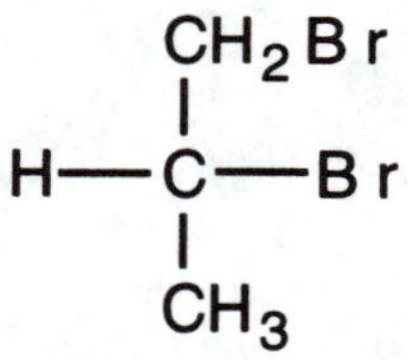

(e)

```
        CH₂OH
         |
         C=O
         |
   H ——— C ——— OH
         |
        CH₂OH
```

(f)

```
        CH₂OH
         |
   H ——— C ——— OH
         |
  HO ——— C ——— H
         |
   H ——— C ——— OH
         |
        CH₂OH
```

**Use the following pairs of compounds to answer questions 6 -7.**

(a)

```
        CHO                            CHO
         |                              |
   H ——— C ——— OH               H ——— C ——— OH
         |                              |
   H ——— C ——— OH              HO ——— C ——— H
         |                              |
  HO ——— C ——— H              HO ——— C ——— H
         |                              |
        CH₂ OH                         CH₂OH
```

(b)

```
        CHO                            CHO
         |                              |
   H ——— C ——— OH              HO ——— C ——— H
         |                              |
   H ——— C ——— OH              HO ——— C ——— H
         |                              |
  HO ——— C ——— H               H ——— C ——— OH
         |                              |
        CH₂OH                          CH₂OH
```

(c)

```
        CH₂OH                          CH₂OH
         |                              |
   H ——— C ——— OH               H ——— C ——— OH
         |                              |
   H ——— C ——— OH              HO ——— C ——— H
         |                              |
  HO ——— C ——— H              HO ——— C ——— H
         |                              |
        CH₂OH                          CH₂OH
```

527

(d)

$$CH_2OH$$
$$H \text{---} C \text{---} OH$$
$$H \text{---} C \text{---} OH$$
$$H \text{---} C \text{---} OH$$
$$CH_2OH$$

$$CH_2OH$$
$$HO \text{---} C \text{---} H$$
$$HO \text{---} C \text{---} H$$
$$HO \text{---} C \text{---} H$$
$$CH_2OH$$

(e)

$$CH_2OH$$
$$H \text{---} C \text{---} OH$$
$$H \text{---} C \text{---} OH$$
$$HO \text{---} C \text{---} H$$
$$CH_2OH$$

$$CH_2OH$$
$$HO \text{---} C \text{---} H$$
$$H \text{---} C \text{---} OH$$
$$H \text{---} C \text{---} OH$$
$$CH_2OH$$

6. Which pairs of compounds are enantiomers?

7. Which pairs of compounds are diastereomers?

8. A pure compound is found to have a specific rotation of +3.3°. Which of the following are statements that can not be made based on this information?
   (a) The compound contains at least one chiral center.
   (b) The compound is said to be chiral.
   (c) The compound has an R configuration.
   (d) The compound has an enantiomer either in nature or one that can be made that has the same melting point or boiling point and same water solubility, but has a specific rotation of -3.3°.
   (e) The mirror image of the compound is superposable.

9. Consider a series of dimethyl cyclobutanes. Which of the following statements are correct?
   (a) 1,1-Dimethylcyclobutane will be achiral.
   (b) *trans*-1,2-Dimethylcyclobutane will exist as a pair of enantiomers.
   (c) *cis*-1,2-Dimethylcyclobutane will exist as a pair of enantiomers.
   (d) *cis*-1,3-Dimethylcyclobutane will be achiral.
   (e) *trans*—1,3-Dimethylcyclobutane will exist as a pair of enantiomers.

10. The compound (-)-carvone is isolated from spearmint oil and (+)-carvone from caraway oil. Which of the following statements are consistent with the structures of the carvones and the information provided?

(-)-carvone

(+)-carvone

(a) The two compounds are enantiomers.

(b) (+) Carvone has an R configuration.

(c) Each of the compounds has only one chiral center.

(d) If the compounds were mixed in a ratio of 1:1 they would constitute a racemic mixture that would not rotate the plane of polarized light.

(e) Humans are able to distinguish between pure samples of these two enantiomers with the tongue.

## Chapter 5 Quiz Answers
(1) **a, b, c, d, e** (2) **c, f** (3) **a, b, c, d, e** (4) **f** (5) **a, b, d, e, f** (6) **b, e** (7) **a** (8) **c, e** (9) **a, b, d** (10) **a, c, d, e**

# CHAPTER 6: QUIZ

*Determine all of the correct answers for each question. Each question may have more than one correct response.*

**Use this reaction to answer questions 1-4.**

$$(CH_3)_3N: \quad + \quad CH_3CHOHCO_2H \longrightarrow (CH_3)_3NH^+ \quad + \quad CH_3CHOHCO_2^-$$

  a                     b                              c                              d

1. Which structure is a Brønsted-Lowry acid?

2. Which structure is a Brønsted-Lowry base?

3. Which structure is a conjugate acid?

4. Which structure is a conjugate base?

5. Place the following compounds in the order most acidic first and least acidic last.

   X) carbonic acid, $H_3CO_3$           $pK_a = 6.36$

   Y) phenol, $C_6H_5OH$              $pK_a = 9.95$

   Z) benzoic acid, $C_6H_5CO_2H$      $pK_a = 4.19$

   (a) XYZ      (b) XZY      (c) YXZ      (d) YZX      (e) ZXY      (f) ZYX

**Use these reactions and the $pK_a$ values to answer questions 6-7.**

(a) $C_6H_5OH \quad + \quad CH_3CH_2O^- \rightleftharpoons C_6H_5O^- \quad + \quad CH_3CH_2OH$

     $pK_a = 9.95$                                $pK_a = 15.9$

(b) $CH_3NH_3^+ \quad + \quad CH_3CO_2^- \rightleftharpoons CH_3NH_2 \quad + \quad CH_3CO_2H$

     $pK_a = 10.64$                              $pK_a = 4.76$

(c) $C_6H_5CO_2H \quad + \quad HCO_3^- \rightleftharpoons C_6H_5CO_2^- + \quad H_2CO_3$

     $pK_a = 4.19$     $pK_a = 10.33$                    $pK_a = 6.36$

(d) $C_6H_5CO_2H \quad + \quad H_2PO_4^- \rightleftharpoons C_6H_5CO_2^- + \quad H_3PO_4$

     $pK_a = 4.19$                              $pK_a = 2.1$

(e) $NH_3 \quad + \quad CH_3CH_2O^- \rightleftharpoons NH_2^- \quad + \quad CH_3CH_2OH$

     $pK_a = 33$                                $pK_a = 15.9$

6. In which reactions is the equilibrium shifted to the right?

7. Could sodium bicarbonate ($NaHCO_3$) be used as a base to react with phenol ($C_6H_5OH$) to make sodium phenoxide ($C_6H_5ONa$)? (a) yes (b) no

**Use this step from a reaction sequence to answer questions 8 and 9.**

$(CH_3)_3C^+$ + $Br^-$ $\longrightarrow$ $(CH_3)_3C{-}Br$
   a        b                                  c

8. Which structure is acting as a Lewis acid?

9. Which structure is acting as a Lewis base?

10. Which of the following statements are correct?
    (a) A Brønsted-Lowry acid is a proton donor.
    (b) A Brønsted-Lowry acid can always be classified as a Lewis acid.
    (c) A Brønsted-Lowry base can always be classified as a Lewis base.
    (d) A carbocation can be classified as both a Lewis acid and a Brønsted-Lowry acid.
    (e) When a Lewis base reacts with a Lewis acid there is a transfer of a pair of electrons from the Lewis base to the Lewis acid.

**Chapter 6 Quiz Answers**
(1) **b** (2) **a** (3) **c** (4) **d** (5) **e** (6) **a, c** (7) **b** (8) **a** (9) **b** (10) **a, b, c**

# CHAPTER 7: QUIZ

*Determine all of the correct answers for each question. Each question may have more than one correct response.*

**Use these compounds to answer questions 1-3.**

(a)

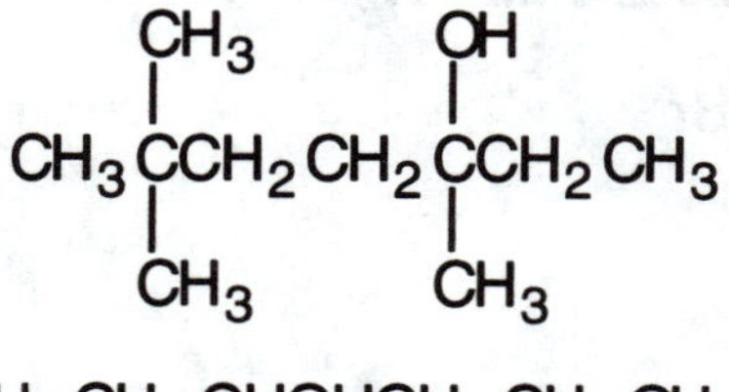

$$CH_3CCH_2CH_2CCH_2CH_3$$

with $CH_3$ and $CH_3$ groups on the first carbon and $OH$ and $CH_3$ on the second substituted carbon.

(b) $CH_3CH_2CHOHCH_2CH_2OH$

(c) 

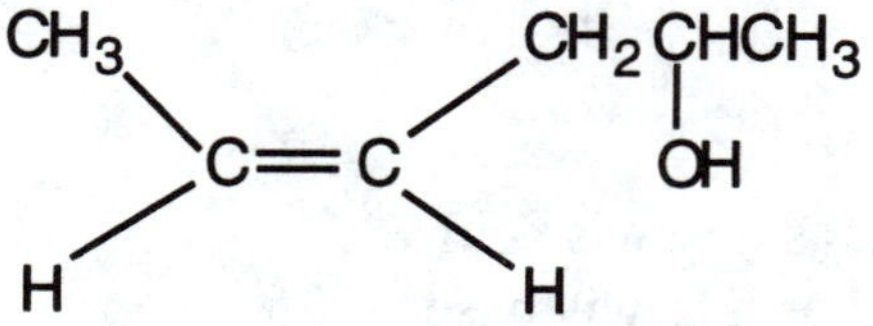

$$CH_3, \quad CH_2CHCH_3$$
$$\backslash C=C \diagup \quad OH$$
$$H \diagup \quad \backslash H$$

(d)

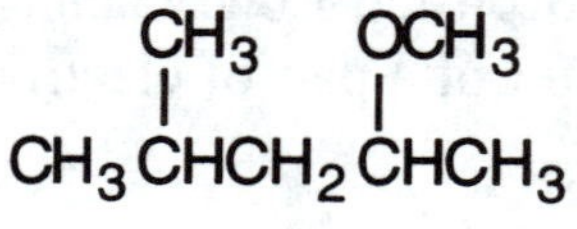

$$CH_3 \quad OCH_3$$
$$CH_3CHCH_2CHCH_3$$

(e)

$$H \quad CH_2CHCH_3$$
$$\backslash C=C \diagup \quad SH$$
$$CH_3CH_2 \diagup \quad \backslash H$$

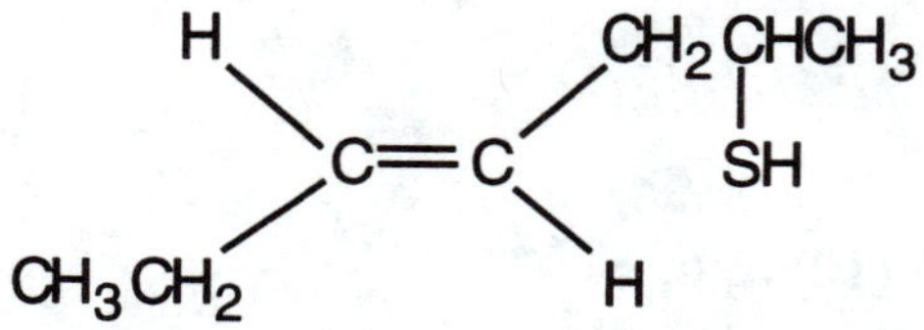

1. Which compound(s) are classified as secondary alcohols?

2. Which of the following names are incorrect? Write the correct name for each one that is incorrect.
   (a) 2,2,5-trimethyl-5-heptanol
   (b) 1,3-pentanediol
   (c) *cis*-2-hexen-5-ol
   (d) 2-methoxy-4-methylpentane
   (e) 4-heptene-2-thiol

3. Which compounds in pure form would exhibit association of its molecules by hydrogen bonding?

**Use these reactions for questions 4-10. Products are not shown.**

(a) $CH_3CHOHCH_3 \quad + \quad Na \longrightarrow$

(b)

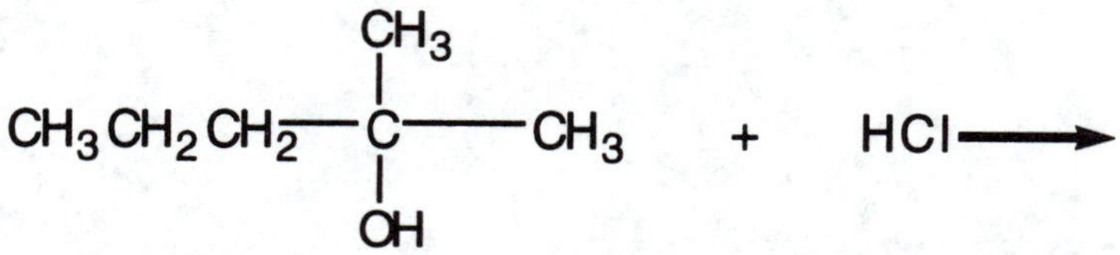

$$CH_3 \\ | \\ CH_3CH_2CH_2 - C - CH_3 \quad + \quad HCl \longrightarrow \\ | \\ OH$$

(c)

$$CH_3CH_2CH_2\overset{\overset{\displaystyle CH_3}{|}}{C}HCH_2OH \qquad + \qquad SOCl_2 \longrightarrow$$

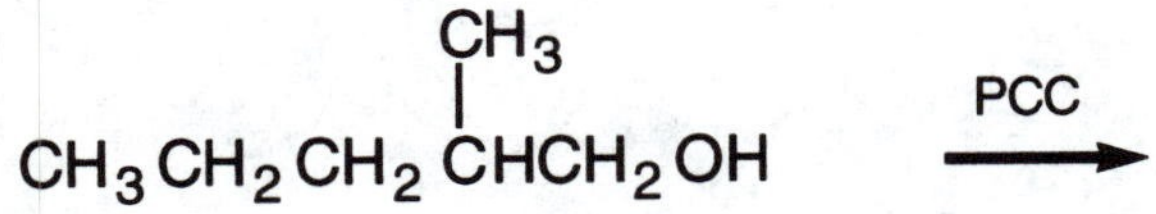

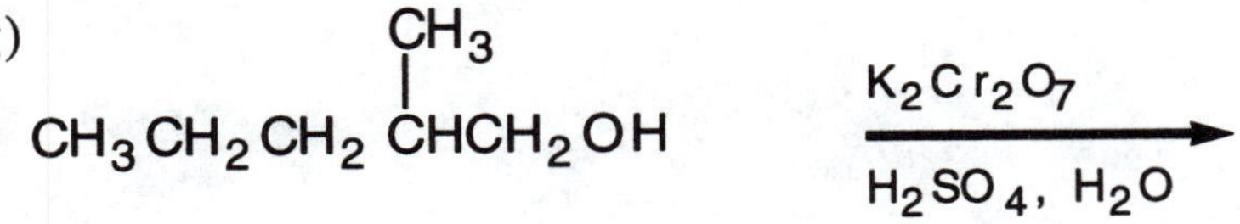

(d)

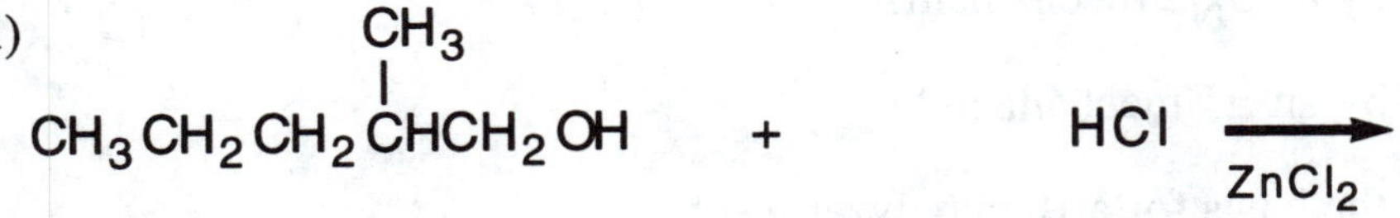

$$CH_3CH_2CH_2\overset{\overset{\displaystyle CH_3}{|}}{C}HCH_2OH \qquad + \qquad HCl \underset{ZnCl_2}{\longrightarrow}$$

(e)

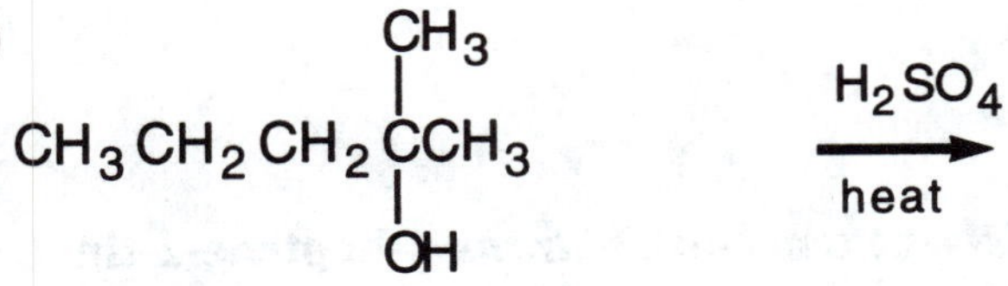

$$CH_3CH_2CH_2\overset{\overset{\displaystyle CH_3}{|}}{\underset{\underset{\displaystyle OH}{|}}{C}}CH_3 \qquad \overset{H_2SO_4}{\underset{heat}{\longrightarrow}}$$

(f)

$$CH_3CH_2CH_2\overset{\overset{\displaystyle CH_3}{|}}{C}HCH_2OH \qquad \overset{PCC}{\longrightarrow}$$

(g)

$$CH_3CH_2CH_2\overset{\overset{\displaystyle CH_3}{|}}{C}HCH_2OH \qquad \overset{K_2Cr_2O_7}{\underset{H_2SO_4,\ H_2O}{\longrightarrow}}$$

(h)

$$\underset{H}{\overset{CH_3CH_2}{}}C=C\underset{CH_3}{\overset{CH_3}{}} \quad + \quad C_6H_5\overset{\overset{\displaystyle O}{\|}}{C}OOH \qquad \underset{CH_2Cl_2}{\longrightarrow}$$

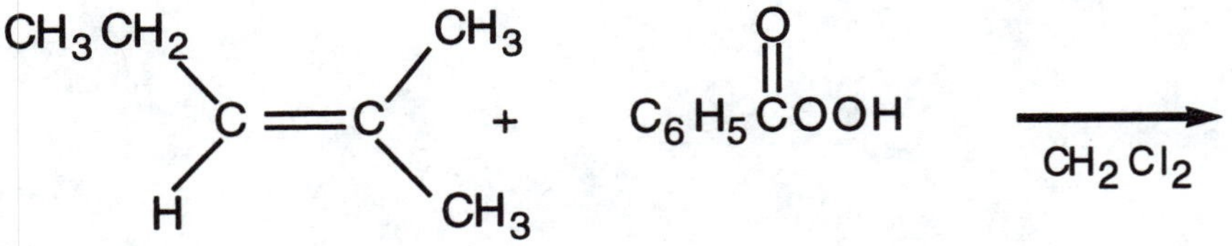

I)

$$\underset{H}{\overset{CH_3CH_2}{}}C\underset{O}{\overset{}{\diagdown}}C\underset{CH_3}{\overset{CH_3}{}} \quad + \quad HOH \overset{H^+}{\longrightarrow}$$

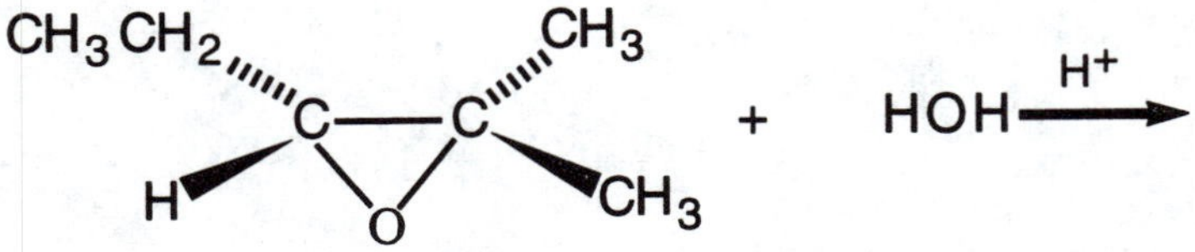

4. Which reaction has the following as the final product?

$$CH_3CH_2CH_2\overset{\overset{\displaystyle CH_3}{|}}{C}HCH_2Cl$$

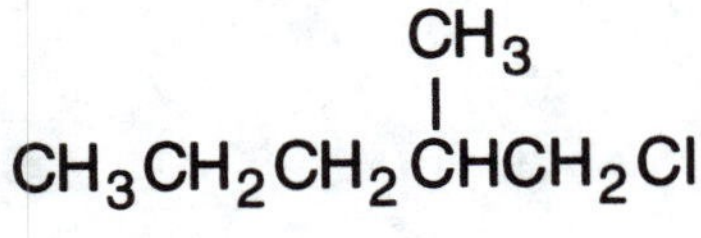

5. Which reaction has the following as the final product?

$$CH_3CH_2CH_2\overset{\overset{\displaystyle CH_3}{|}}{C}HCHO$$

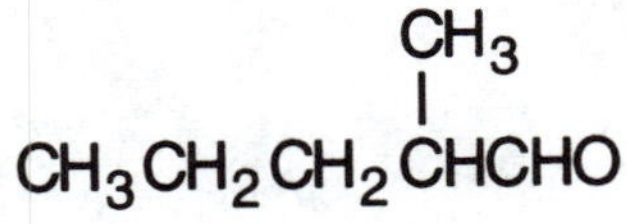

6. Which reaction yields an epoxide as a product?

7. Which reaction proceeds by an $S_N1$ mechanism?

8. Which reaction proceeds by an $S_N2$ mechanism?

9. Which reaction proceeds by an E1 mechanism?

10. Which reaction serves as the basis for a Breathalyzer test?

**Chapter 7 Quiz Answers**
(1) **b, c** (2) **a** (3,6,6-trimethyl-3-heptanol), **c** (*cis*-4-hexen-2-ol), **e** (*trans*-4-heptene-2-thiol) (3) **a, b, c** (4) **c, d** (5) **f** (6) **h** (7) **b** (8) **d** (9) **e** (10) **g**

## CHAPTER 8: QUIZ

*Determine all of the correct answers for each question. Each question may have more than one correct response.*

**Use these compounds to answer questions 1-4 .**

$$CH_3-CH(CH_3)-C(CH_3)(Br)CH_2CH_3 \quad X$$

$$CH_3-CH(CH_3)-CHCH_2CH_2Br \quad Y$$

$$CH_3-CH(CH_3)-CH(CH_3)-CHCH_3(Br) \quad Z$$

1. Place the three compounds in the order: most likely to undergo an $S_N1$ reaction first; least likely to undergo an $S_N1$ reaction last.

   (a) XYZ      (b) XZY      (c) YXZ      (d) YZX      (e) ZXY      (f) ZYX

2. Place the three compounds in the order: most likely to undergo an $S_N2$ reaction first; least likely to undergo an $S_N2$ reaction last.

   (a) XYZ      (b) XZY      (c) YXZ      (d) YZX      (e) ZXY      (f) ZYX

3. Place the three compounds in the order: most likely to undergo an E2 reaction when treated with KOH in water first; least likely to undergo an E2 reaction last when treated with KOH in water.

   (a) XYZ      (b) XZY      (c) YXZ      (d) YZX      (e) ZXY      (f) ZYX

4. Place the three compounds that when heated with KOH in ethanol give in order, the most alkene isomers first and the least alkene isomers last.

   (a) XYZ      (b) XZY      (c) YXZ      (d) YZX      (e) ZXY      (f) ZYX

5. In the synthesis of nerolin by a Williamson ether synthesis which of the following would be useful as the missing reactant in the reaction?

   (naphthalene with $O^- K^+$) + (missing reactant) → (naphthalene with $OCH_2CH_3$) **nerolin**

   (a) $CH_3CH_2OH$      (b) $CH_3CH_2Br$      (c) $CH_3CH_2NH_2$      (d) $CH_3CH_2I$

   (e) $CH_3CO_2CH_2CH_3$

**Use these reactions to answer questions 6-8. Products are not shown.**

(a)

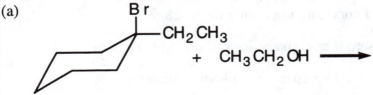

$$+ \quad CH_3CH_2OH \longrightarrow$$

(b)

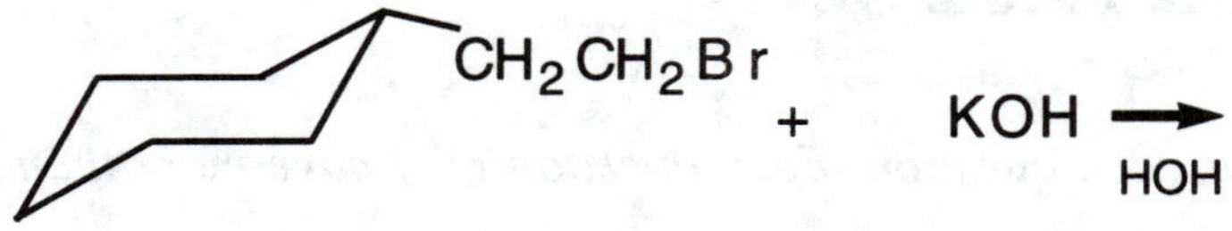

$$CH_2CH_2Br \quad + \quad KOH \xrightarrow{HOH}$$

(c)

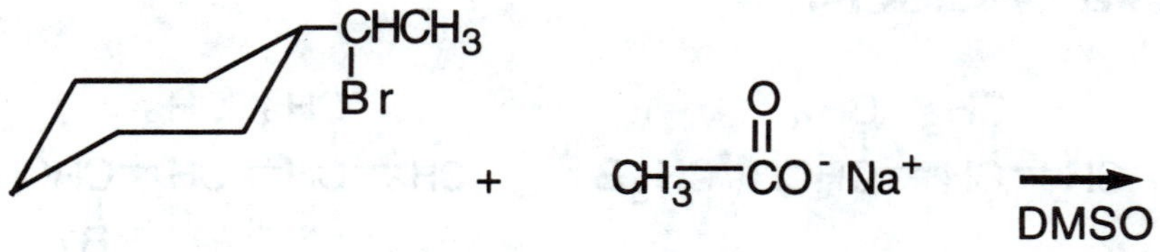

$$\begin{array}{c} CHCH_3 \\ | \\ Br \end{array} \quad + \quad CH_3-\overset{\overset{\displaystyle O}{\|}}{C}O^- Na^+ \xrightarrow{DMSO}$$

(d)

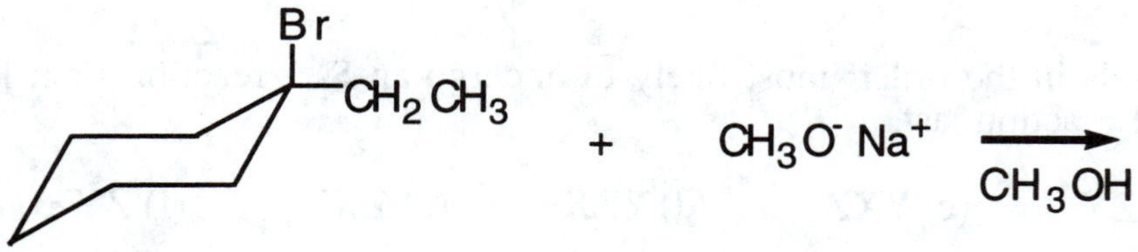

$$\begin{array}{c} Br \\ | \\ CH_2CH_3 \end{array} \quad + \quad CH_3O^- Na^+ \xrightarrow{CH_3OH}$$

(e)

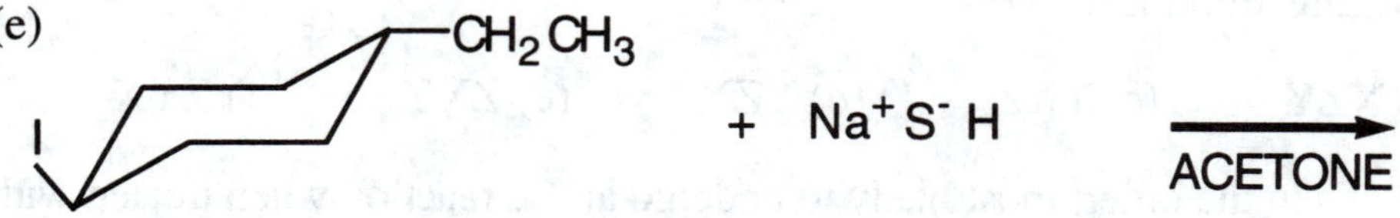

$$CH_2CH_3 \quad + \quad Na^+ S^- H \xrightarrow{ACETONE}$$

(f)

$$\begin{array}{c} CHCH_3 \\ | \\ Br \end{array} \quad + \quad CH_3O^- Na^+ \xrightarrow{CH_3OH}$$

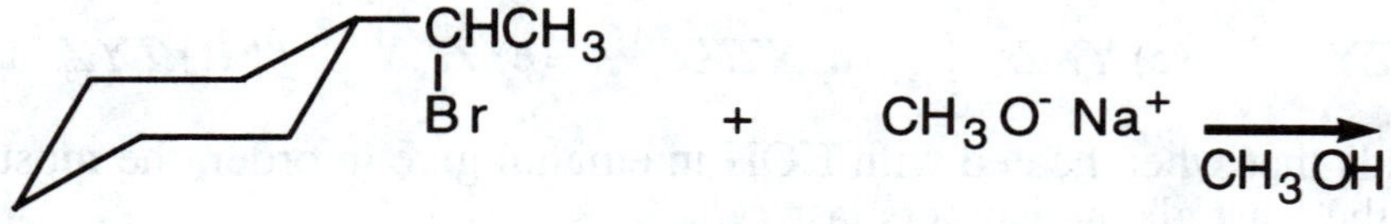

6.  Which reactions would we predict undergo primarily an $S_N2$ reaction?

7.  Which reactions would we expect to have a carbocation intermediate?

8.  In which reactions will the dominant product be an alkene?

9.  Which of the following statements are consistent with the following reaction?

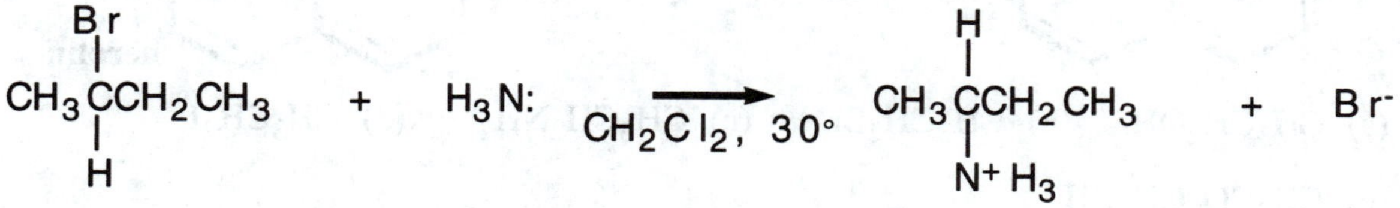

$$\begin{array}{c} Br \\ | \\ CH_3CCH_2CH_3 \\ | \\ H \end{array} \quad + \quad H_3N: \xrightarrow[CH_2Cl_2,\ 30°]{} \quad \begin{array}{c} H \\ | \\ CH_3CCH_2CH_3 \\ | \\ N^+H_3 \end{array} \quad + \quad Br^-$$

(a)  The reaction will go by an $S_N1$ mechanism.

(b)  There will be an inversion of configuration during the reaction.

(c)  There will be a limited amount, if any, alkene formed during the reaction.

(d)  The same method could be used to prepare the following amine.

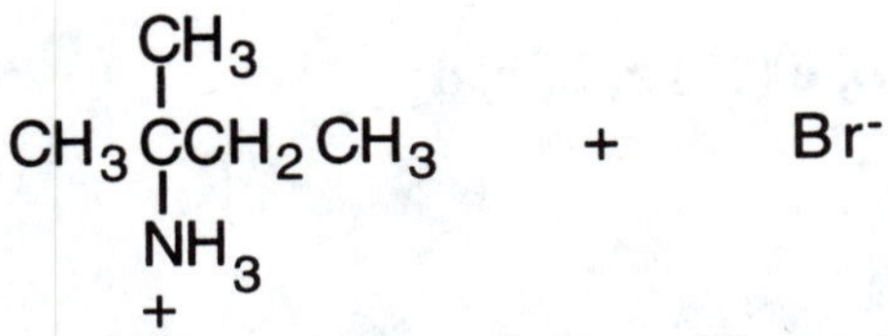

$$\begin{array}{c} CH_3 \\ | \\ CH_3\,\overset{|}{C}CH_2\,CH_3 \\ | \\ \overset{+}{N}H_3 \end{array} \quad + \quad Br^-$$

E) This same reaction would occur using $CH_3O^-$ instead of $Br^-$ as the leaving group.

10. Which of the compounds listed below are potential products for the following reaction?

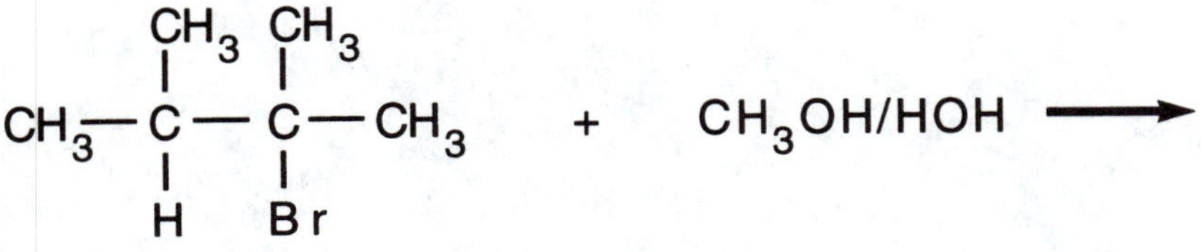

$$\begin{array}{c} CH_3 \quad CH_3 \\ | \qquad | \\ CH_3\!-\!\overset{|}{C}\!-\!\overset{|}{C}\!-\!CH_3 \\ | \qquad | \\ H \qquad Br \end{array} \quad + \quad CH_3OH/HOH \longrightarrow$$

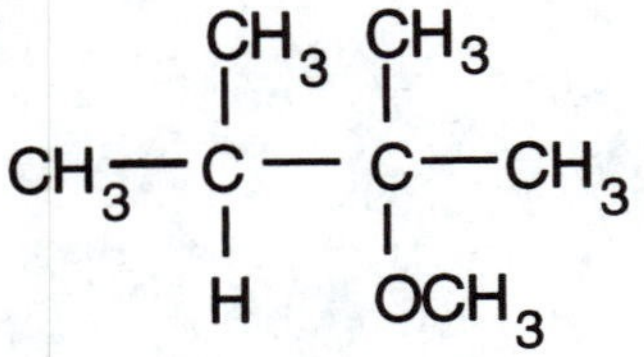

(a)
$$\begin{array}{c} CH_3 \quad CH_3 \\ | \qquad | \\ CH_3\!-\!\overset{|}{C}\!-\!\overset{|}{C}\!-\!CH_3 \\ | \qquad | \\ H \qquad OCH_3 \end{array}$$

(b)
$$\begin{array}{c} CH_3 \quad CH_3 \\ | \qquad | \\ CH_3\!-\!\overset{|}{C}\!-\!\overset{|}{C}\!-\!CH_3 \\ | \qquad | \\ H \qquad OH \end{array}$$

(c)
$$\begin{array}{c} CH_3 \quad CH_2 \\ | \qquad \| \\ CH_3\!-\!\overset{|}{C}\!-\!\overset{}{C}\!-\!CH_3 \\ | \\ H \end{array}$$

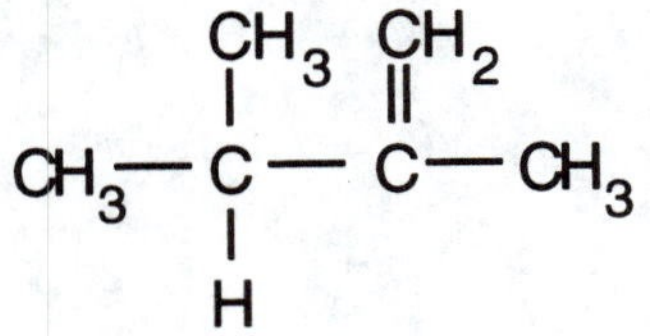

(d)
$$\begin{array}{c} CH_3 \quad CH_3 \\ | \qquad | \\ CH_3\!-\!C\!=\!C\!-\!CH_3 \end{array}$$

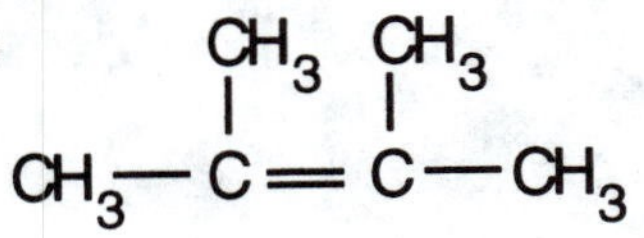

(e)
$$\begin{array}{c} CH_3 \quad CH_3 \\ | \qquad | \\ CH_3\!-\!\overset{|}{C}\!-\!\overset{|}{C}\!-\!CH_3 \\ | \qquad | \\ OH \qquad OCH_3 \end{array}$$

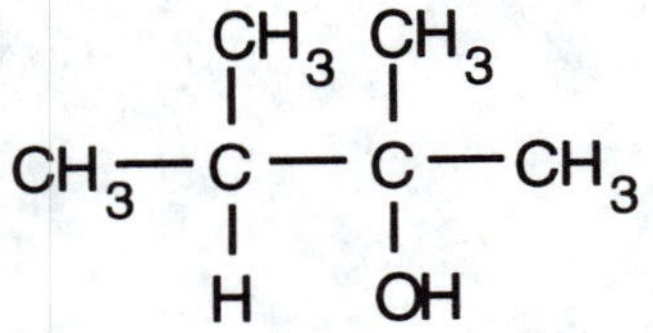

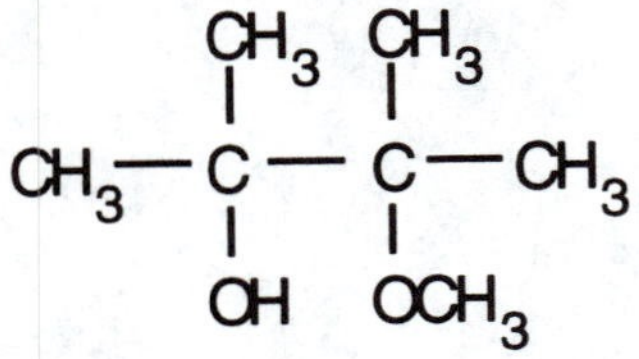

**Chapter 8 Quiz Answers**
(1) **b** (2) **d** (3) **b** (4) **b** (5) **b, d** (6) **b, c, e** (7) **a** (8) **d, f** (9) **b, c** (10) **a, b, c, d**

# CHAPTER 9: QUIZ

*Determine all of the correct answers for each question. Each question may have more than one correct response.*

**Use the compound, aplysin, to answer questions 1-2.**

1. Which of the following are true concerning the aromatic ring in aplysin?
   (a) It contains alternate double and single bonds.
   (b) Each carbon in the aromatic ring exhibits $sp^2$ hybridization.
   (c) The aromatic ring actually exists as a chair conformation.
   (d) Each of the atoms indicated by the letters X, Y, and Z plus the oxygen attached to the aromatic ring lie in the same plane.
   (e) Aplysin has two isomeric forms since a second contributing structure to the resonance hybrid shown can be drawn for the aromatic ring in aplysin.

2. List the location of the three atoms labeled X, Y, and Z in the order *ortho, meta, para* with respect to the oxygen bonded to the aromatic ring as the reference point.

   (a) XYZ     (b) XZY   (c) YXZ     (d) YZX     (e) ZXY     (f) ZYX

3. In the following compound which of the rings would not be classified as aromatic?

4. Place the following compounds in the order most acidic first and least acidic last.

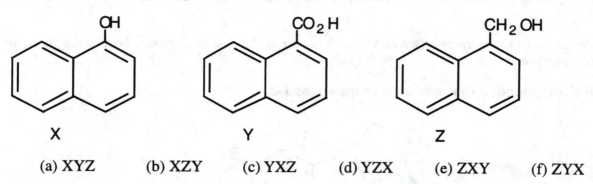

| X | Y | Z |

(a) XYZ      (b) XZY      (c) YXZ      (d) YZX      (e) ZXY      (f) ZYX

5. Place the following phenoxide ions in the order of most stable phenoxide ion first and least stable phenoxide ion last.

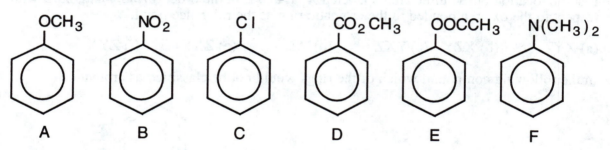

X      Y      Z

(a) XYZ      (b) XZY      (c) YXZ      (d) YZX      (e) ZXY      (f) ZYX

**Use the following compounds to answer question 6-7.**

$OCH_3$    $NO_2$    $Cl$    $CO_2CH_3$    $OOCCH_3$    $N(CH_3)_2$

A      B      C      D      E      F

6. Which of the following compounds, when reacted with bromine ($Br_2$) using $FeBr_3$ as a catalyst would be expected to give primarily a meta substituted product?

7. Which compounds have a substituent that is incapable of participation in the stabilization of the intermediate cation by resonance?

8. To synthesize *meta*-bromoaniline which of the following synthetic procedures can be used?

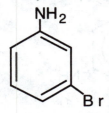

(a) React aniline with $Br_2$ using $FeBr_3$ as a catalyst.
(b) React bromobenzene with ammonia.
(c) React bromobenzene with $HNO_3$ in presence of $H_2SO_4$ and then reduce the nitro group with Fe in HCl, followed by treatment with NaOH.
(d) React nitrobenzene with $Br_2$ using $FeBr_3$ as a catalyst and then reduce the nitro group with Fe in HCl, followed by treatment with NaOH.
(e) React aniline with HBr and then with $Br_2$ using $FeBr_3$ as a catalyst, followed by treatment of product with aqueous NaOH.

9. Place the following compounds in the order reacts fastest with $CH_3COCl/AlCl_3$ first; reacts slowest with $CH_3COCl/AlCl_3$ last.

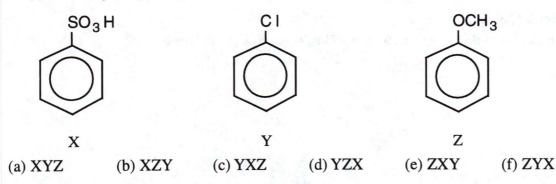

(a) XYZ     (b) XZY     (c) YXZ     (d) YZX     (e) ZXY     (f) ZYX

10. The following series of reactions are performed in the laboratory.
    **Reaction 1:** Toluene is reacted with concentrated nitric acid in the presence of concentrated sulfuric acid.
    **Reaction 2:** The product from reaction 1 is reacted with $Cl_2$ using $FeCl_3$ as a catalyst.
    **Reaction 3:** The product of reaction 2 is reacted with $K_2Cr_2O_7$ in concentrated $H_2SO_4$.
    **Reaction 4:** The product of reaction 3 is reacted with hydrogen gas at 3 atmospheres of pressure using a nickel catalyst.

Which of the following compounds will be the major product from reaction 4?

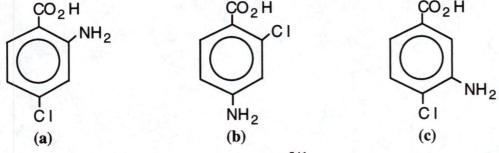

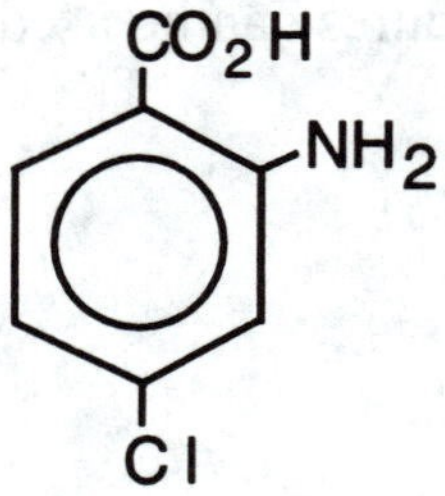

CO$_2$H
NH$_2$
Cl

**(d)**

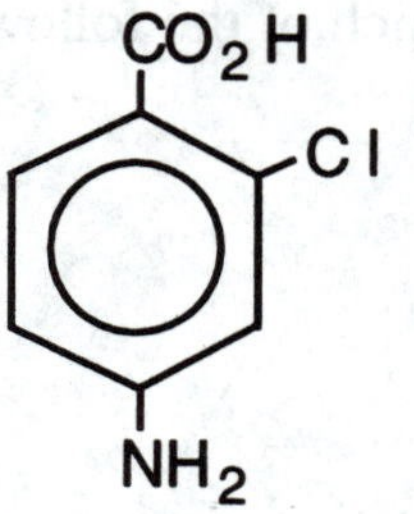

CO$_2$H
Cl
NH$_2$

**(e)**

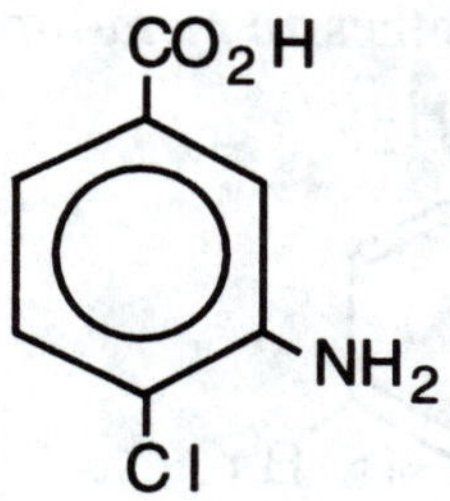

CO$_2$H
NH$_2$
Cl

**(f)**

CO$_2$H
NH$_2$
Cl

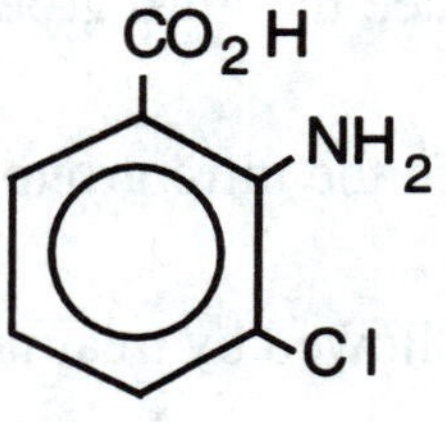

**(g)**

## Chapter 9 Quiz Answers

(1) **b, d** (2) **b** (3) **d** (4) **c** (5) **a** (6) **b, d** (7) **b, d** (8) **d, e** (9) **d** (10) **b**

# CHAPTER 10: QUIZ

*Determine all of the correct answers for each question. Each question may have more than one correct response.*

1. Which of the following amines has a name that is neither an accurate common nor systematic name? Write a correct name for each incorrect one.

(a) [structure: benzene ring with $-NH_2$ and $NO_2$ groups]          nitroaniline

(b) $H_2N$—[benzene ring]—$CO_2H$          *p*-aminobenzoic acid

(c) [benzene ring]—$N(CH_2CH_3)_2$          N-diethylaniline

(d) $(CH_3 CH_2 CH_2 CH_2)_2 NH$          dibutylamine

(e) $(CH_3)_2 CHCH_2$—C—$NH_2$ with $H_3C$ and $H$          4-methyl-2-pentanamine

**Use these amines for questions 2-3.**

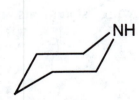

$(CH_3 CH_2)_3 N$

$(CH_3)_2 N$     $N(CH_3)_2$ [naphthalene structure]

(X) piperidine pK$_b$ = 2.88     (Y) triethylamine pK$_b$ = 3.25     (Z) *1,8-bis*(dimethylamino)-naphthalene pK$_b$ = 1.7

2. Place the amines in the order: aliphatic amine; heterocyclic amine; aromatic amine.

(a) XYZ     (b) XZY     (c) YXZ     (d) YZX     (e) ZXY     (f) ZYX

3. Place the amines in the order: most basic first; least basic last.

(a) XYZ     (b) XZY     (c) YXZ     (d) YZX     (e) ZXY     (f) ZYX

4. Place the following three amines in the order: most basic first; least basic last.

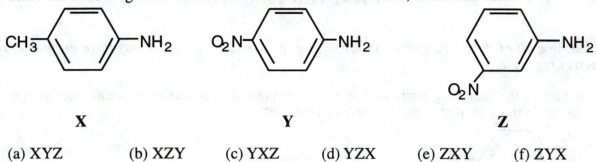

**X**                    **Y**                    **Z**

(a) XYZ          (b) XZY          (c) YXZ          (d) YZX          (e) ZXY          (f) ZYX

5. Place the following three ions in the order: most acidic first; and least acidic last.

$N^+ H_2$
$\|$
$H_2N — C — NH_2$

**X**                    **Y**                    **Z**

(a) XYZ          (b) XZY          (c) YXZ          (d) YZX          (e) ZXY          (f) ZYX

6. Ergotamine tartrate is used for the treatment of migraine headaches. Which of the five nitrogen atoms found in ergotamine tartrate react with tartaric acid to form a salt?

7. Diazepam (Valium) has a $pK_b$ = 10.7. Which of the following statements is consistent with this information?

(a) Diazepam could be considered somewhat equivalent to ammonia in terms of its strength as a base.
(b) Diazepam reacts with concentrated hydrochloric acid to form a salt.
(c) The N in diazepam with the attached methyl group is the one to which the $H^+$ attaches in an ion formed by reaction with an acid.
(d) The $pK_a$ for the conjugated acid of diazepam is 10.7.
(e) If diazepam were reacted with acetic acid, $CH_3COOH$ ($pK_a = 4.76$), the equilibrium for this reaction would favor the formation of the acetate salt of diazepam.

8. Which of the following amines will exhibit hydrogen bonding?

(a) $(CH_3CH_2)_3N$     (b) $(CH_3CH_2)_2NH$     (c)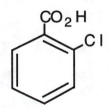

(d) [benzene ring]—$N(CH_3)_2$     (e) [pyrrole ring] NH

9. Place the following chemical reactions in the appropriate order to synthesize *ortho*-chlorobenzoic acid using *ortho*-toluidine as a starting material.

$CO_2H$ ... CI

*ortho*-chlorobenzoic acid

$CH_3$ ... $NH_2$

*ortho*-toluidine

X) React compound with $K_2Cr_2O_7$ in sulfuric acid to convert $-CH_3$ to $-CO_2H$.
Y) Convert $-NH_2$ to a diazonium salt by reaction with $HNO_2$.
Z) React compound with CuCl.

(a) XYZ     (b) XZY     (c) YXZ     (d) YZX     (e) ZXY     (f) ZYX

10. A variety of naturally occurring amines were discussed in the chapter. Which of the following statements pertaining to the amines discussed are accurate?
(a) They can all be classified as naturally occurring Brønsted-Lowry bases.
(b) Most if not all appear to have a human physiological action.
(c) All would be expected to form a salt when reacted with hydrochloric acid.
(d) Synthetic compounds of similar structure to one of the naturally occurring amines are physiologically active in humans.
(e) The naturally occurring amines found in our blood stream or in our cells would normally be expected to exist as an equilibrium mixture of the free base and its conjugate acid.

## Chapter 10 Quiz Answers

(1) *correct names*: **b, d**   *incorrect names*: (a) **3-nitroaniline or *meta*-nitroaniline** (c) **N,N-diethylaniline** (e) **(R)-4-methyl-2-pentanamine** (2) **c** (3) **e** (4) **b** (5) **f** (6) **b** (7) **b** (8) **b, e** (9) **d** (10) **a, b, c, d, e**

# CHAPTER 11: QUIZ

*Determine all of the correct answers for each question. Each question may have more than one correct response.*

1. Which of the following are correct statements for propanal?
   (a) The carbonyl carbon exhibits $sp^2$ hybridization.
   (b) The oxygen exhibits $sp^2$ hybridization.
   (c) The carbonyl carbon is slightly positively charged.
   (d) The oxygen is slightly negatively charged.
   (e) The molecule exhibits association by hydrogen bonding.

2. Which of the following compounds would, upon reaction with phenyl magnesium bromide and then water, yield a secondary alcohol?

   (a) $CO_2$      (b) $CH_3CH_2CH_2CHO$      (c) $CH_3CH_2COCH_3$

   (d) $HCHO$      (e) $CH_3CH_2CHOHCH_3$      (f)

**Use these compounds to answer questions 3-10.**

   (a) $CH_3CH_2CH_2CHO$      (b) $CH_3CH_2COCH_3$      (c) $HCHO$

   (d) $CH_3CH_2CHOHCH_3$      (e) $(CH_3CH_2)_3CCHO$

   (f)

3. Which of the compounds exhibit tautomerism?

4. Which of the compounds will give a positive test with Tollens' reagent?

5. Which of the compounds would not form an acetal or ketal with *cis*-1,2-dihydroxycyclohexanediol in the presence of dry HCl?

6. Which of the compounds would be oxidized by hydrogen peroxide to a carboxylic acid?

7. Which of the compounds would be expected to react with atmospheric oxygen if allowed to stand open to the air for a period of time?

8. Which of the compounds would be reduced by reaction with lithium aluminum hydride to give a secondary alcohol?

9. Which of the compounds would react with 2-propanamine to give an imine?

10. Which of the compounds would react with diethylamine followed by reduction with hydrogen using a nickel catalyst to give a tertiary amine?

**Chapter 11 Quiz Answers**
(1) **a, b, c, d** (2) **b, f** (3) **a, b, f** (4) **a, c, e** (5) **d** (6) **a, c, e** (7) **a, c, e** (8) **b, f** (9) **a, b, c, f, e**
(10) **a, b, c, f, e**

# CHAPTER 12: QUIZ

*Determine all of the correct answers for each question. Each question may have more than one correct response.*

**Use these saccharides drawn in standard Fischer projection to answer questions 1- 4.**

```
        CHO                    CHO                   CH2OH
         |                      |                      |
  HO — C — H            HO — C — H            HO — C — H
         |                      |                      |
  H — C — OH            HO — C — H                  C = O
         |                      |                      |
  H — C — OH            HO — C — H            H — C — OH
         |                      |                      |
  HO — C — H            HO — C — H            H — C — OH
         |                      |                      |
        CH3                   CH2OH           H — C — OH
                                                      |
                                                    CH2OH

         a                      b                      c
```

```
        CHO                                    CH2OH
         |                                       |
  H — C — OH                                  C = O
         |                                       |
  OHC — C — OH                          HO — C — H
         |                                       |
  HO — C — H                           H — C — OH
         |                                       |
        CH3                            H — C — OH
                                                 |
                                        HO — C — H
                                                 |
                                        H — C — OH
                                                 |
                                               CH2OH

         d                                       e
```

1.  Which compound has an L configuration?

2.  Which compound will react with $NaBH_4$ to yield a meso product?

3.  Which compound will give a positive test with a Benedict's solution?

4.  Which compound will form a hemiacetal or hemiketal in solution?

**Use these aldohexoses drawn in standard Fischer projection to answer questions 5-6.**

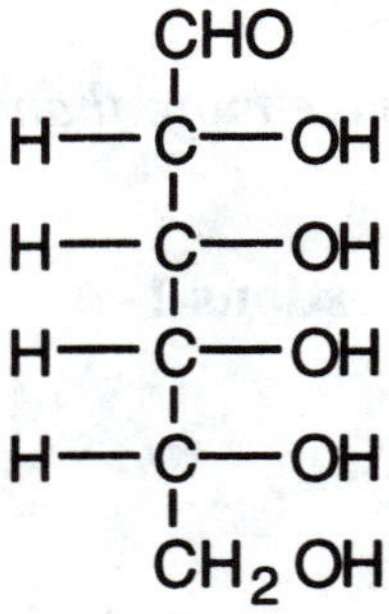

```
      CHO                    CHO                    CHO
  H—C—OH                HO—C—H                 H—C—OH
  H—C—OH                 H—C—OH                HO—C—H
  H—C—OH                 H—C—OH                 H—C—OH
  H—C—OH                 H—C—OH                 H—C—OH
   CH2OH                  CH2OH                  CH2OH

      a                      b                      c


      CHO                    CHO                    CHO
  H—C—OH                HO—C—H                 HO—C—H
  H—C—OH                HO—C—H                  H—C—OH
 HO—C—H                  H—C—OH                HO—C—H
  H—C—OH                 H—C—OH                 H—C—OH
   CH2OH                  CH2OH                  CH2OH

      d                      e                      f


      CHO                    CHO
  H—C—OH                HO—C—H
 HO—C—H                 HO—C—H
 HO—C—H                 HO—C—H
  H—C—OH                 H—C—OH
   CH2OH                  CH2OH

      g                      h
```

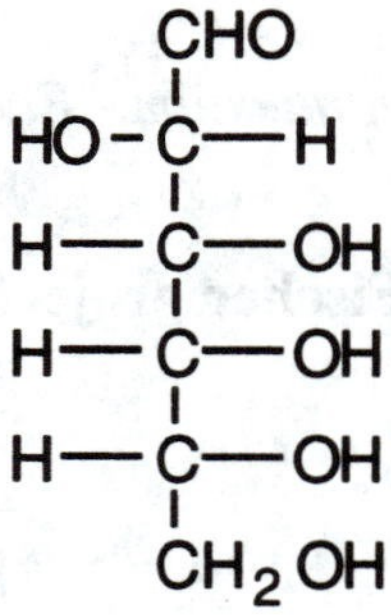

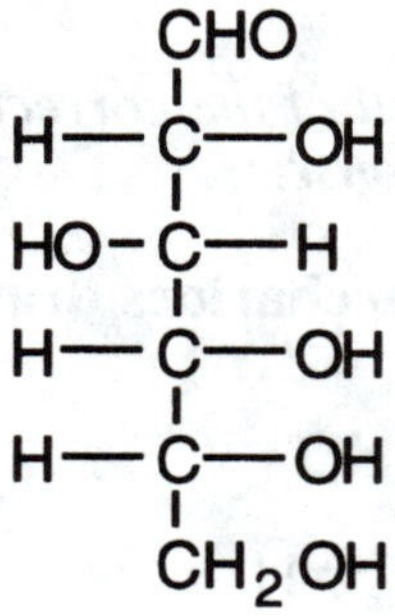

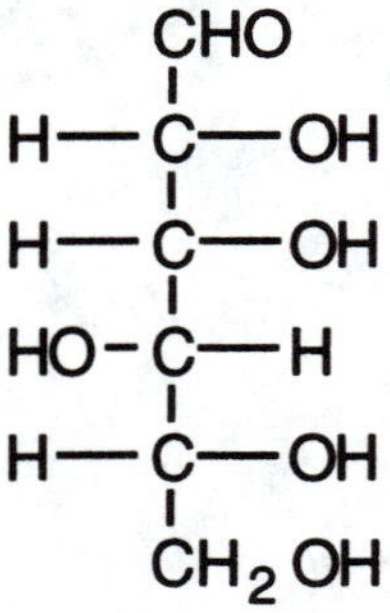

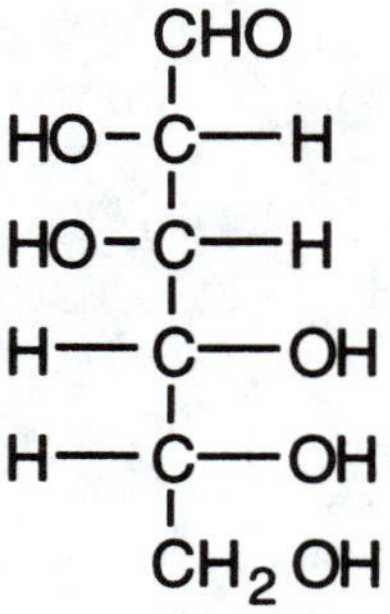

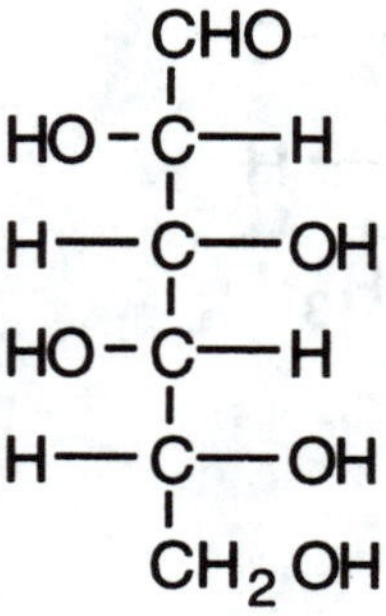

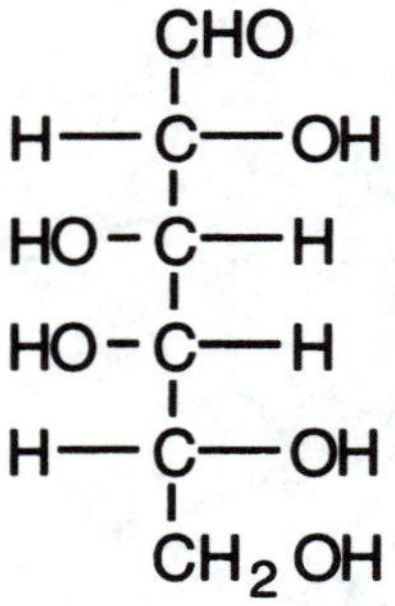

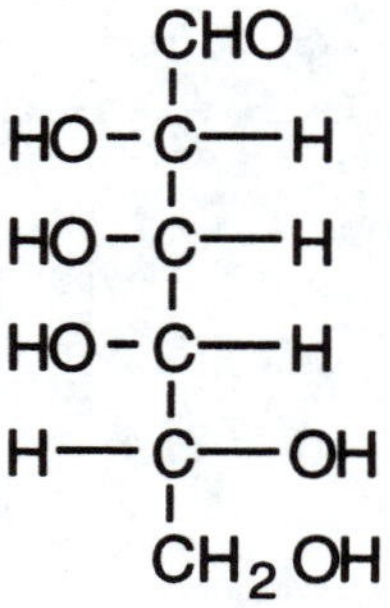

5. Which of the aldohexoses has the following chair conformation?

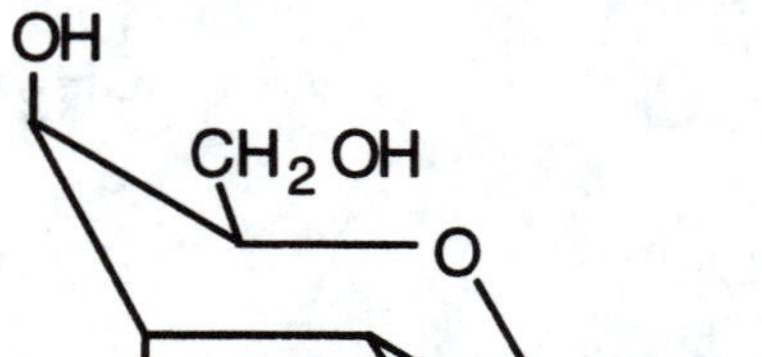

6. Which of the aldohexoses has the following Haworth structure?

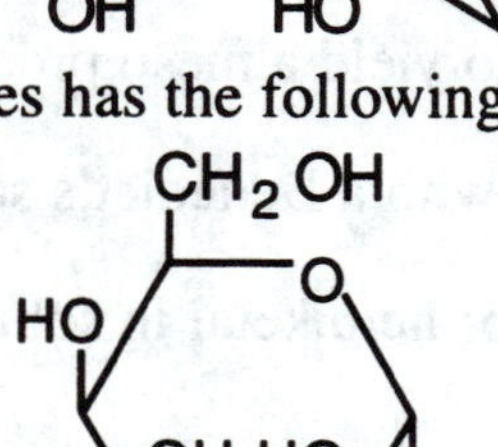

550

**Use these saccharides to answer questions 7-9.**

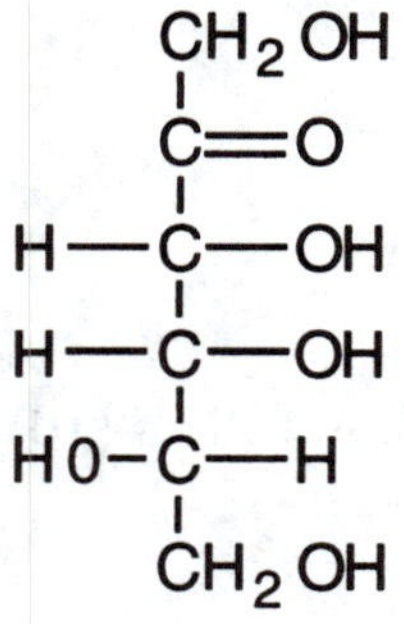

a

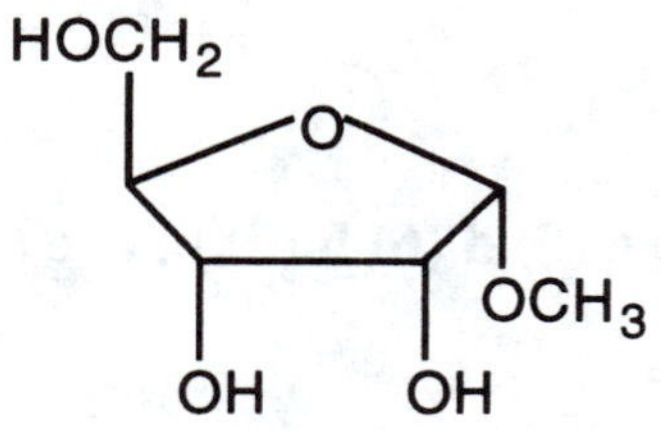

b

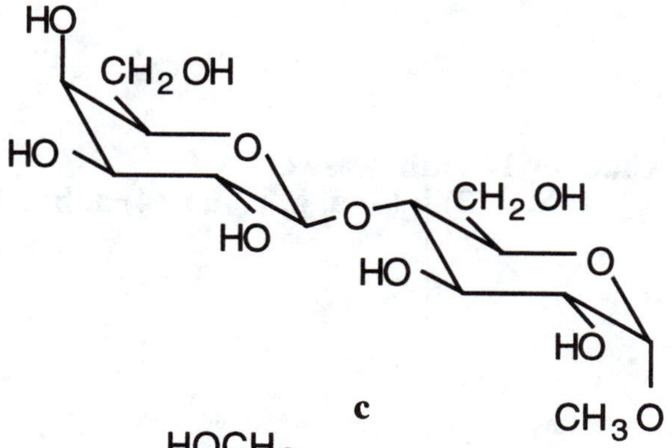

c

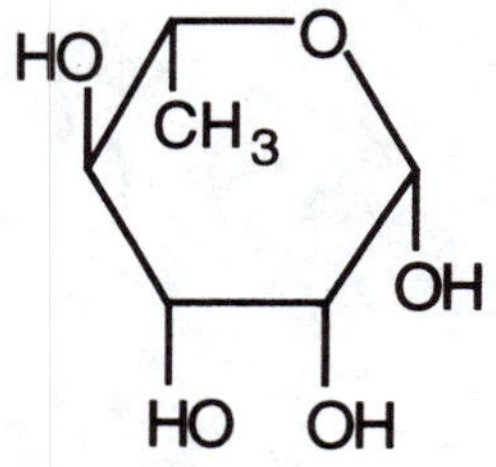

d

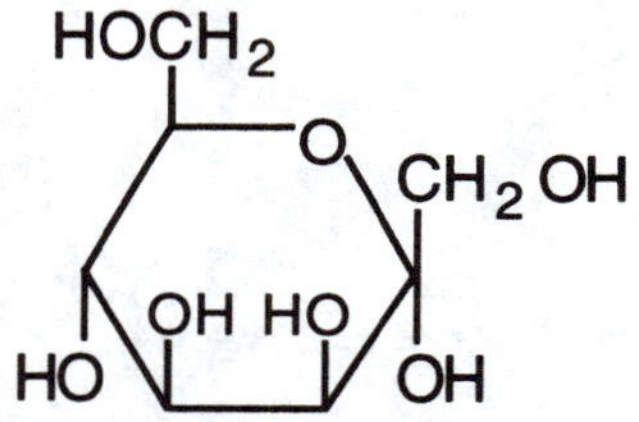

e

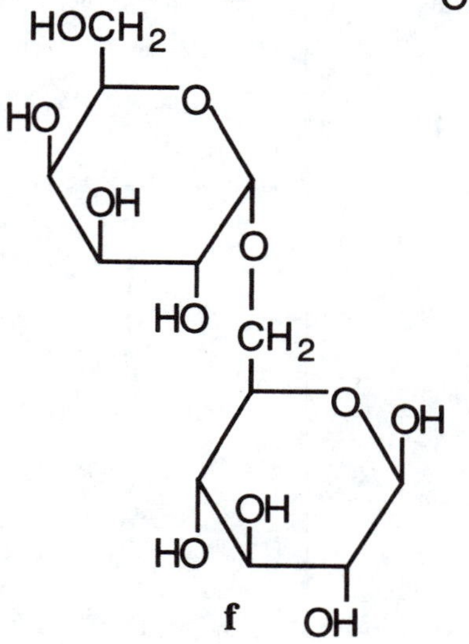

f

7. Which of the saccharides are drawn as the $\alpha$-anomer?

8. Which of the saccharides exhibits mutarotation in a water solution?

9. Which of the following is the proper designation for the glycoside bond between the two monosaccharides units in disaccharide c.
   (a) $\beta_{1\rightarrow 5}$     (b) $\beta_{1\rightarrow 4}$     (c) $\alpha_{1\rightarrow 5}$     (d) $\alpha_{1\rightarrow 4}$     (e) $\beta_{5\rightarrow 1}$     (f) $\beta_{4\rightarrow 1}$
   (g) $\alpha_{5\rightarrow 1}$     (h) $\alpha_{4\rightarrow 1}$

10. Which of the following is not true for glucose?
    (a) Glucose and its derivatives such as 2-aminoglucose would be expected to be a dominant sugar since $\beta$-D-glucose is known to exist in a chair conformation with minimal steric strain in this conformation.
    (b) Since starch containing $\alpha_{1\rightarrow 4}$ glycoside bonds and cellulose containing $\beta_{1\rightarrow 4}$ glycoside are both formed in plants, glucose must exist in two anomeric forms which exhibit mutarotation.
    (c) Humans have the enzymes to cleave $\alpha_{1\rightarrow 4}$ glycoside bonds found in starch but not $\beta_{1\rightarrow 4}$ glycoside bonds found in cellulose.
    (d) Cellulose fibers hold together because of hydrogen bonding between glucose units in the

cellulose strand.

(e) If a sample of urine is tested and it gives a positive test with Benedict's solution the urine sample is known to contain glucose.

## Chapter 12 Quiz Answers
(1) **a, b, d** (2) **b** (3) **a, b, c, d, e** (4) **a, b, c, d, e** (5) **d** (6) **h** (7) **b, c, e** (8) **a, d, e, f** (9) **b** (10) **e**

# CHAPTER 13: QUIZ

*Determine all of the correct answers for each question. Each question may have more than one correct response.*

1. Which of the following carboxylic acids have a correct IUPAC name associated with the structure? Write the correct name for those that are incorrect.

   (a) $CH_3CH_2CH_2CO_2H$          butyric acid

   (b) $CH_3CHOHCO_2H$          α-hydroxypropanoic acid

   (c) $HO_2C(CH_2)_4CHO$          6-oxohexanoic acid

   (d) $HO_2C(CH_2)_{10}CO_2H$          dodecanoic acid

   (e) $CH_3(CH_2)_5$ — $C=C$ — $(CH_2)_8CO_2H$, H, H          9-hexadecenoic acid

**Use these compounds to answer questions 2-3.**

   (X) $CH_3(CH_2)_3CH_2OH$     (Y) $CH_3(CH_2)_3CHO$     (Z) $CH_3(CH_2)_3CO_2H$

2. Place the compounds in the order: exhibiting most hydrogen bonding first; exhibiting least hydrogen bonding last.

   (a) XYZ    (b) XZY    (c) YXZ    (d) YZX    (e) ZXY    (f) ZYX

3. Place the compounds in the order: highest boiling point first; lowest boiling point last.

   (a) XYZ    (b) XZY    (c) YXZ    (d) YZX    (e) ZXY    (f) ZYX

4. Place the following three carboxylic acids in the order: most acidic first; least acidic last.

   (X) $CH_3CO_2H$          $pK_a = 4.72$

   (Y) $BrCH_2CO_2H$          $pK_a = 2.90$

   (Z) $CH_3CH_2CO_2H$          $pK_a = 4.87$

   (a) XYZ    (b) XZY    (c) YXZ    (d) YZX    (e) ZXY    (f) ZYX

5. Place the three carboxylic acids in the order: most acidic first; least acidic last.

   (X) $(CH_3)_3CCO_2H$

   (Y) $CH_3CH_2CH_2CO_2H$

   (Z) $CH_3CHBrCHBrCO_2H$

   (a) XYZ    (b) XZY    (c) YXZ    (d) YZX    (e) ZXY    (f) ZYX

6. Which of the following compounds would react with butanoic acid ($CH_3CH_2CH_2CO_2H$) to yield a salt as the final product?

(a) KOH

(b) $CH_3OH$

(c) $LiAlH_4$ followed by water

(d) $K_2CO_3$

(e) $NaHCO_3$

7. Which of the reagents could be used to reduce only the carbonyl group without removing the double bond or reducing the carboxylic group in the following compound?

$$CH_3(CH_2)_5COCH_2CH=CH(CH_2)_7CO_2H$$

(a) (1) $LiAlH_4$ (2) HOH

(b) $H_2$, Pt, 25°C, 2 atm

(c) (1) $NaBH_4$ (2) HOH

(d) $SOCl_2$

(e) NaOH

8. Which of the following acids would you expect to undergo decarboxylation when heated?

(a)

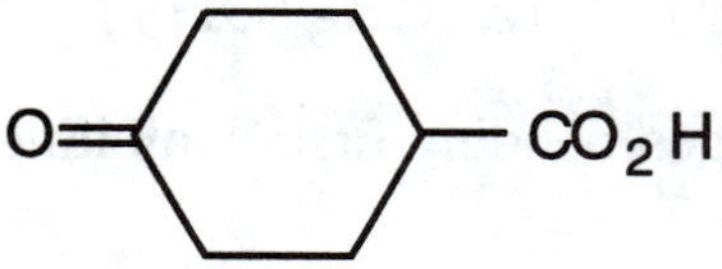

(b)

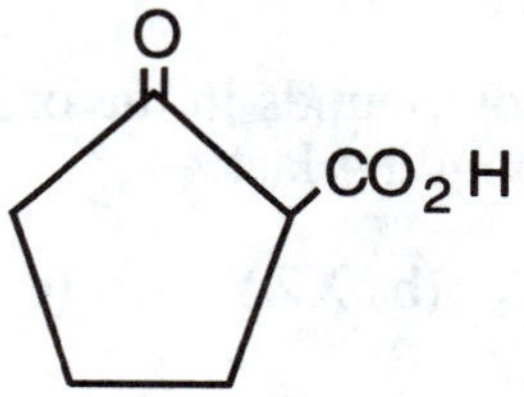

(c)

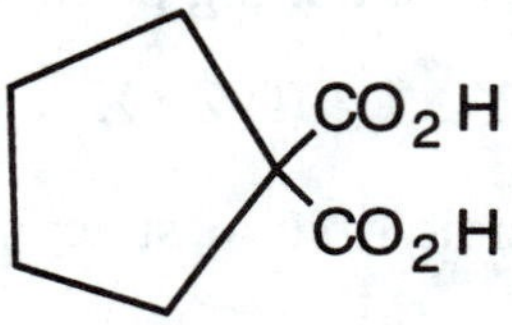

(d)

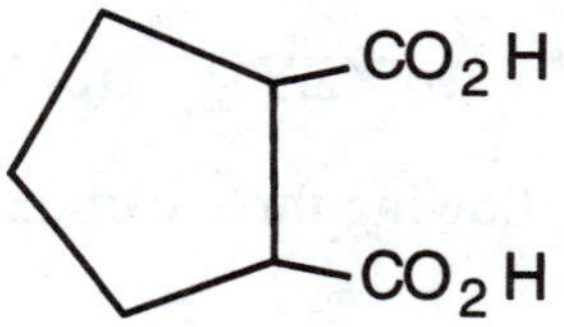

(e)

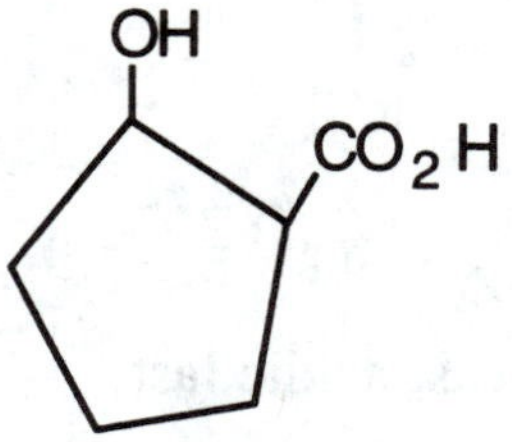

554

9. To obtain a high yield of ester in the following reaction which of the listed procedures can be used?

$$CH_3CO_2H \quad + \quad CH_3CH_2OH \quad \overset{H_2SO_4}{\rightleftharpoons} \quad CH_3CO_2CH_2CH_3 \quad + \quad HOH$$

(a) Use a large concentration of acetic acid compared to the amount of ethanol in the reaction vessel.
(b) Use a large concentration of ethanol compared to the amount of acetic acid in the reaction vessel.
(c) Add water to the reaction vessel.
(d) Remove water from the reaction vessel by doing an azeotropic distillation.
(e) Extend the normal reaction time by a factor of 5 times.

10. Which of the following is not a physiological function associated with the presence of abscisic acid in plants?
(a) Causes closure of stomata to retard water loss during time of drought.
(b) Promotes dormancy in buds protecting plant buds against cold.
(c) Reduces rate of cell growth by inhibition of cell growth hormones.
(d) Promotes abscission (leaf fall) in deciduous trees in the fall.
(e) Promotes formation of potato tubers.

**Chapter 13 Quiz Answers**
(1) **correct names: c; incorrect names: a (butanoic acid), b (2-hydroxypropanoic acid), d (1,12 dodecanoic acid), e (*cis*-9-hexadecenoic acid or (Z)-9-hexadecenoic acid (2) e (3) e (4) c (5) f (6) a, d, e (7) c (8) b, d (9) b, d (10) d**

# CHAPTER 14: QUIZ

*Determine all of the correct answers for each question. Each question may have more than one correct response.*

**Use these physiologically active or commercially important compounds to answer questions 1-2.**

(a)

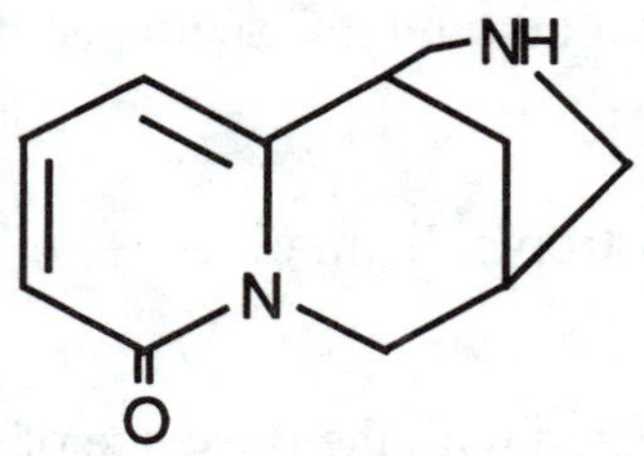

(b)

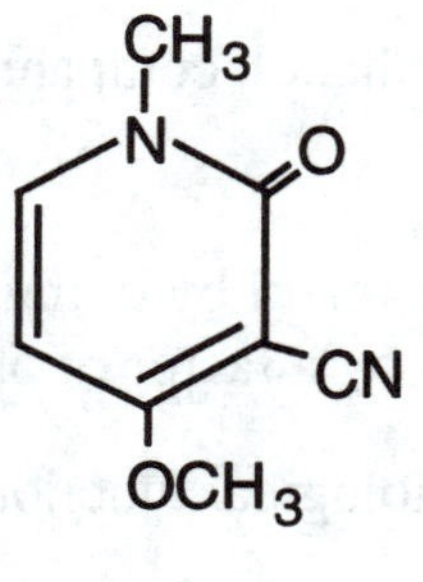

(c)

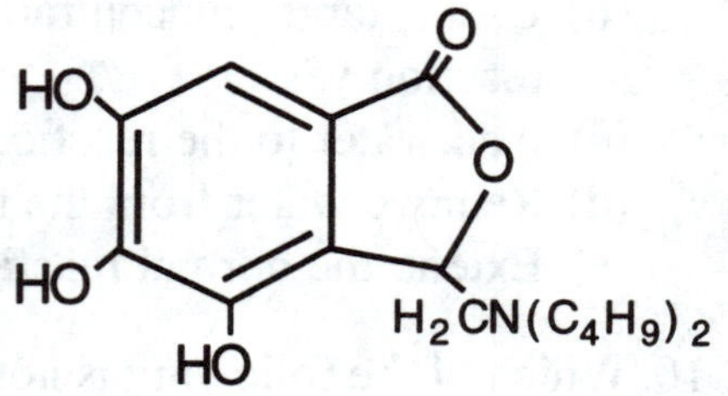

(d)

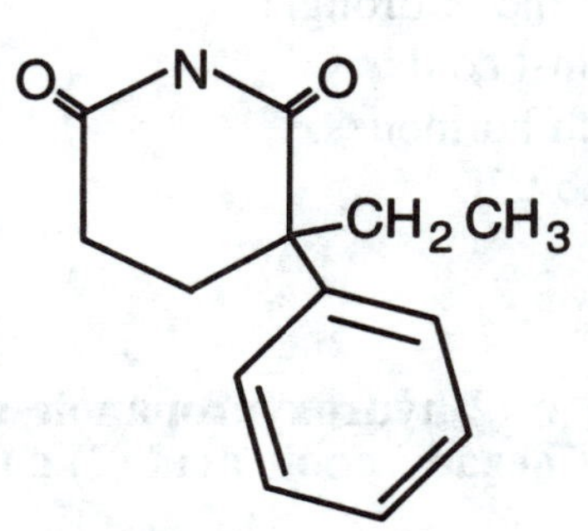

(e)

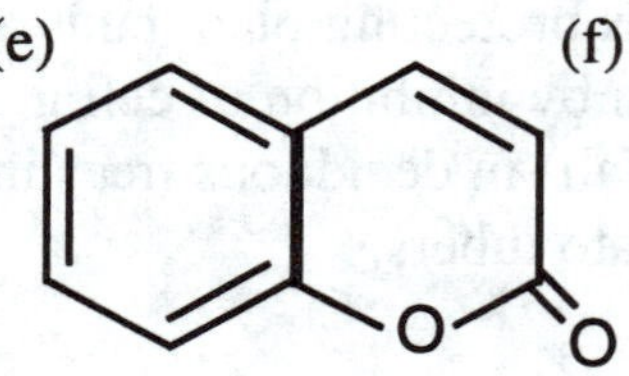

(f)

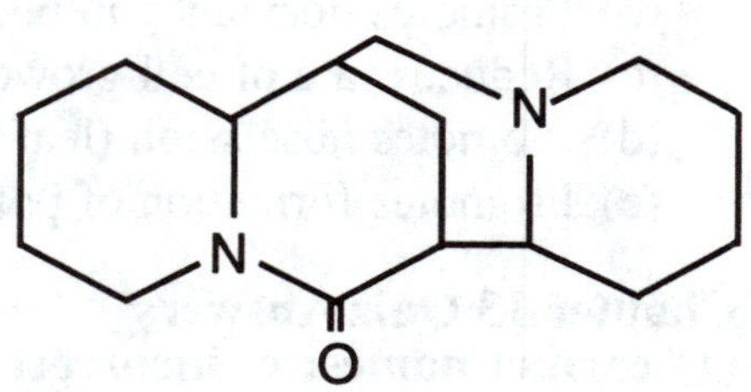

1. Which compounds are lactones?

2. Which compounds are imides?

**Use these compounds to answer questions 3-4.**

(a) $CH_3 CH_2 CO_2 CH_3$     (b) $CH_3 CH_2 COCl$   (c) $CH_3 CH_2 CON(CH_3)_2$

(d) $CH_3 CH_2 \underset{O}{\overset{O}{COCCH_2}} CH_3$     (e)

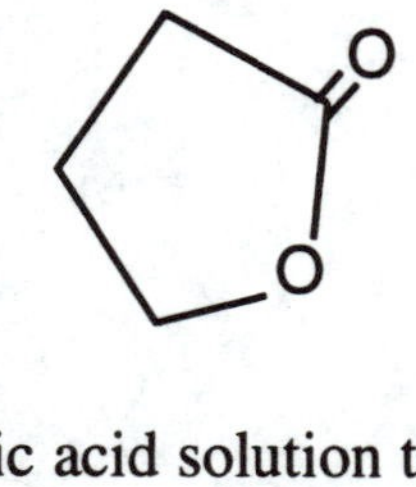

3. Which of the compounds reacts with a hydrochloric acid solution to undergo hydrolysis to form an acid as a product?

4. Which of the above belong to classes of compounds that will readily react with an alcohol to make an ester?

5. These are the structures for salicylic acid, aspirin (acetylsalicylic acid) and oil of wintergreen (methyl salicylate).

salicylic acid          aspirin          oil of wintergreen

Which of the following can be stated?
(a) Both are synthesized from salicylic acid as a starting material.
(b) Aspirin could be made by reaction of salicylic acid with acetic anhydride.
(c) Oil of wintergreen can be made by a Fischer esterification process using methanol and a $H_2SO_4$ as a catalyst.
(d) Aspirin would react with $NaHCO_3$ to form a salt.
(e) In time aspirin in humid climates could undergo a hydrolysis reaction leading to a bottle of aspirin that has a vinegar smell.

6. Which of the following are true when comparing the hydrolysis of an ester using a water/HCl solution versus a water/NaOH solution?
(a) In acid hydrolysis a carboxylic acid is a final product, but in a basic medium a sodium salt of a carboxylic acid is a final product.
(b) An alcohol is a final product in an acid hydrolysis, but in a basic hydrolysis an alkoxide ion is a final product.
(c) The acid and base both act as catalysts in the hydrolysis reaction.
(d) An acid hydrolysis is a reversible process, but a basic hydrolysis is not a reversible process.
(e) More base must be present to do a basic hydrolysis than acid has to be present when doing an acid hydrolysis if all the ester is to be hydrolyzed.

7. For the total hydrolysis of a mole of succinimide which of the following would be true?

(a) An acid hydrolysis using HCl would require two moles of water and one mole of HCl.
(b) In an acid hydrolysis the final product would be succinic acid and ammonium chloride.
(c) A basic hydrolysis using NaOH would require 2 moles of base.
(d) A basic hydrolysis would require two moles of water.
(e) In a basic hydrolysis the final product is succinic acid.

557

8. Put the list of reagents and compounds in the proper order that would be necessary to make $(CH_3)_2NCH_2CH_2N(CH_3)_2$ from $HO_2CCO_2H$.

(X) $(CH_3)_2NH$

(Y) $SOCl_2$

(Z) $LiAlH_4$

(a) XYZ        (b) XZY        (c) YXZ        (d) YZX        (e) ZXY        (f) ZYX

9. Which of the following are features common to nylon 6, nylon 66 and Kevlar?

$-[NH(CH_2)_5CO]_n-$     $-[OC(CH_2)_5CONH(CH_2)_5NH]_n-$     $-[OCC_6H_5CONHC_6H_5NH]_n-$

     nylon 6                   nylon 66                      Kevlar

(a) They are all polyamides.
(b) The are all step-growth polymers.
(c) In the synthetic process water is release as a side product.
(d) Hydrogen bonding occurs between polymer chains.
(e) The aromatic ring in the Kevlar chain weakens the polymer chain.

10. Which of the following are step-growth polymers?

(a) $-[CH_2CHF]_n-$

(b) 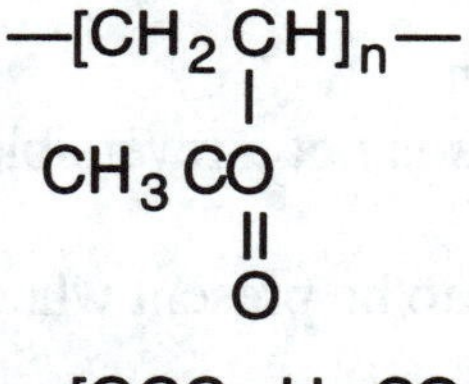 $-[CH_2CH]_n-$
           |
     $CH_3CO$
             ‖
             O

(c) $-[OCC_6H_5CO_2CH_2CH_2O]_n-$

(d) $-[CH_2CCl=CHCH_2]_n-$

(e) 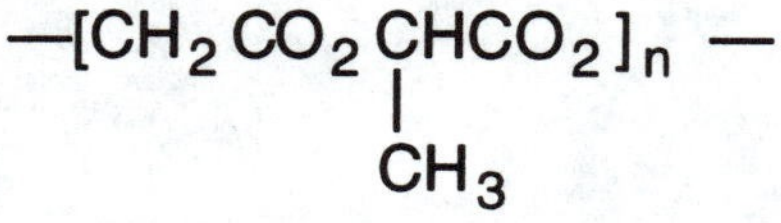 $-[CH_2CO_2CHCO_2]_n-$
               |
            $CH_3$

**Chapter 14 Quiz Answers**
(1) **c, e** (2) **d** (3) **a, b, c, d, e** (4) **b, d** (5) **a, b, c, d, e** (6) **a, d,e** (7) **a, b, c** (8) **c** (9) **a, b, c, d** (10) **c, e**

# CHAPTER 15: QUIZ

*Determine all of the correct answers for each question. Each question may have more than one correct response.*

1. Identify the acidic hydrogens in the compound.

$$CH_3-CO-CH_2-CH_2-CH_2-CH_2-CO_2CH_3$$
   a        b     c     d     e        f

2. Place the following compounds in the order most acidic first and least acidic last.

$CH_3COCH_2CO_2CH_2CH_3$       $CH_3CH_2CH_2CH_2OH$       $CH_3CH_2COCH_3$
       **X**                          **Y**                       **Z**

(a) XYZ       (b) XZY       (c) YXZ       (d) YZX       (e) ZXY       (f) ZYX

3. Place the following three mechanistic steps for the aldo condensation of acetone using KOH in the order step 1, step 2, step 3.

**(X)**

$$CH_3-\overset{\overset{\displaystyle :\ddot{O}:}{|}}{\underset{\underset{\displaystyle CH_3}{|}}{C}}-CH_2-\overset{\overset{\displaystyle O}{||}}{C}-CH_3 \quad \rightleftharpoons \quad CH_3-\overset{\overset{\displaystyle OH}{|}}{\underset{\underset{\displaystyle CH_3}{|}}{C}}-CH_2-\overset{\overset{\displaystyle O}{||}}{C}-CH_3$$

H—OH

**(Y)**

$$HO^- \;+\; H-CH_2-\overset{\overset{\displaystyle O}{||}}{C}-CH_3 \quad \rightleftharpoons \quad HOH + [:CH_2-\overset{\overset{\displaystyle :O:}{||}}{C}-CH_3 \longleftrightarrow CH_2=\overset{\overset{\displaystyle :\ddot{O}:}{|}}{C}-CH_3]$$

**(Z)**

$$CH_3-\overset{\overset{\displaystyle :O:}{||}}{C}-CH_3 \;+\; :CH_2-\overset{\overset{\displaystyle O}{||}}{C}-CH_3 \quad \rightleftharpoons \quad CH_3-\overset{\overset{\displaystyle :\ddot{O}:}{|}}{\underset{\underset{\displaystyle CH_3}{|}}{C}}-CH_2-\overset{\overset{\displaystyle O}{||}}{C}-CH_3$$

(a) XYZ     (b) XZY       (c) YXZ       (d) YZX       (e) ZXY       (f) ZYX

4. Which of the following can be stated concerning the mechanism and the product illustrated in question 3?
   (a) The actual product shown would, with gentle heating in the basic solution, be expected to be an α, ß-unsaturated ketone.
   (b) Reaction of the final product shown in question 3 in the presence of an acid would yield an α, ß-unsaturated ketone.
   (c) The α, ß-unsaturated ketone referred to as the final product in (a) will not have the same structure as the α, ß-unsaturated ketone referred to in (b).

(d) It would be reasonable to assume that an α, ß-unsaturated ketone rather than the product shown in the mechanism would be the more likely product.

(e) The same mechanism would not apply when reacting propanone (acetone) with methanal (formaldehyde).

5. Which of the following compounds would not be logical ones to be used as the second reactant with $C_6H_5CH_2CHO$ in a mixed aldo condensation?

(a) $C_6H_5COCH_3$

(b) $HCHO$

(c) $C_6H_5COC_6H_5$

(d) $(CH_3)_3CCHO$

(e) $(CH_3)_3CCOCH_2CH_3$

6. Which of the following esters would not be logical ones for use as the second reactant with $C_6H_5CH_2CO_2CH_2CH_3$ in a crossed Claisen condensation?

(a) $C_6H_5CO_2CH_2CH_3$

(b) $(CH_3)_2CHCO_2CH_2CH_3$

(c) $HCO_2CH_2CH_3$

(d) $CH_3O_2CCO_2CH_3$

(e) $CH_3CH_2O_2CCH_2CO_2CH_2CH_3$

7. Which of the following compounds when treated with $CH_3CH_2O^-\ Na^+$ will yield one or more ring compounds as product?

(a) $C_6C_5COCH_2CH_2CH_2CH_2CH_2CHO$

(b) $C_6C_5COC(CH_3)_2CH_2CH_2CH_2C(CH_2)_3CHO$

(c) $CH_3CH_2O_2CCH_2CO_2CH_2CH_3$

(d) $CH_3CH_2O_2C(CH_2)_5CO_2CH_2CH_3$

(e) $C_6C_5COCH_2CHO$

8. Place the following three mechanistic steps for a crossed Claisen condensation of ethyl propanoate and ethyl methanoate in the order: step 1, step 2, step 3.

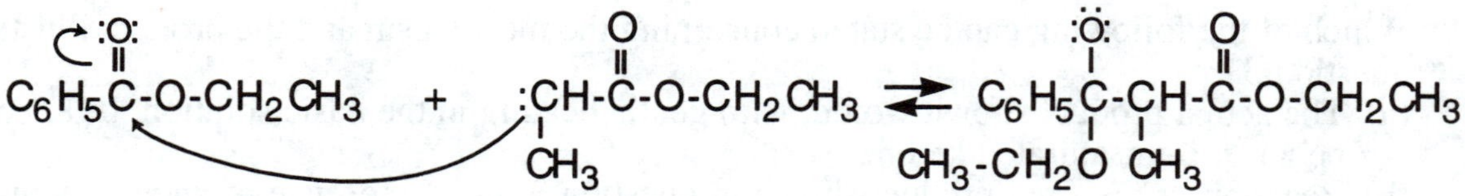

(Y)

$$CH_3CH_2O^- \quad + \quad CH_3\text{-}CH\text{-}CO_2CH_2CH_3 \quad \rightleftharpoons$$

$$CH_3CH_2OH \quad + \quad [\: :CH\text{-}\overset{\overset{\displaystyle :O:}{\|}}{C}\text{-}O\text{-}CH_2CH_3 \longleftrightarrow \overset{\overset{\displaystyle :\ddot{O}:}{|}}{CH}=C\text{-}O\text{-}CH_2CH_3 \:]$$

(Z)

$$C_6H_5\overset{\overset{\displaystyle :\ddot{O}:}{|}}{C}\text{-}CH\text{-}\overset{\overset{\displaystyle O}{\|}}{C}\text{-}O\text{-}CH_2CH_3 \quad \rightleftharpoons \quad C_6H_5\overset{\overset{\displaystyle O}{\|}}{C}\text{-}CH\text{-}\overset{\overset{\displaystyle O}{\|}}{C}\text{-}O\text{-}CH_2CH_3$$

(a) XYZ     (b) XZY     (c) YXZ     (d) YZX     (e) ZXY     (f) ZYX

9. Which of the following are true concerning the crossed Claisen condensation used in question 8?

    (a) The product shown would actually be present in very low yield at the conclusion of the reaction unless an aqueous solution of HCl is added after condensation is complete.

    (b) Prior to adding aqueous HCl solution, the major component in the reaction vessel is an anion.

$$C_6H_5\overset{\overset{\displaystyle O}{\|}}{C}\text{-}\overset{\overset{\displaystyle ..}{C}}{\underset{\underset{\displaystyle CH_3}{|}}{}}\text{-}\overset{\overset{\displaystyle O}{\|}}{C}\text{-}O\text{-}CH_2CH_3$$

    (c) The following ketone $C_6H_5COCH_2CH_3$ can be made relatively easily from the final β-ketoester.

    (d) The following β-ketoester would be expected as a side product.

$$CH_3CH_2\text{-}\overset{\overset{\displaystyle O}{\|}}{C}\text{-}\overset{\underset{\underset{\displaystyle CH_3}{|}}{C}H}{}\text{-}\overset{\overset{\displaystyle O}{\|}}{C}\text{-}O\text{-}CH_2CH_3$$

    (e) The yield of the desired cross condensation product will be the major product.

10. Which of the following are not true concerning the Claisen condensation of acetyl coenzyme A?

$$CH_3-\overset{\overset{\displaystyle O}{\|}}{C}-SCoA$$

    (a) The reaction will occur spontaneously to form a ß-ketothioester in the cell of a living organism.

    (b) After two molecules of acetyl-CoA have condensed the ß-ketothioester will undergo an additional Claisen condensation with a third of the molecules of acetyl-CoA in the presence of an enzyme.

(c) The stereochemical center formed when the ß-ketothioester undergoes a Claisen condensation reaction will be mixture of S and R configuration.

(d) The Claisen condensation reactions of acetyl-CoA are the first reactions leading to the formation of isopentenyl pyrophosphate.

(e) Terpenes and steroids would still be present in living organisms in the absence of the Claisen condensation of acetyl coenzyme A.

## Chapter 15 Quiz Answers
(1) **a, b, e** (2) **b** (3) **d** (4) **a, b, d** (5) **a, e** (6) **a, c, d** (7) **a, d** (8) **c** (9) **a, b, c, d** (10) **a, b, d**

# CHAPTER 16: QUIZ

*Determine all of the correct answers for each question. Each question may have more than one correct response.*

**Use these fatty acids to answer questions 1-2.**

(a) $CH_3(CH_2)_{14}COOH$

(b) $CH_3(CH_2)_{15}COOH$

(c) $CH_3(CH_2)_{28}COOH$

(d) 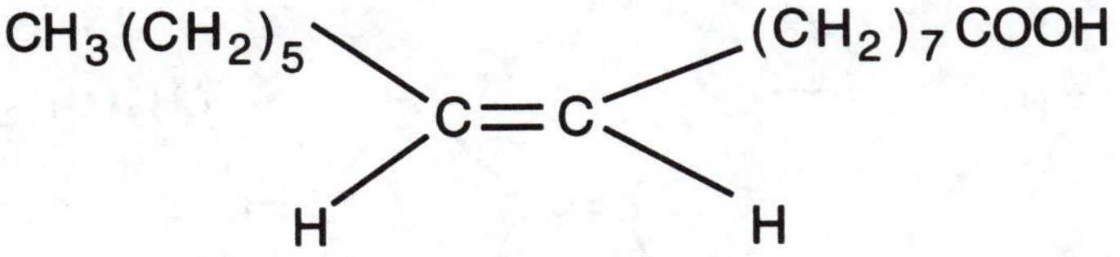 $CH_3(CH_2)_5$ $\diagdown$ $(CH_2)_7COOH$

$C=C$

$H$ $H$

(e) 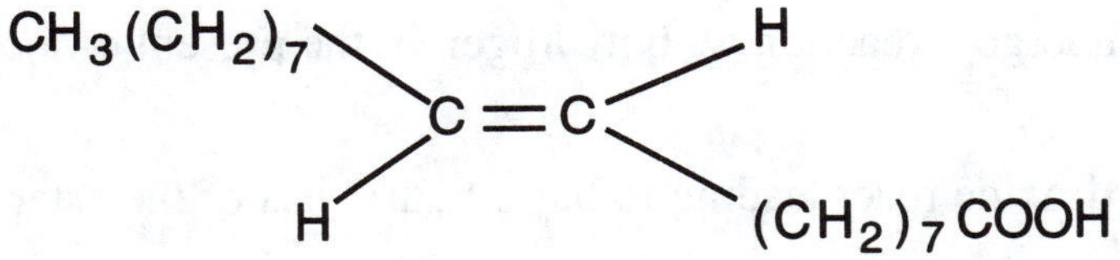 $CH_3(CH_2)_7$ $\diagdown$ $H$

$C=C$

$H$ $(CH_2)_7COOH$

(f) 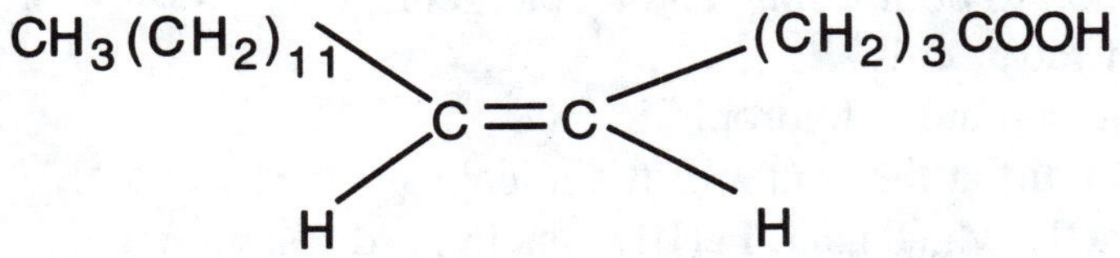 $CH_3(CH_2)_{11}$ $\diagdown$ $(CH_2)_3COOH$

$C=C$

$H$ $H$

(g) 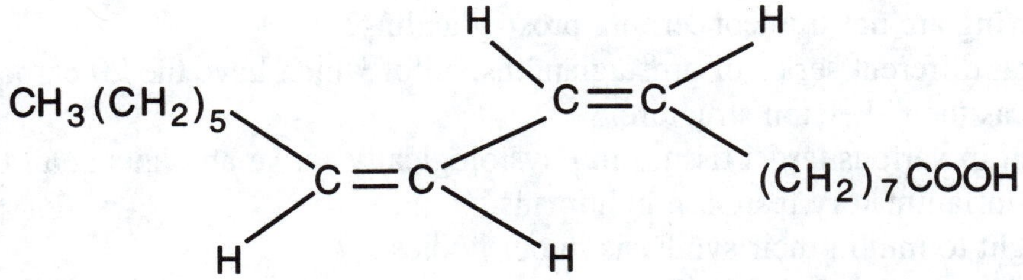 $CH_3(CH_2)_5$ $\diagdown$ $H$ $H$

$C=C$ $C=C$

$H$ $H$ $(CH_2)_7COOH$

(h) $CH_3(CH_2)_4$ $CH_2$ $(CH_2)_7COOH$

$C=C$ $C=C$

$H$ $H$ $H$ $H$

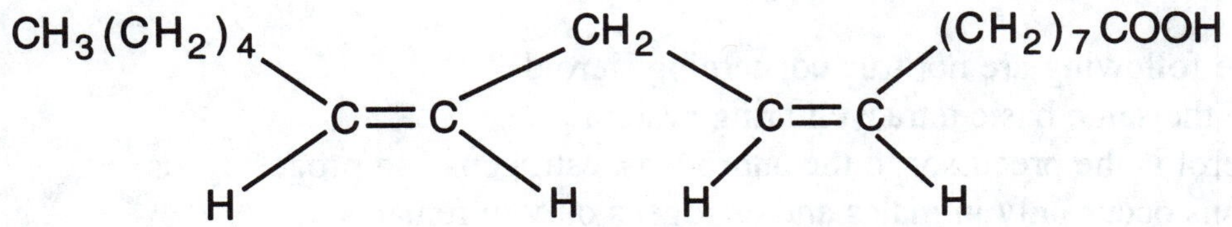

1. Which of the fatty acids would you expect to be commonly found as components of glycerides found in many living organisms?

2. Which of the fatty acids would you expect to be more commonly found as components of glycerides found in the plant kingdom as compared to those found in the animal kingdom?

**Use this compound for question 3-4.**

3. Which carbon is most likely to undergo a reaction with hydrogen in the present of a catalyst such as Raney nickel?

4. At which carbon does sodium hydroxide react leading to basic hydrolysis of the ester and a soap being formed as the product?

5. Which of the following are common to both a soap and a detergent?
   (a) They are both sodium salts of fatty acids.
   (b) They both have a hydrophobic end and a hydrophilic end.
   (c) They both form a micelle with dirt at the center of the micelle.
   (d) They both precipitate with Ca(II), Mg(II) and Fe(III) ions in hard water.
   (e) They are both sold in approximately equal amounts in the market place.

6. Which of the following are not true concerning prostaglandins?
   (a) There are several different series of prostaglandins, all of which have the 20 carbon acid, prostanoic acid as their skeleton structure.
   (b) They are present in various target tissues in physiologically active amounts at all times.
   (c) Some cause an inflammatory response in humans.
   (d) Aspirin is thought to inhibit their synthesis in our bodies.
   (e) A synthetic analog is used as a abortifacient.

7. Which of the following are not true concerning steroids?
   (a) All have the same basic tetracyclic ring system.
   (b) Cholesterol is the precursor to the androgens, estrogens and progestagens.
   (c) Androgens occur only in males and estrogens only in females.
   (d) The synthetic anabolic steroids have no androgenic function in the body.
   (e) A combination of a synthetic progestagen and estrogen is now used in some birth control pill.

564

8. Which of the following are not true concerning the phospholipids?
   (a) They contain two fatty acids attached to a glycerol backbone.
   (b) The name refers to the fact that all contain a phosphate group having a low molecular weight alcohol attached.
   (c) They form a bimolecular cellular membrane with nonpolar ends of the fatty acid chains on the inside.
   (d) The bimolecular membrane, once formed, is a rigid structure.
   (e) Snake venom phospholipase enzyme causes hydrolysis at carbon-2 of phospholipids.

9. Which of the following are not true concerning vitamin A?
   (a) B-carotene serves as a precursor to vitamin A which in turn is a precursor for the formation of 11-*cis*-retinal.
   (b) Rhodopsin is an imine formed in the rods of the retina by the reaction of 11-*cis*-retinal with opsin, a protein.
   (c) The formation of rhodopsin occurs both day and night.
   (d) The process of vision involves a cis to trans isomerization of a double bond in the 11-*cis*-retinal portion of rhodopsin.
   (e) Only the 11-*cis*-retinal portion of rhodopsin is affected during the process in which our optic nerve is stimulated.

10. Which of the following are not true concerning the other fat soluble vitamins in humans.
    (a) Vitamin D is formed in the skin by ultraviolet irradiation of a steroid.
    (b) Vitamin D is the actual physiologically active ingredient in the human body.
    (c) Vitamin E may retard the aging process.
    (d) Vitamin E leads to the formation of peroxy radicals in the body.
    (e) Vitamin K has a fundamental role in blood clotting.

**Chapter 16 Quiz Answers**
(1) **a, d, h** (2) **d, h** (3) **e** (4) **b** (5) **b, c** (6) **b** (7) **c, d** (8) **d** (9) **c, e** (10) **b, d**

# CHAPTER 17: QUIZ

*Determine all of the correct answers for each question. Each question may have more than one correct response.*

**Use these amino acids drawn in Fischer projection to answer questions 1-2.**

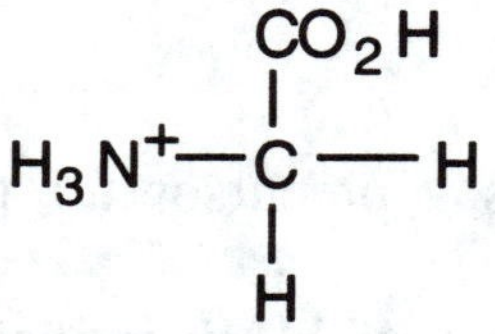

A 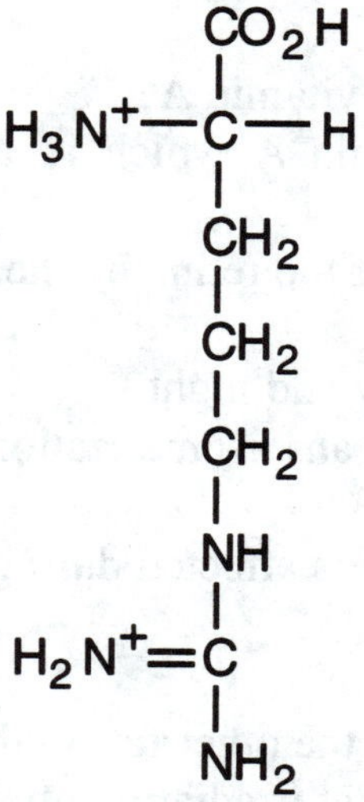       B 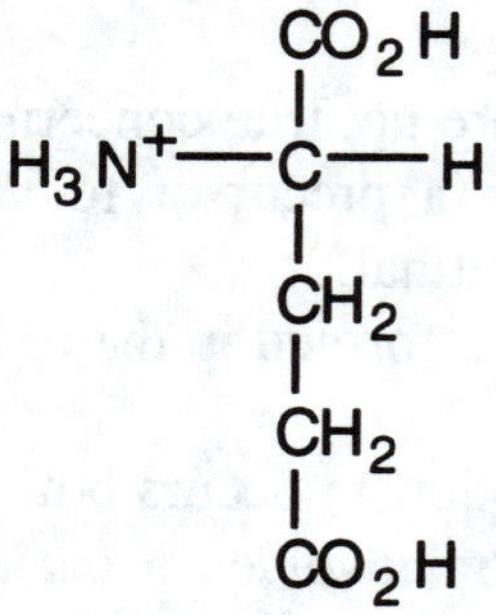       C 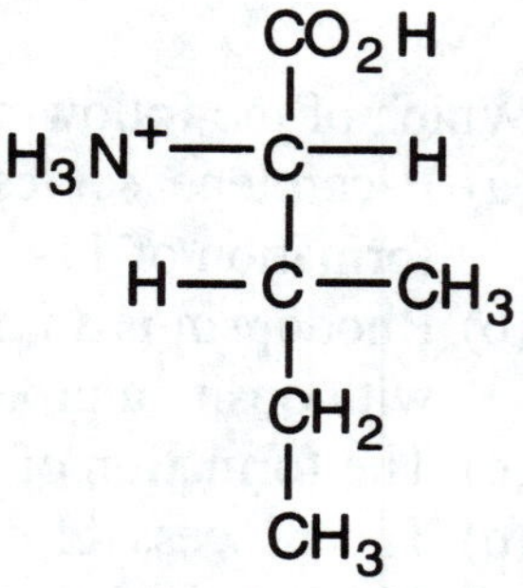       D

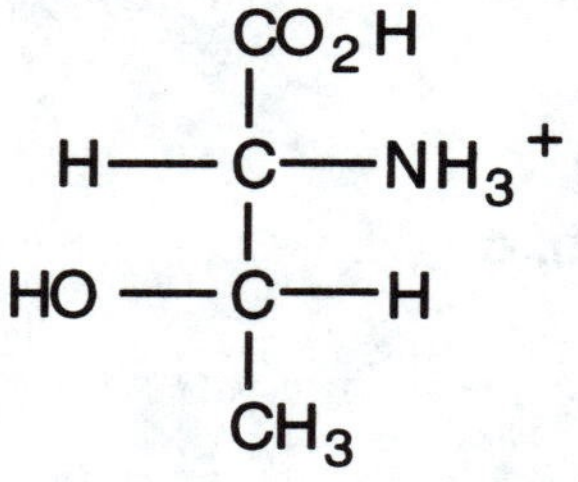

E 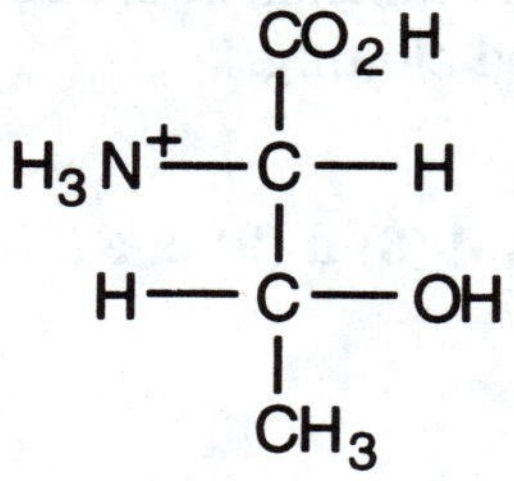       F 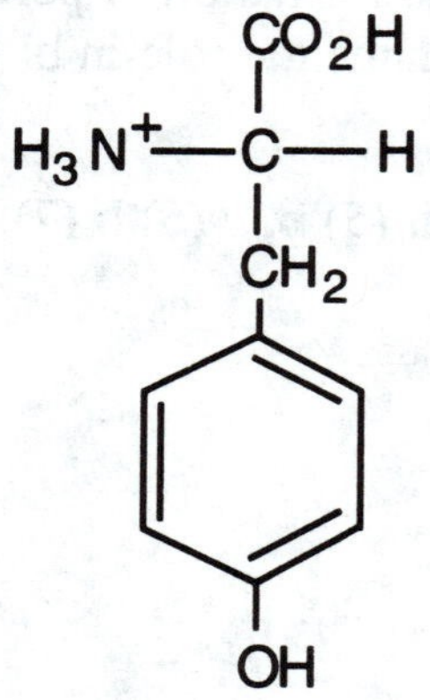       G 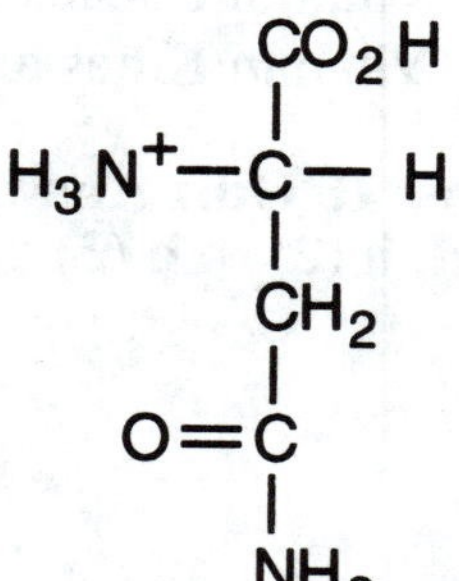       H

1. Which amino acids have an L configuration?

2. Which amino acids have an acidic side chain?

**Use the following information and structures for the amino acid, lysine, to answer question 3-6.**

(a) $H_3N^+ CH_2 CH_2 CH_2 CH_2 CHCO_2H$
$\qquad\qquad\qquad\qquad\quad |$
$\qquad\qquad\qquad\qquad\quad NH_3^+$

$\qquad\qquad\qquad\qquad$ pK$_a$ of $\alpha$-CO$_2$H = 2.18

(b) $H_3N^+ CH_2 CH_2 CH_2 CH_2 CHCO_2^-$
$\qquad\qquad\qquad\qquad\quad |$
$\qquad\qquad\qquad\qquad\quad NH_3^+$

$\qquad\qquad\qquad\qquad$ pK$_a$ of $\alpha$-NH$_3^+$ = 8.95

(c) $H_3N^+ CH_2 CH_2 CH_2 CH_2 CHCO_2^-$
$\qquad\qquad\qquad\qquad\quad |$
$\qquad\qquad\qquad\qquad\quad NH_2$

$\qquad\qquad\qquad\qquad$ pK$_a$ of side chain = 10.53

(d) $H_2NCH_2 CH_2 CH_2 CH_2 CHCO_2^-$
$\qquad\qquad\qquad\qquad\quad |$
$\qquad\qquad\qquad\qquad\quad NH_2$

3. Which structure is the zwitterion?

4. At what pH value in an aqueous solution will the concentration of zwitterion be at a maximum value?

   (a) 1.18    (b) 2.18    (c) 3.18    (d) 5.565    (e) 7.00    (f) 7.95

   (g) 8.95    (h) 9.53    (i) 9.74    (j) 10.53    (k) 11.53

5. In an aqueous solution what pH value will have to be exceeded before the concentration of structure d will be greater than its conjugate acid?

   (a) 1.18    (b) 2.18    (c) 3.18    (d) 5.565    (e) 7.00    (f) 7.95

   (g) 8.95    (h) 9.53    (i) 9.74    (j) 10.53    (k) 11.53

6. At what pH value in an aqueous solution will structures b and c be present in equal amounts?

   (a) 1.18    (b) 2.18    (c) 3.18    (d) 5.565    (e) 7.00    (f) 7.95

   (g) 8.95    (h) 9.53    (i) 9.74    (j) 10.53    (k) 11.53

7. Use the following information to determine the sequence of amino acids in a pentapeptide listed in order starting with the N-terminal end.

| Experimental Procedure | Amino Acid Composition |
| --- | --- |
| Hydrolysis of pentapeptide | gly, leu, lys, thr, tyr |
| Edman degradation | gly |
| Hydrolysis catalyzed by trypsin | Fragment A: thr |
|  | Fragment B: gly, leu, lys, tyr |
| Hydrolysis catalyzed by chymotrypsin | Fragment A: gly, tyr |
|  | Fragment B: leu, lys, thr |

(a) gly-leu-lys-tyr-thr   (b) gly-lys-tyr-leu-thr   (c) gly-tyr-leu-lys-thr

(d) gly-leu-tyr-lys-thr   (e) gly-lys-leu-tyr-thr   (f) gly-tyr-lys-leu-thr

(g) thr-leu-lys-tyr-gly   (h) thr-lys-tyr-leu-gly   (i) thr-tyr-leu-lys-gly

(j) thr-leu-tyr-lys-gly   (k) thr-lys-leu-tyr-gly   (l) thr-tyr-lys-leu-gly

8. Which of the following are true concerning the primary structure of a protein and peptide bond?
   (a) The primary structure refers to the sequence of amino acids in a protein.
   (b) This representation of the peptide bond between two glycine molecules is totally consistent with our knowledge of peptide bonds found in proteins.

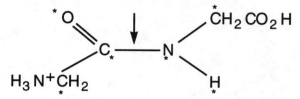

   (c) The carbon-nitrogen bond indicated by an arrow in the peptide bond allows relatively free rotation of the protein structure.
   (d) All the atoms identified by an asterisk in the peptide bond lie in the same plane.
   (e) All the bond angles shown in the peptide bond are approximately 120°.
   (f) A cis arrangement rather than the trans arrangement shown for the α-carbons is just as prominent in protein structure.

9. Serine will form hydrogen bonds with which of the following amino acids?
   (a) threonine   (b) proline   (c) lysine   (d) tyrosine   (e) aspartic acid
   (f) glutamine   (g) valine   (h) methionine

10. Which of the following are associated with the secondary structure of a protein?
    (a) The sequencing of amino acids in the chain.
    (b) The amino acid chain assumes the shape of an α-helix.
    (c) The amino acid chain assumes the shape of a ß-pleated sheet.
    (d) Disulfide bonds between side chains of cysteine.
    (e) Hydrogen bonding between the N-H in one peptide bond and the C=O of another peptide bond.
    (f) Hydrophobic effect between nonpolar amino acid side chains.
    (g) Salt linkage or electrostatic attraction between oppositely charged -$NH_3^+$ and negatively charged -$CO_2^-$ groups on the side chains of amino acids.

**Chapter 17 Quiz Answers**
(1) **b, c, d, f, g, h** (2) **c, g** (3) **c** (4) **i** (5) **j** (6) **g** (7) **c** (8) **a, d, e** (9) **a, c, d, e, f** (10) **b, c, e**

# CHAPTER 18: QUIZ

*Determine all of the correct answers for each question. Each question may have more than one correct response.*

**Use these statements associated with determination of the structure of DNA to answer questions 1-2.**

   (a) Determination of the 4 different nitrogen bases in DNA.

   (b) Determination that DNA had a 2-deoxyribose/phosphate backbone.

   (c) Determination that the amount of guanine equals cytosine and the amount adenine equals thymine in DNA.

   (d) Determination of the genetic code leading to formation of specific amino acid sequences in proteins.

   (e) Determination that the base pairs of adenine/thymine and guanine/cytosine maximized hydrogen bonding thereby leading to paired bases having almost identical dimensions and therefore an $\alpha$-helix structure.

   (f) Determination of the process of replication of DNA.

   (g) Determination that DNA was probably composed of 2 strands.

   (h) Determination that 2-deoxyribose was the sugar in DNA.

1. For which structural feature determination did Watson and Crick receive a Nobel prize?

2. Which information could have been experimentally determined only after Watson and Crick completed their efforts?

3. Which of the following is true concerning ATZ (azidothymidine) used for treatment of people diagnosed as having AIDS.

azidothymidine                      deoxythymidine

   (a) It has a structure that mimics deoxythymidine and thus is incorporated into the HIV-1 retrovirus DNA strand.

   (b) The lack of a 3´-OH group in AZT prevents additional nucleotides from being added to the DNA strand.

   (c) The base in AZT will not be able to hydrogen bond to adenine on the complimentary DNA strand.

   (d) Unfortunately AZT binds equally well to human DNA polymerase and to viral DNA polymerase.

   (e) AZT is actually an inhibitor for DNA polymerase.

4. Which of the following are true statements that can be made in contrasting DNA and RNA?
   (a) There is only one type of DNA but three types of RNA.
   (b) Both exist as two stranded molecules.
   (c) Both have a sugar phosphate ester backbone containing ß-D-ribose.
   (d) The RNA structure pairs uracil rather than thymine with cytosine.
   (e) There is approximately as much RNA as DNA in cells.

5. Which of the following act as a site in a cell where protein synthesis takes place?
   (a) DNA       (b) rRNA     (c) tRNA     (d) mRNA     (e) none of the nucleic acids

6. Which of the following carry the coded genetic information on protein amino acid sequence to the site where protein synthesis occurs?
   (a) DNA       (b) rRNA     (c) tRNA     (d) mRNA     (e) none of the nucleic acids

7. Which of the following are true concerning the genetic code?
   (a) All condons contain exactly three nucleotides.
   (b) Each amino acid has only one specific codon that codes for its presence in a protein.
   (c) There are three codons (UAA, UAG, and UGA) that code chain termination procedures.
   (d) To date, the code has been found to be the same for essentially all organisms on the planet.
   (e) During transcription the following mRNA strand would have been synthesized from the DNA strand shown.

   mRNA ⟶ 5'-UUA-CAG-UCG-ACG-GAC-GUG-CGU-AGU-3'
   DNA ⟶ 3'-AAT-GTC-AGC-TGC-CTG-CAC-GCA-TCA-5'

8. Which of the restriction endonucleases will cleave the following DNA strand?

   5'-AAACCTGGGCCCGTTGTGTACAAGCTTTATCACGAGCTCGTC-3'

   (a) AluI    AG↓CT       (b) BalI    TGG↓CCA   (c) FnuDII     CG↓CG

   (d) HeaIII CG↓CC       (e) HpaII   C↓CGG      (f) MboI        ↓GATC

   (g) NotI   GC↓GGCCGC   (h) SasI    GAGCT↓C

9. Which of the following DNA fragments will undergo an excision reaction when reacted first with dimethyl sulfate and then with piperidine in NaOH?
   (a) 5'-ATATCCCTATACCAC
   (b) 5'-AATGTTCTACAAACT
   (c) 5'-CATTTTATTCTCCGT
   (d) 5'-TCTCACAAAACCTTA
   (e) 5'-AGCCCTCGCGACACA

10. Arginine is coded on mRNA by the six codons CGU, CGC, CGA, CGG, AGA, AGG. Which of the following statements are consistent with this information for the coding of arginine by mRNA?
    (a) If the first two bases in a codon are CG then the third base is irrelevant as far as coding goes.

(b) If the first two bases in a codon are AG then the third base is irrelevant as far as coding goes.

(c) In all the codons for arginine the middle base is irrelevant as far as coding is concerned.

(d) AGU and AGC are not viable codons for another amino acid.

(e) It would be impossible for each amino acid to have six separate codons which coded for it on a mRNA strand.

## Chapter 18 Quiz Answers

(1) e (2) d, f (3) a, b (4) a, d (5) b (6) d (7) a, d, e (8) a, h (9) b, c, e (10) a, e

# CHAPTER 19: QUIZ

*Determine all of the correct answers for each question. Each question may have more than one correct response.*

1. When an individual drinks an alcoholic beverage the following reaction occurs in the human body.

$$CH_3CH_2OH + NAD^+ \xrightarrow[\text{dehydrogenase}]{\text{alcohol}} CH_3CHO$$

Which of the following statements concerning this metabolic process are correct?
(a) The $NAD^+$ is acting as an oxidizing agent.
(b) The $NAD^+$ accepts an $H^+$ from a carbon plus 2 electrons in a separate step.
(c) The reaction would occur even without the presence of an enzyme.
(d) The alcohol dehydrogenase removes an $H^+$ from the hydroxyl group.
(e) A hydride ion is added to the $NAD^+$ in the process to give $NADH$.

2. Place the following three metabolic reactions in the proper sequence leading to the formation of a carbon-carbon double bond in a fatty acid.

(X) $RCH_2CH_2COSCoA + FAD \longrightarrow RCH=CHCOSCoA + FADH_2$

(Y) $RCH_2CH_2CO_2^- + ATP \longrightarrow RCH_2CH_2CO_2^-AMP + P_2O_7^{4-}$

(Z) $RCH_2CH_2CO_2^-AMP + CoA-SH \longrightarrow RCH_2CH_2COSCoA$

(a) XYZ      (b) XZY      (c) YXZ      (d) YZX      (e) ZXY      (f) ZYX

3. Which of the following reactions occurring during fatty acid metabolism are oxidation reduction reactions?
(a) $RCH_2CH_2CO_2^- + ATP \longrightarrow RCH_2CH_2CO_2^-AMP + P_2O_7^{4-}$
(b) $RCH_2CH_2CO_2^-AMP + CoA-SH \longrightarrow RCH_2CH_2COSCoA$
(c) $RCH_2CH_2COSCoA + FAD \longrightarrow RCH=CHCOSCoA + FADH_2$
(d) $RCH=CHCOSCoA + HOH \longrightarrow RCHOHCH_2COSCoA$
(e) $RCHOHCH_2COSCoA + NAD^+ \longrightarrow RCOCH_2COSCoA + NADH$
(f) $RCOCH_2COSCoA + CoA-SH \longrightarrow RCOSCoA + CH_3COSCoA$

4. Which of the following is true for the fatty acid metabolism cycle?
(a) The fatty acid chain is shortened by two carbons with each completion of the cycle.
(b) Each reaction in the cycle requires a different enzyme.
(c) Two acetyl-CoA's are formed during the completion of each cycle.

(d) The total metabolism of stearic acid, $CH_3(CH_2)_{16}CO_2H$ would require a total of 8 $NAD^+$, and 8 $FAD$.

(e) A total of 8 $ATP$ would be required for the complete metabolism of stearic acid.

5. If the sugar consumed is fructose rather than glucose which of the following is true concerning glycolysis?

(a) $ATP$ is still required in the initial step of the process.

(b) The net production of $ATP$ in the total process will be 3 $ATP$ rather than 2 $ATP$.

(c) A total of 2 $NADH$ will still be produced in the total process.

(d) The final product of the process is now lactate, rather than pyruvate.

(e) Fructose provides the body with more energy than glucose through the process of glycolysis.

6. During rapid physical activity which of the following is true concerning the glycolysis process?

(a) The net amount of $ATP$ formed is less.

(b) The net amount of $NADH$ remaining at the end of the process is less.

(d) An additional step, the reduction of pyruvate to lactate occurs.

(e) Additional $H^+$ ions are formed.

(f) The process is an anaerobic process eventually leading to muscle fatigue.

7. Which of the reactions in the glycolysis constitute an oxidation reduction reaction and thus leads to the production of NADH?

(a) glucose $\longrightarrow$ glucose 6-phosphate

(b) glucose 6-phosphate $\longrightarrow$ fructose 6-phosphate

(c) fructose 6-phosphate $\longrightarrow$ fructose 1,6-diphosphate

(d) fructose 1,6-diphosphate $\longrightarrow$ dihydroxyacetone + glyceraldehyde 3-phosphate

(e) glyceraldehyde 3-phosphate $\longrightarrow$ 1,3-bisphosphoglycerate

(f) 1,3-bisphosphoglycerate $\longrightarrow$ 3-phosphoglycerate

(g) 3-phosphoglycerate $\longrightarrow$ 2-phosphoglycerate

(h) 2-phosphoglycerate $\longrightarrow$ phosphoenolpyruvate

(i) phosphoenolpyruvate $\longrightarrow$ pyruvate

8. Positron emission tomography is a new medical analysis procedure based on examination of the site at which glycolysis is occurring in the body. To do the analysis a patient is given an injection of glucose where the hydroxyl group on the C-6 position has been replaced with a positron emitting fluorine. At which of the reactions will the glycolysis process stop or be discontinued for the positron emitting glucose molecule so a build up of a positron emitting compound will occur?

(a) glucose $\longrightarrow$ glucose 6-phosphate

(b) glucose 6-phosphate $\longrightarrow$ fructose 6-phosphate

(c) fructose 6-phosphate $\longrightarrow$ fructose 1,6-diphosphate

(d) fructose 1,6-diphosphate $\longrightarrow$ dihydroxyacetone + glyceraldehyde 3-phosphate

(e) glyceraldehyde 3-phosphate $\longrightarrow$ 1,3-bisphosphoglycerate

(f) 1,3-bisphosphoglycerate $\longrightarrow$ 3-phosphoglycerate

(g) 3-phosphoglycerate $\longrightarrow$ 2-phosphoglycerate

(h) 2-phosphoglycerate $\longrightarrow$ phosphoenolpyruvate

(i) phosphoenolpyruvate $\longrightarrow$ pyruvate

9. Which of the following sugars, which can occur in our diet can enter directly into glycolysis without first undergoing some other enzymatic reaction?
   (a) sucrose, a disaccharide composed of a fructose and a glucose unit
   (b) galactose, a monosaccharide
   (c) mannose, a monosaccharide
   (d) isomaltose, composed of two glucose units
   (e) fructose

10. Which is not true concerning the fermentation process leading to the formation of ethanol as a final product?
    (a) Initially, glucose undergoes glycolysis.
    (b) The final product of glycolysis, pyruvate, undergoes decarboxylation leading to the formation of ethanal.
    (c) Ethanal undergoes a reduction using NADH as the reducing agent.
    (d) Carbon dioxide is a byproduct of the process.
    (e) The process requires ATP.

**Chapter 19 Quiz Answers**
(1) **a, d, e** (2) **d** (3) **c, e** (4) **a, b, d** (5) **a, c** (6) **b, d, e, f** (7) **e** (8) **a** (9) **e** (10) **e**

# CHAPTER 20: QUIZ

*Determine all of the correct answers for each question. Each question may have more than one correct response.*

**Use these compounds to answer questions 1-6.**

(a)

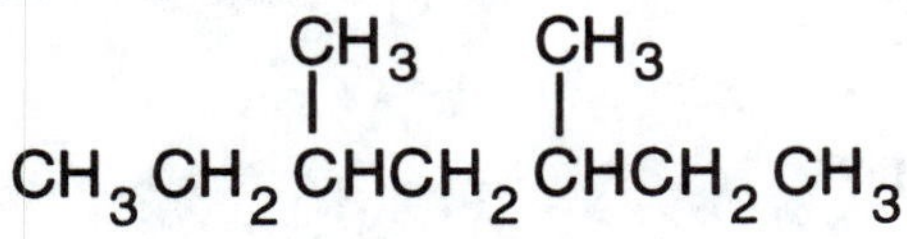

$$CH_3CH_2\overset{\overset{\displaystyle CH_3}{|}}{C}HCH_2\overset{\overset{\displaystyle CH_3}{|}}{C}HCH_2CH_3$$

(b) 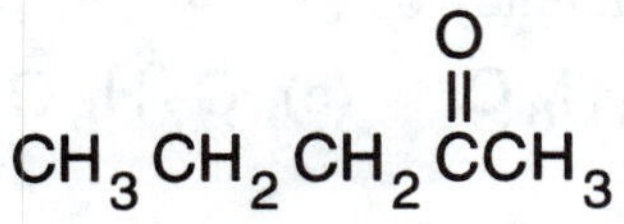 $CH_3CH_2CH_2\overset{\overset{\displaystyle O}{\|}}{C}CH_3$

(c) $BrCH_2CH_2CH_2Br$

(d)

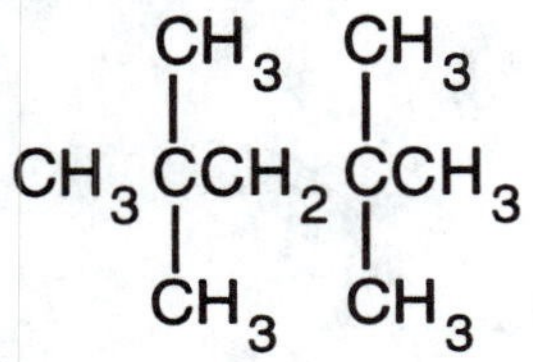

$$CH_3\overset{\overset{\displaystyle CH_3}{|}}{\underset{\underset{\displaystyle CH_3}{|}}{C}}CH_2\overset{\overset{\displaystyle CH_3}{|}}{\underset{\underset{\displaystyle CH_3}{|}}{C}}CH_3$$

(e)

$$\begin{array}{c} CH_3 \quad\quad\quad \overset{\displaystyle CH_3}{|} \\ \quad\quad C\text{-}O\text{-}CH_3 \\ C=C \quad CH_3 \\ H \quad\quad\quad H \end{array}$$

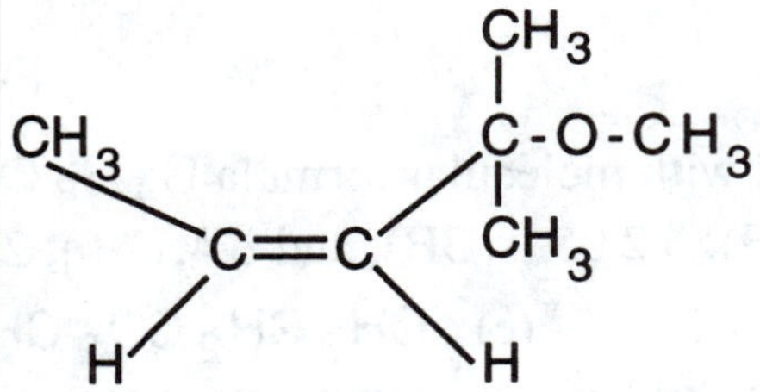

(f) $CH_3CH_2O_2CCH_2CH_2CO_2CH_2CH_3$

(g)

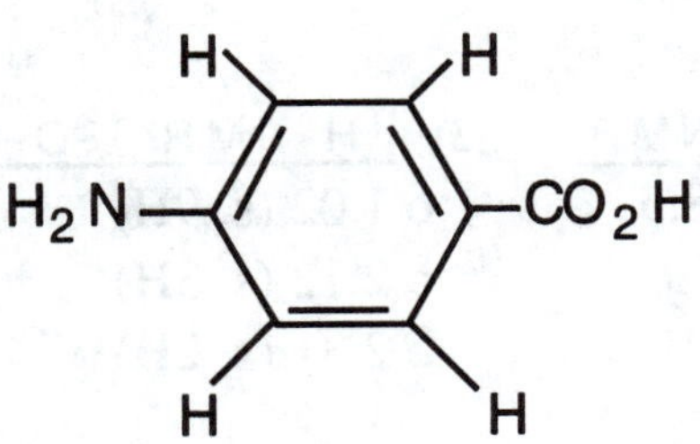

$$H_2N-\!\!\!\!\!\!\bigcirc\!\!\!\!\!\!-CO_2H$$

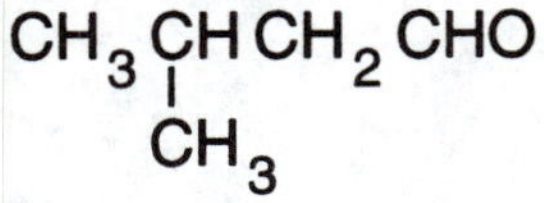

(h) $CH_3\overset{}{\underset{\underset{\displaystyle CH_3}{|}}{C}}HCH_2CHO$

1. Which compounds will give four $^1$H-NMR signals in the NMR?

2. Which compounds will give two $^1$H-NMR signals with an integrated signal heights ratio of 9:1?

3. Which compound would be expected to have $^1$H-NMR signals with chemical shifts in the three ranges δ2.2-δ2.9, δ6.5-δ8.5, and δ10-δ13?

4. Which compound would have $^1$H-NMR signals that exhibit no splitting of the signal by nonequivalent neighboring hydrogens?

5. Which compound would have three $^1$H-NMR signals in the order: a triplet, a singlet; and a quartet with a ratio of 3:2:2?

6. Which compounds would give four $^{13}$C-NMR signals?

7. Which of the following compounds has an index of hydrogen deficiency of 3?

   (a) $C_4H_4Br_2$    (b) $C_4H_6O_2$    (c) $C_5H_7N$    (d) $C_6H_8O$    (e) $C_7H_8O$
   (f) $C_7H_7NO_2$

8. A compound of molecular formula $C_4H_9Br$ has the following spectral data.

   | $^1$H-NMR | $^{13}$C-NMR |
   |---|---|
   | δ1.02 (d, 6H) | 21 |
   | δ1.97 (m, 1H) | 32 |
   | δ3.30 (d, 2H) | 42 |

   Which of the following is the structure of the compound?

   (a) $BrCH_2CH_2CH_2CH_3$       (b) $CH_3CHBrCH_2CH_3$

   (c) $BrCH_2CH(CH_3)_2$         (d) $(CH_3)_3CBr$

9. Which of the following is the structure for a compound with molecular formula $C_4H_8O_2$ and an $^1$H-NMR spectrum having peaks at δ 1.25 (t, 3H), δ 2.05 (s, 3H), and δ 4.12 (q, 2H)?

   (a) $CH_3CH_2CH_2CO_2H$       (b) $(CH_3)_2CHCO_2H$       (c) $CH_3CH_2CO_2CH_3$
   (d) $CH_3CO_2CH_2CH_3$        (e) $HCO_2CH_2CH_2CH_3$     (f) $HCO_2CH(CH_3)_2$

10. Following are the $^1$H-NMR and the $^{13}$C-NMR spectra for three isomeric ketones with molecular formula $C_7H_{14}O$.

| (1) $^1$H-NMR | $^{13}$C-NMR | (2) $^1$H-NMR | $^{13}$C-NMR | (3) $^1$H-NMR | $^{13}$C-NMR |
|---|---|---|---|---|---|
| δ .91 (t, 6H) | 15 | δ 1.08 (d, 6H) | 18.5 | δ 1.02 (s, 9H) | 30.5 |
| δ 1.6 (sex, 4H) | 18 | δ 2.79 (sep, 2H) | 39 | δ 2.12 (s, 3H) | 32 |
| δ 2.39 (t, 4H) | 44 | | 219 | δ 2.31 (s, 2H) | 34 |
| | 212 | | | | 208 |

Match the following structures with the data given in the above tables.

(X) $CH_3-CO-CH_2C(CH_3)_3$

(Y) $(CH_3)_2CH-CO-CH(CH_3)_2$

(Z) $CH_3CH_2CH_2-CO-CH_2CH_2CH_3$

**Chapter 20 Quiz Answers**
(1) **b, g, h** (2) **d** (3) **g** (4) **d** (5) **f** (6) **b, f, h** (7) **c, d** (8) **c** (9) **d** (10) **x=3, y=2, z=1**

# CHAPTER 21: QUIZ

*Determine all of the correct answers for each question. Each question may have more than one correct response.*

**Use these compounds to answer questions 1-6.**

(a) $CH_3 CH_2 CH_2 CH_2 CH_2 CH_2 CH_2 CHO$

(b) $CH_3 CH_2 CH_2 CH_2 OH$

(c) $CH_3 CH_2 CH_2 CH=CH_2$

(d) $CH_3 CH_2 CH_2 CH_2 CH_2 CO_2 CH_3$

(e) $CH_3 CH_2 CH_2 CH_2 NHCH_3$

(f) $CH_3 CH_2 - CO - CH_3$

(g) $H-\overset{\overset{\textstyle O}{\|}}{C}-N(CH_2 CH_3)_2$

(h) $CH_3 O$ —⟨benzene ring⟩— $CO_2 H$

1. Which compounds will have infrared absorption in the range 1630 $cm^{-1}$ to 1780 $cm^{-1}$?

2. Which compounds will have an infrared absorption in the range 3200 $cm^{-1}$ to 3550 $cm^{-1}$?

3. Which compounds will have infrared absorption in the range 1050 $cm^{-1}$ to 1150 $cm^{-1}$?

4. Which compounds will have infrared absorption in the range 3010 $cm^{-1}$ to 3095 $cm^{-1}$ plus absorption in the range 1620 $cm^{-1}$ to 1680 $cm^{-1}$?

5. Which compounds will have a set of infrared absorptions that fall within the following ranges: 1050 $cm^{-1}$ to 1150 $cm^{-1}$, 1475 $cm^{-1}$ to 1600 $cm^{-1}$, 1710 $cm^{-1}$ 1780, 2500 $cm^{-1}$ to 3000 $cm^{-1}$?

6. A compound with a molecular formula of $C_8H_8O_2$ has a strong, very broad infrared absorption centered at 3040 $cm^{-1}$ and another strong absorption at 1699 $cm^{-1}$. The $^1$H-NMR has signals at δ 6.05 (s, 2H), δ 7.29 (m, 5H) and at δ 11.98 (s, 1H). Which of the following statements concerning its structure are not consistent with this information?

(a) The index of hydrogen deficiency is 5.

(b) It contains a C=O.

(c) It is an ester.

(d) It contains a benzene ring which has only one substituent attached.

(e) It can not be some type of phenol/ketone structure.

(f) It can not be some type of phenol/aldehyde structure.

7. Which of the following statements concerning the infrared, $^1$H-NMR, and $^{13}$C-NMR spectra are not consistent with the structure of ethyl propanoate?

$$CH_3-CH_2-CO_2-CH_2-CH_3$$
$$\quad w \qquad x \qquad\quad y \qquad z$$

(a) The infrared has a strong absorption at 1740 cm$^{-1}$.
(b The $^{13}$C-NMR contains 3 peaks.
(c) The hydrogens labeled $w$ and $z$ will both be triplets.
(d) The signal for hydrogens labeled $w$ and $z$ will have the same chemical shift value.
(e) The hydrogens labeled $x$ and $y$ will both be two separate quartets each of which integrates to 2 H's.

8. A compound of molecular formula $C_4H_8O$ has a strong infrared absorption at 1740 cm$^{-1}$ but no absorption in the 3200 cm$^{-1}$ to 3600 cm$^{-1}$ range. The $^1$H-NMR signals are δ 1.06 (t, 3H), δ 2.04 (s, 3H) and δ 2.48 (q, 2H). Which of the following is the structure of the compound?

(a) $CH_2=CH-CH_2-CH_2OH$

(b) $CH_2=CH-CH(OH)-CH_3$

(c) $CH_2=COH-CH_2-CH_3$

(d) $CH_3-CH=CH-CH_2OH$

(e) $CH_3-CH=COH-CH_3$

(f) $CH_3-CH_2-CH_2-CHO$

(g) $CH_3-CH_2-CO-CH_3$

9. A compound of molecular formula $C_5H_{12}O$ has a strong, broad infrared absorption at 3315 cm$^{-1}$. The compound gives the following NMR spectral data.

| $^1$H-NMR | $^{13}$C-NMR |
|---|---|
| δ .91 (d, 6H) | 24 |
| δ 1.56 (q, 2H) | 26 |
| δ 1.71 (sep, 1H) | 43 |
| δ 2.24 (s, 1H) | 62 |
| δ 3.66 (t, 2H) | |

Which of the following is the structure of the compound?
(a) $CH_3-CH_2-CH_2-CH_2-CH_2OH$
(b) $CH_3-CH_2-CH_2-CH(OH)-CH_3$
(c) $CH_3-CH_2-CH(OH)-CH_2-CH_3$
(d) $(CH_3)_2CH-CH_2-CH_2OH$
(e) $(CH_3)_2CH-CH(OH)-CH_3$
(f) $(CH_3)_2COH-CH_2-CH_3$
(g) $HOCH_2-CH(CH_3)-CH_2-CH_3$
(h) $(CH_3)_3CCH_2OH$

10. In a $40,000 grand larceny case, the owners of a gasohol company became suspicious that the supplier of the alcohol was trying to substitute absolute methanol for absolute ethanol. The owners of the gasohol company tested the density of the shipment of alcohol. They asked a chemist to analyze a tanker load of the alcohol they had just purchased. Time was also a necessary consideration in analyzing the alcohol. Considering that the chemist may be called on to testify in court and explain to a lay person that the supplier had indeed substituted methanol for ethanol which of the following would be a most useful course for the chemist to follow?

(a)   Decide that the measurement of density alone would be quite sufficient to convince a jury that the tanker load was methanol and not ethanol and no spectral data need be gathered.
(b)  Take an infrared spectrum of the alcohol in the tanker.
(c)  Take a $^1$H-NMR of the sample.
(d)  Take a $^{13}$C-NMR of the sample.
(e)  Take an infrared spectra of samples of known methanol and ethanol for comparison with the sample.
(f)  Take a $^1$H-NMR of samples of known methanol and ethanol for comparison with the sample.
(g)  Take a $^{13}$C-NMR of samples of known methanol and ethanol for comparison with the sample.

**Chapter 21 Quiz Answers**
(1) **a, d, f, g, h** (2) **b, e** (3) **b, d, h** (4) **c** (5) **h** (6) **c** (7) **b, d** (8) **g** (9) **d** (10) **c, f**